Separations Division

Presentations at the 2008 AIChE Annual Meeting

Philadelphia, Pennsylvania
16 - 21 November 2008

Volume 1 of 2

ISBN: 978-1-61567-217-2

Printed from e-media with permission by:

Curran Associates, Inc.
57 Morehouse Lane
Red Hook, NY 12571

Some format issues inherent in the e-media version may also appear in this print version.

Printed by Curran Associates, Inc. (2009)

For permission requests, please contact AIChE
at the address below.

AIChE
3 Park Avenue
New York, NY 10016-5991

Phone: (203) 702-7660
Fax: (203) 775-5177

www.aiche.org

Additional copies of this publication are available from:

Curran Associates, Inc.
57 Morehouse Lane
Red Hook, NY 12571 USA
Phone: 845-758-0400
Fax: 845-758-2634
Email: curran@proceedings.com
Web: www.proceedings.com

TABLE OF CONTENTS

Volume 1

Volume 2

Electrokinetic Flotation of Paint Solids in a Continuous Flow Tank

Julie Q. Shang
Yanqing Xu

Department of Civil and Environmental Engineering
The University of Western Ontario
London, Ontario
Canada
N6A 5B9

Introduction

The treatment of water from paint spray booths in automobile assembly plants typically involves applications of large quantity of chemicals as coagulants and flocculants. Suspended paint solids are brought to water surface in sludge pits after these chemicals are added. The suspended paint particles are then mechanically collected and removed as sludge for further treatment and disposal. There are several disadvantages of the existing floatation process, including:

1. The cost of chemicals;
2. The dosage (feed rate) and type of chemicals need to be closely monitored and routinely tested to ensure the performance of the system. This increases the operation costs.
3. Unsatisfactory performance of the system may not be discovered and corrected immediately. This causes sludge build-up in the sludge pit and recirculation system, leading to additional maintenance costs.
4. Chemicals added to the sludge pit are non-recoverable and increase the mass and volume of the sludge generated in the process. This leads to higher disposal costs.

A study was carried out in an effort to explore alternative solutions for treating the circulating water in paint operation systems. Electrokinetic flotation (EKF hereafter) was tested for removing suspended paint solids from water. The EKF is based on principles of electrostatics and electrochemistry, including (1) directional movement of electrically charged particles in an electric field (electrophoresis and dielectrophoresis), (2) oxygen and hydrogen gas bubbles generated by an electric current through conducting electrodes (electrolysis) and (3) flocculation generated by changes of surface charges of suspended solids under the influence of the electric filed. As a result, when a DC field is applied via submerged electrodes in the circulating water with suspended paint solids, electrophoresis and dielectrophoresis generates movement of paint particles, the electrical field generates flocculation of paint solids to form larger flocs, and electrolysis generates gas bubbles at the both cathode and anode that carry paint particles and flocs to the water surface. The influencing parameters of EKF treatment include the water electrical conductivity (EC), electrode type and layout, type of paints, paint solid loading, and applied DC current. The results showed that the EKF process generated quick floatation of paint particles without using any flocculants. The suspended solids (SS) in the sludge water are significantly reduced along with significant color and COD removal. This paper presents the results of a study using a lab-scale continuous flow tank that simulates the paint spray operation in an automatic assembly plant.

Background

Flotation is commonly used in water and wastewater treatment and is most effective for the separation of low-density suspended solids and oil *(Lafrance et al., 1995; Chen & Horan, 1998; Huang & Liu, 1999; Vaughan et al., 2000; Manjunath et al., 2000)*. The first attempt to use electrokenetic flotation (also known as electroflotation) was by Elmore *(1904)* who used the technique to separate minerals from ore. EKF has been an attractive alternative because of characteristics such as high separating efficiency with shortened process duration, simple operation, small and compact equipment, and requiring few accessories. The EKF method has shown to be effective in removing suspended solids, oil and grease, organic pollutants, as well as sludge from wastewater. Experiments have shown that smaller gas bubbles are more efficient in the flotation of fine particles *(Bennett et al., 1958; Ahmed & Jameson, 1985; Ketkar et al., 1991; Burns et al., 1997)*. EKF generates essentially uniform and fine gas bubbles. The gas bubble flux can be adjusted by varying the current density on electrodes. The performance of an EKF system for treating oily wastewater has been proven to be better than either dissolved air flotation, sedimentation, or impeller flotation (IF) *(Il'in & Sedashova, 1999; Chen, 2004)*.

A major challenge in EKF treatment is the electrode system design. The common electrode materials for EKF include iron, aluminum, stainless steel and coated titanium anode (Ti/IrO_x). The former three materials are cheap, readily available, but are consumable during an EKF process. The Ti/IrO_x anode represents the state of the art anode development, which is non-consumable and stable.

Continuous Flow Tank for EKF Study

The design of the lab-scale continuous flow tank is based on the paint spray operation in an automobile assembly plant (Fig.1). The water circulates between the paint spray booth and sludge pit. In the spray booth, the paint over- spray is forced into the underlying circulating channel by drawn-down air. The circulating water carries the paint solids to the sludge pit, where they are separated from water by chemical floatation.

The continuous flow tank has geometrical dimensions proportional to those in the plant, and all tests performed in the tank are designed based on the similar hydraulic retention time (HRT) (6-8 minutes). The facility consists of a treatment tank (simulating the sludge pit in the plant) and a mixing tank. The water used in the experiments was prepared from a concentrate (Q_1), which was fed to the treatment tank (Q_3) and returned to the mixing tank (Q_2), as shown in Fig. 2. The recirculation pump keeps water circulating between the mixing tank and the treatment tank, and the feeding pump controls the paint solid loading in the treatment tank by adjusting the feeding rate of the concentrated paint sludge water.

The electrode designed for the continuous flow EKF tests is made of fine grain (0.0004") graphite. The electrode module consists of 13 graphite plates, 70 mm in width, 15 mm in height and 6 mm in thickness, as shown in Fig. 3(a). The face-to-face spacing of the graphite plates is 18.5 mm. The electrode placement in the treatment tank is shown in Fig. 3(b). The module can be removed from the tank without shutdown water circulation. The position of the electrode module in the tank is adjustable for optimum treatment effects. The water-borne base coat sludge water was prepared by spraying the paint with 80 kPa air pressure into a bucket containing tap water. The TS concentration was measured and adjusted based on the design target.

Results

Two types of EKF tests were carried out in the continuous flow tank, i.e. the close-loop test and open-loop test.

The close-loop test was performed with a constant initial suspended solid (SS) concentration. There was no additional paint solid loading during the close-loop EKF process. As a result, the SS of the paint sludge water reduced with time and reached the minimum. Figure 4 shows the results of a close loop test, presented as the SS versus treatment time. The Figure shows SS removal by EKF and chemical coagulant is comparable and much higher than the control test under the same initial SS concentration.

The open-loop EKF tests were designed to simulate the operation of paint spray booth and sludge water treatment and circulation. As shown in Fig. 2, the concentrate is fed to the mixing tank using a variable-speed tubing pump, where it is mixed with circulating water. The SS concentration of the mixed water simulates the water coming from the paint spray booth. The mixed water flows into the treatment tank to undergo EKF treatment. The treated water (the circulating water) is pumped back to complete the loop. The open-loop EKF tests included a control test, a chemical coagulant floatation test and an EKF test. A water-borne paint was used to prepare the concentrate. Three tests were carried out under the same SS loading rate and hydraulic retention time (HRT). The initial SS concentration of the water was zero (tap water). The SS of water increased with time to simulate the operation of the paint spray booth. The coagulant dosage rate was 0.21 mg/L/min in the chemical floatation test, and the applied DC current was 0.5 A in the EKF test. Fig. 5 shows the SS concentration vs. treatment time in the open-loop tests. In the control test, the SS concentration increased with time to 170 mg/L after 28 hours. In the coagulant and EKF test, on the other hand, the SS concentrations remained below 20 mg/L during the entire testing period of 48 hours. It is also noticed that the SS concentrations reached equilibrium after 32 hours in both tests.

The mass balance of the paint sludge after the chemical and EKF tests was analyzed and shown in Fig. 6. The results indicate that in the EKF test, 61.4 % of the suspended paint solids became the floated sludge, on the other hand, in the coagulant test, only 1.9 % of the total paint solids sprayed floated to the top of the sludge tank. It was also noted that the EKF test generated much less sediment than the coagulant test. The result indicated that floatation in the EKF process is generated by hydrolysis induced gases whereas coagulant floatation relies on air bubbles from flow turbulence. The flow in the lab continuous flow tank is laminar with low a Reynolds number, which explains the negligible floatation in the coagulant test.

Figure 7 presents the water quality during the EKF test. It is noted that water quality remained stable during EKF testing. It is also found that water temperature change was not significant (Fig. 7(b)), indicating the EKF test did not generate significant heating effect under the lab-scale conditions.

Summary and Conclusions

The results of the laboratory continuous flow tests on EKF have shown the benefits of EKF treatment of paint sludge water, including

1) EKF generated significantly better paint sludge flotation than that by the chemical coagulant;
2) The EKF process eliminated coagulant and other chemicals, which lead to less sludge;
3) The water quality in the EKF treatment is comparable to chemical coagulant treatment;
4) The EKF requires DC current and the power consumption depends on the water electrical conductivity
5) Heating in EKF treatment is insignificant (< 5°C)
6) Water remained neutral water pH during the EKF operation

The EKF technology has other features such as COD removal, simple implementation, retrofitting to existing sludge treatment systems, easy operation and control, as well as no adverse environmental impacts.

References

Ahmed, N. & Jameson, G. J. 1985. The effect of bubble size on the rate of flotation of fine particles. International journal of mineral processing, 14 (1985), pp195-215.

Bennett, A. J. R.; Chapman, W. R. & Dell, C. C. 1958. Studies in froth flotation of coal. Third international coal preparation congress, Brussels-Liege, June, Paper E2.

Burns, S. E.; Yiacoumi, S. & Tsouris, C. 1997. Microbubble generation for environmental and industrial separations. Separation and purification technology, 11 (1997), pp221-232.

Chen, Guohua. 2004. Electrochemical Technologies in Wastewater Treatment. Separation and Purification Technology. Vol. 38, 2004. pp11-41.

Chen, W.; Horan, N. 1998. The treatment of a high strength pulp and paper mill effluent for wastewater re-use. III. Tertiary treatment options for pulp and paper mill wastewater to achieve effluent recycle, J. Environ. Technol. Vol. 19, 1998, pp173-182.

Elmore, F. E. 1904. A process for separating certain constituents of subdivided ores and like substances, and apparatus therefore, British Patent No. 13,578.

Huang, C. J.; Liu, J. C. 1999. Precipitate flotation of fluoride-containing wastewater from a semiconductor manufacturer, Water Res. Vol. 33, 1999, pp3403-3412.

Il'in, V.I. and Sedashova, O.N. 1999. An electroflotation method and plant for removing oil products from effluents, Chem. Petrol. Eng. 35 (7–8) (1999) 480–481.

Ketkar, D. R.; Mallikarjunan, R. & Venkatachalam, S. 1991. Electroflotation of quartz fines. International journal of mineral processing. 31 (1991), pp127-138.

Lafrance, P.; Grasso, D. 1995. Trajectory modelling of non-Brownian particle flotation using an extended Derjaguin–Landau–Verwey–Overbeek approach, Environ. Sci. Technol. Vol. 29, 1995, pp1346- 1352.

Manjunath, N. T.; Mehrotra, I.; Mathur, R. P. 2000. Treatment of wastewater from slaughter-house by DAF-UASB system, Water Res. Vol. 34, 2000, pp1930-1936.

Tchobanoglous, G.; Burton, F.L.; Stensel, H.D. 2003. Wastewater Engineering Treatment and Reuse (Fourth Edition). Metcalf & Eddy, Inc.. McGraw-Hill Series in Civil and Environmental Engineering. 1819pp.

Vaughan, R. L.; Reed, B. E.; Roark, G. W.; Masciola, D. A. 2000. Pilot-scale investigation of chemical addition-dissolved air flotation for the treatment of an oily wastewater, Environ. Eng. Sci. Vol. 17, 2000, pp267-277.

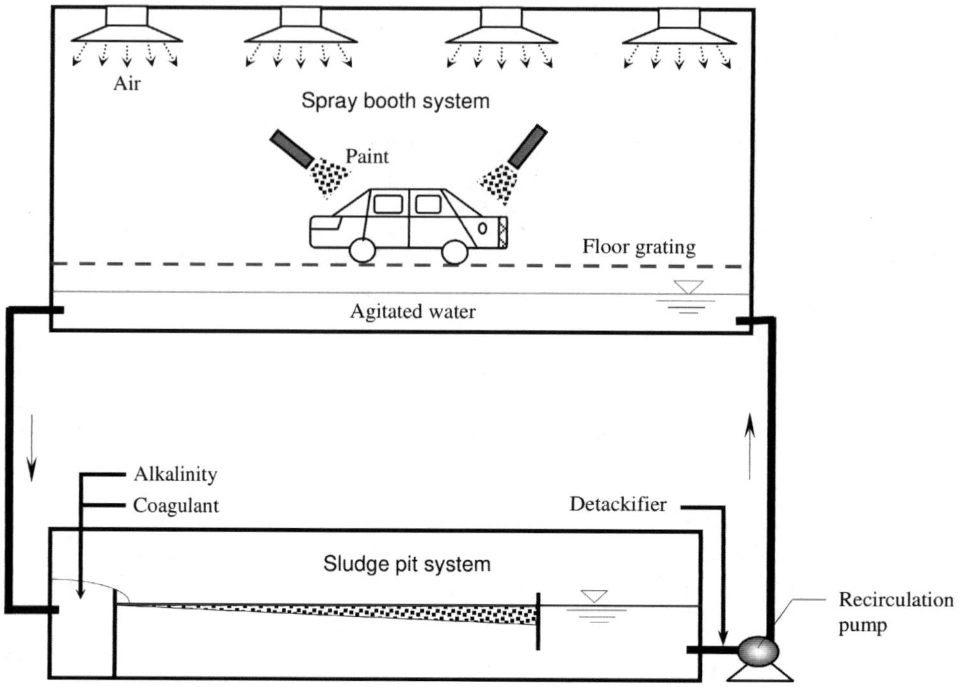

Fig. 1 Schematic of the paint spray booth and sludge pit in an automobile assembly plant.

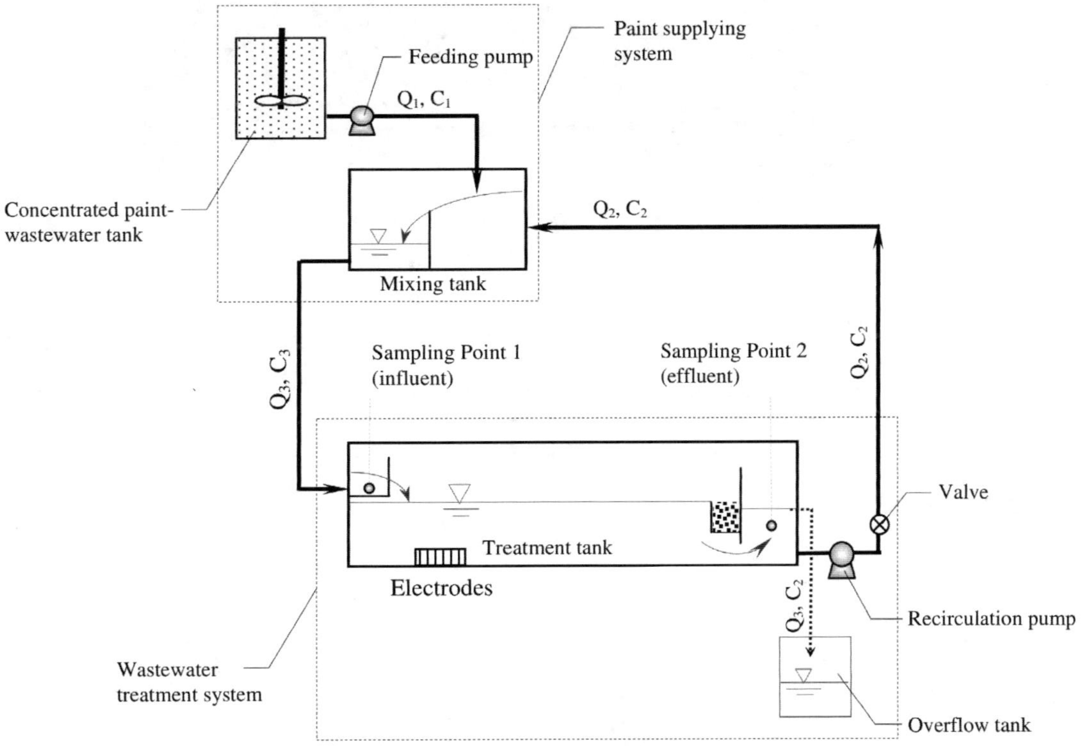

- Effective volume of treatment tank: 38.4 L;
- Capacity of recirculation pump: 0 to 14 L/min;
- Capacity of feeding pump: 0 to 67.5 mL/min
- Minimum Hydraulic retention time of treatment tank: 2.7 min;

Fig. 2 Schematic of the lab-scale continuous flow system

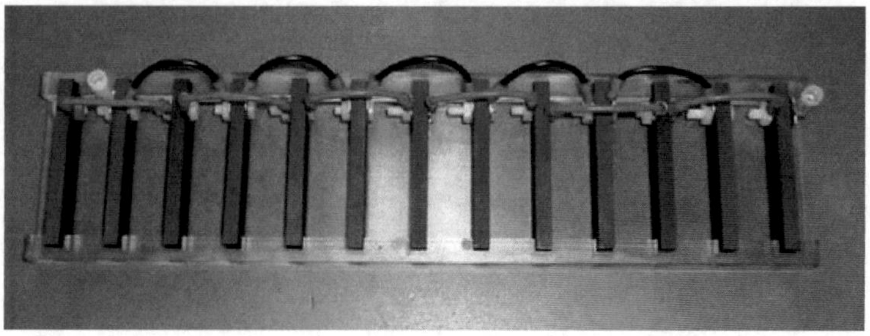

(a)

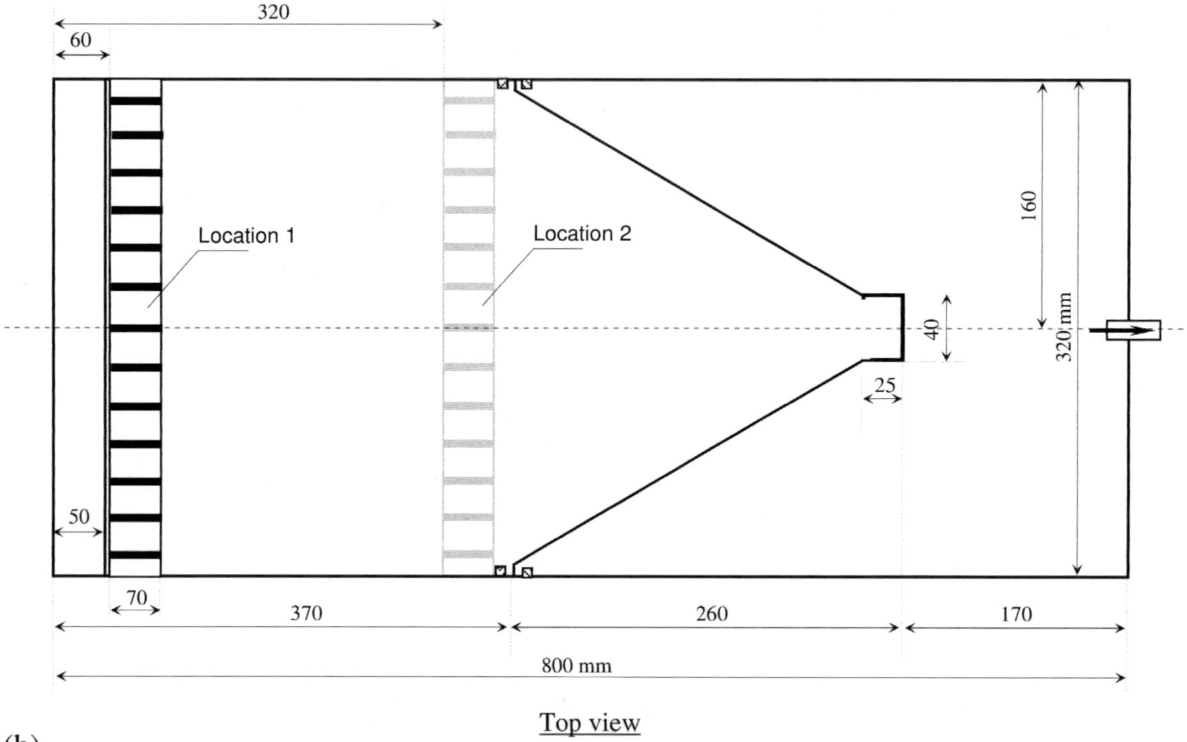

(b)

Fig. 3 Electrodes in the continuous flow tank, (a) electrode; (b) electrode location in the tank.

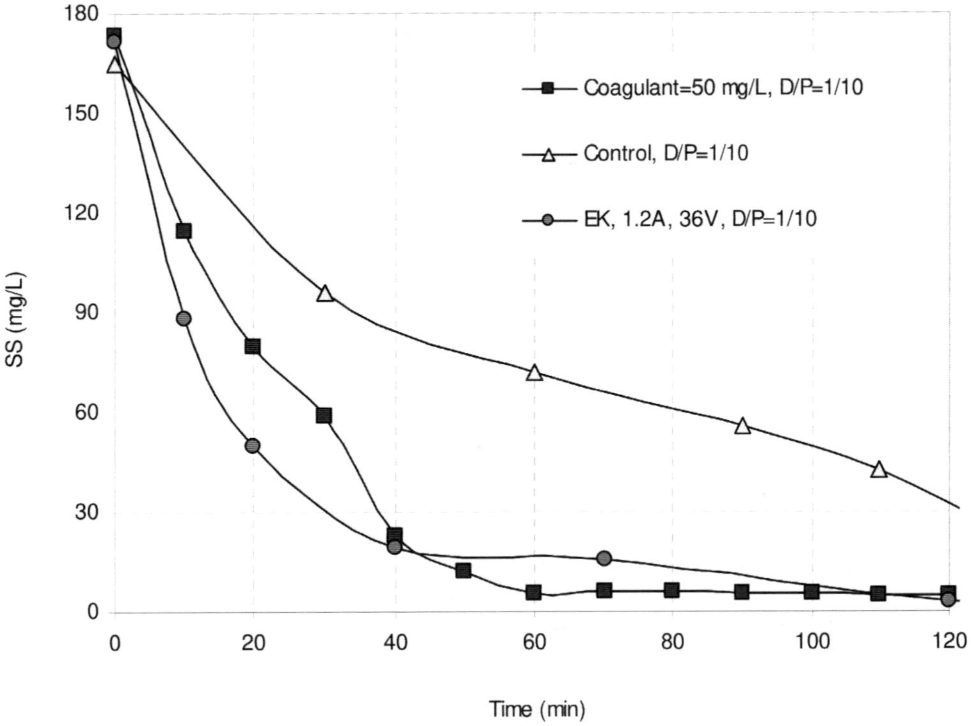

Fig. 4 Close-loop EKF test: SS concentration versus time in control, EKF and coagulation tests

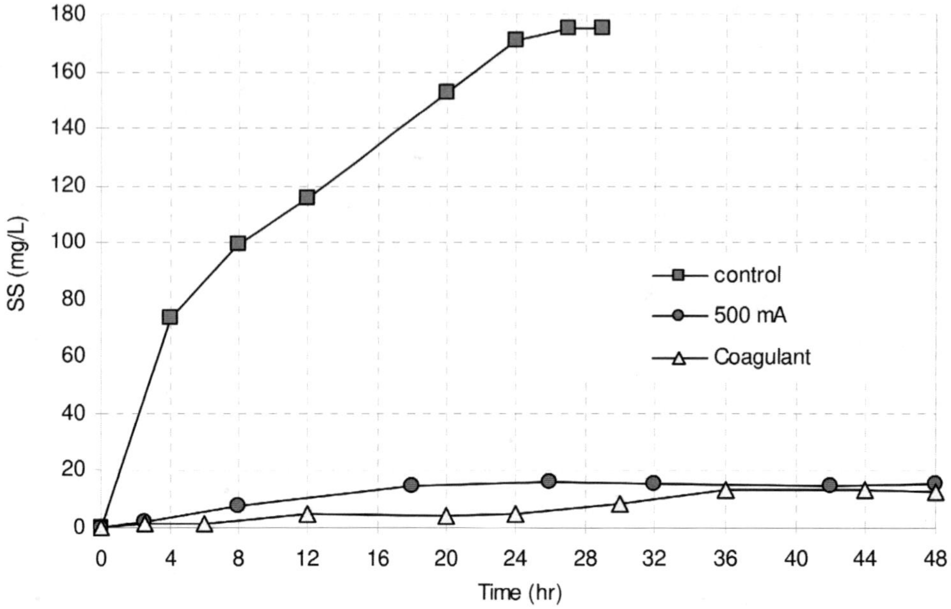

Fig. 5 Open-loop EKF test: SS concentration in control, EKF and coagulation tests

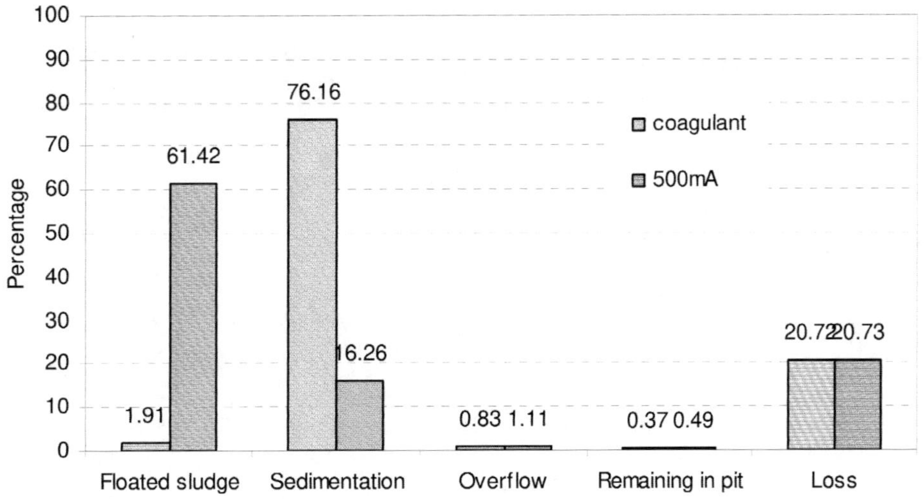

Fig. 6 Sludge mass balance after the open-loop test.

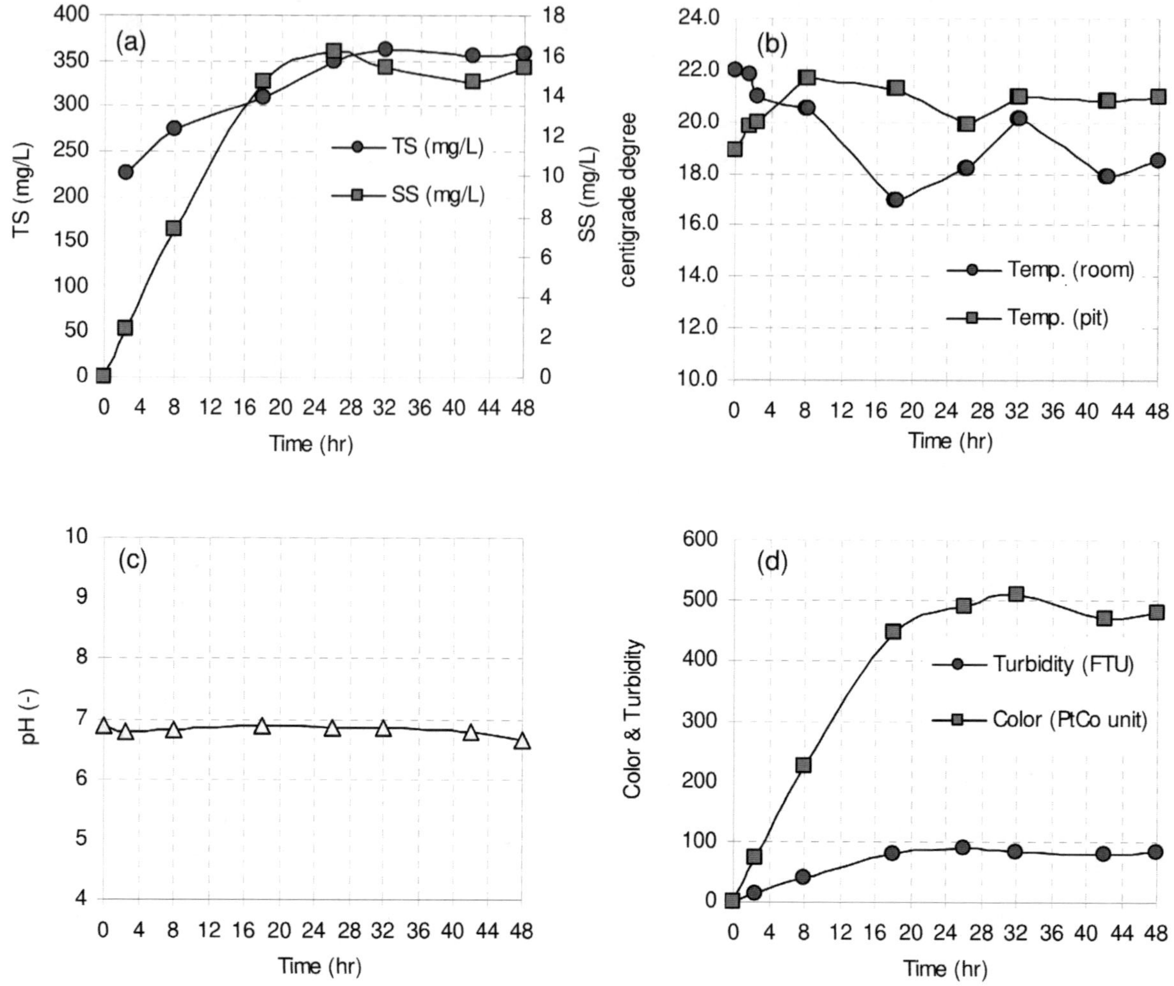

Fig. 7 Results of the open-loop EKF tests: (a) TS and SS; (b) temperature; (c) pH, (d) Color and turbidity.

11b The Effect of Flow-Field Heterogeneity on the Aggregation and Breakage of Colloidal Suspensions Under Turbulent Conditions

Lyonel Ehrl[1], Miroslav Soos[2], Hua Wu[2], and Massimo Morbidelli[1]. (1) Institute for Chemical and Bioengineering, Department of Chemistry and Applied Biosciences, ETH Zurich, Wolfgang-Pauli-Str. 10, HCI F 133, Zurich, 8093, Switzerland, (2) Institute for Chemical and Bioengineering, Department of Chemistry and Applied Biosciences, ETH Zurich, Wolfgang-Pauli-Str. 10, Zurich, 8093, Switzerland

Separation of fine particles from liquid is commonly achieved by a coagulation step, where the particles, after destabilization through addition of coagulant, are aggregated to from larger clusters. In a stirred vessel, besides Brownian aggregation, which is the controlling mechanism for sub-micron particles or clusters, shear-induced aggregation is an additional mechanism promoting cluster growth. As the forces acting on the clusters increase with their size, when the clusters become large enough, eventually, breakage sets in and ceases further growth. It is well known that the rates of both processes, aggregation and breakage, are nontrivial functions of the shear rate [1]. Therefore, process dynamics as well as steady-state characteristics will denpend strongly on the shear rate distribution inside a coagulation unit. In order to investigate the effect of flow field heterogeneity, we have investigated the aggregation of a model system consisting of surfactant-free polystyrene particles with diameter equal to 610nm in a stirred tank and three different types of Taylor-Couette device, which exhibit distinctly different shear rate distributions as calculated by computational fluid dynamics [2]. The time evolution of the cluster mass distribution, characterized by the root-mean-square radius of gyration, $\langle R_g \rangle$, and the zero-angle intensity of scattered light, $I(0)$, was measured by in-loop small-angle static light scattering. When these measured integral quantities, $\langle R_g \rangle$ and $I(0)$, were plotted as a function of the volume average shear rate, they were significantly smaller for the stirred tank than for the Taylor-Couette-type devices, proving the direct impact of the shear rate distribution, which differs substantially for the given devices [2], on aggregation and breakage processes. Additionally, it was found that the aggregate structure as analyzed by image analysis is independent of the unit type. Present results confirm our previous consideration that the steady-state aggregate size cannot be simulated just by the volume average shear rate even for dilute conditions and complete information about shear rate distribution has to be included in the analysis.

References:

[1] D.L. Marchisio, M. Soos, J. Sefcik, M. Morbidelli, Role of turbulent shear distribution in aggregation and breakage processes. AIChE J. 2006, 52, 158-173.

[2] M. Soos, H. Wu, M. Morbidelli, Taylor-Couette unit with a lobed inner cylinder cross section. AIChE J. 2007, 53, 1109-1120.

11c Development of a Rotary Microfilter to Treat Radioactive Waste

Michael Poirier and David T. Herman. Savannah River National Laboratory, Savannah River Site, Aiken, SC 29808

Development of a Rotary Microfilter to Treat Radioactive Waste

11d Waste Coal Beneficiation with High Efficient Flotation Process

Fan Shi, Department of Chemical Engineering, University of Michigan, 2300 Hayward Avenue, 3230 H. H. Dow Building, Ann Arbor, MI 48109-2136 and Shaio-hung Chiang, Chemical Engineering Department, University of Pittsburgh, 1249 Benedum Hall, Pittsburgh, PA 15261.

The vast quantity of waste coal in the United States presents us an opportunity to recover it as an alternative energy source that otherwise would not only be unused but also be a threat to the environment.

In this paper a newly developed multi-stage loop-flow flotation column (the MSTLFLO Process) has been tested as an effective means for recovering waste coal from disposal sites and to produce a clean coal product for electricity generation. Tests conducted under different operating conditions were carried out with two different waste coal samples. The results have demonstrated that the MSTLFLO process is capable of producing low ash (less than 10 % ash) clean coal product with greater than 90 % heating value recovery. The cost of producing clean coal product from waste coal using the MSTLFLO process in a 2,000 ton/day plant is estimated as low as $2.38/ton. These findings indicate that the MSTLFLO process has the potential to be developed for commercial applications in producing utility quality clean coal products from waste coal deposits at a competitive cost.

EXPERIENCE FROM WORLD'S LARGEST SEAWATER FILTRATION PLANT FOR OIL RESERVOIR INJECTION

Mohammed H. Al-Ghamdi, Seawater Injection Department, Saudi Aramco, P.O. Box 5770, Dhahran, Saudi Arabia
Nicos P. Isaias, Consulting Services Department, Saudi Aramco, P.O. Box 9981, Dhahran 31311, Saudi Arabia

ABSTRACT

Oil reservoir injection water should be non-corrosive, non-scaling, disinfected and free of particulate matter to minimize reservoir damage. Plant # 1 – a Saudi Aramco facility – with a present design capacity of 9.5 million barrels per day, is the world's largest seawater filtration treatment plant for oil reservoir injection. The design capacity is to be further increased to 14 million barrels per day in 2008. The present filtration system consists of 76 horizontal deep bed pressure filters using granular filter media. These downflow pressure filters are the largest of this type that have ever been built. Each filter is capable of handling 125,000 barrels of water per day. The filters are designed so that suspended solids in treated water should not exceed 0.2 mg/l, with no more that 10% of the solids exceeding the 2 micron size.

The objective of this paper is to give an overall overview of this unique seawater filtration facility and to share some of the challenges encountered, as well as the improvements done to this system.

KEYWORDS

Seawater, Filtration, Filter Media, Deep Bed Filtration, Reservoir Injection

1. Introduction

Plant # 1 – a Saudi Aramco facility in the Arabian Gulf is the world's largest seawater filtration treatment plant for oil reservoir injection. It has present design capacity of 9.5 million barrels per day of treated seawater which is presently being expanded to 14 million barrels per day. The treated water is injected into producing reservoirs in order to support oil production and to obtain optimum oil recovery. As a rule of thumb one barrel of oil is produced for every barrel of water injected.

To minimize any formation damage to the reservoir, the seawater is filtered at Plant # 1 prior to reservoir injection using granular deep bed filters. These downflow horizontal pressure filters are the largest of this type that have ever been built. The latest filters installed at this facility are of 11 feet (3.35 m) diameter x 40 feet (12.19 m) length.

The injection of water into the reservoir rock is necessary to maintain oil production from oil reservoirs. This method is called water-flooding. Water-flooding increases the oil production in two ways:

1. Water injected at the outer edges of the reservoir pushes the oil toward the producing wells located within the reservoir.

2. Reservoir pressure can be held constant by injecting a barrel of water for each barrel of oil produced. This helps to maintain the oil production rate because the reservoir pressure is maintained.

Seawater is the preferred source of injection water as compared to aquifer water because it is widely available and cheap. Seawater is usually relatively clean and of a general uniform composition. In addition, the seawater salinity is such that clay swelling is avoided. The general disadvantages of seawater are that it requires de-aeration to reduce corrosion related problems, disinfection for bacterial control and perhaps the greatest drawback of seawater being its scaling tendency. For the injected water to do its job it must move through the tiny pores of the reservoir rock to displace the oil. To ensure constant supply of water into the reservoir rock over a period of time it is critical that the water be non-corrosive, non-scale forming, disinfected and free of particulate matter. Any appreciable levels of corrosion products, scale, suspended solids, microorganisms, and other impurities could result in plugging of the small pores of the reservoir rock. This will cause a reduction in oil production. Therefore, the filtered water must meet stringent water quality requirements. Table 1 details the major contaminants of seawater that require treatment as well as the final treated seawater quality specifications used at Plant #1 of Saudi Aramco filtration facility.

Impurity	Symbol	Untreated Seawater	Treated Seawater
Total dissolved solids, (mg/l)	TDS	57,000	57,000
Total suspended solids (mg/l)	TSS	2 - 3	<0.2
pH, (standard units)	pH	8.1 - 8.2	7.2 - 7.3
Oxygen, (mg/l)	O_2	6.0 - 8.0	0.0
Sulfur dioxide, (mg/l)	SO_2	----	3.0 - 5.0
Residual chlorine, (mg/l)	Cl_2	----	0.0
Particle size (microns)			Less than 10% of the particles shall be larger than 2.0 microns

Table 1. Injection water specifications produced by Saudi Aramco Plant #1

Suspended solids at Plant #1 are removed by screening and media filtration. Growth of microorganisms is controlled by injecting sodium hypochlorite generated by on-site hypo-chlorinators and periodic use of non-oxidizing biocides. Oxygen is removed by nitrogen stripping and the residual oxygen is scavenged using sulfurous acid. Both nitrogen and sulfurous acid are manufactured on site. Removal of oxygen minimizes contamination of the water with corrosion products. The sulfurous acid injection causes the seawater pH to decrease slightly and this prevents calcium carbonate scale formation. However, a side effect of the sulfurous acid addition is the removal of the chlorine and therefore, periodically other biocides are injected into the system.

The objective of this paper is to give an overall overview of this unique facility, where the world's largest filtration process takes place, and to share some of the challenges and improvements done to this system.

2. Description of Saudi Aramco Plant #1

Plant #1 receives seawater from the Arabian Gulf with high salinity of approximately 57,000 mg/l in total dissolved solids (TDS). Arabian Gulf seawater contains an enormous amount of living organisms ranging from very small bacteria and plants to large fish. These species can cause fouling, plugging, and corrosion. Shellfish, barnacles, jellyfish, etc., can cause fouling of the intake structure, screens, pipes, and canal walls which can lead to the decrease in the flow capacity to the plant. Figure 1, which is the process flow diagram of Plant #1, shows that the first step is for the seawater to be disinfected by injecting sodium hypochlorite. Sodium hypochlorite is generated in-situ using the electrolytic cell process.

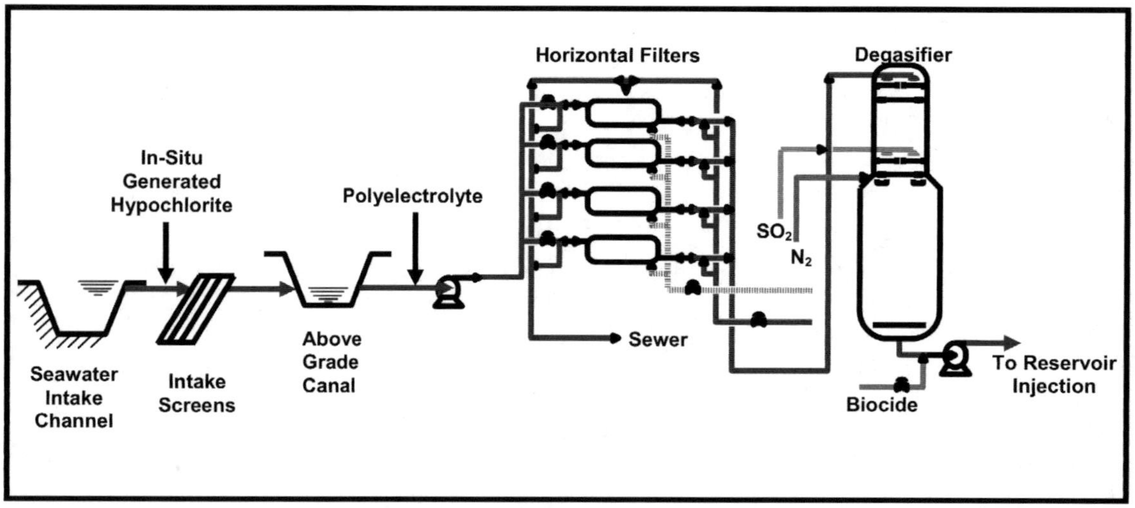

Figure 1. Saudi Aramco Plant #1 Process Flow Diagram

Sodium hypochlorite is added to the seawater at the intake structure to prevent larger fish from reaching the plant intake. All the marine organisms (large fish, plankton, and algae) are removed from the seawater by the use of mechanical screens, sodium hypochlorite, and filter media. Failure to remove the microorganisms from the system will cause problems such as:

- Fouling of the canal, piping, filters and de-aerators
- Fouling of the filters by the growth of plankton and algae in the filters
- Plugging of the injection wells downstream of the filters
- Corrosion of stainless steel piping and other system components
- Generation of hydrogen sulfide which in turn will cause corrosion and plugging problems

The seawater is then passed through bar and drum screens to remove any seaweed, marine organisms and other debris. The seawater is then pumped to an above grade canal and sent to 19 treatment modules. Each module has a design capacity of 0.5 million barrels per day and consists of four filters and a deaerator column for oxygen removal.

Filtration aid (polyelectrolyte) is added to agglomerate the seawater colloids and enhance the filtration performance. This chemical treatment is absolutely necessary to achieve the required high suspended solids removal efficiency.

The filter modules and the degasifiers are located on either site of a central canal from which they are fed. The filters are designed so that the residual suspended solids in treated water should not exceed 0.2 mg/l and no more that 10% of the suspended solids should be larger than 2 microns in size to avoid oil reservoir formation damage.

The filtrate from the horizontal filters is fed to degasifiers for de-aeration. Oxygen needs to be removed from seawater for corrosion control. The oxygen content in seawater is approximately 6 and 8 mg/l for summer and winter respectively. After oxygen removal in the degasifier by nitrogen stripping the residual oxygen level is of the order of 10 ppb. Sulfurous acid is then added to scavenge this residual oxygen and to achieve a complete oxygen removal. The residual chlorine is removed in the de-aerator by the sulfurous acid addition. Once the de-aerated water leaves the de-aerator, there is no residual biocidal capability in the treated water. It is therefore necessary to inject a second biocide into the treated water after the de-aerator. The injection of biocide is generally injected in one to three slugs per week rather than being added continuously. The treated water is pumped through pipelines as far as 300 km away for reservoir injection.

Sometimes, the desired injection water quality specifications are difficult to meet because of several challenges. The filters are operated at a high flux rate in order to cope with the high filtrate demand requirement, single chemical dosage may not be capable of consistently achieving the required specifications, and filter bed upsets occur due to inconsistent backwash practices. Furthermore, the filters are arranged in modules (each module consisting of 4 filters). A small number of large filters may be more convenient as less valves, actuators, connections etc are required, but can lead to operational control problems. The increase in flow rate when one filter is out of service (under backwash) is the factor n/n-1. Thus for a module of 4 filters the factor is 4/3, i.e. 1.33. Such an increase in flow rate can cause a spike of turbidity from the three filters still in service.

3. Plant #1 Filtration Improvements

Twenty (20) filters out of the 76 filters are new installed filters. Several modifications such as lowering of flux rate, change in filter media configuration, and change in chemical dosing were made to the new filters in order to improve their filtration efficiency.

Reduction of Flux Rate:

The flux rate (filtration velocity) is inversely proportional to the filtrate quality. To lower the flux rate, the size of filters installed was increased from 10 ft (3.05 m) diameter x 36 ft (10.97 m) tan-to-tan length to 11 ft (3.35 m) diameter x 40 ft (12.19 m) tan-to-tan length. This has caused decrease if flux rate while maintaining the same output of 125,000 barrels per filter per day. The decrease in flux rate improved the filtrate water quality as shown in figure 2.

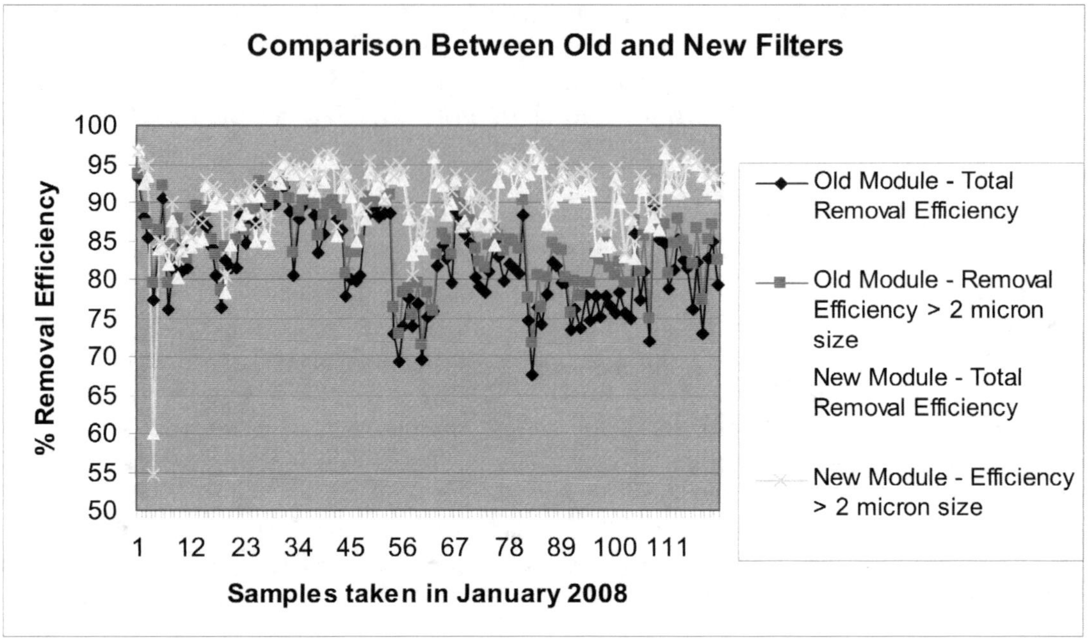

Figure 2. Comparison of filtration efficiency between old and new filters

Filter Media Configuration:

The old filter media configuration consisted of gravel support layers, coarse sand and fine sand on top. The new filter media configuration consists of fine sand and anthracite on top. The gravel support layers were removed and this has enabled the installation of deeper layers of fine sand and anthracite. The removal of the support layers was made possible by the installation of special wire wound laterals whose nozzles cannot be blocked by the sand medium. Anthracite was installed to enable the use of ferric sulfate $Fe_2(SO_4)_3$ coagulant. Special care was taken in matching the filter media sand and anthracite in relation to their size and specific gravity to avoid intermixing and interpenetration. Filter media of narrow size distribution were used (uniformity coefficients of <1.3 and <1.5 for sand and anthracite respectively) to sustain long filter runs, avoid media interpenetration, be able to backwash the filter bed with the minimum of free board while avoiding media loses.

Chemical Dosing:

To further improve the filtration efficiency of the new filters apart from the flocculent addition it was also decided to add ferric sulfate coagulant. This change is presently in progress. Since ferric sulfate is anticipated to produce a voluminous floc anthracite layer has been installed on top of the sand to retain the voluminous floc and to achieve acceptable filter runs between back washings. Ferric sulfate coagulant was chosen instead of ferric chloride because ferric sulfate is expected to have a better performance at the natural pH of seawater of pH 8.1 - 8.2 than ferric chloride. The isoelectric point (zero point charge) of ferric sulfate floc is around pH 8.0 – 8.5 whereas the isoelectric point of ferric chloride floc is around pH 6.0.

Modification of Feed Water Distribution Showers:
Initially the feed to the horizontal filter was taking place through an assembly of showers. The nozzles of these showers were bigger in size than the size of the filter media, thus filter media were getting lost during backwashing upsets. This shower feed system has been replaced with a lateral type of assembly with opening sizes smaller than the media size. The new feed system helped to make the distribution of the feed water to the filter more uniform and have eliminated filter media loss during backwash upsets.

Siphon Arrangement (to enhance air scouring efficiency and water backwash):
The use of a coagulant such as ferric sulfate can result in the build up of a surface "cake" on the anthracite layer. This "cake" needs to be broken during air scouring and flushed out of the filter during water backwash. The filter is drained to about 3 inches above the filter media surface prior to commencing the air scour. This is very important in achieving the braking of the "cake" since most of the agitation during air scouring should take place where the "cake" is and not above or below the "cake" location. Failure to drain the water to the correct level prior to the start of the air scouring will result in insufficient breaking of the surface "cake" which can lead to mud-ball formation. The draining of the filter to the correct level prior to air scouring is accomplished by the installation of a siphon system instead of having to rely on drainage time which is a function of the filter's degree of blockage.

4. Conclusion

Plant # 1 is unique in terms of size and process. Treating large volumes of sea water is very challenging and meeting the water quality specifications for oil reservoir injection is even more challenging. Many design improvements such as filters' size, filtration media, filtration aids (flocculants and coagulants) feed water distribution lateral assembly, and siphon arrangement for filter water drainage prior to air scouring have led to significantly increasing the filtration efficiency.

5. References

1. Nicos P. Isaias, Ahmed S. Al-Rammah, Saudi Aramco Engineering Report (SAER #5606), March 1998

2. Degrémont, Water Treatment Handbook, Chapter 9, Filtration Through a Filter-Bed, 5th English ed; Published by Halsted Press, 1979, pp254-274.

3. Quaye B. A., Isaias N. P., "Contact Flocculation-Filtration of Low Turbidity, Highly Colored Acid Moorland Water" Journal of the Institution of Water Engineers and Scientists Vol. 39 N0. 4 August 1985 pp. 325-340.

4. Purchas, D. B., Wakeman, R.J., Solid/Liquid Separation Equipment Scale-Up, Uplands Press, London, 2nd Edition, 1986: pp. 497-508.

5. Saudi Aramco Materials System Specification (32-SAMSS-034) "Granular Filtering Materials"

MODEL OF MICROFIBER COALESCENCE FILTERS AUGMENTED WITH NANOFIBERS

G.G. Chase* and S. Dharmanolla
Department of Chemical and Biomolecular Engineering
The University of Akron

ABSTRACT

The effect of adding nanofibers to a non-woven microfiber filter media is determined theoretically. The filter performance is characterized by a Relative Quality Factor and the amount of nanofiber added to the filter is measured as the ratio of surface area of nanofibers per surface area of microfibers. Calculations show the Relative Quality Factor passes rises rapidly with addition fo small amounts of nanofiber, reaches a maximum, and gradually decreases as more nanofibers are added to the medium. This shows that the amount of nanofibers mixed into the microfiber can be selected to optimize filter performance.

* Corresponding author: PHONE 330.972.7943, EMAIL gchase@uakron.edu

INTRODUCTION

A model of the performance of mixed micro and nanofiber media in coalescence filter media is presented in this paper. Similar models may be applied for filtration of solid particles. Typical coalescence filters are vacuum formed using glass fibers with average diameters of about 2 to 5 microns. The glass fibers are held together with a binder material. The performance of the filter is measured by its capture efficiency and pressure drop. The capture efficiency and pressure drop are combined together to define the quality factor (Brown, 1993). The greater the quality factor the better the filter performance.

Coalescing filter media augmented with nanofibers having diameters in the range of 100 to 200 nm are known to improve filter performance. The presence of the nanofiber increases the surface area available for capture of the droplets but because the diameters are so small the gas phase "slips" around the fibers and the corresponding increase in pressure drop is not as large as when micron fibers are used. The net effect is an increased quality factor with the use of nanofibers.

This paper describes a theoretical model to determine the amount of nanofiber added to the microfiber filter media to optimizes the filter performance.

THEORY DEVELOPMENT

A volume averaged continuum theory is used to model the process (Chase, 2002) To simplify the model equations the following assumptions are made:

1. Media fibers are rigid and stationary.
2. The media is incompressible.
3. The process is isothermal.
4. The flow is constant and unidirectional in the z direction (see Figure 1).
5. The liquid droplets are captured by the fibers. Capture of droplets by other droplets on the fibers or by binder are neglected.

6. The permeability of media is uniform and isotropic.

7. The porosity and saturation of the medium are uniform and constant.

8. The coalesced droplets do not re-entrain into the air stream.

9. The concentration of liquid droplets in the inlet stream is low enough to assume the bulk density of inlet air stream is equal to density of air at that temperature.

At steady state the amount of liquid holdup (saturation) in the medium and the pressure drop are steady. The mass rate of droplets in the inlet stream that are captured in the filter is equal to the rate of liquid that drains from the filter medium.

The conservation of mass of each phase (gas phase and liquid captured on the fibers), the conservation of species (air, liquid droplets), and momentum balance of the gas phase are the starting equations for the model. The mass and species balances in the model are used to determine the outlet droplet concentration. The gas phase momentum balance is used to determine the pressure drop across the media. At steady state the rate of droplet capture by the fibers must equal the rate of liquid drainage from the medium. Hence, a model applying the mechanisms of droplet capture, along with the assumed average saturation value, is sufficient for closure of gas phase equations for the purpose of determining whether there is an optimum amount of nanofibers.

The full set of conservation equations are too complex to model directly. The above assumptions are applied to simplify the volume averaged continuum equations. The simplified gas phase equations are:

Oil Species Balance

$$\frac{\partial\left((1-S)\varepsilon w_0{}^G \rho^G v_z^G\right)}{\partial z} + I_0^G = 0 \qquad [1]$$

Momentum Balance (z component)

$$(1-S)\varepsilon\frac{\partial P}{\partial z} + F_Z{}^G = 0 \qquad [2]$$

The species balance is solved by introducing a constitutive equation for the mass transfer term $I_0^{\ G}$. The momentum balance is solved by introducing a constitutive relation for the drag force, $F_z^{\ G}$.

CALCULATION OF OUTLET CONCENTRATION

To determine the outlet concentration Eq.[1] is rearranged to obtain

$$\frac{d}{dz}(w_0^{\ G}) = \frac{-I_0^{\ G}}{(1-S)\varepsilon\rho^G v_z^G} \qquad [3]$$

where the interphase mass transfer term is related to the filter coefficient, α, by

$$I_0^{\ G} = \alpha w_0^{\ G} \rho^G (1-S)\varepsilon v_z^G \qquad [4]$$

Substituting [9] into [8] and canceling terms gives,

$$\frac{d}{dz}(w_0^{\ G}) = -\alpha w_o^{\ G} \qquad [5]$$

On integrating equation [5] from 0 to z,

$$w_0^{\ G}(z) = w_0^{\ G}(0)\exp(-\alpha z) \qquad [6]$$

and the outlet concentration is

$$w_{OUT} = w_{IN} \exp(-\alpha L) \qquad [7]$$

The overall filter coefficient, α is related to the single fiber efficiency by

$$\alpha = E_i \alpha_i \qquad [8]$$

$$\alpha_i = \frac{4\varepsilon_i}{\pi d_i} \qquad [9]$$

where E_i is the capture efficiency and ε_i is the volume fraction of the filter media occupied by fibers of diameter d_i. For a mixture of fibers the filter coefficient is given by

$$\alpha = \sum_i E_i \alpha_i \qquad [10]$$

Correlations for the single fiber capture efficiencies, E_i, are available from literature (Brown, 1993). For submicron sized droplets the dominant mechanisms are direct interception and Brownian diffusion. Other mechanisms such as inertial impaction are present but are considered negligible.

CALCULATION OF PRESSURE DROP

The total force acting on all fibers, F_{tot}, is related to the pressure drop by

$$F_{tot} = A\Delta P \qquad [11]$$

We define the force per unit length of fiber as f_i. For a filter of one fiber diameter the pressure drop is related to the force per unit length by

$$A\Delta P = L_i f_i \qquad [12]$$

where L_i is the total length of fibers of diameter d_i within a differential volume element of the filter. For a medium with multiple fiber diameters,

$$A\Delta P = \sum L_i f_i \qquad [13]$$

Integration of Eq.[2], assuming uniform properties over the filter thickness, the pressure drop is related to the drag force by

$$\Delta P = \frac{F_z^G L}{(1-S)\varepsilon} \qquad [14]$$

where F_z^G is the total drag force and L is the thickness of media.

Combining equations we get the expression for the drag force to be

$$F_z^G = \frac{(1-S)\varepsilon \sum L_i f_i}{AL} \quad [15]$$

We can relate the length to the volume fraction of the fibers as

$$L_i = \frac{4AL\varepsilon_i}{\pi d_i^{\,2}} \quad [16]$$

Substituting equation [16] in [15],

$$F_z^G = \frac{4\varepsilon(1-S)}{\pi} \sum\left(\frac{\varepsilon_i}{d_i^{\,2}} f_i\right) \quad [17]$$

For a medium of just one fiber size, combining Eqs.[16] and [12] gives

$$f_i = \frac{\Delta P}{L} \frac{\pi d_i^{\,2}}{4\varepsilon_i} \quad [18]$$

If we have correlation for the pressure drop for a filter medium of fibers of one size then by inspection with Eq.[18] we can deduce an expression for f_i. The correlations for f_i should account for continuum, slip, or molecular flow phenomena depending on the diameters of the fibers. The flow regimes are distinguished by the Knudsen number.

For $Kn < 0.25$ (for continuum and slip flow of the microfibers), from Brown (1993), we get a correlation for the pressure drop to be

$$\Delta P = \frac{16\mu L v_Z^{\,G}(1+1.996Kn)\varepsilon_i}{d_i^{\,2}\left[\begin{array}{l}(-0.5\ln(\varepsilon_i) - 0.75 + \varepsilon_i - \dfrac{\varepsilon_i^{\,2}}{4} \\ +1.996Kn\left(-0.5\ln(\varepsilon_i) - 0.25 + \dfrac{\varepsilon_i^{\,2}}{4}\right)\end{array}\right]} \quad [19]$$

By inspection we derive the expression for f_i to be

$$f_i = \frac{4\mu\pi\upsilon_Z{}^G(1+1.996Kn)}{\left[\begin{array}{l}(-0.5\ln(\varepsilon_i)-0.75+\varepsilon_i-\frac{\varepsilon_i{}^2}{4} \\ +1.996Kn\left(-0.5\ln(\varepsilon_i)-0.25+\frac{\varepsilon_i{}^2}{4}\right)\end{array}\right]} \qquad [20]$$

For Kn >10 (for molecular flow with very small nanofibers) we have

$$\Delta P = \frac{4.58\mu\varepsilon_i L v_z^G}{d_i \lambda} \qquad [21]$$

hence

$$f_i = \frac{2.29\pi\mu\upsilon_Z^G}{K_n} \qquad [22]$$

There is no correlation for the flow conditions between the slip flow regime ($Kn < 0.25$) and free molecular regime ($Kn > 10$). For fibers and conditions resulting in a Knudsen number in the intermediate range $0.25 \leq Kn \leq 10$ the model uses a linear interpolation to estimate the f_i.

These equations allow us to calculate the pressure drop for a medium with multiple fiber diameters, by the equation

$$\Delta P = \frac{4L}{\pi}\sum\frac{\varepsilon_i}{d_i^2}f_i \qquad [23]$$

CALCULATION OF QUALITY FACTOR

The Quality Factor (also known as figure of merit) is given by Brown (1993) as

$$QF = \frac{-\ln\left(\frac{w_{OUT}}{w_{IN}}\right)}{\Delta P} \qquad [24]$$

which represents the ratio of the capture performance to the power required to achieve the capture performance. In general, media with higher quality factors perform better than media with lower quality factors. A filter with a larger quality factor has a greater capture efficiency than a filter with a lower quality factor when operated at the same pressure drop. Substitution of Eqs.[7] and [23] into [24] gives the working equation,

$$QF = \frac{\alpha}{\frac{4}{\pi}\Sigma(\frac{\varepsilon_i}{d_i^2} f_i)} \qquad [25]$$

for calculating the Quality Factor. Of most interest is how much change occurs in the Quality Factor due to the presence of the nanofibers. We define the Relative Quality Factor, RQF as

$$RQF = \frac{QF}{QF_{Basis}} \qquad [26]$$

where the basis quality factor is the quality factor for the filter medium without nanofibers.

MODEL RESULTS

The model is easily programmed into a spreadsheet for calculations. As an example the filter is modeled at room temperature, atmospheric pressure, with air face velocity of 2.1 m/s, and constant, uniform 10% saturation. All filters are 1 cm thick. Porosity is determined from the volumes of the fibers present in the filter.

The amount of nanofibers in the filter is calculated as the ratio of the external surface area of the nanofibers divided by the external surface area of the microfibers in the filter medium. An area ratio of zero means no nanofibers are present in the filter.

Figure 2 shows the RQF curves for three droplet sizes for 3 micron diameter microfibers and 300 nm diameter nanofibers. The RQF increases rapidly with small additions of nanofiber, reaches a maximum at an area ratio of about 1 to 3, then slowly declines. The rapid increase in RQF is due to the significant increase in surface area available for capture of particles. However, as more nanofibers are added to the filter the pressure drop due to the added fibers increases faster than the corresponding increase in particle capture and the RQF passes through a maximum and then declines. Figure 2 also shows that the RQF is sensitive to the size of the particles being captured.

Figure 3 is similar to Figure 2, but in Figure 3 three sizes of microfibers are compared. The 5 micron fiber curve shows the largest RQF because these fibers have the largest difference in size between the micro and nanofibers. If the magnitudes of the quality factors were compared instead of the relative quality factors the 1 micron fibers have the highest quality factors. This plot shows the filters made of larger microfibers will show the greatest increase in RQF. These plots show the same trend of a rapid rise in RQF with a maximum occurring with area ratios of about 1 to 3.

Figure 4 shows a similar plot for different sizes of nanofibers. For the 150 nm droplet size the 100 nm fibers have the highest RQF of those compared. Maximums in the RQF curves occur in the range of about 1 to 3 area ratios.

The plots in Figures 2 through 4 show the RQF has a maximum relative to the amount of nanofibers added to the filter medium. The RQF also varies with fibers sizes and particle sizes. These variations suggest that it may be possible to optimize the filter design for the capture of particular drop sizes.

The improvement in filter performance may not be obvious from the Quality Factor calculations. Comparisons of filter performances in Figures 5 and 6 show how the filter properties change with addition of nanofibers. Both figures have a common base case of a 6 cm diameter filter 1 cm thick made of 3 micron fibers occupying 28.3% of the filter volume; the air is at room temperature, atmospheric pressure, contains 150 nm droplets, and has at a face velocity of 2.1 m/s (flow rate of 0.00603 m3/s).

In Figure 5, to maintain a constant pressure drop (21.4 kPa) across the filter, the filter thickness is reduced as the area ratio of nanofibers to microfibers increases from 0 to 6. As nanofiber is added to the microfiber medium the capture efficiency significantly increases as shown by the rapid decrease in outlet concentration. The relative outlet concentration is the the concentration for a given condition divided by the corresponding outlet concentration for a medium with zero nanofibers. As more and more nanofibers are added to the medium the effect on the outlet concentration diminishes and the relative outlet concentration approaches a constant value. This is due to most of the particles being captured near the filter inlet surface. This result is consistent with the Quality Factor plots in Figures 1-3, a small amount of nanofiber added to the filter results in a large change in the filter thickness.

Figure 6 shows a similar plot as in Figure 5 but here the outlet concentration is held constant (at 38.2% of the inlet concentration) and the pressure drop is allowed to vary. The outlet concentration is maintained constant for the varying amounts of nanofiber by adjusting the medium thickness. Here we see that to maintain the same outlet concentration the medium thickness must be significantly reduced. The reduction in the medium thickness also contributes to a decrease in the relative pressure drop (normalized to the pressure drop required for a medium with no nanofibers).

These model results demonstrate the advantages of adding nanofibers to microfiber filter media. Validation of the model through experiments is the subject on-going work.

CONCLUSION

A model was developed to calculate the Relative Quality Factor for filter media mixed with micro and nanofibers. The model was based on assumptions such as uniform saturation and porosity to obtain a tractable set of equations. The mechanisms of drag force and droplet capture apply a linear summation of single fiber mechanisms to predict filter performance. The mechanisms account for the effects of slip and molecular flow phenomena.

The model results show a significant increase in the Relative Quality Factor with the addition of small amounts of nanofibers. The RQF plots pass through a maximum and gradually decline as greater amounts of nanofibers are added to the filters. The dependence of the RQF on the fiber amounts, fiber sizes, and drop sizes suggest that it may be possible to optimize the filter design.

ACKNOWLEDGMENTS

This work was supported by the Coalescence Filtration Nanomaterials Consortium: Ahlstrom Paper Group, Donaldson Company, Cummins Filtration, Hollingsworth and Vose, and Parker Hannifin.

REFERENCES

Chase, G.G, "Transport Phenomena in Porous Media," in Fluid Flow Handbook, J. Saleh ed., McGraw Hill, New York, 2002.

Brown, R.C., Air Filtration: An Integrated Approach to the Theory and Applications of Fibrous Filters, Pergamon Press, New York 1993.

NOTATION

A =Area

d_i = fiber diameter

E_i = Single fiber efficiency

F_z^G =Drag force

F_{tot} = Total force acting on the filter

f_i = Drag force per unit length of fiber of diameter d_i

I_0^G = mass flux of oil droplets captured from gas phase onto fibers

Kn =Knudsen number

L = Filter thickness

L_i = Total length of fibers of diameter d_i within elemental volume of the filter medium

P, ΔP = Gas phase pressure, pressure drop across filter

QF = Quality Factor

RQF = Relative quality factor (QF of the medium/QF of medium of only microfibers)

S = saturation, volume fraction of the void space occupied by liquid

v_z^G = intrinsic gas phase velocity in the z direction

w_0^G = Mass fraction of oil in gas phase

w_{in}, w_{out} = Inlet and outlet mass fractions of droplets in gas phase

α = Overall Filter Coefficient for the filter medium

α_i = Filter Coefficient of a100% efficient fiber

ε = porosity of the media

μ = viscosity of gas phase

ρ^G = gas phase density

ρ_0^G = mass concentration of liquid oil droplets in the gas phase, kg/m3

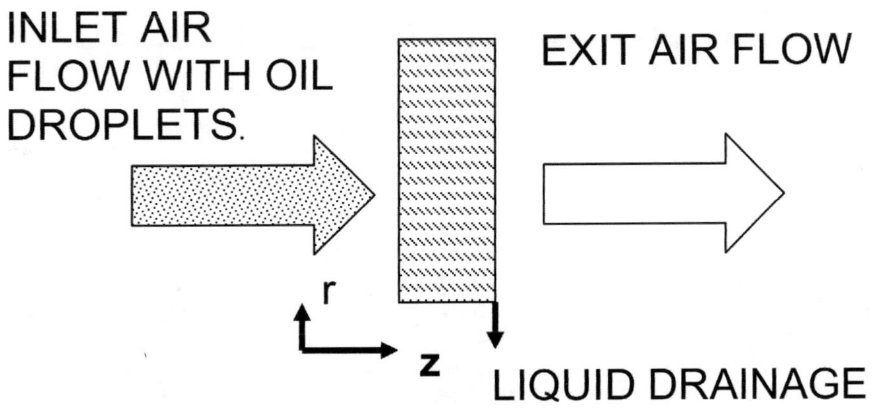

Figure.1 Diagram of a coalescing filter medium. The inlet gas stream enters the medium, droplets are captured and coalesce into large drops, and the coalesced drops drain from the medium by gravity.

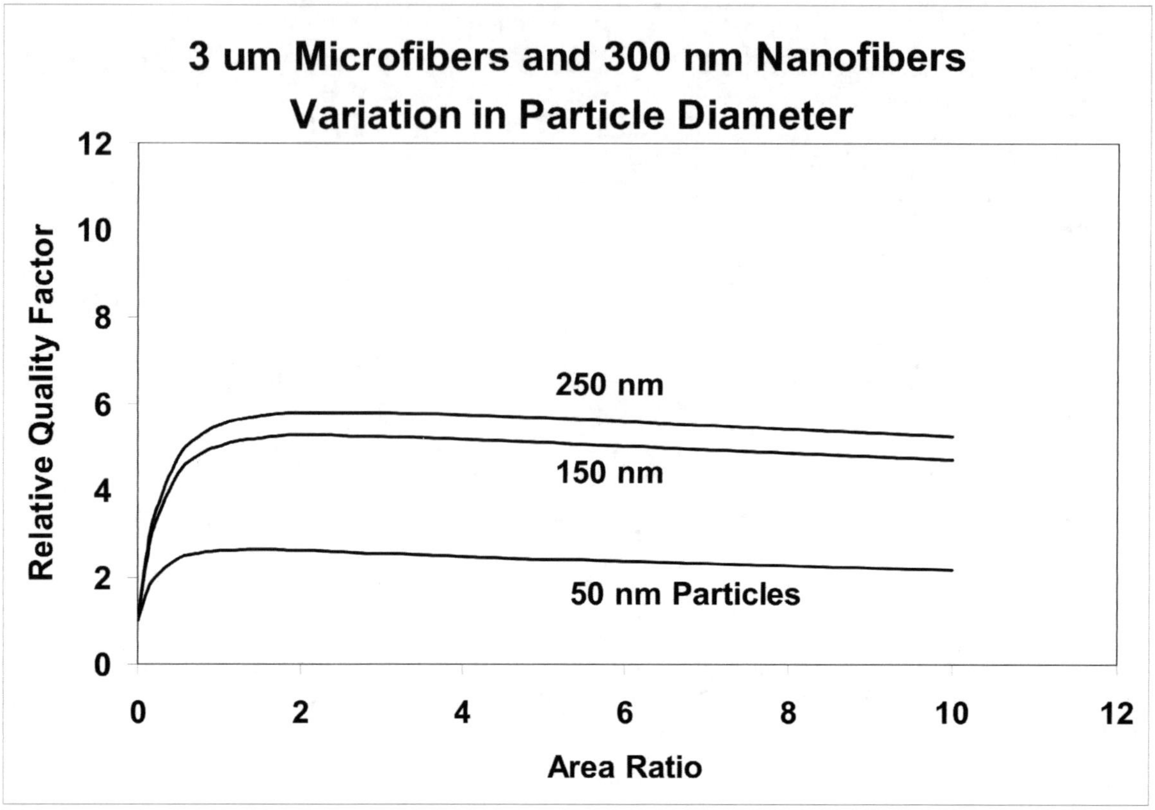

Figure 2. Relative quality factors for different sizes of droplets.

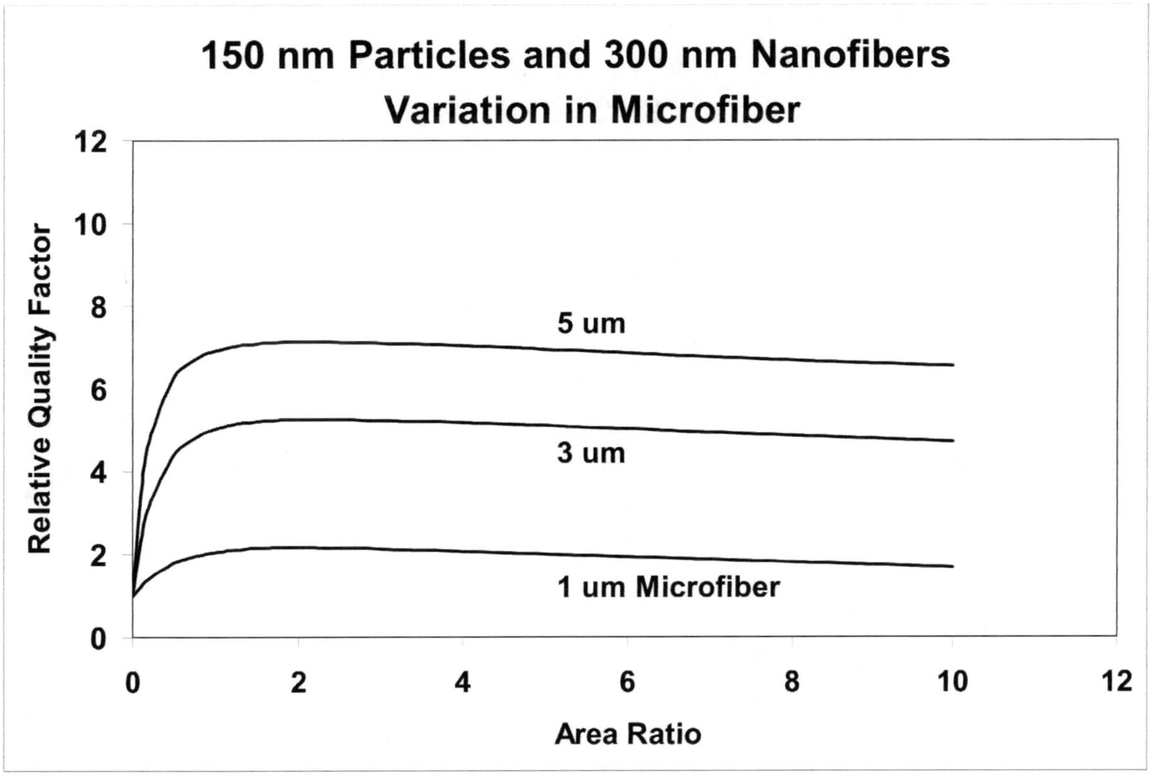

Figure 3. Relative Quality Factors for different size micro fibers.

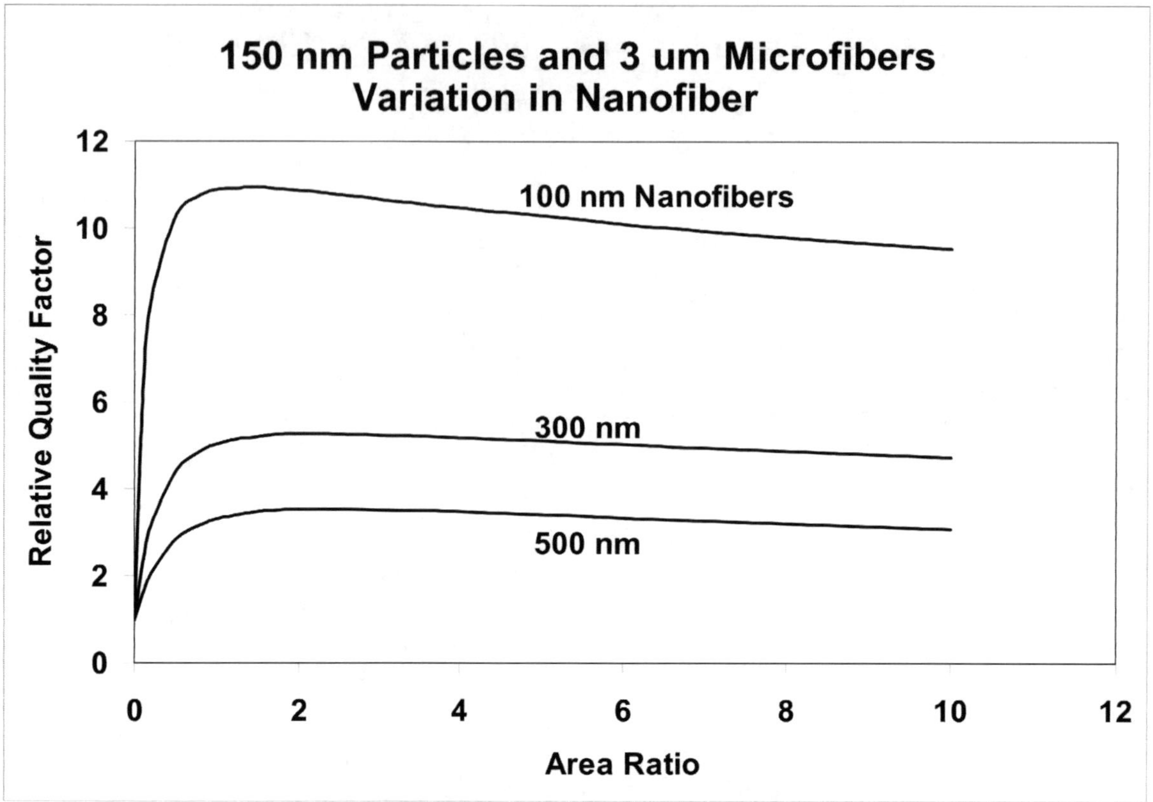

Figure 4. Relative Quality Factor for different size nanofibers.

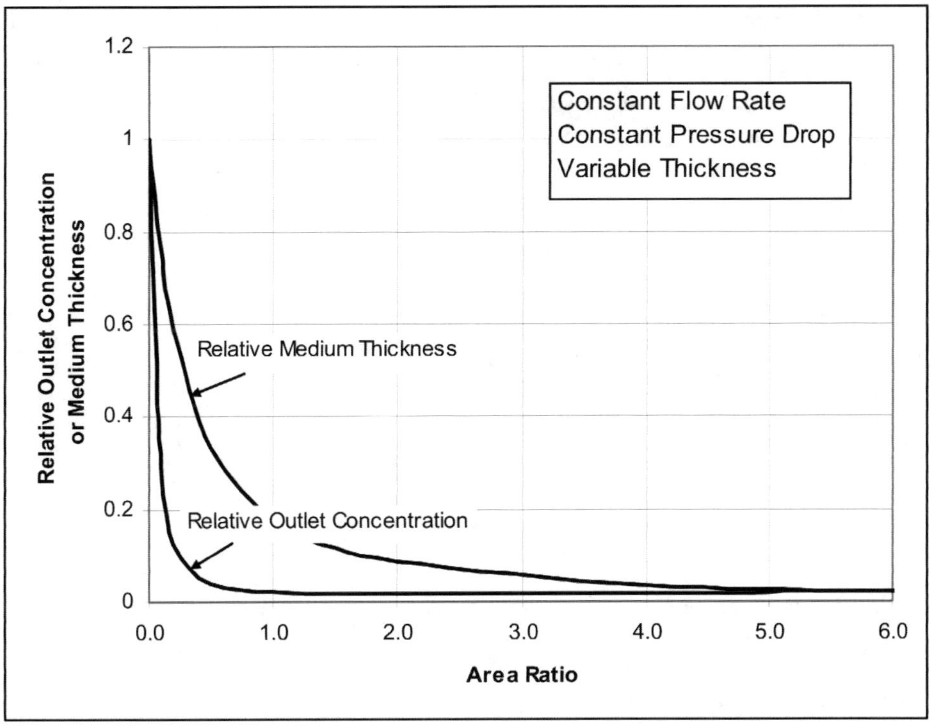

Figure 5. Comparison of relative filter outlet concentration and relative thickness with addition of nanofibers holding flow rate and pressure drop constant. The relative outlet concentration is the ratio of the outlet concentration divided by the outlet concentration for the base case of no nanofibers. The relative medium thickness is the medium thickness divided by the thickness of the medium with no nanofibers (ie, 1 cm).

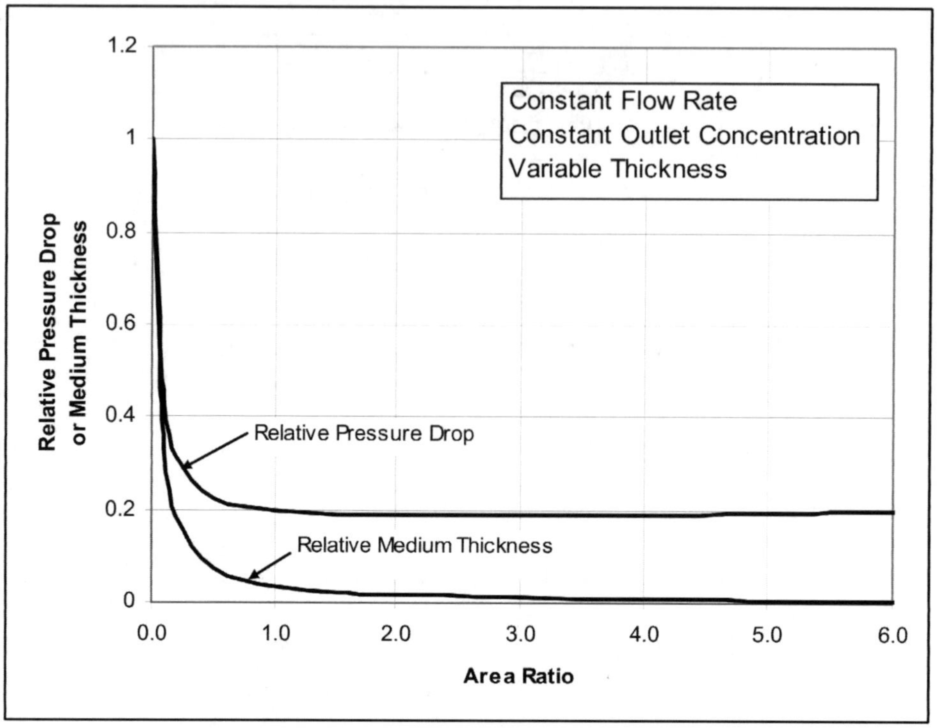

Figure 6. Comparison of relative filter pressure drop and relative thickness with addition of nanofibers holding flow rate and outlet concentration constant (at 38.2% of the inlet concentration). The relative pressure drop is the ratio of the pressure drop divided by the pressure drop for the base case of no nanofibers. The relative medium thickness is the medium thickness divided by the thickness of the medium with no nanofibers (ie, 1 cm).

30a Humidity Effects on CO_2 Removal by Adsorption In Zeolites: Equilibria and Kinetics Study

Yu Wang and M. Douglas LeVan. Department of Chemical and Biomolecular Engineering, Vanderbilt University, VU Station B 351604, 2301 Vanderbilt Place, Nashville, TN 37235-1604

The removal of water vapor and carbon dioxide from gases is important in air separation, environmental protection applications, and some purifications. Zeolites are selective adsorbents for carbon dioxide, water vapor, and other impurities and are used in gas processing broadly. However, reliable data for coadsorption of carbon dioxide and water vapor are still largely lacking, as is knowledge of how adsorbed water influences mass transfer rates of carbon dioxide. This work provides a systematic experimental study on both equilibrium and mass transfer rates for water, carbon dioxide, and their binary mixtures in zeolites 5A and 13X. A volumetric apparatus was designed and used to measure adsorption isotherms over a wide temperature range, from -45 C to 175 C. Pure CO_2 and water isotherms are described well by a multi-temperature Toth model. Binary data were measured at three different temperatures for three different water loadings. Mass transfer rates were measured by frequency response methods. A pressure-swing frequency response method (PSFR) was used to measure pure CO_2 adsorption rates on both zeolites. The mass transfer mechanism is best described using a non-isothermal nanopore diffusion model. Water mass transfer rates were measured by a concentration-swing frequency response method (CSFR) and are described well by a linear driving force model. Binary diffusion of water and CO_2 was studied by CSFR to investigate how the interactions between the molecules affect diffusion rates.

30b Experimental Study of Water and Ethanol Adsorption and Desorption in Adiabatic Fixed Bed on a-Type Zeolites

Marian Simo[1], Siddharth Sivashanmugam[1], Christopher J. Brown[2], and Vladimir Hlavacek[1]. (1) Chemical and Biological Engineering, University at Buffalo, 303 Furnas Hall, Buffalo, NY 14260, (2) Thermal Kinetics Systems, LLC, 667 Tifft Street, Buffalo, NY 14220

Kinetic study for water and ethanol adsorption-desorption process was carried out in near adiabatic fixed bed. The range of experimental conditions corresponded to those encountered in commercial fuel ethanol PSA process. Adsorption isotherms obtained from adsorbent manufacturer were compared with data from fixed bed runs. Several models for single components in inert carried were formulated. The validity of particular models was tested in numerical calculation. Mathematical model for pilot scale fixed bed assumed axially dispersed plug flow with variation of velocity, heat exchange between gas phase and solid matrix and negligible pressure drop. Experimental as well as numerical results will be presented and discussed.

30c Concentration Dependence of Diffusion of Condensable Vapors In Single Adsorbent Particles Measured Via Concentration-Swing Frequency Response

T. Grant Glover, Yu Wang, and M. Douglas LeVan. Department of Chemical and Biomolecular Engineering, Vanderbilt University, VU Station B 351604, 2301 Vanderbilt Place, Nashville, TN 37235-1604

This paper extends a concentration-swing frequency response method to examine mass transfer rates and concentration dependence for adsorption of condensable vapors in single adsorbent particles. The new approach can be used to measure equilibria and rates on new adsorbent materials, for which only a small research sample may be available. Mass transfer mechanisms can be easily identified. We consider some common systems as examples of implementation of the method. The adsorption kinetics of water and hexane in BPL activated carbon and water in silica gel are determined at several different concentrations. The mechanism that best describes the adsorption of water in BPL activated carbon is nanopore diffusion, with the diffusivity having a clear minimum near P/Po = 0.5. The concentration dependence of the diffusion data is not described well by the Darken relationship. Both nanopore diffusion and the Glueckauf linear driving force model can be used to describe the diffusion of hexane in BPL activated carbon, and the dependence of the diffusivity on concentration can be qualitatively described using the Darken relationship. The diffusion of water in silica gel cannot be described by the nanopore diffusion model and is best characterized by the linear driving force.

30d Estimation of Competitive Langmuir Isotherm Parameters. the Retention Time Method for Binary Injections

Arvind Rajendran and Wenda Chen. School of Chemical and Biomedical Engineering, Nanyang Technological University, Nanyang Drive 62, Singapore, Singapore

The estimation of adsorption isotherm parameters is vital to the design, scale-up and optimization of adsorption and chromatographic separation processes. Based on the equilibrium theory of chromatography a method to estimate competitive Langmuir isotherm parameters by measuring the retention time of the two shock fronts from pulse injections of binary mixtures has been developed. This method offers several advantages compared to the available techniques for isotherm measurements. Firstly, only the retention times of the shock fronts are required, and hence this method effectively eliminates the need for detector calibration. Secondly, by using injections of binary mixtures, it eliminates the need for the availability of pure components. Thirdly, since the experiment involves a pulse injection it reduces the amount of solute needed for isotherm determination. Finally, since the estimation procedure involves the calculation of the retention times of the shock fronts alone, i.e. without calculating the entire elution profile, the computational time required is significantly reduced allowing for rapid parameter estimations. The method has been tested on several published results and it has been found that parameters obtained from this "retention time" method are comparable to those estimated from more rigorous techniques such as the inversion method. The effectiveness of this method in rapid determination of isotherm parameters is highlighted, and its limitations with respect to aspects such as column efficiency are discussed.

30e Conformation of Cationic Polyacrylamide on Mica Studied by Single Molecule "Pulling" Technique

Yulin Deng and Brett Brotherson. School of chemical and biomolecular engineering, Georgia Institute of Technology, 500, 10th st. N.W., Atlanta, GA 30332

In this work, an improved "pulling" method using scanning probe microscopy (SPM) has been used to gain information about the conformation of an adsorbed polymer. The sequential adsorption/desorption of cationic polyacrylamide on freshly cleaved mica in deionized water was directly measured using these modified SPM experiments. Our method uses a specially etched tip which ensures only one polymer molecule is adsorbed on the cantilever tip allowing more accurate interpretation of the results. The effect of cantilever tip retraction velocity from the surface was shown to be insignificant to the measured loop length distribution. The effects of site-clocking polymer and nanoparticles on the conformation of post-adsorbed polymer were investigated. It was found that the loop length of absorbed polymer remains almost constant, but the tail length increases significantly in the presence of site blocking agents. For the first time, SPM was used to show the effects of dwell time on adsorbed polymer conformation.

Web Page: www.chbe.gatech.edu/fac_staff/faculty/deng.php

30f Extraframework Cations In Faujasite Adsorbents : Analytical Methods and General Distribution Trends

Philibert Leflaive, Tom Frising, Anne-Agathe Quoineaud, and Christophe Pichon. Institut Francais du Pétrole, BP 3, Vernaison, 69390, France

Understanding the factors that control separation or catalytic properties in zeolites requires a detailed knowledge of the structure of the material. In aluminosilicate zeolites, the presence of aluminum atoms introduces negative charges, which are compensated with nonframework cations. Faujasites are characterized by the fact that the number of potential cation sites in the structure exceeds the number of cations needed to ensure electroneutrality of the material.

Knowledge concerning the positions of the extraframework cations is critical to the understanding of the adsorption, separation, and catalytic properties of cation-exchanged zeolitic materials. The nature and positions of the cations will control the electrostatic field present within the zeolite pores, which strongly influences the adsorption and reactivity of the sorbed molecules.

However, the precise location of cations by conventional diffraction (XRD) techniques is rendered difficult because of partial occupancies affecting low symmetry sites. Moreover, if the system is complex, i.e. if there are different extraframework cations or if an adsorbed phase is present, XRD determination is no longer unambiguous.

The use of more sophisticated techniques such as anomalous X-ray diffraction, neutron diffraction and Solid State Magic Angle Spinning (MAS) and Multi Quanta MAS (MQMAS) Nuclear Magnetic Resonance may then be required to bring additional information for complete understanding.

The interest and drawbacks of such methods will be illustrated and discussed.

Protein Recovery from Enzyme-Assisted Aqueous Extraction of Soybean Oil

Kerry A. Campbell, Ramon Morales Chabrand, and Charles E. Glatz, Department of Chemical and Biological Engineering, Iowa State University

Introduction

Because of the high transportation cost of bio-based feedstocks, the transition from a petroleum-based to a bio-based economy will require the development of small-scale distributed production centers rather than large centralized facilities seen in industry today. In vegetable oil production, the use of hazardous petroleum solvents drives up capital costs, resulting in larger processing plants and, therefore, increased transportation costs. In contrast, enzyme-assisted aqueous extraction processing (EAEP) uses only water as a solvent and is potentially less capital intensive.

One EAEP process developed at the Center for Crops Utilization Research (CCUR) at Iowa State University uses extrusion to disrupt cells of soy flakes, from which oil and water soluble components are extracted with the assistance of proteases, resulting in three fractions: residual solids, skim (an aqueous fraction with emulsified oil stable toward creaming), and cream (emulsified oil) [1].Protein values, which potentially account for 65% of total EAEP revenues, are in the solid and skim fractions. The skim fraction differs from the traditional protein-containing fractions and characterization and recovery of this protein is critical to implementation of this alternative process.

We have investigated methods to recover the hydrolyzed proteins of the skim fraction of the CCUR process with an additional benefit of reduction of antinutritional factors, such as oligosaccharides and trypsin inhibitors [1]. Traditional protein concentration methods such as pH precipitation, ultrafiltration, and ion exchange chromatography are complicated not only by the increased solubility and small molecular weight of hydrolyzed polypeptides, but also by the presence of emulsified oil in the skim fraction. Results from various recovery methods will be presented.

Materials and Methods

Skim was produced following a process developed by [1]. Extruded pellets were provided by the Center for Crops Utilization and Research. Two hundred grams of extruded flake pellets were added to 2L of deionized water with 0.5% (w/w flour) endoprotease Protex 6L, (Genencor Int., Rochester, NY) in a 2L jacketed reactor (Model 4742, Chemglass Inc., Vineland, NJ) held at 50 °C by a circulating water bath and agitated at 500 rpm by a stirrer (Model BDC 3030, Caframo, Ltd., Wiarton, Ontario) with a 1-inch, 3-bladed screw impeller. Constant pH 9 was maintained by an autotitrator (Model 718 Stat Titrino, Metrohm, Ltd., Herisau, Switzerland) by adding 2M sodium hydroxide. Following extraction, the slurry was centrifuged at 3000 g, 15 min, 20 °C to remove the solids. Skim and cream fractions were separated overnight in a separatory funnel at 4 °C.

Ultrafiltration of skim was carried out with a YM3 3 kDa nominal molecular weight cut off (NMWCO) regenerated cellulose membrane, (Millipore, Inc. Billerica, MA) in an Amicon 50 ml, 43mm diameter stirred cell (also by Millipore) at 40 psi. For this experiment, skim fractions were produced under various hydrolysis conditions: 0.5% (w/w solid) Protex 7L (P7L 0.5) for lower hydrolysis, 0.5% Protex 6L (P6L 0.5) for medium hydrolysis, and 1.0% Protex 6L (P6L 1.0) for greater hydrolysis. The three treatments resulted in molecular weight profiles having 28%, 14%, and 10% of the peptides of a molecular weight greater than 10 kDa, respectively, as determined by size exclusion chromatography [1]. Fifty grams of skim were weighed into the membrane cell. Permeate was collected, weighed, and aliquots were withdrawn for saccharide analysis. Retentate composition was calculated as the difference between material content in the permeate and the content of the starting material. Moisture content was determined by weight loss after freeze drying. Protein content of the resulting solids was based on total nitrogen analyzed according to AOAC method 993.13 [2] using a RapidN III combustion analyzer (Elementar Americas, Inc. Mt Laurel, NJ) and a nitrogen to protein conversion factor of 6.25.

Protein recovery by pH precipitation was carried out by placing 50 ml skim in a beaker, adjusting to desired pH by adding 2M hydrochloric acid and agitating for one hour with a magnetic stirrer (pH varied by less than 0.1 pH unit over the course of the experiments). Skim was then transferred to pre-weighed 50 ml centrifuge tubes and centrifuged at 4500 g, 30 min, 20 °C. Supernatant was collected and dried at 130 °C in pre-weighed containers to determine solids content. Protein content was based on total nitrogen content, determined as above. Supernatant protein content was based on the difference between protein present in precipitate and protein present in the untreated skim.

Batch adsorption experiments were carried out by adding 2 to 13.7 g of Streamline Q XL (GE Healthcare, Chalfont St. Giles, United Kingdom), a strong anion-exchange EBA resin, to 50 ml skim in a beaker and agitating 40 min with a magnetic stirrer (such agitation was only used for initial screening as this resin is prone to shear damage from the stir bar). To avoid size segregation, resin aliquots had been obtained by pouring resin slurry into a sintered glass vacuum funnel, rinsing several times with deionized water, draining liquid, and transferring full-depth portions of the cake into a beaker with a spatula. After adsorption, resin was allowed to settle, and the supernatant was collected, dried, and analyzed as above for solids and protein content. Protein concentration was adjusted by diluting skim with a phosphate buffer of pH and conductivity matching that of the skim. Adsorbed protein was calculated by difference from the initial and final liquid phase protein concentrations. Resin was regenerated by adding 1M sodium chloride solution, agitating several minutes, decanting liquid, and repeating until the supernatant was clear.

Breakthrough experiments were conducted with a Streamline 25 expanded bed adsorption column from GE Healthcare using between 20 and 30 g Streamline Q XL resin. The column was equilibrated by pumping a phosphate buffer at a pH and conductivity equal to that of the skim through the column at 15 ml/min (3 cm/min), resulting in an expanded bed height of 12-16 cm. After feed loading, the column was rinsed with equilibration buffer, and then proteins were eluted with 1M sodium chloride solution. Resin was cleaned and regenerated by recycling a 1M sodium chloride, 1M sodium hydroxide solution (after discarding the first 3 column volumes of effluent) for several hours, and then rinsing for three column volumes with a 20% ethanol solution. Effluent fractions were collected into 15 ml centrifuge tubes. Because of the size of the samples collected, total nitrogen determination was not practical. Protein content was therefore determined by measuring the absorbance of samples at 215 nm with an Ultrospec 4000 UV/visible spectrophotometer (Pharmacia Biotech, now GE Healthcare). Samples were diluted with phosphate buffer to have an absorbance of less than 1 AU/cm, which also reduced sample opacity to negligible levels.

Protein fractions were characterized by size exclusion chromatography (SEC) using a 300 mm x 7.8 mm Biobasic SEC 120 column (BioRad Laboratories, Ltd.) following methods described previously [1] Raffinose and stachyose concentrations were determined by cation-exchange chromatography, also described previously [1].

Results and Discussion

Optimum selectivity between protein and non-protein components, in particular the stachyose/protein ratio, is desired. Of the hydrolysis treatments, the P6L 0.5 treatment gave the best protein-stachyose separation (Figure 1). This treatment is milder hydrolysis than the P6L 1.0 skim [1] providing a larger size difference of selective filtration. The P7L skim has still lower hydrolysis but also lower initial protein concentration that was not offset by any selectivity advantage.

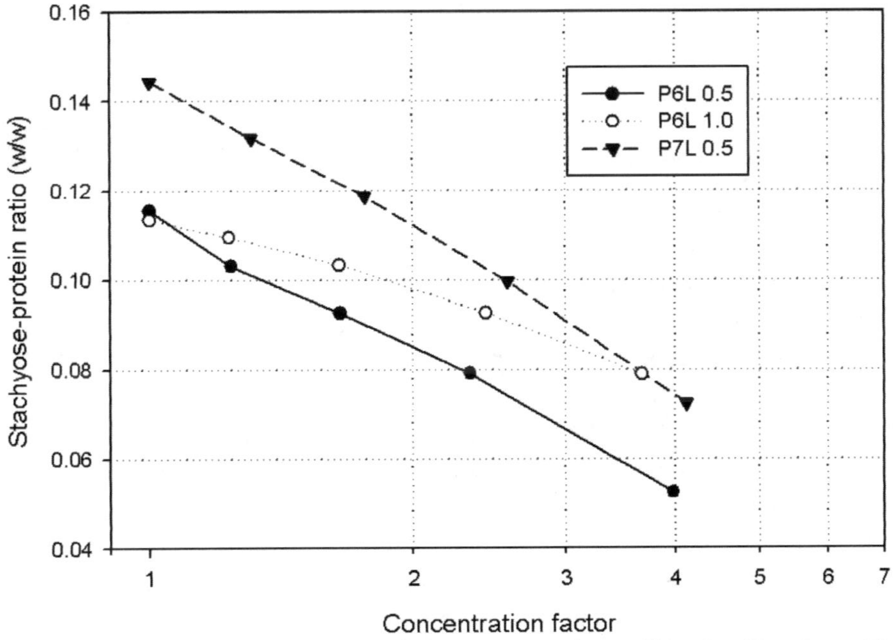

Figure 1- Stachyose-protein ratio of the retentate from skim ultrafiltration with a 3,000 Da NMWCO membrane up to a concentration factor of 4 for skims obtained with three different hydrolysis conditions.

For skim from P6L 0.5, at a concentration factor of four, the dry-basis protein concentration of the retentate had increased from 55% to 70%, while stachyose decreased from 6% to about 2%, shown in Figure 2. Along those lines, ultrafiltration also achieved significant dewatering with total solids of the retentate increasing from 5% to 14%. The major drawback of this treatment is the loss of valuable protein, which for P6L 0.5 skim was 26%. Experiments with a 1,000 Da NMWCO resulted in decreased protein losses, but the stachyose retention increased as well, resulting in no substantial protein-stachyose separation (data not shown).

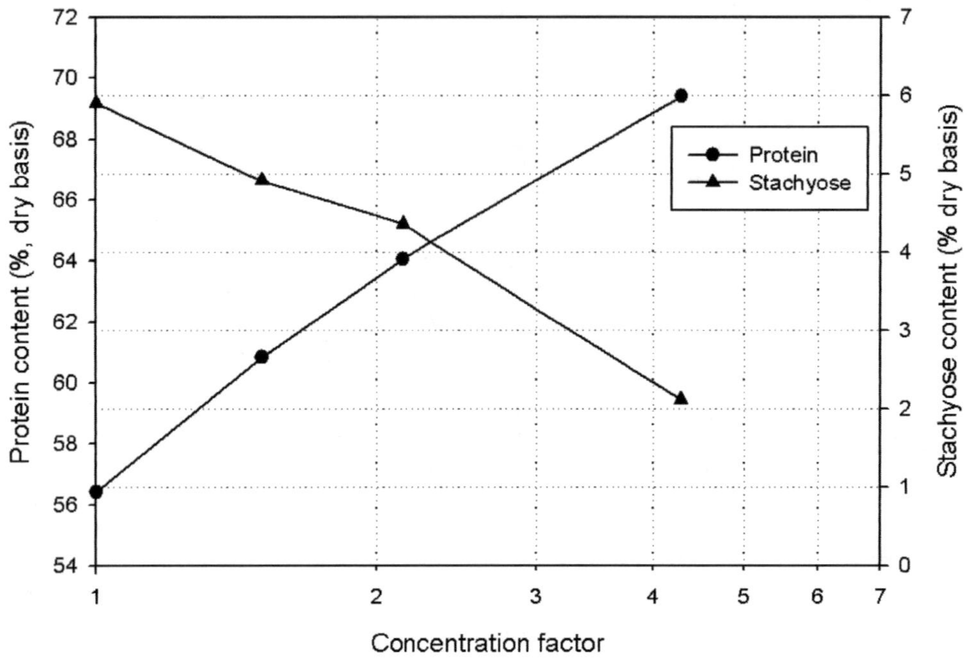

Figure 2- Dry-basis protein and stachyose composition of skim retentate from P6L 0.5 extraction with a 3,000 Da NMWCO membrane.

The solubility profile of skim over a range of pH for skim containing 26 mg/ml and 50 mg/ml protein is shown in Figure 3. Unhydrolyzed soy protein aqueous extracts from soy flours and extrudates have very low solubility (less than 10%) at or near the pI of soy proteins, between pH 4 and 5 [3-5]. The hydrolyzed skim proteins in this case appear to have a minimum solubility at the same pH; however, the percentage of the protein that remains soluble has been dramatically increased, with 70% of the skim proteins remaining in solution at pH 4. This is high compared to other studies of hydrolyzed soy protein. The minimum solubility of a 10 mg/ml extruded soy protein aqueous extract with a degree of hydrolysis of 4% was reported to be about 30% [5]. The reason for the increased solubility appears to be entirely attributable to the hydrolysis. Molecular weight profiles, as determined by SEC (Figure 4), show that no polypeptide with a molecular weight greater than 30 kDa remains in solution at a pH of 4. The large peak to the left on Figures 4 A and B represents large, presumably unhydrolyzed, proteins of molecular weights greater than 30 kDA [1]. While this demonstrates that pH precipitation would be a beneficial first step for a protein recovery process mirroring traditional SPI or SPC production, recovery of the remaining 70% is still necessary for economic success. Furthermore, the high solubility at low pH of the hydrolyzed fraction would be a potentially valuable functional property setting this apart from traditional SPI.

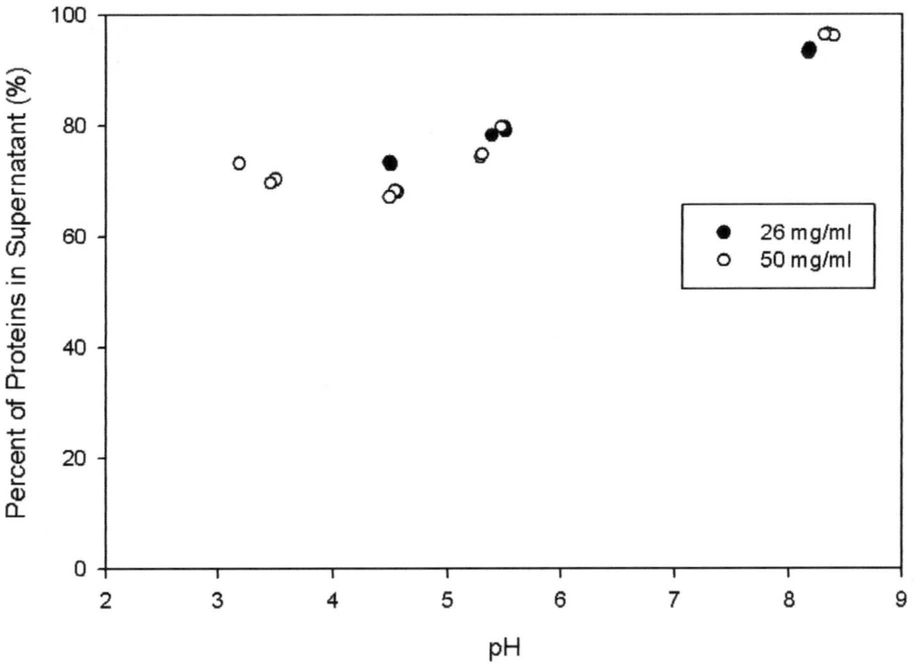

Figure 3- pH solubility profile of skim proteins for trials using 26 mg/ml and 50 mg/ml protein in skim.

It would be logical to suspect that many polypeptides resulting from hydrolysis would have surface charge properties similar to the intact proteins. Since soy proteins have an acidic isoelectric point, and therefore have a net negative charge at high pH, anion exchange chromatography was investigated as a means of recovering a large portion of the hydrolyzed proteins. For batch adsorption experiments, very good repeatability was achieved where the same mass concentration of resin was used, while the specific protein adsorption increased as the resin concentration decreased, as shown in Figure 5. This can be attributed to different binding constants of different protein fractions in the skim. A linear adsorption isotherm model with two different adsorption coefficients fits the data quite well in the linear region of the adsorption isotherms. With low resin concentrations, specific protein adsorption as high as 100 mg/ml was achieved.

The idea of protein fractions of varying adsorption coefficients is supported by chromatography breakthrough curves, shown in Figure 6. Initial experiments used the column in reverse flow (packed bed) mode, and the breakthrough appears very similar to expanded bed mode. During the protein elution, however, severe channeling was observed, apparently caused by resins forming a compact plug during application. Operating the column in expanded bed mode during application and elution seemed to eliminate this problem. Although some early breakthrough is evident, the tracer profile shows that column flow approaches plug flow. Hence, deviation from this behavior for protein breakthrough indicates either selective binding of only portions of the peptides or slow-binding kinetics. The latter should not be the case given the size of the proteins, the resin, and the flow rate. Overall protein balances have closures of around 95%, with

19% and 17% of total applied protein remaining bound after rinsing for 2 and 4.5 bed volumes applied, respectively. Based on this, the specific protein adsorption after rinsing was 24 mg/g and 46 mg/g for the two respective trials, much lower than would be expected based on batch adsorption data. Protein and resin conditions used in these two EBA trials correspond with those used in two of the batch trials, which resulted in 44 mg/g and 74 mg/g adsorbed protein, respectively. In other words, had these EBA experiments been conducted as batch adsorption experiments with the same masses of skim and resin, the specific protein adsorption would have been almost double the mass adsorbed under plug flow operation.

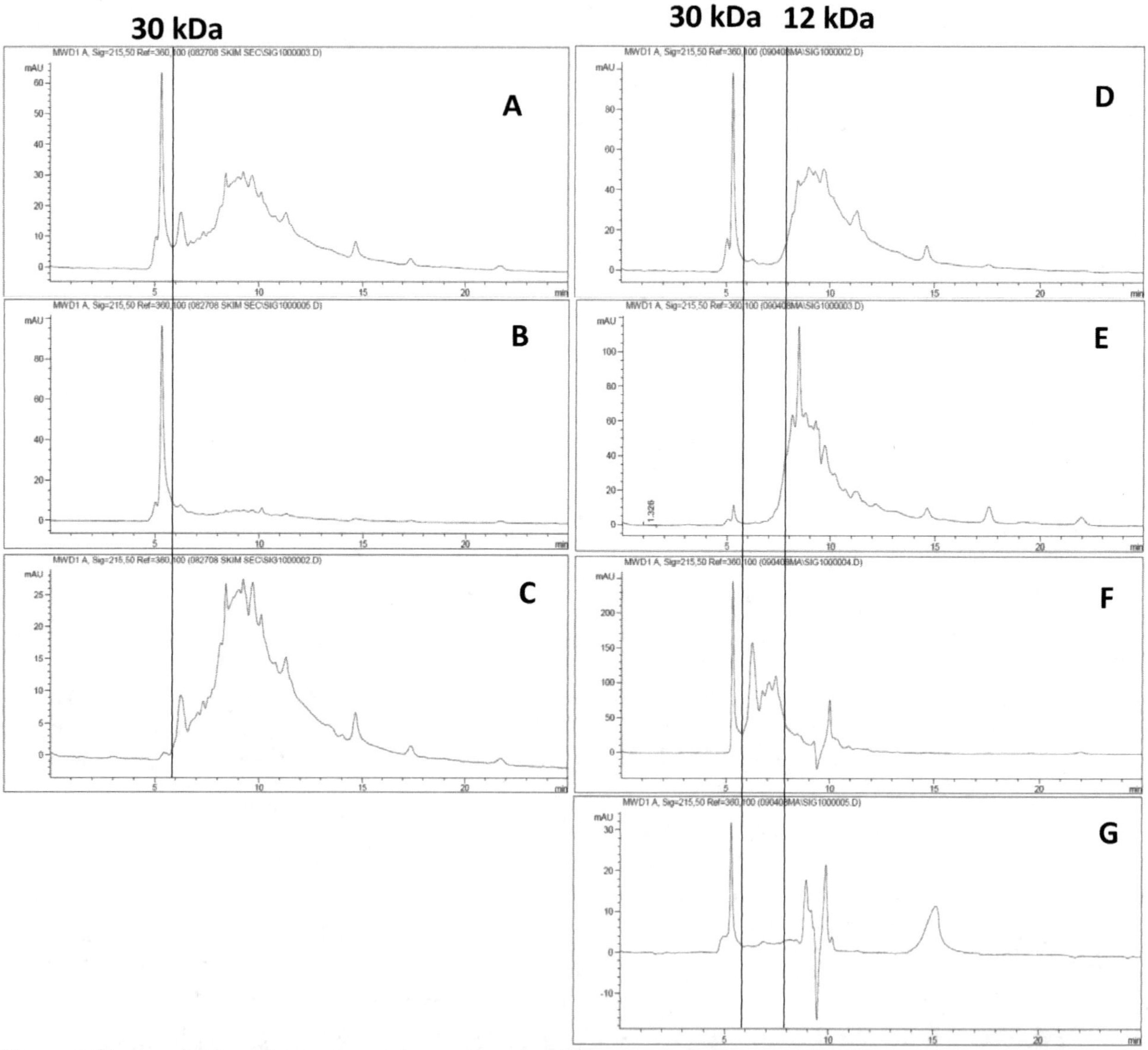

Figure 4-Size exclusion chromatography profiles for (A) untreated skim at pH 9, (B) proteins precipitated at pH 4.0, (C) supernatant proteins at pH 4, (D) EBA column breakthrough at applied volume of 4 column volumes, (E) proteins rinsed from column using equilibration buffer (40mM phosphate buffer, pH9), (F) proteins eluted from column using 1M NaCl, and (G) proteins eluted from column during cleaning cycle using 1M NaOH with 1M NaCl. Elution times corresponding to molecules of 30 kDa and 12 kDa molecules are indicated. Sample dilution and injection volumes were all such that approximately 10 μg of protein were injected for each chromatogram.

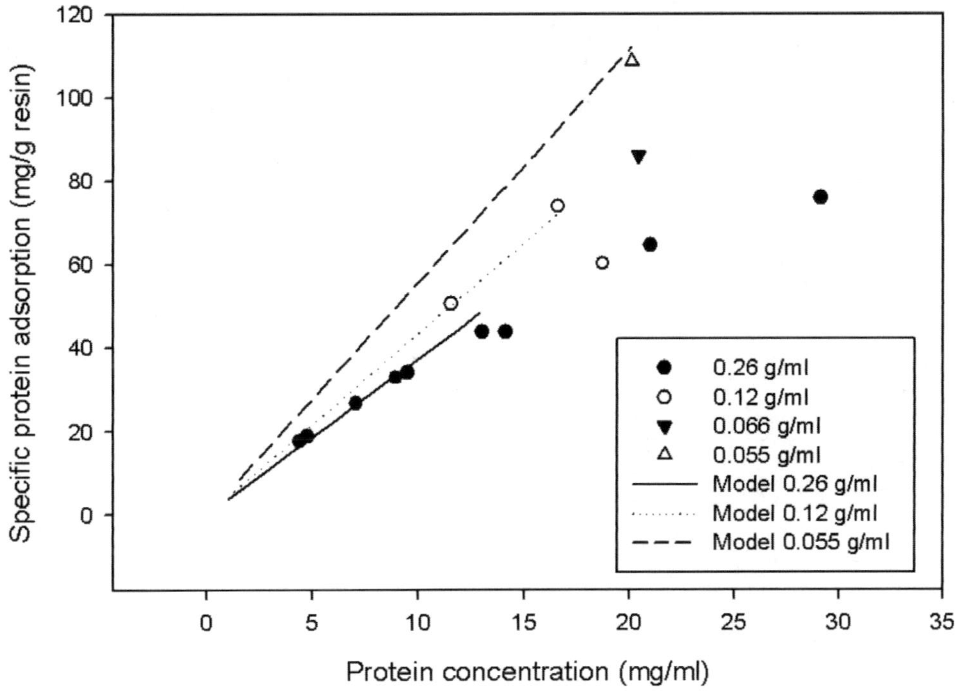

Figure 5- Adsorption isotherm for Streamline Q XL resin with skim using different resin concentrations for batch adsorption experiments with preliminary model fit.

SEC profiles of selected EBA eluent fractions are shown in Figure 4 D through G. The profile for column effluent collected after 4 column volumes had been applied appears to be practically identical to the profile of the initial skim (Figure 4 A), except for polypeptides between 12 and 30 kDa, which bind strongly to the resin with no breakthrough. Even though the >30 kDa proteins appear in the breakthrough chromatograph, they do not appear in the rinse fraction (Figure 4 E), indicating strong adsorption. The presence of these proteins in the breakthrough indicates that either adsorption sites specific to these have become saturated or their binding equilibrium coefficients are less than the 12-30 kDa fraction, or both. Another fraction of strongly binding proteins is seen between 3.9 kDa and 7.4 kDa, in Figures 4 F and G.

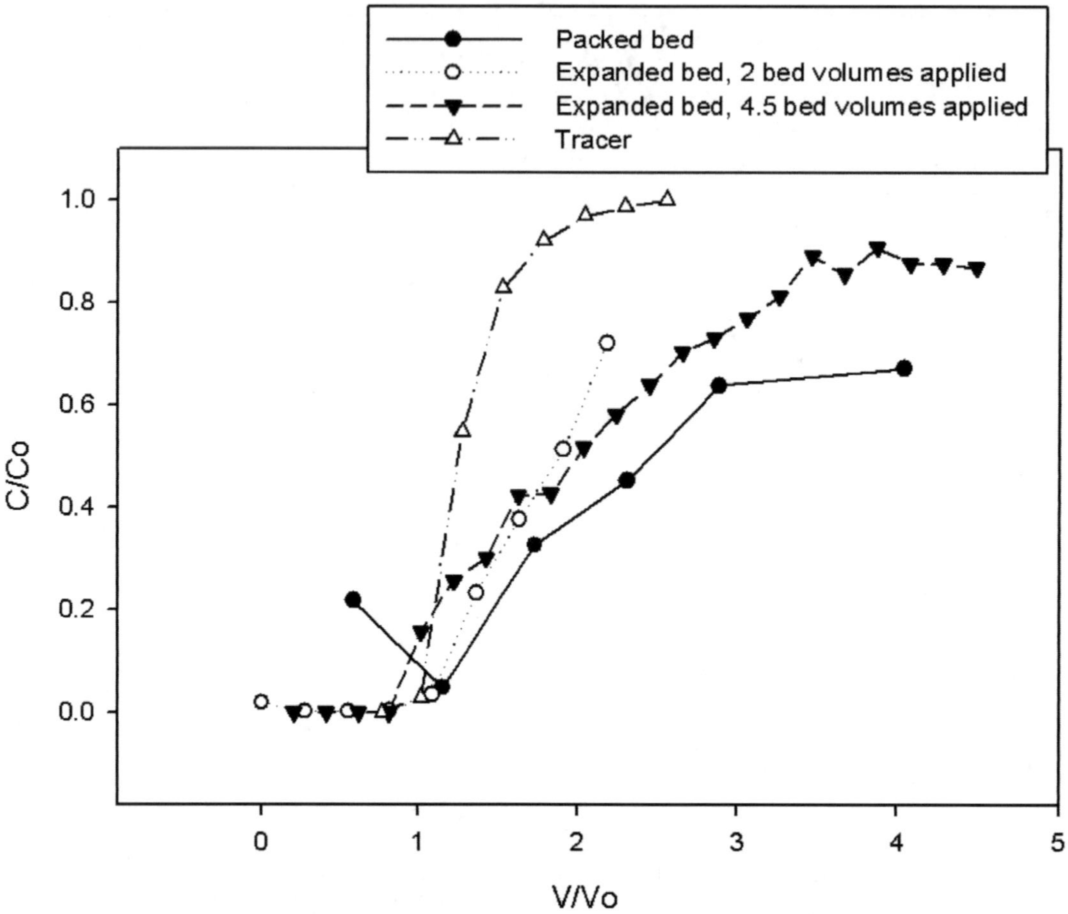

Figure 6- Breakthrough profiles for expanded bed adsorption and packed bed adsorption compared to tracer breakthrough profile. X-axis is volume applied (V) divided by the column volume (Vo).

Conclusions

Hydrolysis complicates the concentration and purification of protein from the soy EAEP skim fraction. Ultrafiltration may work, however it will result in protein losses of at least 25%. Traditional isoelectric precipitation allows the recovery of large, presumably unhydrolyzed proteins. Expanded bed anion exchange chromatography shows promising results from an operational point of view and may provide some interesting polypeptide selectivity; however, high protein recovery yields were not attained. An approach more likely of success will be a multistage process, incorporating isoelectric precipitation, with staged anionic and/or cationic chromatographic separation and this is being investigated.

References

1. de Moura, J., et al., *Enzyme-Assisted Aqueous Extraction of Oil and Protein from Soybeans and Cream De-emulsification.* Journal of the American Oil Chemists' Society, 2008. **85**(10): p. 985-995.
2. AOAC, *Method 960.13*. 16th ed ed. Official Methods of Analysis of Association of Official Analytical Chemists. 1995, Arlington, VA.
3. Shen, J.L., *Solubility profile, intrinsic viscosity, and optical rotation studies of acid precipitated soy protein and of commercial soy isolate.* J. Agric. Food Chem., 1976. **24**(4): p. 784-788.
4. Rosenthal, A., D.L. Pyle, and K. Niranjan, *Simultaneous aqueous extraction of oil and protein from soybean: mechanisms for process and design.* Trans. Food and Bioproducts Processing, 1998. **76**: p. 224-230.
5. Lamsal, B.P., S. Jung, and L.A. Johnson, *Rheological properties of soy protein hydrolysates obtained from limited enzymatic hydrolysis.* LWT - Food Science and Technology, 2007. **40**(7): p. 1215-1223.

31b High Power Ultrasonics In the Chemical, Food & Bio Industry: From Laboratory to Commercial Production

Alex Patist, Technology Development Center, Cargill, Inc., Crescent Ridge II, 10900 Wayzata Blvd. St. 2-164, Minnetonka, MN 55305

High power ultrasound has only recently become an efficient tool for large scale commercial food and non-food applications, such as extraction, emulsification, homogenization, crystallization, dewatering, low temperature pasteurization, degassing, defoaming, activation and inactivation of enzymes, milling and viscosity reduction. More recently, researchers have started to explore the technology for its ability to provide unique product functionality. High power ultrasound is not only safe and environmentally friendly in its application, but efficient and economical as well. This can be attributed to improved equipment design and higher efficiencies of large-scale continuous flow-through systems. It offers a number of advantages over other conventional technologies including low energy and maintenance costs, and equipment that is robust and durable which can generally be retrofitted into existing facilities due to the relative small size. Like most innovative processing technologies, high power ultrasonics is not an off-the-shelf technology and therefore needs to be developed and scaled up for each application (for example using a Stage-Gate approach). This implies that the process from concept testing to commercial launch can take up to several years and involves collaboration between vendors, institutes or experts in the field. The advantage however, is that the specific design and application can often be kept as trade-secret or protected through patents or exclusivity agreements. The objective of the present paper is to present examples of ultrasonic applications that have made it to commercialization, its underlying fundamental mechanism, and to share some key learning's involving scale-up of an innovative process technology in general.

31c Extraction of Proteins from Corn Dry Mill Co-Product by Bio-Based Solvents

Saurav Datta[1], Y. J. Lin[1], M. C. Negri[1], R. Datta[2], Bryan Bals[3], Bruce E. Dale[3], and Seth W. Snyder[1]. (1) Argonne National Laboratory, 9700 S. Cass Avenue, C-205/362, Lemont, IL 60517, (2) Vertec BioSolvents, Downers Grove, IL, (3) Chemical Engineering and Material Science, Michigan State University, 3246 Engineering Building, East Lansing, MI 48824

Distiller's grains and solubles (DGS) is the primary co-product in dry grind ethanol production. DGS are a suitable feed for ruminants such as cattle. DGS suffers from two disadvantages: 1) Due to high fiber content, DGS are not suitable for monogastrics such as hogs and poultry; 2) Drying DGS to produce dried distiller's grains and solubles (DDGS) is energy intensive and requires about 1/3 of the energy consumer in dry mills. Drying is required so that DDGS can be transported from the corn ethanol belt to the southern cattle belt. To address these disadvantages, we are developing a process to recover the high value protein and oils from DGS. In specific, we employ food-grade biobased solvents to extract the nutritional components.

Typical DGS contain about 30 % protein and 50-60 % moisture. The extracted protein could be used as a high value feed to a wider variety of animals. The energy requirement for drying would also be significantly reduced. The separated fiber could be used as a captured cellulosic stream to be used either to produce additional ethanol or as a fuel source for the mill. This process will also help build markets for biobased products. Overall, the process economics of the corn dry mill could be improved dramatically by reductions in energy costs and increased value of the animal feed co-products.

In this study, effects of different process parameters, such as, time, temperature and ratio of DGS to solvent, on the recovery of proteins have been examined. Hydrophilic, hydrophobic and mixtures of both types of solvents have been used for the protein recovery and their effects on the recovery have also been studied. We report that solvent extraction using bio-based solvents were able to recover significant amount of proteins from DGS. We also report that hydrophobic solvents are more efficient for the protein recovery than to hydrophilic solvents.

31d Separation of Chemicals Produced by Acidogenic Fermentation

Emilie Fockedey[1], Juray De Wilde[1], and Patrick A. Gerin[2]. (1) Materials and Process Engineering Dept. (IMAP), Universite catholique de Louvain (UCL), Reaumur, Place Sainte Barbe, 2, Louvain-la-Neuve, 1348, Belgium, (2) Unit of Bioengineering (GEBI), Universite catholique de Louvain (UCL), Croix du Sud, 2 bte 19, Louvain-la-Neuve, 1348, Belgium

Introduction

Decreasing oil reserves and environmental concerns have motivated the use of alternative and renewable resources, such as biomass, for the production of energy and chemicals. Acidogenic fermentation of biomass for the production of ethanol and acetic, propionic, butyric and lactic acids offers three main advantages: (i) germ free conditions are not required; (ii) very diversified, less pretreated and cheap biomass can be used; (iii) the molecules produced and their esters are consumed on a large scale by industries.

Whereas many studies address the transformation of biomass, few report the product recovery, separation and concentration steps downstream of the biomass transformation. These steps remain, in fact, a major challenge for the expansion of green chemistry in a sustainable development context and in particular for the application of acidogenic fermentation which produces very dilute (concentrations of only a few percent) and complex aqueous solutions of acids and ethanol.

A number of separation processes can be potentially used and the process which is technically, energetically and environmentally the most efficient has not been identified or developed yet. In the present work, the separation step following the acidogenic fermentation of biomass is investigated. A novel extraction - re-extraction process is presented for the recovery of acids and ethanol produced by acidogenic fermentation. The process is based on the transfer of the acids and ethanol to a glycerol phase via an intermediate solvent phase and it is schematically presented in Figure 1. Tri-n-octylamine based solvents are chosen in the first step of the process for their preferential extraction of acids. In the second, re-extraction step, the acids are extracted from the intermediate solvent with glycerol. Glycerol being a by-product of the growing biodiesel production, its offer largely exceeds its demand and its market value has decreased over the years. New markets and applications of glycerol are expected to increase its value[1] as well as the profitability of the biodiesel industry. The novel glycerol based extraction – re-extraction process for the recovery of acids opens perspectives for the transformation of glycerol into short chain esters, directly using the acids recovered.

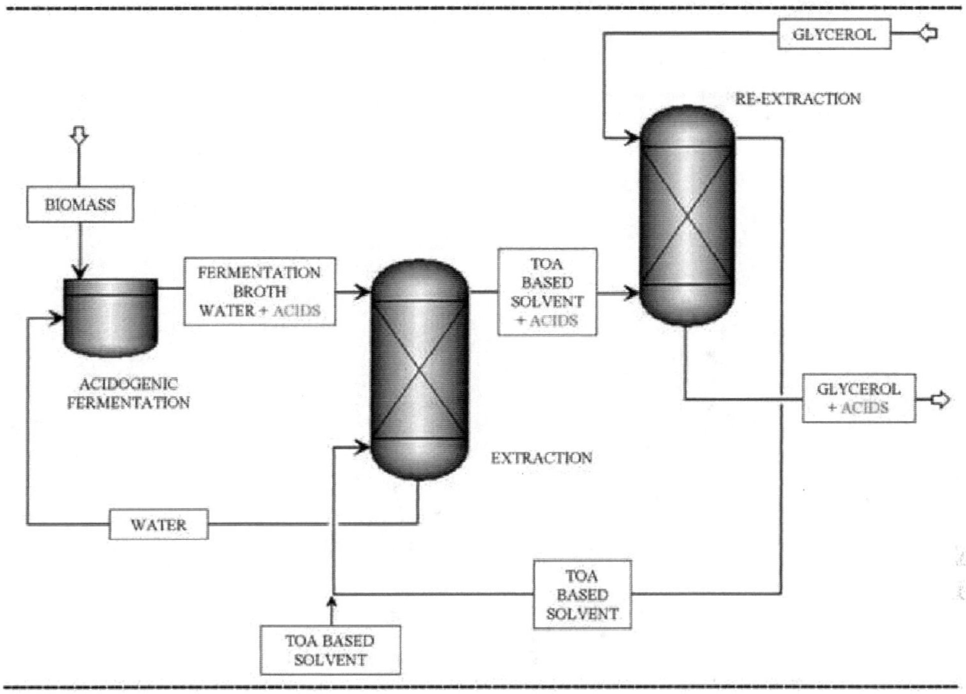

Figure 1: TOA-based extraction – glycerol-based re-extraction process for the separation of volatile fatty acids and ethanol produced by acidogenic fermentation of biomass

Experimental set-up and results

Experiments were carried out in order to obtain mass based distribution coefficients. Pure tri-n-octylamine (Acros, 98%) was used as intermediate solvent. First, butyric acid aqueous model solutions were tested. Then, the behavior with real fermentation broths was measured. Solutions were stirred to reach equilibrium and then centrifuged. The concentrations of butyric acid in the aqueous phase before and after extraction were determined by gas chromatography. Before the extraction, the pH was not adjusted except when measuring the influence of the pH on the distribution coefficients.

The extraction – re-extraction performance was shown to depend highly on the operating conditions. In particular, the influence of the pH, the temperature and the composition of the aqueous phase was studied.

The distribution coefficient of butyric acid for the first extraction as a function of the pH of the aqueous solution is presented in Figure 2. The mass distribution coefficient is seen to increase significantly with decreasing pH. Moreover, the pH after the first extraction is linearly related to the pH before the first extraction.

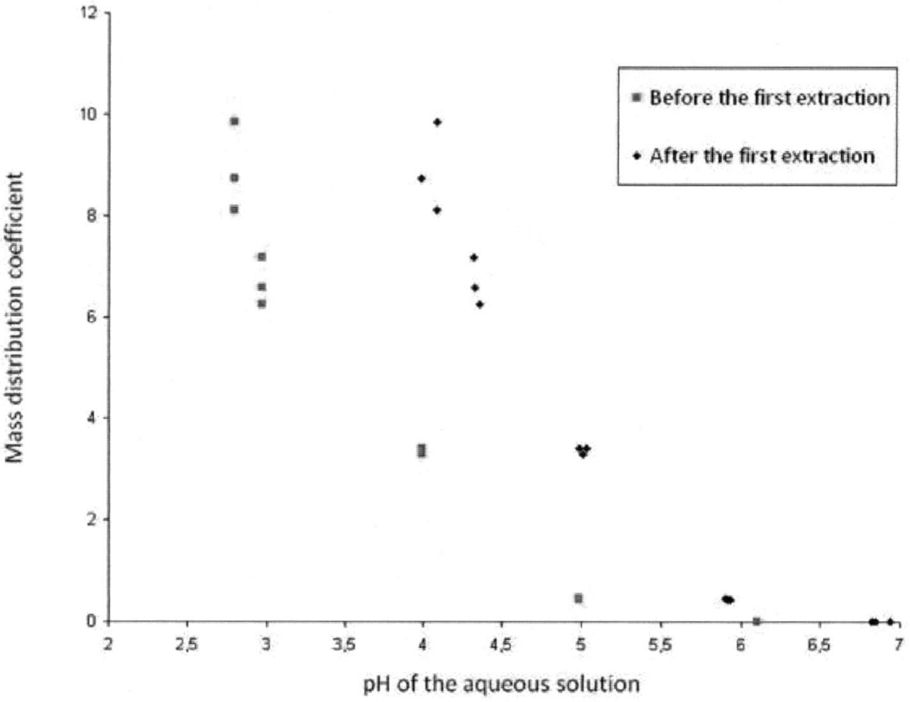

Figure 2: The mass distribution coefficient as a function of the pH of the aqueous solution before and after the first extraction

For the first extraction step (water/TOA system), the distribution coefficient was shown to increase when decreasing temperature. Inversely, for the second extraction step (TOA/glycerol system), the extraction was more efficient when the temperature increased.

The simultaneous presence of butyric and acetic acids did not affect the extraction of the individual acids.

Conclusion

A novel TOA-based extraction – glycerol-based re-extraction process for the recovery of ethanol and acetic, propionic, butyric and lactic acids from the fermentation broth was experimentally studied. First, the distribution coefficients in simple cases were determined. Next, more complex solutions, approaching fermentation broths, were studied. The extraction re-extraction performance was shown to depend highly on the solvent and on the operating conditions. The influence of the most important process parameters, such as the pH, the temperature, the composition of the aqueous phase was analyzed. The influence of the composition of the organic phase, i.e. different modifiers and diluents, is to be further investigated.

Acknowledgements

The authors would like to acknowledge the Belgian Fonds pour la formation à la Recherche dans l'Industrie et dans l'Agriculture (F.R.I.A.) for the financial support.

References

1. C. O'Driscoll (2007) Biofuels, Bioprod., Bioref., 1, 6–7

Equilibrium Study on Reactive Extraction of Lactic Acid with Tri-*n*-Butyl Phosphate in *n*-Hexane

Panut Mulyono and Anita Pardah
Department of Chemical Engineering, Faculty of Engineering, Gadjah Mada University
Jl. Grafika No. 2, Kampus UGM, Yogyakarta 55281, Indonesia.
Tel. +62-274-6492171, Fax. +62-274-6492170, E-mail: pmulyono@chemeng.ugm.ac.id

Abstract

Extraction of carboxylic acids from dilute aqueous solution using traditional solvents such as ketones, alcohols, ethers, and ester is inefficient because the distribution ratio is rather low. Reactive extraction which exploits reversible chemical complexation is an effective separation process for extraction of carboxylic acids from aqueous streams such as fermentation broths and wastewaters. In the extraction process, selection of the solvent is an important aspect to be considered. Considering its solubility in water, cost and availability, tri-*n*-butyl phosphate (TBP) seems to be an attractive solvent for the extraction of lactic acid from aqueous solution.

The purpose of this experiment is to study the equilibrium of the reactive extraction of lactic acid in aqueous solution with TBP in *n*-hexane. The parameters studied in this experiment were initial concentration of lactic acid in the aqueous phase, TBP concentration in *n*-hexane phase, and the extraction temperature. The experiments at ambient temperature were carried out using a separatory funnel, while the experiments at other than ambient temperature were carried out using erlenmeyer flask and water bath shaker to adjust the temperature. In this experiment, the initial concentration of lactic acid was varied from 0.1 to 0.5 gmol/dm^3. The range of initial TBP concentrations in *n*-hexane was 0.1 to 1.0 gmol/dm^3 and the extraction temperature range was 283 to 313 K.

The experimental results showed that the higher the initial concentration of lactic acid in aqueous solution, the higher the distribution ratio for a fixed TBP concentration and extraction temperature. For a fixed initial concentration of lactic acid in aqueous solution and extraction temperature, the distribution ratio of lactic acid is increased by increasing TBP concentration. The overall equilibrium constants (K_{pq}) for the experiments using TBP concentration ranging from 0.1 to 1.0 gmol/dm^3 at the extraction temperature of 293 K are calculated to be 0.0668 to 0.5144. K_{pq} for the experiments at the temperature ranging from 283 to 313 K at the initial concentration of lactic acid of 0.2 gmol/L are found to be 0.0122 to 0.8856. The K_{pq} as a function of temperature (T) in K can be expressed as $\ln K_{pq} = 10{,}596/T - 38.08$ with sum of square of error of 0.14.

Keywords: Reactive Extraction, Equilibrium, Lactic Acid, TBP, *n*-Hexane.

Introduction

Production of carboxylic acids by fermentation usually produces low concentrations of carboxylic acids in water which is sometimes less than 10% (w/w).[1] Separation, purification and even to concentrate the fermentation products are very difficult because affinities between acids and water is very strong. Separation of acids and water by distillation is energy consumable to evaporate water.[1] The classical process to recover a carboxylic acid from a wide variety of dilute aqueous solutions is based on precipitation of the calcium salt upon addition of calcium hydroxide to the acid aqueous solution.[2] This method of recovery results low yield of acid crystallization, costly, and unfriendly to the environment.[3, 4]

To recovery carboxylic acids from dilute acid solutions can be conducted by extraction method. Traditional solvents which are usually used are ketone, alcohol, ether and ester. Using traditional solvent is not very effective because its distribution ratio is very low. To increase the distribution ratio, reactive extraction is one of the method used to recovery carboxylic acids from dilute solutions.[5] Many researchers have investigated reactive extraction of carboxylic acids using some extractants. Juang and Huang[5] has studied the equilibrium of reactive extraction of lactic acid with an amine extractant. Synergistic extraction of lactic acid with tri-*n*-octylamine and TBP has been studied by Matsumoto *et al.*[6]

This paper presents the equilibrium of reactive extraction of lactic acid with TBP in *n*-hexane. TBP reacts with lactic acid to form complexes, while *n*-hexane functions to dilute TBP. TBP has an advantage as solvent for use in aqueous extraction due to its negligible solubility in water.[7]

Experimental

Reagent and Solutions

Distilled water with the density of 0.9959 g/cm^3 and viscosity of 0.0089 g/(cm·s) was used in this experiment. Lactic acid, TBP, *n*-hexane, and other chemicals were supplied by Merck Co. as analytical reagent grade, and all were used without further purification.

Experimental Procedure

The experiments were done for the extraction of lactic acid with pure *n*-hexane and for the extraction of lactic acid with TBP diluted by *n*-hexane. For measuring the distribution ratio of lactic acid in *n*-hexane, the aqueous solution of lactic acid with a fixed concentration and *n*-hexane of equal volume of 50 cm^3 were mixed in a flask and shaken at a fixed temperature for one hour to attain the extraction equilibrium. After two phases were separated, lactic acid concentration in the aqueous phase was determined by titration with known concentration of NaOH solution using phenolphthalein as indicator. The concentration of lactic acid in organic phase was calculated by total mass balance of lactic acid. The experimental parameters were temperature (283, 293, 303, and 313 K) and lactic acid initial concentration (0.1, 0.2, 0.3, 0.5, and 0.5 mol/dm^3).

The experiments for the extraction of lactic acid with TBP in *n*-hexane were carried out by shaken of equal volume (50 cm^3) of aqueous solution of lactic acid with solution of TBP in *n*-hexane. The experimental procedure was similar to that explained above, with the experimental parameters of temperature (283, 303, and 313 K), lactic acid initial concentration (0.1, 0.2, 0.3, 0.5, and 0.5 mol/dm^3), and TBP concentrations (0.1, 0.2, 0.5, and 1.0 mol/dm^3).

Mathematical Modeling

Extraction of Lactic Acid with Pure *n*-Hexane

Most carboxylic acids such as lactic acid mainly exist as dimers in organic phase due to the strong intermolecular hydrogen-bonding, especially in the non-polar or lowly polar solvents.[5] On the contrary, in the aqueous phase, they exist mainly as monomers because the intermolecular hydrogen bonding between the acids is destroyed owing to their preferential hydrogen bonding with water molecules.[3, 5]

The reactions in this extraction as the aqueous pH is far less than the dissociation constant of lactic acid (pK_a =3.86) are as follows:[5]

Distribution of lactic acid monomer from the aqueous phase to the organic phase.

$$\text{HA} \Leftrightarrow \overline{\text{HA}};\ K_d = \frac{[\overline{HA}]}{[HA]} \tag{1}$$

Dimerization of lactic acid in the organic phase.

$$2\overline{\text{HA}} \Leftrightarrow \overline{(\text{HA})_2};\ K_2 = \frac{[\overline{(HA)_2}]}{[\overline{HA}]^2} \tag{2}$$

Dissociation of lactic acid in aqueous phase.

$$\text{HA} \Leftrightarrow \text{H}^+ + \text{A}^-;\ K_a = \frac{[H^+][A^-]}{[HA]} \tag{3}$$

The overbar refers to the organic phase. The concentrations of lactic acid in organic phase and aqueous phase at equilibrium are calculated from the following equations:

$$[\overline{HA}]_t = [\overline{HA}] + 2[\overline{(HA)_2}] \tag{4}$$

$$[HA]_t = [HA] + [A^-] \tag{5}$$

The distribution ratio (D_0) of lactic acid in the absence of TBP is defined by Eq. (6).

$$D_0 = \frac{[\overline{HA}]_t}{[HA]_t} \tag{6}$$

where $[\overline{HA}]_t$ is the total concentration of lactic acid in the organic phase and $[HA]_t$ is the total concentration of lactic acid in the aqueous phase at equilibrium.

The concentration of lactic acid in aqueous phase at equilibrium is determined by titration using NaOH solution with phenolphthalein as indicator, while its concentration in the organic phase is calculated by mass balance as follows:

$$[HA]_t V_0 = [\overline{HA}]_t V_{org} + [HA]_t V_1 \tag{7}$$

Substitution of $[\overline{HA}]_t$ from Eq. (4) and $[HA]_t$ from Eq. (5) to Eq. (6) we have the following equation:

$$D_0 = \frac{[\overline{HA}] + 2[\overline{(HA)_2}]}{[HA] + [A^-]} = \frac{\dfrac{[\overline{HA}]}{[HA]} + \dfrac{2[\overline{(HA)_2}]}{[HA]}\dfrac{[\overline{HA}]^2}{[\overline{HA}]^2}\dfrac{[HA]}{[HA]}}{\dfrac{[HA]}{[HA]} + \dfrac{[A^-]}{[HA]}\dfrac{[H^+]}{[H^+]}} \tag{8}$$

or $$D_0 = \frac{K_d + 2K_d^2 K_2 [HA]}{1 + \dfrac{K_a}{[H^+]}} \tag{9}$$

The value of $[HA]$ at Eq. (9) can be obtained from Eq. (5) and at the measured equilibrium pH, that is $[A^-] = [H^+]$, so that:

$$[HA] = [HA]_t - [H^+] \tag{10}$$

Eq. (9) can be arranged as:

$$D_0\left(1 + \frac{K_a}{[H^+]}\right) = K_d + 2K_d^2 K_2 [HA] \tag{11}$$

By plotting $D_0\left(1 + \frac{K_a}{[H^+]}\right)$ versus $[HA]$, we get the value of K_d as the intercept of the graph and $2K_d^2K_2$ as its slope. The values of K_d and K_2 can also be obtained by linear regression method.

Extraction of Lactic Acid with TBP

The extraction of lactic acid with TBP in *n*-hexane at pH far less than pK_a can be expressed by the following stoichiometry:[5]

$$pHA + q\overline{TBP} \Leftrightarrow \overline{(HA)_p(TBP)_q} \ ;\ K_{pq} \tag{12}$$

where K_{pq} is the extraction equilibrium constant, which is given by Eq. (13).

$$K_{pq} = \frac{\left[\overline{(HA)_p(TBP)_q}\right]}{[HA]^p\left[\overline{TBP}\right]^q} \tag{13}$$

The total concentration of lactic acid in the organic phase can be expressed by Eq. (14) or Eq. (15) as follows:

$$\left[\overline{HA}\right]_t = \left[\overline{HA}\right] + 2\left[\overline{(HA)_2}\right] + \sum_p\sum_q p\left[\overline{(HA)_p(TBP)_q}\right] \tag{14}$$

$$\left[\overline{HA}\right]_t = K_d[HA] + 2K_2K_d^2[HA]^2 + \sum_p\sum_q p\left[\overline{(HA)_p(TBP)_q}\right] \tag{15}$$

Eq. (15) shows that the amount of lactic acid complexed with TBP is corrected by subtracting that physically extracted by pure *n*-hexane from the measured $\left[\overline{HA}\right]_t$. The values of K_d and K_2 in Eq. (15) are calculated from the experimental data of the extraction of lactic acid using pure *n*-hexane. The value of $[HA]$ is calculated by Eq. (10) and the value of $\left[\overline{HA}\right]_t$ is calculated from the total mass balance. So the value of $\left[\overline{(HA)_p(TBP)_q}\right]$ can be calculated as follows:

$$\left[\overline{(HA)_p(TBP)_q}\right] = \frac{1}{p}\left[\left[\overline{HA}\right]_t - K_d[HA] - 2K_2K_d^2[HA]^2\right] \tag{16}$$

The value of $\left[\overline{TBP}\right]$ in Eq. (13) is obtained from the TBP mass balance as expressed by Eq. (17).

$$\left[\overline{TBP}\right] = \left[\overline{TBP}\right]_O - \sum_p\sum_q q\left[\overline{(HA)_p(TBP)_q}\right] \tag{17}$$

Knowing the values of $\left[\overline{(HA)_p(TBP)_q}\right]$, $[HA]$, and $\left[\overline{TBP}\right]$, then the value of K_{pq} in Eq. (13) can be obtained. In this calculation, both number of lactic acid molecules involved in the complex (*p*) and number of TBP molecules involved in the complex (*q*) are assumed to be 1.

The complexation reaction between lactic acid and TBP in the organic phase is expected to be exothermic. Formation of complex makes the system more ordered and thus decreases the entropy (ΔS). Therefore, as the temperature (*T*) is increased, the amount of lactic acid extracted decreases. If the enthalpy (ΔH) and entropy of reaction are assumed to be constant over the temperature range, the following expression:

$$\ln K_{pq} = \frac{-\Delta H}{RT} + \frac{\Delta S}{R} \tag{18}$$

indicates that a plot of ln K_{pq} vs $1/T$ gives a straight line.[1] The enthalpy of reaction is obtained from the slope, and the intercept is proportional to the entropy.

Results and Discussion

Figure 1 shows the distribution ratio of the extraction of lactic acid with pure *n*-hexane as a function of initial lactic acid concentration in the aqueous solution at four different temperatures. It appears in **Figure 1** that at a fixed temperature, the distribution ratio increases as a function of the initial concentration of lactic acid in the aqueous solution. It can be seen from **Figure 1** that the distribution ratio decreases as the temperature increases for the experiments at the same initial concentration of lactic acid in the aqueous solution.

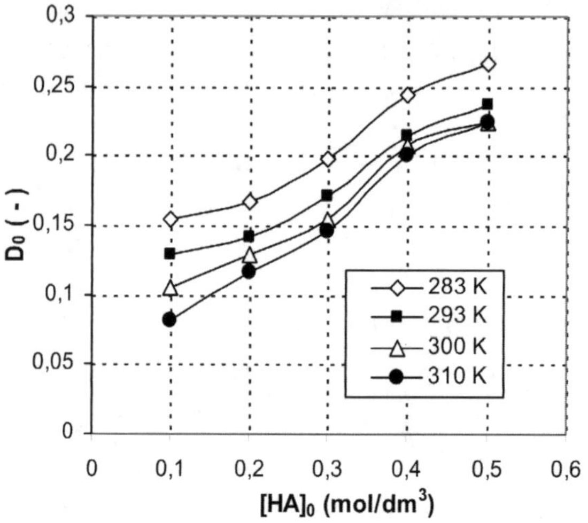

Figure 1. Effects of initial lactic acid concentration and temperature on the distribution ratio of the extraction of lactic acid with pure *n*-hexane.

From the experimental data, then the value of K_a was calculated by Eq. (3) and the values of K_2 and K_d were obtained based on Eq. (11). The values of K_a, K_2, and K_d as a function of temperature are shown in **Figure 2**. **Figure 2** depicts that K_a increases as a function of the temperature. K_a attains the maximum value of 33.3 x 10^{-5} mol/dm^3 at 300 K, then decreases to the value of 17.8 x 10^{-5} mol/dm^3 at 310 K. The value of K_2 increases as a function of temperature, while the value of K_d is decreased by increasing the temperature.

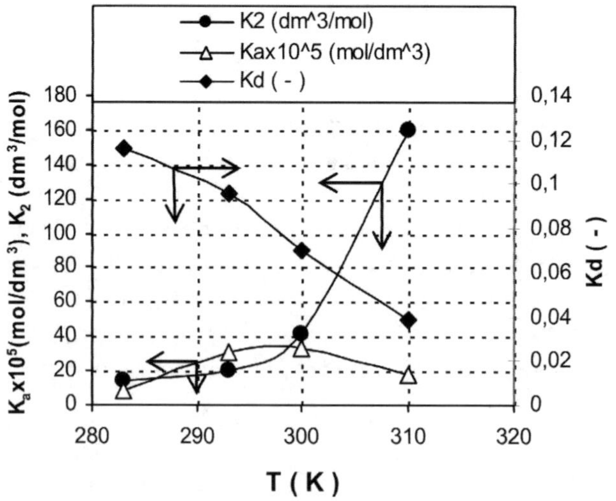

Figure 2. Values of K_a, K_2, and K_d as a function of temperature.

Figure 3 illustrates the distribution ratio of the extraction of lactic acid with TBP in *n*-hexane as a function of initial lactic acid concentration in the aqueous solution at different TBP concentration in the organic phase at 293 K. It is found that at a fixed TBP concentration, the distribution ratio tends to increase with increasing the initial lactic acid concentration in the aqueous solution. It is also shown in **Figure 3** that at the same initial acid concentration, the distribution ratio increased with increasing TBP concentration.

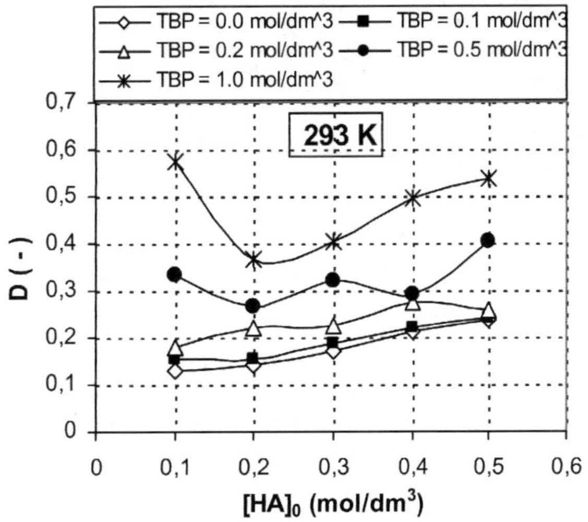

Figure 3. Effects of initial lactic acid and TBP concentrations on the distribution ratio of the extraction of lactic acid with TBP in *n*-hexane at 293 K.

The effect of temperature on the distribution ratio of extraction of lactic acid with TBP in *n*-hexane is shown in **Figure 4**. The complexation reaction of lactic acid and TBP in the organic phase involve proton transfer or hydrogen bond formation and are therefore expected to be exothermic. Also, formation of a complex makes the system more ordered and therefore decreases the entropy. Thus, as the temperature increases, the amount of lactic acid extracted decreases.[1, 5]

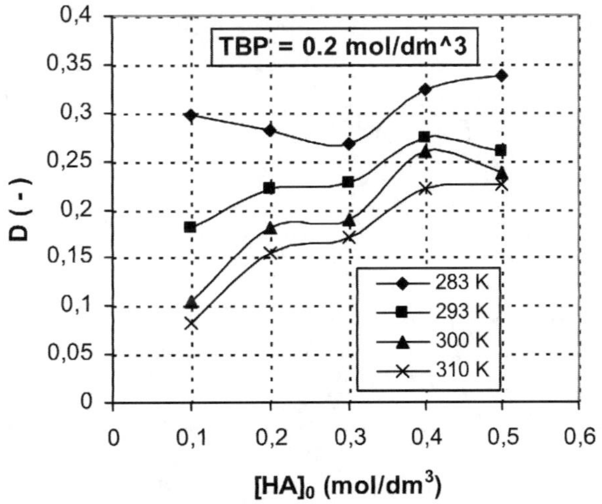

Figure 4. Effect of temperature on the distribution ratio of extraction of lactic acid with TBP in *n*-hexane at the TBP concentration of 0.2 mol/dm^3.

Figure 5 reveals the temperature dependence of the equilibrium constant K_{pq}. It can be seen at **Figure 5** that the extraction of lactic acid with TBP is exothermic. The enthalpy of reaction (ΔH) is obtained from the slope of the graph and it was found to be -88,1 kJ/mol, and the intercept is proportional to the entropy (ΔS) and it was found to be -316.6 J/(mol·K).

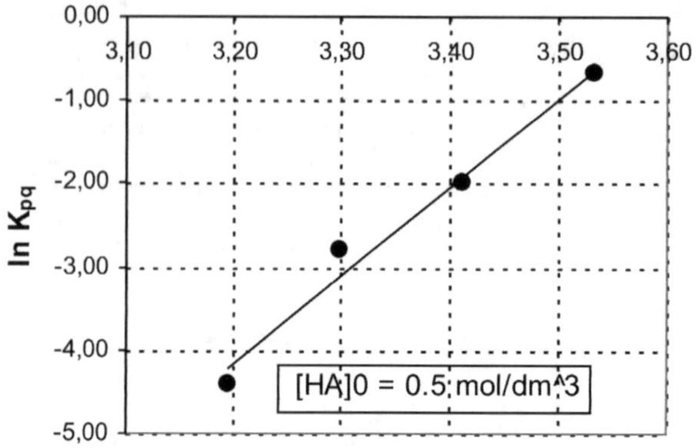

Figure 5. Temperature dependence of the equilibrium constant K_{pq}.

Conclusions

The reactive extraction of lactic acid from aqueous solution with TBP in *n*-hexane has been studied. Based on the results of this study, the following conclusions are made.

1. The higher the initial concentration of lactic acid in aqueous solution, the higher the distribution ratio for a fixed TBP concentration and extraction temperature.
2. For a fixed initial concentration of lactic acid in aqueous solution and extraction temperature, the distribution ratio of lactic acid is increased by increasing TBP concentration.
3. The overall equilibrium constants (K_{pq}) for the experiments using TBP concentration ranging from 0.1 to 1.0 gmol /dm^3 at the extraction temperature of 293 K are found to be 0.0668 to 0.5144.
4. K_{pq} for the experiments at the temperature ranging from 283 to 313 K at the initial concentration of lactic acid of 0.2 gmol/dm^3 are found to be 0.0122 to 0.8856.
5. The values of ΔH and ΔS are found to be -88,1 kJ/mol and -316.6 J/(mol·K), respectively. The K_{pq} as a function of temperature (T) in K can be expressed as $\ln K_{pq} = 10{,}596/T - 38.08$ with sum of square of error of 0.14.

Nomenclature

D = distribution ratio of lactic acid by TBP (dimensionless)
D_0 = distribution ratio of lactic acid by pure *n*-hexane (dimensionless)
ΔH = apparent enthalpy change (kJ/mol)
K_2 = dimerization constant of lactic acid in the organic phase (dm^3/mol)
K_a = dissociation constant of lactic acid in the aqueous phase (mol/dm^3)
K_d = distribution constant of the monomeric acid defined in Eq. (3) (dimensionless)
K_{pq} = overall equilibrium constant defined in Eq. (13) ($(\text{mol/dm}^3)^{1-p-q}$)
p = number of lactic acid molecules involved in the complex
q = number of TBP molecules involved in the complex
R = universal gas constant (8.314 J/(mol·K))
ΔS = apparent entropy change (J/(mol·K))
T = temperature (K)
[ ] = molar concentration of species in the brackets (mol/dm^3)

Subscripts

l	=	liquid phase
org	=	organic phase
t	=	total value at equilibrium
0	=	initial

References

1. Tamada, J.A. and King, C.J. (1990), "Extraction of Carboxylic Acids with Amine Extractants. 3. Effect of Temperature, Water Coextraction, and Process Considerations," *Ind. Eng. Chem. Res.*, 29(7), 1333-1338.
2. Malmary, G., Faizal, M., Albet, J., and Molinier, J. (1997), "Liqui-liquid Equilibria of Acetic, Formic, and Oxalic Acids Between Water and Tributyl Phosphate + Dodecane," *J. Chem. Eng. Data*, 42(5), 985-987.
3. Juang, R.S. and Huang, W.T. (1994), "Equilibrium Studies on The Extraction of Citric Acid from Aqueous Solutions with Tri-n-Octylamine," *J. Chem. Eng. Japan*, 27(4), 498-504.
4. Poposka, F.A., Nikolovski, K., and Tomovska, R. (1997), "Equilibria and Mathematical Models of Extraction of Citric Acid with Isodecanol/n-Paraffins Solutions of Trioctylamine," *J. Chem. Eng. Japan*, 30(5), 777-785.
5. Juang, R.S. and Huang, W.T. (1997), "Equilibrium Studies on The Extraction of Lactic Acid with an Amine Extractant," *Chem. Eng. Journal*, 65, 47-53.
6. Matsumoto, M., Yuba, S., and Kondo, K. (1998), "Synergetic Extraction of Lactic Acid with Tri-n-Octylamine and Tri-n-Butylphosphate," *J. Chem. Eng. Japan*, 31(6), 996-998.
7. Shah, D.J. and Tiwari, K.K. (1981), "Recovery of acetic acid from Dilute Aqueous Stream Using Liquid-liquid Extraction with Tri-n-Butyl Phosphate as Solvent," *J. Sep. Process Technol.*, 2(4), 1-6.

31f Extraction and Purification of Isoflavones from Soybeans and Characterization of Their Estrogenic Activities

Elizabeth J. Zhang[1], **Ka M. Ng**[2], and Kathy Q. Luo[1]. (1) Bioengineering Graduate Program, Department of Chemical Engineering, Hong Kong University of Science and Technology, Clear Water Bay, Kowloon, Hong Kong, China, (2) Department of Chemical Engineering, Hong Kong University of Science and Technology, Clear Water Bay, Kowloon, Hong Kong, China

The objective of this study was to develop a purified soy isoflavone product which is potentially efficacious for relieving estrogen-deficient diseases (i.e. menopausal symptoms). In this study, a simple method was developed for producing isoflavones from soybean flour including only three steps extraction, hydrolysis and crystallization. Then the extraction and purification parameters were optimized to give a high yield of total isoflavones, about 0.66 mg aglycones/g soybean flour, which is more than two times of the initial yield. HPLC analysis and MTT cell proliferation assay using MCF-7 cells revealed that the product thus obtained not only contained a high content of isoflavone aglycones but also had estrogenic activity. MTT data also revealed that both genistein and daidzein exhibited estrogenic effects at lower concentrations and anti-proliferative effects at higher concentrations, and 1 μM genistein and 10 μM daidzein exerted significant estrogenic activities which were not more than that of endogenous level of 17beta-estradiol (E2). The production method developed can be used as guideline for manufacturing soy isoflavones, and MTT assay was demonstrated to be suitable for quality control on isoflavone products, and the results on the estrogenic properties of isoflavones can be used as reference data for their effective and safe usages in estrogenic therapy.

31g Separation and Purification of Schisandrin B from *Fructus Schisandrae*

Ka F. Luk[1], Robert Kam M. Ko[2], and Ka M. Ng[1]. (1) Department of Chemical Engineering, Hong Kong University of Science and Technology, Clear Water Bay, Kowloon, Hong Kong, China, (2) Department of Biochemistry, Hong Kong University of Science and Technology, Clear Water Bay, Kowloon, Hong Kong, China

Schisandrin B (Sch B) is the most abundant and biologically active dibenzocyclooctadiene lignan present in *Fructus Schisandrae* (FS). In the past two decades, the pharmacological activities of Sch B have been extensively investigated. There is a growing body of evidence suggesting its health benefits in the prevention and treatment of a number of diseases including protection against oxidative liver, heart and brain damage. Recently, Sch B was also found to be a potential adjuvant for cancer chemotherapy. Thus, it is a highly desirable ingredient for a dietary supplement.

While Sch B has been isolated from FS in the laboratory, the experimental procedure was usually lengthy and the yield was relatively low. The objective of this study is to develop a process that is amenable to large-scale recovery of Sch B from FS. The process starts with solid-liquid extraction of FS by petroleum ether to yield a Sch B-containing extract, which is then purified by removing the relatively non-polar substances, mostly fatty oils, by an adsorption-desorption step. This is followed by a chromatographic process to isolate the Sch B-enriched fraction that is then used to yield Sch B by crystallization.

32a Defect-Free Composite Pd Membranes with High Temperature Long-Term Stability

M. Engin Ayturk, Dept. of Chemical Engineering, Worcester Polytechnic Institute, 100 Institute Road, Worcester, MA 01609 and Yi Hua Ma, Worcester Polytechnique Institute, Chemical Enginering, Worcester, MA 01609.

The membrane selectivity is among the key determinants that affect the overall H_2 separation efficiency and the long-term stability of the composite Pd and Pd/alloy membranes. To mitigate high temperature leak growth and to improve the long-term stability of the membrane selectivity, a novel synthesis procedure was implemented during the preparation of the composite Pd membrane supported on a media-grade 0.1 μm Inconel support, which was oxidized at 800°C for 12 hours to generate the intermetallic diffusion barrier. The successful synthesis of the 7.6 μm thick pure-Pd/Inconel membrane with a special emphasis on the formation of leak resistant deposition layers and saturated (stable) grain boundaries, led to an excellent long-term stability at 450°C with essentially infinite H_2/He ideal selectivity. At 450°C and at a ΔP of 35 psi (P_{High}=50 psia & P_{Low}=15 psia), the He leak of the membrane was undetectable and the membrane maintained an excellent H_2/He selectivity stability over a period of ~845 hours with numerous H_2/He cyclings. Upon the completion of the long-term H_2 permeation characterization (~1250 hours within 300-450°C), the high pressure testing of the 7.6 μm Pd /Inconel membrane was conducted at a feed side pressure as high as 116 psia (7.9 bara). At 400°C and a ΔP of 101 psi (P_{High}=116 psia & P_{Low}=15 psia), the stable H_2 flux for the 7.6 μm Pd/Inconel membrane was as high as 150 scfh/ft^2, exceeding the 2007 DOE target under similar operating conditions. Furthermore, the He leak after the high-pressure testing remained undetectable. The long-term selectivity stability is a significant achievement towards the successful utilization of Pd- and/or Pd/alloy-based catalytic membrane reactors for the production of high-purity H_2 suitable for fuel cell applications in industrial steam reformers and/or high temperature water-gas shift reactors integrated into the downstream of the industrial coal gasification units.

32b High Purity Hydrogen Generation from Methanol Steam Reforming Using Pd-Based Membrane Reactors

Sameer H. Israni and Michael P. Harold. Department of Chemical Engineering, University of Houston, 4800 Calhoun, Houston, TX 77204

Pd membrane based reactors have the potential to generate high purity H2 in a single unit for stationary and mobile applications spanning power stations, soldier-power, and vehicles. Previous studies [1,2] from our group have analyzed in some detail the use of methanol reforming in membrane reactors as a way of intensifying the reaction and separation/purification into a single unit. In the current study methanol steam reforming was carried out using Cu/ZnO/Al2O3 catalyst. The methanol conversion, hydrogen productivity, hydrogen utilization and outlet CO/CO2 ratio for a packed bed reactor (PBR) and a packed bed membrane reactor (PBMR) were compared at different pressures and temperatures. 5 micron thick Pd-Ag 'nanopore' membranes [3] were used in the PBMR. Our previous studies [4,5] have highlighted that slow radial diffusion of hydrogen is one of the major factors limiting productivity & utilization in the membrane reactor. In order to further study this effect we have used membrane reactors with different radial diffusion lengths. The radial diffusion lengths have been varied by changing the reactor diameters while correspondingly changing the ratio of catalyst to inert SiC, to maintain the catalyst volume to membrane surface area ratio constant. A 2 dimensional non-isothermal reactor model has been developed to simulate the results and to elucidate the rate limiting processes. Scale-up principles derived from the experiments and model will be discussed.

References

1. Harold, M.P., B. Nair, and G. Kolios, "Hydrogen Generation in a Pd Membrane Fuel Processor: Assessment of Methanol-Based Reaction Systems," Chemical Engineering Science, 58, 2551-2571 (2003).

2. Lattner, J.R., and M.P. Harold, "Comparison of Methanol Based Fuel Processors for PEM Fuel Cell Systems," Appl. Catalysis B. Environmental, 56, 149-169 (2005).

3. Nair, B., and M.P. Harold, "Pd Encapsulated and Nanopore Hollow Fiber Membranes: Synthesis and Permeation Studies," J. Membrane Sci., 290, 182-195 (2007).

4. Nair, B., and M.P. Harold, "Experiments and Modeling of Transport in Composite Pd and Pd/Ag Coated Alumina Hollow Fibers," J. Membrane Sci., doi:10.1016/j.memsci.2007.11.034 (2008).

5. Israni, S.H., B. Nair and M. P. Harold, "Hydrogen Generation and Purification in a Composite Pd Hollow Fiber Membrane Reactor: Experiments and Modeling," Catalysis Today, to appear (2008).

32c Steam Methane Reforming in a Membrane Reactor for Hydrogen Production

Ashok S. Damle and Jim Acquaviva. Pall Corporation, 2200 Northern Boulevard, East Hills, NY 11548

Economical generation of pure hydrogen represents a critical technology component for power generation by PEM fuel cells in a variety of mobile and stationary power applications. Hydrogen is conventionally produced by steam reforming of hydrocarbon fuels followed by a two-step water gas shift (WGS) reaction and hydrogen separation and purification by pressure swing adsorption (PSA). Combining the WGS reactor and hydrogen purification steps into a single membrane reactor, has the potential to significantly reduce the capital and operating costs of producing hydrogen, and consequently, reduce the price of hydrogen to the consumer. Due to simultaneous reaction and product separation, it is possible to increase the efficiency of the overall process by shifting the equilibrium and thereby producing more hydrogen than would be possible using the conventional approach, providing additional benefit to the consumer. High performance, high-temperature hydrogen separation membranes thus represent a key enabling technology for efficient hydrogen production using synthesis gas derived from a variety of feedstocks.

Palladium-alloy foils and extruded tubes are known to be completely selective for hydrogen separation, however, they are a relatively expensive option for large scale industrial applications. These product formats exhibit low hydrogen flux rates due to the thickness necessary for structural stability. By depositing thin palladium-alloy film on a porous substrate, hydrogen flux and structural stability of the composite Pd-alloy membrane is increased while reducing the membrane costs. Both the thin metal film deposition process and the porous substrate characteristics influence development of successful composite membranes for hydrogen separation application. Pall Corporation has successfully developed high flux and selectivity hydrogen separation membranes by depositing thin Pd-alloy films on tubular ceramic/porous stainless steel composite (AccuSep® Inorganic media) substrates. These composite membranes have demonstrated performance approaching U.S. DOE's year 2010 hydrogen flux target with structural stability against thermal cycling necessary for their commercial application.

Utilization of the Pd-alloy composite membranes in a membrane reactor configuration for conducting steam methane reforming (SMR) reaction allows operation at temperatures much lower than conventional SMR temperatures with simultaneous hydrogen separation for producing high purity hydrogen. This paper will present current status of Pd-alloy membrane technology at Pall Corporation. Results of model simulations to determine the effect of various operating conditions of steam to carbon ratio, space velocity, membrane temperature, and feed pressure on the potential benefits of conducting the SMR reaction in a membrane reactor configuration are discussed.

The composite membrane technology will reduce the capital cost of equipment required for hydrogen production by combining the hydrogen generation and purification steps. The U.S. DOE has set year 2010 targets of 250 scfh/ft2 for hydrogen flux and $1000/ft2 for membrane module cost. Combining these goals into a single goal of $4/scfh of hydrogen capacity sets a relevant, challenging goal for commercialization. This paper will also present the techno-economic evaluation of the membrane reactor process for hydrogen production and approaches for meeting the hydrogen production cost target.

32d a Novel Membrane Reactor for Electrochemical Preferential Oxidation (ECPrOx) of CO

Saurabh A. Vilekar, Chemial Engineering Department, Worcester Polytechnic Institute, 100 Institute Road, Worcester, MA 01609 and Ravindra Datta, Chemical Engineering, Worcester Polytechnic Institute, 100 Institute Road, Worcester, MA 01609.

Steam reforming constitutes a major source of hydrogen production. A typical reformate from a steam reformer followed by water-gas-shift (WGS) contains roughly 75% H2, 24% CO2, and 5000-10000 ppm CO. However, CO present in the reformate poisons platinum electro-catalyst and adversely affects the fuel cell performance. Thus, preferential oxidation (PrOx) is used to selectively oxidize CO. Despite being universally adopted in fuel reformers, the current PrOx technology used to reduce the CO is in fact, cumbersome and bulky. Another limitation is the poor selectivity of the PrOx reactor for CO oxidation over H2.

An entirely novel membrane reactor with the configuration of a fuel cell, namely ECPrOx, i.e., electrochemical preferential oxidation of CO is being developed in our laboratory. This membrane-based process exploits the oscillatory CO electro-oxidation by utilizing electric potential to cleanse CO from reformate. Here selective CO electro-oxidation is achieved at the anode by rendering the process electrochemical. In addition, supplemental power is produced with no H2 wasted.

The electrochemical device has direct application in on-board reforming process for fuel cell applications and the potential to replace both WGS reaction and PrOx.

32e Advanced Membrane Technologies for Shipboard and Apu Fuel Cell Systems

John Heinzel, NSWCCD Code 985, Naval Sea Systems Command - Philadelphia, 5001 S. Broad St, Philadelphia, PA 19112, Donald Hoffman, Code 331, Office of Naval Research, 875 N. Randolph St., Arlington, VA 22203, and John Kuseian, Code 985, Naval Sea Systems Command - Philadelphia, 5001 S. Broad St, Philadelphia, PA 19112.

The U.S. Navy, through the Office of Naval Research, has invested in compact fuel processing technologies for some time. Central to the lessons learned from various development and demonstration efforts has been the need for compact, selective means of gas separation. Such devices are critical for use in removing diluents and poisons from the gas stream to ensure long life. Additionally, high temperature membrane technologies are necessary, to enable simpler process integration for systems where robustness and power density are the key drivers.

This paper and presentation discuss the current programs that the U.S. Navy is investing in, to advance technologies necessary for use in notional shipboard and vehicle APU systems. The work will discuss system drivers and requirements that present the needs for specific types of membrane separation units, as well as how their notional usage plays into certain system designs. Further, this work describes efforts being performed with respect to fuel cell membrane characterization, particularly under use in Naval environments. The overall intent is to inform the viewer of the needs of membrane-based separation operations, from the aspect of system integration, for Navy applications. Real data from various demonstrations and equipment-based models will be presented to augment the points and information shown.

32f Water Transport in Nafion

Jay B. Benziger[1], Paul M. Majsztrik[2], M. Barclay Satterfield[3], and Andrew B. Bocarsly[2]. (1) Department of Chemical Engineering, Princeton University, Princeton, NJ 08544, (2) Chemistry, Princeton University, Princeton, NJ 08544, (3) Chemical Engineering, Princeton University, Princeton, NJ 08544

The permeation of water through Nafion membranes has been measured as a function of film, temperature and water activities. Water permeation coefficients increased with water content in Nafion. For water activity in the range 0 - 0.8, permeation coefficients increased linearly with water activity and scaled inversely with membrane thickness. The permeation coefficient was independent of membrane thickness when the membrane was in contact with liquid water (aw=1). Water transport is shown to be consistent with water diffusion through the hydrophilic phase of Nafion. The volume fraction of the hydrophilic phase increases with water activity, which correlates with the increase in the permeation coefficient. As aw approaches 1, the hydrophilic domain volume increases sufficiently whereby the limiting resistance to water transport changes from diffusion to interfacial transport across the membrane/gas interface. The mechanical and transport properties are linked to structural changes in the microphases in Nafion induced by water absorption.

Web Page: pemfc.princeton.edu

43a Separation of Plasmid DNA Isoforms Using Ultrafiltration

David R. Latulippe and Andrew L. Zydney. Chemical Engineering, Pennsylvania State University, 219 Fenske Laboratory, University Park, PA 16802

Recombinant plasmid DNA can be used as a novel therapeutic molecule for gene therapy and DNA vaccination. One of the challenges in plasmid DNA production is the separation of the desired supercoiled isoform from host-cell related impurities including genomic DNA, RNA, proteins, as well as the undesirable linear and open-circular plasmid DNA isoforms formed by irreversible breaks in the DNA backbone. Current FDA regulations specify that at least 90% of the plasmid DNA in the final product must be in the supercoiled form. Due to the very large size of plasmid DNA, traditional chromatographic strategies suffer from a number of limitations including limited resolution, very low binding capacity, mass transfer limitations, and high processing costs. Membrane-based processes have great potential for large-scale plasmid DNA purification. The objective of this work was to obtain quantitative data on the effects of plasmid isoform structure on the transmission of plasmid DNA during ultrafiltration.

Experiments were conducted with Ultracel composite regenerated cellulose membranes and a 3000 base pair pBluescript plasmid. The plasmid as supplied was predominantly in the supercoiled form (>90%). Open-circular and linear isoforms were prepared by incubating the plasmid with appropriate restriction enzymes. Plasmid concentrations were determined using PicoGreen DNA fluorescent reagent allowing accurate detection of concentrations as low as 250 ng/L. Stirred-cell ultrafiltration experiments were conducted over a range of filtrate flux and solution ionic strength using different molecular weight cut-off membranes. The different DNA isoforms were analyzed using agarose gel electrophoresis.

Experiments with the individual isoforms showed that transmission was a strong function of both the filtrate flux and solution ionic strength. The strong flux-dependence was due to the elongation of the DNA in the converging flow field into the membrane pores. For all conditions tested, the linear DNA had a higher sieving coefficient than the supercoiled and open-circular forms due to differences in conformation and flexibility of the DNA isoforms. Experiments with a mixture of plasmid isoforms confirmed that it was possible to purify the different isoforms by proper selection of the membrane pore size, filtrate flux, and solution ionic strength. For example, very good separation of the linear form from the supercoiled and open-circular forms was achieved with the largest pore size membrane by operating at a filtrate flux slightly above the critical flux for transmission of the linear DNA. This work demonstrates that high-resolution purification of supercoiled DNA can be achieved using membrane ultrafiltration by exploiting differences in the conformational flexibility of the different plasmid isoforms.

43b Ultrafiltration of Highly Aggregating Proteins

Gopinath V. Annathur[1], Sa .V Ho[1], and Sami Kawas[2]. (1) Global Biologics, Pfizer Global R&D, 700, Chesterfield Pkwy west, Chesterfield, MO 63017, (2) Chemical Engineering, University of Texas, Austin, Austin, TX 000

Ultrafiltration (UF) and Diafiltration (DF) processes are typically performed at constant trans-membrane pressures (TMP) or with increasing TMP to maintain certain flux rate. For some proteins, such operations, however, can cause uncontrolled material buildup on the membrane wall, potentially leading to disruptive membrane fouling as well as product loss. Operation at constant, controlled wall concentration has been proposed by Van Reis et. al that can both help avoid these complications and achieve optimal performance of the process. Their outlined procedure for optimizing UF operation involves one single experimental run to generate sufficient data allowing determination of the parameters required for constant Cw operation, optimization of Cw, and accurate prediction of the respective process runtime. We tested this constant wall concentration (Cw) approach for three different classes of proteins: a monoclonal antibody from cell culture and two microbially-derived proteins -- a highly soluble one and a highly aggregating protein. We also gathered diffusional kinetic data for various proteins to explore the effect of protein concentration on its aggregation behavior. It was found that the constant Cw approach is effective for highly soluble proteins but encounters difficulty in predicting UF performance of proteins that have a strong tendency to aggregate in a concentration dependent manner. An alternative treatment for those proteins is being investigated.

43c Microfiltration and Thermodynamic Investigations of Proteins on Anopore Nanocomposite Membranes

Tahmina Imam, Chemical Engineering, Texas A&M University, College Station, TX 77843, **Seunguk Yeu**, Department of Chemical Engineeing, Texas A&M University, TAMU 3122, College Station, TX 77843-3122, Zivko L. Nikolov, Biological and Agricultural Engineering, Texas A&M University, 2117 TAMU, College Station, TX 77845, and Daniel F. Shantz, Department of Chemical Engineering, Texas A&M University, TAMU 3122, College Station, TX 77843-3122.

Over the past few decades, there has been a great interest in exploring alternatives to conventional packed-bed chromatography for protein purification. Although membrane filtration offers a potentially attractive alternative, membrane fouling has been an impediment to practical use. To address this problem, we are exploring the potential of using organic-inorganic hybrid membranes. This presentation will summarize our recent work using functionalized AnoporeTM alumina membranes with macropores of 200nm diameter (Whatman). The fouling behavior before functionalization, and after a variety of surface modifications will be reported for bovine serum albumin (BSA) (pI 4.7) and lysozyme (pI 11.0) using a stirred cell filtration system to determine the fouling and the flux decline during filtration. In short, the results show that the fouling behavior can be modified through a judicious choice of surface functionalization. These experimental results will be summarized. More recently we have investigated the equilibrium adsorption properties of these composite membranes and observe clear trends between the equilibrium adsorption of protein (BSA) and the flux enhancement. Finally, preliminary work will be reported investigating the competitive adsorption of IgG1 and BSA on these nanocomposite membranes.

43d Tangential Flow Ultrafiltration of Virus Containing Feed Streams

David Grzenia[1], Jonathan O. Carlson[2], and Ranil Wickramasinghe[1]. (1) Colorado State University, Department of Chemical Engineering, Fort Collins, CO 80523-1370, (2) Microbiology, Immunology and Pathology, Colorado State University, Department of Chemical Engineering, Fort Collins, CO 80523-1370

Purification of virus particles and viral vectors for viral vaccines and gene therapy applications is a major large scale separations challenge. Purification of parvovirus particles such as adeno-associated virus, the leading candidate for gene therapy applications, is particularly challenging given their small size, typically 18-26 nm. We have investigated the use of ultrafiltration for purification of Aedes aegypti densonucleosisvirus, a mosquito parvovirus and minute virus of mice, and FDA recommended model parvovirtus..

Four ultrafiltration membranes with molecular weight cut offs of 30, 50, 100 and 300 kD have been tested in tangential flow mode. Experiments have also been conducted where part of the permeate is pumped co-current to the feed. This results in a more constant transmembrane pressure drop along the entire membrane. Maintaining a more constant transmembrane pressure results in improved passage of host cell proteins and smaller virus particles.

Currently chromatographic separations such as ion exchange are used to remove DNA and host cell proteins. Our results indicate that by carefully selecting the molecular weight cut off of the membrane and maintaining a constant transmembrane pressure along the membrane, virus particles may be purified and concentrated by ultrafiltration. Reducing the impurity load to subsequent chromatographic steps will lead to the use of smaller columns resulting in cost savings.

43e Membrane Chromatography: Protein Purification Using Newly Developed, High-Capacity, Weak Anion-Exchange Membranes

Bharatkumar V. Bhut[1], Ranil Wickramasinghe[2], and Scott Husson[1]. (1) Chemical and Biomolecular Engineering, Clemson University, 112 Earle Hall, Dept. of Chemical and Biomolecular Engineering, Clemson University, Clemson, SC 29634-0909, (2) Colorado State University, Department of Chemical Engineering, Fort Collins, CO 80523-1370

Considering that the total cost of protein therapeutics is shifting from cell culture to downstream purification, high productivity and high resolution separation techniques are in demand by the biopharmaceutical industry. Membrane chromatography offers several advantages over resin-based media, such as low pressure drop and facile scale up and set up. With the advent of biotechnology, worldwide demand for protein therapeutics is increasing rapidly. Therefore, high dynamic capacity membranes are essential to meet productivity demands. The objective of this research was to investigate dynamic adsorption capacities and protein fractionation behavior of newly developed adsorptive membranes. High dynamic capacity (>50 mg/mL BSA) ion-exchange membranes were produced by grafting functional polymer nanolayers from commercially available regenerated cellulose membranes using atom transfer radical polymerization. Separation parameters and dynamic adsorption capacities were measured using polymerization time as independent variable. Influence of ionic strength and volumetric flow effects on dynamic binding capacities were studied using an Akta purifier. A set of experiments was performed to determine efficiency and recovery of proteins from mixtures using membrane chromatography.

43f Evaluation of Anion Exchange Membranes for the Removal of Trace Impurities from Mammalian Cell Culture Derived Protein Solutions

Justin R. Weaver[1], Scott Husson[2], and Ranil Wickramasinghe[1]. (1) Chemical and Biological Engineering, Colorado State University, Fort Collins, CO 80523-1370, (2) Clemson University, 126 Earle Hall, Clemson, SC 29634

The single largest cost incurred during development/manufacturing of a biopharmaceutical protein product (such as a Monoclonal Antibody or "mAb") occurs during the purification process. Purification of a mAb usually involves three chromatography steps: a capture step from cell culture medium (such as Protein A affinity); an intermediate purification step to remove product-associated impurities and other major contaminates; and a polishing step to remove trace impurities such as virus, DNA, and other low concentration impurities. Owing to the negative charge present on the surface of DNA and most viruses, as well as many Host Cell Proteins (HCP), at neutral pH, anion-exchange (AEX) chromatography is employed often as the predominant polishing step for such impurities, operating under conditions that the engineered protein of interest does not bind but "flows thru." Historically, the biopharmaceutical industry has used strong quaternary amine (Q) AEX resins in their processes for removal of trace impurities. The advantage of membrane adsorbers over traditional resins are convective flow to all the binding ligands enabling binding at higher flow rates (not limited by pore diffusion), low pressure drops across the membrane, one-time use disposability, and increased virus access to binding ligands (virus particles only bind to resin bead surface, too large to diffuse into pores). Therefore, we seek to evaluate the potential of two commercially available AEX membrane adsorbers (Sartorius Sartobind® Q and Pall Mustang® Q) and one beta AEX membrane from Millipore (ChromaSorb™ Prototype) to remove specific impurities: Minute Virus of Mice (MVM), HCP, and DNA. Results to date, with MVM as the only impurity, show maximum MVM removal factors of > 3.5 log for all membranes. The Sartobind Q shows a significant ($p < 0.05$) effect on LRV (log virus removal) with pH and NaCl concentration changes, while the Mustang Q shows a significant effect on LRV with changes in NaCl concentration only. Interestingly, the Millipore ChromaSorb™ Prototype membrane showed no significant effect on LRV with changes in pH, NaCl conc., or flow rate. Ongoing/future work will demonstrate the trace impurity removal capabilities of all three membranes with and without the presence of a "protein of interest" present, i.e. a monoclonal antibody. In addition, efforts will be made to develop a mathematical model to explain the removal potential of the membranes under different solution conditions and differing levels of binding solutes.

43g Cephalexin Recovery from Aqueous Solutions by Supported Liquid Membranes with Strip Dispersion

Michael E. Vilt, Chemical and Biomolecular Engineering Department, The Ohio State University, 125 Koffolt Laboratories, 140 West 19th Ave., Columbus, OH 43210 and W.S. Winston Ho, Department of Chemical and Biomolecular Engineering, The Ohio State University, Department of Materials Science and Engineering, 2041 College Road, Columbus, OH 43210-1178.

Cephalexin is an important and widely used semisynthetic cephalosporin. Cephalosprorins along with penicillins are Beta-lactam antibiotics, which account for the majority of the antibiotic world market. Cephalexin is traditionally produced by a ten-step chemical synthesis. An enzymatic synthesis for Cephalexin has been developed, and offers several advantages over the classical route. The enzymatic synthesis reduces energy and solvent waste, but has been used in industrial production on a limited basis due to its difficult separation. Liquid membranes, in particular supported liquid membranes (SLMs), are a promising solution to the separation. SLMs are not used commercially, as they are still plagued with problem of long term instability. SLM with strip dispersion has been a recent development to solve the issue of stability.

SLMs with strip dispersion are formed when an aqueous strip solution is dispersed in an organic membrane solution by a mixer, and passed onto one side of a membrane support. When a microporous hydrophobic support is used, the organic phase of the dispersion becomes imbedded in the pores of the support, forming a stable SLM. Stability is maintained by having a constant supply of organic membrane solution to the pores. An aqueous feed solution is passed on the other side of the membrane support, and the target species is extracted into the organic solution by a selective carrier. The target species is then stripped by the aqueous strip solution. For final recovery of the target species, the mixer is turned off or a settler is used, and the dispersion quickly separates.

In this study, Cephalexin has been separated and concentrated from an aqueous solution using SLMs with strip dispersion. The following key parameters were investigated: feed and strip dispersion flowrate, strip dispersion mixing rate, carrier concentration, counter ion concentration, pH, and the volume of aqueous strip solution. Experiments used a Liqui-Cel® hollow fiber module as a microporous support. The organic membrane solution of the SLM consisted of Aliquat 336, Isopar L (isoparaffinic hydrocarbon solvent), and 1-decanol. The aqueous strip solution was composed of potassium chloride and citrate buffer. An enrichment factor of 3.2 was observed in the aqueous strip solution while achieving over 99% extraction and 96.2% total recovery. In this case, the aqueous feed solution of 5500 ppm (15 mM) was lowered to 30 ppm when using an organic membrane solution containing 2.5% Aliquat 336. The resulting overall mass transfer coefficient was 1.6 x 10-5 cm/sec. The mass flux of Cephalexin for this system was found to be independent of aqueous feed and strip dispersion flowrates, suggesting a major mass transfer resistance due to chemical reaction kinetics, which is supported by calculated individual mass transfer resistances. The pH of the aqueous strip phase was found to play a more significant role when trying to achieve higher enrichment ratios. It was observed that the highest stripping efficiency occurs when the pH of the aqueous strip phase is between the values of 5 and 6.

63a Interactions of SO2 and H2S with Modified Amorphous Carbon Films

W. Michalak, E. Broitman, J.B. Miller, and A.J. Gellman. Chemical Engineering, Carnegie Mellon University, Pittsburgh, PA 15213

In addition to their potential for adverse environmental impact, sulfur-compounds commonly encountered in processing of fossil fuels can poison catalysts and foul process equipment. Therefore, there is significant interest in development of efficient processes for their capture and conversion. Activated carbons are well known for their ability to adsorb sulfur-compounds; both morphology and surface chemistry contribute to their performance. However, because control and characterization of activated carbon's surface properties are difficult, the interactions between sulfur-compounds and carbon surfaces are not understood at a fundamental level that enables rational design of new materials.

In this work, we use ultra-high vacuum (UHV) techniques to carefully prepare and thoroughly characterize amorphous carbon (a-C) thin films as models of activated carbon sorbent-catalysts. Films with modified surface chemistries were prepared by post-deposition oxidation of a sputtered carbon film (a-CO_x) and by sputtering in the presence of N_2 (for a-CN_x) or methane (for a-CH_x). Temperature programmed desorption (TPD) and x-ray photoelectron spectroscopy (XPS) were used to study H_2S and SO_2 surface chemistry on the films. For comparison, we performed similar experiments on a highly oriented pyrolytic graphite (HOPG) surface.

SO_2 and H_2S readily adsorb on the unmodified a-C surface at low temperature in UHV. In TPD experiments, both desorb with first-order kinetics and desorption energies (E_{des}) that are higher than their respective heats of sublimation. The chemically modified a-CO_x surface adsorbs less H_2S and SO_2 than a-C, but desorption kinetics and energetics are unaffected, suggesting a simple site blocking mechanism. In contrast, modification as a-CN_x and a-CH_x significantly alters the nature of the interactions between the adsorbed molecules and the carbon surface. SO_2 continues to adsorb in significant quantities on these surfaces, but desorbs from both with zero order kinetics, indicative of adsorption as islands or droplets on the surfaces. H_2S does not adsorb onto either a-CN_x or a-CH_x in appreciable quantities, mirroring the behavior of the inactive HOPG surface.

These results illustrate the potential for chemical modification of carbon surfaces for altering adsorption/ reaction pathways.

63b Adsorption Separation of Carbon Dioxide, Methane and Nitrogen on H- and Na-Exchanged β-Zeolite

Xiaoqin Liu[1], Xiaoliang Xu[1], Xingxiang Zhao[1], and **Xiaohua Lu**[2]. (1) State Key Laboratory of Materials-oriented Chemical Engineering, Nanjing University of Technology, Nanjing, Xinmofan Road 5#, Nanjing, China, (2) State Key Laboratory of Materials-oriented Chemical Engineering, Nanjing University of Technology, Xinmofan Road 5#, Nanjing, China

In recent years, due to the sharp increase in the price of petroleum and the stringent environmental regulations, there has been significant increase in the production of natural gas as an efficient and environmentally clean fuel supply. However, a large amount of the natural gas reserves as well as aging natural gas wells, which are available worldwide, have been not discovered to date because of containing significant concentrations of contaminating gases such as nitrogen and carbon dioxide. To meet "pipeline-quality" methane, the natural gases field that contain significant amounts of nitrogen need to be upgraded to meet the pipeline quality for minimum heating value specifications. The maximum amount of carbon dioxide cannot exceed 2% because the existence of carbon dioxide reduces the conversion rate and energy content of natural gas. Furthermore, carbon dioxide is acidic and corrosive in the presence of water within the transportation and/or storage system. Recently, adsorption technology has been considered as an alternative to carbon dioxide absorption with liquid amines and cryogenic distillation for nitrogen removal.

In this paper, static volumetric measurements of the adsorption of carbon dioxide, methane and nitrogen onto H- and Na-β zeolite were performed in the pressure range of 0-1 atm at (273, 303)K. The adsorption capacity for carbon dioxide, methane and nitrogen increased on sodium-exchanged β-zeolite compared to Hβ. The magnitude of the increase in the adsorption capacities for carbon dioxide was much higher than that of methane and nitrogen. The carbon dioxide over methane as well as carbon dioxide over nitrogen selectivities in the Henry's law region increased upon Naβ. The limiting heat of adsorption data showed stronger interactions of carbon dioxide molecules with sodium cations in Naβ. These observations had been explained in terms of larger electrostatic interaction of carbon dioxide with extra-framework zeolite cations. However, Naβ showed a decrease in the selectivity of methane over nitrogen. This decrease could be ascribed to the moderate affinity of nitrogen to Naβ. By using the Clausius-Clapeyron equation, the surface energetic character of the adsorbents was analyzed. All the isotherms could be described appropriately with Langmuir-Freundlich model.

63c Promoted ZnO Sorbents for Low Temperature H_2s Removal in PEM Fuel Cell

Priyanka Dhage, Vivekanand Gaur, Hongyun Yang, and Bruce Tatarchuk. Department of Chemical Engineering, Auburn University, Center for Microfibrous Materials Manufacturing, Auburn, AL 36849

The present work is focused on the modification of ZnO based sorbents by promoting with varied molar ratios of transition metals for enhancing their H_2S removal capacity at low temperatures. Silica supported sorbents with formulation M_xZnO_{1-x}/SiO_2, (M = Mn, Fe, Co, Ni, Cu) were prepared by incipient wetness method and tested in 1 vol% flowing H_2S/H_2 in packed bed reactor in both dry and moist conditions. $Cu_{0.05}ZnO_{0.95}/SiO_2$ showed the highest capacity of 77 mg S/g sorbent and almost 90% theoretical metal utilization followed by Fe > Co > Ni = Mn at same doping levels. The structural modification of ZnO surface with Cu-promoter (5 mol%) showed up to 30% increase in saturation capacity over the un-promoted ZnO/SiO_2 and 58% increase over commercial ZnO sorbent, examined under identical conditions. The experimental results on the effect of varying Cu concentration level in Cu_xZnO_{1-x}/SiO_2, where $0 \leq x \leq 1$, revealed that $Cu_{0.2}ZnO_{0.8}/SiO_2$ showed the highest capacity of 106 mg S/g sorbent. The bulk and surface properties of the sorbents before and after promotional modification were characterized by X-ray diffraction and N_2 adsorption-desorption isotherms. The sulfided sorbents were thermally regenerated under airflow for repeated use in multiple adsorption- desorption cycles. To investigate the influence of the structural modification on the regenerability of Cu_xZnO_{1-x}/SiO_2, the sorbents promoted with distinct Cu-contents were tested at wide temperature range for multiple cycles.

Furthermore, to develop novel microfibrous entrapped sorbents and enhance the H_2S removal efficiency for fuel cell application, composite bed design in which glass fiber entrapped sorbent (particle size – 150-250 μ m) is used as the polishing layer at the downstream of a packed bed, was comparatively examined. This new and pioneer design resulted a dual advantage of lower pressure drop and enhancement in sulfur adsorption capacity at the same time.

Equilibrium Calculation of Gaseous Reactive Systems with Simultaneous Species Adsorption

Mingheng Li *

Abstract

This work focuses on the calculation of chemical equilibrium in a gaseous reactive system with simultaneous single or multiple species adsorption under isothermal and isobaric conditions. Two different algorithms are developed following the minimization of Gibbs free energy and the concept of equilibrium constant, respectively. In either case the problem formulation is converted to a set of nonlinear algebraic equations solved using the Newton-Raphson scheme. An example of steam reforming of ethanol with simultaneous CO_2 adsorption is used to illustrate the proposed approaches. It is shown that at $T = 500$ °C and $P = 5$ bar, the CO_2 removal ratio should exceed 40% in order to achieve a decent enhancement in hydrogen production and purity. An integrated process that combines the endothermic reforming and the exothermic combustion of CH_4 from the off-gas supplemented with simultaneous CO_2 adsorption in the reforming process yields a theoretical maximum overall conversion rate of 86.3% (the corresponding H_2 purity out of the reformer is 89.4% on wet basis or 96.2% on dry basis) with little or no external heat supply. The analysis in this work is potentially useful in the design and optimization of adsorption enhanced reforming reactors for hydrogen generation and other applicable reactive systems.

Keywords: Equilibrium, Adsorption, Optimization, Hydrogen, Adsorption Enhanced Reforming

1 Introduction

Chemical equilibrium in a reactive system is a state where the chemical activity of each species does not have a net change over time.[1] The calculation of chemical equilibrium provides the thermodynamic limit of a chemical process and has been traditionally used in the design and analysis of chemical equipment such as turbines and engines etc.[2,3] The advances in chemical equilibrium algorithms have been significantly facilitated by the ever-increasing computational power of modern computers. One milestone in chemical equilibrium calculation is the development of the NASA chemical equilibrium and applications (CEA) code, which is capable of tracking hundreds of species in a gaseous reactive system and has been used in various applications such as rockets, incident/reflected shocks and Chapman-Jouguet detonations etc.[4] The so-called Gibbs reactor module that is based on the minimization of Gibbs free energy has also been incorporated in various modern engineering solvers including Pro/II®

*Department of Chemical and Materials Engineering, California State Polytechnic University, Pomona, CA 91768 USA. Tel: +1-909-869-3668. Fax: +1-909-869-6920. E-mail: minghengli@csupomona.edu.

and Aspen Hysys® to predict equilibrium compositions.[5,6] Nowadays thermodynamic analyses based on chemical equilibrium calculations have been found in a variety of applications such as material interfacial phenomena,[7,8] biological processes,[9,10] and reacting nozzle flows[11–14] to name a few.

More recently, researchers in various groups have shown a great interest in the application of chemical equilibrium to processes related to hydrogen generation and fuel cells.[15–23] As reducing the demand on fossil resources has been a public concern, hydrogen, being a potential carrier of clean energy, is now an important topic that may lead into a new era of energy research. The generation of hydrogen can be realized through various routes such as steam reforming of hydrocarbons,[24] gasification of coal,[25] chemical looping combustion of coal,[26] water splitting by thermochemical cycle[27] and electrolytic decomposition of water[28] etc. In particular, hydrogen generation from biomass has caught lots of attention as it is a potentially viable, renewable and carbon-neutral (or even carbon-negative in conjunction with sequestration) process.[29] Most of these processes involve water gas shift reaction ($CO + H_2O = CO_2 + H_2$). If CO_2 can be simultaneously removed as the water gas shift reaction proceeds, the thermodynamic limitation can be circumvented and therefore, the chemical equilibrium shifts to the right, resulting in an enhancement in the extent of the forward reaction and the hydrogen yield.

This work focuses on the development of computational algorithms for chemical equilibrium calculations in gaseous reactive systems with simultaneous single or multiple species adsorption. The algorithms are then applied to the thermodynamic analysis of steam reforming of ethanol. Implications to an integrated autothermal process for hydrogen generation from steam reforming of hydrocarbons are discussed at the end of this work.

2 Chemical equilibrium calculation through nonlinear optimization

For a closed system under constant temperature and pressure, the Gibbs free energy decreases as equilibrium is approached. Therefore, one approach to equilibrium calculation in an isothermal and isobaric reactive system is to minimize its total Gibbs free energy. The mathematical formulation generally leads to a nonlinear optimization problem which can be solved using numerical methods. This approach is advantageous over the equilibrium constant method when simultaneously tracking a large number of species is necessary because it circumvents specifying a set of reactions a priori. To this end, it is worth noting that even though the NASA CEA package is capable of handling various types of equilibrium calculations, no species adsorption is explicitly accounted for in the current code.[4]

The focus of this work is to calculate the chemical equilibrium of a reactive gaseous system with species adsorption, which is a common problem to adsorption enhanced reactive systems.[30–35] In the problem formulation, the following assumptions are made: (1) the reactive system is maintained at isothermal and isobaric conditions; (2) the gaseous phase obeys the ideal gas law; (3) there is no heterogeneous reaction on the surface of the adsorbent; and (4) there is no condensed species in the entire reactive system. Consider such a reactive system with one kilogram of reactants, the Gibbs free energy (G) of the entire reactive system is:

$$G = \sum_{j=1}^{s}(\mu_j^{(g)} n_j^{(g)} + \mu_j^{(ads)} n_j^{(ads)}) \tag{1}$$

where s is the total number of species, n_j is the amount of species j in mole, μ is the chemical potential, and the superscripts $^{(g)}$ and $^{(ads)}$ stand for properties related to species in the gas phase and on the surface, respectively. When equilibrium is reached, the following criteria should be satisfied:

$$\mu_j^{(ads)} = \mu_j^{(g)} \quad (j = 1, ..., s) \tag{2}$$

For a gaseous species that obeys ideal gas law, its chemical potential can be calculated using the following equation:[1]

$$\mu_j^{(g)} = \mu_j^{0(g)} + RT \ln \frac{P}{P^0} + RT \ln \frac{n_j^{(g)}}{n_T^{(g)}} \quad (j = 1, ..., s) \tag{3}$$

where T and P are temperature and pressure of the system, respectively, R is the gas constant (8.314 J/mol/K), ^{0} stands for properties under standard conditions ($P^0 = 1$ bar), and $n_T^{(g)} = \sum_{j=1}^{s} n_j^{(g)}$. A detailed expression for $\mu_j^{(ads)}$ might not be necessary because the relationship between the gaseous species and the adsorbed species can sometimes be experimentally obtained. Based on the Langmuir isotherm widely used in the analysis of adsorption phenomena[36] or experimental observations (e.g., equilibrium chemisorption isotherms of CO_2 on hydrotalcites[18,35]), the adsorption/desorption equilibrium typically follows the relationship below:

$$n_j^{(ads)} = S\frac{n_{j,sat}^{(ads)} \beta_j P_j}{1 + \sum_{j=1}^{s} \beta_j P_j} = S\frac{n_{j,sat}^{(ads)} \beta_j \frac{n_j^{(g)}}{n_T^{(g)}} P}{1 + \sum_{j=1}^{s} \beta_j \frac{n_j^{(g)}}{n_T^{(g)}} P} \quad (j = 1, ..., s) \tag{4}$$

where $n_j^{(ads)}$ is the actual adsorbed amount of species j in mole, $n_{j,sat}^{(ads)}$ is the saturated surface concentration in mole/m^2, S is the total surface area of the adsorbent, β_j is a constant for species j under isothermal conditions, P_j is the partial pressure of species j in the gas phase. Based on Eq. 4, the ratio of $n_j^{(ads)}$ to $n_j^{(g)}$ is expressed as follows:

$$r_j = \frac{n_j^{(ads)}}{n_j^{(g)}} = \frac{S n_{j,sat}^{(ads)} \beta_j P}{n_T^{(g)} + \sum_{j=1}^{s} \beta_j n_j^{(g)} P} \quad (j = 1, ..., s) \tag{5}$$

There are two ways to solve the relationship between the equilibrium composition and the amount of adsorbent. One could start with the surface area of the adsorbent and include Eq. 4 in the optimization problem to solve the equilibrium composition. Alternatively, one could start with r_j and solve the equilibrium composition from the optimization problem. The surface area of the adsorbent is then calculated based on Eq. 5 and the equilibrium composition. The latter approach is used in this work because the number of constraints can be significantly reduced in the optimization problem. However, it is important to note that r_j are not independent if multiple species are adsorbed simultaneously. In such a case, r_j should be specified based on the following relationship:

$$\frac{r_i}{r_j} = \frac{n_{i,sat}^{(ads)}\beta_i}{n_{j,sat}^{(ads)}\beta_j} \tag{6}$$

To minimize the Gibbs free energy of the entire system described by Eq. 1, the mass balance should be satisfied at the same time. This is achieved by the so-called atomic balance equations.[37] Let a_{ij} be the number of chemical element i contained in species j, the atomic balance equation is:

$$\sum_{j=1}^{s} a_{ij}n_j^{(g)} + \sum_{j=1}^{s} a_{ij}n_j^{(ads)} = \sum_{j=1}^{s} a_{ij}n_j^{(g)}(0) \quad (i = 1, ..., \epsilon) \tag{7}$$

where ϵ is the total number of chemical elements in this reactive system. $b_i(0) = \sum_{j=1}^{s} a_{ij}n_j^{(g)}(0)$ is the number of moles of elements i per kilogram reactants. It is worth noting that the effective number of equations in Eq. 7 is equal to the rank of the atomic matrix, which might be less than ϵ occasionally. In such a case, Eq. 7 should be replaced by its minimum realization. Interested readers are directed to open literature for more details.[37] However, the optimization algorithms can still be developed in a similar manner.

Based on the above analysis, the calculation of equilibrium compositions is formulated as the following nonlinear optimization problem:

$$\begin{aligned}
\min_{n_j^{(g)},n_j^{(ads)}} G &= \sum_{j=1}^{s}(\mu_j^{(g)}n_j^{(g)} + \mu_j^{(ads)}n_j^{(ads)}) \\
s.t. & \\
0 &= \sum_{j=1}^{s} a_{ij}n_j^{(g)} + \sum_{j=1}^{s} a_{ij}n_j^{(ads)} - b_i(0) \quad (i = 1, ..., \epsilon) \\
\mu_j^{(ads)} &= \mu_j^{(g)} \quad (j = 1, ..., s) \\
\frac{n_j^{(ads)}}{n_j^{(g)}} &= r_j \quad (j = 1, ..., s) \\
\mu_j^{(g)} &= \mu_j^{0(g)} + RT\ln\frac{P}{P^0} + RT\ln\frac{n_j^{(g)}}{n_T^{(g)}} \quad (j = 1, ..., s) \\
n_T^{(g)} &= \sum_{j=1}^{s} n_j^{(g)}
\end{aligned} \tag{8}$$

Eq. 8 can be simplified by canceling the terms related to the adsorbed species. The resulting problem is:

$$\begin{aligned}
\min_{n_j^{(g)}} G &= \sum_{j=1}^{s}(1+r_j)\mu_j^{(g)}n_j^{(g)} \\
s.t. & \\
0 &= \sum_{j=1}^{s}(1+r_j)a_{ij}n_j^{(g)} - b_i(0) \quad (i=1,...,\epsilon) \\
\mu_j^{(g)} &= \mu_j^{0(g)} + RT\ln\frac{P}{P^0} + RT\ln\frac{n_j^{(g)}}{n_T^{(g)}} \quad (j=1,...,s) \\
n_T^{(g)} &= \sum_{j=1}^{s} n_j^{(g)}
\end{aligned} \tag{9}$$

Strictly speaking, inequity constraints $n_j \geq 0$ $(j = 1, ..., s)$ should be included in the optimization problem described by Eq. 8 or 9. This is handled by solving $\ln n_j$ instead of n_j in the detailed solution procedure[4] . Even when this strategy is applied, sometimes n_j can still be very close (or theoretically equal) to zero during the iterations and singularity problems occur when taking logarithms of extremely small positive numbers. These can be avoided by purposely setting the minimum of n_j to a very small number (e.g., 10^{-100}). During the iterations, any n_j that is smaller than its minimum will be replaced by this lower limit. Such a modification does not have an effect on the accuracy of the solution because the contribution of $n_j\mu_j$ of these components to the total Gibbs free energy would be very close to zero (Note that $\lim_{n_j\to 0} n_j \ln n_j = -\lim_{n_j\to 0} n_j = 0$ based on L'Hopital's rule). With this in mind, the optimization problem described in Eq. 9 is then converted to a set of nonlinear algebraic equations using the lagrange multiplier method. First we define $f = \sum_{j=1}^{s}(1+r_j)\mu_j^{(g)}n_j^{(g)} + \sum_{i=1}^{\epsilon}\lambda_i\left[\sum_{j=1}^{s}(1+r_j)a_{ij}n_j^{(g)} - b_i(0)\right]$, where λ_i are the lagrange multipliers, the optimal solution to the optimization problem of Eq. 9 should be determined by solving the following nonlinear algebraic equations:

$$\begin{aligned}
0 &= \frac{\partial f}{\partial n_j^{(g)}} = (1+r_j)\mu_j^{(g)} + \sum_{i=1}^{\epsilon}(1+r_j)a_{ij}\lambda_i \quad (j=1,...,s) \\
0 &= \frac{\partial f}{\partial \lambda_i} = \sum_{j=1}^{s}(1+r_j)a_{ij}n_j^{(g)} - b_i(0) \quad (i=1,...,\epsilon) \\
\mu_j^{(g)} &= \mu_j^{0(g)} + RT\ln\frac{P}{P^0} + RT\ln\frac{n_j^{(g)}}{n_T^{(g)}} \quad (j=1,...,s) \\
n_T^{(g)} &= \sum_{j=1}^{s} n_j^{(g)}
\end{aligned} \tag{10}$$

The variables to be determined are $n_j^{(g)}$ $(j = 1, ..., s)$, λ_i $(i = 1, ..., \epsilon)$ and $n_T^{(g)}$. These $s + \epsilon + 1$ variables in Eq. 10 can be solved using the decent Newton-Raphson method. The central idea of the Newton-Raphon method is to apply multi-variable Taylor series expansion to a nonlinear vector function and then use truncated terms that contain only the first order derivatives to build a linear iterative formula. The formula can be then used to compute the solution with a given initial guess close to the solution; interested readers may refer to specialized books[38] for details. Following a method employed by the NASA CEA package,[4] the iterative variables are chosen to be $\Delta \ln n_j^{(g)}$ ($j = 1, ..., s$), $\Delta \ln n_T^{(g)}$ and $\pi_i = -\lambda_i/RT$ ($i = 1, ..., \epsilon$) in order to avoid taking the logarithm of negative numbers in the iteration procedure. Note here $\pi_i = -\lambda_i/RT$ is solved directly at each iteration step.

With the definition of π_i, the first equation in Eq. 10 can be converted to a dimensionless form as follows:

$$(1+r_j)\frac{\mu_j^{(g)}}{RT} - \sum_{i=1}^{\epsilon}(1+r_j)a_{ij}\pi_i \quad = \quad 0 \quad (j = 1, ..., s) \tag{11}$$

which has the following Newton-Raphson iterative formula:

$$\frac{(1+r_j)\mu_j^{(g)}}{RT} - \sum_{i=1}^{\epsilon}(1+r_j)a_{ij}\pi_i - \frac{\partial\left[\frac{(1+r_j)\mu_j^{(g)}}{RT}\right]}{\partial \ln n_T}\Delta \ln n_T^{(g)} + \frac{\partial\left[\frac{(1+r_j)\mu_j^{(g)}}{RT}\right]}{\partial \ln n_j}\Delta \ln n_j^{(g)} \quad = \quad 0 \quad (j = 1, ..., s) \tag{12}$$

or

$$\Delta \ln n_j^{(g)} \quad = \quad -\frac{\mu_j^{(g)}}{RT} + \sum_{i=1}^{\epsilon} a_{ij}\pi_i + \Delta \ln n_T^{(g)} \quad (j = 1, ..., s) \tag{13}$$

Similarly, the Newton-Raphson iterative formulas for the other equations can be determined as follows:

$$\sum_{j=1}^{s}(1+r_j)a_{ij}n_j^{(g)} - \sum_{j=1}^{s}a_{ij}n_j^{(g)}(0) + \sum_{j=1}^{s}(1+r_j)a_{ij}n_j^{(g)}\Delta \ln n_j^{(g)} \quad = \quad 0 \quad (i = 1, ..., \epsilon) \tag{14}$$

and

$$\sum_{j=1}^{s}n_j^{(g)} - n_T^{(g)} + \sum_{j=1}^{s}n_j^{(g)}\Delta \ln n_j^{(g)} - n_T^{(g)}\Delta \ln n_T^{(g)} \quad = \quad 0 \tag{15}$$

In order to expedite the calculation, especially when a large number of species are present in the reactive system, the so-called "reduced Gibbs iteration" scheme has been proposed to significantly reduce the number of variables in the iteration steps.[4] The idea is to cancel $\Delta \ln n_j^{(g)}$ in Eqs. 14 and 15 using Eq. 13. Consequently, Eqs. 14 and 15 are converted to

$$\begin{aligned} &\sum_{k=1}^{\epsilon}\left[\sum_{j=1}^{s}(1+r_j)a_{kj}a_{ij}n_j^{(g)}\right]\pi_k + \left[\sum_{j=1}^{s}(1+r_j)a_{ij}n_j^{(g)}\right]\Delta \ln n_T^{(g)} \\ &= \left[\sum_{j=1}^{s}a_{ij}n_j^{(g)}(0) - \sum_{j=1}^{s}(1+r_j)a_{ij}n_j^{(g)} + \sum_{j=1}^{s}(1+r_j)a_{ij}n_j^{(g)}\frac{\mu_j^{(g)}}{RT}\right] \quad (i = 1, ..., \epsilon) \end{aligned} \tag{16}$$

and

$$\sum_{k=1}^{\epsilon}\left[\sum_{j=1}^{s}a_{kj}n_j^{(g)}\right]\pi_k + \left[\sum_{j=1}^{s}n_j^{(g)} - n_T^{(g)}\right]\Delta \ln n_T^{(g)} \quad = \quad \left[n_T^{(g)} - \sum_{j=1}^{s}n_j^{(g)} + \sum_{j=1}^{s}n_j^{(g)}\frac{\mu_j^{(g)}}{RT}\right] \tag{17}$$

After the above transformations, the number of variables reduces from $s + \epsilon + 1$ to $\epsilon + 1$, which is favorable when a large amount of species are to be tracked simultaneously. It can be readily verified that the algorithm reduces to the standard NASA CEA one when there is no adsorption, or $r_j = 0$ (for $j = 1, ..., s$).[4]

The procedure of the calculation is to calculate the terms on the right hand side of Eqs. 16 and 17 with initial guesses of $n_j^{(g)}$ and $n_T^{(g)}$, and then solve $\Delta \ln n_T^{(g)}$ and π_k $(k = 1, ..., \epsilon)$ using these $\epsilon + 1$ linear equations. Subsequently, $\Delta \ln n_j^{(g)}$ $(j = 1, ..., s)$ are calculated using Eq. 13 based on the value of $\Delta \ln n_T^{(g)}$. The following formulas

$$\begin{aligned} \left(n_j^{(g)}\right)^{(m+1)} &= \exp\left[\left(\ln n_j^{(g)}\right)^{(m)} + \alpha^{(m)}\Delta(\ln n_j^{(g)})^{(m)}\right] \quad (j = 1, ..., s) \\ \left(n_T^{(g)}\right)^{(m+1)} &= \exp\left[\left(\ln n_T^{(g)}\right)^{(m)} + \alpha^{(m)}\Delta(\ln n_T^{(g)})^{(m)}\right] \end{aligned} \tag{18}$$

are then applied to update $n_j^{(g)}$ and $n_T^{(g)}$ to be used in the next iteration step. When the composition is far from equilibrium, a positive α that is smaller than one should be chosen to avoid divergence. When it is close to equilibrium, α is set to be 1. The readers may refer to the NASA CEA technical report[4] for more detailed discussions.

In the above equations, the thermodynamic data of each species at temperature T, such as the heat capacity, enthalpy, entropy and chemical potential are calculated as functions of temperature:[4]

$$\begin{aligned} \frac{c_p^0(T)}{R} &= \frac{a_1}{T^2} + \frac{a_2}{T} + a_3 + a_4T + a_5T^2 + a_6T^3 + a_7T^4 \\ \frac{H^0(T)}{RT} &= -\frac{a_1}{T^2} + \frac{a_2}{T}\ln T + a_3 + \frac{a_4}{2}T + \frac{a_5}{3}T^2 + \frac{a_6}{4}T^3 + \frac{a_7}{5}T^4 + \frac{a_8}{T} \\ \frac{S^0(T)}{R} &= -\frac{a_1}{2T^2} - \frac{a_2}{T} + a_3\ln T + a_4T + \frac{a_5}{2}T^2 + \frac{a_6}{3}T^3 + \frac{a_7}{4}T^4 + a_9 \\ \frac{\mu^{0(g)}}{RT} &= \frac{H^0(T)}{RT} - \frac{S^0(T)}{R} \end{aligned} \tag{19}$$

where $a_1, \ldots, a_9$ are constant for a given species.

Remark 1: Regarding the assumptions made in the problem formulation, if the ideal gas law is not adequately accurate (e.g., under very high system pressures), the partial pressure should be replaced by fugacity when the chemical potential is calculated. The formation of condensed species (e.g., the formation of carbon under low steam/ethanol ratios) is not accounted for in the current work and a more comprehensive algorithm is being developed to have this capability.

Remark 2: When the Gibbs minimization problem is solved using the lagrange multiplier method, only the first-order (necessary) optimality conditions are used. At an optimal solution $n_j^{(g)}$ such that $\frac{\partial f}{\partial n_j^{(g)}} = 0$ $(j = 1, ..., s)$, it can be verified that $\frac{\partial^2 f}{\partial {n_j^{(g)}}^2} = RT(1 + r_j)\left(\frac{1}{n_j^{(g)}} - \frac{1}{n_T^{(g)}}\right) > 0$ $(j = 1, ..., s)$. Given this fact, the computed optimal solution is indeed a local minimum. Interested readers are directed to open literature for global optimization techniques in the calculation of liquid-gas phase and chemical equilibriums with possible multiple minima and maxima in the Gibbs free energy.[39]

In the solution procedure, the initial estimate of $n_T^{(g)}$ is set to be 0.1 and those of $n_j^{(g)}$ are set to be uniformly distributed. This initial guess corresponds to an average molecular weight of 10, which follows the NASA CEA algorithm.[4] With these estimates, the optimization converges successfully within a small number of iterations in all cases studied in this work.

3 Chemical equilibrium calculation using the equilibrium constant method

Different from the Gibbs minimization method presented in the previous section, the Gibbs energy of the adsorbed species is not required if the amount ratios r_j are available. However, a set of gaseous reactions should be specified before the concept of equilibrium constant can be applied. A reaction scheme can be determined using a set of independent reactions such that any gaseous reaction in this reactive system can be expressed by a linear sum of these basis reactions.[40] The minimum number of independent chemical reactions in a reactive system is equal to the number of species minus the rank of the atomic matrix, provided that the former is greater than the latter.[37] For example, if the rank of the atomic matrix is e, the number of independent reactions is $s-e$. In most cases $e=\epsilon$. Occasionally $e<\epsilon$ when not all the row vectors in the atomic matrix are independent.

Consider a small change in the Gibbs free energy in the gaseous phase under isothermal and isobaric conditions:

$$\delta G^{(g)} = \sum_{j=1}^{s} \mu_j^{(g)} \delta n_j^{(g)} = \sum_{j=1}^{s} \mu_j^{(g)} \sum_{i=1}^{s-e} \nu_{ij} \delta \xi_i \tag{20}$$

where ν_{ij} is the stoichiometric coefficient of species j and ξ_i is the reaction extent in the i-th reaction, $\delta G^{(g)}=0$ for any $\delta\xi_i$ at chemical equilibrium implies:

$$\sum_{j=1}^{s} \mu_j^{(g)} \nu_{ij} = 0 \quad (i=1,...,s-e) \tag{21}$$

A combination of Eqs. 3 and 21 yields:

$$\sum_{j=1}^{s} \nu_{ij} \left[\frac{\mu_j^{0(g)}}{RT} + \ln \frac{P}{P^0} + \ln \frac{n_j^{(g)}}{n_T^{(g)}} \right] = 0 \quad (i=1,...,s-e) \tag{22}$$

or

$$\prod_{j=1}^{s} \left[\frac{P n_j^{(g)}}{P^0 n_T^{(g)}} \right]^{\nu_{ij}} = \exp\left[\sum_{j=1}^{s} \left(-\nu_{ij} \frac{\mu_j^{0(g)}}{RT} \right) \right] = K_{p_i} \quad (i=1,...,s-e) \tag{23}$$

which is just the so-called equilibrium constant method.[1]

The Newton-Raphson iterative formula for Eq. 22 is:

$$\sum_{j=1}^{s} \nu_{ij} \Delta \ln n_j^{(g)} - \left(\sum_{j=1}^{s} \nu_{ij} \right) \Delta \ln n_T = -\sum_{j=1}^{s} \nu_{ij} \frac{\mu_j^{(g)}}{RT} \quad (i=1,...,s-e) \tag{24}$$

As a result, the chemical equilibrium can be solved based on Eqs. 14, 15 and 24 (Note that the effective number of equations in Eqs. 14 is e if $e < \epsilon$. Therefore, the number of variables is equal to the number of equations). The iterative variables are $\Delta \ln n_j^{(g)}$ $(j = 1, ..., s)$ and $\Delta \ln n_T^{(g)}$. Even though more iterative variables are involved in the equilibrium constant method, a solution procedure similar to the one presented in section 2 can be followed.

4 A case study - steam reforming of ethanol with simultaneous CO_2 adsorption

A reactive system consisting of steam and ethanol is studied in this section. The steam reforming of biomass-generated ethanol would be a potentially viable and renewable process for hydrogen production.[29,41,42] Experimental studies have shown that for a system of hydrocarbon (e.g., methane, methanol or ethanol) and steam, the yield and purity of hydrogen can be significantly improved if CO_2 is adsorbed.[31,32,43–46] Calculations in this section will provide a quantitative analysis of the underlying physicochemical behavior.

In this reactive system, $T = 500$ °C and $P = 5$ bar and the feed ratio of C_2H_5OH to H_2O is 1/3, which are from open literature.[47] The species to be tracked in the systems are CH_4, CO, CO_2, C_2H_4, C_2H_5OH, CH_3CHO, H_2, H_2O and O_2. The formation of byproducts such as C_2H_4, CH_3CHO and CH_4 has been observed in the ethanol reforming processes under certain conditions.[29,47] However, only CH_4 is shown to have a differentiable concentration under the operating conditions in this work. CO_2 is assumed to be the only species that is adsorbed on the surface (i.e., only r_{CO_2} can be a nonzero number) even though the adsorption of multiple species can be readily handled with little extra computational effort. A list of the thermodynamic coefficients, the atomic matrix, and the reaction stoichiometric coefficient matrix are shown in Tables 1, 2 and 3 respectively. Note that the reaction scheme is required only in the equilibrium constant method. In Table 3, a set of 6 independent reactions are representative of the entire system because the number of species is 9 and the rank of the atomic matrix is 3.

The conversion rate of ethanol to hydrogen with simultaneous CO_2 adsorption is solved using the algorithms presented in sections 2 and 3. Computer simulations demonstrate that both algorithms end up with the same results. Typically, it takes only seconds to finish hundreds of simulations. For steam reforming of ethanol, the ideal overall reaction would be $C_2H_5OH + 3\ H_2O = 6\ H_2 + 2\ CO_2$, which corresponds to a 100% conversion of C_2H_5OH to H_2. When CO_2 is adsorbed under different $CO_2^{(ads)}/CO_2^{(g)}$ ratios, the H_2 yield is shown in Figure 1. Note that the horizontal axis is chosen to be $r_{CO_2} = CO_2^{(ads)}/CO_2^{(g)}$ instead of $CO_2^{(ads)}$ because the former is roughly proportional to the amount of adsorbent, and therefore, the size of the reactor (See Figure 2 for a plot generated based on Eq. 5 and the results from the equilibrium calculation. β_{CO_2} under the operating conditions in this work is estimated to be 22 bar^{-1} based on the experimental data of promoted hydrotalcites[48]). It is seen that with no or little CO_2 adsorption, the conversion rate of C_2H_5OH to H_2 is only 15.4%, which is consistent with the NASA CEA code.[4] When CO_2 is adsorbed, the conversion can be significantly enhanced. For example, at $CO_2^{(ads)}/CO_2^{(g)} = 10^2$, the conversion rate of C_2H_5OH to H_2 is 38.9%, or enhanced by a factor greater than 2 (A list of the composition under these two conditions is shown in Table 4). However, a further improvement in the conversion of C_2H_5OH to H_2 requires a significant increase in $CO_2^{(ads)}/CO_2^{(g)}$, or the amount of adsorbent. This can be explained by Figure 3, in which it is shown that as CO_2 is adsorbed on the surface, a further removal of CO_2 from the gas phase becomes more and more difficult.

The gas composition in the product mixture under different $CO_2^{(ads)}/CO_2^{(g)}$ ratios is shown in Figure 4. Under the operating conditions of interest, none of C_2H_4, CH_3CHO, C_2H_5OH or O_2 has a differentiable fraction in the product mixture. As $CO_2^{(ads)}/CO_2^{(g)}$ increases, the fraction of $CO_2^{(g)}$ rapidly decreases to a minimal level (at $CO_2^{(ads)}/CO_2^{(g)} = 10^{1\sim2}$), beyond which a significant reduction in the $CO_2^{(g)}$ concentration is not obvious. The low partial pressure of CO_2 in the gas phase makes it difficult for an effective adsorption. In fact, at $CO_2^{(ads)}/CO_2^{(g)} > 10^1$ the primary carrier of carbon in the gaseous phase is CH_4 instead of CO or CO_2. When CO_2 is gradually adsorbed on the surface, the forward reaction in the reversible water gas shift reaction $CO + H_2O = CO_2 + H_2$ is favored. As a result, part of the CH_4 is converted to CO through the methane reforming reaction $CH_4 + H_2O = CO + 3\ H_2$. In this sense, the entire system behaves like a buffer solution. Due to the coupling of multiple equilibrium phenomena, a complete removal of CO_2 is very difficult.

An interesting phenomenon is also observed in this reactive system regarding the specific volume of the product mixture. Note that the calculation is based on a fixed amount of mass (i.e., 1 kilogram) of the reactant mixture, and therefore, the volume of the product is proportional to the total amount $n_T^{(g)}$. As shown in Figure 5, the volume of the gas mixture expands about 1.37 times at equilibrium if there is no CO_2 adsorption. According to Le Chatelier's principle, a low system pressure is preferred for a high conversion of C_2H_5OH to H_2 at chemical equilibrium, which is consistent with experimental observations.[47] When CO_2 is gradually adsorbed on the surface, the volume of the product shrinks. After reaching a minimum at a $CO_2^{(ads)}/CO_2^{(g)}$ ratio around 14, the volume of the product increases even as CO_2 is adsorbed. This is because the effect of volume expansion due to methane reforming $CH_4 + H_2O = CO + 3\ H_2$ is greater than the shrinking effect caused by CO_2 adsorption.

Figure 6 shows the relationship between the conversion rate of C_2H_5OH to H_2 as a function of the CO_2 removal ratio, defined as the absolute amount of CO_2 adsorbed on the surface divided by its theoretical upper limit. It is seen that the enhancement in the H_2 production is not significant when the CO_2 removal ratio is below 40% (corresponding to a $CO_2^{(ads)}/CO_2^{(g)}$ around 13). This is because the methane reforming reaction and the water gas shift reaction shift only a little to the right, which is reflected by the fact that the total amount drops significantly (see Figure 5) and that the fraction of CH_4 and H_2O might even increase (see the peaks in CH_4 and H_2O fractions in Figure 4). As the CO_2 removal ratio exceeds 40%, the H_2 production increases linearly (A similar phenomenon is observed in H_2 purity). In such a case, there is little CO_2 or CO in the gas phase (Figure 4), and therefore, the overall reaction in the gas phase is $CH_4 + 2\ H_2O = CO_2 + 4\ H_2$. Because the CO_2 generation rate is roughly balanced by its adsorption rate, 4 mole of H_2 will be generated when 1 mole of CO_2 is removed from the gas phase by adsorption. If the result is interpreted in terms of the H_2 production and CO_2 removal ratio, the slope is 4/3 because the ideal reaction is $C_2H_5OH + 3\ H_2O = 6\ H_2 + 2\ CO_2$.

The reactive system in the above analysis is based on a setting similar to a single stage batch reactor, in which the gas phase has a uniform composition and is in equilibrium with the adsorbent. If a reforming reactor is divided into several zones (or stages) and a local chemical equilibrium is maintained in each zone, the overall hydrogen yield at the end of the reactor could be higher. This is because CO_2, once being adsorbed on the surface, does not have an effect on the reactive system downstream, which is true in a real reforming reactor. In the next case study, the CO_2 is adsorbed using multiple stages in series. This is realized by setting the gas composition at the outlet of one stage to be the inlet composition at the next stage. In addition, the gaseous reactive system at each stage is in local equilibrium with the adsorbent. Four cases are studied in which the stage $CO_2^{(ads)}/CO_2^{(g)}$ ratios are different. However, the product of $CO_2^{(ads)}/CO_2^{(g)}$ and the number of stages is fixed which makes the total adsorbent area roughly the same. A smaller $CO_2^{(ads)}/CO_2^{(g)}$ at each equilibrium stage corresponds to a shorter time scale for reaction and adsorption equilibrium. In each case, the cumulative adsorbent surface area and the H_2 yield are solved as functions of number of stages times $CO_2^{(ads)}/CO_2^{(g)}$ and the results are shown in Figures 7 and 8. It is seen that the cumulative adsorbent surface area is a linear function of the number of stages and the total surface areas are indeed roughly the same in all cases. However, the conversion rate to hydrogen might be higher if the reaction and adsorption equilibrium are faster, even though the difference becomes smaller at an adequately large adsorbent area. In all cases, the long tails in the hydrogen yield profiles suggest that an improvement in the hydrogen yield becomes more and more challenging and a trade off between the adsorbent size and the hydrogen yield should be made.

Based on this finding, and the fact that the overall gas phase reaction is endothermic at high hydrogen yields, an integrated process design could encompass the heat exchange between the endothermic reforming and exothermic combustion of remaining CO and CH_4 in the off-gas downstream of a H_2 separation unit. See Figure 9 for a schematic of this process. C_2H_5OH and H_2O are preheated by the flue gas before entering the reformer, where H_2 is generated and a major part of CO_2 is adsorbed. The product mixture passes through a hydrogen separation unit (e.g., a pressure swing adsorption unit, or PSA), where H_2 is separated from unconverted CH_4, H_2O as well as some CO and CO_2 in trace amount. The CH_4 from the H_2 separation unit is sent to a combustor where heat is supplied to the reformer as well as the feed streams (C_2H_5OH and H_2O) through heat exchange (HE) units. Interested readers might find more discussions in autothermal processes in open literature.[24,29,49–51] If the heat released from the CO_2 adsorption is not taken into account (since it will be released in subsequent CO_2 desorption), a plot of the heat demand in the reformer and the heat supply in the combustor v.s. the CO_2 removal ratio is given in Figure 10. Under the current operating conditions, the gaseous reaction could be exothermic (i.e., negative heat demand) at low CO_2 removal ratios where the H_2 yields are also low (see Figure 6). However, as the H_2 yield increases, the reforming reaction becomes endothermic and the waste heat provided by the combustion of un-reacted CH_4 and CO decreases. A heat neutral point occurs at a CO_2 removal ratio around 90%. In the above calculation both reactions occur at 500 °C. Therefore, it does not account for pre-heating of feed streams to 500 °C or cooling of flue gases to room temperature or the energy required for the H_2 separation unit. Moreover, a 100% efficiency of heat transfer between these two processes is assumed. This rough calculation indicates that an 86.3% theoretical maximum overall conversion of C_2H_5OH to H_2 (the corresponding H_2 purity in the product mixture out of the reformer is 89.4% on wet basis or 96.2% on dry basis, see Table 5 for the composition of other components.) can be achieved with little or no external heat supply in the reforming process if combined with simultaneous CO_2 adsorption. The reactions in the reformer reactor and the combustor as well as the overall reaction of the entire process (T = 500 °C) are summarized as follows:

$$
\begin{array}{lrcl}
(1) & C_2H_5OH + 3\ H_2O & = & 0.2052\ CH_4 + 5.1790\ H_2 + 0.4105\ H_2O + 1.7947\ CO_2\ (ads) \\
(2) & 0.2052\ CH_4 + 0.4105\ O_2 & = & 0.2052\ CO_2 + 0.4105\ H_2O \\
\hline
(sum) & C_2H_5OH + 3\ H_2O + 0.4105\ O_2 & = & 0.2052\ CO_2 + 5.1790\ H_2 + 0.8210\ H_2O + 1.7947\ CO_2\ (ads)
\end{array}
$$

Note that the gaseous CO and CO_2 in trace amount are not listed above. The net energy demand of the combined reforming and combustion process is zero at 500 °C if the heat released from CO_2 adsorption is not taken into account ($\Delta H_1 = -\Delta H_2 = 1.64 \times 10^5$ J/mol. In Figure 10 the heat demand and supply are both 1.64×10^6 J because they are based on 1 kilogram of the ethanol and steam mixture, which contains 10 moles of ethanol and 30 moles of steam). The overall H_2 yield could be less if the heat transfer efficiency and the operating cost are taking into account in this integrated process. A more detailed engineering calculation, which includes regeneration of CO_2, will be presented in a separate work.

It should be noted that even though the equilibrium calculation can provide a mean-field analysis and predict the overall trend of a reforming process, it is not adequate to mimic the real operation. Due to the continuous nature of the adsorption enhanced reforming process, a full description requires kinetic models (including the kinetics of reactions and adsorption/desorption as well). Interested readers are directed to open literature[30,32] for research efforts in this direction.

5 Concluding remarks

Two approaches for the calculation of chemical equilibrium in a gaseous reactive system combined with simultaneous species adsorption are developed in this work following algorithms similar to the NASA CEA code. Both algorithms are very effective in solving gaseous reactive systems with simultaneous species adsorption. Whereas the equilibrium constant method requires specifying a set of independent reactions and solving more nonlinear equations, these reactions are useful in the analysis of various coupled equilibrium phenomena in the entire reactive system.

The algorithms are applied to an enhanced ethanol reforming system where CO_2 is partially or fully removed by adsorption. It is shown that at T = 500 °C and P = 5 bar, the CO_2 removal ratio should exceed 40% in order to achieve a decent enhancement in hydrogen production and purity. Whereas the hydrogen production increases linearly with the CO_2 removal ratio if the latter is higher than 40%, the amount of adsorbent (and therefore, the reactor size) required for CO_2 increases exponentially. A trade off between the H_2 production and the reactor size can be achieved by burning CH_4 in the off-gas from the H_2 separation unit to provide heat for the endothermic reforming process. It is shown that a theoretical maximum overall conversion of C_2H_5OH to H_2 around 86.3% can be achieved with little or no external heat supply in the reforming process if combined with simultaneous CO_2 adsorption (the corresponding H_2 purity in the product mixture out of the reformer is 89.4% on wet basis or 96.2% on dry basis).

Even though the adsorption of a single species is discussed in the case study, the algorithms can be potentially applied to multiple species adsorption in a similar manner.

Acknowledgements

This work is partly supported by Intelligent Energy, Inc. (through a DOE subcontract) and the Faculty Center for Professional Development at California State Polytechnic University, Pomona. Industrial feedback from Dr. Durai Swamy and coworkers at Intelligent Energy, Long Beach, CA and discussions with Drs. T. K. Nguyen and W. Dong are gratefully acknowledged.

References

1. Sandler, S. I. *Chemical and Engineering Thermodynamics*, 3rd ed.; Wiley: New York, 1999.

2. Huzel, D. K.; Huang, D. H. *Design of Liquid Propellant Rocket Engines*; Technical Report NASA SP-125, 1971.

3. Tran, D. H.; Snyder, C. A. *Effects of Chemical Equilibrium on Turbine Engine Performance for Various Fuels and Combustor Temperatures*; Technical Report NASA TM-105399, 1992.

4. Gordon, S.; McBride, B. J. *Computer Program for Calculation of Complex Chemical Equilibrium Compositions and Applications*; Technical Report NASA RP-1311, 1996.

5. Hysys 4.0 User Guide; Aspen Technology, Inc., 2004.

6. Pro/II 8.0 User Guide; Invensys Systems, Inc., 2006.

7. Rao, Y. K. Chemical vapour deposition of WSi_2 thin films: Equilibrium W-Si-H-Cl-Ar system. *J. Alloys Compd.* **2008**, *452*, 116–121.

8. Won, Y. S.; Varanasi, V. G.; Kryliouk, O.; Anderson, T. J.; McElwee-White, L.; Perez, R. J. Equilibrium analysis of zirconium carbide CVD growth. *J. Cryst. Growth* **2007**, *307*, 302–308.

9. Lay-Son, M.; Drakides, C. New approach to optimize operational conditions for the biological treatment of a high-strength thiocyanate and ammonium waste: pH as key factor. *Water Res.* **2008**, *42*, 774–780.

10. Smirnov, V. V.; Brinkley, D. W.; Lanci, M. P.; Karlin, K. D.; Roth, J. P. Probing metal-mediated O_2 activation in chemical and biological systems. *J. Mol. Catal. A: Chem.* **2006**, *251*, 100–107.

11. Li, M.; Christofides, P. D. Multi-scale Modeling and Analysis of HVOF Thermal Spray Process. *Chem. Eng. Sci.* **2005**, *60*, 3649–3669.

12. Li, M.; Shi, D.; Christofides, P. D. Diamond Jet Hybrid HVOF Thermal Spray: Gas-phase and Particle Behavior Modeling and Feedback Control Design. *Ind. Eng. Chem. Res.* **2004**, *43*, 3632–3652.

13. Li, M.; Christofides, P. D. Computational study of particle in-flight behavior in the HVOF thermal spray process. *Chem. Eng. Sci.* **2006**, *61*, 6540–6552.

14. Li, M.; Christofides, P. D. Modeling and Analysis of HVOF Thermal Spray Process Accounting for Powder Size Distribution. *Chem. Eng. Sci.* **2003**, *58*, 849–857.

15. Assabumrungrat, S.; Laosiripojana, N.; Pavarajarn, V.; Sangtongkitcharoen, W.; Tangjitmatee, A.; Praserthdam, P. Thermodynamic analysis of carbon formation in a solid oxide fuel cell with a direct internal reformer fuelled by methanol. *J. Power Sources* **2005**, *139*, 55–60.

16. Assabumrungrat, S.; Pavarajarn, V.; Charojrochkul, S.; Laosiripojana, N. Thermodynamic analysis for a solid oxide fuel cell with direct internal reforming fueled by ethanol. *Chem. Eng. Sci.* **2004**, *59*, 6015–6020.

17. Colpan, C. O.; Dincer, I.; Hamdullahpur, F. Thermodynamic modeling of direct internal reforming solid oxide fuel cells operating with syngas. *Int. J. Hydrogen Energy* **2007**, *32*, 787–795.

18. Ding, Y.; Alpay, E. Equilibria and kinetics of CO_2 adsorption on hydrotalcite adsorbent. *Chem. Eng. Sci.* **2000**, *55*, 3461–3474.

19. Lutz, A. E.; Bradshaw, R. W.; Bromberg, L.; Rabinovich, A. Thermodynamic analysis of hydrogen production by partial oxidation reforming. *Int. J. Hydrogen Energy* **2004**, *29*, 809–816.

20. Lutz, A. E.; Bradshaw, R. W.; Keller, J. O.; Witmer, D. E. Thermodynamic analysis of hydrogen production by steam reforming. *Int. J. Hydrogen Energy* **2003**, *28*, 159–167.

21. Matelli, J. A.; Bazzo, E. A methodology for thermodynamic simulation of high temperature, internal reforming fuel cell systems. *J. Power Sources* **2005**, *142*, 160–168.

22. Nikulshina, V.; Hirsch, D.; Mazzotti, M.; Steinfeld, A. CO_2 capture from air and co-production of H_2 via the $Ca(OH)_2$-$CaCO_3$ cycle using concentrated solar power-Thermodynamic analysis. *Energy* **2006**, *31*, 1715–1725.

23. Posada, A.; Manousiouthakis, V. Hydrogen and dry ice production through phase equilibrium separation and methane reforming. *J. Power Sources* **2005**, *156*, 480–488.

24. Hershkowitz, F.; Berlowitz, P. J.; Socha, R. F.; Marucchi-Soos, E.; Frederick, J. W. Hydrogen Production via Steam Reforming in a Reverse-Flow Reactor. *Proceedings of AIChE Annual Meeting*, New York, 2005.

25. Sharma, A.; Takanohashi, T.; Morishita, K.; Takarada, T.; Saito, I. Low temperature catalytic steam gasification of HyperCoal to produce H_2 and synthesis gas. *Fuel* **2008**, *87*, 491–497.

26. Gupta, P.; Velazquez-Vargas, L. G.; Li, F.; Fan, L. S. Chemical looping combustion of coal. *Proceedings of AIChE Annual Meeting*, New York, 2005.

27. Steinfeld, A. Solar Thermochemical Production of Hydrogen - A Review. *Sol. Energy* **2005**, *78*, 603–615.

28. Scholz, W. H. Processes for industrial production of hydrogen and associated environmental effects. *Gas Sep. Purif.* **1993**, *7*, 131–139.

29. Deluga, G. A.; Salge, J. R.; Schmidt, L. D.; Verykios, X. E. Renewable Hydrogen from Ethanol by Autothermal Reforming. *Science* **2004**, *303*, 993–997.

30. Ding, Y.; Alpay, E. Adsorption-enhanced steam-methane reforming. *Chem. Eng. Sci.* **2000**, *55*, 3929–3940.

31. Hufton, J. R.; Mayorga, S.; Sircar, S. Sorption-enhanced reaction process for hydrogen production. *AIChE J.* **1999**, *45*, 248–256.

32. Xiu, G.; Li, P.; Rodrigues, A. E. New generalized strategy for improving sorption-enhanced reaction process. *Chem. Eng. Sci.* **2003**, *58*, 3425–3437.

33. Yong, Z.; Mata, V. G.; Rodrigues, A. E. Adsorption of Carbon Dioxide on Chemically Modified High Surface Area Carbon-Based Adsorbents at High Temperature. *Adsorption* **2001**, *7*, 41–50.

34. Lee, K. B.; Beaver, M. G.; Caram, H. S.; Sircar, S. Chemisorption of carbon dioxide on sodium oxide promoted alumina. *AIChE J.* **2007**, *53*, 2824–2831.

35. Lee, K. B.; Beaver, M. G.; Caram, H. S.; Sircar, S. Novel Thermal-Swing Sorption-Enhanced Reaction Process Concept for Hydrogen Production by Low-Temperature Steam-Methane Reforming. *Ind. Eng. Chem. Res.* **2007**, *46*, 5003–5014.

36. Langmuir, I. The adsorption of gases on plain surfaces of glass, mica and platinum. *J. Am. Chem. Soc.* **1918**, *40*, 1361–1403.

37. Li, M. Synthesis of Thermodynamically Feasible Reaction Pathways. *Proceedings of AIChE Annual Meeting*, New York, 2004.

38. Press, W. H.; Teukolsky, S. A.; Vetterling, W. T.; Flannery, B. P. *Numerical recipes in C++: the art of scientific computing*, 2nd ed.; Cambridge University Press: New York, USA, 1997.

39. McDonald, C. M.; Floudas, C. A. Global optimization for the phase and chemical equilibrium problem: application to the NRTL equation. *Comput. Chem. Eng.* **1995**, *19*, 1111–1139.

40. Aris, R.; Mah, R. H. S. Independence of Chemical Reactions. *Ind. Eng. Chem. Fund.* **1963**, *2*, 90–94.

41. Arteaga, L. E.; Peralta, L. M.; Kafarov, V.; Casas, Y.; Gonzales, E. Bioethanol steam reforming for ecological syngas and electricity production using a fuel cell SOFC system. *Chem. Eng. J.* **2008**, *136*, 256–266.

42. Florin, N. H.; Harris, A. T. Enhanced hydrogen production from biomass with in situ carbon dioxide capture using calcium oxide sorbents. *Chem. Eng. Sci.* **2008**, *63*, 287–316.

43. Stevens, J. F.; Krishnamurthy, B.; Atanassova, P.; Spilker, K. *Development of 50 kW Fuel Processor for Stationary Fuel Cell Applications*; DOE Technical Report DOE/GO/13102-1, 2007.

44. Ortiz, A. L.; Harrison, D. P. Hydrogen production using sorption-enhanced reaction. *Ind. Eng. Chem. Res.* **2001**, *40*, 5102–5109.

45. Wang, Y. N.; Rodrigues, A. E. Hydrogen production from steam methane reforming coupled with in situ CO_2 capture: Conceptual parametric study. *Fuel* **2005**, *84*, 1778–1789.

46. Koumpouras, G. C.; Alpay, E.; Stepanek, F. Mathematical modelling of low-temperature hydrogen production with in situ CO_2 capture. *Chem. Eng. Sci.* **2007**, *62*, 2833–2841.

47. Rampe, T.; Hubner, P.; Vogel, B.; Heinzel, A. Hydrogen Generation from Ethanol by Allothermal Reforming. *Proceedings of the First World Conference on Biomass for Energy and Industry*, 2001; pp 1889–1892.

48. Lee, K. B.; Verdooren, A.; Caram, H. S.; Sircar, S. Chemisorption of carbon dioxide on potassium-carbonate-promoted hydrotalcite. *J. Colloid Interface Sci.* **2007**, *308*, 30–39.

49. Baldea, M.; Daoutidis, P. Dynamics and control of autothermal reactors for the production of hydrogen. *Chem. Eng. Sci.* **2007**, *62*, 3218–3230.

50. Tiemersma, T. P.; Patil, C. S.; van Sint Annaland, M.; Kuipers, J. A. M. Modelling of packed bed membrane reactors for autothermal production of ultrapure hydrogen. *Chem. Eng. Sci.* **2006**, *61*, 1602–1616.

51. Iordanidis, A. A.; Kechagiopoulos, P. N.; Voutetakis, S. S.; Lemonidou, A. A.; Vasalos, I. A. Autothermal sorption-enhanced steam reforming of bio-oil/biogas mixture and energy generation by fuel cells: Concept analysis and process simulation. *Int. J. Hydrogen Energy* **2006**, *31*, 1058–1065.

List of Tables

Table 1: Coefficients for thermodynamic properties.[4]

	CH_4	CO	CO_2
a_1	-1.766850998E+05	1.489045326E+04	4.943650540E+04
a_2	2.786181020E+03	-2.922285939E+02	-6.264116010E+02
a_3	-1.202577850E+01	5.724527170E+00	5.301725240E+00
a_4	3.917619290E-02	-8.176235030E-03	2.503813816E-03
a_5	-3.619054430E-05	1.456903469E-05	-2.127308728E-07
a_6	2.026853043E-08	-1.087746302E-08	-7.689988780E-10
a_7	-4.976705490E-12	3.027941827E-12	2.849677801E-13
a_8	-2.331314360E+04	-1.303131878E+04	-4.528198460E+04
a_9	8.904322750E+01	-7.859241350E+00	-7.048279440E+00
	C_2H_4	CH_3CHO	C_2H_5OH
a_1	-1.163605836E+05	-1.373904369E+05	-2.342791392E+05
a_2	2.554851510E+03	2.559937679E+03	4.479180550E+03
a_3	-1.609746428E+01	-1.340470172E+01	-2.744817302E+01
a_4	6.625779320E-02	5.922128620E-02	1.088679162E-01
a_5	-7.885081860E-05	-6.240006050E-05	-1.305309334E-04
a_6	5.125224820E-08	3.703324410E-08	8.437346400E-08
a_7	-1.370340031E-11	-9.342697410E-12	-2.234559017E-11
a_8	-6.176191070E+03	-3.318731310E+04	-5.022229000E+04
a_9	1.093338343E+02	1.007417652E+02	1.764829211E+02
	H_2	H_2O	O_2
a_1	4.078322810E+04	-3.947960830E+04	-3.425563420E+04
a_2	-8.009185450E+02	5.755731020E+02	4.847000970E+02
a_3	8.214701670E+00	9.317826530E-01	1.119010961E+00
a_4	-1.269714360E-02	7.222712860E-03	4.293889240E-03
a_5	1.753604930E-05	-7.342557370E-06	-6.836300520E-07
a_6	-1.202860160E-08	4.955043490E-09	-2.023372700E-09
a_7	3.368093160E-12	-1.336933246E-12	1.039040018E-12
a_8	2.682484380E+03	-3.303974310E+04	-3.391454870E+03
a_9	-3.043788660E+01	1.724205775E+01	1.849699470E+01

Table 2: Atomic matrix.

	CH_4	CO	CO_2	C_2H_4	CH_3CHO	C_2H_5OH	H_2	H_2O	O_2
(C)	1	1	1	2	2	2	0	0	0
(H)	4	0	0	4	4	6	2	2	0
(O)	0	1	2	0	1	1	0	1	2

Table 3: Reaction stoichiometric matrix*.

Reaction No.	CH_4	CO	CO_2	C_2H_4	CH_3CHO	C_2H_5OH	H_2	H_2O	O_2
(1)	1	1	0	0	0	-1	1	0	0
(2)	-1	1	0	0	0	0	3	-1	0
(3)	0	-1	1	0	0	0	1	-1	0
(4)	0	2	-2	0	0	0	0	0	1
(5)	0	0	0	0	1	-1	1	0	0
(6)	-2	0	0	1	0	0	2	0	0

* (used only in the equilibrium constant method)

Table 4: Stoichiometric coefficients of the gas products at equilibrium (the reactants are 1 C_2H_5OH + 3 H_2O).

$CO_2^{(ads)}/CO_2^{(g)}$	CH_4	CO	CO_2	C_2H_4	CH_3CHO	C_2H_5OH	H_2	H_2O	O_2
0	1.2570	0.0489	0.6941	0.0000	0.0000	0.0000	0.9231	2.5629	0.0000
100	0.9773	0.0021	0.0101	0.0000	0.0000	0.0000	2.0888	1.9566	0.0000

Table 5: Product composition under different conditions.

	Post-CO_2 adsorption (wet)	CO_2-H_2O free (dry)
H_2	89.4%	96.2%
CH_4	3.5%	3.8%
H_2O	7.1%	

List of Figures

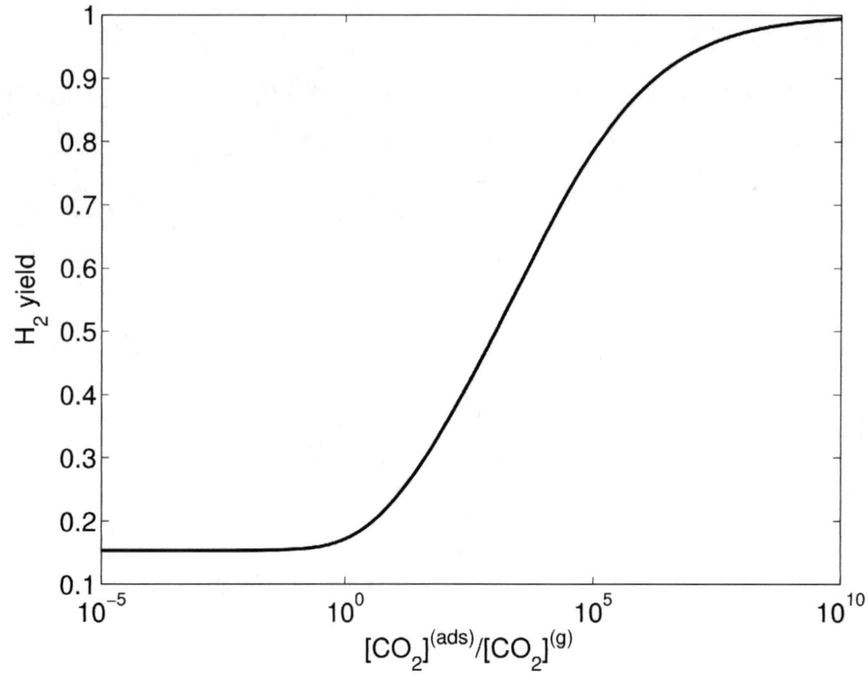

Figure 1: Conversion of C_2H_5OH to H_2 as a function of $CO_2{}^{(ads)}/CO_2{}^{(g)}$ (feed mole ratio = 1/3, P = 5 bar and T = 500 °C).

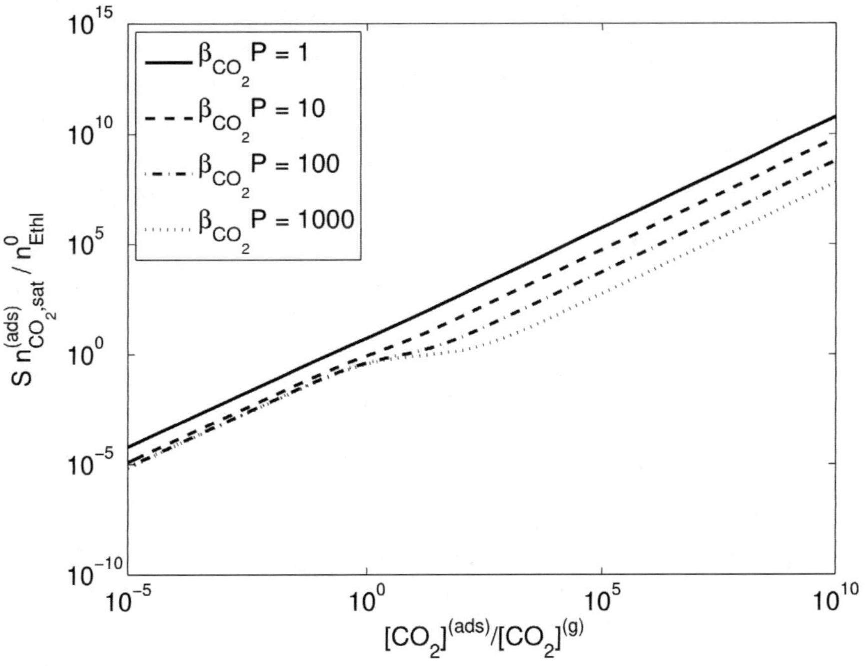

Figure 2: Relationship between ${CO_2}^{(ads)}/{CO_2}^{(g)}$ and the surface area of adsorbent (feed mole ratio = 1/3, $P = 5$ bar and $T = 500$ °C).

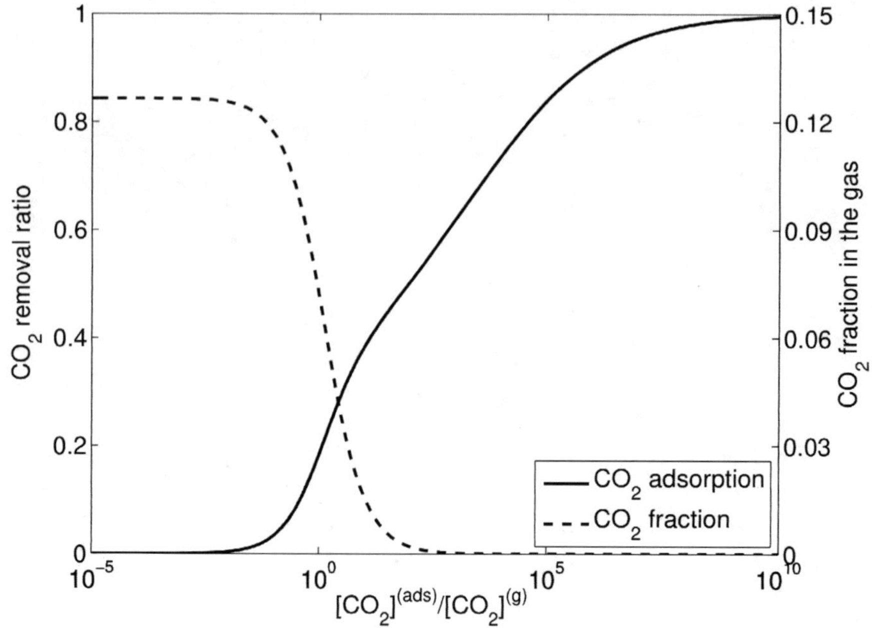

Figure 3: CO_2 removal ratio and CO_2 fraction in the gas as a function of ${CO_2}^{(ads)}/{CO_2}^{(g)}$ (feed mole ratio = 1/3, P = 5 bar and T = 500 °C).

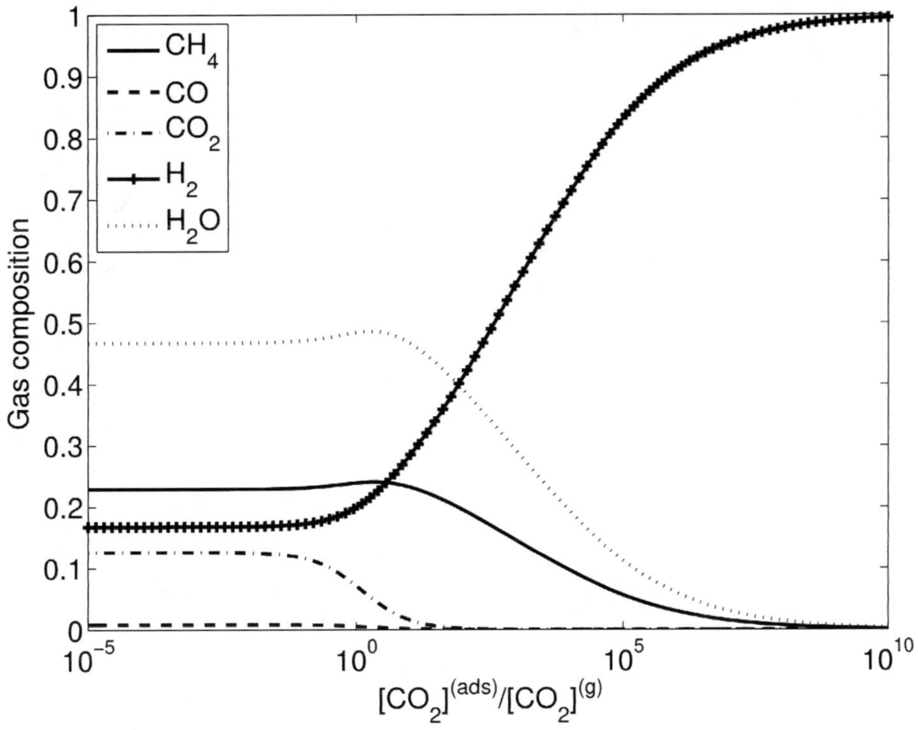

Figure 4: Gas composition at equilibrium as a function of $CO_2{}^{(ads)}/CO_2{}^{(g)}$ (feed mole ratio = 1/3, P = 5 bar and T = 500 °C).

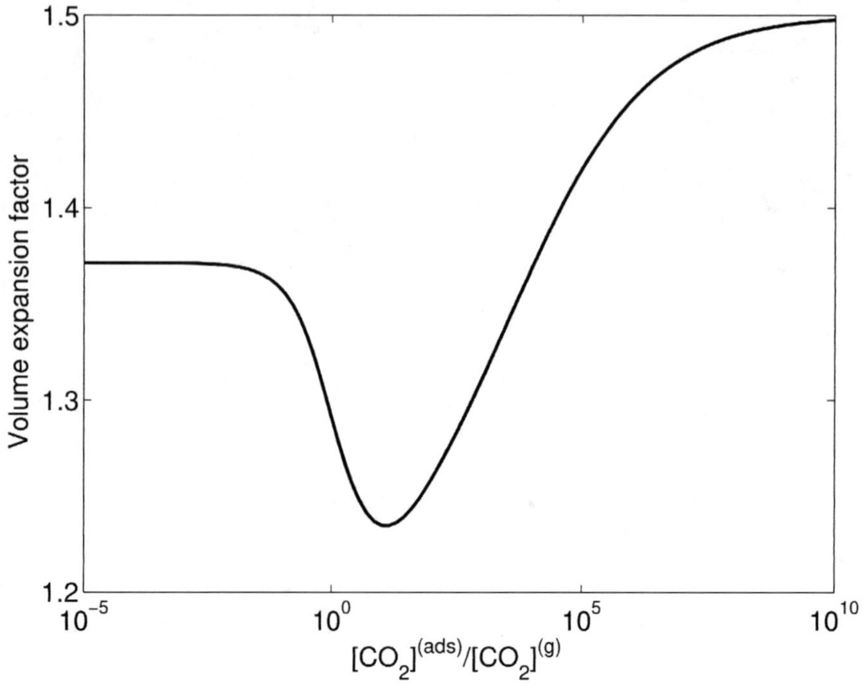

Figure 5: Volume expansion (the volume ratio of the product mixture to the reactant mixture) with respect to the reactant mixture (feed mole ratio = 1/3, P = 5 bar and T = 500 °C).

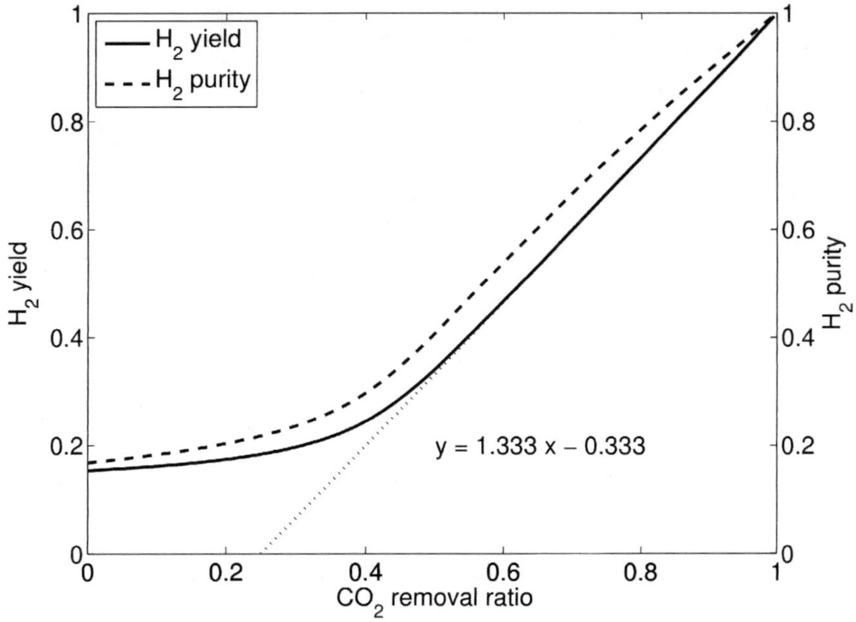

Figure 6: H_2 yield and purity as functions of the CO_2 removal ratio (feed mole ratio $= 1/3$, $P = 5$ bar and $T = 500$ °C).

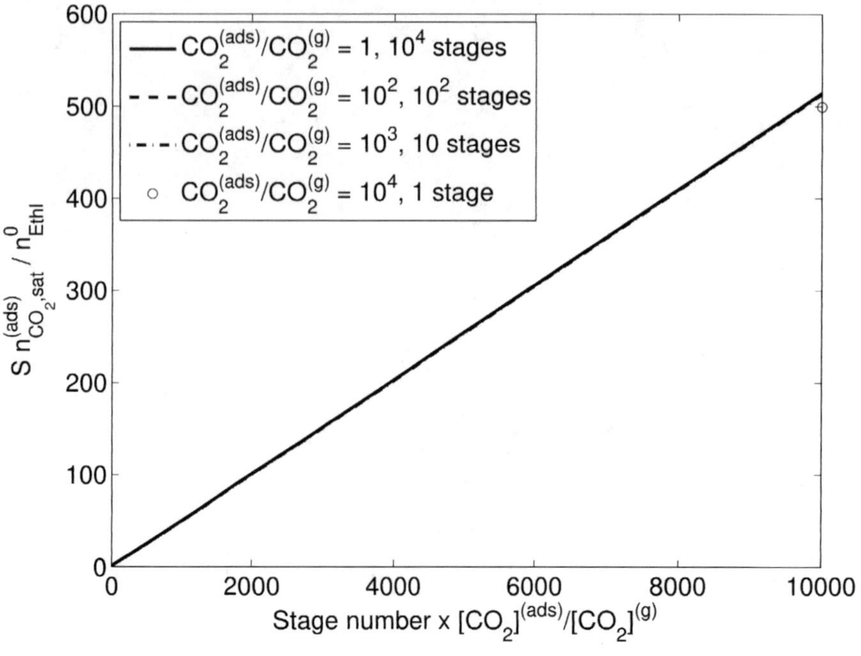

Figure 7: Cumulative adsorbent area as a function of number of equilibrium stages and ${CO_2}^{(ads)}/{CO_2}^{(g)}$ ratios (feed mole ratio $= 1/3$, $P = 5$ bar and $T = 500$ °C, $\beta_{CO_2} = 22\ \text{bar}^{-1}$).

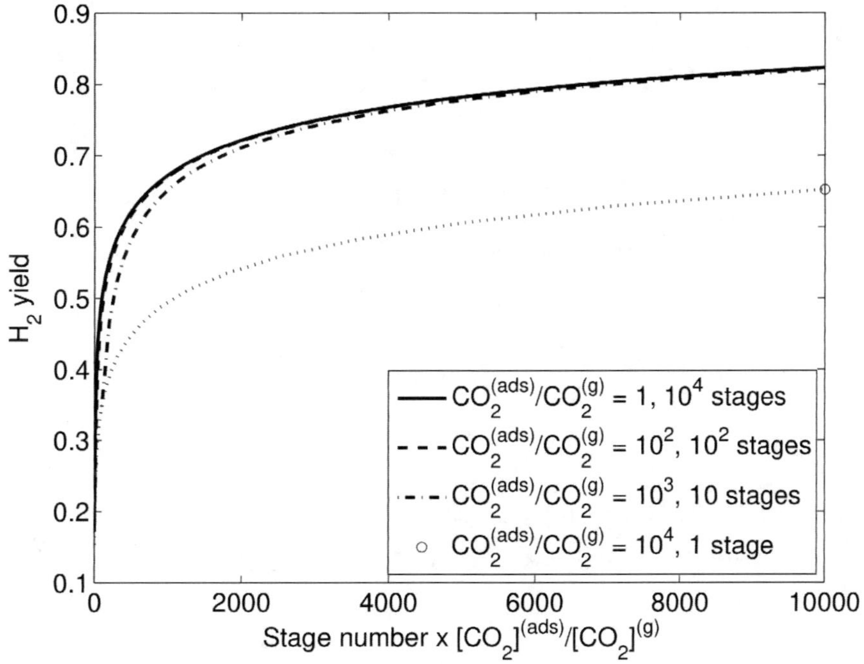

Figure 8: Conversion of C_2H_5OH to H_2 as a function of number of equilibrium stages and ${CO_2}^{(ads)}/{CO_2}^{(g)}$ ratios (feed mole ratio = 1/3, $P = 5$ bar and $T = 500$ °C).

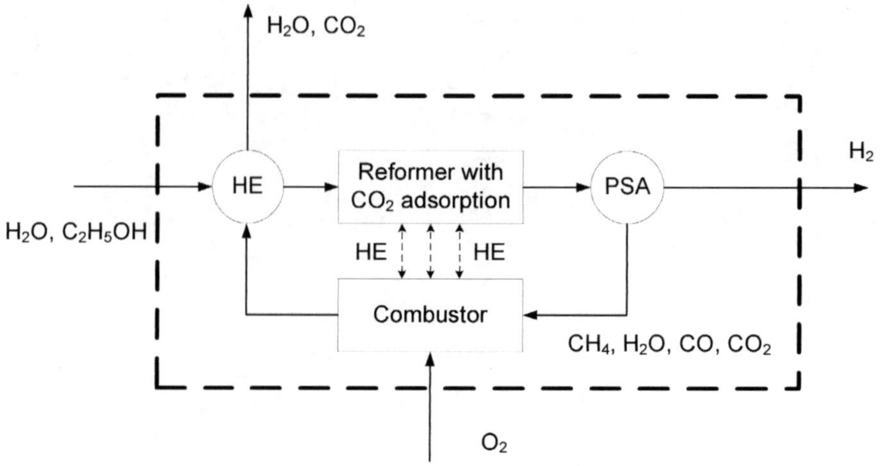

Figure 9: Integrated endothermic reforming and the exothermic combustion of unconverted CH_4 with simultaneous CO_2 in the ethanol reforming process.

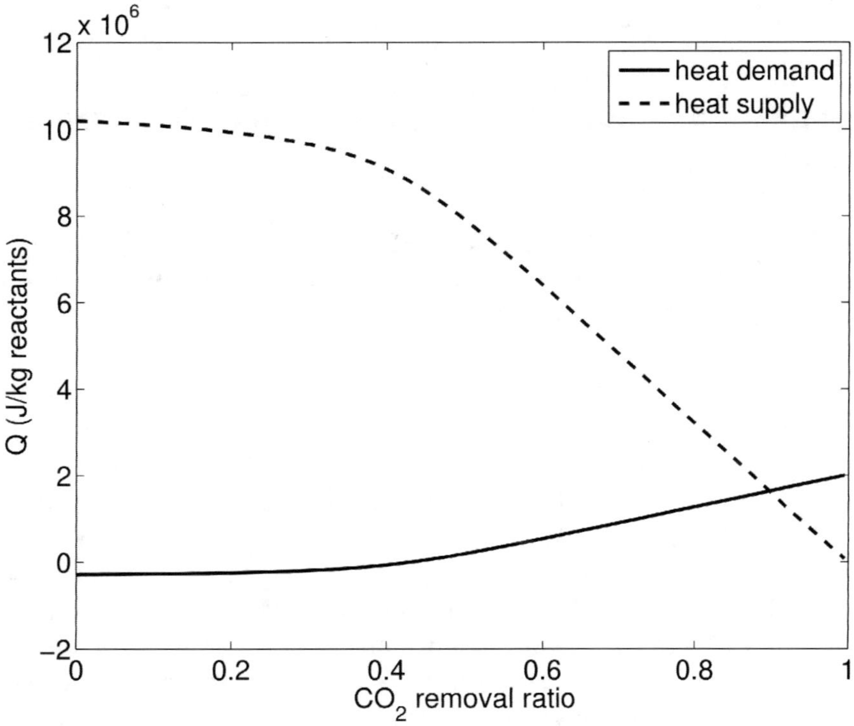

Figure 10: Heat demand by ethanol reforming and heat supply through combustion of exhausting gases from the H_2 separation unit as a function of removal ratio of carbon in the form of CO_2 (feed mole ratio = 1/3, P = 5 bar and T = 500 °C).

63e Two-Dimensional Equation-of-State Modeling of Adsorption of Coalbed Methane Gases

Sayeed Mohammad[1], Zhejun Pan[1], R.L. Robinson[1], and K. a. M. Gasem[2]. (1) Chemical Engineering, Oklahoma State Univ., Stillwater, OK 74078, (2) School of Chemical Engineering, Oklahoma State University, 423 Engineering North, Stillwater, OK 74078

Adsorption equilibrium models are essential for optimizing coalbed methane (CBM) production and carbon dioxide (CO2) sequestration processes. Although a number of frameworks are available for describing the adsorption phenomenon, two-dimensional (2-D) equations of state (EOS) offer distinct advantages in modeling supercritical, high-pressure adsorption systems.

Current applications of 2-D EOS typically involve regressions of data on individual adsorption isotherms to determine the temperature-dependent EOS parameters (fÑ, f", k). For the EOS to have maximum utility, generalized correlations for these parameters are needed. Such generalizations would facilitate (a) correlation of adsorption data over a range of operating temperatures, and (b) a priori predictions of adsorption behavior. Accordingly, we have focused on developing new correlations for the 2-D EOS parameters that yield precise representations and accurate predictions of high-pressure, supercritical, pure-gas adsorption encountered in CBM recovery and CO2 sequestration. Further, we have extended the 2D-EOS to adsorption from gas mixtures by incorporating "mixing rules" to describe the composition dependence of the model parameters.

In this work, we have used the 2-D Peng-Robinson (PR) EOS to illustrate the proposed method for determining the EOS pure-fluid parameters and to demonstrate the 2-D EOS capability to represent and predict pure-gas adsorption of CBM gases (CH4, CO2, and N2) on carbon adsorbents. Experimental adsorption measurements, including both activated carbons and coals (both dry and wet), were used to evaluate the efficacy of this approach.

The new correlations for the 2-D EOS parameters appear effective in modeling pure-gas adsorption on carbon matrices at supercritical and near-critical conditions. The 2-D PR EOS, using the new parameter correlations, can represent adsorption on activated carbon and coals within their expected experimental uncertainties. Specifically, the 2-D EOS parameter correlations can represent the pure-gas adsorption over a range of temperature with 2.4% absolute average deviation (AAD) for activated carbons and 4.4% AAD for coals. Further, the new correlations, which are generalized in terms of accessible adsorbate and adsorbent properties, can predict (a) pure-gas adsorption on activated carbons with 9% AAD (within three times the expected experimental uncertainties), and (b) binary and ternary gas adsorption within three times the experimental uncertainties, on average.

Adsorption of carbon dioxide adsorption on mesoporous porous silica materials containing amine-like motifs

Guoying Zhao, Baroz Aziz, Niklas Hedin
Material Chemistry Research Group, Department of Physical Inorganic and Structural Chemistry, Arrhenius Laboratory, Stockholm University, SE-106 91 Stockholm, Sweden

Introduction

CO_2 is separated from gas mixtures in various industrial applications, e.g. in enhanced oil recovery[1], natural gas processing, and in specialized CO_2 production plants. In the context of carbon capture, three different capture tactics are discussed and studied: post combustion, pre combustion, and oxyfuel or chemical looping capture. Different separation may be employed depending on the thermodynamic properties of the mixtures, the size of the CO_2 point source, and economical factors. In post combustion carbon capture, rather low partial pressures of CO_2 in a N_2 stream, alkanolamines are used in aqueous solutions. This use is because of their high reactivity, ease of reuse, and low absorption of hydrocarbons.[2] There are many disadvantages with this method, high energy consumption, severe corrosion of amine aqueous solution, and the re-pollution of the gas stream with water and amine vapor. This has limited its applicability from both an environmental and separation efficiency point of view.[3] More advanced solvents (ionic liquids), membrane technology and adsorption mediated separation have been proposed for low cost CO_2 separation. Porous adsorbents, with a comparably low heat capacity, high surface area and high stability, have been developed in industrial settings for similar gas separation applications.[4] CO_2 has a large electric quadrupole moment and physisorb easily to most surfaces. Porous materials have show large potential for CO_2 separation. Microporous materials may have a high selectivity for gas separation [5]; however, their small pores may slow down the cycle times in swing adsorption process. In this study we focused on the study of modifications of mesoporous (pore sizes 2-50 nm) materials, including their chemical modification and capacity for CO_2 uptake. These chemical modifications could enhance carbon dioxide separation efficiency with an easy regeneration of the adsorbents.

Functional mesoporous silica materials can be synthesized in different ways and we employ post-synthetically modifications of the materials with organo-silica groups[6]. This method is appealing for screening of different groups and other synthetic variables. To our knowledge, no studies have been published for silica materials modified with alkanolamine or amidine groups. Amidine molecules are reported to bind CO_2 readily at low CO_2 pressure in liquid solutions, and the chemical reaction is reversible at temperatures around 65 ºC.[7] We explored the potential for CO_2 separation in hybrid material, with amidine groups on a solid support. Silica materials with primary amine modification have been reported to enhance CO_2 adsorption.[8] In this study, we extended those by careful optimizing the conditions for the chemical reactions, and by comparing the capacity for carbon dioxide adsorption for these "amine-like" modifications.

The adsorption capacity (n) of CO_2 (dry and wet), and N_2 (dry and wet), on the organically modified mesoporous silica materials are characterized as a function of pressure (p) and temperature (T) conditions using a slightly modified commercial volumetric adsorption apparatus. The thermodynamic (*npT*) behavior for the solid-gas system is analyzed and compared for the samples with amine, alkanolamine or amidine moities. A certain propyl amine modified sample showed significantly large carbon dioxide adsorption capacity than the others. The propyl amine modified samples showed, consistently, a higher carbon dioxide adsorption capacity. In addition, a high selectivity of carbon dioxide adsorption was shown for the primary amines on a mesoporous solid support.

2. Materials and Methods

In this study, Davisil™ (LC60A pore size 40-63 nm), was used as substrate for the post-synthesis "grafting" reactions. Thee chemicals were used as received from Aldrich. Gases (helium, carbon dioxide and nitrogen) were supplied by the Linde gas company. The modification of Davisil™ with aminopropyl groups was carried out according reported procedures and further optimized using the principles of factorial design. A series of final materials with a surface coverage of aminopropyl groups ranging from 1.56 to 4.16 groups/ nm^2 was obtained. The loading of organics were estimated from thermogravimetric analysis (TGA). The material with propyl bis(2-hydroxyethyl)amine was post-synthesized using similar procedure. The product with pending propyl N,N-dimethylacetamidine moieties, was prepared from the Davisil™ with aminopropyl groups. In this synthesis we employed the Scoggins' procedure for acetamidines.[9] A range of characterization methods (such as thermogravimetric analysis, nitrogen adsorption-desorption isotherms BET analysis, solid-state NMR spectroscopy) were used to study the composition and micro-structure of the materials.

The CO_2 and N_2 uptake capacity of the derived materials were studied under various thermodynamic conditions using a slightly modified Micromeritics Gemini™ 2375 device (schematically shown in Figure 1). Before the experimentation, each sample was pretreated or regenerated under a flow of dry N_2 gas at a temperature of 140 °C for an extended amount of time. Humid gas (containing around 1.9% water) was produced to examine the effect of moisture on the adsorption capacity of carbon dioxide and nitrogen. Dry gas components (nitrogen or carbon dioxide, the partial pressure kept at 1.5bar) were bubbled through liquid water (kept in a temperature controlled room (23°C)). The temperature for the adsorption uptake experiment was controlled by a circulating bath (Huber ministat 230) and a Dewar (siloxane oligomer as heat transfer medium).

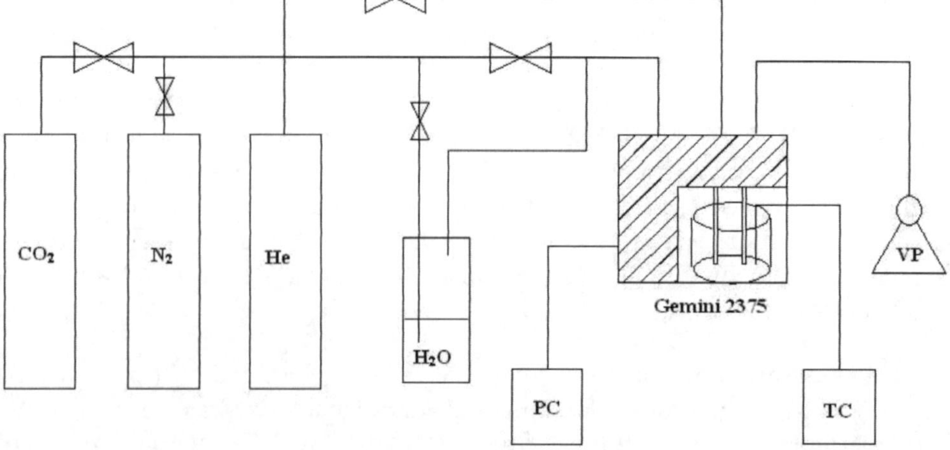

Figure 1: A representation of the volumetric adsorption device used to measure temperature and pressure dependent gas uptake. CO_2, N_2, and He denote gas bottles with pressure regulators. H_2O is liquid water, and Gemini 2375 a slight modified Micrometrics Gemini 2375 device. PC, TC, and VP, means pressure controller, temperature controller and vacuum pump.

3. Results and Discussions

3.1 Material synthesis

We have two major aims in modifying the mesoporous (2 – 50 nm) silica materials. One aim is to increase the degree of surface modification using multi parameter optimization. For this we employed Experiment of Design (the details of that optimization will be presented elsewhere). A second aim is to compare the binding efficiency for CO_2 adsorption and try to relate that to the density of organic fragments. Three amine-like moieties were chosen. The propyl amine modification have been studied by other researchers and a significant carbon dioxide uptake have been found[6,7,8]. Therefore we employed the multi parameter modification for this group. The other two surface modifications, propyl N,N-dimethylacetamidine and propyl bis(2-hydroxyethyl)amine, are new to our knowledge. In the liquid state these latter moieties have significant and interesting interactions with carbon dioxide.[1, 9]

The amount of organic material in the different materials was established by TGA. We are well aware about the potential for systematic errors associate with this methodology. In Table 1 the degree of modification is presented with the under the label loading. We achieved a surface coverage of 2.4 – 4.2 molecular amine fragments per nanometer squared. These values are comparable, if not higher, than those presented by others.[7]

Table 1. Presented are the surface areas, pore volumes, loadings and carbon dioxide uptake for modified mesoporous silica materials. The five last rows in the table refer to various propyl amine materials. The acetamidine sample is synthesized from the sample msf12.

Chemical Modification	**S_{BET} (m^2/g)**	**Pore size (nm)**	**Pore volume (cm^3/g)**	**Loading ($1/nm^2$)**	**CO_2 adsorption efficiency ($mmol/g_{ads}$)**		**N(efficiency)***	
					No H_2O		No H_2O	
Unmodified Davsil™	540	63	0.87					
Propyl amine (Amd16)	199	67	0.33	4.16	1.40	2.13	0.45	0.69
bis(2-hydroxyethyl)-amino propyl	284	55	0.39	2.38	0.33	0.35	0.20	
N,N-dimethyl-	309	84	0.65	3.45	0.28		0.10	

* The number N(efficiency) relates to a hypothesized case where the carbon dioxide adsorption is presumed to occur only on the amine moieties.

acetamidine propyl								
Amd7	284	62	0.44	3.55	1.34		0.49	
Amd13	361	62	0.56	2.49	0.89		0.44	
DTAS107	394	57	0.57	2.54	0.89	0.98	0.44	0.48
DTAS90	370	60	0.56	2.84	0.87	0.99	0.39	0.44
Msf12	347	75	0.65	2.70	1.25	0.90	0.56	0.40

All the samples were analyzed with nitrogen adsorption. As expected the BET surface decreased with an increased degree of modification. This effect is in particular visible for the various propyl amine modifications, presented in Table 1 as propyl amine (amd16), amd7, amd13, DTAS107, DTAS90, and Msf12. The amd16 modification has the smallest specific BET area and the highest loading of amine groups. For all the samples, the average pore size does not change significantly after modification but Msf12 and the sample with amidine, which were pretreated in boiling water for three hours.

The local arrangements of various molecular fragments were studied with solid-state NMR spectroscopy, under magic angle spinning conditions. The degree of condensation among the organosiloxanes was studied (for the aminopropyl modified samples) with solid-state ^{29}Si NMR spectroscopy, using direct and cross-polarization detection. The degree of condensation was found to be high. In addition it was realized (using solid-state ^{1}H and ^{13}C NMR spectroscopy) that the samples contained a small amount of non hydrolyzed ethoxy groups. The propyl N,N-dimethylacetamidine modification posed challenges to characterization; from TGA data and supported by NMR spectroscopy, we noted that only 30% of the amine functions where converted to the amidine functionality. On the other hand, it was concluded that the functionalization with both propyl amine and propyl bis(2-hydroxyethyl)amine was successful. In the next section, the carbon dioxide uptake on these samples are presented, besides for the less defined amidine sample.

3.2 Carbon dioxide uptake

The pressure dependency for the carbon dioxide uptake was measured under isothermal conditions (temperature between 0 – 70 °C, with or without water present). A volumetric adsorption device was used for this (Figure 1). Besides of being able to screen for an enhanced capacity for carbon dioxide uptake, these detailed measurements might inform about specific thermodynamic properties of the adsorbents. This combined pT information is important and interesting for swing adsorption, in particular for temperature swing adsorption (TSA) or thermally coupled PSA[5] mediated carbon dioxide gas separation.

The materials was screened by two factors: the gas (carbon dioxide, or carbon dioxide and water) uptake capacity per gram and the N(efficiency), according to the gas uptake at standard condition (25 °C and 1 atm). The uptake capacity of the propyl amine modified samples are significant, in particular the amd16 sample takes up significant amount of gas (see table 1). The uptake for the bis(2-hydroxyethyl)-amino propyl and N,N-dimethyl-acetamidine propyl modified samples were not promising and, hence, we did not optimize the synthetic conditions any further. The amount 1.4 mmol/ g corresponds to 6.2 wt% carbon dioxide and with "wet" carbon dioxide even more gas is adsorbed (2.1 mmol/ g). These values should be compared with some alternative microporous materials, which have capacities of 2-3 mmol/g at such standard pressures.[10]

In Figure 2 the carbon dioxide adsorption isotherms at 25 °C is presented. The propyl amine samples (amd16,13, 7; msf12; dats107, 90) all have higher uptake capacities than the bis(2-hydroxyethyl)amino propyl and a N,N-dimethylacetamidine propyl modified samples at all pressures.

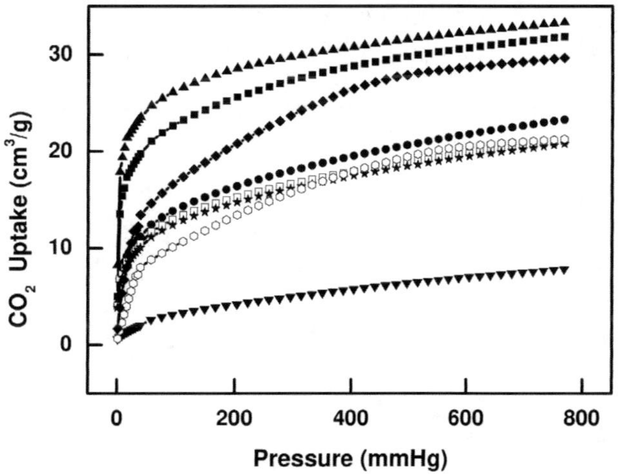

Figure 2. Carbon dioxide uptake on organically modified mesoporous silica materials, with different amine motifs at 25 ℃ without water present (propyl amine: ▲Amd16;■amd7;◆msf12;●dats107;□amd13;★dats90. amidine: ◇. bis-ethanol amine: ▼).

The carbon dioxide uptake in presence of water vapor was studied and is presented in Figure 3. For most samples the effect of water vapor is negligible. however the gas adsorption capacity of amd 16 is enhanced significantly with water present, which could be very well related to different degrees of hydrophobicity.

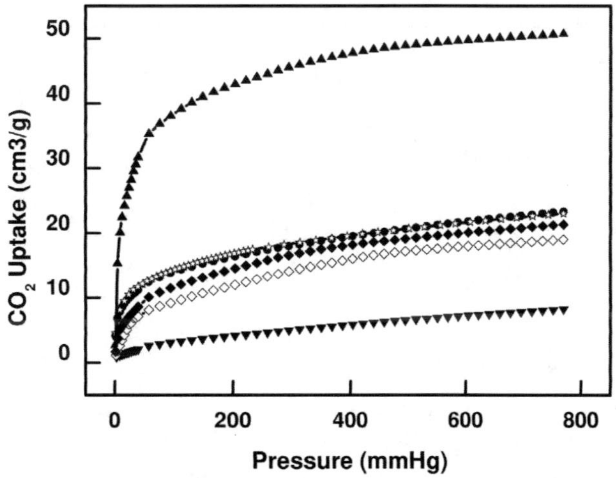

Figure 3. Gas uptake on organically modified mesoporous silica materials, with different amine motifs at 25 ℃ with water vapor present (propyl amine: ▲ Amd16;◆msf12;●dats107;☆ dats90. amidine: ◇. bis-ethanol amine: ▼).

The temperature dependent carbon dioxide adsorption isotherms for the bis(2-hydroxyethyl)amino propyl and propyl amine (amd16) sample are presented in Figure 4 and Figure 5. From these two sets of curves, clearly, the adsorption is much higher for the sample modified with propyl amine groups. The largest differences are in the low pressure regime, where the adsorption increases significantly in Figure 4. In a very simplified dual site

Langmuir model, the adsorption energies for the carbon dioxide on the propyl amine modified samples were estimated. This model did not describe the initial low pressure dependencies well; however, the dependencies at higher pressures ("the second step") is better described and corresponds to an adsorption energy of roughly 6 kcal/mol (10 kT) - a typical values for physisorption.

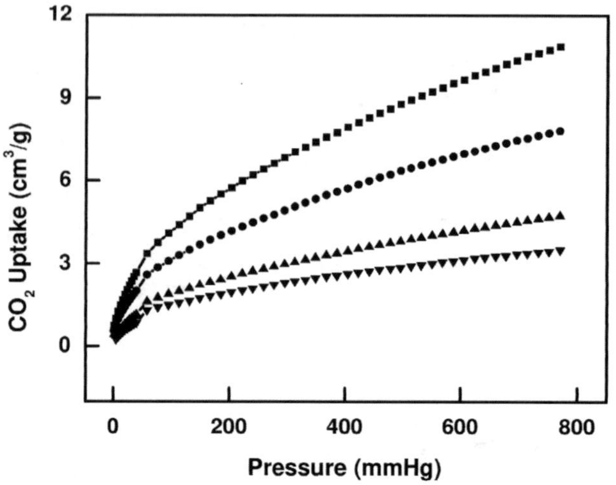

Figure 4. Carbon dioxide uptake in mesoporous silica materials modified with bis-ethanol amine groups at different temperature conditions, without water vapor present (■ 0℃;●25℃;▲50℃;▼70℃)

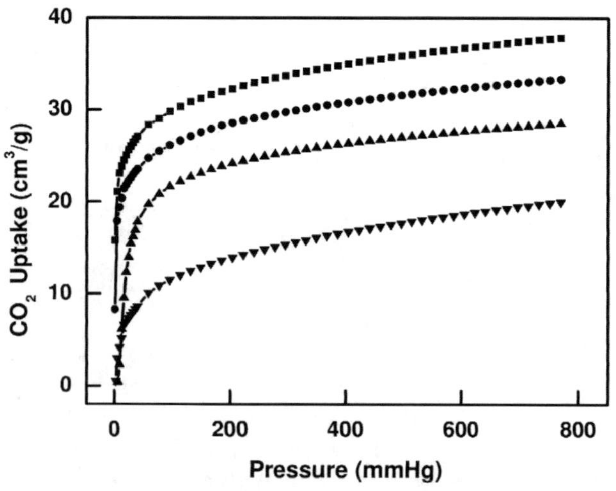

Figure 5. Temperature dependent carbon dioxide uptake capacity for propyl amine modified mesoporous silica materials (amd16) without water vapor present. (■ 0℃;●25℃;▲50℃;▼70℃)

The selectivity of carbon dioxide over nitrogen adsorption is very important for gas separation of flue gases. The propyl amine modified materials have significant selectivity. At a carbon dioxide pressures of 1 atmosphere the selectivity of CO_2 is between 30-50 depending on the temperature (at a higher temperature the selectivity is higher); at lower pressures, where both gases adsorb in a Henry's law manner, this selectivity is larger. The high selectivity for the amine modified materials is significantly higher than for zeolites or active carbons. Zeolites adsorbents show carbon dioxide selectivity of about 15 and active carbons of about 5-10 [ref] at an atmospheric pressure of either carbon dioxide or nitrogen gas.[16] It is not fully clear for us if the high selectivity is a reflection of a "chemisorption" nature on these amine modified materials.

4. Conclusions

A range of propyl amine modified mesoporous silica materials have been synthesized and studied. By careful optimization a specific loading of over 4 propyl amine per nm^2 was achieved. In addition, propyl bis(2-hydroxyethyl)amine and propyl N,N-dimethylacetamidine moieties were attached to mesoporous silica materials, for the first time.

The carbon dioxide uptake was significantly higher on propyl amine modified samples than on the amidine and bis-ethanol amine modified samples. The lower carbon dioxide uptake indicates that amidine and bis-ethanol modifications are less useful in adsorbents for carbon dioxide separation from flue gases.

On a silica adsorbent modified with propyl amine groups 1.4 mmol/ g carbon dioxide can be adsorbed at standard conditions of 25 °C and 1 atmosphere. This value is similar to those for microporous materials (zeolites or metal organic frameworks) or active carbons. Speculatively, the larger pores enable faster molecular diffusion, which could allow for faster adsorption cycles. The carbon dioxide selectivity (over nitrogen) was determined to be comparably high on these materials. The high capacity, large pore size, and significant carbon dioxide selectivity, all combined indicates that these or similar materials are important candidates for carbon dioxide adsorbents.

Acknowledgement

The Berzelii Center EXSELENT (VR/ Vinnova) and the Swedish research council (VR) supported this study. We thank Professor Aatto Laaksonen for stimulating discussions.

References

1. Isaacs EE, Otto FD, Mather AE, Journal of Chemical and Engineering Data, 1980, 25(2):118-120;Blunt M, Fayers FJ, Orr FM, Energy Conversion and Management, 1993, 34(9-11):1197-1204.

2. Donaldson TL, Nguyen YN, Industrial and Engineering Chemistry Fundamentals, 1980, 19(3):260-266.

3. Bates ED, Mayton RD, Ntai I, Davis JH, Journal of the American Chemical Society, 2002, 124(6):926-927;Mago BF, West CW, 1976, US Patent 3959170.

4. Crittenden B, Thomas WJ, Adsorption Technology and Design, Butterworth-Heinemann, Oxford, 1998.

5. Ruthven DM, Reyes SC, Microporous and Mesoporous Materials, 2007, 104(1-3):59-66.

6. Stein A, Melde BJ, Schroden RC, Advanced Materials, 2000, 12(19):1403-1419; Angeletti E, Canepa C, Martinetti G, Venturello P, Journal of the Chemical Society-Perkin Transactions 1, 1989, (1):105-107.

7. Liu YX, Jessop PG, Cunningham M, Eckert CA, Liotta CL, Science, 2006, 313(5789):958-960.

8. Hicks JC, Drese JH, Fauth DJ, Gray ML, Qi GG, Jones CW, Journal of the American Chemical Society, 2008, 130(10):2902-2903; Huang HY, Yang RT, Chinn D, Munson CL, Industrial & Engineering Chemistry Research, 2003, 42(12):2427-2433; Kim S, Ida J, Guliants VV, Lin JYS, Journal of Physical Chemistry B, 2005, 109(13):6287-6293.

9. Knowles GP, Graham JV, Delaney SW, Chaffee AL, Fuel Processing Technology, 2005, 86(14-15):1435-1448;Scoggins MW, Journal of Chromatographic Science, 1975, 13(3):146-148.

10. Siriwardane RV, Shen MS, Fisher EP, Poston JA, Energy & Fuels, 2001, 15(2):279-284.

11. Bae YS, Mulfort KL, Frost H, Ryan P, Punnathanam S, Broadbelt LJ, Hupp JT, Snurr RQ, Langmuir, 2008, 24(16):8592-8598.

12. Wang B, Cote AP, Furukawa H, O'Keeffe M, Yaghi OM, Nature, 2008, 453(7192):207-211.

63g Characterization of the Adsorption of CO_2 In Metal-Organic Framework and Zeolitic-Imidazolate Framework Materials by Density Functional Theory Calculations

Rees B. Rankin, Department of Chemical and Petroleum Engineering, University of Pittsburgh (and National Energy Technology Laboratory I.A.E.S.), Benedum Hall #1123, University of Pittsburgh, Pittsburgh, PA 15261 and J. Karl Johnson, Department of Chemical and Petroleum Engineering, University of Pittsburgh, 1242 Benedum Hall, Pittsburgh, PA 15261.

Recent years have seen an increased attention in the development of advanced materials for the separation and storage of gas mixtures. Specific focus to date has been placed on novel groups of crystalline nanoporous materials such as Metal-Organic Frameworks (MOFS) and Zeolitic-Imidazolate Frameworks (ZIFS). These materials offer potentially high surface area, high purity, and can be functionally-tailored by selection and modification of the metal, ligand, or linker groups in the material structure, all at a comparatively low cost. Although MOF and ZIFS can be synthesized at relatively low cost, an experimental-screening based approach for performance characterization can be overly daunting; due to the combinatorial potential of materials that can form MOFs/ZIFs, the number of possible MOF/ZIF like materials is immense. One critical component in the design of novel MOF/ZIF like materials for selected applications is an atomic level understanding of the interactions between the framework and the gaseous adsorbate(s); specifically, the nature of changes in adsorption site and adsorption energy between similar framework materials remains largely unknown. Much attention has been recently paid to theoretical modeling of H_2/MOF interactions in the literature; results reported reveal weak binding energies of 1-5 kcal/mol. However, it is suggested experimentally that some MOFs and ZIFs can potentially bond much more energetically with other gaseous adsorbates such as CO_2.

To date there has significantly little research focus in the literature to apply quantum-mechanics based theoretical calculations and models to quantify the energy-structure dependence in the bonding of CO_2 with these materials. In this work we present fully-periodic DFT calculations performed to quantitatively characterize and contrast the bonding of CO_2 in the materials CuBTC, ZIF-68, ZIF-69, and ZIF-70. We began by calculating the fully relaxed bulk crystal structure of these materials based on the proposed structures from experimental crystallography. In most cases, minor expansions of the lattice constant, on the order of 0.5-1.0% are observed, accompanied by minor (measured by RMS analysis) atomic relaxations away from the crystallographic positions. We have also examined the effects induced on the crystal structure by defect/vacancy formation in the ideal material. We then performed calculations to examine the nature of CO_2 bonding sites and bonding energy in these materials. We present a comparison of the results of these calculations to help determine the relationships between CO_2 bonding in these materials associated with material purity, structure, and composition.

104a Desulfurization of Diesel Fuel by Extraction with Lewis-Acidic Ionic Liquid

Hongshuai Gao, **Jianmin Xing**, Yuguang Li, Xiaochao Xiong, Wangliang Li, Qinfeng Liu, and Huizhou Liu. State Key Laboratory of Biochemical Engineering, Key Laboratory of Green Process and Engineering, Institute of Process Engineering, Chinese Academy of Sciences, Beier Tiao No.1, Zhongguan Cun, Haidian District, Beijing, China, Beijing, China

In the last decade, much attention has been paid to the deep desulfurization of fuels due to more stringent environmental regulations.[1]Although hydrodesulfurization (HDS) is highly efficient in removing thiols, sulfides, and disulfides. It is difficult to reduce refractory sulfur-containing compounds such as dibenzothiophene (DBT) and its derivatives to an ultra-low level. Therefore, the development of alternative ultra-deep desulfurization processes, such as adsorption, biodesulfurization, oxidation, is desired. Ionic liquids (ILs) have attracted intensive interest because of their unique chemical and physical properties[2]such as negligible vapor pressure, low toxicity, high chemical and thermal stabilities, and ability to dissolve a wide range of organic and inorganic compounds. Ionic liquids have been used as alternatives to classical molecular solvents in a range of fundamental researches and applications, including synthesis, catalysis, separation and so on. Extraction of fuels using ionic liquids to remove sulfur compounds has been reported. [3,4] Most of them are neutral ionic liquids, and their efficiencies of sulfur removal are rather low. Although $AlCl_3$-based Lewis-acidic ionic liquids have excellent desulfurization results, they are sensitive to moisture and air.

In this work, a kind of Lewis-acidic ionic liquid 1-butyl-3-methylimidazolium tetrahalogenoferrate(III) ([BMIM][$FeCl_4$]) and other two ionic liquids 1-butyl-3-methylimidazolium hexafluorophosphate ([BMIM][PF_6]) and 1-butyl-3-methylimidazolium tetrafluoroborate ([BMIM][BF_4]), are studied for the extraction sulfur compounds from model diesel and diesel fuel, and the feasibility of their regeneration is also investigated. Unlike AlCl3-based ionic liquids, [BMIM][$FeCl_4$] is air and moisture stable, and is immiscible with water. The electronic configuration of $Fe^{3+}$is$1s^2 2s^2 2p^6 3s^2 3p^6 3d^5 4s^0$. According to π-complexation mechanism the cations with empty s-orbitals and the electron density can form the usual σ bonds with their s-orbitals and, in addition, their d-orbitals can back-donate electron density to the anti-bonding π-orbitals of the sulfur rings. Fe^{3+} can form π-complexation bonding with aromatic sulfur compounds. [BMIM][$FeCl_4$] demonstrated effective for removal of aromatic sulfur compounds from diesel. It can be regenerated through reextraction by hexane, and might be used as a promising solvent for the desulfurization of diesel by an extractive desulfurization process.

Keywords: desulfurization, extraction, ionic liquid, Lewis-acidic, diesel

References

[1] Song C S. An overview of new approaches to deep desulfurization for ultra-clean gasoline, diesel fuel and jet fuel. Catal. Today, 2003, 86: 211-263.

[2] Welton T. Room-temperature ionic liquids: Solvents for synthesis and catalysis. Chem. Rev., 1999, 99: 2071-2084.

[3] Bösmann A, Datsevich L, Jess A, Lauter A, Schmitz C, Wasserscheid P. Deep desulfurization of diesel fuel by extraction with ionic liquids. Chem. Commun., 2001, 23: 2494-2495.

[4] Alonso L, Arce A, Franciso M, Rodr•guez O, Soto A. Gasoline desulfurization using extraction with [C8mim][BF_4] ionic liquid. AIChE Journal, 2007, 53: 3108-3115.

104b Liquid-Liquid Extraction of Alcohols from Water Using Ionic Liquids

Alexandre Chapeaux, Luke D. Simoni, Mark A. Stadtherr, and Joan F. Brennecke. Department of Chemical and Biomolecular Engineering, University of Notre Dame, 182 Fitzpatrick Hall, Notre Dame, IN 46556

Separation of organics from aqueous solutions, as is frequently encountered in recovery of products from fermentation, usually requires energy intensive distillation. Here we explore the potential use of ionic liquids to remove alcohols from aqueous solutions using ionic liquids. Ionic liquids are low melting salts that exhibit negligible volatility at ambient conditions. Therefore, the recovered alcohol could be removed from the ionic liquid extraction solvent in a single evaporation stage. To evaluate various ionic liquids for this application, this study details the phase behavior of binary and ternary systems of ionic liquids with water and alcohols. The mutual solubilities of various ionic liquids with water were determined in order to study the effect of the anion, and the alkyl chain length on the cation and substitution on the cation. As expected, an increase in the alkyl chain length or in the substitution decreases the mutual solubility of the ionic liquid with water. Of the ionic liquids tested, the ones with the tetrafluoroborate anion have the largest mutual solubility with water, followed by ionic liquids with tetracyanoborate anion, hexafluorophosphate anion, bis-(trifluoromethylsulfonyl)imide anion, methide anion, and tris(pentafluoroethyl)trifluoro-phosphate anions.[1]

The second component of our work is to optimize ionic liquids for various liquid-liquid separations. The general feasibility of separating water and alcohols using ionic liquids has been shown previously.[2-4] Ternary diagrams are necessary in order to design liquid-liquid separations. With the binary data as starting points, ternary diagrams of systems with water and alcohols for various ionic liquids were developed, measuring all three components of each phase. We also test the ability of excess Gibbs energy models to predict these ternary data over the entire composition range, using only parameters determined from binary and pure component data. Further the Hunter-Nash method was used to determine the number of stages that would be required in order to separate water and alcohols using a liquid-liquid extractor.[5]

Replacing distillation with liquid-liquid extraction could reduce energy consumption and make the production of alcohols cheaper and more environmentally sound. To that end, this study also measured enthalpies of mixing for mixtures of ionic liquids and alcohols. With this data we use a thermodynamic cycle to calculate the energy required to evaporate the alcohol from the ionic liquid and compare this to energy requirements for conventional distillation.

References

[1] A. Chapeaux, L. D. Simoni, M. A. Stadtherr and J. F. Brennecke, J. Chem. Eng. Data, 2007, 52, 2462-2467.

[2] M. M. M. Fadeev A. G., Chem. Comm., 2001, 295-296.

[3] X. S. Hu, J. Yu and H. Z. Liu, J. Chem. Eng. Data, 2006, 51, 691-695.

[4] V. Najdanovic-Visak, L. P. N. Rebelo and M. N. da Ponte, Green Chem., 2005, 7, 443-450.

[5] A. Chapeaux, L. D. Simoni, M. A. Stadtherr and J. F. Brennecke, Green Chem., 2008, Submitted.

104c Residual Monomer Reduction In Polymer Latexes Using Supercritical Carbon Dioxide

Marijke Aerts[1], R.J. Kemp[1], M.F. Kemmere[1], J. Meuldijk[1], J.T.F. Keurentjes[1], and F. Tumakaka[2]. (1) Department of Chemical Engineering and Chemistry /Process Development Group, Eindhoven University of Technology, P.O. Box 513, Eindhoven, 5600MB, Netherlands, (2) Lehrstuhl für Thermodynamik, Universitat Dortmund, Emil-Figge-Strasse 70, Dortmund, 44227, Germany

An important issue in the production of polymeric materials is the amount of residual monomer that remains in the product after polymerization. Existing techniques to reduce the amount of residual monomer are energy-intensive, time consuming and not able to meet future requirements for the level of residual monomer in polymer based products. As an alternative a (sc)CO_2 based extraction process of the monomer from the polymer is evaluated.

Carbon dioxide is chosen as extraction medium for its many advantageous characteristics over other solvents. Carbon dioxide is a so called „green solvent" due to its environmentally benign nature and its low toxicity and flammability. Moreover, carbon dioxide at supercritical conditions (T_c=31°C and P_c=74 bar) is liquid-like in density and gas-like in viscosity and diffusion coefficient of solutes. As the critical conditions are relatively mild, easy separation of low molecular species is obtained by simply tuning pressure and temperature.

In the (sc)CO_2 based extraction column for the reduction of residual monomer in latex products, latex droplets (~250 mm) are dispersed in (sc)CO_2. The latex droplets contain submicron (~100 nm) polymer paricles. Monomer diffuses from the polymer via the water phase into the (sc)CO_2 phase. A parallel mass transfer mechanism is diffusion of the monomer from the polymer particles directly to the (sc)CO_2 phase. This occurs when a particle resides at the water-(sc)CO_2 interface and is in direct contact with the (sc)CO_2 phase. This mass transfer mechanism is also reffered to as the shuttle effect. **Figure 1** illustrates the mass transfer mechanisms in the extraction column.

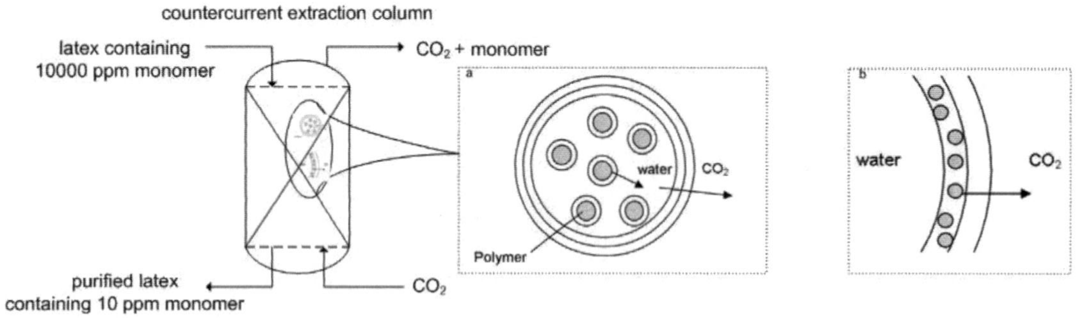

Figure 1 Mass transfer in series (a) and the shuttle effect (b).

For the shuttle effect to dominate mass transfer, only a small fraction of the (sc)CO_2-water interface has to be occupied with polymer particles due to the plasticizing effect of (sc)CO_2. The plasticizing effect of (sc)CO_2 results in a tremendous increase of the monomer diffusion coefficient in the polymer. Our work demonstrates that (sc)CO_2 based extraction to reduce residual monomer in polymer products is a very promising and a sustainable alternative to existing techniques, e.g. steam stripping.

References

1. M. F. Kemmere, M. H. W. Cleven, M. A. van Schilt, J. T. F. Keurentjes, „Process design for the removal of residual monomer from latex products using supercritical carbon dioxide“, Chemical Engineering Science, 2002, 57, 3929-3937.

2. M. Gornert, G. Sadowski, „Phase Equilibrum Measurements of the Polystyrene/Styrene/Carbon Dioxide Ternary System at elevated Pressures using ATR-FTIR Spectroscopy“, Macromolecular Symposia, 2007, 259, 236-242.

104d Cfd-Supported Design In Liquid-Liquid Extraction and Experimental Validation of Simulation Results

Thomas Gurker and Rolf Marr. Department of Chemical Engineering and Environmental Technology, Graz University of Technology, Inffeldgasse 25/C/II, Graz, A-8010, Austria

Axial dispersion of the continuous phase was investigated for single phase flow and dual phase flow in a lab size Rotating Disk Contactor (RDC) column. The column diameter was 100 mm. The inner diameter of the stator rings was 70 mm. The rotor disc diameter was 60 mm and the diameter of the rotor axis was 35 mm. Disc and stator ring distance was 36 mm. The design of the column except the disc and stator ring distance is based on the Misek design recommendations.

Residence time distribution was determined with the tracer NaCl by detecting the conductivity on two points along the extraction column. Flow rate of both phases and the rotor revolution were varied. The conductivity probes were installed in a distance of 1118 mm. Axial dispersion was determined for open-open boundary and expressed in terms of the dispersion number.

Based on these experimental results the flow pattern of the investigated RDC was then modelled by CFD-simulation with Fluent®. For that purpose, different mesh-geometries and turbulence models were investigated. This was followed by the validation of single phase flow in 2D-axisymmetric and 3D applications with a simulation area up to 10^6 cells.

Results:

The realizable k-ε model matches most favourably with experimentally obtained data. 2D-axisymmetric configuration sufficed simulation of the complex flow pattern in the RDC without notable loss of accuracy.

The calculation of the axial dispersion number of the continuous phase $E_{ax,c}$ was carried out by applying the Discrete Phase Model on the simulated flow field. Subsequent comparison of simulation results with experimentally obtained data showed an acceptable correlation.

104e Computational and Experimental Analysis of the Flow In An Annular Centrifugal Contactor

Kent E. Wardle, Chemical Sciences and Engineering Division, Argonne National Laboratory, 9700 S. Cass Avenue, Argonne, IL 60439

The annular centrifugal contactor has been developed for solvent extraction processes used in the reprocessing of spent nuclear fuel. The compact size and high efficiency of these contactors have made them the choice for advanced reprocessing schemes and the central equipment for a future advanced fuel cycle facility. While a sufficient base of experience exists to facilitate successful operation of current contactor technology, a more complete understanding of the fluid flow within the contactor would enable further advancements in design and operation of future units and greater confidence for use of such contactors in a variety of other solvent extraction applications. This research effort has coupled computational fluid dynamics modeling with experimental measurements and observations to provide a valid detailed analysis of the flow within the centrifugal contactor. CFD modeling of the free surface flow in the annular mixing zone using the Volume of Fluid (VOF) volume tracking method combined with Large Eddy Simulation (LES) of turbulence was found to have very good agreement with the experimental measurements and observations. From this, detailed study of the flow and mixing for different housing vane geometries was performed. Flow within the separation zone was also modeled providing a useful description of the flow in this region and identifying critical design features which are the cause of specific experimentally observed effects. It is anticipated that this work will form a foundation for additional efforts at improving the design and operation of centrifugal contactors and provide a framework for progress towards simulation of solvent extraction processes.

104f Porous Membranes In Liquid Membrane Permeation

Marlene Fritz, Hannes Noll, and Matthäus Siebenhofer. Department of Chemical Engineering and Environmental Technology, Graz University of Technology, Inffeldgasse 25/C/II, Graz, A-8010, Austria

Supported liquid membrane separation is a very efficient technology [1]. Until now supported liquid membrane separation has not established in industrial scale because of membrane bleeding.

To overcome bleeding, stabilization of the membrane is subject of ongoing research activities. Improvement has been expected from feeding the solvent phase directly in PE- membranes cross currently to the mass transfer direction.

Therefore membrane carrier modules needed to be characterized considering pore size, voidage, membrane thickness, phase flow rate and bleeding. Membrane thickness was varied in a range of 1 to 5 mm, pore size was 7 to12 μm and 20 to 60 μm and porosity was 37%. The best compromise between membrane bleeding, pore size distribution and flow rate was obtained for membrane modules with a pore size distribution of 20-60 μm and 2 mm thickness.

After hydraulic characterization mass transfer was investigated with the test system Zn2+/DEHPA/ H2SO4. DEHPA was dissolved in ShellSoll-T, a hydrocarbon blend. Mass transfer experiments included transfer of constituents and water transfer. In a first step mass transfer was measured without any solvent flux, in a second step the organic solvent phase was pumped through the membrane. Mass transfer showed positive dependence on flow of the organic phase perpendicular to the mass transport direction. Enhancement was limited. Extended solvent flux perpendicular to the mass transport direction did not rise mass transfer.

[1] N.M.Kocherginsky; Q.Yang; L.Seelam Recent advances in supported liquid membrane technology, Seperations and Purification Technology 2007, 53 171-177

REDROP – AN EFFICIENT MONTE-CARLO METHOD FOR SIMULATING EXTRACTION COLUMNS

Murat Kalem, AVT - Thermal Process Engineering, RWTH Aachen University, Aachen, Germany
Andreas Pfennig, AVT - Thermal Process Engineering, RWTH Aachen University, Aachen, Germany

Introduction

Due to increase in the price of fossil fuels and the decrease in their reserves, renewable raw materials are industrially used more and more frequently. Therefore extraction, as an energy-saving unit operation has gained additional importance. Precise design methods for technical columns are required for the future. So far, pilot-plant experiments are carried out as a basis for equipment design. These cost-intensive experiments can be minimized or even avoided by simulating the behavior of extraction columns. The simulations have to account for drop movement, axial dispersion, mass transfer, coalescence and drop breakage. The key to a successful simulation is solving the drop-population balances. A large number of researchers have tried to numerically solve the system of equations. A great disadvantage of this procedure is the increase in numerical efforts if additional properties of the drop population have to be regarded. An alternative approach solving the population balances is using the Monte-Carlo method. ReDrop (Representative Drops) is a simulation program which can successfully predict the behavior of pilot-plant columns.[1,2] Because different internal types (random packing, regular packing, sieve tray) can be modeled with ReDrop, simulations can reveal which kind of internal is the best for the interested material system. Also the dimension of the internals as well as the operating conditions can be optimized If the design of the technical column is based on pilot-plant experiments, a type of internal has to be chosen in a early design of the design procedure. A later change of the internals is cost-intensive. In this paper, ReDrop is explained in detail and results are compared to experimental values in literature. Finally, an outlook is given on the extensions of ReDrop which are currently treated.

ReDrop

The basic idea of the ReDrop model is to follow a discrete number of drops through their total life-time in an extraction column. Every effect which influences each drop is taken into account. The structure of the code is module-based. So any new phenomena which have to be considered can be easily implemented. The ReDrop algorithm is shown in Fig. 1. Before starting the simulation, the user has to supply raw data like mass flows, initial concentrations or type of the internals. Every time step of the simulation starts with feeding of new drops into the column. In the drop loop, all effects which act on every single drop are calculated. For a given diameter, the sedimentation velocity of the drop is calculated. The new concentration of the solute in the drop is computed. Models are used to determine the probability of each drop to coalesce or to break. After all drops are worked through the program focuses on the continuous height elements. Regarding back mixing the new concentration of the transfer component is calculated for each height element. After determining the holdup for each height element the simulation continues with the next time step. The core of the ReDrop algorithm are the single-drop models which can predict the sedimentation and the mass transfer of the drop. These models include system-specific parameters which have to be obtained in simple-lab scale experiments. Thus experiments are unavoidable because even small amounts of impurities can change the drop behavior drastically. But compared to pilot-plant scale column experiments, the required amount of substance is negligible for the lab-scale measurements. After the model parameters are successfully fitted to single- drop experiments, they are input for ReDrop. For the calculations, the influence of other drops and of the internals is included.

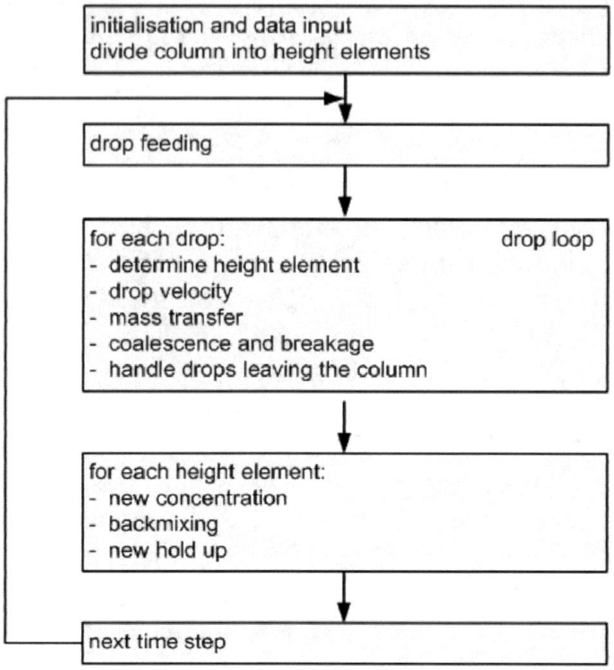

Figure 1. The ReDrop algorithm.

Usually 1000 drops per meter column are considered. Simulation time on a PC with a Pentium IV processor is typically 10 minutes to reach steady state. The output of the simulation can be followed on a Windows console (Fig. 2). Menu bars at the top of the window allows the user to immediately change operational parameters as mass flows. Thus dynamic simulations are allowed for.

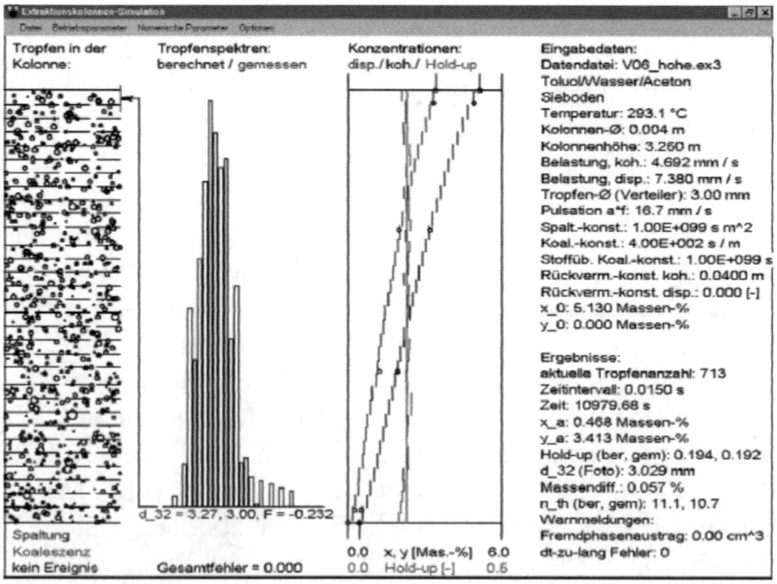

Figure 2. The ReDrop program

Single-Drop Experiments

For any new material system, sedimentation experiments in a lab-scale cell have to be carried out in order to run ReDrop. In Fig. 3, the sketch of such a cell is shown. The cell is divided into sections in order to determine the sedimentation velocity as a function of the residence time of the drop. In order to keep a constant temperature, the cell is temperature controlled. A drop is produced by a computer-driven syringe and a nozzle. Due to the density difference, the drop will rise through the stagnant continuous phase. The sedimentation of the drop is recorded by a video camera. By evaluating the videos, the sedimentation velocity of the drop can be determined. The experiments are performed for at least 15 drop diameters.

The obtained experimental values are then with the model of Henschke.[1] This model combines 4 single models for rigid drops, drops with totally mobile interface, oscillating drops and deformed drops. This combination results in one model which is valid over the complete range of drop diameter. The model has in total 5 parameters, two of which are fixed to typical characteristic values while three have to be fitted to the results of the single-drop experiments. In Fig. 4 such a fit is shown with the two limiting cases of an ideally-mobile and a rigid interface. The influence of internals and holdup is included in the ReDrop program.

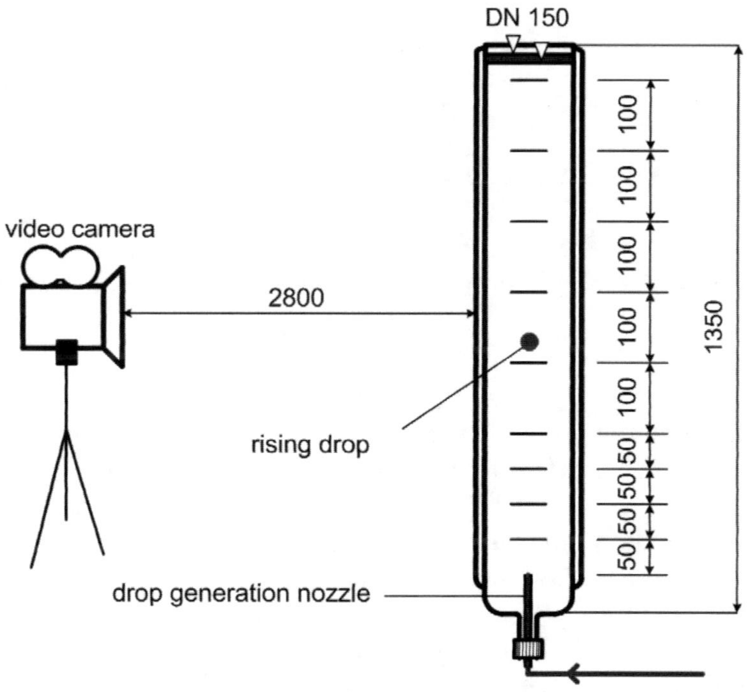

Figure 3. The sedimentation cell.

As for sedimentation, single-drop experiments deliver the necessary information about the dependency of mass transfer from drop diameter, contact time and concentration of the components in both phases.[1] A mass-transfer model with one additional parameter will be used to fit the experimental values. The mass-transfer model is based on the Handlos-Baron model.[3] The Handlos-Baron model assumes always the same effective diffusion coefficient for the same material system. In reality, the effective diffusion coefficient depends on the intensities of interfacial eruptions and on the presence of contaminants. Henschke and Pfennig extended the Handlos-Baron model in such a way that eruptions and contaminants are considered by an adjustable parameter.[4] Experiments with packing or sieve-tray will reveal the influence of the geometry of the chosen internal. These experiments are necessary because it has been seen that internals can have different influence for different material systems. While for physical extraction a sieve-tray improved the mass transfer this effect has not been seen for the reactive standard test system zinc + D2EHPA.[5]

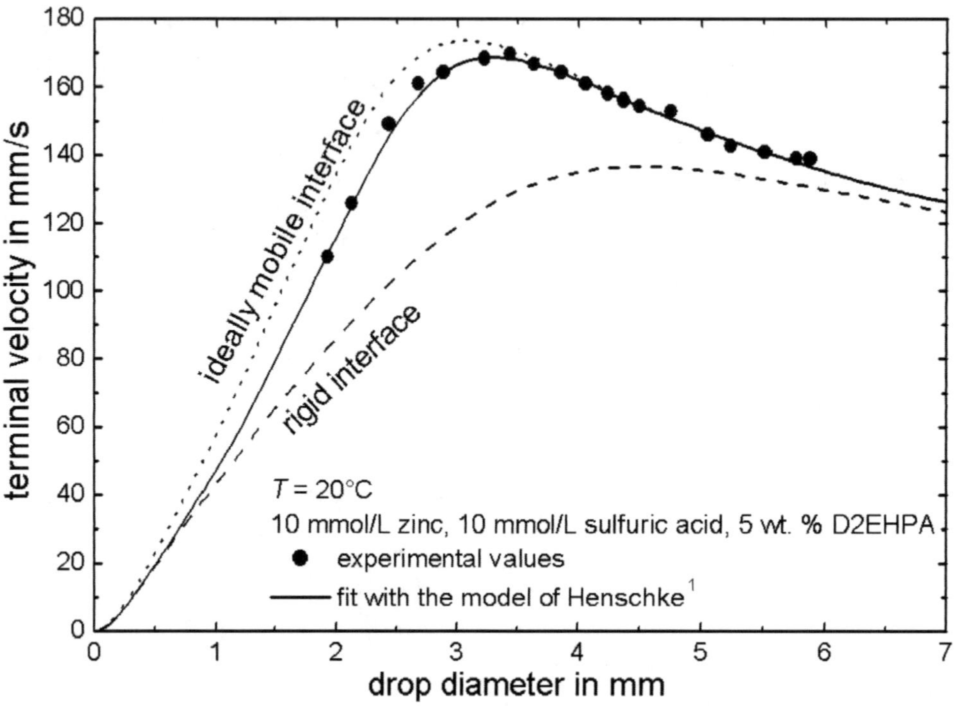

Figure 4. Experimental values and fit of the sedimentation model.

Population models

Single-drop sedimentation as well as mass transfer are calculated by a deterministic model. Drop breakage and coalescence are determined stochastically for each drop. For the drop breakage, the diameter of the drop is compared to a stable diameter.[1] The stable diameter depends on the type of internal, material properties, drop diameter and drop velocity. Drops with a diameter smaller than the stable diameter pass an internal without being broken. If the drop diameter is bigger than the stable diameter, the breakage probability will be calculated. By comparison of the probability with a random number, it is decided if the drop will truly break. For a drop which will split, the number of daughter drops and as well as their drop size distribution are generated again based on distribution functions and random numbers.

Coalescence is the second effect which is based on stochastic calculations. For each drop, a coalescence probability p_k is calculated, depending on holdup, drop diameter, viscosity of the continuous phase, density difference and surface tension.[1] A random variable between 0 and 1 is generated. If this random number is smaller than the coalescence probability, the drop is marked for coalescence. If a second drop (close to the first one) is also marked, the coalescence of both drops is carried out. A detailed description of the models is given by Henschke.[1]

Results

ReDrop simulations have been validated for the physical test systems toluene + water + acetone and butyl acetate + water + acetone.[1] Due to the numerical approach, ReDrop is not only able to simulate the concentration profiles, but also of the predetermination of dynamic limits of the column as flooding. In a recent project with industrial partners ReDrop has proven his applicability for technical systems.[6] The predicted concentration profiles and flooding limits agree within 10 % with experimental points.

Actual work

Reactive Equilibrium and Kinetics

The next goal is to extend ReDrop to reactive extraction. Contrary to physical extraction, chemical reactions have to be considered. So modules are added which are capable of calculating an arbitrary number of reaction equilibria. For slow reactions, reaction kinetics are the rate-determining factor and, therefore, must be modeled and can be quantified in a modified Lewis cell.[7] If diffusion or convection is rate-determining, a multi-component mass transfer model for reactive systems must be developed. Today investigations and validation of ReDrop for reactive extraction are carried out with the reactive standard test system.[8] The reaction equation for this test system is:

$$Zn^{2+} + 1.5\overline{R_2H_2} \leftrightarrow \overline{ZnR_2RH} + 2H^+ \qquad (1)$$

D2EHPA ($\overline{R_2H_2}$) is a cation exchanger which extracts zinc from the aqueous phase by an interfacial reaction. Bars are indicating components in the organic phase. Activity coefficients are used to include the non-ideality in both phases.[9] Currently, the Pitzer model for aqueous phases and the Hildebrand-Scott model are used for reactive equilibria. In future, an IK-CAPE interface will allow to use any G^E model. In Fig. 5, calculated equilibria versus own experimental data for the reactive standard test system are shown. The equilibrium module is programmed in such way that an arbitrary number of reactive equations can be solved simultaneously.

Single-Drop Experiments With A Reactive System

Sedimentation experiments with the standard test system have been carried out. For physical test systems it was known that mass transfer slows drops down. This is due to the destruction of the internal circulation of the drop by mass-transfer induced convections. However, for the reactive system, mass transfer accelerates drops. In case of missing mass transfer, the interfacially active D2EHPA can accumulate at the interface, the drop will thus show rigid behavior. The extraction of zinc by D2EHPA will lead to partly free areas on the interface where more momentum can transferred between the phases. The different behavior of both types of extraction systems is illustrated in Fig. 6.

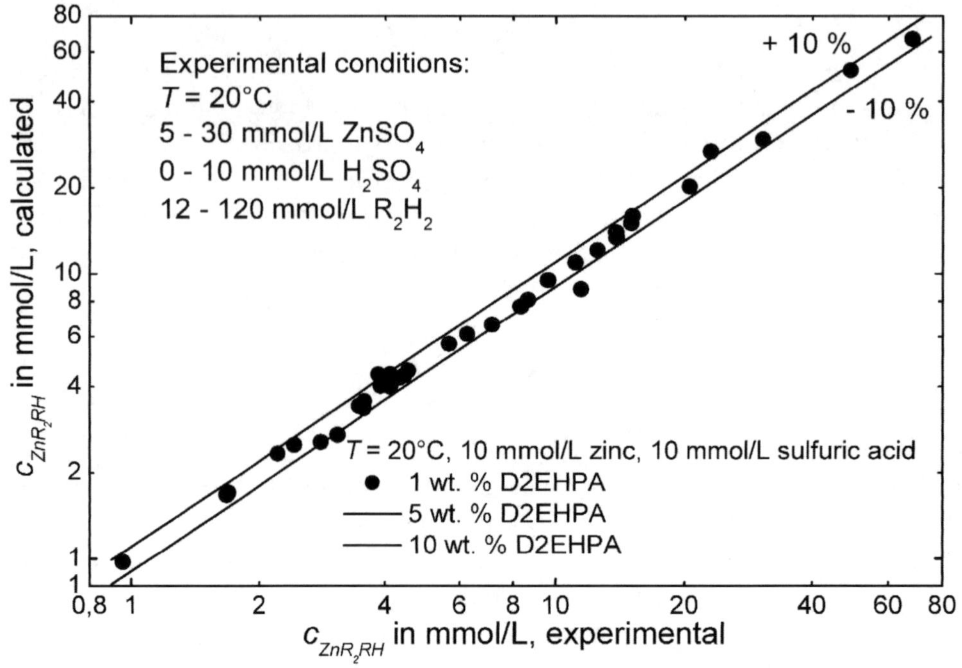

Figure 5. Reactive equilibria for the standard test system of reactive extraction zinc + D2EHPA.

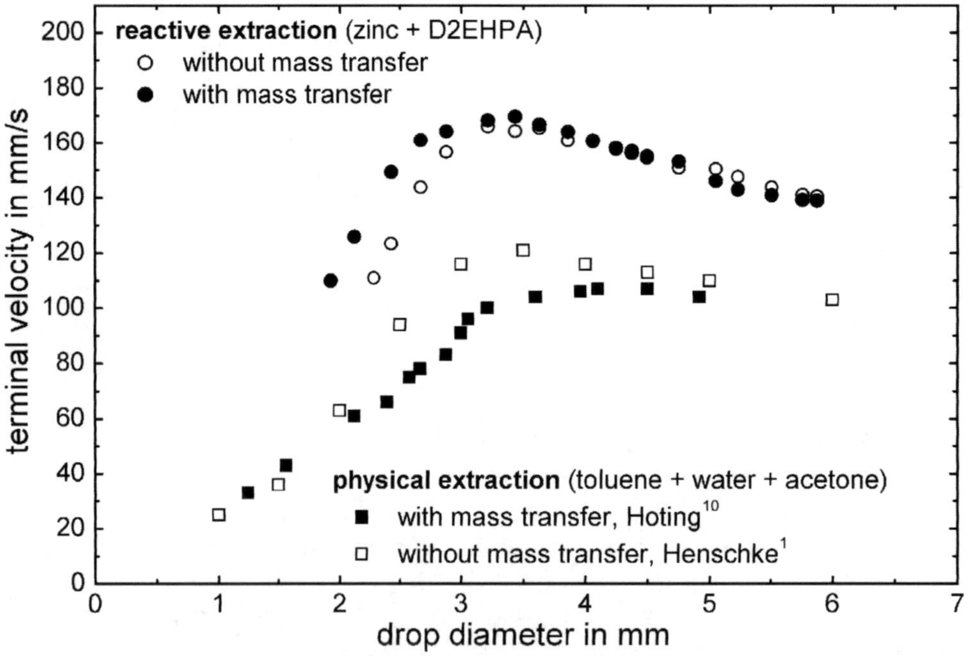

Figure 6. Influence of sedimentation on mass transfer.

Summary

ReDrop is an efficient Monte-Carlo method to predict the behavior of pulsed pilot-plant extraction columns. Necessary single-drop model parameters are determined in lab-scale experiments. For physical systems (even for technical) ReDrop has proved its applicability. By designing columns with the ReDrop program, strong savings in time and costs are possible. Actual work is on the extension of ReDrop to reactive extraction. So modules for calculating reactive equilibriums and mass transfer are being developed.

Acknowledgements

The authors would like to thank Dr. Martin Henschke, Dipl.-Ing. Tobias Grömping and Dipl.-Ing. Mehmet Yücel Altunok for their help in the preparation of this paper..

References

1. Henschke, M. (2004), "Auslegung pulsierter Siebboden-Extraktionskolonnen", Shaker, Aachen.
2. Pfennig, A. (2008), "Reliable design of physical and reactive extraction based on simple lab-scale measurements", Proceedings of the International Solvent Extraction Conference 2008.
3. Handlos, A.E. and Baron,T. (1957), "Mass and heat transfer from drops in liquid-liquid extraction", *AIChE J.*, 3 (1), pp. 127 - 136.
4. Henschke, M., Pfennig, A. (1999), "Mass-transfer enhancement in single-drop extraction experiments", *AIChE J.*, 45(10), pp. 2079 - 2086.
5. Kalem, M. and Pfennig, A (2007), "Hydrodynamik und reaktiver Stofftransport an Einzeltropfen", *Chem. Ing. Tech.*, 79(9).
6. Pfennig, A. and Grömping, T. (2008), "Reaktivextraktion effizient ausgelegt", *Chem. Ing. Tech.*, 79 (9).
7. Bertakis, E., Kalem, M. and Pfennig, A. (2008), "Model-based geometry optimization on a Nitsch cell using the Fisher information matrix, *Chem. Eng. Sci.*, doi:10.1016/j.ces.2007.07.043.
8. Bart, H.-J. and Slater, M.J. (2001), "Standard test system of reactive extraction", European Federation of Chemical Engineering: Working Party on Distillation, Absorption and Extraction.
9. Klocker, H. (1996), "Multikomponentenstoffaustausch bei der Reaktivextraktion", Dissertation, TU Graz, Austria.
10. Hoting, B. (1996), "Untersuchung zur Fluiddynamik und Stoffübertragung in Extraktionskolonnen mit strukturierten Packungen", Nr. 439 in Fortsch.-Ber. VDI, Reihe III, VDI.

108a Crystallization: Its Place In the History and Future of Separation and Purification Processes

Ronald W. Rousseau, School of Chemical & Biomolecular Engineering, Georgia Institute of Technology, 311 Ferst Drive, Atlanta, GA 30332-0100

Crystallization is unique among separation and purification processes, for it alone must meet criteria on morphology as well as purity and yield. The use of crystallization from solutions and melts, using technologies involving stirred tanks, circulating pumps, falling films, impinging jets, and fluidized beds, is ubiquitous throughout the chemical industry. It is found in the production of commodity and specialty chemicals, and it plays an important role in the formulation and efficacy of pharmaceutical compounds. Indeed, it could be argued that crystallization is the equal of distillation in being the workhorse of the chemical industry.

The prominence and utility crystallization has is due to the production of a solid with specific characteristics. There was little sophistication in its early use: e.g. salt crystals were simply harvested from brines and sea water whose evaporation left behind one of the staples of life. The purpose of this presentation is to review the developments that have contributed to bringing crystallization science and technology to its present status. Moreover, how crystallization may contribute to meeting future strategic industrial and societal needs will be examined.

108b Past and Future In Water Desalination – the Cheapest Most Valuable Product on Earth

Rafi Semiat, The Wolfson Chemical Engineering Department, Technion - Israel Institute of Technology, Technion City, Haifa, Haifa, 32000, Israel

The world is facing drought. Many countries are suffering from lack of water or polluted water. Desalination techniques can convert seawater, brackish or polluted water into drinkable water. Solutions based on evaporation techniques and membrane processes can be applied, yet the cost is still high for those with economies that are based on simple agriculture. Proximity to the ocean facilitates seawater desalination. Desalination of brackish water and the treatment of sewage water and industrial wastewater is the key for solving environmental problems while helping to resolve the water shortage situation. Those solutions are based on long evolution of the appropriate separation processes, a key issue in Chemical Engineering.

More R&D is needed in order to further reduce the cost of producing water in order to enable wide spread and affordable usage of desalination. Two main desalination directions survived the long evolution process: reverse osmosis and evaporation processes. Heat and mass transfer beside fluid dynamics are playing a significant role in the evolution of these processes. Competition and similarities exist between the different processes: product cost and market penetration, problems associated with concentration polarization, techniques used to improve the transfer mechanisms, and finally, directions for future research and development. The presentation will highlight some of the important achievements and future directions in water desalination.

108c Historical Overview of the Science and Technology of Separations in Supercritical Fluids

Charles A. Eckert[1], Vittoria M. Blasucci[2], Ryan Hart[2], and Charles Liotta[3]. (1) School of Chemical and Biomolecular Engineering, Georgia Institute of Technology, 2206 Environmental Science and Technology Building, 311 Ferst Drive, NW, Atlanta, GA 30332-0100, (2) School of Chemical & Biomolecular engineering, Georgia Institute of Technology, 2277 Peachtree Rd NE Unit 601, Atlanta, GA 30309, (3) School of Chemistry and Biochemistry, Georgia Institute of Technology, 311 Ferst Drive NW, Atlanta, GA 30332-0100

Although the phenomenon of supercritical fluids (SCFs) has been known for more than a century, it was not until the early 70's that chemical engineers began to appreciate the potential of SCFs for chemical processes. After an initial burst of enthusiasm where SCFs were hailed as the new panacea, the community settled down to learning their strengths and limitations and began discovering an ever-growing host of useful applications. The primary advantage of the SCF state is that it couples efficient solvation (never as good as in the liquid) with transport properties far superior to liquids (but never as good as those of gases). The earliest work primarily involved extractions with SCF CO2, with food products and environmental applications leading the way, and in each of these the exceptional ability of a solvent to penetrate a matrix made unique processes possible.

In the ensuing years, a plethora of other applications were explored, and many of these are now in commercial use. Perhaps the first application proposed other than extraction was SCF water oxidation of waste streams, which has been thoroughly investigated but has never been widely implemented. However, subsequent work on other reactions in SCFs has generated broader use. Very often the advantage of using SCFs in the reaction system is the vastly improved mass transfer. Other early applications spanned the range from the rehabilitation of old books to miscible flooding for tertiary oil recovery.

A bit later a variety of methods using SCFs for particle formation have been proposed, and many are now in widespread use for applications ranging from drug delivery to nanoparticle formation. In addition there are many applications in coating and cleaning, including such diverse targets as microelectronic and commercial dry cleaning. Other applications abound, such as SCF drying, chromatography, processing of polymers, impregnation and dyeing of polymers, and the processing of renewable resources.

The region in which SCFs display unusual, and very useful properties, is generally where both the density and the compressibility are large (typically TR = 1.01 – 1.1 and PR = 1 – 2). The fluid density must be relatively high for it to be a good solvent, but also the compressibility must be high so that small pressure changes give the large density changes allowing for selective solution or dissolution. The SCF does not dissolve insoluble compounds; rather it enables higher concentrations of dilute solutes by a combination of physical and chemical solvation forces. Typical enhancements for low solubility compounds may be 103-106, and in rare cases, effects as large as 1010or even 1015 have been reported.

The majority of the applications have used SCF CO_2, which has a readily accessible critical point at 31ºC and 74 bars, and is inexpensive, nontoxic, and nonflammable. Unfortunately, it is a rather poor solvent, but it can be substantially enhanced by additions of small amounts of cosolvents or modifiers – generally polar or protic compounds. However, a wide range of SCF solvents and cosolvents are available, and will be discussed with the merits and limitations of each. Also we shall discuss the physical and chemical intermolecular forces operating in such systems, and cite examples of chemical engineers taking advantage of these for specific applications.

Finally, the talk will focus on the choice of when to select the use of SCF methods. Generally these are cases where more conventional separations will not serve the purpose, and often where cost alone is not the issue. Rather it could be used where there is the need for a safe solvent, for ease of solvent removal, for the need for enhanced mass transport, for treating a thermally labile product, or even to create a unique form of product.

111a Membrane Module Design

G. Glenn Lipscomb, Chemical & Environmental Engineering, University of Toledo, Mail Stop 305, Toledo, OH 43606

Membrane module designs today are very similar to those developed 30 years ago for the first desalination units. However, our understanding of the underlying physics of transport within modules has progressed significantly. Professor Sirkar has made seminal contributions to this body of knowledge and provided a unifying framework for analyzing performance based on the nature of the contacting phases and flow profiles within the module. The impact of Professor Sirkar's work on the future of module design is discussed.

111c A New Hybrid Ion Exchange-Nanofiltration (HIX-NF) Separation Process for Energy Efficient Desalination: Process Concept and Laboratory Evaluation

Sudipta Sarkar and **Arup K. Sengupta**. Lehigh University, 13 E. Packer Avenue, Bethlehem, PA 18015

The study presents the results of a new energy-efficient hybrid ion exchange-nanofiltration (HIX-NF) process to desalinate brackish and sea water. From a scientific viewpoint, the HIX-NF process is unique in its approach for it partially alters the feed water chemistry prior to the membrane treatment. Specifically it converts monovalent chloride, the predominant anion in salt water, into divalent sulfate through a reversible anion exchange process without requiring any external regenerant. As a result, reverse osmosis (RO) membrane can be replaced altogether by nanofiltration (NF) membrane resulting in a marked reduction in energy requirement. Also, conversion of chloride to sulfate reduces the osmotic pressure of the feed water by nearly thirty percent. The reject stream from nanofiltration, rich in sulfate, is used to regenerate the anion exchanger. Laboratory results validate that sodium sulfate can be desalinated with NF membranes at much lower transmembrane pressure than with RO membranes, all other conditions remaining identical. The reversible sulfate-chloride anion exchange and, in particular, sulfate/chloride selectivity plays a central role for the success of the proposed process. The most significant finding is that a single anion exchange resin cannot sustain the HIX-NF process for different feed water concentrations. The investigation, however, shows that by changing the size of the amine functional group (e.g., quaternary-, tertiary-, secondary- and mixed amine), the sulfate/chloride selectivity of the anion exchange resin can be tailored for a specific feed water salinity allowing optimum reversibility of the anion exchange process. Laboratory studies have validated the basic premise of the hybrid process including greater than two times lesser energy requirement than RO process for the same feed water. Future proto-type experimental studies are strongly recommended to confirm the overall viability of the HIX-NF process.

Key Words: Desalination, Brackish Water, Sea Water, Nanofiltration, Hybrid Ion Exchange, sulfate-chloride ion exchange, Regenerability, Coupled transport.

111d Recent Advances and Future Opportunities for Solvent Resistant Nanoporous Membranes

Ramesh Bhave, Materials Science & Technology Division, Oak Ridge National Laboratory, Oak Ridge, TN 37831-6116

During the past decade, there has been a growing interest in the development of membrane-based processes, using solvent resistant nanofiltration membranes, as alternatives to distillation, crystallization and liquid-liquid extraction to improve efficiency of many chemical and petrochemical applications. This paper will review recent advances in the development and application of polymeric and inorganic solvent resistant nanoporous membranes in these critically important organic separations when used as standalone or in combination with existing chemical/petrochemical operations.

111e New Membranes for Hydrogen Purification and Proton Transport for Fuel Cells

He Bai and **W.S. Winston Ho**. Department of Chemical and Biomolecular Engineering, The Ohio State University, 125A Koffolt Labs, 140 West 19th Avenue, Columbus, OH 43210-1180

This presentation covers two types of new membranes for fuel cells: (1) carbon dioxide-selective membranes for hydrogen purification and (2) proton-exchange membranes (PEMs). On hydrogen purification for fuel cells, the membranes have been synthesized by incorporating amino groups into polymer networks. These membranes are selective to carbon dioxide and hydrogen sulfide preferentially versus hydrogen since the acid gases permeate through the amine-containing membranes via the facilitated transport mechanism due to their reversible reactions with the amine. The membranes synthesized have shown high carbon dioxide and hydrogen sulfide permeabilities and selectivities vs. hydrogen up to 170oC. Using the membrane synthesized, we have obtained <10 ppm carbon monoxide in the hydrogen product in water gas shift membrane reactor experiments via carbon dioxide removal. The data have been in good agreement with modeling prediction. Since hydrogen sulfide has much higher reaction rate with the amine than carbon dioxide, hydrogen sulfide can permeate through the membrane much faster than carbon dioxide. Our initial experiments have shown a nearly complete removal of hydrogen sulfide from 50 ppm in the synthesis gas feed to about 10 ppb in the hydrogen product, which is good for fuel cell applications. For high temperature PEMs, we have synthesized sulfonated polyimide copolymers containing hydrophilic soft segments to increase the water retention of the membranes at high temperatures and low humidities. In fuel cell performance testing, the new membrane showed similar performance as Nafion® 112 at 70oC and 80% RH, but much better performance than Nafion® 112 at 120oC and 50% RH. Recently, we have also synthesized new sulfonated polybenzimidazole (SPBI)-based membranes. The membrane has exhibited a very high conductivity (> 0.1 S/cm) at high temperatures (> 120oC) and low humidities (even anhydrous). All these new PEMs should be much more cost-effective than Nafion®.

132a Application of Do & Do Models to Single-Walled Carbon Nanotubes: Insight into Water Adsorption In Microporous Carbons

Pyoungchung Kim, Civil and Environmental Engineering, University of Tennessee, Knoxville, 70 Perkins Hall, Knoxville, TN 37996-2010 and Sandeep Agnihotri, Civil and Environmental Engineering, University of Tennessee, 73A Perkins Hall, Knoxville, TN 37996-2010.

This paper describes the application of a water-activated carbon isotherm model to study the effect of oxidation and operating temperature on water adsorption data collected for single-walled carbon nanotubes (SWNTs). Water adsorption isotherms in carbon adsorbents are commonly curve-fitted by several adsorption isotherm equations. Our previous study showed that of these isotherm equations the Do and Do equation modified by Marban et al. (2004) was the most suitable equation for interpreting the experimental water isotherms collected for single-walled carbon nanotubes (SWNTs). This model is able to deconvolute the total experimentally-observed isotherm with great accuracy into two components: micropore filling and adsorption on primary sites. We applied the modified Do model to better understand water adsorption mechanisms on SWNTs exposed to ozonation and different temperatures. Water adsorption isotherms were obtained using a customized high sensitivity gravimetric balance technique at different temperatures from 5 to 55 oC. SWNT samples exposed to ozone were analyzed using x-ray photospectroscopy (XPS) and presented the gradually increasing oxygenated functional groups with increasing ozone concentrations. Water adsorption isotherm of the ozone treated SWNT samples showed the water adsorption by functional groups at lower relative pressure (P/Po) gradually increased with increasing ozone concentrations. Parameters of curve-fitting by modified Do model showed that functional groups (So) also increased as ozone concentration increased which was identical to XPS result whereas water cluster size (m) diffusing into micropore decreased. Overall, the oxidation of nanocarbons enhanced the water uptake at lower vapor pressures, and reduced the size of water cluster that grows on primary sites and the water cluster that migrates into the micropores. Water adsorption isotherms at different temperatures on SWNT samples showed that as temperature increased the relative pressure at which water uptake abruptly increased. Parameters of curve-fitting by modified Do model showed that as temperature increased the functional group concentrations were almost constant, micropore volume decreased and the size of cluster size penetrating into micropore increased which is due to the thermal expansion of water clusters. Exploring the applicability of water-activated carbon models to water-carbon nanotubes is a goal unique to our research. This work should help identify those physical and chemical properties of microporous carbons that are most influencing in uptake of water vapor from air and process streams.

132b Adsorption and Diffusion of Gases In Mof-5, Mof-177 and Zeolite 5A

Dipendu Saha and Shuguang Deng. Chemical Engineering Department, New Mexico State University, P.O. Box 30001, MSC 3805, Las Cruces, NM 88003

A comparative study was carried out to evaluate adsorption properties of two metal organic framework adsorbents (MOF-5 and MOF-177) and a conventional zeolite adsorbent (5A) for hydrogen purification and other gas separation processes. Three different greenhouse gases, carbon dioxide (CO2), methane (CH4) and nitrous oxide (N2O) and three other potential trace components in hydrogen, carbon monoxide (CO), hydrogen sulfide (H2S) and ammonia (NH3) are selected to investigate both adsorption equilibrium and kinetics these adsorbents. The adsorption equilibrium and kinetic data were measured in a volumetric unit at various temperatures. It is observed that zeolite 5A has a much higher CO2 capacity than MOF-5 and MOF-177 at ambient conditions. The three adsorbents have a similar adsorption capacity for other gases. However, MOF-5 and MOF-177 have much faster adsorption kinetics. Diffusion time constants and isosteric heat of adsorption will also be reported.

Predictions of Protein Adsorption and Desorption in Liquid-Solid Circulating Fluidized Bed Ion Exchange System

Jahirul I. Mazumder, Jesse Zhu, Amarjeet S. Bassi and Ajay K. Ray
Department of Chemical & Biochemical Engineering, The University of Western Ontario, 1151 Richmond St., London, ON N6A 5B9, Canada

Introduction

The interest in recovery of various functional proteins from large volume of industrial broths and biological wastewater streams has increased in recent years, due to the advancement in genetic engineering, and concerns about recycling limited resources and environmental protection (Lan, 2001). Ion exchange adsorption and chromatography have long been used for downstream processing in the biotechnology industry because of its high binding capacity, high resolving power, and relatively inexpensive (Chang and Chase, 1996). As the protein concentration in the feedstock is fairly low, large volumes must be treated to recover proteins in an industrial scale. Therefore, a highly efficient continuous ion exchange process is necessary for the industrial scale protein purification.

Lan et al. (2002) had developed a Liquid-Solid Circulating Fluidized bed (LSCFB) ion-exchange system for the continuous recovery of protein from unclarified broth. LSCFB ion-exchange system is used as an integrated reactor and regenerator system; two different operations (adsorption and desorption) are carried out simultaneously in two separate columns (downcomer and riser respectively) in continuous mode (with continuous circulation of ion exchange particles between two columns). Again, with the shift into the circulating fluidization regime from convention fluidization, liquid and solid phase dispersion is significantly reduced. Moreover, LSCFB has some obvious advantages in addition to the advantages of conventional fluidized bed such as high liquid-solid contact efficiency, higher throughput and enhanced mass transfer due to higher velocity, reduced back mixing of phases and uniform flow pattern (Zhu et al., 2000).

Detailed modeling of the hydrodynamics, mass transfer and kinetics of adsorption and desorption of protein in the LSCFB ion-exchange system is fundamental and crucial for better understanding of the adsorption and desorption behaviors, the design and scale up of the LSCFB system, and optimization of the operating parameters. A number of models have been developed to describe the protein adsorption behaviors in packed bed, expanded bed and fluidized beds considering various types and degrees of approximation to physical reality. Several steps are involved in the process of adsorption of proteins onto the adsorbents: convective and diffusion mass transfer from liquid phase to adsorbent surface, diffusion through the pore of the ion exchange resins and the surface reactions. The surface adsorption process is sufficiently rapid compared to the former two steps and is not considered as a limiting step. Lan et al. (2000) developed a model for continuous protein recovery in LSCFB ion-exchange system assuming the process is surface reaction limited.

In this work, we present a model to predict the continuous adsorption and desorption of BSA protein onto and from anion exchange resin (Diaion HPA25®) in the LSCFB ion-exchange system taking into account the fluidization behaviors in the LSCFB, overall mass transfer for adsorption process and surface desorption kinetics. The model was validated by comparing the predicted results with experimental data reported by Lan et al. (2000; 2002).

LSCFB Ion Exchange System and Continuous Protein Recovery

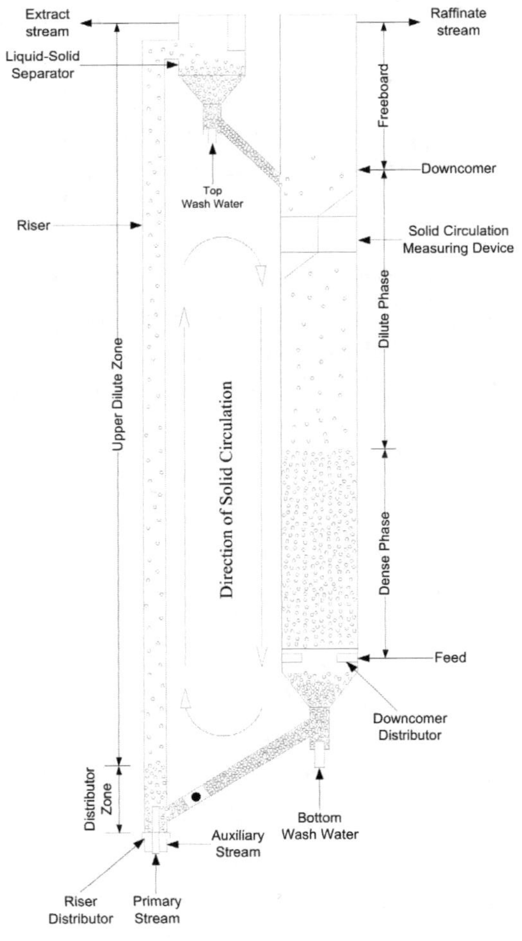

Figure 1: Schematic diagram of the Liquid-solid Circulating Fluidized bed (LSCFB) ion exchange system (Lan et al., 2002).

LSCFB is composed of two interconnected columns namely: a riser and a downcomer. Lan et al. (2002) used LSCFB for simultaneous adsorption and desorption of BSA protein using anion exchange resins Diaion HPA25® in the two separate columns of the LSCFB. *Figure 1* shows a schematic diagram of the LSCFB ion-exchange system developed by Lan et al. (2002). Adsorption of protein from the BSA solution was carried out in the downcomer of the system and the adsorbed protein were extracted from the ion exchange resins in the riser. The configurations of the LSCFB ion exchange system and continuous protein recovery in the LSCFB unit were described by Lan et al. (2002).

Hydrodynamic Modeling

Fluidization in the Downcomer:

The liquid and solids contact counter-currently in the downcomer of the liquid solid circulating fluidized bed: the liquid moves upward and the solids downward. Lan et al. (2002) observed three different operating zones in the downcomer which differ in solids holdup: the dense phase zone, the dilute phase zone and the freeboard zone. The section from the top to the solids entrance of the downcomer is the freeboard region which is necessary to prevent the loss of ion exchange resins through the raffinate stream. The dilute phase is in between the freeboard and the dense phase. The protein concentration is very low in the dilute phase and the freeboard zone. Solids holdup in the dense phase zone is much higher than that of the other zones and contains most of the ion-exchange particles. Hence, the dense phase zone is the most important zone for protein adsorption in the downcomer (Lan et al., 2002). The dense-phase zone operates as a conventional fluidized moving bed (Kwauk, 1992). Ricardson-Zaki equation is modified for the counter current arrangement to calculate the bed-voidage in the dense-phase zone:

$$U_{ld} + U_{sd}\frac{\varepsilon_d}{1-\varepsilon_d} = U_i {\varepsilon_d}^n \tag{1}$$

The bed expansion index (*n*) can be determined from the following correlation:

$$n = (4.4 + 18\frac{d_p}{D_c})\mathrm{Re}_t^{-0.01} \tag{2}$$

$$\text{where, } \mathrm{Re}_t = \frac{U_t d_p \rho}{\mu} \tag{3}$$

U_i, the superficial liquid velocity at $\varepsilon = 1$, which can be determined using following empirical equation (Khan and Richardson, 1989):

$$\frac{U_i}{U_t} = 1 - 1.15(\frac{d_p}{D_c})^{0.6} \tag{4}$$

Hence, ε_d can be calculated from Eq.1 for a given U_{ld} and U_{sd} and determining n and U_i from Eq. 5 and Eq. 7 respectably.

Fluidization in the Riser:

The riser of the LSCFB ion-exchange system is operated in circulating fluidization regime: liquid velocity in the riser is greater than the terminal settling velocity of the solid particles. As discussed earlier, the riser distributor separates the liquid flow into two portions: primary liquid flow and auxiliary liquid flow. Because of the arrangement of the riser distributor, two distinct zones, namely a distributor zone and a upper dilute zone, are observed along the riser based on their solids holdup. Solids holdup in the distributor zone is much higher than the upper dilute zone and the distributor zone can be considered to be operated as conventional fluidization. Therefore, the voidage in the distributor section can be approximated using the correlations for homogeneous conventional fluidization. The modified Richardson–Zaki equation for the distributor zone of the co-current up flow riser can be written as:

$$U_{lr} - U_{sr}\frac{\varepsilon_{r1}}{1-\varepsilon_{r1}} = U_i \varepsilon_{r1}^{\ n} \tag{5}$$

The bed expansion index (n) and the superficial liquid velocity (U_i) at $\varepsilon = 1$ in the distributor zone of the riser can be calculated using Eqs. (2)-(4).

The upper dilute zone is operated in circulating fluidization regime and solids holdup in this zone is nearly uniform. Solids holdup in the riser is function of both the liquid velocity in the riser and solids circulation rate. Lan et al. (2002) had reported data for the solids holdup in upper dilute phase of the riser (ε_{sr2}) as a function of liquid velocity and solids circulation rate. Based on these data a correlation has been developed:

$$\varepsilon_{sr2} = 2.64\times10^{-14} U_{lr}^{\ -5.343} + 2.57\times10^{-5} G_s U_{lr}^{\ -1.578} \tag{6}$$

where, U_{lr} is the superficial liquid velocity in the riser in m/s and G_s is the solids circulation rate in kg/m^{2}s.

Modeling of LSCFB Ion Exchange System

Design equation for the downcomer:

$$\mathrm{U_{ld}}\frac{\partial \mathrm{C_d}}{\partial \mathrm{Z_d}} + \psi k_f a (C_d - C_s)(1-\varepsilon_d) = 0 \tag{7}$$

where, C_d is the protein concentration in the bulk liquid phase of the downcomer; U_{ld} is the superficial liquid velocity in the downcomer; Z_d is the axial distance from the bottom of the downcomer; ε_d is the voidage in the downcomer dense phase; ψ is a constant factor which includes the effects of intra-particle diffusion and liquid phase axial dispersion; a is the specific surface area of the ion-exchange resins. C_s is the equilibrium liquid phase protein concentration at liquid-solid interface predicted using the Langmuir isotherm:

$$C_s = \frac{K_d q_d}{q_m - q_d} \tag{8}$$

where, q_m is the maximum adsorption capacity of the ion-exchange particles, K_d is the dissociation constant, q_d is the solids phase protein concentration in the downcomer.

$$q_d = \frac{U_{ld}}{U_{sd}}(C_d - C_{ed}) + q_{od} \tag{9}$$

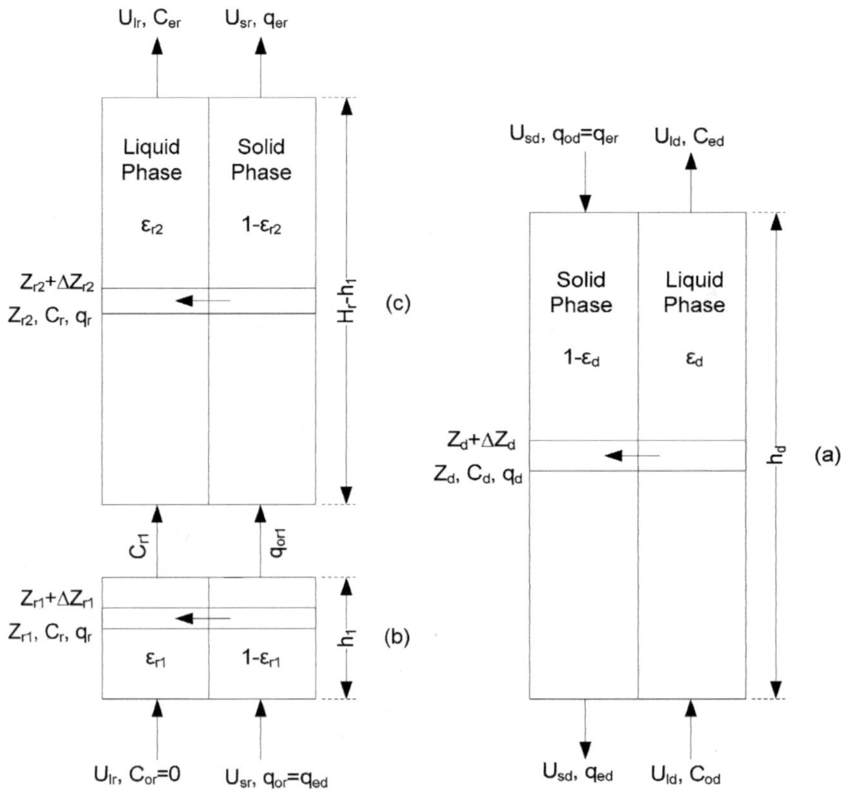

Figure 2: Flow patterns in (a) the downcomer, and (b) distributor zone and (c) upper dilute zone of the riser.

k_f is the film mass transfer co-efficient (k_f) in the downcomer dense phase calculated as a function of solids holdup (ε_{sd}) and particles Reynolds number (Re_p) using the correlation reported by Fan et al. (1960):

$$k_f = \frac{D_m}{d_p}(2+1.03(\varepsilon_{sd}\,\mathrm{Re}_p)^{1/2}(Sc)^{1/3} \quad (10)$$

where, $\mathrm{Re}_p = \frac{d_p U_{slip} \rho}{\mu}$; $Sc = \frac{\mu}{\rho D_m}$; $u_{slip} = \frac{U_{ld}}{\varepsilon_d} + \frac{U_{sd}}{1-\varepsilon_d}$ (11)

The value of D_m for BSA solution was estimated using the following correlation reported by Young et al. (1980):

$$D_m = 8.34 \times 10^{-8} (\frac{T}{\mu M^{1/3}}) \quad (12)$$

Design equation for the riser:

Distributor zone of the riser, $U_{sr} \frac{\partial q_r}{\partial Z_{r1}} + k_{r1} q_r \varepsilon_{sr1} = 0$ (13)

Boundary Conditions:

At $Z_{r1} = 0$ $\quad C_r = 0$; $q_r = q_{or}$ (14a)

At $Z_{r1} = h_1$ $\quad C_r = C_{r1}$; $q_r = q_{or1}$ (14b)

Upper dilute zone of the riser, $U_{sr} \frac{\partial q_r}{\partial Z_{r2}} + k_{r2} q_r \varepsilon_{sr2} = 0$ (15)

Boundary Conditions:

At $Z_{r2} = h_1$ $\quad C_r = C_{r1}$; $q_r = q_{or1}$ (16a)

At $Z_{r2} = 3$ $\quad C_r = C_{er}$; $q_r = q_{er} = q_{od}$ (16b)

where, k_{r1} and k_{r2} are desorption rate constants in the distributor zone and the upper dilute zone respectively.

Simulation

The design equations for the downcomer, the distributor zone and upper dilute zone of the riser were solved in succession using FORTRAN (IMSL subroutine). Non-linear algebraic equations and ordinary differential equations were solved using NEQNF and IVPAG (Gear's BDF method) subroutines of the IMSL library respectively. The computation flowchart for the simulation of the LSCFB ion-exchange system is outlined in *Figure 3*.

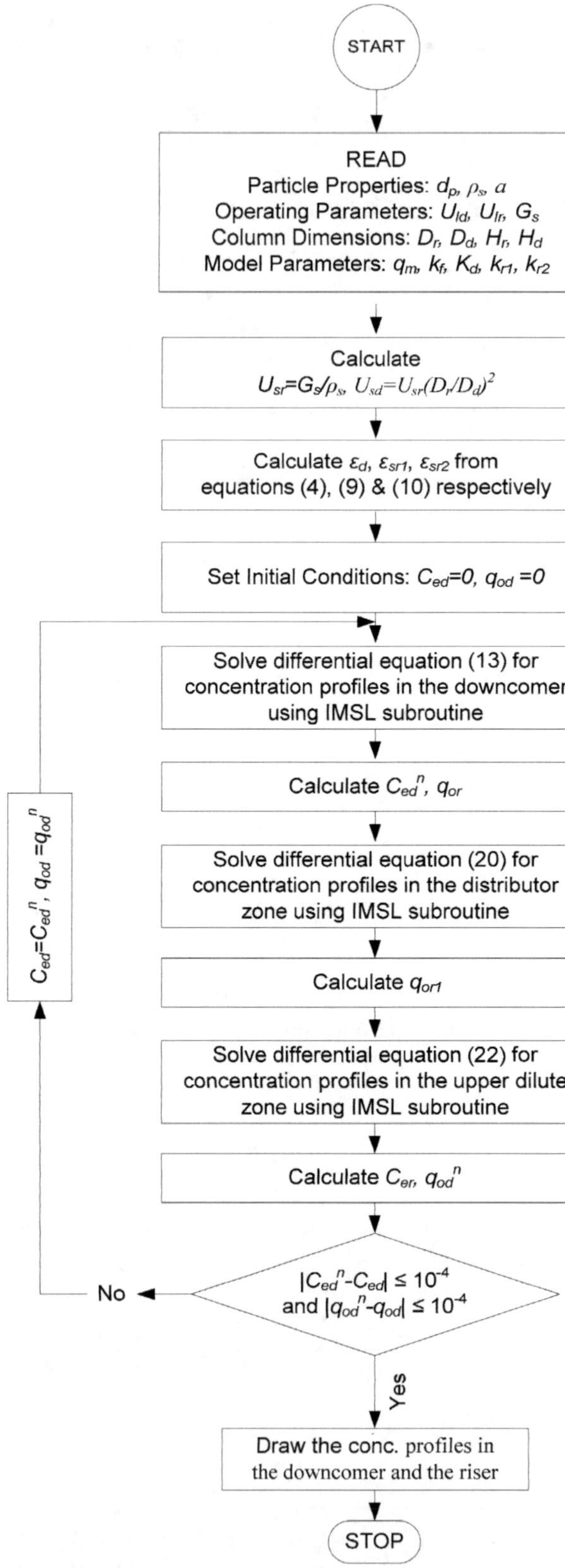

Figure 3: Algorithm for simulation of the LSCFB ion-exchange system.

Parameter Estimation and Model Validation

As the values of some of the parameters for this particular system are not available in the literature, these parameters were tuned by fitting the model predicted results with the experimental results (Lan et al., 2001). The downcomer and the riser are coupled by the circulation of the ion-exchange resins and change in any parameters affect both the adsorption and the desorption process. Hence, predicted results for both the variation of liquid phase concentration profile in the downcomer and the variation of liquid phase concentration profile in the riser with operating conditions were fitted with the experimental data simultaneously.

Predicted results for the liquid phase protein concentration profile in the downcomer under different solid circulation rate and the liquid phase protein concentration profile in the riser under different superficial liquid velocity were fitted with the experimental data simultaneously as shown in *Figure 4* and *Figure 5* respectively. Estimated values of dissociation constant (K_d), constant lumped parameter related to diffusion and dispersion (ψ), desorption rate constant at the distributor zone (k_{r1}) and desorption rate constant at the upper dilute zone of the riser, (k_{r2}), are listed in *Table I.*

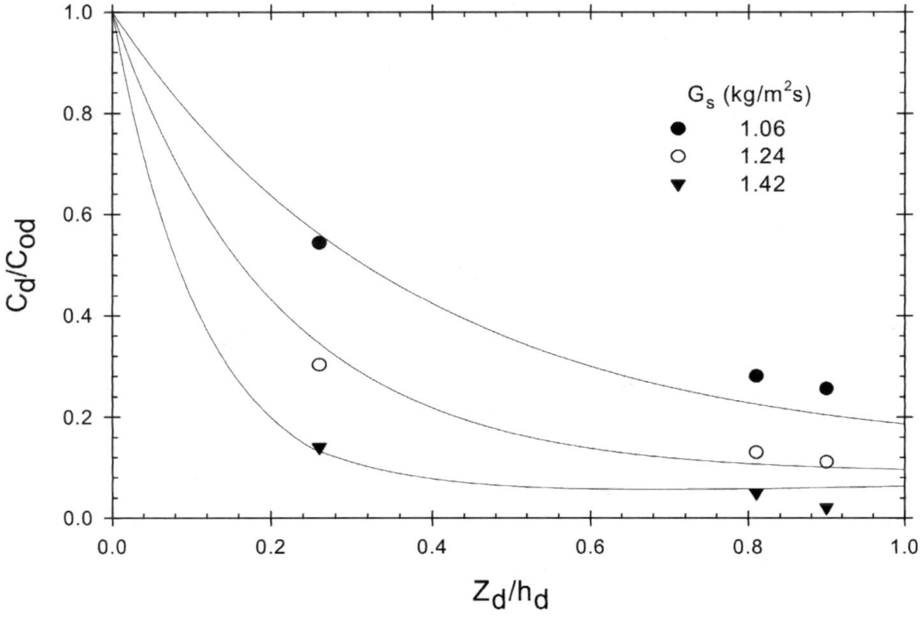

Figure 4: Liquid phase protein concentration profile in the downcomer under different solids circulation rate G_s (experimental data from Lan et al. (2000) and solid lines: model predicted results using the values of the parameters as listed in Table IV, C_{od} = 2 kg/m^3, U_{ld}= 0.6 cm/s, U_{lr}= 11.3 cm/s, 3 kg dry resins).

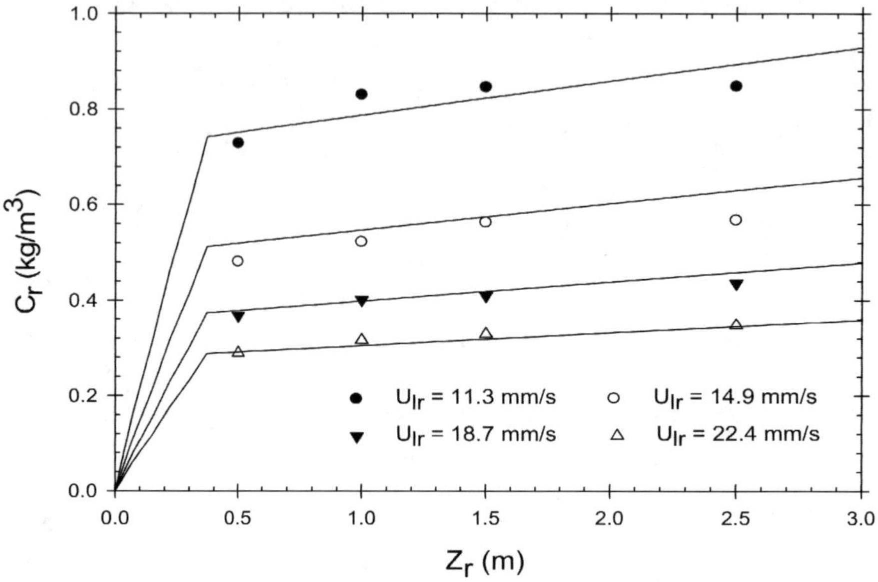

Figure 5: Liquid phase protein concentration profile in the riser under different superficial liquid velocity in the riser U_{lr} (experimental data from Lan et al. (2000) and solid lines: model predicted results using the values of the parameters as listed in Table IV, $C_{od} = 2$ kg/m^3, $G_s = 1.24$ kg/m^{2}s, $U_{ld} = 0.6$ cm/s, 3 kg dry resins).

Table I: Tuned parameters obtained based on the experimental data using NSGA II

K_d (kg/m^3)	K_{r1} (1/s)	K_{r2} (1/s)	ψ
0.25	0.005253	0.0006	0.68

To validate the model and the values of estimated model parameters, the predicted results for the liquid phase protein concentration profile under different superficial liquid velocity in the downcomer are compared with the experimental data (Lan et al. 2000; 2002) as shown in *Figure 6*. It is seen that the model resulted in quite good prediction of the variation of adsorption capacity of the downcomer with the change in superficial liquid velocity.

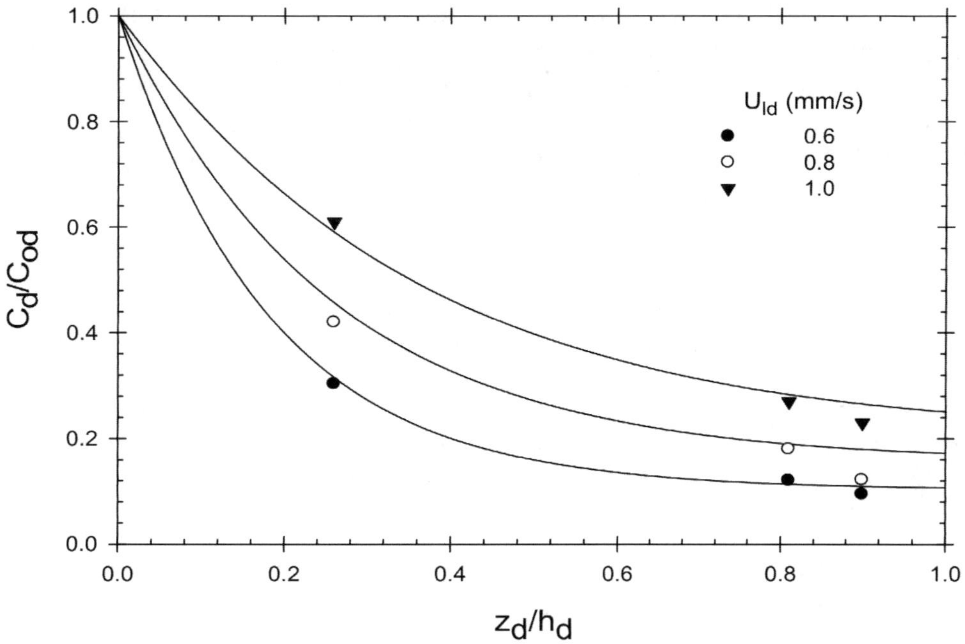

Figure 6: Liquid phase protein concentration profile in the downcomer under different superficial liquid velocity in the downcomer U_{ld} (experimental data from Lan et al. (2000; 2002) and solid lines: model predicted results using the values of the parameters as listed in Table IV, $C_{od} = 2$ kg/m^3, $G_s = 1.24$ kg/m^{2}s, $U_{lr} = 11.3$ cm/s, 3 kg dry resins).

Conclusions

The interaction between the hydrodynamics and the kinetics makes the adsorption and desorption behavior in the liquid-solid circulating fluidized bed quite complex. The change in the operating conditions in one column influences the performance of another column of the LSCFB. A experimentally verified model was developed to predict adsorption and the desorption performance of the LSCFB ion-exchange system and the model showed good agreement with the experimental data available. As LSCFB is being applied in an ever increasing number of new applications, this model can be used as a tool for predicting the performance of LSCFB for many more applications.

References

Chang, Y.K., Chase, H.A., "Development of Operating Conditions for Protein Purification Using Expanded Bed Techniques: The Effect of the Degree of Bed Expansion on Adsorption performance" *Biotechnol. Bioeng.* 49, 512–526 (1996)

Fan, L-T., Yang, Y-C., Wen, C-Y., "Mass Transfer in Semifluidized Beds for Solid-Liquid System" *AICHE J.* 6(3), 482-487 (1960)

Lan, Q.D., "Development of Continuous Liquid-Solid Circulating Fluidized Bed Ion Exchange System and its Application in Protein Purification" *Ph.D. Thesis, U of Western Ontario* (2001)

Lan, Q.D., Zhu, J.X., Bassi, A.S., Margaritis, A., Zheng, Y., Rowe, G.E., "Continuous Protein Recovery Using a Liquid-Solid Circulating Fluidized Bed Ion Exchange System: Modeling and Experimental Studies" *Can. J. Chem. Eng.* 78, 858–866 (2000)

Lan, Q.D., Zhu, J.X., Bassi, A.S., Margaritis, A., "Continuous Protein Recovery with a Liquid-Solid Circulating Fluidized Bed Ion Exchanger" *AICHE J.* 48, 252-261 (2002)

Young, M.E., Carroad, P.A., Bell, L.R., "Estimation of Diffusion Coefficients of Proteins" *Biotechnol. Bioeng.* 22, 947-955 (1980)

Zhu, J-X., Zheng, Y., Karmanev, D., Bassi A.S., "(Gas-) Liquid – Solid Circulating Fluidized Beds and their Potential Applications to Bioreactor Engineering" *Can. J. Chem. Eng.* 78, 82-94 (2000)

132e Effect of Metallic Nanoparticles Supported on Polymeric Matrix on Water Contaminants Adsorption

Vladimir A. Escobar-Barrios[1], René Rangel-Mendez[1], **Guillermo Andrade-Espinosa**[1], and Juan M. Arce-Ramos[2]. (1) Ciencias Ambientales, Instituto Potosino de Investigación Científica y Tecnológica (IPICYT), Camino a la presa San Jose # 2055, Lomas 4a. Sección, San Luis Potosi, San Luis Potosi, 78216, Mexico, (2) Chemical Engineering, Universidad Autónoma de San Luis Potosi, San Luis Potosi, 78114, Mexico

There are several places where the drinking water is naturally rich in fluoride and/or arsenic with high concentrations of such contaminants, usually the content is higher than the allowed limit by health organizations. The presence of such contaminants, at high concentrations, can cause negative effects on human health.

In order to remove fluoride or arsenic from drinking water, it is used adsorption methods. However, studies looking for new and effective adsorbents continue. Nanosilver or nano zinc oxide dispersions are used, conventionally, for medical applications and the studies regarding their use for water treatment are incipient. The key point in the use of such dispersions is their immobilization or supporting into materials with mechanical resistance to water treatment conditions and capable to interact with such dispersions without affecting the adsorption characteristics.

Polyurethane prepolymers were polymerized using metallic nanoparticles dispersions. Such dispersions are silver and oxide zinc based, and are effective to catalyst such prepolymers. Variables like initial polymerization temperature, nanoparticles dispersion quantity and nanoparticles dispersion types, were evaluated. The dispersion content affects both the polymerization rate and also the foaming of the final product. The results show that nanocomposites based on silver nanoparticles were able to adsorb water contaminants like fluoride.

As the silver content increases into polymeric matrix the adsorption capacity increases for the fluoride.

132f Metal Organic Framework Supported Molybdenum Carbide Sorbents for Sulfur Removal

Fan Shi[1], Katie Cychosz[2], Antek-Wong Foy[2], Adam J. Matzger[2], and Levi T. Thompson[1]. (1) Department of Chemical Engineering, University of Michigan, 2300 Hayward Avenue, 3230 H. H. Dow Building, Ann Arbor, MI 48109-2136, (2) Department of Chemistry and Macromolecular Science and Engineering, University of Michigan, 930 N. University, Ann Arbor, MI 48109-1055

New emissions restrictions limit the sulfur content in transportation fuels. Sulfur in these fuels not only leads to SO_xemissions, but severely deactivates most catalysts including those used for automotive emissions control and in fuel cells. Sulfur levels need to be reduced to less than 1 ppm to protect these catalysts. Compared to traditional hydrodesulfurization technologies operated at high temperature and high hydrogen partial pressure, selective adsorption of organosulfur compounds directly from the liquid fuel is very attractive strategy for desulfurization due to its low energy requirements (ambient operation conditions) and low operation costs (no hydrogen consumption).

In this paper we describe the synthesis and performance characteristics of a new type of sorbent consisting of early transition metal carbide domains dispersed onto very high surface area metal organic framework (MOF) supports. A wet impregnation method was used to produce these sorbents. The resulting materials were characterized using x-ray diffraction, sorption analysis, elemental analysis, and electron microscopy. Surface areas for the sorbents were greater than 2000 m^2/g or more than 70% of the surface area of the fresh MOF-5 support, and Mo loadings were as high as 10 wt%. The incorporation of MoC_x domains increased the adsorption capacities as shown in Table 1. The liquid feed contained organosulfur compounds including thiophene, dibenzothiophene, and/or 4, 6-dimethyl dibenzothiophene, in iso-octane or benzene solvents.

Total and breakthrough dibenzothiophene adsorption capacities for a feed containing 35 $ppmw_S$ ranged up to 2.5 and 0.8 $mgS/g_{sorbent}$, respectively. These values are comparable to or superior to those reported for other types of sorbents including those based on zeolite and transition metal oxide supports. The results indicate that functionalized MOFs are promising adsorbents for the organosulfur compounds typically found in transportation fuels.

Table 1. Adsorption results for MOF-5 and MoC_x/MOF5 sorbents at room temperature.

Sorbents	**LHSV** (hr^{-1})	**Mo loading (wt%)**	**Total Capacity** (mgS/g)	**Breakthrough Capacity**	
				(mgS/g)	(mgS/gMo)
MOF-5	120	0	1.1	0.15	-
MoC_x/MOF-5	120	1.2	1.4	0.33	28
MoC_x/MOF-5	1.6	1.2	1.1	0.8	68
3MoC_x/MOF-5	120	3.3	2.5	0.70	21

*Corresponding author: Levi T. Thompson: ltt@umich.edu

Deep desulfurization of diesel by adsorption over Ni/ZSM-5

K.K.Sarda and K.K.Pant*
Department of Chemical Engineering
Indian Institute of Technology Delhi-110016, India
*Corresponding Author: E.mail: kkpant@chemical.iitd.ac.in

Introduction

Environmental restriction regarding the quality of transportation fuels and the emissions from them are the most important issues. Because of more stringent regulations and fuel specifications regarding development of better method for deep-desulfurization of fuels are important. There has been recent interest in developing adsorbents for selective desulfurisation by adsorption. Mercaptans, thiophenes, sulfides that are present in diesel are linear compounds and hence easy to remove by adsorptive desulfurisation process but benzothiophenes, dibenzothiophene, 4-MDBT, 4,6-DMDBT are cyclic compounds, are difficult to remove due to very high sterric hinderence. [1-5]

Desulfurization by adsorption is based on the ability of a solid sorbent to selectively adsorb organosulfur compounds from refinery streams. The pi complexation bonds are stronger than those formed by Vander Waals interactions, but they are also weak enough as to be broken by traditional engineering means, such as increasing temperature and/or decreasing pressure. Therefore, this provides opportunities for tailoring and developing new adsorbents for processes where selective adsorption is needed, such as in the case of sulfur removal from fuels. [6-7]

Although research has been carried out using various sorbents, there is not a single sorbent identified for the removal of sulfur compounds. It was reported in literature that Cu, Ni, Mo and Ag based sorbents have given satisfactory results for desulfurization. Howover details of sorbent preparation and characterization are not revealed. There is also a need to understand the mechanism of adsorption of various compounds of sulfur on the surface.

Experimental

The adsorbents prepared in this work were based on Nickel over ZSM-5 support. Incorporation of Ni into the zeolite framework was done using liquid-phase ion-exchange method at room temperature. Different amount of Nickel was loaded on to the support. Ni(I)-ZSM-5 adsorbent was prepared by ion-exchanging ZSM-5[(Si/Al)=20] with $Ni(NO_3)_2$ aq. solution for 24hours.This solution is heated in oven for 18 hours at 110 °C and then calcined at 550 °C for 4 hrs.

Results & Discussion

To study the effectiveness of the adsorbent on sulfur removal experiments were carried out in batch. Desulfurization process depends on the overall activity of metal as well as support material. Effect of Si/Al ratio of ZSM -5 on sulfur removal is given in Table 1. As the silica to alumina ratio increases the acidity of ZSM-5 decreases. Also the pore size of the catalyst is different for different silica to alumina ratio. These two factors affect the desulfurization process. The maximum % sulfur removal was observed for the zeolite with silica to alumina ratio (SAR) of 20. As silica to alumina ratio decreases the acidity of ZSM increases. Therefore it can take the lone pair of sulfur compounds easily and hence more the sulfur removal. Also for the ZSM 20 the constraint index is 0.5 that is less than 2 hence it has large pore size hence it can remove variety of sulfur compounds present in diesel sample according to their size. As there were many types of sulfur compounds present in diesel hence it is not feasible to adsorb all kind of sulfur compounds. Hence oxidative desulfurization was used to further remove the sulfur compounds from the diesel. Removal by oxidation was also studied for diesel in a batch process.

Table 1: Effect of varying Silica to Alumina ratio(SAR) in ZSM on sulfur removal [T= 27 °C, P= 1 atm, Initial sulfur content = 350 ppm]

Weight of Sorbent (gm.)	% Removal of Sulfur for ZSM (SAR20)	% Removal of Sulfur ZSM SAR 34	% Removal of Sulfur ZSM SAR 45	% Removal of Sulfur ZSM SAR 200	Amount of sulfur adsorbed per gm of sorbent (mg/gm)			
					ZSM (SAR 20)	ZSM (SAR 34)	ZSM (SAR 45)	ZSM (SAR 200)
2	30	25	25	20	0.52	0.43	0.43	0.35
2.5	35	30	27.5	23.75	0.49	0.42	0.38	0.33
3	40	32.5	32.5	30	0.46	0.37	0.37	0.35
3.5	42.5	40	38.75	35	0.42	0.4	0.38	0.35
4	42.5	40	38.75	35	0.37	0.35	0.33	0.3

For the ZSM (SAR20) the constraint index is 0.5 that is less than 2 that indicates that it has large pore size hence it can remove variety of sulfur compounds present in diesel sample according to there size. ZSM (Si/Al (SAR) = 20) has shown highest sulfur removal because of its acidic in nature. The pore size of the SAR 20 catalyst is high as its constraint index is low 0.5. (<2) and its surface area is 500 m^2/g.[15] Hence ZSM (SAR=20) was used for further investigation studies in batch process.

Effect of wt % of Ni on ZSM (SAR20) and H_2O_2 dosing:

Removal of sulfur was highest (76%) at 8 wt % of Ni . Removal of sulfur first increased and then decreased because initially when the amount of Ni was increased in the sorbent the adsorption sites also increased. As shown in the Figure 1, after 8 wt. % of Ni there was a decrease because the desulfurization process was dependent on the overall activity of the catalyst as a whole. Highest sulfur removal was obtained with Ni (8 wt % of Ni).

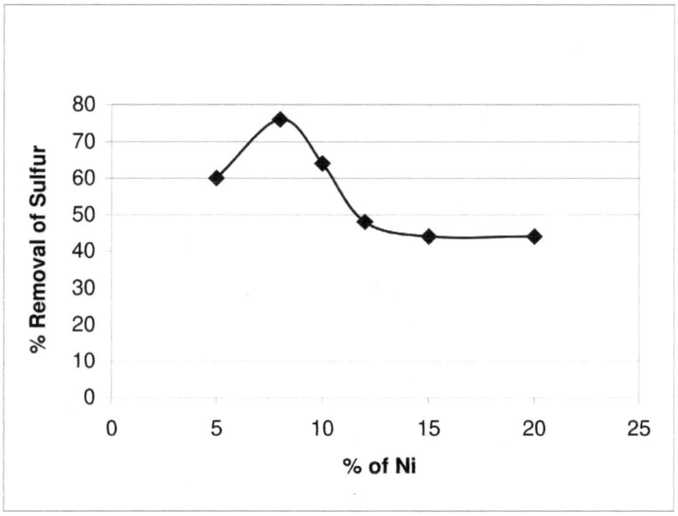

Fig. 1: % Removal of sulfur from commercial diesel v/s wt % of Ni ON ZSM 5

Effect of H_2O_2 concentration is also presented in Fig. 2. In a batch process, 1.2 gm of sorbent is added in 10 ml of model diesel sample. Hydrogen peroxide (6 % w/v solution) concentration in the sample was 1.0 ml per 10 ml of the diesel sample. Experiments were carried out at 27 °C and pressure. Initially when the H_2O_2 dose increased sulfur removal increased and than it reached to a constant value. This was because the amount of sulfur compounds present in the diesel sample that reacted with H_2O_2 was constant. When initially the amount of H_2O_2 increased then there were more chances for sulfur compounds to react with H_2O_2. But after this amount of sulfur compounds present in the diesel is the limiting factor for this reaction.

To perform these experiments model diesel was prepared using calculated amount of thiophene, dibenzothiophene, benzyl mercaptan and dibutylsulfide in isoctane. Amount of above mentioned sulfur compounds was taken in such a way that finally they accounted for 400 ppm of sulfur in the final model sample. Maximum removal of sulfur compounds with Ni-ZSM and H_2O_2 is observe because Ni removes the hindered sulfur compounds such as thiophene, benzothiophene whereas H_2O_2 is effective in removing linear compounds such as mercaptan and dibutyl sulfides. To study the effect of H_2O_2 30% w/v solution of H_2O_2 is used and the results are given in Fig. 9.

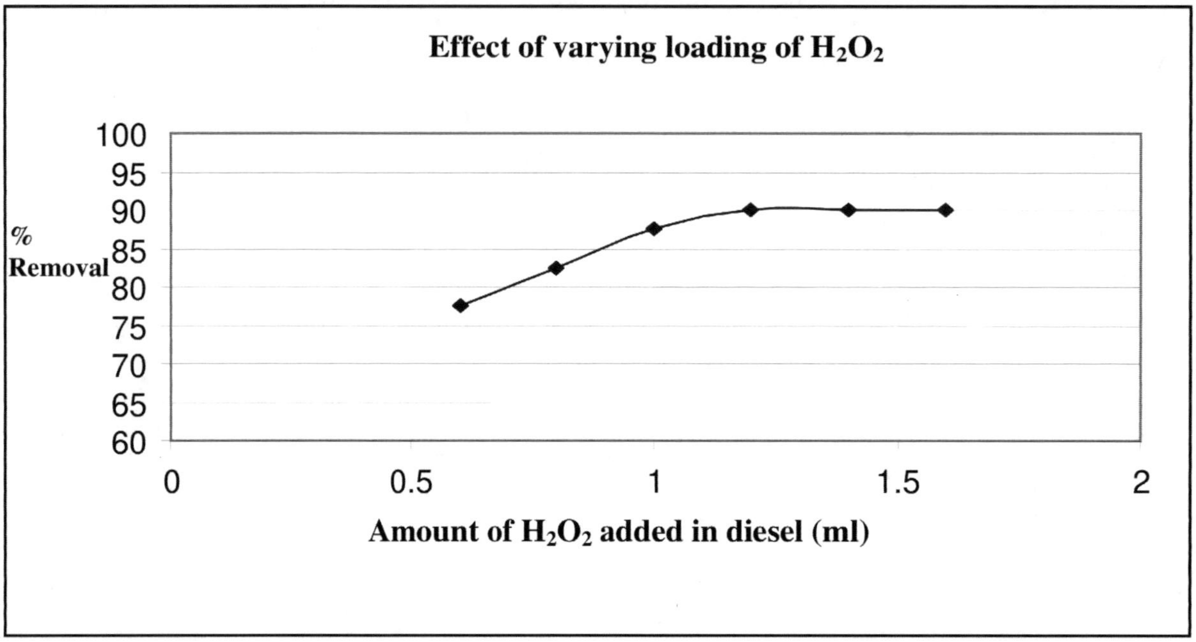

Fig. 2: Effect of H_2O_2 dose with constant loading of Ni-ZSM

The removal observed with 3.0 gm of Ni-ZSM (8%) was approximately 76%. Sulfur removal increased as the concentration of H_2O_2 up to a constant value. This is because the amount of sulfur compounds present in the diesel sample that can react with H_2O_2 is constant. The combination is effective because Ni has strong molecular orbital that can attract the hindered sulfur compounds and H_2O_2 can remove the linear sulfur compounds by converting them to sulfones.

Column Experiments

Effect of bed height on sulfur removal:

Batch process has given satisfactory results with approximately 90% sulfur removal with Ni-ZSM (8 wt % of Ni) and Hydrogen peroxide (6% w/v solution). Column experiments were carried out with varying amount of weight of sorbent in the column. The experimental conditions were as follows: sorbent: Ni-ZSM-5 (SAR20),adsorbent dose 2-8 g; T:27 °C; flow rate 0.2-2.0 ml/min.; Initial sulfur in diesel: 350 ppm.

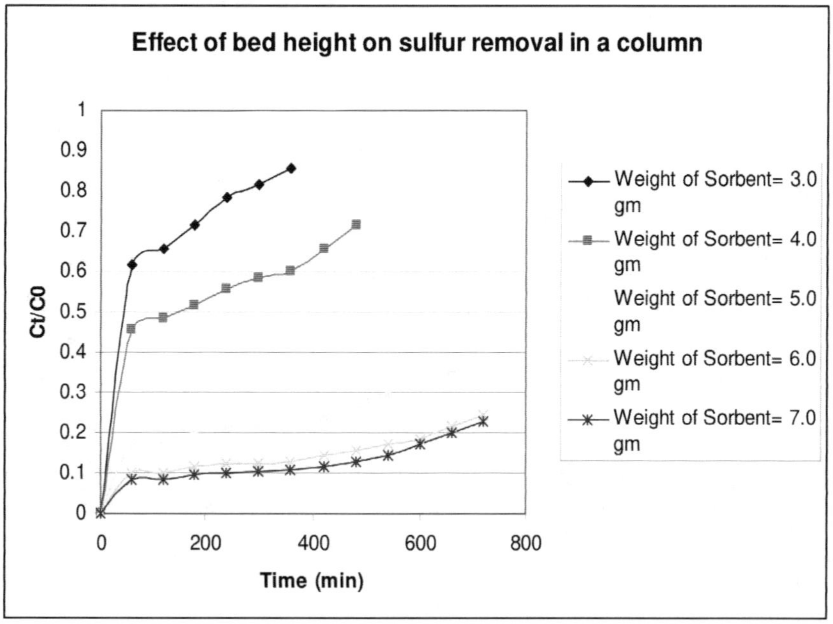

Fig 3: Effect of bed height on sulfur removal in column [H_2O_2 = 5 ml, Flow Rate = 0.2 ml/min, T= 27 °C, P= 1 atm, Initial sulfur content = 350 ppm]

It can be seen from Figure 3 that as the bed height increases the sulfur removal increase and than it reached to a maximum value after that there was no effect of further increase in the bed height. This was because as the sorbent amount increased the sorption sites increased. Hence the sulfur removal increased. But after the maximum there was no effect of increase of sorption sites as sorption capacity of the sorbent comes into picture.

Effect of flow rate on sulfur removal:

Column experiments were carried out with increasing the flow rate of diesel in the column. Ni-ZSM (8 wt % of Ni and Si/Al =20) was used as sorbent and flow rate was varied from 0.2 ml/min to 2.0 ml/min. It can be seen from the Figure 4 that as the flow rate increases the sulfur removal decrease. This was because as the flow rate increased the contact time between the diesel and sorbent decreased. Hence the sulfur was not able to come in contact with diesel and hence the adsorption decreased.

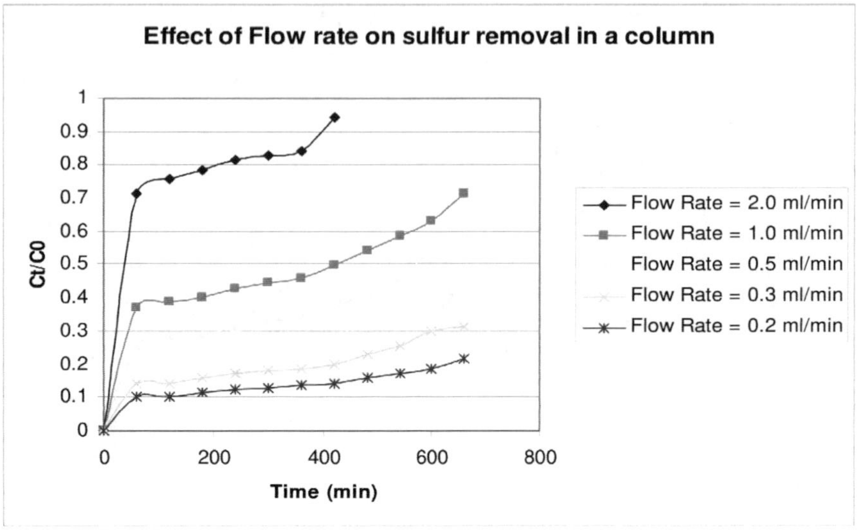

Fig. 4 Effect flow rate of diesel on sulfur removal in column [Weight of Ni-ZSM = 6 gm, H_2O_2 = 5 ml, T= 27 °C, P= 1 atm, Initial sulfur content = 350 ppm]

Experiments were also carried out with increasing the flow rate of diesel in the column. Catalyst bed was initially saturated with H_2O_2. Ni-ZSM (8 wt % of Ni and Si/Al =20) was used as sorbent and flow rate was fixed at 0.3 ml/min. On increasing the concentration of H_2O_2 the % sulfur removal increased. This was because as the concentration of H_2O_2 increased the amount of oxidant present in the sorbent increased and hence there was more oxidation of the linear sulfur compounds.

Conclusion

Among the various ZSM-5 catalysts with varying silica to alumina ratio, ZSM (Si/Al =20) was found to be most effective. A detailed investigation of Ni-ZSM [(Si/Al) =20] has confirmed high adsorption capacity (4.8mg S/g adsorbent) of sulfur. Ni-ZSM [(Si/Al) =20] was found to be very effective on the removal of sulfur compounds from commercial diesel. In the batch study Ni-ZSM and H_2O_2 has also shown approximately 90% removal of sulfur. 8%Ni-ZSM with H_2O_2 has shown more than 90% removal of sulfur from commercial diesel in the column study.

References

1. Babich V., Moulijn J. A.; '*Deep desulfurisation of oil refinery stream*' Applied Sciences, 136, (1); 2002, 262-268
2. C. Song ;'*Clean Fuels and Catalysis Program*'; Applied Catalysis; 41, (1-2); 207-238
3. Molinari D.; 'Oxidation and Functionalization: Classical and Alternative Routes and Sources'; DGMK-Conference, Milan; 2005
4. Bhandari V. M., Chang H. K., Park U. G., Han S. S., Cho H. S.and Kim J. N.; '*Desulfurisation of diesel using ion exchanged zeolite*'; Chemical Engineering Science; 61,(8); 2006; 2599-2608,
5. Richardeau D., Joly C., Canaff C., Magnoux P., Guisnet M. and M. Thomas; '*Adsorption and reaction over HFAU zeolites of thiophene in liquid hydrocarbon solutions*'; Applied Catalysis; A 263; 2004; 49–61.
6. Xue M., Chitrakar R., Sakane K., Hirotsu T., Ooi K., Yoshimura Y., Toba M. and Qi F.; ' *Preparation of cerium-loaded Y-zeolites for removal of organic sulfur compounds from hydrodesulfurizated gasoline and diesel oil* '; Journal of Colloid and Interface Science; 298 (2006); 535-542
7. Song C. , Sun L. , Xiaoliang M.; ' *Active carbon based desulfuriser*'; Catalysis Today; 77,(1); 2002; 107-116
8. Arturo J. Herna´ndez-Maldonado, Frances H. Yang,Gongshin Qi, Ralph T. Yang; 'Desulfurization of transportation fuels by p-complexation sorbents; Cu(I)-, Ni(II)-, and Zn(II)-zeolites'; Applied Catalysis; l 56 (2005); 111–126
9. A. Marafi , A. Al-Hindi, A. Stanislaus; '*Deep desulfurization of full range and low boiling diesel streams from Kuwait Lower Fars heavy crude*'; Fuel Processing Technology; 88 (2007); 905–911
10. Jinbo Gao, Shouguo Wang, Zongxuan Jiang, Hongying Lu, Yongxing Yang, Fei Jing, Can Li; '*Deep desulfurization from fuel oil via selective oxidation using an amphiphilic peroxotungsten catalyst assembled in emulsion droplets*'; Journal of Molecular Catalysis; 258 (2006); 261–266
11. Farhan Al-Shahrani , Tiancun Xiao , Simon A. Llewellyn , Sami Barri c, Zheng Jiang , Huahong Shi , Gary Martinie , Malcolm L.H. Green; '*Desulfurization of diesel via the H_2O_2 oxidation of aromatic sulfides to sulfones using a tungstate catalyst*'; Applied Catalysis; 73 (2007); 311–316
12. G. McKay, H.S. Blair and J.R. Gardner; '*Adsorption of dyes on chitin Equilibrium studies*'; J. Applied Polymer Science; 27 (1982); pp. 3043–3047

175a Synthetic Membrane Filtration: Breakthroughs and Needs

Georges Belfort, Chemical and Biological Engineering, Rensselaer Polytechnic Institute (RPI), 110 8th St., Troy, NY 12180

Bioseparations - past, present and future developments

175b Adsorption Processes - History and Recent Developments

James A. Ritter, Department of Chemical Engineering, University of South Carolina, 301 Main St., Columbia, SC 29208

Adsorption processes are as obscure as they are ubiquitous. They are rarely introduced in the undergraduate chemical engineering curriculum; and thus, they remain very ambiguous to the practicing engineer. Yet, as a phenomenon or as a unit operation, adsorption processes show up just about everywhere. They are readily found in the chemical plant drying air, cleaning water, purifying hydrogen, producing oxygen or nitrogen from air, and separating linear from branched hydrocarbons, to name just a few applications. Adsorption processes are also found in the private sector as in-home water purification, medical oxygen and odor removal systems. They are even used in automotive applications for odor removal, and aerospace applications for inerting fuel systems, producing oxygen for pilots, and removing water vapor, carbon dioxide and trace contaminates from spacecraft cabins. The list of applications for adsorption processes goes on and on, with new ones appearing continually.

All of these applications have one thing in common. They are all based on the decades, if not centuries, old notion that the surface of a solid material may have a different affinity for different substances, species, molecules or atoms when contacted with a fluid phase, thereby effecting separation of certain species from each other in the fluid phase. It is this simple affinity difference concept imparted by the solid phase that has allowed the adsorption process to evolve into one of the most important separation processes that exist today for purifying or separating components from either the gas or liquid phase.

Although adsorption processes have been around seemingly forever, it has been only recently, maybe within the past 40 or 50 years, that widespread commercial applications have flourished. What is truly remarkable about the prosperity of these commercial adsorption processes is that most, if not all of them, have relied on just a few different types of adsorbent materials. Just four classes of adsorbent materials have lead the way for huge commercial successes, including activated carbons, activated aluminas, silica gels and aluminosilicates. This presentation will provide a brief history of adsorbent and adsorption process technology and also discuss some of the recent developments in the area.

179a Membrane Extraction for Acetic Acid and Lignin Removal from Biomass Hydrolysates

David Grzenia[1], Dan Schell[2], and **Ranil Wickramasinghe**[1]. (1) Colorado State University, Department of Chemical Engineering, Fort Collins, CO 80523-1370, (2) National Renewable Energy Laboratory, National Bioenergy Center, 1617 Cole Blvd., Golden, CO 80401

A major obstacle to the large scale industrial use of biobased products and biofuels is the lack of efficient, cost-effective separation methods. Separations operations currently account for 60-80% of the processing costs of most mature chemical processes. Here we focus on the development of membrane extraction as a low cost, robust separation process in future biorefineries. As membrane extraction is non-dispersive it overcomes all of the disadvantages of conventional extraction.

Acetic acid is produced during thermochemical pretreatment of lignocellulosic biomass. It is a weak acid that is strongly inhibitory to microorganisms used for bioconversion of sugars. Removal of acetic acid could be essential for increasing ethanol yields during fermentation.

We have conducted experiments using dilute sulfuric acid pretreated corn stover. Acetic acid, in its protonated form, was extracted into an organic phase consisting of octanol and Alamine 336, a tertiary amine, containing 8-10 carbon aliphatic chains. Importantly, acetic acid removal is most efficient at pH values below 4.8, the pKa of acetic acid, thus no pH adjustment is required after pretreatment. Further as sulfuric acid is co-extracted the pH of the hydrolysate increases during extraction. Our results indicate co-extraction of furfural, hydroxymethylfurfural, acid soluble lignin and other phenolic compounds. Thus addition of membrane extraction to remove acetic acid may simplify and/or eliminate current hydrolysate detoxification technologies such as overliming. Development of a practical membrane extraction process for removal of weak acids such as acetic acid depends on carefully choosing the organic diluent and extractant (octanol and Alamine 336 used here).

179b Perfluoropolymer Membranes for Dehydration of Ethanol

Sudip Majumdar, Donald Stookey, and Stuart Nemser. Compact Membrane Systems, Inc, 335 Water Street, Wilmington, DE 19804

Bio-based ethanol is a renewable energy source. Renewable bioethanol from agricultural sources has many potential advantages including development of fuel independence and reduction in greenhouse gas generation. Ethanol as derived through fermentation from biomass contains a significant amount of water. Dehydration is a capital and energy intensive process step that is complicated by the ethanol–water azeotrope.

Compact Membrane Systems (CMS) is currently investigating highly permeable, thermally and chemically resistant perfluoropolymer membranes to selectively remove water from water–ethanol mixtures. While the focus is on ethanol, the concept is a platform technology and can be used in many other applications to make high purity chemicals. The broad platform technology is to make concentrations of organics by this process where the material that is being removed is water. Basic data in combination with preliminary economic and engineering analysis show that a process scheme that includes CMS membranes can improve the overall economics for dewatering and producing fuel-grade ethanol. This presentation will identify a number of potential bioethanol streams that can be enhanced by introduction of high water flux – high water/ethanol selectivity membranes. Presentation will focus on analyzing one of these streams.

179c VOC Capture & Recovery Based on Hollow Fiber Membranes: Evolution from Benchtop to Pilot Scale Systems

Stephen Conover, Applied Membrane Technology,Inc., 11558 Encore Circle, Minnetonka, MN 55343, Gordana Obuskovic, Otto York Department of Chemical Engineering, New Jersey Institute of Technology, 161 Warren St, Newark, NJ 07102, Kamalesh K. Sirkar, Chemical Engineering, New Jersey Institute of Technology, Membrane Seperations, 161 Warren st, Newark, NJ 07102, and Yingjie Qin, Chembrane Research & Engineering,Inc., NJIT Enterprise Development Center III, 211 Warren St., Newark, NJ 07103.

Hollow Fiber based VOC capture & recovery membrane modules offer several advantages such as compact system designs, lower energy demands, and less concern over handling concentrated explosive vapors. Professor Sirkar's work on VOC separations utilizing hollow fibers has evolved a series of operating modes which can be tailored to suit not only a variety of VOCs but also a range of VOC loadings and feed flow rates associated with different industrial gas streams.

This presentation will trace some of the journey from benchtop studies and modeling in Professor Sirkar's labs at NJIT to the pilot scale skid based systems currently undergoing field demonstration studies under NASA sponsorship. Applied Membrane Technology Inc., has worked with Dr. Sirkar since the 1990's to evolve these concepts incl

uding modules utilizing plasma polymerized composites, with and without immobilized liquids trapped in the pores, as well as, by combining such composites with liquid absorbents for enhanced extraction rates. Economic analyses and recent field study data from a skid based system employing over 600,000 hollow fibers will be presented.

179d Functionalized Membranes for Water Treatment: An Overview (Invited Keynote Speech)

D. Bhattacharyya[1], Jian Xu[2], Leonidas Bachas[3], Saurav Datta[3], A. Hollman[1], and Scott R. Lewis[2]. (1) Chemical and Materials Engineering, University of Kentucky, 177 F. Paul Anderson Tower, Lexington, KY 40506, (2) Chemical & Materials Engineering, University of Kentucky, 177 F. Paul Anderson Tower, Lexington, KY 40506, (3) University of Kentucky, 177 F. Paul Anderson Tower, Lexington, KY 40506

Functionalization of membranes with appropriate macromolecules or nanoparticles provides water treatment applications such as, tunable water flux and separations at low pressure, toxic metal capture, toxic organics destruction, etc. Traditionally, microfiltration membranes have been used for filtration of suspended solids, bacteria, viruses, etc. However, microfiltration membranes (eg, cellulosics, silica, polysulfone, polycarbonate, alumina) can be functionalized with a variety of reagents. Depending on the types of functionalized groups (such as, chain length, charge of groups, biomolecule, etc.) and number of layers, these types of membranes could be used in applications ranging from environmental applications to biocatalysis. In addition, electrostatic self assembly in pores (layer-by-layer, LBL) can also be achieved through alternate adsorption of cationic and anionic poly-aminoacids under convective flow conditions. Non-stoichiometric immobilization of charged multilayers within a confined pore geometry leads to an enhanced volume density of ionizable groups in the membrane phase. For example, the use of polypeptides with helix-coil transitions allows nano-domain interactions in membrane pores for selective environmental separations (using layer-by-layer nano-assembly in pores), and for the capture of various toxic metals. Multilayer assemblies of polyelectrolytes provide excellent platform for protein/enzyme immobilization by providing reusability and high reactivity. The presentation will include the role of nano-domain interactions for selective separations, polypeptide assembly in membrane pores for water treatment applications, in-situ synthesis methods of nano-structured bimetallic systems in membrane domain for chloro-organic detoxification from water. The use of nanostructured metals immobilized in membrane phase is expected to have significant positive impact on pollution remediation through compact and flexible dechlorination technology development with high reaction rates at room temperature, significant reduction of metals usage, and improvement in water quality. The authors would like to thank NIEHS-SBRP, US EPA, and NSF-IGERT program for funding various aspects of this work.

179e Gas Separation Membranes at Air Liquide/medal

Uttam Shanbhag[1], Edgar S. Sanders[2], S. S. Kulkarni[3], I. C. Roman[3], T. Li[3], O. M. Ekiner[3], C. Anderson[2], S. Karode[2], and C. Kim[2]. (1) AL-DRT, 200 GBC Drive, Newark, DE 19702, (2) MEDAL - Air Liquide, 305 Water St., Newport, DE 19804-2410, (3) AL-DRTC, 200 GBC Drive, Newark, DE 19702

MEDAL is the Air Liquide subsidiary responsible for manufacture of hollow fiber membranes, membrane bundles and complete system skids. MEDAL's core membrane technology is unique in the industry in that it is able to employ cheap polymers to provide structural support and a "relatively" expensive functional polymer to achieve gas separation

Since the 1980s, MEDAL focused on H2 recovery applications with its polyaramid product, a chemically robust and physically durable membrane product able to operate at high pressure. Polyimide products were developed in the 1990s to serve the N2 and inert gas generation market as well as the natural gas purification markets.

Recently, in response to low-sulfur regulations and increased need for hydroprocessing in refineries, Air Liquide has introduced a more productive H2 recovery product. Additionally, with higher energy prices, there are growing applications in landfill gas purification and under-balanced oil-field drilling.

This presentation will discuss the various topics outlined above.

195a Donnan Membrane Principle: It's Application In Developing Hybrid Nanosorbents

Luis H. Cumbal, Sudipta Sarkar, and Arup SenGupta. Lehigh University, 13 E.Packer Avenue, Bethlehem, PA 18015

Donnan membrane equilibrium arises from the inability of ions to diffuse out from one phase into the other in a heterogeneous system. This principle was judiciously exploited to formulate a new class of hybrid sorbents developed at Lehigh University. Hybrid anion exchange resins (HAIX) dispersed with nanoparticles of iron oxide, popularly known by the brand name ArsenXnp, use the Donnan membrane principle to achieve high sorption capacity for environmentally important ligands like arsenic. As a result of a counter-ion inclusion effect that may be derived from the principle, concentration of anions inside an anion exchanger far exceeds those in the bulk solution. Hydrated ferric oxide (HFO) nanoparticles have high selectivity towards dissolved arsenic species which predominantly remain as oxy-anions in natural waters. However, the nanoparticles are needed to be encapsulated in a suitable host material before their field application. The hybrid anion exchange resins not only provide support as a host material for the HFO nanoparticles dispersed inside, it also provide a synergistic effect of concentrating the target arsenic anions inside the resin phase so that the effective overall adsorption capacity of the hybrid sorbent is greatly enhanced. Moreover, the hybrid sorbents are regenerable and reusable, making the process of arsenic removal cost-effective. The arsenic selective hybrid nanosorbents have been successfully used in Indian subcontinent and USA. Field results of sustainable arsenic removal in remote villages in India shall also be presented.

195b Preparation of High-Capacity, Weak Anion-Exchange Membranes for Protein Separations Using Surface-Initiated Atom Transfer Radical Polymerization

Bharatkumar V. Bhut[1], Ranil Wickramasinghe[2], and Scott Husson[1]. (1) Chemical and Biomolecular Engineering, Clemson University, 112 Earle Hall, Dept. of Chemical and Biomolecular Engineering, Clemson University, Clemson, SC 29634-0909, (2) Colorado State University, Department of Chemical Engineering, Fort Collins, CO 80523-1370

When used as chromatography media, synthetic microporous or macroporous membranes offer advantages over resin-based media, such as low pressure drop, high production rate, and facile scale up and set up. In this presentation, we will describe how to surface modify commercially available regenerated cellulose membrane by atom transfer radical polymerization to produce high-capacity (>50 mg/mL) anion-exchange membranes for protein chromatographic separations. The monomer 2-dimethylaminoethyl methacrylate was polymerized from the pore surface of cellulose membranes to convert them into weak anion-exchange membranes. Physicochemical properties of surface-modified membranes were studied as a function of polymerization time with various analytical measurement techniques that include scanning electron microscopy, atomic force microscopy, and attenuated total reflectance FTIR. Performance properties that were measured include buffer permeability and static protein adsorption capacities. Kinetic studies of surface-initiated ATRP using model flat substrates were performed to approximate three dimensional polymer nanolayer thickness evolution during graft polymerization from membrane surfaces.

195c Pervaporation Membrane Analysis for Logistics Fuel Separation for Navy Platforms

John Heinzel[1], Ian Peek[2], Donald Hoffman[3], and John Kuseian[2]. (1) NSWCCD Code 985, Naval Sea Systems Command - Philadelphia, 5001 S. Broad St, Philadelphia, PA 19112, (2) Code 985, Naval Sea Systems Command - Philadelphia, 5001 S. Broad St, Philadelphia, PA 19112, (3) Code 331, Office of Naval Research, 875 N. Randolph St., Arlington, VA 22203

The U.S. Navy's Office of Naval Research (ONR) is sponsoring advanced research in a variety of component areas, with the goal of obtaining a three to four time improvement in overall system power density for logistics fuel-based fuel cell generators. The technology development programs ongoing as part of the ONR advanced fuel cell program are addressing numerous areas of process improvements. One of the most critical is the ability to remove sulfur from F-76, JP-5 and JP-8 Navy logistics fuel, without the use of HDS reactors and similar equipment. Such removal of refractory sulfur compounds is a challenge due to the nature of the molecules themselves, as well as their high concentrations, up to 5000 ppm of sulfur.

The use of fractionization/distillation has been well established in laboratory demonstrations, coupled to desulfurization sorbent beds. Work performed by the Navy and its partnering institutions has shown the benefits of utilizing a majority light cut fraction from a parent fuel, with respect to sorbent capacity. This paper shows the results of the fractionation of various Navy logistics fuels, containing various sulfur content, through a pervaporation-based membrane separation process. Discussion of the process, as well as the separation behavior and speciation of components of interest is provided. Additionally, an introduction to the practical sorbent performance implications is also provided, without going into substantial detail regarding adsorbent formulations and that program overall.

195d Ion Exchange Membranes from Sulfonated Polysulfone through Phase Inversion: Influence of Ion Exchange Capacity on Membrane Morphology

Cuiming Wu[1], Jiuhua Cheng[2], and Tai-Shung Chung[1]. (1) Department of Chemical and Biomolecular Engineering, National University of Singapore, 10 Kent Ridge Crescent, Singapore 119260, Singapore, Singapore, (2) Chemical & Biomolecular Engineering Department, National University of Singapore, 10 Kent Ridge Crescent,, Singapore, 119260, Singapore

Many theories and experimental data have been developed and confirmed for the relationship between membrane morphology and the difference in solubility parameter for the formation of non-charged membranes. However, this relationship between membrane morphology and solubility parameter difference is not well established and clear for the formation of charged membranes. In this paper, polysulfone and sulfonated polysulfone (SPSf) with different ion exchange capacities (IEC) were used as the starting materials for membrane preparation through a wet phase inversion process, while different non-solvents, i.e., water and isopropanol were employed as coagulants. The resultant charge membranes have diverse membrane morphology that includes finger-like, sponge-like, and dense-like structures. Efforts were given to understand the thermodynamic and dynamic driving forces for the inversion processes and to correlate with membrane morphology with them. It was found that as the IEC value of SPSf increases, the phase inversion process in isopropanol becomes slow. Such phenomenon is in contrary to the predication by the Hildebrand solubility parameter theory. In other words, as the calculated Hildebrand solubility parameter of SPSf increases with increasing IEC value, the difference in Hildebrand solubility parameters between SPSf with IPA would also increase and favor a rapid phase inversion. Calculations suggest that the Hansen solubility parameter (dd , dp , dh) theory gives a consistent relationship with experimental observation between membrane morphology and solubility parameter difference for the formation of charged membranes.

195e Separation of Lignosulfonate from Aqueous Solution Using Liquid Membrane

Kabita Chakrabarty, K. Vamsy Krishna, Prabirkumar Saha, and **Aloke Kumar Ghoshal**. Chemical Engineering, Indian Institute of Technology Guwahati, Guwahati, India

The pulp and paper industry consumes a large volume of water and subsequently discharges enormous amount of effluent containing organic and inorganic pollutants such as lignosulfonate, sucrose, polyose, cooking chemicals and silica [1]. Lignosulfonate is a polydispersed polyelectrolyte that has a spherical macromolecular structure [2]. Extensive uses of commercial grade lignosulfonate include production of vanillin, animal feed pellet binders, dispersant for dye, pesticides, carbon black etc. It has applications in water treatment and industrial cleaning operations too. Attempts to utilize the paper mill waste stream as a source of lignosulfonate requires an efficient separation technology which will ensure effective separation on one hand and clear reusable water on the other.

Howard process is a well known industrial unit for separation of lignosulphonate from aqueous solution however it uses huge quantity of lime for the separation. Several other methods exist such as ultra filtration, ion-exchange, dialysis, electro dialysis and extraction with amines. Nevertheless, they are still in their lab-scale operation stage and are yet to be economically viable for industrial applications. Membrane technology has emerged out as a practical alternative for concentration and purification of macromolecular species in aqueous solutions [1]. Although its application in paper industries is still not possible, many researchers are putting their consistent effort to develop an efficient membrane separation process [3-8].

The liquid membrane technology, on the other hand, has so far found its application mainly in the field of metal ions separation. It is a separation method in which an organic phase, immiscible with water, containing a complexing agent (carrier) selective towards the target element is interposed between two aqueous phases i.e. source and receiving phases. The complex element is formed in the source phase, diffuses across the liquid membrane (organic phase) due to concentration gradient and releases the target element in the receiving phase [9]. Thus, target element can be extracted and stripped simultaneously in a single step. Very few researchers applied this concept of liquid membrane to pulp and paper mill effluent. Kontturi et al [2] studied liquid-liquid extraction of lignosulfonate from paper mill effluent using various solvents such as alcohols, ketones and cyclohexane and found that the process was not suitable for large scale application due to difficulties in separating the components and loss of solvent. Later on they developed a supported liquid membrane comprising of Decanol-Trilaurylamine and polytetrafluoroethylene and also explained their experimental results with a simple mass transfer model [10]. They observed that the membrane was stable for about 48 hours but the flux of lignosulfonate was too low for practical applications.

In this present work an effort has been made to look for a more efficient liquid membrane for the separation and recovery of lignosulfonate from its aqueous solution. Out of various liquid membrane configurations, bulk liquid membrane is the simplest one and it has been chosen to test the feasibility of the desired application. Two phase equilibrium study has been conducted first to choose a suitable membrane and also to study the effect of various parameters such as pH, temperature etc. on the equilibrium distribution of lignosulfonate. Dichloroethane has been found to be a good organic solvent that can be used as the liquid membrane along with tri-octylamin as a carrier agent. With the membrane chosen based on equilibrium study, three phase separation process has been carried out and the effects of various parameters such as stirring speed, carrier concentration, temperature, feed concentration, etc. have been investigated.

It has been observed that stirring of feed and organic phases and an increase of temperature enhanced the transport of lignosulfonate, whereas stirring of stripping phase and initial feed concentration below 1000 mg/l have no appreciable effect on the separation of lignosulfonate. The increase of rate of lignosulfonate transport with increase in stirring speed of the feed phase implies that the extraction process is basically controlled by diffusion. Stripping process is not affected much by stirring, but it is influenced by the chemical reaction at the membrane-strip interface. The feed concentration has no impact on the transport of lignosulfonate at lower concentration, which is a merit for this separation process. The separation achieved under the experimental conditions studied is around 90-98%, but recovery is only 5-10%. Further optimization of the strip phase condition, specially the chloride concentration, would enhance the recovery. Further research is going on in this direction.

References

[1] Ganguli Liu, Yangsheng Liu, Jinren Ni, Hanchang Shi,Yi Qian. Treatability of Kraft spent liquor by micro and ultrafiltration, Desalination 160 (2004) pp.131-141.

[2] Kontturi.A.K., Kontturi.K, Niinikoski.P, Sundholm.G. Extraction of a polyelectrolyte using SLM –I, Acta chemical Scandinavica, 44 (1990) 879-882.

[3] F.G.Neytzell De Wilde, Recovery of lignosulfonate from a calcium bisulphite pulp mill effluent by ultra filtration, Desalination, 67(1987) 495-505.

[4] S.Sridhar, P.K. Bhattacharya, Limiting flux phenomena in ultra filtration of Kraft black liquor, journal of membrane sci., 57(1991) 187-206.

[5] B. Chakravorty, A.S. Srivastava, Application of membrane technology for recovery of water from pulp and paper mill effluent, Desalination 67 (1987) 363-369.

[6] I.K.Bansal, A.J.Wiley, Application of reverse osmosis in the pulp and paper industry, Technology and Engineering (1977) 459-480.

[7] B.Chakravorty, R.N Mukherjee, S. Basu, Synthesis of ion exchange membrane for electrodialytic treatment of bleaching plant effluent, Indian j. Technology 19 (1981) 102-105.

[8] B.Chakravorty, R.N Mukherjee, S. Basu, Studies on synthesis of ion exchange membrane for electrodialytic treatment of bleaching plant effluent, journal of Environmental studies, 27(1986) 173-181.

[9] M. Mulder, Basic Principles of Membrane Technology, Kluwer Academic Publishers, Dordrecht, 1991.

[10] Kontturi.A.K. Kontturi.K, Niinikoski.P, Sundholm.G, Extraction of a polyelectrolyte using SLM – II extraction and fraction of lignosulfonate, Acta chemical Scandinavica 44 (1990) 883-891.

195f Membrane Based Purification Process for Lactic Acid Production

Weihong Xing, Weixing Li, Nanping Xu, and **Xiaohua Lu**. State Key Laboratory of Materials-oriented Chemical Engineering, Nanjing University of Technology, Xinmofan Road 5#, Nanjing, China

Lactic acid has a wide range of applications in the food and pharmaceutical industries. Especially, lactic acid is being used in chemical and plastic industries nowadays due to its renewable characteristics. Fermentation by adding $CaCO_3$ is the general method to produce lactic acid with reacidification by a strong acid [1]. Therefore, some impurities including residual sugar, bacteria, protein and pigment exist in the fermentation broth, which resulted in the poor quality of the final product. Furthermore, the process was high energy and substance cost. To develop a new clean method, the membrane separation (microfiltration and nanofiltration) and electrodialysis were investigated.

In the new production of lactic acid, the acidity of the fermentation broth was neutralization by alkali (NaOH). Then, the sodium lactate fermentation was filtrated by ceramic membrane with pore size of 200 nm. The permeate from the filtration of ceramic membrane was clarified by nanofiltration to remove the residual sugar and multivalent metal ions. At last, the obtained clean sodium lactate was treated by electrodialysis with bipolar membranes to prepared the lactic acid and alkali. In this work, the membrane processes including ceramic membrane filtration, nanofiltration and electrodialysis with bipolar membranes were studied.

The ceramic membranes with four pore sizes were used to clarify the fermentation broth and the effects of velocity on the permeate flux using the four ceramic membranes were obtained. The ceramic membrane with pore size of 200nm had the better performance and the crossflow velocity had little influence on the filtration. The tubidity data of the permeates using ceramic membranes with pore sizes of 20, 50, 200 and 500nm were 0.06, 0.1, 0.1 and 0.3, respectively. Therefore, the ceramic membrane with pore size of 200 nm was more suitable to clarify the fermentation broth. The total sugar in the fermentation system decreased from 2.5% to 2.0%. And then, the permeate was fed to the nanofiltration unit. The nanofiltration membrane with the molecular weight cut-off of 200 Da was run at the pressure of 2.5MPa and the temperature of 30 ^oC. After nanofiltration, the permeate was purified sodium lactate solution with the concentration of 12%. In the permeate solution, the content of Ca^{2+}/Mg^{2+}, Fe^{2+} and Zn^{2+} were 3, 2 and 1 mg/L, respectively.

In the electrodialysis process, the three-apartment equipment was used to investigate the function of the bipolar membranes. The yield and current efficiency increased with the increasing voltage from 27 to 40 V. The conversion rates of sodium lactate by bipolar membrane electrodialysis were over 99%. Under the voltage of 40 V and the current of 8 A, the current efficiency of the bipolar membrane electrodialysis arrived at the maximum (61%).

The experimental results of the membrane based purification process for lactic acid production showed the good functions for clarification of fermentation broth, purification and conversion of soldium lactate by the integrated membrane processes including ceramic membranes, nanofiltration and bipolar membranes. The new clean process for lactic acid producing is promising in the future.

Keywords: Microfiltration; Fermentation; electrodialysis with bipolar membrane; Ceramic membrane

Acknowledge: This work is supported by Program for Changjiang Scholars and Innovative Research Team in University (PCSIRT 0732) , National Hi-tech Research Development Program of China (No. 2007AA03Z437) .

References

[1] S. Kascak, J. Kominek, M. Roehr, H.J. Rehm, G. Reed, A. Puhler, P. Stadler(Eds.), Biotechnology, Products of Primary Metabolism,Vol.6, VCH, Weinheim Germany, 1996, Chapter 8.

195g Selectivity Between Potassium, Sodium and Calcium Ions In Synthetic Media and Juice Media Using Wafer Enhanced- Electrodeionization (WE-EDI)

Thang Ho[1], Jamie A. Hestekin[2], and Anju Kurup[2]. (1) Chemical Engineering, University of Arkansas, 3202 Bell Engineering Center, Fayetteville, AR 72701, (2) Ralph E. Martin Department of Chemical Engineering, University of Arkansas, 3202 Bell Engineering Center, Fayetteville, AR 72701

Selectivity between Potassium, Sodium and Calcium Ions in Synthetic Media and Juice Media Using Wafer Enhanced- Electrodeionization (WE-EDI)

Wafer enhanced electrodeionization (WE-EDI) is a new technology that has been shown to removal ions from fermentation broths to very dilute levels (Arora et al., 2007). The novelty in the process is producing unique wafers, that transport ions, and incorporating these into an electrodialysis stack. Although the technology has been shown to be viable for dilute ion separations, areas such as selective separations, wafer enhancement, and ion exchange bead selection have not been explored. This paper is focused on the removal of sodium ion in the present of other competing ions such as potassium and calcium. The purpose is to produce low sodium juice for health purpose especially for low sodium tolerance patients. Using WE-EDI technology would allow controlling the selectivity of the ions. Moreover, WE-EDI will provide and attractive alternative to the use of bipolar membranes in electrodialysis. The WE-EDI technology has been shown to increase the performance of the membranes by increasing the transport of ions through the system. For instance, early studies of removal of sodium and potassium show up to a 40% reduction of power under certain conditions. WE-EDI also may increase the life time of the membrane especially in high complex media such as juice by allowing water dissociation on the surfaces of the resin beads instead of on the surface of the membranes, which is violent and hard on the surface layer.

This paper addresses the characteristics of wafer: porosity and capacity with different composition. We also propose a mathematical model that provides a good prediction for product quality, especially for low sodium juice production. The selectivity difference between sodium, potassium, and calcium will be evaluated in order to optimizing the process performance. The mathematical model based on the diffusion model of ions transport at the surface of the resins beads. The mathematical model will provide the fundamental understanding of ions transport in WE-EDI.

213a Towards Further Contribution to Reduction of CO2 Emissions by An Ideal Heat Integrated Distillation Column (HIDiC)

Masaru Nakaiwa, Research Institute for Innovation in Sustainable Chemistry, National Institute of Advanced Industrial Science and Technology, AIST Tsukuba Central 5, 1-1-1, Higashi, Tsukuba-shi, Ibaraki, 305-8565, Japan, Koichi Iwakabe, Research Institute for Innovation in Sustainable Chemistry, National Institute of Advanced Industrial Science and Technology (AIST), Energy-Efficient Chemical Systems Group, AIST Tsukuba Central 5, 1-1-1, Higashi, Tsukuba-shi, Ibaraki, 305-8565, Japan, and Keigo Matsuda, Department of Chemistry and Chemical Engineering, Graduate School of Science and Engineering, Yamagata University, 4-3-16, Jonan, Yonezawa, Yamagata, 992-8510, Japan.

(not provided at time of submission request)

The Effect of Outlet Weir Height on Sieve Tray Performance

A. Shariat*, T. J. Cai

Fractionation Research, Inc., Stillwater, OK 74076, USA

Presented in the Distillation Honors I of the 2008 AIChE Annual Meeting

Philadelphia, Pennsylvania

November 16-21, 2008

UNPUBLISHED

THE EFFECT OF OUTLET WEIR HEIGHT ON SIEVE TRAY PERFORMANCE

A. Shariat*, T. J. Cai, Fractionation Research, Inc., Stillwater, OK

Abstract

The effect of the outlet weir height on sieve tray performance is described in this paper. The studies were made to investigate the effect of the outlet weir height on sieve tray mass transfer efficiency and hydraulic performance. Tests were conducted in the 1.22 m diameter section of the Fractionation Research, Inc. (FRI) low-pressure column. Sieve trays with nominal hole areas of 14%, 13%, and 8% of the bubbling area and hole diameters of 12.7 mm were used. The weir heights used were 0, 25, 51, and 102 mm, with operating pressures ranging from deep vacuum with ortho/para xylene (o/p Xylene) at approximately 20 mmHg absolute to 11.38 bar with iso-butane/n-butane (iC_4/nC_4). A third system, cyclo-hexane/n-heptane (C_6/C_7) at 0.276 bar and 1.65 bar was also used. Experimental results, including tray capacity, mass transfer efficiency, pressure drop, and entrainment are presented and discussed. The results indicate that the effect of weir height very much depends on the system properties and open area.

Key Words: Distillation, sieve tray, open area, weir height, capacity, efficiency, entrainment, pressure drop

*Corresponding author. E-mail: shariat@fri.org , Phone: 1-405-707-8649,
Fax: 1-405-385-0357

Introduction

Trays with downcomers have been widely used in petrochemical industries for many years. The performance of trays depends on the uniformity of the liquid flow and the vapor liquid contact on the tray. Vapor from tray below, flows through tray perforations, forming bubbles that go through the liquid that is flowing across the tray. At tray exit a dam, called an overflow or outlet weir, maintains a pool of liquid. In general, outlet weirs are segmental and can be extension of the inner section of segmental downcomer. The outlet weir height typically is 51 mm, this is the height that has been used in obtaining most of the FRI data with the trays with downcomer. In vacuum distillation, operating in the spray regime, typical weir height is 0 to 25 mm, while absorbers and strippers may use 102 mm weir height(1). FRI has tested trays with weir heights ranging from 0 to 150 mm. For weir heights exceeding 15% of the tray spacing, one must be aware of capacity decrease due to the reduction in the effective tray spacing(2).

Care must be taken in installation of weir, an out of level weir promotes uneven liquid flow over the tray deck, which is detrimental to the tray performance. An out of level weir is a source of vapor-liquid channeling. The accepted tolerance on weir height ranges from ±1.5 to ±3 mm(2). Adjustable outlet weirs, ranging from 25 to 51 mm, used in rare occasions to provide flexibility interchanging efficiency and capacity; these are fabricated from slotted bars that can be bolted to the edge of the segmental downcomer at various heights. However, the possibility of mal-distribution due to out of levelness after adjustments far outweighs the gain, and the practice is no longer recommended[2].

The weir length is an important design criterion that determines the weir loading. The length may range from 50 to 80 percent of column diameter. For 1-pass and split flow trays, it is recommended to use a weir length of 60 to 70% of the tray diameter, while for 2-pass trays, a length of 50 to 60% of the tray diameter is more appropriate[3]. It is a normal practice to design the weir length to achieve a crest height (i.e., height of liquid over the weir) of 6 to 12 mm. A reasonable design requires a weir loading of 18 $m^3/(h.m)$[4], with a minimum weir load of 2 and maximum of 60 $m^3/(h.m)$[5]. At low liquid loads, where maintaining the liquid crest height between 6 to 12 mm may be difficult, triangular notch or picket fence weirs are used[2].

Inlet weirs are used on occasions to have a more uniform liquid flow from under the downcomer to the tray active area. Inlet weirs provide a barrier to the high inertia liquid flowing under downcomer, reducing the hydraulic jump and weeping at the inlet section of the tray. However, at high liquid rates, the inlet weir may cause weeping by projecting the high momentum liquid[1]further downstream.

This paper intends to present experimental data measured in a commercial size column using a range of systems and pressures. The purpose of this work is not to compare the experimental data to those predicted by the current correlations available in the open literature. It is hoped that having accurate experimental data from a commercial scale unit, would enable the developers to improve the existing models for the sieve tray designs. **Figure 1** is a photograph of a typical sieve tray tested by FRI.

Experimental Equipment

Plant Description

Figure 2 shows the low-pressure column and the auxiliary support system of the FRI experimental unit, as was configured for the measurement of entrainment rate in these tests. The current FRI experimental unit consists of two commercial size distillation columns and their support systems. For most operation modes only one column is used. The 1.22 m inside diameter high-pressure column is 8.4 m tall from bottom head seam to the top flange and rated for pressures up to 37 bar. The low- pressure column is rated from deep vacuum to 11.4 bar and consists of two sections. The lower section is essentially identical to the high-pressure column but is topped with a 3.66 m tall transition zone. The upper section has a 2.44 m inside diameter and is 6.7 m tall. Each column has a flanged head and clean inner wall design, which allows installation of hardware at any location in the column. Sight windows are strategically located to provide visual observation points inside the column. Couplings are available every 152 mm along the shell, which permits temperatures and pressures to be measured and samples to be withdrawn. The description of the FRI experimental unit, the procedure for obtaining and analyzing these data have been detailed previously[6][7][8]. For tests included in this paper, the entrainment collection tray was only installed for the deep vacuum operations at approximately 20 mmHg with o/p Xylene system.

Tray Design

Figure 3 is a schematic drawing of the test tray, with 14% hole area over bubbling area, as was installed in the column. **Table 1** shows the design details of this tray that had three weir heights of 0, 51, and 102 mm. **Table 2** shows the dimensions for the 8% hole area tray, and **Table 3** includes the dimensions for the tray used in the deep vacuum operation for o/p Xylene at 20 mmHg system. Trays in **Table 1** and **Table 2** were used for cyclohexane/n-heptane (C_6/C_7) system at 0.276 and 1.65 bar, and isobutane/n-butane (iC_4/nC_4) system at 11.38 bar.

For the 14% hole area, nine trays with smooth side of hole facing the vapor flow with a weir height of 51 mm, and eight trays with the sharp edge facing the vapor flow with weir heights of 0 and 102 mm were used. The trays had a bubbling area of 74% of the column area. The hole diameter was 12.7 mm on 30.2 mm equilateral triangular centers. The downcomer was a stepped type with a 102 mm recess height above the tray deck. A seal pan 102 mm deep and 184 mm wide was used. The bottom of the downcomer was 6 mm above the tray deck. A hold down bar, 25 mm wide and 6 mm thick, at the outlet weir position was considered as a negligible weir height. The detail dimensions of this set of tray are in **Table 1.**

Table 2 displays the detail dimensions of the set of trays with 8% nominal hole area as percent of the bubbling area. A hold down bar 6 mm high and 25 mm wide at outlet weir location, was considered to be the 0 mm weir height. Ten trays for the outlet weir heights of 0 and 102 mm, and nine trays for the 51 mm weir height were installed in the column. In all cases, the sharp edge of 12.7 mm hole diameter on 38.1 mm equilateral centers holes were facing the vapor. Straight downcomers with 38 mm downcomer clearance and 940 mm chordal length were used with the 51 and 102 mm weir heights. A seal pan 184 mm wide with a depth of 102 mm, under a stepped downcomer with a 102 mm recess height and 6 mm deck clearance, was used with the 0 mm weir height. Additional details are in **Table 2**.

The deep vacuum studies were conducted using seven trays, with 13% hole area over the bubbling area, at 610 mm tray spacing, with 12.7 mm hole diameter on 30.2 mm equilateral triangular spacing, with the sharp side of holes facing the vapor flow. Weir heights of 0, 25, and 102 mm with a weir length of 762 mm were used. The small downcomer occupied about 5% of the column area. The top tray was used as an entrainment collection tray, and was identical to the other trays except there was no downcomer. A 102 mm thick demister was located about 1550 mm above the entrainment tray to return any residual entrainment from the vapor stream to the entrainment tray. Gravity flow was used to meter and conduct the entrainment liquid collected in a 102 mm recessed seal pan to the reflux accumulator. To ensure that no liquid is backed up in the withdrawal line, causing the entrainment tray flooding, a liquid level guage was installed in the line at the same elevation of the entrainment collection tray and was monitored during entrainment measurement runs. Rather than using a seal pan, as in the previous cases, for 0 mm weir height a 25 mm high inlet weir was installed at the downcomer opening. The column setup is shown in **Figure 2**, and the tray dimensions are given in **Table 3**. To measure mass transfer efficiency and the operating conditions, liquid samples were withdrawn from bottom of each downcomer as the outlet sample of the tray attached to the downcomer.

Test Mixtures and Test Procedures

Systems

The results for three different test systems are discussed in this paper. The pressure range was from high vacuum with o/p Xylene system at approximately 20 mmHg absolute to high pressure with iC_4/nC_4 at 11.38 bar pressure. The third system used in these tests is cyclohexane/n-heptane (C_6/C_7) at 0.276 and 1.65 bar. The average approximate physical properties for the systems tested at the indicated pressures are given in **Table 4**.

Capacity

Except for the high vacuum runs, the capacity runs were made at total reflux and several constant liquid rate loads. The maximum attainable rates under stable conditions, defined as incipient flood by FRI, were determined for sieve trays by procedure previously described[6].

Efficiency

All efficiency studies were conducted using the total reflux operation. To calculate the efficiency, liquid samples were withdrawn from the bottom of each downcomer. The analyzed liquid samples were the liquid composition at the outlet of the tray connected to the downcomer. The compositions were obtained using vapor phase chromatography for the o/p Xylene and the iC_4/nC_4 mixtures, and by refractive index for the C_6/C_7 mixture. The analysis were within 1% band width accuracy. The composition profiles for the trays were plotted in terms of log[x/(1-x)] versus the tray location. The overall tray efficiencies were calculated, after dropping the outlier points and smoothing the profiles. The Fenske-Underwood[9]equation was used to obtain the total reflux efficiency from

$$N_t = \frac{\log\left[\frac{x_T(1-x_B)}{x_B(1-x_T)}\right]}{\log\alpha}$$

And

$$E_o = N_t / N_a$$

Pressure Drop

Total reflux pressure drop were measured for single trays, as well as multiple trays in sections of the column. The sections included all the trays, top half, and the bottom half. The reported pressure drops, are measurements averaged per tray basis.

Entrainment

The entrainment rates were measured at total reflux and several constant liquid rates. The following standard procedure was followed to obtain the constant liquid load entrainment measurement:

1. For the specified liquid load, the vapor rate was increased incrementally till a high entrainment rate was observed.

2. The conditions in the column were held steady until equilibrium was reached, after which all the measurements and flow rates were recorded.
3. The procedure was repeated by decreasing the vapor rate in stepwise manner, obtaining the entrainment for each vapor rate with a fixed liquid load.

The intervals were selected in such a way to obtain three to five entrainment rates for a specified liquid load.

Results for 14% & 13% Hole Areas

Capacity

The effect of weir height on capacity of 14% hole area sieve trays with 0, 51, and 102 mm outlet weirs for C_6/C_7 at 0.276 and 1.65 bar and that for iC_4/nC_4 at 11.38 bar displayed in **Figures 4 – 6** respectively. From these results, no significant effect on the capacity of sieve tray with 14% hole area and 610 mm tray spacing is observed. One expects to have the 51 mm weir height capacity in between the 0 and 102 mm weir height capacity, but this was not the case specifically for C_6/C_7 at 0.276 bar. This difference may be attributed to the actual difference in the trays tested.

Efficiency

Figures 7 – 9 show the effect of outlet weir heights of 0, 51, and 102 mm on the efficiency of 14% sieve trays for C_6/C_7 at 0.276 and 1.65 bar and that for iC_4/nC_4 at 11.38 bar. For C_6/C_7 at both pressures the efficiency increases with the increase in the weir height. In general, for butane system, except at a capacity factor of 0.077 m/s and 51 mm weir height with the highest capacity, the weir height from 0 to 102 mm does not affect the efficiency of iC_4/nC_4 at 11.38 bar.

Figure 10 shows the effect of outlet weir heights of 0, 25, and 102 mm on overall tray efficiency of the 13% sieve tray with 25 mm holes with o/p Xylene at approximate 20 mmHg absolute pressure. The increase in outlet weir height to 102 mm resulted in increased tray efficiency, removal of the outlet weir resulted in decreased efficiency.

Pressure Drop

Figures 11 – 13 compare the tray pressure drop, at total reflux condition, of the 14% sieve trays for outlet weir heights of 0, 51, and 102 mm with C_6/C_7 at 0.276 and 1.65 bar and that with iC_4/nC_4 at 11.38 bar. As expected, the pressure drop per tray in all cases increases with an increase in the weir height

The effect of weir height on the total reflux tray pressure drop of 13% sieve tray at total reflux on o/p Xylene at 20 mmHg absolute pressure, with 0, 25, and 102 mm weir height, is displayed in **Figure 14**. The tray with the 0 mm weir exhibits the lowest pressure drop, while the tray with 102 mm weir had the heighest pressure drop.

The pressure drop per theoretical stage is shown in **Figure 15.** There is no essential difference among the three weir heights.

Entrainment

Figure 16 shows the entrainment characteristics of the 13% hole area sieve tray having 0, 25, and 102 mm high outlet weirs with the o/p Xylene system at approximately 20 mmHg absolute. It can be observed that the trays with 0 and 102 mm outlet weirs had about the same entrainment characteristics, while the tray with 25 mm outlet weir had an abnormally high entrainment rate. This may be attributed to the operating pressure difference between the two sets of trays, with the test of the tray with 25 mm weir being conducted at 16 mmHg, and for 0 and 102 mm weir height at 20 mmHg.

There are conflicting views on the effect of liquid load on entrainment in the literature. Various researchers have found that increasing the liquid load either increases entrainment(10, 11), decreases entrainment(12), or both(13, 14). The effect of weir height with the constant liquid load of 18 and 25 $m^3/(h.m)$ on entrainment are shown in **Figures 17** and **18**.

Results for 8% Hole Area

Capacity

Figures 19 – 21 show the effect of weir height on capacity of 8% hole area sieve trays with 0, 51, and 102 mm high outlet weirs for C_6/C_7 at 0.276 and 1.65 bar and that for iC_4/nC_4 at 11.38 bar. For C_6/C_7 at 0.276 and 1.65 bar, **Figures 19** and **20**, no significant difference in capacity was observed between the 0, 51, and 102 mm weirs at any liquid rate. For butane system at 11.38 bar pressure, **Figure 21**, there was little difference in capacity for the 0 and 51 mm weirs. The 102 mm outlet weirs reduced the capacity for the system at very high liquid rate.

Efficiency

In **Figures 22 - 24**, the overall tray efficiencies for three weir heights are compared. For C_6/C_7 system at 0.276 bar pressure, **Figure 22**, the 0 mm outlet weir was more efficient at low vapor rates than were the higher weirs. For 1.65 bar pressure C_6/C_7 system, **Figure 23**, the weir height had only small effect on the efficiency.

For the butane system, **Figure 24**, the highest efficiency was obtained with the 51 mm weir, while the lowest efficiency was obtained with 0 mm weir. However, there are indication from additional tests of this tray that the seal pan were not properly installed and excessive leakage was responsible for the low efficiency of the butane system.

Pressure Drop

The effect of weir height on total reflux pressure drop measurements are shown in **Figure 25** for C_6/C_7 system at 0.276 bar and in **Figure 26** and for 1.65 bar, and in **Figure 27** for butane system at 11.38 bar. All these figures indicate, the contribution of weir height to tray pressure drop is a function of vapor rate. The contribution is the highest at low vapor rate, becoming negligible at higher vapor rates.

Conclusions

Performance characteristics of a sieve tray with 12.7 mm holes, weir heights ranging from 0 to 102 mm, with three nominal hole areas of 14%, 13%, and 8% of bubbling area, with systems that cover a wide range of properties have been obtained in a large scale commercial installation. Although the results from these studies are not typical of all sieve tray designs, certain findings such as the effect of system properties, liquid loadings, and weir height on capacity, efficiency, pressure drop, and entrainment can be important to hardware design and optimum operating conditions.

The results are summarized below:

1. Weir heights of 0 to 102 mm have no significant effect on the capacity of the sieve trays with hole area of 8% or greater at tray spacing of 610 mm.
2. The effect of weir height on the efficiency of sieve tray is not conclusive. The following results were obtained from these studies with 14% hole area tray:

System	Comparative Performance of 0, 51, and 102 mm Weirs
C_6/C_7, 0.276 bar	Efficiency increases with weir height

System	Comparative Performance of 0, 51, and 102 mm Weirs
C_6/C_7, 1.65 bar	Efficiency increases with weir height
iC_4/nC_4, 11.38 bar	Highest efficiency with 51 mm weirs, 0 and 102 mm same

In contrast, the results on an 8% sieve tray were as follows:

System	Comparative Performance of 0, 51, and 102 mm Weirs
C_6/C_7, 0.276 bar	Efficiency about the same with all three
C_6/C_7, 1.65 bar	Highest efficiency with 0 mm weirs, 51 and 102 mm same
iC_4/nC_4, 11.38 bar	Highest efficiency with 51 mm weirs, 0 and 102 mm same

These results suggest that the effects of weir height are functions of system properties and hole area.

Nomenclature

C_b = superficial capacity factor based on tray bubbling area, (m/s)
C_6 = ciclo-hexane
C_7 = normal heptanes
iC_4 = iso-butane
nC_4 = normal butane
E_o = overall tray efficiency
N_a = actual number of trays
N_t = theoretical number of trays
o/p = ortho/para
x = mole fraction of more volatile component in liquid phase

x_B = mole fraction of more volatile component in liquid phase at bottom of section of trays
x_T = mole fraction of more volatile component in liquid phase at top of section of trays

Greek Letters
α = relative volatility

Bibliography

1. Lockett, M.J., "Distillation Tray Fundamentals", Cambridge University Press, 1986.
2. Kister, H.Z., "Distillation Operation", McGraw-Hill, Inc., 1990.
3. Van Winkle, M., "Distillation", McGraw-Hill, Inc. 1967.
4. Lieberman, N. P., and Lieberman,E. T., "Working Guide to Process Equipment", McGraw-Hill Professional, 2003
5. Stichlmair, J.G., and Fair, J.R., "Distillation Principles and Practice", Wiley-VCH, 1998.
6. Silvey, F. C. and Keller, G. J. Chem. Eng. Prog., 1966, 62, No. 1, p.68.
7. Silvey, F. C. and Keller, G. J. in "Distillation – 1969"; Institute of Chemical Engineering: London, 1969; p4:18.
8. Yanagi, T., Paper presented in the AIChE Annual Meeting, Detroit, Michigan, December 4-8, 1966.
9. Underwood, A. J. V. Trans. Inst. Chem. Eng. 1932, 10, 112.
10. Holbrook, G. E., and Baker, E.M., Trans AIChE, 1932, 30, 520.
11. Pyott, W.T., Jackson, C.A., and Huntington, R. L., Ind. Eng. Chem. 1935, 27, 821.
12. Bain, J. L. and Van Winkle, M., AIChE Journal, 1961, 7, 363.
13. Atteridg, P.T., Lemieux, E. J., Schreiner, W. C., and Sundback, R. A., AIChE Journal 1956, 2, 3.
14. Friend, L., Lemieux, E. J., and Schreiner, W. C., Chem. Eng., 1960, 67, No. 22, 101.

Table 1. Design Detail of Trays with 14% Hole Area

Column Diameter, mm	1213
Tray Spacing, mm	610
Perforated Sheet, Material	316 SS
Perforated Sheet, Thickness, mm	1.59
Outlet Weir Length, mm	940
Outlet Weir Height, mm	0[*], 51, 102
Clearance under Downcomer, mm	6[!] ,19.1
Effective Bubbling Area, m^2	0.860
Nominal Downcomer Area at Top, m^2	0.107
Nominal Downcomer Area at Bottom, m^2	0.049[!!] , 0.107
Hole Diameter and Spacing, mm x mm	12.7 x 30.2
Number of Holes per Tray	930
Nominal Hole Area, % Bubbling Area	14
Hole Area, m^2	0.118
Edge of Hole Facing Vapor Flow	Sharp**

*6 mm Hold Down Bar
! Distance above a 184 mm x 101 mm Seal Pan
!! Stepped Downcomer with 101 mm Recess
**For 51 mm Weir Height Smooth edge was used

Table 2. Design Detail of Trays with 8% Hole Area

Column Diameter, mm	1213
Tray Spacing, mm	610
Perforated Sheet, Material	316 SS
Perforated Sheet, Thickness, mm	1.59
Outlet Weir Length, mm	940
Outlet Weir Height, mm	0[*], 51, 102
Clearance under Downcomer, mm	6[!] ,19.1
Effective Bubbling Area, m^2	0.860
Nominal Downcomer Area at Top, m^2	0.107
Nominal Downcomer Area at Bottom, m^2	0.049[!!] , 0.107
Hole Diameter and Spacing, mm x mm	12.7 x 38.1
Number of Holes per Tray	565
Nominal Hole Area, % Bubbling Area	8
Hole Area, m^2	0.072
Edge of Hole Facing Vapor Flow	Sharp

*6 mm Hold Down Bar
! Distance above a 184 mm x 101 mm Seal Pan
!! Stepped Downcomer with 101 mm Recess Height

Table 3. Design Detail of Trays for Deep Vacuum Operation

Column Diameter, mm	1213
Tray Spacing, mm	610
Perforated Sheet, Material	316 SS
Perforated Sheet, Thickness, mm	1.59
Outlet Weir Length, mm	762
Outlet Weir Height, mm	0, 25, 102
Inlet Weir Length, mm	762
Inlet Weir Height, mm	25*
Edge of Hole Facing Vapor Flow	Sharp
Downcomer Clearance, mm	19.1
Effective Bubbling Area, m^2	1.041
Nominal Downcomer Area at Top, m^2	0.049
Nominal Downcomer Area at Bottom, m^2	0.049
Hole Diameter and Spacing, mm x mm	12.7 x 30.2
Number of Holes per Tray	1066
Nominal Hole Area, % Bubbling Area	13
Hole Area, m^2	0.135

*For 0 mm Outlet Weir studies

Table 4. Average Physical Properties of the Test Systems under Operating Conditions

System						
	Unit	Cyclo-hexane/n-Heptane			o/p Xylene	Iso-butane/n-Butane
Pressure	bar	0.276	0.35	1.65		11.38
	mm Hg				16 - 20	
Vapor Density	kg/m^3	1.210	1.128	5.032	0.124	28.27
Liquid Density	kg/m^3	707.9	705.6	659.1	845.6	493.4
Liquid Viscosity	mPa/s	0.43	0.41	.25	.52	0.091
Surface Tension	mN/m	19.4	19.1	14.2	26.4	5.27
Relative Volatility		1.82	1.81	1.57	1.29	1.24

Figure 1. 14% Hole Area Commercial Sieve Tray

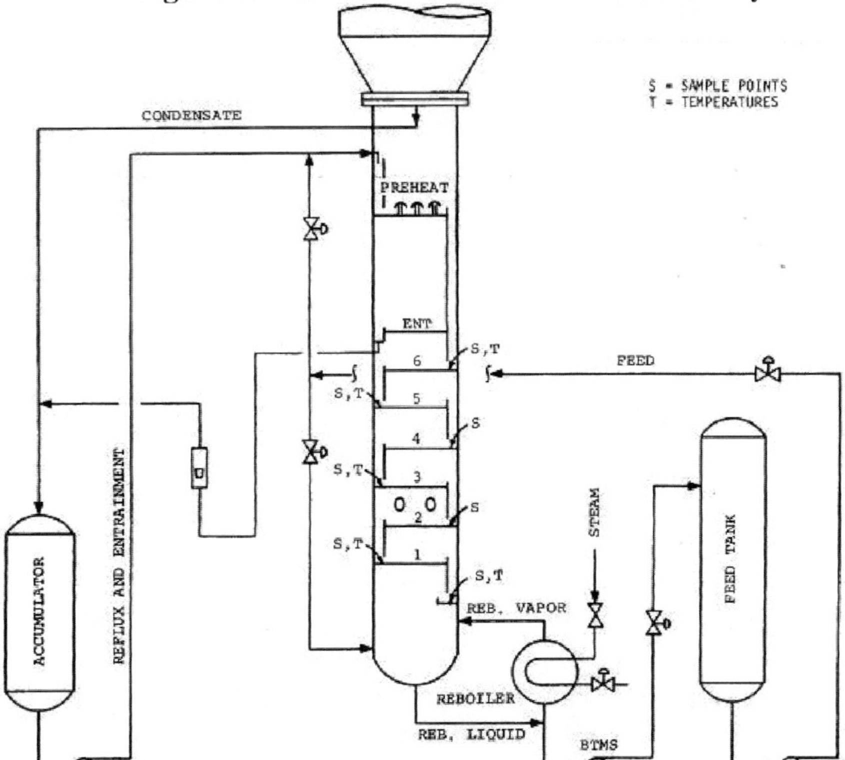

Figure 2. FRI Distillation Unit Setup with Entrainment Capture Tray

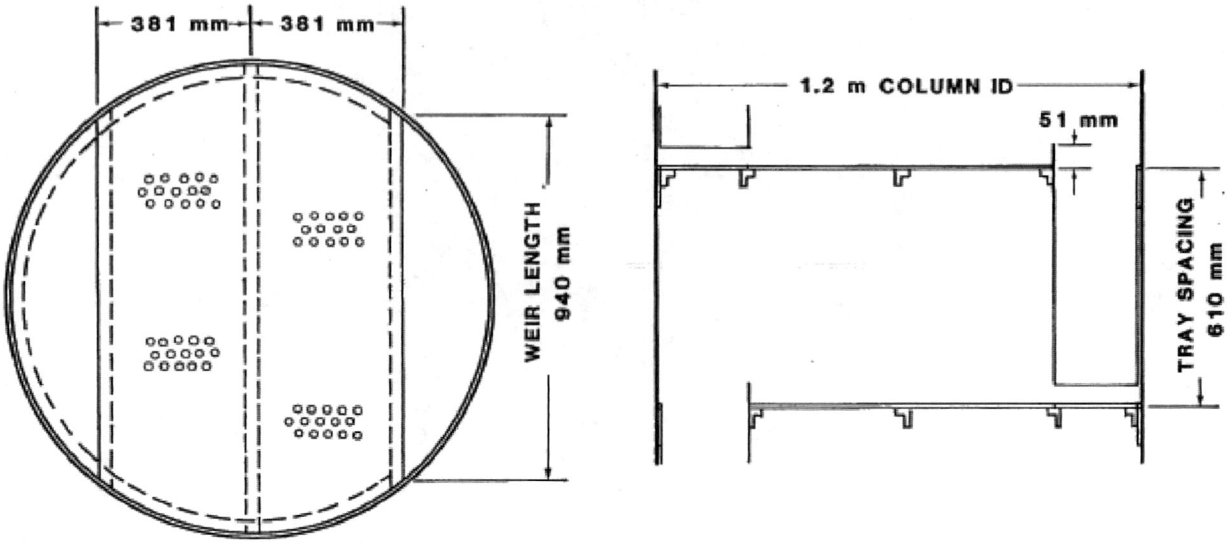

Figure 3. Diagram of Test tray with 14% Hole Area

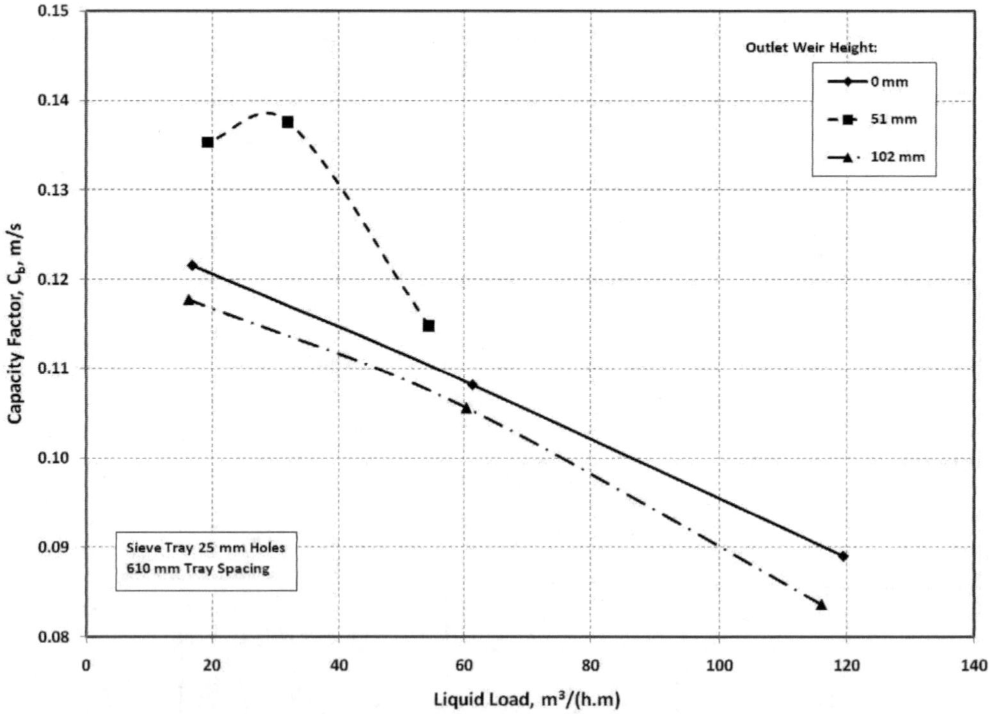

Figure 4. Effect of Weir Height on Capacity for 14% Hole Area Tray (C_6/C_7 at 0.276 bar Pressure)

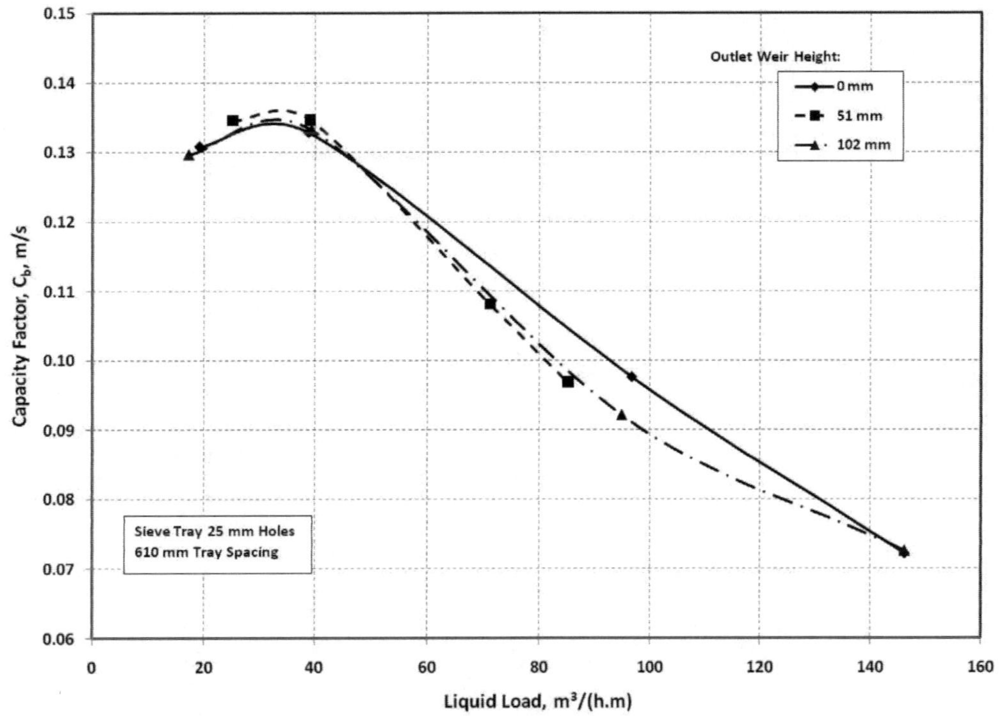

Figure 5. Effect of Weir Height on Capacity for 14% Hole Area Tray (C_6/C_7 at 1.65 bar Pressure)

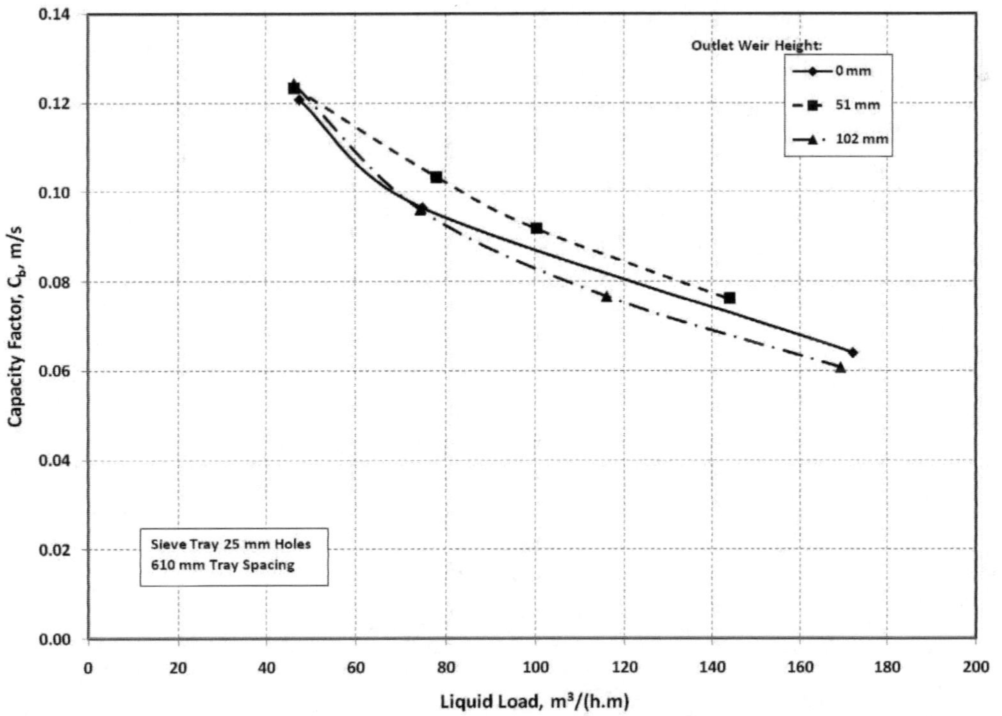

Figure 6. Effect of Weir Height on Capacity for 14% Hole Area Tray (iC_4/nC_4 at 11.38 bar Pressure)

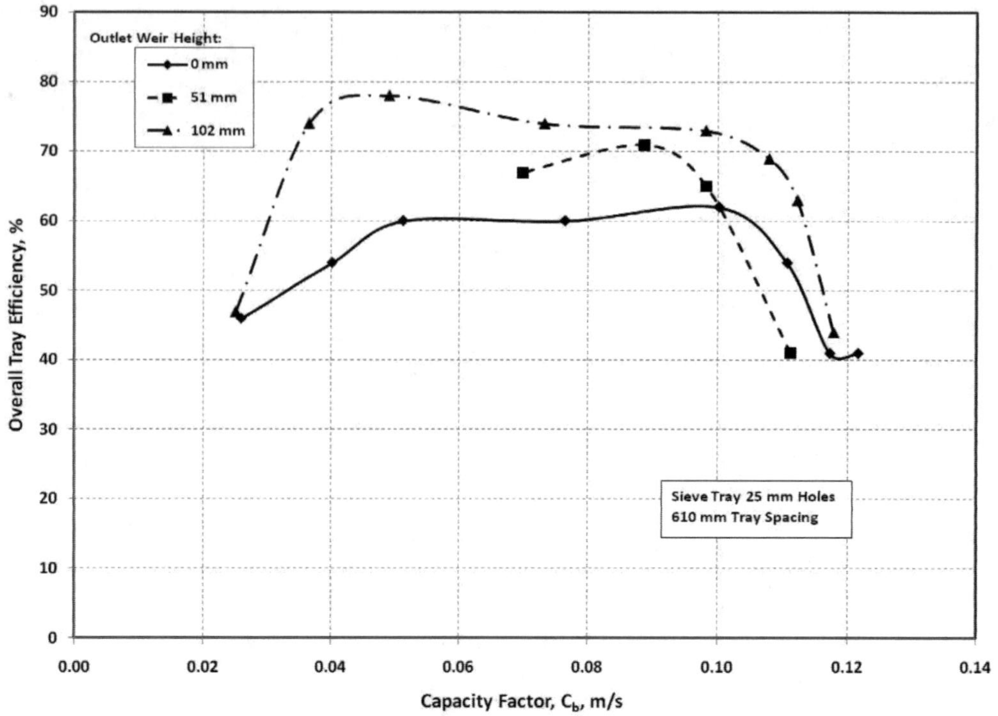

Figure 7. Effect of Weir Height on Efficiency for 14% Hole Area Tray (C_6/C_7 at 0.276 bar Pressure)

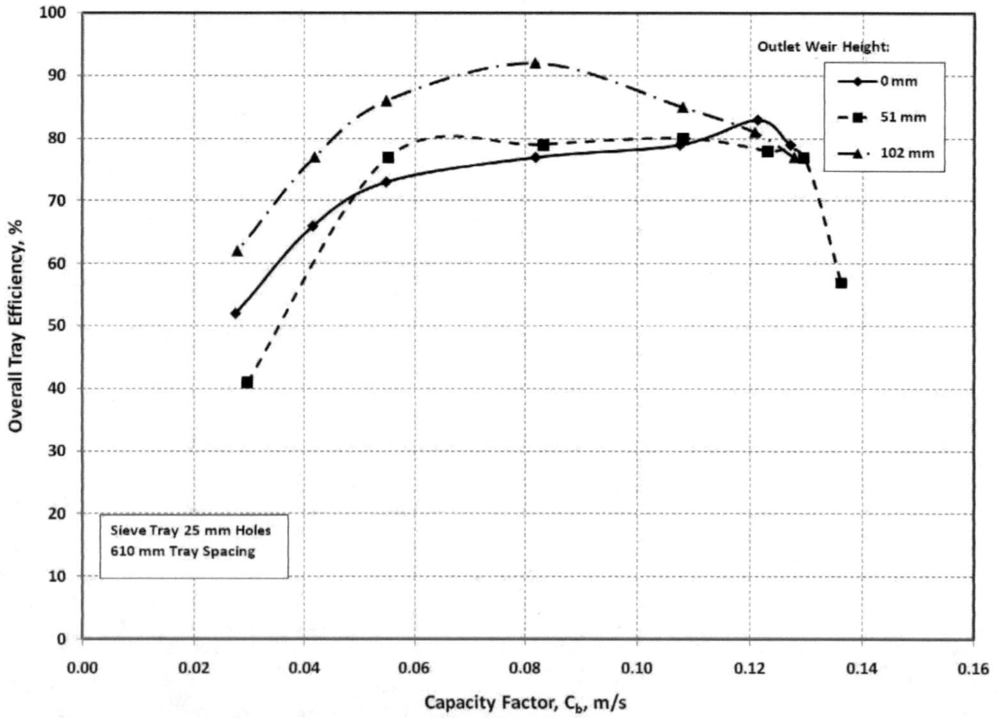

Figure 8. Effect of Weir Height on Efficiency for 14% Hole Area Tray (C_6/C_7 at 1.65 bar Pressure)

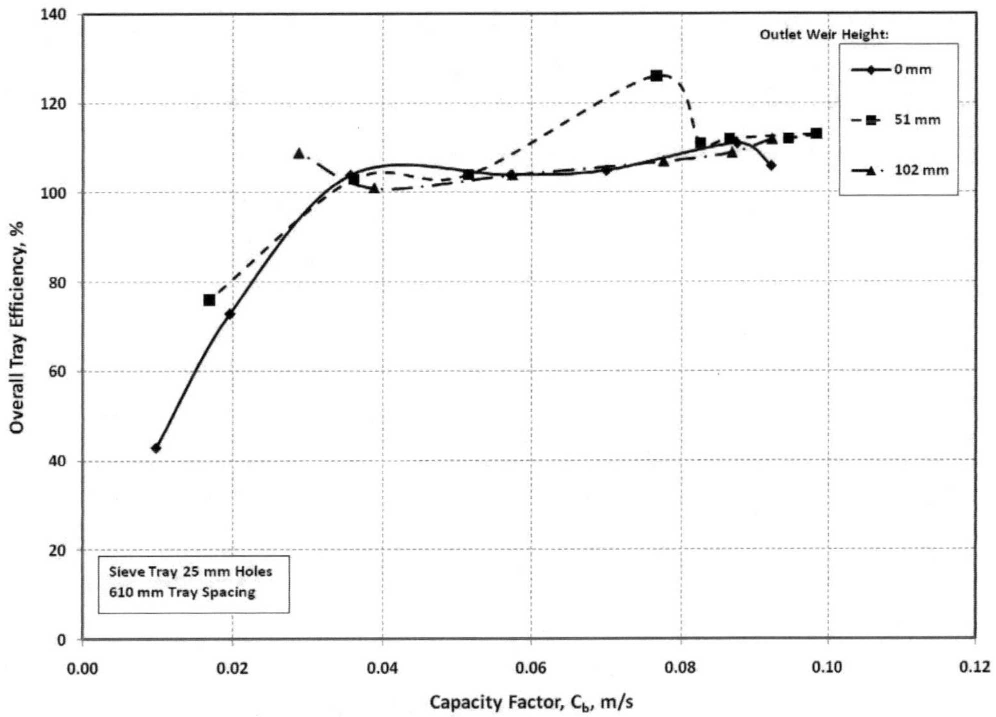

Figure 9. Effect of Weir Height on Efficiency for 14% Hole Area Tray (iC_4/nC_4 at 11.38 bar Pressure)

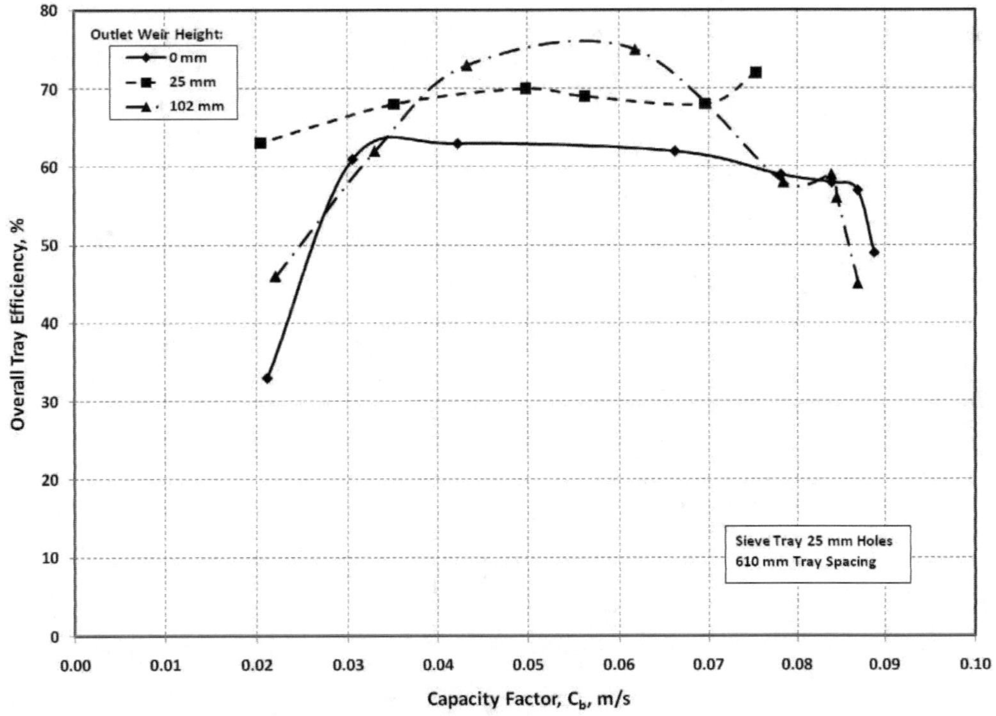

Figure 10. Weir Height Effect on Efficiency with 13% Hole Area Tray (o/p Xylene at 20 mmHg Absolute Pressure)

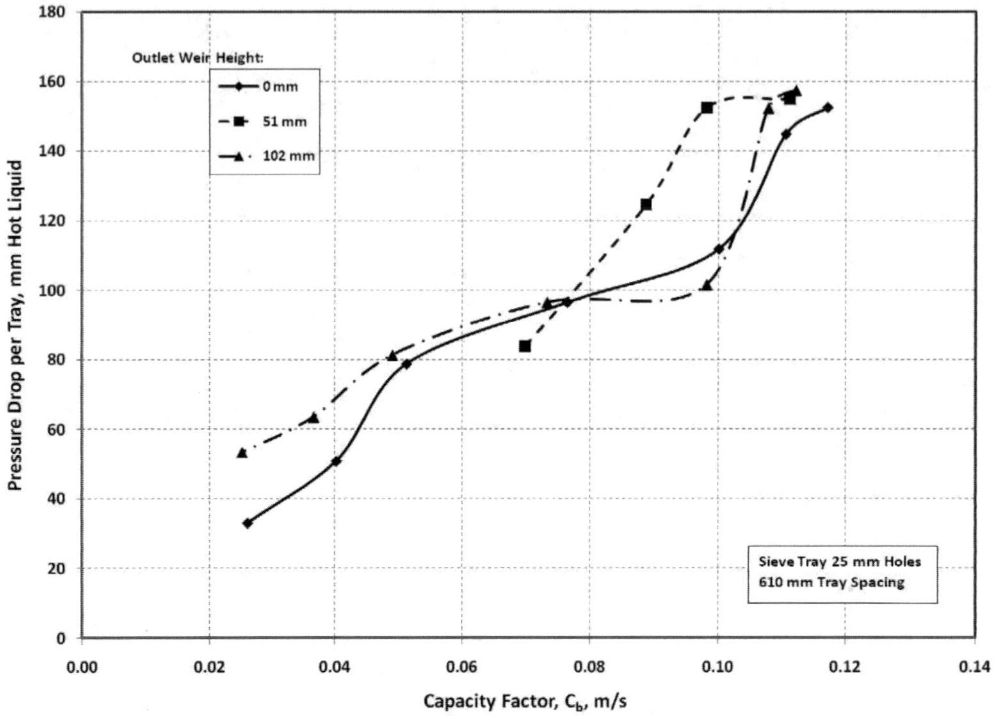

Figure 11. Effect of Weir Height on Pressure Drop for 14% Hole Area Tray (C_6/C_7 at 0.276 bar Pressure)

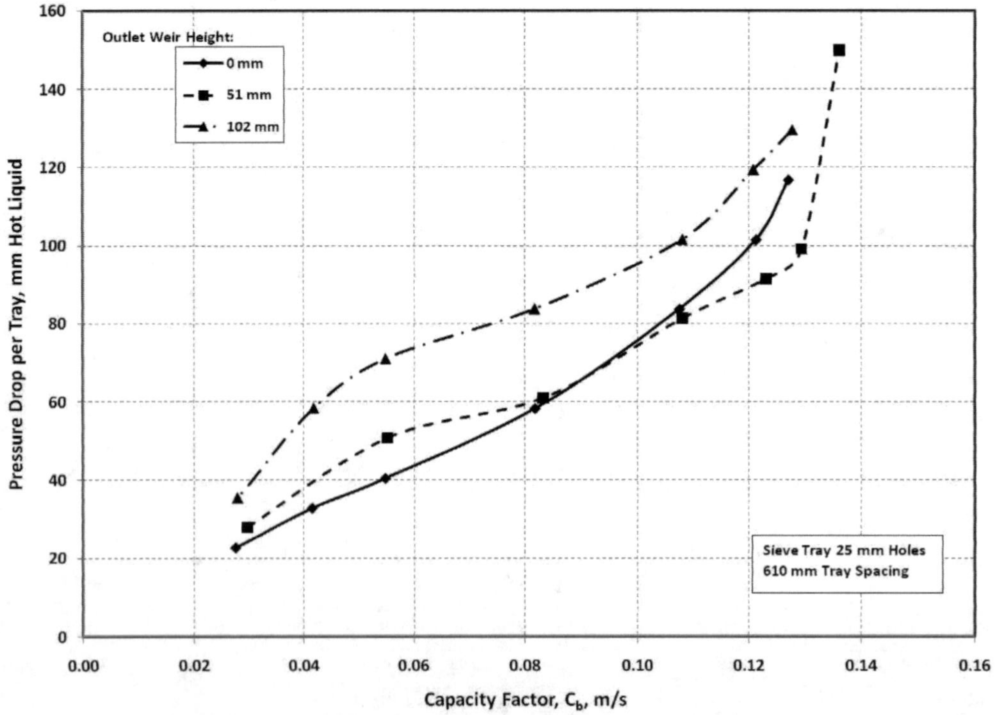

Figure 12. Effect of Weir Height on Pressure Drop for 14% Hole Area Tray (C_6/C_7 at 1.65 bar Pressure)

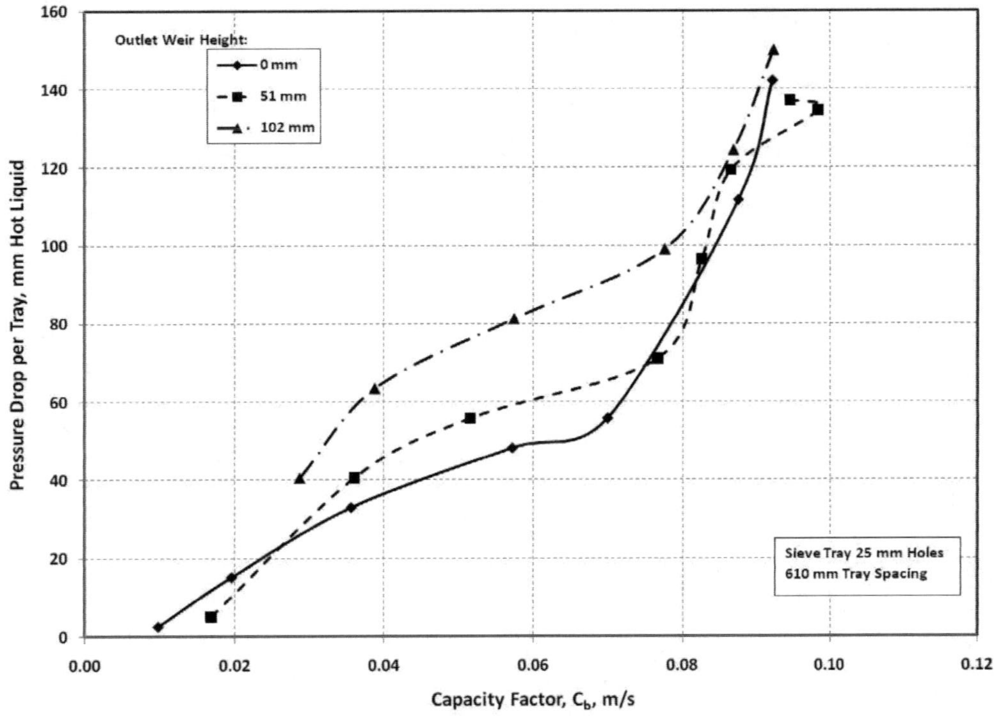

Figure 13. Effect of Weir Height on Pressure Drop for 14% Hole Area Tray (iC$_4$/nC$_4$ at 11.38 bar Pressure)

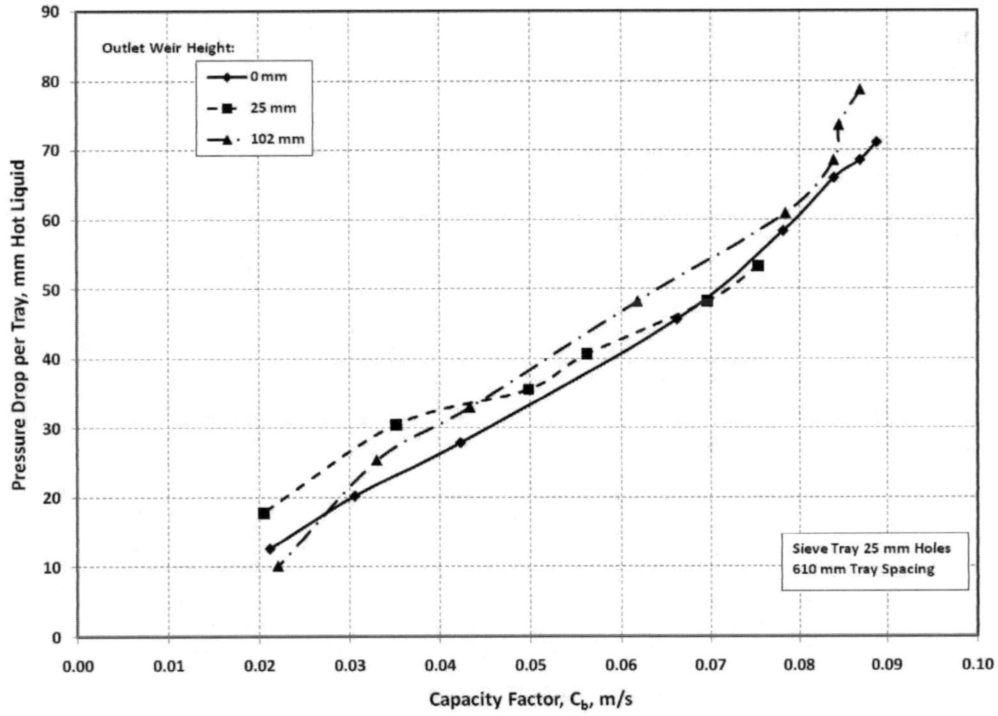

Figure 14. Effect of Weir Height on Pressure Drop for 13% Hole Area Tray (o/p Xylene at 20 mmHg Absolute Pressure)

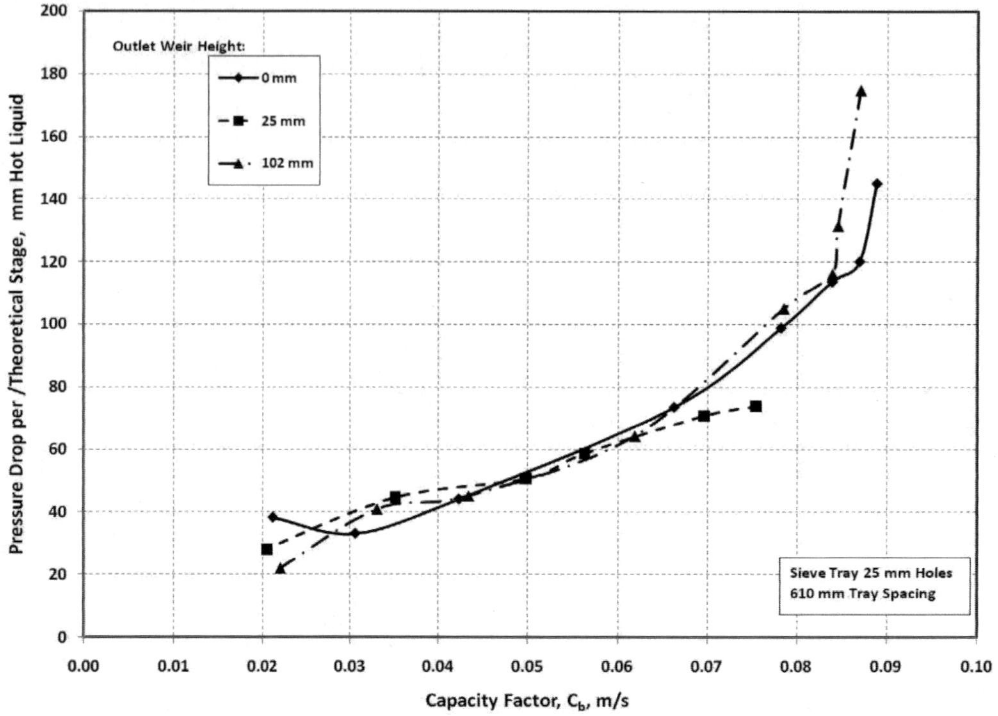

Figure 15. Effect of Weir Height on Pressure Drop per Stage for 13% Hole Area Tray (o/p Xylene at 20 mmHg Absolute Pressure)

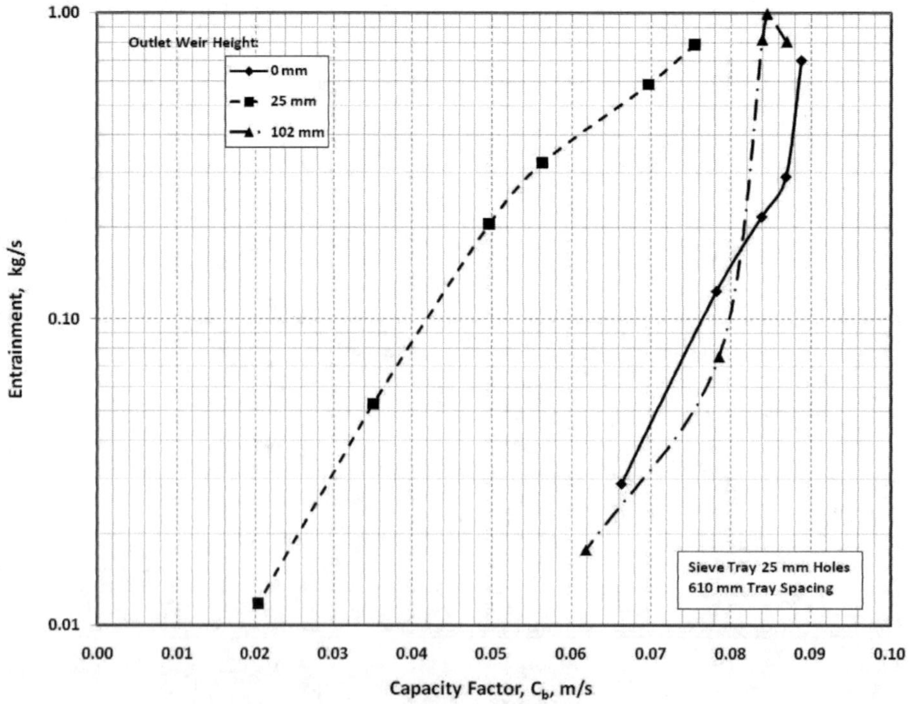

Figure 16. Effect of Weir Height on Total Reflux Entrainment 13% Hole Area Tray (o/p-Xylene at 20 mmHg Absolute Pressure)

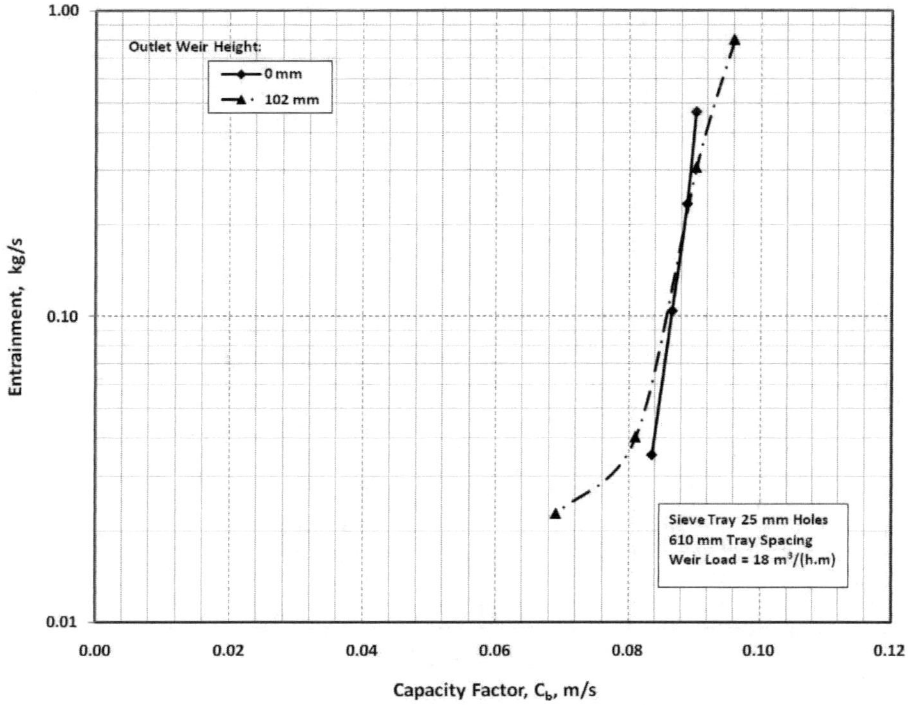

Figure 17. Constant Weir Load (18 m^3/(h.m)) Entrainment, 13% Hole Area Tray (o/p Xylene at 20 mmHg Absolute Pressure)

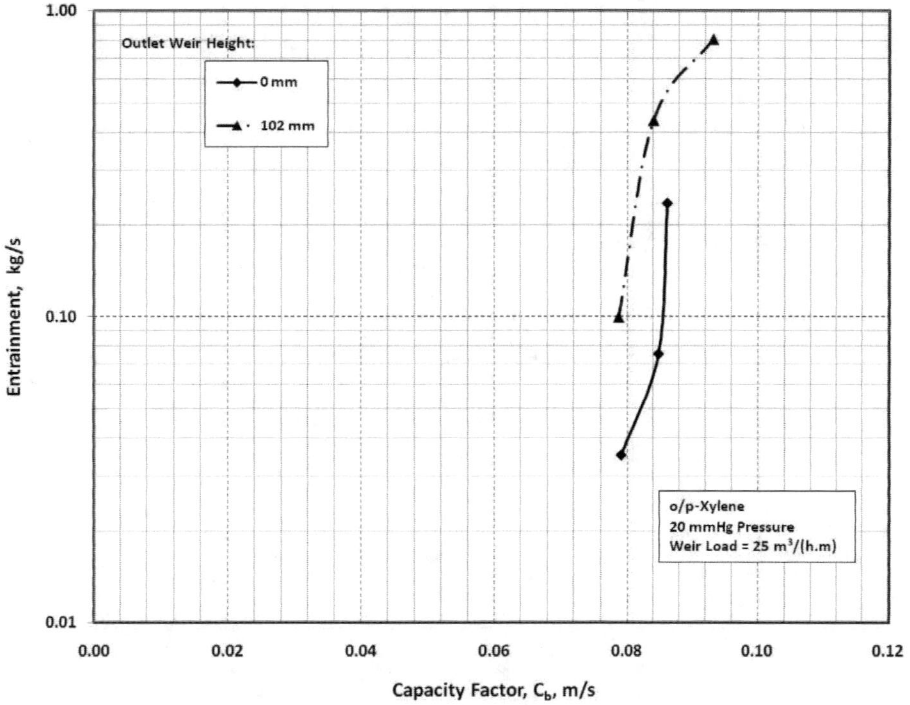

Figure 18. Constant Weir Load (25 m^3/(h.m)) Entrainment, 13% Hole Area Tray (o/p Xylene at 20 mmHg Absolute Pressure)

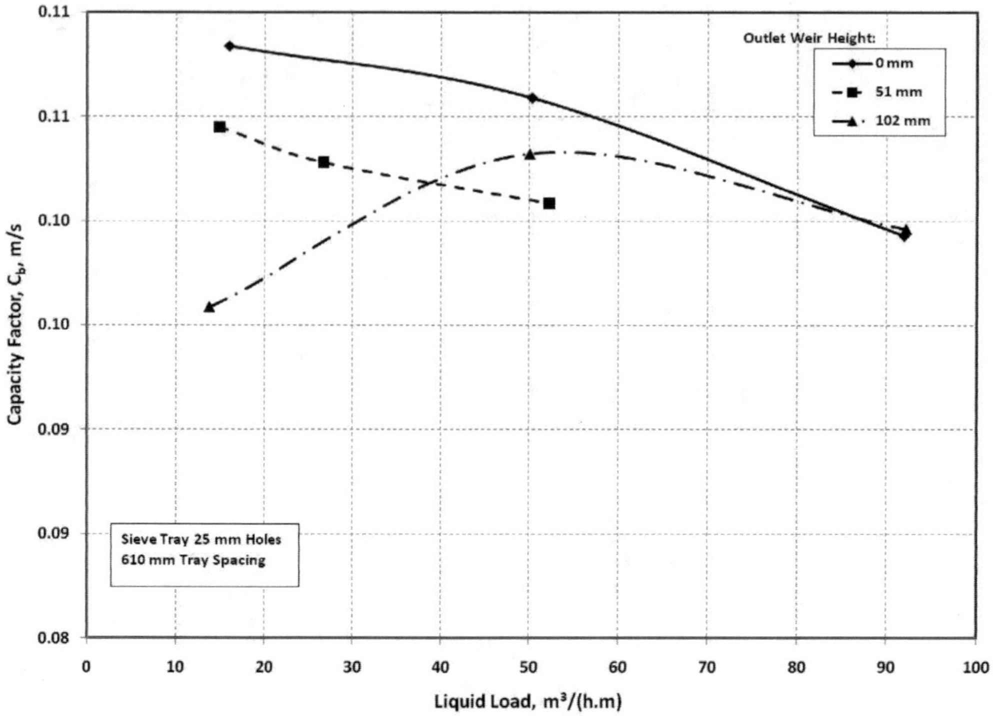

Figure 19. Figure Effect of Weir Height on Capacity for 8% Hole Area Tray (C_6/C_7 at 0.276 bar Pressure)

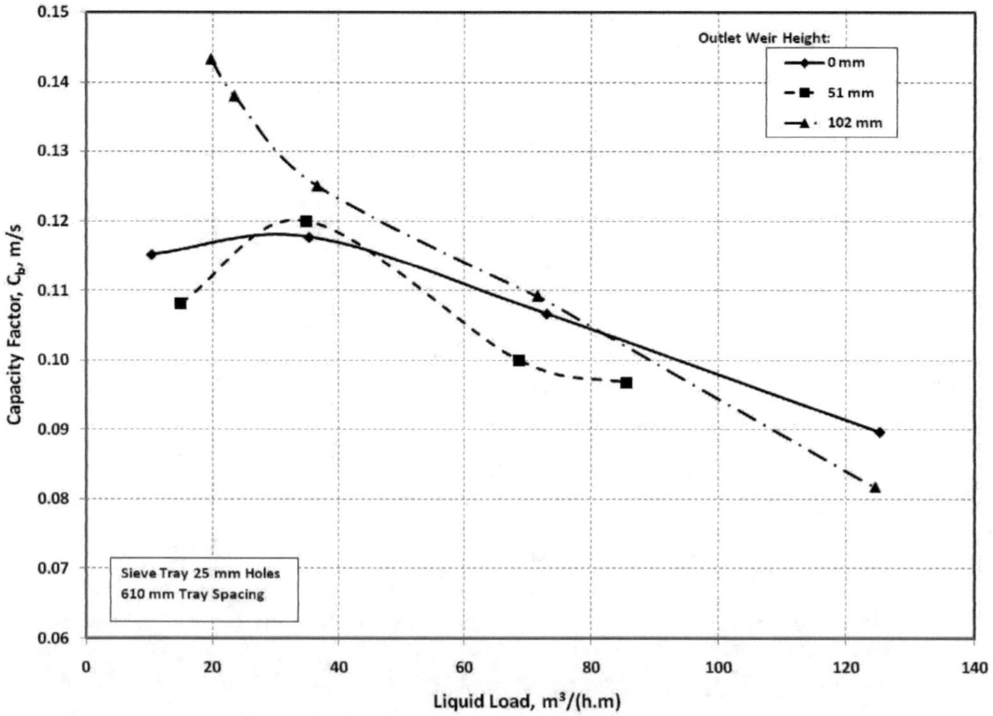

Figure 20. Effect of Weir Height on Capacity for 8% Hole Area Tray (C_6/C_7 at 1.65 bar Pressure)

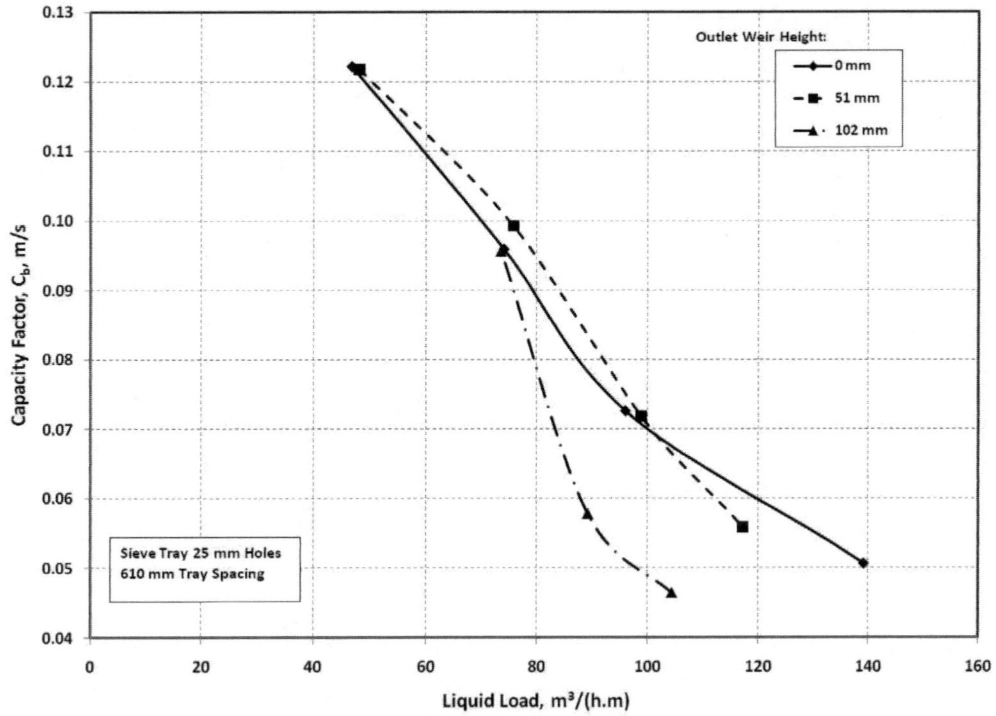

Figure 21. Effect of Weir Height on Capacity for 8% Hole Area Tray (iC_4/nC_4 at 11.38 bar Pressure)

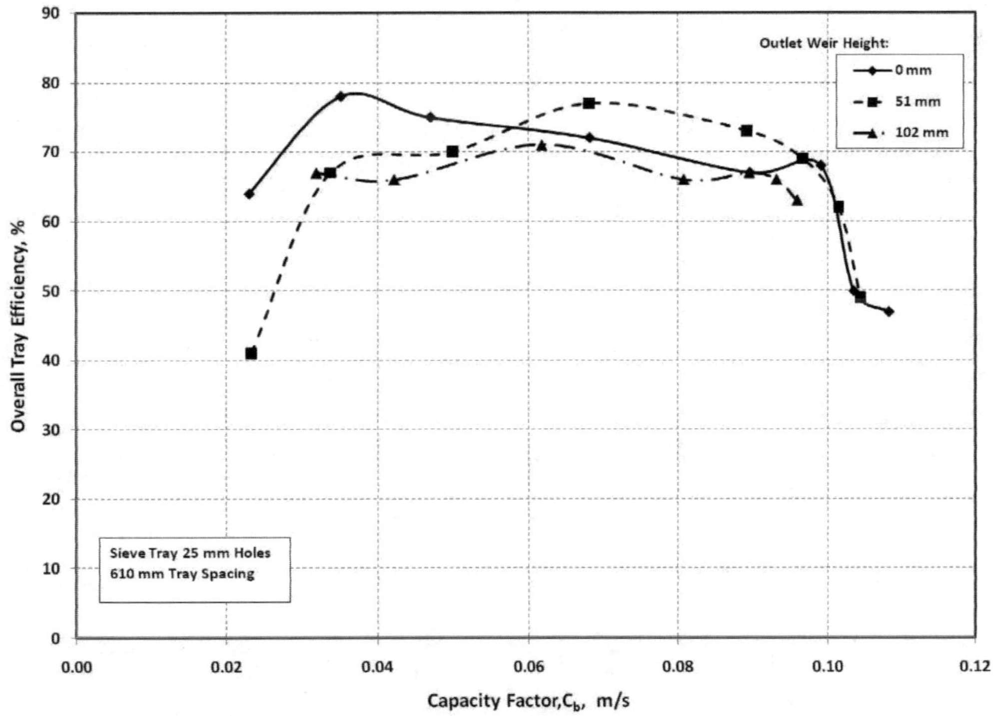

Figure 22. Effect of Weir Height on Efficiency for 8% Hole Area Tray (C_6/C_7 at 0.276 bar Pressure)

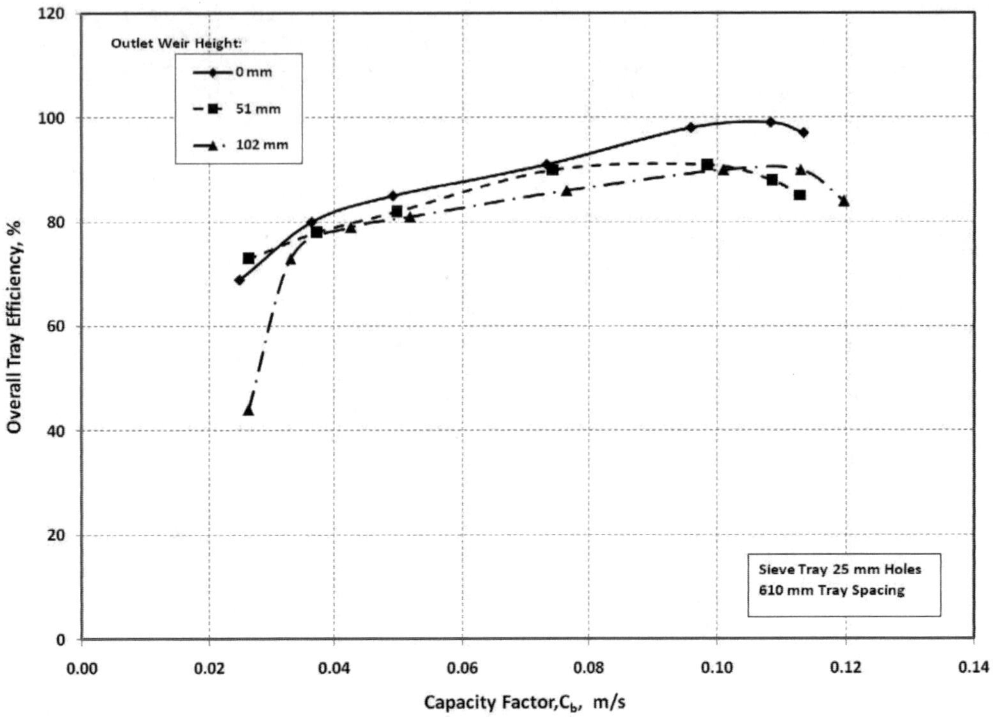

Figure 23. Effect of Weir Height on Efficiency for 8% Hole Area Tray (C_6/C_7 at 1.65 bar Pressure)

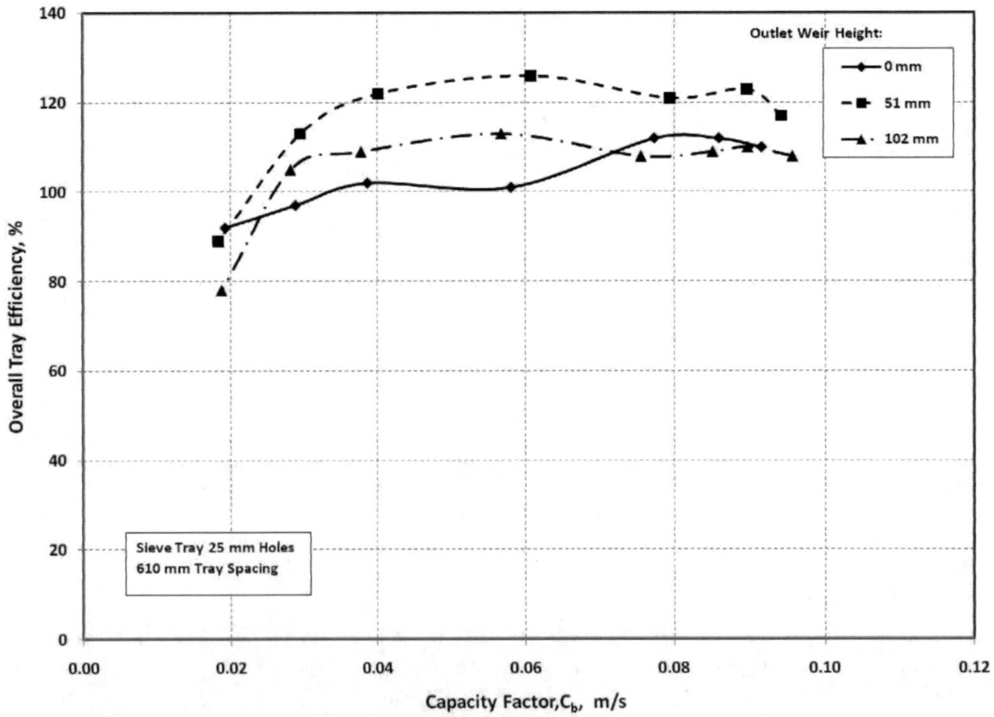

Figure 24. Effect of Weir Height on Efficiency for 8% Hole Area Tray (iC_4/nC_4 at 11.38 bar Pressure)

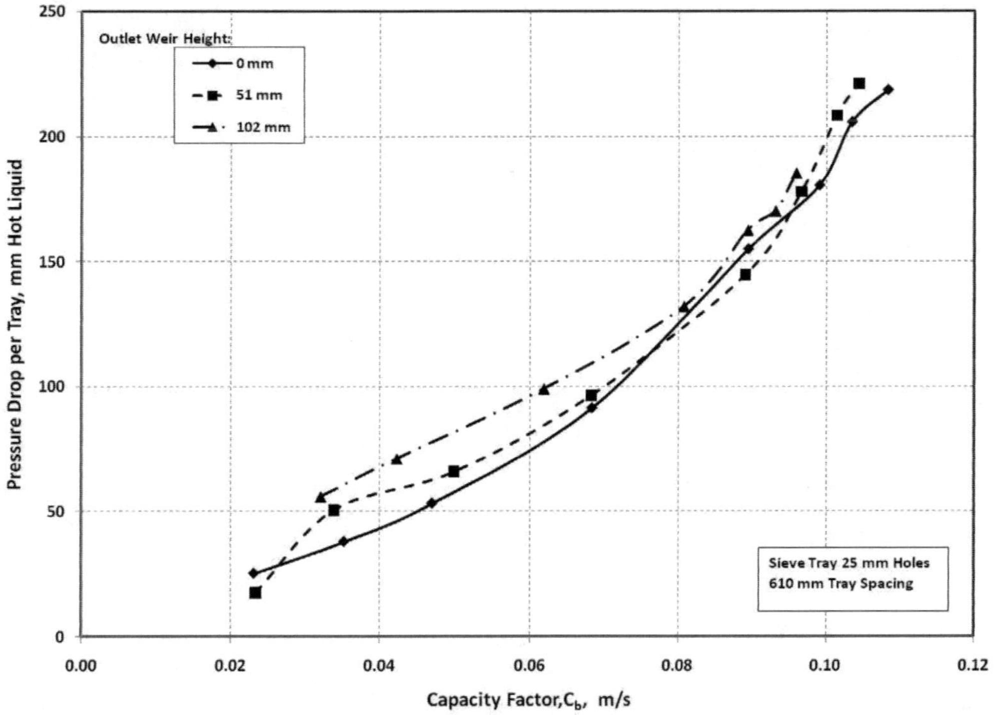

Figure 25. Effect of Weir Height on Pressure Drop for 8% Hole Area Tray (C_6/C_7 at 0.276 bar Pressure)

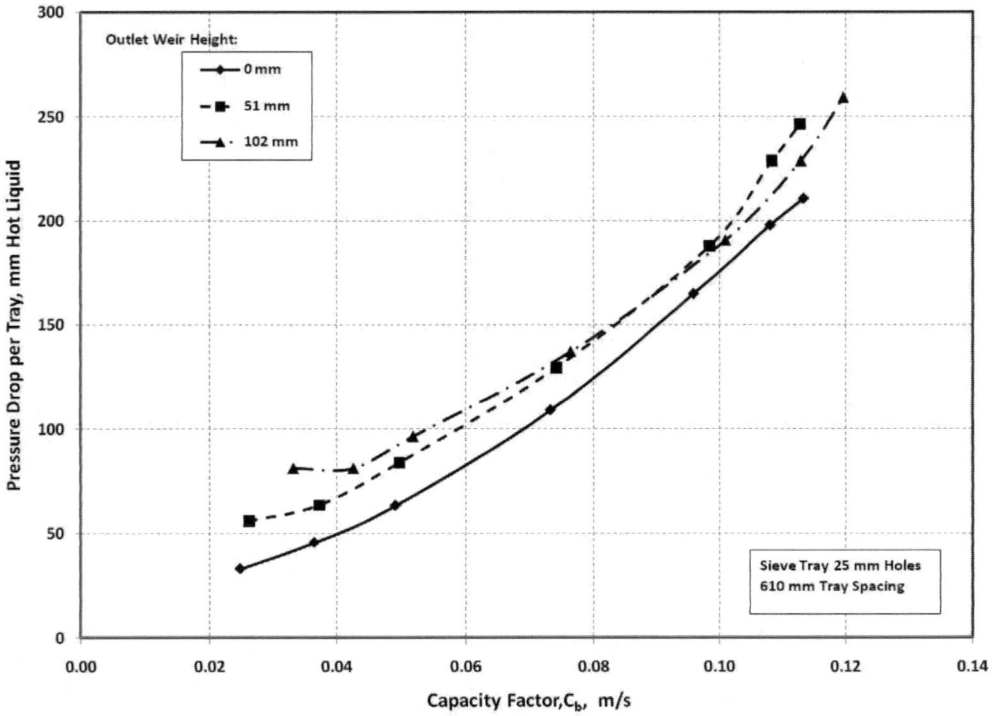

Figure 26. Effect of Weir Height on Pressure Drop for 8% Hole Area Tray (C_6/C_7 at 1.65 bar Pressure)

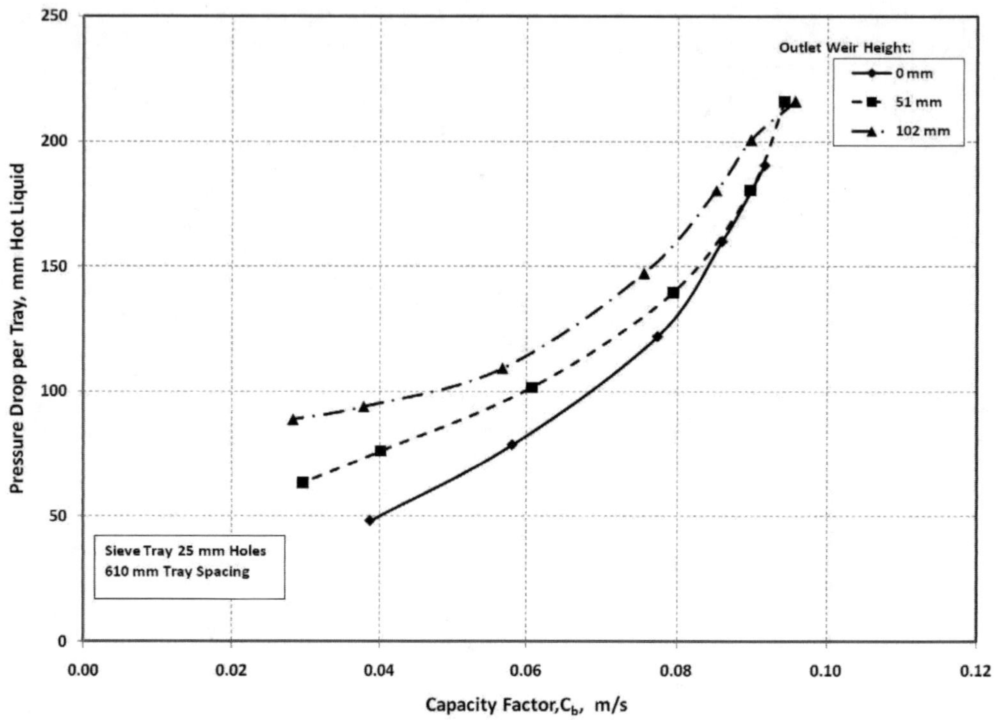

Figure 27. Effect of Weir Height on Pressure Drop for 8% Hole Area Tray (iC_4/nC_4 at 11.38 bar Pressure)

213c Analysis of Operation Results of the Internally Heat Integrated Distillation Column (HIDiC) Pilot Pant In Japan

Koichi Iwakabe[1], Kinpei Horiuchi[2], Keigo Matsuda[3], Toshinari Nakanishi[4], Masahiro Ikeda[5], Satoshi Taniguchi[1], Takuji Yamamoto[1], Sho Kataoka[1], Takao Ohmori[1], and Masaru Nakaiwa[6]. (1) Energy-Efficient Chemical Systems Group, Research Institute for Innovation in Sustainable Chemistry, National Institute of Advanced Industrial Science and Technology (AIST), AIST Tsukuba Central 5, 1-1-1, Higashi, Tsukuba-shi, Ibaraki, 305-8565, Japan, (2) Japan Society for Safety Engineering, List Kannai Bldg. 3F, 4-47, Onoecho, Nakaku, Yokohama, Kanagawa, 231-0015, Japan, (3) Department of Chemistry and Chemical Engineering, Graduate School of Science and Engineering, Yamagata University, 4-3-16, Jonan, Yonezawa, Yamagata, 992-8510, Japan, (4) R&D Department, Kimura Chemical Plants Co. Ltd., 2-1-2, Kuiseterajima, Amagasaki,, Hyogo, 660-8567, Japan, (5) Institute of Industrial Science, The University of Tokyo, 4-6-1, Komaba, Meguro-ku, Tokyo, 153-8505, Japan, (6) Research Institute for Innovation in Sustainable Chemistry, National Institute of Advanced Industrial Science and Technology, AIST Tsukuba Central 5, 1-1-1, Higashi, Tsukuba-shi, Ibaraki, 305-8565, Japan

Our honoree Professor Dr. Shuzo Ohe was the chairman of the evaluation committee of the Japanese NEDO HIDiC project conducted in 2002-2006. Not only he gave us a lot of suggestion on our project, but also he has introduced our HIDiC technology in AIChE meetings. He has presented the application of the HIDiC technology for the separation of azeotropic mixtures. We would like to express our sincere appreciation to his generous works on distillation technologies by reporting the analysis of operation results of the HIDiC pilot plant. The HIDiC pilot plant was designed by Kimura Chemical Plants Co., Ltd., constructed in Chiba factory of Maruzen Petrochemical Co., Ltd in 2005. Among the operation results obtained in the NEDO HIDiC project, those with the zero external reflux condition have been analyzed in this study. The overall heat transfer coefficients, which is one of the most important properties to design the HIDiC, were calculated by the operation results under an assumption of the constant average enthalpy of the internal reflux. The results were correlated to the flow rate of the internal reflux. Based on the correlation, the steady-state simulations were carried out. The simulation results showed the better reproduction of operation results than those from the previous model with the constant overall heat transfer coefficient.

213d Mass Transfer Performance of Packed Beds Containing a Modular Catalytic Structured Packing: Model Vs. Experiment

Zarko Olujic, Process & Energy, Delft University of Technology, Leeghwaterstraat 44, Delft, 2628 CA, Netherlands and Marcel Behrens, Packaged Gases Technology, COE, Air Products Management S.A./N. V., Houtemsesteenweg 20, Vilvoorde, 1800, Belgium.

A comparative experimental study has been carried out to establish the hydraulic behaviour of a new modular catalytic packing (Sulzer's Katapak®-SP 12) and the hybrid packed bed consisting of a catalytic packing section placed in between two sections comprising elements of a high capacity structured packing (MellapakPlus 752.Y). Total reflux distillation experiments were carried out at atmospheric pressure using a packed column with an internal diameter of 0.45 m. As expected, the pressure drop of combined bed were in between those measured for beds consisting purely of the catalytic and structured packings. However, unlike the two reference beds the combined bed exhibited a clear upper gas load limit due to a pronounced liquid build-up at the transition from the structured packing to the top element of catalytic packing section. Also it appeared that the Delft model, with appropriate packing geometry modifications is capable of predicting reliably the preloading region pressure drop and efficiency of a hybrid packed bed containing Katapak-SP.

Basic design of a double-pipe unit type internally heat integrated distillation column (HIDiC)

Toshinari Nakanishi, Naoaki Adachi and Noriyuki Nishida, Kimura Chemical Plants Co., Ltd, Amagasaki, Hyogo, Japan
Koichi Iwakabe, Prime Polymer Co., Ltd., Ichihara, Chiba, Japan
Kinpei Horiuchi, Japan Society for Safety Engineering, Yokohama, Kanagawa, Japan
Masaru Nakaiwa, National Institute of Advanced Industrial Science and Technology (AIST), Japan, Tsukuba, Ibaraki, Japan

1. Introduction

Distillation is not only the most energy consumer but also the most frequently-adopted separation operation in chemical industries since it separates a mixture only by the difference among vapor pressures (boiling points) of components. So distillation is often called as the king of separation processes. Therefore, research and development of technologies on energy savings of distillation is an extremely important and eternal mission for chemical engineers.

In 1978, Secondary Reflux Vaporization (SRV) system, that is an ingenious technology of energy saving for the distillation, was proposed by Dr. Mar in USA. In the SRV system, the rectifying section is pressurized to elevate its temperature level to be higher than that for the stripping section, and these two sections are contacted with each other like a heat exchanger. This structure makes the rectifying section a heat source or a side heater for the stripping section. If the temperature difference and the contact between these two sections are appropriate, a great energy saving can be expected by the SRV system. However, this system has not been put into practical use at that time and R&D had not been continued.

In the nearly time in Japan, Heat Integrated Distillation Column (HIDiC), that is almost equivalent to the SRV system, has been investigated by numerical simulations. A lot of papers for the HIDiC have been reported by Prof. Takamatsu and Dr. Nakaiwa et al. until now. In these papers, some possibilities of operation of HIDiC without a reboiler and a condenser, that is called an ideal HIDiC operation, are shown to obtain the best energy saving from thermodynamic point of view. After long R&D period, a pilot plant of the HIDiC was built and operated at last in 2005 [2)]. Since the feed flow rate of the HIDiC pilot plant is about 2000 kg/h, it is rather a commercial scale column. The consecutive operation of the pilot HIDiC for 1000 hours has already been achieved. The energy consumption of the pilot HIDiC has been saved at less than 40% of a conventional distillation column with the same scale.

In this paper, the structure and the detail of the pilot HIDiC are shown, and its basic design strategy is explained.

2. Principle of HIDiC

A distillation column separates a mixture by the difference of boiling points. Vapor and liquid phases appear in distillation columns, so it is a gas-liquid contacting unit. In conventional distillation columns, liquid is usually produced by condensation at external coolers (condensers), and vapor is produced by vaporization at external heaters (reboilers). It is a great waste of energy since cooling and heating are concurrent at different external heat exchangers. The Vapor Recompression (VRC) system, that is a well-known heat pump technology, can recover the energy from the top vapor by reusing it as a heat medium in the heater at the bottom of the column (Figure 1). Figure 2 shows a concept of the HIDiC. In the HIDiC system, a heat pump is used not to the top and the bottom of the column but to the whole column, that is, the cooling of the rectifying section is directly carried out by contacting the rectifying section with the stripping section. In an ideal situation, both cooling and heating outside the column are no longer required since both vapor and liquid are produced inside the column.

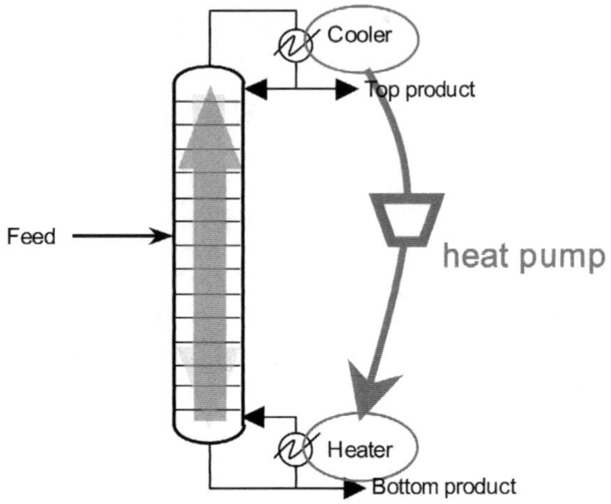

Figure 1 Concept of VRC

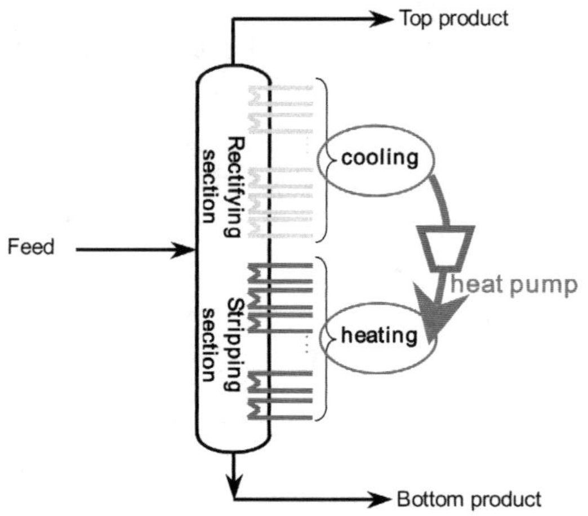

Figure 2 Concept of HIDiC

According to the concept of the HIDiC, the structure must allow the heat transfer from the rectifying section at high temperature to the stripping section at low temperature (Figure 3). Therefore, a compressor should be arranged between the rectifying section and the stripping section to pressurize the rectifying section and make its temperature higher than that in the stripping section.

By this contact between these two sections, the liquid and vapor flow rate profile of the HIDiC is very different from those for the conventional column. It is considered that liquid and vapor flow rate are nearly constant in the conventional distillation column. However, the flow rate profile changes in the HIDiC since the internal heat transfer causes condensation in the rectifying section and evaporation in the stripping section. As shown in Figure 3, both the liquid and vapor flow rates increase downward in the rectifying section, and upward in the stripping section. The condenser and the reboiler duty are significantly reduced since the vapor and liquid flows needed for separation are produced not in the external heat exchangers but inside the column.

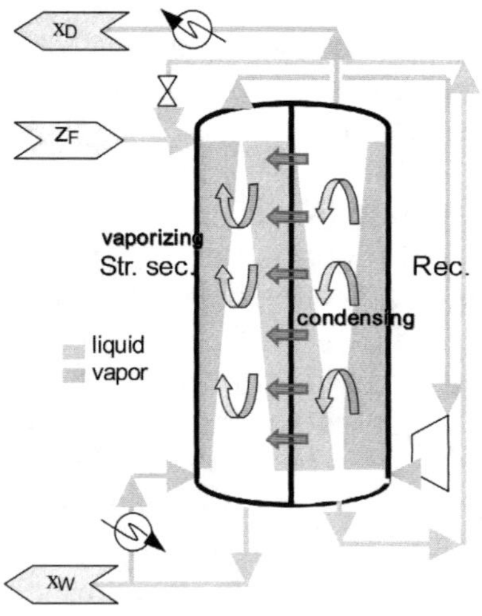

Figure 3 Liquid and vapor flow rate in HIDiC

3. Structure and Design of HIDiC

Figure 4-(a) shows a structure for HIDiC named as Double Pipe Unit. The inside and the outside of the inner pipe function as the rectifying section and the stripping section respectively, and the wall surface of the inner pipe is the heat transfer area. Structured packings are filled both in the rectifying and stripping sections. The heat transfers from the inner pipe to the outer pipe. As mentioned above, the liquid and vapor flow rates change in the HIDiC, so the diameter of the inner pipe must be altered in the axial direction. Broken lines in Figure 4-(b) show liquid loads when the diameter of inner pipe is constant and solid lines show liquid loads when the diameter of the inner pipe is changed.

To realize the structure, four items must be designed; diameters of the inner and outer pipes, the height at which the diameter of the inner tube is changed, the number of double pipe units for a HIDiC pilot plant, and the packing height. To design these items, performance of distillation and heat transfer such as shown in Figure 4-(c) were required. It is important to see a range of design, that is, minimum and maximum loads.

The performance data was obtained by a test equipment in Kimura Chemical Plants, in which packing heights of both inner and outer pipe sides are 1 m each. The test equipment was operated at flow rates and temperatures that correspond to those in the actual pipe unit of the HIDiC pilot plant.

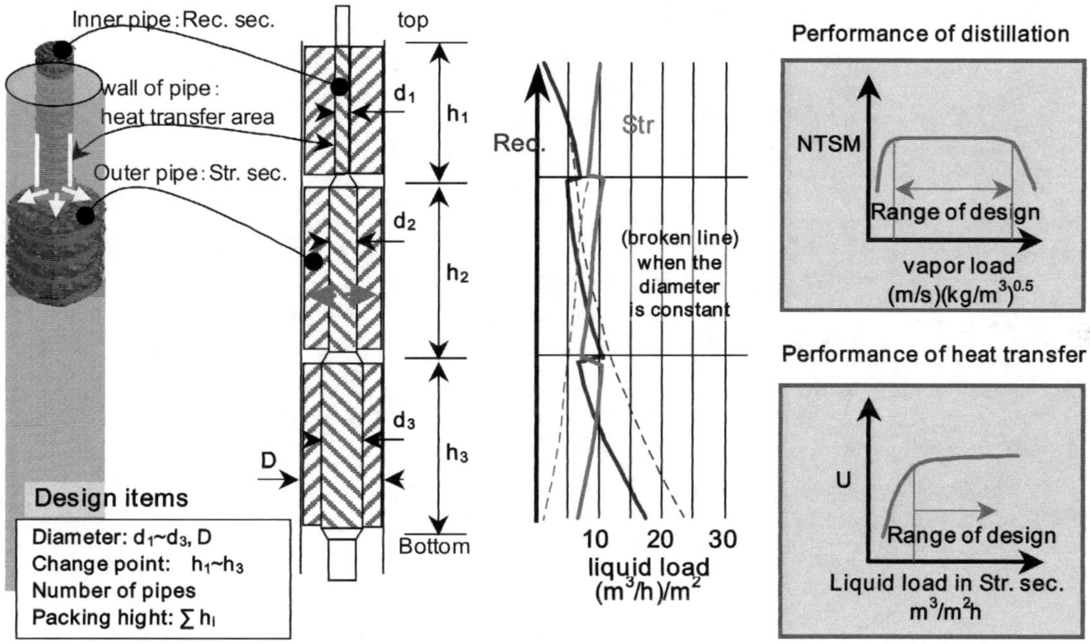

a) Structure b) Flowrate in the column c) Performance data

Figure 4 Design of double pipe-unit

4. Detail of Pipe-unit

The performance of the heat exchange between the rectifying and stripping sections is very important since that indeed is the way of energy saving in the HIDiC. To obtain a good performance, the following three requirements must be satisfied.

1) Liquid condensed on the heat transfer surface in the rectifying section should be steadily removed to packings in order to make the heat transfer efficiency of condensation higher.

2) No dry spots should be generated on the heat transfer surface in the stripping section in order to keep the heat transfer area.
3) Vapor and liquid must flow uniformly both in the rectifying and the stripping section.

As one of structures that satisfies above three requirements, the Double Pipe-Unit was proposed. Its cut model is shown in Figure 5. In the figure, "wire spiral" is a device that is coiled on the wall of the outside surface of the inner pipe (heat transfer surface of the stripping section side). It allows the liquid flow to stick around the surface to ensure the heat transfer, and at the same time, to spread appropriate amount of liquid to the packing section. It also makes the heat transfer area larger, and good contact between the wall of pipe and packing.

Figure 5 Cut model of double pipe-unit

5. Results of experiments

Figure 6 shows the result of distillation performance, and Figure 7 shows the result of heat transfer performance. In Figure 6, the circle and the square plots show Number of Theoretical Stage per Meter (NTSM) in the inner side and the outer side, respectively. It was found that NTSM in the outer side decreases sharply over about 2.5 of F-factor, which is the product of vapor velocity and square root of the vapor density. It is thought as the flooding point. In Figure 7, it was found that the overall heat transfer coefficient (U) is about 1100 W/(m^2·K) if the liquid load in the outer side is more than 5.

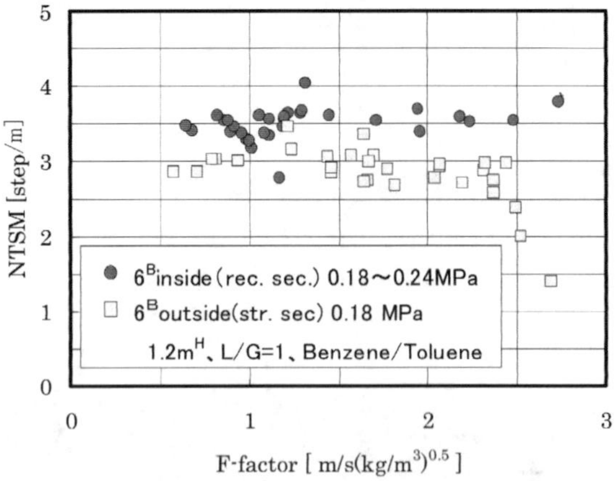

Figure 6 performance of distillation

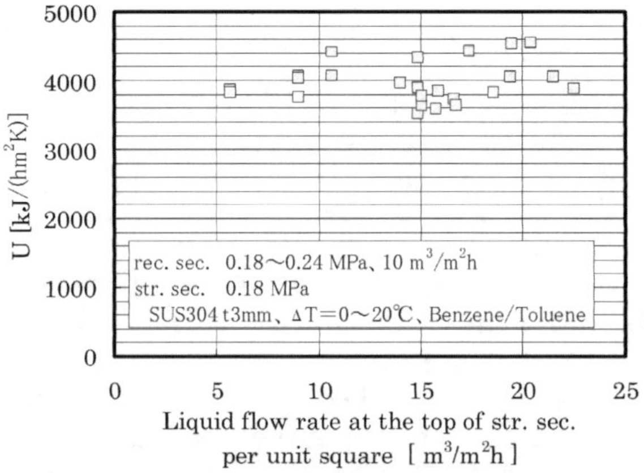

Figure 7 performance of heat transfer

6. Pilot HIDiC

The pilot HIDiC has been built and operated in 2005 at Chiba factory of Maruzen Petrochemical Co., Ltd. It is a cyclopentane stripper from the feed of 12-component gasoline distillate. The pilot HIDiC is 27 [m] of height and 1.4 [m] of diameter. It consists of seven Double Pipe Units (Figure 8). The diameter of the inner pipe of each Double Pipe Unit was altered three times (165 mm, 216 mm and 267 mm), and the diameter of the outer pipe is fixed at 406 mm. This change in the ratio of the cross sectional areas of the rectifying and the stripping section makes constant vapor velocity.

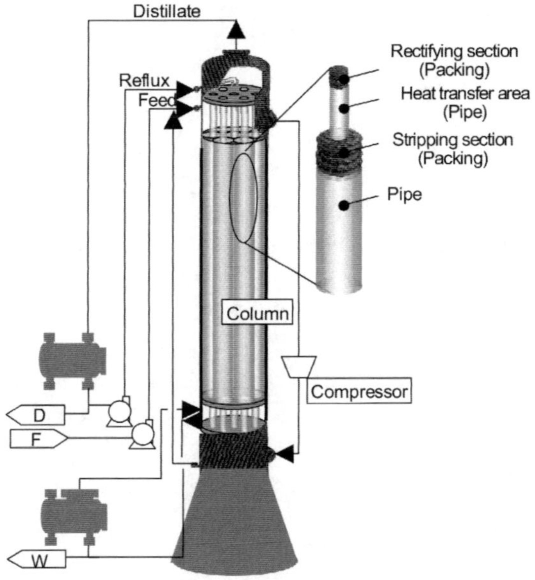

Figure 8 Structure of the pilot HIDiC

Figure 9 Photograph of the pilot HIDiC

7. Conclusions

In this paper, R&D for the first pilot HIDiC in Japan was explained. The principles of VRC and HIDiC systems were introduced and compared with each other. The structure and the design strategy of the HIDiC was explained. Double Pipe Unit for the HIDiC needs no redistribution device in the column since the wire spiral on the outer surface of the inner pipe also functions as a distributor.

The experimental results obtained in a test double pipe unit showed that NTSM decline abruptly at 2.3 $m/s(kg/m^3)^{0.5}$ in the stripping section, suggesting the flooding occurred at the point. Since this flooding point in the rectifying section was not found, additional test would be required to brush up the design strategy of the HIDiC. It seems that the overall heat transfer coefficient U will drop if the liquid flow rate per unit cross sectional area in the stripping section decreases less than 5 $m^3/(m^2h)$ because of the increase in dry spots. Consequently the cross sectional areas of the rectifying and the stripping sections should be designed in the above mentioned F-factor (<2.3 $m/s(kg/m^3)^{0.5}$) and the liquid flow rate (>5 $m^3/(m^2h)$) by changing the diameter of inner pipe. These results shown in Figure 6 and 7 might be dependent on the size of unit (the inner pipe = 6", the outer pipe = 12"). If the diameter of the outer pipe increases up to 20" for example, the liquid load might be less than about 15 $m^3/(m^2h)$.

According to the results obtained, the pilot HIDiC has been built and operated very successfully. The authors wish that the HIDiC technology become common in chemical industries to realize our sustainable world.

Acknowledgements

This work was supported by Ministry of Economy, Trade and Industry (METI), Japan, New Energy and Industrial Technology Development Organization (NEDO), Japan and National Institute of Advanced Industrial Science and Technology (AIST), Japan. Authors would like to express sincere appreciation to Prof. Shuzo Ohe of Science University of Tokyo for worthy advices on our HIDiC development project.

References

1. Nakaiwa, M., K. Huang, A. Endo, T. Ohmori, T. Akiya, T. Takamatsu; "Internally heatintegrated distillation columns: A review," Chem. Eng. Res. Des., 81, 162-177(2003)
2. Horiuchi, K., M. Yamamoto, K. Yanagimoto, K. Kataoka, M. Nakaiwa; "A Pilot Plant of Heat Integrated Distillation Column for Multi-Component Petroleum Distillation," The 7th International Conference on Separation Science and Technology, pp171,Yeongju, Korea(2005)
3. Nakanishi, T., Aso, K., Matsuda, K., Nakaiwa, M., Hasebe, S., Takamatsu, T., "Method of Design for Packed Column Type HIDiC", Distillation & Absorption 2006 Conference, IChemE

231a Magnetically Responsive Membrane for Water Treatment

Xinying Wang, Department of Chemical And Biological Engineering, Colorado State University, Fort Collins, CO 80523-1370, Xianghong Qian, Department of Mechanical Engineering, Colorado State University, Fort Collins, CO 80523-1374, Scott Husson, Clemson University, 126 Earle Hall, Clemson, SC 29634, and Ranil Wickramasinghe, Colorado State University, Department of Chemical Engineering, Fort Collins, CO 80523-1370.

Water treatment is drawing increasing attention worldwide since water is a highly valuable natural resource. Pressure driven membrane separation processes provide many advantages for water treatment. Membrane fouling is a major concern. In this research program, we will modify commercially available nano-filtration membranes by growing polymer brushes from the membrane surface and then attaching super-paramagnetic nano-particles to the brush ends resulting in magnetically responsive, anti-fouling membranes for water treatment.

We have grafted polyacrylic acid (PAA) brushes to the surface of composite polyamide nanofiltration membranes (NF 270) and attached amine group coated super-paramagnetic particles (SPMPs) to PAA brush ends. Membrane surfaces were characterized by field emission scanning electron microscopy (FESEM), Atomic Force Microscope (AFM) and attenuated total reflection Fourier transform infrared (ATR-FTIR) spectroscopy. The results confirm our success at grafting PAA brush and attaching SPMPs. Attachment of SPMPs is also shown using a vibrating sample magnetometer (VSM) which indicates that there is no hysteresis in the magnetization curve as expected for SPMPs. Some preliminary oily water separation data were obtained using our modified membrane under an alternating magnetic field. The results indicate that movement of the magnetically active brushes suppresses fouling.

231b Fabrcation of Ultra High Flux, Highly Concentric and Macrovoid-Free Pvdf Hollow Fibers for Membrane Distillation Process

Sina Bonyadi, Chemical & Biomolecular Engineering, National University of Singapore, 10 Kent Ridge Crescent, Singapore, Singapore and **Tai Shung Chung**, Department of Chemical and Biomolecular Engineering, National University of Singapore, 10 Kent Ridge Crescent,, Singapore, 119260, Singapore.

Highly permeable and macrovoid-free PVDF hollow fibers with a thin wall and a big inner diameter are highly desirable for membrane distillation processes. However, fabrication of these fibers is usually hindered by instabilities such as nascent fiber collapse or deformation during the spinning process. In this communication, we demonstrate a new approach to overcome such obstacles by means of the dual layer spinning and material engineering of PVDF with hydrophobic clay particles in the outer layer and a highly shrinkable polymer such as PAN in the inner layer of the fibers. The function of the inner support layer is to prevent the collapse of the outer nascent layer, and to produce a highly concentric and uniform macrovoid-free thin wall. The resultant single layer PVDF fibers were highly porous (80% porosity), macrovoid-free with a small wall thickness of 120 μm and a big inner diameter of 860 μm. The fabricated fibers showed an outstanding performance during desalination of water by direct contact membrane distillation. A 100% salt rejection and fluxes as high as 70 kg/m2hr, were achieved at 86 °C, which is a much higher flux compared to all the data reported in the literature for hollow fiber membranes so far.

231c Characterization of Commercial Reverse Osmosis Membrane Performance

Elizabeth M. Van Wagner[1], Alyson C. Sagle[1], Benny D. Freeman[1], and Mukul M. Sharma[2]. (1) Chemical Engineering, The University of Texas at Austin, Center for Energy & Environmental Resources, 10100 Burnet Road Building 133, Austin, TX 78758, (2) Petroleum & Geosystems Engineering, The University of Texas at Austin, Cockrell School of Engineering, 1 University Station C0300, Austin, TX 78712

Thin-film composite reverse osmosis (RO) membranes have been studied for nearly fifty years, gradually evolving to the high water flux, high salt rejection (typically >98%) materials used today. However, the high throughput and selectivity that make RO membranes viable candidates for desalination also make measuring their properties difficult. This study was undertaken to identify some important variables responsible for measured performance values (water flux and salt rejection) of commercial RO membranes. Crossflow filtration experiments were performed to characterize the water flux and NaCl rejection of three commercial polyamide reverse osmosis (RO) membranes (LE and XLE from Dow FilmTec and AG from GE Water and Process Technologies). Thorough cleaning of the crossflow system, along with following the recommended pretreatment and test conditions (including pressure, flowrate, temperature, feed pH, and feed filtration) of the manufacturer resulted in measured performance values in agreement with manufacturer benchmarks. Correction for the effect of concentration polarization also proved important. The influence of feed pH and continuous feed filtration on water flux and salt rejection was also determined. While rejection was strongly affected by feed pH, water flux was unaffected. Continuous filtration of the feed led to higher water flux and lower salt rejection than was measured in experiments with unfiltered feed, suggesting fouling of the membrane surfaces by the unfiltered feed. The flux and rejection of these three membranes obeyed a general tradeoff relation: membranes that exhibited higher flux had lower rejection and vice versa.

231d Surface Functionalization of Polybenzimidazole Nanofiltration Membranes for Forward Osmosis

Deepthi Mettu[1], **Isabel Escobar**[1], Maria Coleman[2], and T.S. Chung[3]. (1) Department of Chemical and Environmental Engineering, University of Toledo, 3801 W. Bancroft St, Toledo, OH 43606, (2) Chemical and Environmental Engineering, The University of Toledo, 2801 West bancroft, Toledo, OH 43606, (3) Chemical and Biomolecular Engineering, National University of Singapore, Kent Ridge, Singapore, 119260, Singapore

Membrane selection and operation are critical issues in design of forward osmosis (FO) processes. For example, hydrophilicity and surface negative charge can be combined with thin film composite membrane configuration to enhance both water flux and salt rejection. Recent studies by Chung et.al. indicate that polybenzimidazole (PBI) nanofiltration membranes are efficient for FO which is a ground-breaking technique as a nano-filtration membrane has been incorporated in FO for the first time. The ideal nanofiltration membranes for forward osmosis would be highly negatively charged at most pH values. Since PBI is neutrally charged at a pH of 7.0, there is a decrease in electrostatic repulsion between the membrane surface and electrolytes to be rejected which would adversely affect the separations process. Modification of the membrane surface with negatively charged hyrophylic groups can be used to enhance the FO separations and extend range of application of this membrane system.

The presentation will focus on a joint project with Dr. Chung of National University of Singapore to functionalize the surface of PBI based nanofiltration membranes with short hydrophilic charged groups. Specifically, the membrane functionalization and characterization of surface chemistry using a variety of techniques will be discussed. Preliminary studies of impact of surface functionalization on the steady state flux for pure water and electrolyte solution will be discussed.

Web Page: www.sciencedirect.com/science?_ob=ArticleURL&_udi=B6TGK-4NWCGV0-9&_user=10&_coverDate=08%2F15%2F2007&_alid=737888

231e Biofilm Resistant Polypropylene Feed Spacers

Richard T. Hausman, Isabel C. Escobar, and Tilak Gullinkala. Chemical Engineering, University of Toledo, 2801 West Bancroft, Toledo, OH 43606

Biofouling, which is the accumulation of microorganisms onto the membrane surface and on the feed spacer, is a very prominent problem in the use of reverse osmosis (RO) membranes. In the past, research and development of biofouling prevention has commonly focused on the pretreatment of the feed water, the improvement of cleaning solutions and procedures, and modifications to the membranes themselves. This research focuses on developing anti-biofouling polypropylene (PP) feed spacers. This is particularly significant because membrane replacement due to fouling is the single largest operating cost in water separation (Escobar et al. 2005).

PP was functionalized via the addition of a spacer arm with metal chelating ligands. These ligands were charged with copper or silver ions to allow for slow release of metals into the feed water in the membrane systems. Copper and silver ions have been previously used to disinfect water against microbial life because the ions interfere with enzymes involved in cellular respiration, and they bind DNA at specific sites. This functionalization was chosen as the focus because these chelating ligands are quite stable and easily synthesized, operate over a diverse range of conditions, have easily controlled binding affinities, and are well suited for model studies. In this research, the spacer arm was glycidyl methacrylate (GMA) which was polymerized to the PP using benzoyl peroxide as a radical initiator. Iminodiacetic acid (IDA) was then added to the end of the GMA spacer arm and to allow for the chelatation of the copper and silver ions.

Characterization of the PP modification was performed using Fourier transform infrared spectroscopic (FITR), atomic force microscopy (AFM), scanning electron microscopy (SEM) and mechanical strength measurements. The effects on the biological content of water in the presence of the modified spacer were tested using a synthetic water matrix of sterile phosphate buffer solution containing Pseudomonas fluorescens P17 and Spirillium volutans strain NOX. Both modified and unmodified spacers were contacted with the feed water for 2 weeks to allow for growth and then placed in stomacher bags where the biofilm was removed. R2A Agar spread plates will then be made for each specimen and bacterial growth were quantified using a Model 920A Colony Counter.

References:

Escobar I., Hoek, E., Gabelich, C., F. DiGiano, Y. Le Gouellec, Berube, P., K. Howe, J. Allen, K. Atasi, M. Benjamin, P. Brandhuber, Brant, J.A., Chang, Y., Chapman, M., A. Childress, W. Conlon, T. Cooke, I. Crossley, G. Crozes, P. Huck, S. Kommineni, J. Jacangelo, A. Karimi, J. Kim, D. Lawler, Q. Li, L. Schideman, S. Sethi, J. Tobiason, T. Tseng, S. Veerapaneni, and A. Zander, "American Water Works Association Membrane Technology Research Committee Report: Membrane Fouling – Recent Advances and Research Needs", Journal American Water Works Association, 97 (8), 2005, pages 79-89.

Boron Removal from Aqueous Solution by Bipolar Electrodialysis

Hiroki Nagasawa[a, *], Atsushi Iizuka[a], Akihiro Yamasaki[b] and Yukio Yanagisawa[a]

[a] *Department of Environment Systems, Institute of Environmental Studies, School of Frontier Sciences, The University of Tokyo, Chiba, Japan*

[b] *Department of Materials and Life Science, Seikei University, Tokyo, Japan*

Abstract

Boron removal process using electrodialysis method combined with bipolar membrane water splitting reaction was investigated. By using bipolar membrane electrodialysis, pH value of boron contained solution will immediately increase and the dissociation reaction of boric acid into borate ions was enhanced. Therefore, boron can be effectively removed from boron contained solution. Experimental results confirm that bipolar membrane electrodialysis has an advantage over existing electrodialysis. Effect of initial concentration was investigated. It was confirmed that energy requirement for boron removal increase when the initial boron concentration increase.

Keywords: Boron; Electrodialysis, Bipolar membrane; Water

Introduction

Boron compounds in water at high concentration are known to be toxic to humans and plants. The World Health Organization (WHO) has limited boron concentration in drinking water to less than 0.5 mg-B/L. Because of the increased use of desalinated seawater and underground water for drinking water and large amount of boron emissions via wastewater from the manufacturing process, the removal of boron from aqueous solutions has been an important issue in recent years.

In aqueous solutions, boron is mainly present in the form of boric acid. According to the dissociation reaction, boric acid will form into borate ion at alkaline condition.

$$H_3BO_3 + OH^- \leftrightarrow B(OH)^- \quad pKa = 9.1 \quad (1)$$

There are several methods investigated for removing boron from aqueous solutions, such as precipitation [1], adsorption [2], solvent extraction [3] and ion-exchange [4]. However, the precipitation and the adsorption methods are efficient only under high boron concentration. In the case of the solvent extraction and the ion-exchange, post regeneration process causes cost increases.

Electrodialysis is one of the considerable options for boron removal from aqueous solution [5]. Using electrodialysis, borate ion will be efficiently removed from aqueous solution. However, borate ion only exist at alkaline condition, existing electrodialysis method cannot be utilized to acidic solutions.

* Corresponding author. E-mail: hiroki_nagasawa@yy.k.u-tokyo.ac.jp

To resolve this problem, we focus on water splitting reaction caused by bipolar membrane. Combining the water splitting reaction and electrodialysis, the dissociation reaction of boric acid into borate ions will be enhanced and boron removal will become available in acidic solutions.

In this study the boron removable process using electrodialysis method combined with bipolar membrane water splitting reaction was investigated.

Process description

In a bipolar membrane electrodialysis (BPED) system, a number of bipolar membranes and anion exchange membranes are stacked in an alternating pattern between a cathode and an anode (Fig. 1). The repeating unit is called a cell pair, which consists of a bipolar membrane, a compartment filled with a boron contained solution (feed cell), and an anion exchange membrane and a compartment filled with an electrolyte solution (concentrate cell).

When an electrical potential is applied across the system, water molecules in the bipolar membranes will be split into protons and hydroxyl ions. According to the electrical field applied across the system, hydroxyl ions will be transported to the feed compartment. Due to the increase of hydroxyl ions, pH of the boron contained solution will increase and the dissociation of boric acid into borate ions will be enhanced. Generated borate ions will be transported from feed compartment to concentrate compartment via the anion exchange membrane by function of the electrical field. Thus boron will be removed from boron contained solution and concentrated in the electrolyte solution.

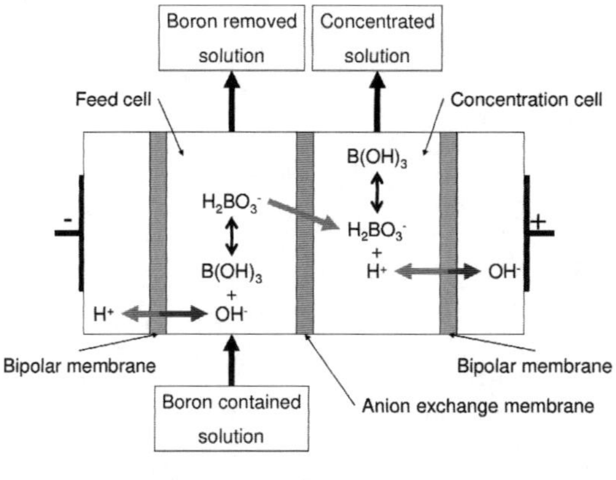

Fig. 1. Schematic of the bipolar membrane electrodialysis system for boron removal.

Experimental

An experimental study was conduced in a laboratory-scale electrodialysis apparatus (AGC). This apparatus contains three cell pairs between two electrodes at the both end. Neosepta BP-1E (Tokuyama) was employed as a bipolar membrane and Selemion AMV (AGC) was employed as an anion exchange membrane. The membrane effective area was 117.5 cm^2 and the distance between two membranes was 1.5 mm.

Solution for feed cell (feed solution, 1 L), solution for concentrate cell (concentrate solution, 1 L) and electrode rinse solution (0.1 M Na_2SO_4 *aq*, 1L) was circulated using magnetic pumps. The flow rates of three solutions were all set at 1.0 L/min. The electrodialysis experiments were operated with a constant electrical potential at 10 V.

Results and discussion

Comparison of BPED and existing electrodialysis

Comparison study of BPED and exsisting electrodialysis (ED) system was performed. For

ED apparatus, cation exchange membrane, CMV (AGC), was employed instead of bipolar membrane.

$Na_2B_4O_7$ solution having 100 mg-B/L was employed for the feed solution. Initial pH values of the feed solution were set at 2.6 and 9.2 by adding HCl and NaOH (the concentration of Cl^- in the feed solution was controlled at 0.01 M). The concentrate solution was 0.01 M NaOH solution.

When the electrical potential was applied, the current value increased with time in BPED cases. While in ED cases, current value decreased with time and reached 0 in 30 minutes. This is due to the increase in the electrical resistance in the feed solution. Since ED process employed cation exchange membrane instead of bipolar membrane, the ion concentration in the feed solution decreased with time (Fig. 2a).

The pH value of the feed solution increased immediately in BPED cases, while the change of pH value was smaller in ED cases (Fig. 2b). The pH value of the concentrate solution decreased immediately in BPED cases, while the change of pH value was smaller in ED cases.

The decrease of boron concentration was observed in all cases (Fig. 2c). In BPED cases, more than 90% of boron in the feed solutions was removed in both acidic and alkaline conditions. This result suggests that dissociation reaction of boric acid into borate ion was enhanced by water splitting reaction by bipolar membrane. On the other hand, in ED case, 50% of boron was removed in alkaline condition, 42% of boron was removed in acidic condition. This difference in boron removal would be due to difference of pH value in the feed solutions, which fix the degree of

a)
b)
c)

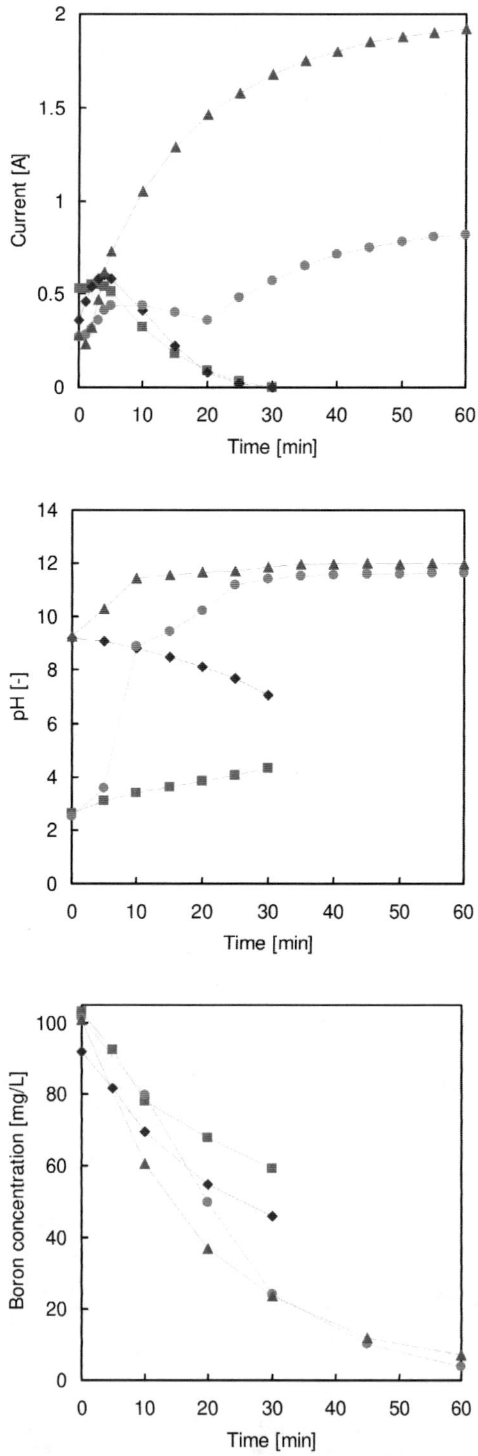

Fig. 2. Comparison of BPED and ED. a) Current, b) pH value of feed solution, c) concentration of boron. ● BPED pH 2.6, ▲ BPED pH 9.2, ■ ED pH 2.6, ◆ ED pH 9.2.

ionization of boric acid.

Effect of initial boron concentration

The effect of initial boron concentration in the solution was investigated. $Na_2B_4O_7$ solution having 50, 100, 200 mg-B/L was employed for the feed solution and concentrate solution.

In case of 50 mg-B/L, 90% of boron was removed in 30 minutes and boron concentration in the feed solution reached less than 1.0 mg-B/L in 60 minutes.

In case of 100 mg-B/L, 90% removal in 60 minutes and <1.0 mg-B/L in 120 minutes.

In case of 200 mg-B/L, boron concentration in the feed solution did not reach below 20 mg/L in this experiment.

The effect of initial concentration on the energy requirement for boron removal calculated from the experiment was shown in Fig. 4. It suggests that energy requirement for boron removal increases with the initial boron concentration increase.

Conclusions

Boron removable process using bipolar membrane electrodialysis was investigated. Water splitting reaction in bipolar membrane enhanced the dissociation reaction of boric acid into borate ion. Therefore, boron in the feed solution was effectively removed and concentrated.

More than 90% of boron was removed in 30 minutes and boron concentration in the feed solution reached less than 1.0 mg-B/L in 60 minutes in the best case studied.

It was confirmed that energy requirement will increase when the initial boron concentration increased.

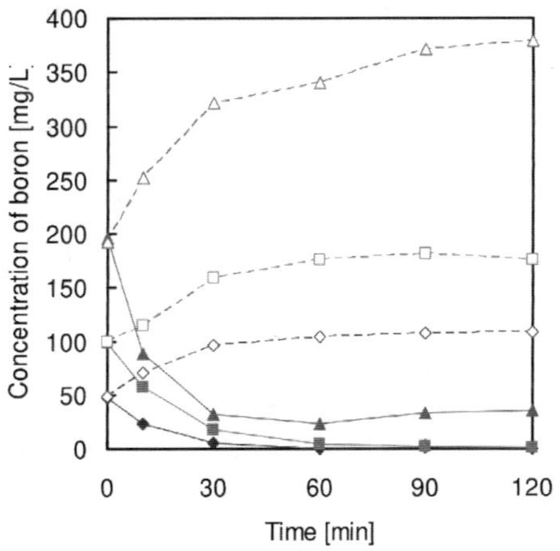

Fig. 3. Effect of initial boron concentration on boron removal. ◆feed 50 mg-B/L, ◇ concentrate 50 mg-B/ L, ■ feed 100 mg-B/L, □ concentrate 100 mg-B/L, ▲ feed 200 mg-B/L, △ concentrate 200 mg-B/L.

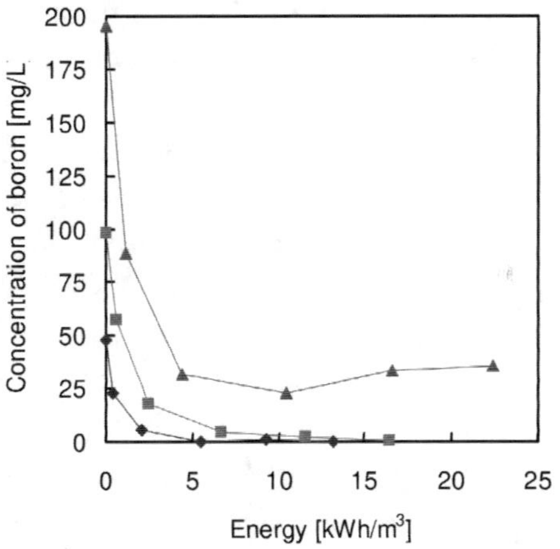

Fig. 4. Energy requirement for boron removal. ◆feed 50 mg-B/L, ◇ concentrate 50 mg-B/ L, ■ feed 100 mg-B/L, □ concentrate 100 mg-B/L, ▲ feed 200 mg-B/L, △ concentrate 200 mg-B/L.

References

[1] N. Ozturk *et al.*, *Adsorption* **2004**, *10*, 245 – 257. [2] W. Bouguerra *et al.*, *Desalination* **2008**, *223*, 31 – 37. [3] N. Bical *et al.*, *Separation Science and Technology* **2003**, *38*, 165 – 177. [4] M. Simonnot *et al.*, *Water Research* **2000**, *34*, 109 – 116. [5] N. Kabay *et al.*, *Desalination* **2008**, *223*, 49 – 56.

EFFECT OF MEMBRANE STRUCTURE ON THE PERFORMANCE OF NANOFILTRATION HOLLOW FIBER MEMBRANE USED IN FORWARD OSMOSIS

Kai Yu Wang, Qian Yang, Tai-Shung Chung, Raj Rajagopalan

National University of Singapore, Singapore

Introduction

Along with the global water scarcity, less and less access to clean water has become a world-wide issue afflicting people, especially in the developing and industrialized nations. More and more effort has been put into evaluating the potential techniques to recover fresh water from wastewater and sea water/brine desalination with low capital costs. Even though the reverse osmosis (RO) process has replaced conventional thermal distillation processes as the dominant desalination technology after 40 year development, the high oil prices have shadowed its future potential because high hydraulic pressures must be provided in the RO process to overcome the osmotic pressure of feed solutions. In addition, the low water recovery by RO is another disadvantage during sea water desalination. Membrane separation technology, important for non-thermal separation, has been becoming more attractive to avoid thermodynamically imposed efficiency limitations on heat utilization. Recently, forward osmosis (FO) has drawn much attention by researchers and its applications have been developed in various fields, such as wastewater treatment [1], pharmaceutical and juice concentration [2-4], desalinating seawater [5, 6], even in the power generation [7-9] and the potable water reuse in space [10, 11]. Forward osmosis process, which is a natural process, utilizes the osmotic pressure gradient generated by a highly concentrated solution (draw solution) to allow water to diffuse through a semi-permeable membrane from a saline feedwater, which has a relatively low concentration. By employing the osmotic pressure as a driving force, forward osmosis can operate without high hydraulic pressures which are necessary in the reverse osmosis process. Therefore, less energy is required for the FO process compared to the RO process. In addition, FO may offer the advantages of high rejection of a wide range of contaminants and lower membrane-fouling propensities than traditional pressure-driven membrane processes. Because the osmotic forces from draw solutions in FO are greater than the trans-membrane pressures used in RO, potentially higher water productivity and recoveries may be achievable.

Membrane, as the barrier between the draw solution and the feed solution to transport water and reject solute, play an important role on the realization of forward osmosis. However, most available membranes used in the forward osmosis are either from dense semi-permeable membranes available in the reverse osmosis industry or cellulose triacetate based FO membrane developed by Hydration Technologies Inc. (Albany, OR) through coating cellulose triacetate on the non-woven fabric [5]. In addition, almost all FO studies were mainly based on the flat membrane configuration. As compared to flat sheet membrane, hollow fiber membranes have been widely used for liquid phase separation due to their high surface area to volume ratio and ability to self-support. Because nanofiltration (NF) membranes have much higher rejections to di-valent ions but partially permit the mono-valent ions get through due to the Donnan electrostatic exclusion, they can be employed to fractionate electrolyte mixtures [12]. Nanofiltration membranes have been widely used for the separation of small neutral and charged molecules in aqueous solutions through their active charged nano-scale pores, with the estimated pore sizes of around 0.5 ~ 2 nm in diameter. As a pressure-driven process, nanofiltration has become an effective means for the removal of heavy metals ions and divalent anions from wastewater [13]. It is possible to develop NF membrane with applicable nano-scale pore size and narrow pore size distribution in the forward osmosis to obtain desirable permeation flux and high separation performance [14].

Polybenzimidazole is a generic name of a class of heterocyclic polymers, and the most commonly concerned is the poly-2, 2′ - (m-phenylene)-5, 5′ -bibenzimidazole (PBI). PBI are a class of heterocyclic polymers that were commercially developed by the Celanese Corporation in 1983 [15, 16], In addition to higher moisture affinity and high glass transition temperature (Tg, 425 ~ 435°C), PBI was known to possess outstanding thermal, physical and chemical stability over a wide range of pH. Since hollow fiber configuration has the advantage of self-support and provides a higher ratio of membrane area over unit module volume [17], fabricating PBI NF hollow fiber membranes may have greater potential in the water recovery with high productivity. Moreover, the relatively low cost of fabricating the hollow fibers make it of interest for large-scale industrial applications.

In this study, hydrophilic polybenzimidazole hollow fiber membranes through dry-jet wet phase inversion and its modified patterns were fabricated with different structures, for instances, wall thickness, microstructure and pore size in order to investigate the effects of membrane morphology on the membrane performance during FO process. It has been acknowledged that the support layer structure have the important effect on the water transport due to the serious internal concentration polarization in the porous support layer.

Experimental

The first batch PBI hollow fiber (A) was produced from the polymer dope of PBI/DMAc/LiCl (22.6/75.6/1.8 wt%). The selective layer was formed on the shell side of hollow fiber. After immersing in tap water for three days in order to remove the residual DMAc and LiCl, the fabricated hollow fiber membranes were soaked in a 50 wt% glycerol aqueous solution for 48 hrs and then air-dried at room temperature. SEM images illustrate the hollow fiber structure which consists of four types of morphologies, namely: asymmetric outer selective skin layer, finger-like macrovoids, sponge-like substructure, and porous inner skin layer. The outer and inner diameters of fiber are 273 and 137 microns, respectively. During PBI hollow fiber spinning, water is applied as the external coagulant to form the nano-porous outer selective layer. The formed sponge-like substructure near inner skin is due to the delayed phase separation induced by the solvent enriched bore fluid of DMAc/ethylene glycol mixture. This kind of microstructure makes PBI hollow fiber mechanically strong to sustain high trans-membrane pressure during nanofiltration except that the substructure may produce internal concentration polarization and seriously influence the water recovery during forward osmosis. The second batch PBI hollow fiber (A) was produced from the polymer dope of PBI/DMAc/LiCl (22.0/76.3/1.7 wt%). The as-spun PBI hollow fiber (B) was dipped in fresh methanol for three times to replace water. Then, the membrane was dipped in the solution of substituting agent *p*-Xylylene dichloride (1.0 wt%) in ethanol for 2.0 and 9.0 hours. After taking out of *p*-Xylylene dichloride solution, the modified PBI membrane was rinsed by ethanol again, and then kept in water [18]. The solute transport experiments were conducted to characterize the pore size distribution. 200 ppm solutions containing glycerol, glucose, saccharose or raffinose with known Stokes radius r_s were used to measure the solute rejection.

Membrane characterization

SEM images illustrate the hollow fiber structure which consists of four types of morphologies, namely: asymmetric outer selective skin layer, finger-like macrovoids, sponge-like substructure, and porous inner skin layer, as shown in Fig. 1. The outer and inner diameters of fiber are 273 and 137 microns, respectively. During PBI hollow fiber spinning, water is applied as the external coagulant to form the nano-porous outer selective layer. The formed sponge-like substructure near inner skin is due to the delayed phase separation induced by the solvent enriched bore fluid of DMAc/ethylene glycol mixture. This kind of microstructure makes PBI hollow fiber mechanically strong to sustain high trans-membrane pressure during nanofiltration except that the substructure may produce internal concentration polarization and seriously influence the water recovery during forward osmosis. The pure water permeability is 0.5 liter m^{-2} bar^{-1} hr^{-1} measured at 10 bar. From the solute transport method [19], the mean effective pore radius of PBI NF hollow fiber membrane was 0.32 nm.

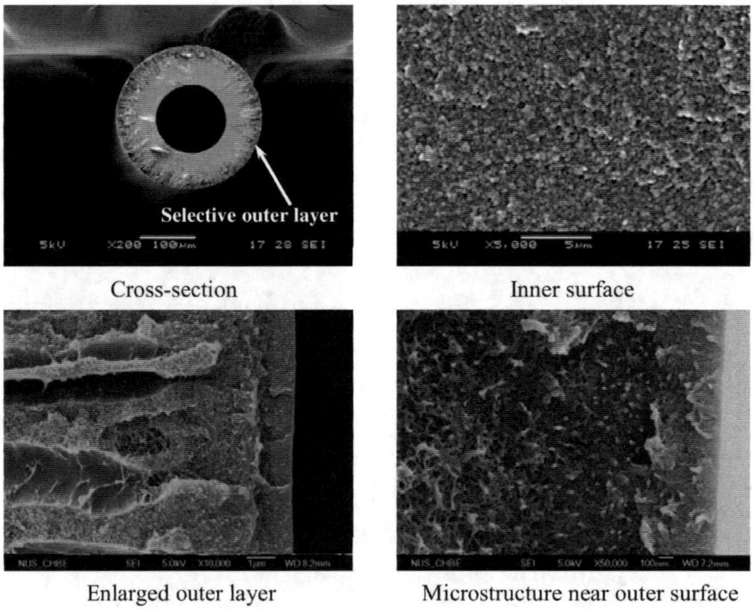

Figure 1 Morphology of PBI hollow fiber A

The PBI hollow fiber B from second batch has the same structure as the PBI fiber A, as shown in Fig. 2. The wall thickness was designed to be thin as 40 μm to reduce the mass transport resistance. The outer and inner diameters of the formed fiber B are 293 and 213 microns, respectively. Following table 1 lists the pure water permeability, mean pore size and MWCO of the polybenzimidazole membrane as a function of modifying period in the *p*-Xylylene dichloride/ethanol solution. After the p-xylylene dichloride modification, the PBI membrane has a decreased mean pore size and a narrowed pore size distribution which lead to an enhanced selectivity for the separation of solutes, as shown in Fig. 3. This suggests that the membrane structure and subsequent separation performance can be effectively controlled through the chemical modification.

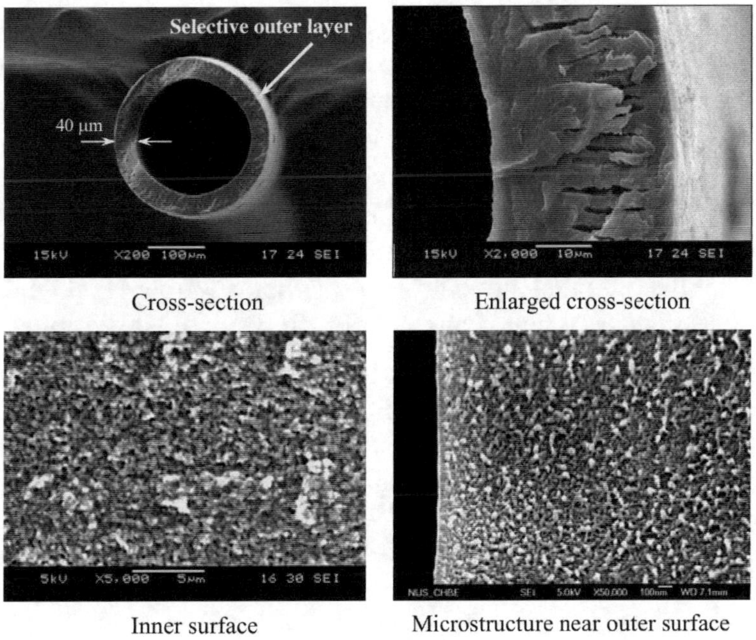

Figure 2 Morphology of PBI hollow fiber B

Table 1 properties of PBI fiber B vs. *p*-xylylene dichlororide modification

PBI NF hollow fiber membrane (B)	PWP ($l\ m^{-2}\ bar^{-1}\ hr^{-1}$)	r_p (nm)	MWCO (Da)
Original membrane	2.43	0.41	993
Modified membrane (2 hour)	1.53	0.34	528
Modified membrane (9 hour)	1.25	0.29	354

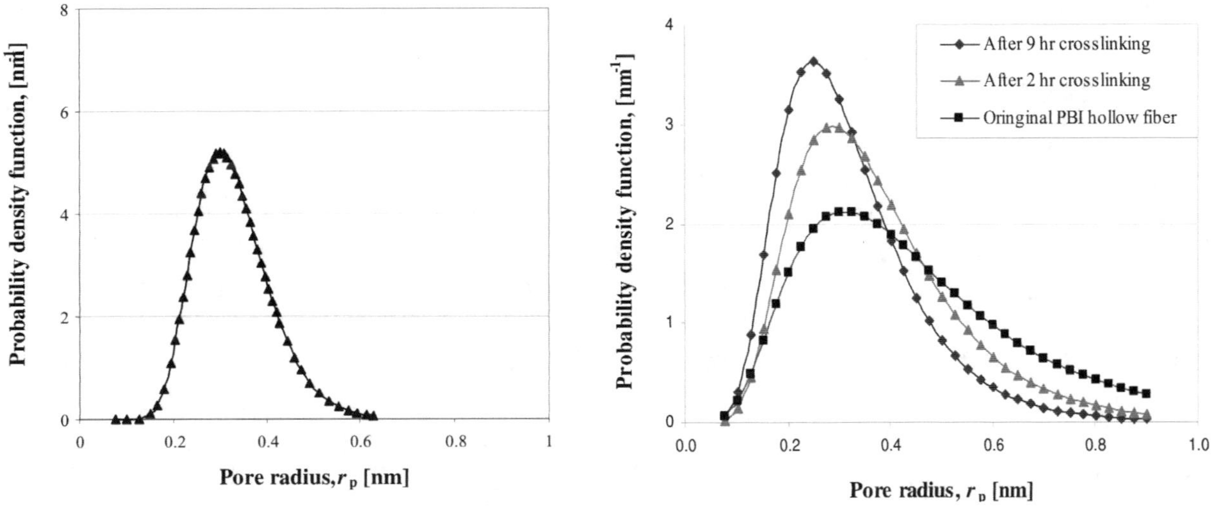

Figure 3. Pore size distribution probability density of the PBI hollow fiber A (left) and B (right)

Forward osmosis based on the PBI NF membrane

Forward osmosis experiments were conducted on a lab-scale circulating filtration unit, as shown in Fig. 4. In addition, membrane orientations have an obvious influence on the water permeation flux resulting from effects of membrane structure and concentration polarization; namely, the pressure retarded osmosis (PRO) mode for when the draw solution flows against the selective layer, and the forward osmosis (FO) mode for when the draw solution flows against the porous support layer. When draw solution flows against the outer selective layer of the hollow fiber (PRO mode), the water permeation is much higher than the other way that draw solution flows against the inner porous support layer of the hollow fiber (FO mode) when two streams are under laminar flow. This is attributed to the lack of internal concentration polarization when DI water flows against the inner porous layer in PRO mode. On the other hand, when draw solution flows against the porous support layer, the dilutive internal concentration polarization due to the porous support structure of asymmetric membrane seriously decreases the net driving force. This leads to the reduced water permeation flux although there is no external concentration polarization because the feed is the DI water in FO mode.

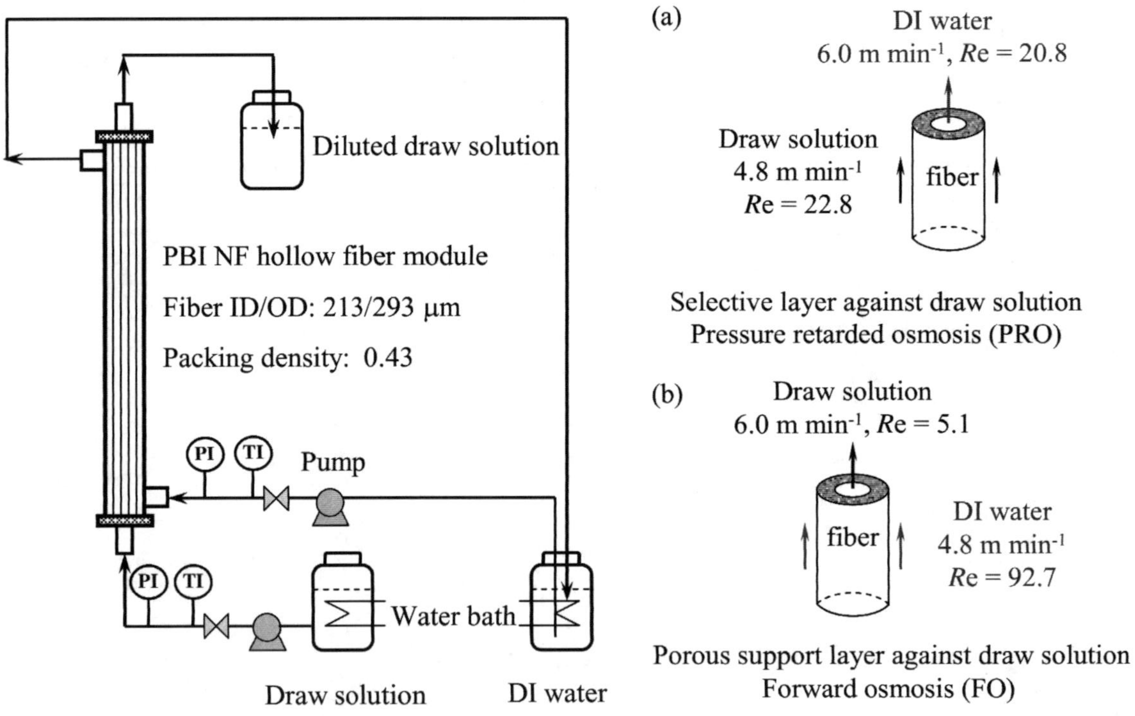

Figure 4. Schematic of forward osmosis set-up and membrane orientations

The dependence of water permeation flux on draw solution concentration for two different membrane orientations inside the membrane module is plotted in Fig. 5 and 6. The water transport through the PBI membrane goes up with an increase in draw solution concentration. The draw solution concentration dominates the membrane performance because the osmotic pressure is the driving force in the forward osmosis process.

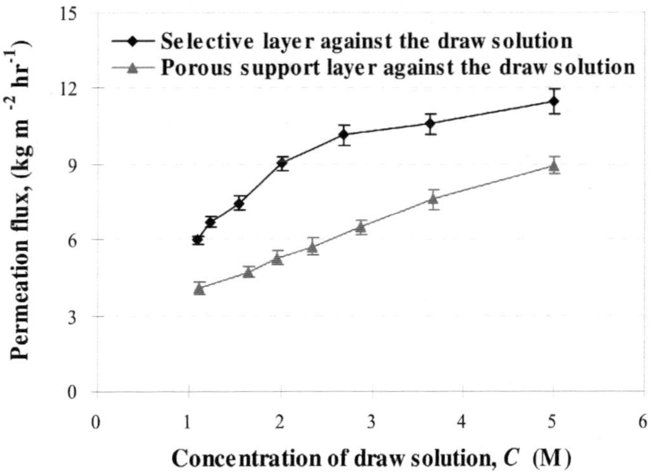

Figure 5 Effect of $MgCl_2$ draw solution concentration on the water flux (fiber A)

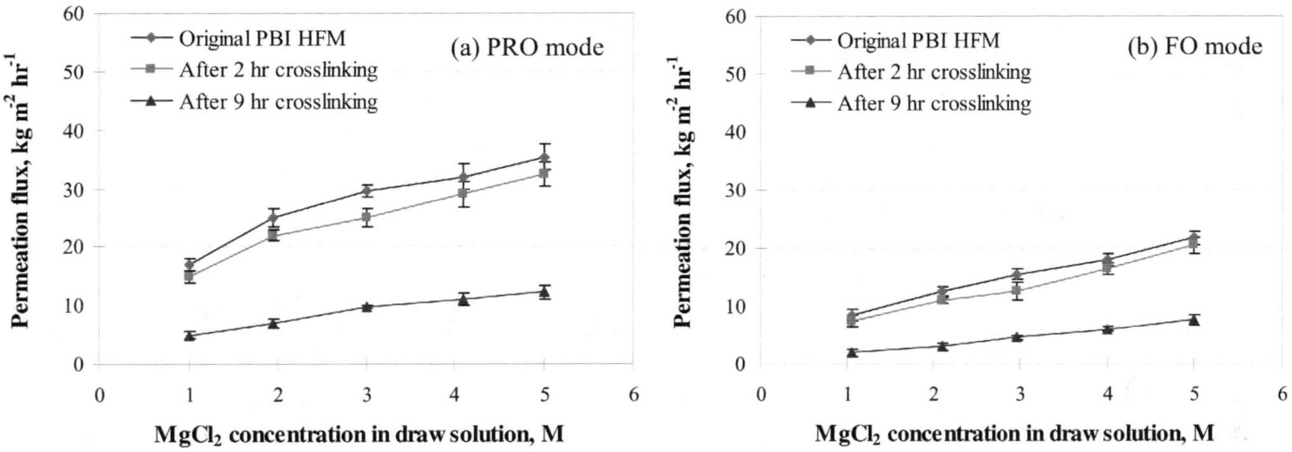

Figure 6 Effect of $MgCl_2$ draw solution concentration on the water flux (fiber B)

Compared to PBI fiber A, the flux of the newly developed PBI NF membranes B is almost double because of the thin wall effect. Flux slightly decreases for the 2-hr chemical modification sample even though its reject rates for ions improve significantly. The 9-hr chemical modification may not only narrow the pore size but also increase substructure resistance; as a result, it has the lowest flux.

Conclusions

This study has demonstrated that through *p*-xylylene dichloride modification the asymmetric PBI nanofiltration hollow fiber membranes can adjust the mean effective pore size to a level which can be used for forward osmosis. The 2-hr modified PBI nanofiltration membrane with high permeation flux and improved salt selectivity may be used for water recovery from wastewater whereas a longer cross-linking time may make the PBI membrane applicable for the seawater desalination. The PBI NF hollow fiber designed with a thin wall has an improved water permeation flux because of reduced transport resistance. Compared with the PBI fiber with thick wall, the water permeation flux is greatly increased. The internal concentration polarization has a serious influence on the water permeation flux when the draw solution flows against the membrane porous support layer. As a result, the pore size, membrane structure, draw solution concentration, membrane orientation and operating conditions will influence the water transport flux and salt back-diffusion through the membrane. Future studies will further optimize the nanofiltration hollow fiber structure to decrease the internal concentration polarization and investigate their performance in the seawater desalination.

References:

1. Cath, T. Y., Childress, A. E., Elimelech, M., (2006). Forward osmosis: principles, and recent developments, Review. Journal of Membrane Science, 281, 70-87.
2. Wright, J. C., Johnson, R. M., Yum, S. I., (2003). DUROS® osmotic pharmaceutical systems for parenteral & site-directed therapy, Drug Delivery Technol. 3, 64.
3. Jiao,B., Cassano, A., Drioli, E., (2004). Recent advances on membrane processes for the concentration of fruit juices: a review, J. Food Eng. 63, 303.
4. Petrotos, K. B., Lazarides,H. N., (2001). Osmotic concentration of liquid foods, J. Food Eng. 49, 201.
5. McCutcheon, J. R., McGinnis, R. L., Elimelech, M., (2005). A novel ammonia–carbon dioxide forward (direct) osmosis desalination process, Desalination 174, 1.
6. Kessler, J.O., Moody, C.D. (1976). Drinking water from sea water by forward osmosis, Desalination 18, 297.
7. Lee, K. L., Baker, R. W., Lonsdale, H. K., (1981). Membranes for power generation by pressure-retarded osmosis, J. Membr. Sci. 8, 141.
8. Seppălă, A., Lampinen, M. J., (1999). Thermodynamic optimizing of pressure retarded osmosis power generation systems, J. Membr. Sci. 161, 115.
9. Loeb, S. (2001). One hundred and thirty benign and renewable megawatts from Great Salt Lake? The possibilities of hydroelectric power by pressure retarded osmosis, Desalination 141, 85.
10. Cath, T. Y., Gormly, S., Beaudry, E. G., Flynn, M. T., Adams, V. D., Childress, A. E. (2005). Membrane contactor processes for wastewater reclamation in space Part 1. Direct osmosis concentration as pretreatment for reverse osmosis. J. Membr. Sci. 257, 85.
11. Cath, T. Y., Adams, V. D. Childress, A. E. (2005). Membrane contactor processes for wastewater reclamation in space II. Combined direct osmosis, osmotic distillation, and membrane distillation for treatment of metabolic wastewater. J. Membr. Sci. 257, 111.
12. Raman, L. P., Cheryan, M., Rajagopalan, N. (1994). Consider nanofiltration for membrane separations, Chem. Eng. Progr. 90(3), 68.
13. Schäfer, A. I., Fane, A. G., Waite, T. D. (2005). Nanofiltration: principles and applications, New York: Elsevier Advanced Technology.
14. Wang, K. Y., Chung, T. S., Qin, J. J. (2007). Polybenzimidazole (PBI) nanofiltration hollow fiber membranes applied in forward osmosis process, J. Membr. Sci. 300, 6.
15. Choe, E. W., Choe,D. D. (1996). In Polymeric Materials Encylopedia. Salamone, J. C., Ed.; CRC Press: New York.
16. Chung, T. S. (1997). Polybenzimidazoles, Handbook of thermoplastics, O. Olabisi ed., Marcel Dekker, New York, 701.
17. Strathmann, H. (2001). Membrane separation processes: Current relevance and future opportunities, AIChE J. 47(5), 1077.
18. Wang, K. Y., Xiao, Y., Chung, T. S. (2006). Chemically modified polybenzimidazole nanofiltration membrane for the separation of electrolytes and cephalexin, Chem. Eng. Sci. 61, 5807.

19. Singh, S., Khulbe, K., Matsuura, T., Ramamurthy, P. (1998). Membrane characterization by solute transport and atomic force microscopy, J. Membr. Sci. 142, 111.

235a The Role of Solid and Liquid Phase Nonidealities In Determining the Solubility and Purity of Isomorphic Compounds

Angel Olivera-Toro and Amyn S. Teja. Georgia Institute of Technology, 311 Ferst Drive NW, Atlanta, GA 30332-0100

The correlation and prediction of crystal solubility and purity in mixtures of hydrocarbons and amino acids is described, with emphasis on systems in which solid solution formation is the dominant mechanism that leads to impure crystals. The crystal phase exhibits significant nonidealities even when the dominant mechanism that leads to impure crystals is lattice substitution by isomorphic or near-isomorphic compounds. Nonidealities in the liquid phase are also significant and are described using predictive UNIFAQ and COSMO-RS models. Surprisingly, solute-solute interactions in the liquid phase are shown to play a significant role in isomorphic systems.

AIChE 2008 Annual Meeting

Abstract

Chiajen Lai*, Bill Bartels, Justin Burt, Jessica Defreese, Prashant Deshpande, Junying Fan, Chenchi Wang

Particle Engineering of a Pharmaceutical Intermediate via Crystal Agglomeration

The filtration of a pharmaceutical intermediate was historically challenging due to the intrinsic needle-like morphology of the crystals. Attempts were made in the past to change the crystal morphology to facilitate filtration. Micronized seeds, either jet or wet-milled, were found to be efficient in growing larger particles with improved filtration rate, and a process based on this approach has been demonstrated in multiple pilot plant scale batches. However, implementation of the process requires a solvent swap from acetonitrile to ethanol, and use of 5% wet-milled seeds to achieve desirable filtration rate. An alternative crystallization was recently developed that eliminates the need for distillation and wet-milled seeds. Crystal agglomeration, enabled by a specific by-product of the reaction and triggered by solvent composition during antisolvent charge, significantly changes the filtration characteristics of the material. This crystallization offers particle engineering opportunities to optimize the process by reducing the number of unit operations and solvent consumption. Additional washing is needed to remove the by-product and other impurities to ensure product quality. The design of this crystallization through particle agglomeration, and the challenges encountered for efficient washing will be discussed in this presentation.

*presenting author

235c Novel Chiral Separation Via Integration of Membrane Separation with Cooling Crystallization: Dl-Glutamic Acid Resolution

Apichit Svang-Ariyaskul, William J. Koros, and Ronald W. Rousseau. School of Chemical & Biomolecular Engineering, Georgia Institute of Technology, Atlanta, GA 30332-0100

Chiral separation of racemic mixtures is essential in the production of many pharmaceutical compounds, and the objective of here is to establish an alternative for existing chiral separation processes. Resolution of DL-glutamic acid is used as model system in testing the integration of membrane separation with cooling crystallization. The proposed process utilizes two crystallization chambers that are separated by a membrane that prevents transport of crystals from one chamber to the other. It is critical that system conditions are controlled so that only one isomer crystallizes in each of the chambers. This is done by appropriate addition of seed crystals to each chamber and by restricting the formation of new crystals to secondary nucleation mechanisms. The seed crystals introduced to each of the chambers may grow or participate in secondary nucleation, but conditions must be controlled so as to prevent primary nucleation, which would result in the formation of both crystal species in each chamber. Experiments with different amount of seed crystals were conducted to determine operating conditions that produce the highest product yield and purity. Results were promising as the product purity was over 94% (separation factor of 16) and the product yield was increased by at least 30% of the cooling crystallization process without the membrane.

235d Simultaneous Separation of Fullerene C_{60} and C_{70} Solids Using a Crystallization-Based Process

Kui S. Kwok, Yik C. Chan, and Ka M. Ng. Department of Chemical Engineering, Hong Kong University of Science and Technology, Clear Water Bay, Kowloon, Hong Kong, China

The unique properties of fullerenes and their derivatives have inspired extensive research on their potential applications. However, the high cost of pure fullerenes hinders further development of such applications into products. While mixed fullerenes can be bought at US$ 4 per gram, C_{60} and C_{70} solids of 99% or higher purity cost several hundreds US dollars. This situation is attributed to the lack of an economical technique to separate the fullerene mixtures. Chromatography, the conventional method for fullerene isolation, suffers from the drawback of low product throughput and high capital cost arising from the requirement of large amounts of stationary and mobile phases. Crystallization, on the other hand, because of its ease to be scaled up, is a cost-effective alternative. Although previous studies had shown the capability of crystallization to separate fullerene C_{60} and C_{70}, these works were only carried out on a trial-and-error basis in bench-scale and no account was given on the principle of separation. It is thus unclear how the process can be scaled up for the mass production of pure fullerenes.

This work presents a crystallization-based process that can simultaneously produce pure fullerene C_{60} and C_{70} solids in a pilot scale using *o*-xylene as the solvent. We began by examining the solid-liquid phase behavior of the ternary system C_{60}-C_{70}-*o*-xylene and measured its solid-liquid equilibrium phase diagram at several temperatures. Based on this scientific information, the flowsheet of a crystallization process was then synthesized to separate a C_{60}-C_{70} fullerene mixture by temperature swing. To verify the practicability of the process, pilot-scale experiments were performed according to the generated flowsheet to separate a commercially available fullerene mixture. C_{60} and C_{70} solids of purity higher than 98% were finally obtained.

235e Saxs and PFG-NMR Investigations of Zeolite Synthesis Mixtures

Daniel F. Shantz, Department of Chemical Engineering, Texas A&M University, TAMU 3122, College Station, TX 77843-3122

Despite intense research efforts a truly molecular description of the role of the organocation in high-silica zeolite nucleation and growth has remained elusive. In this talk I will summarize recent efforts from my laboratory that have applied two very different, yet complimentary, methods to interrogate zeolite synthesis mixtures. In the first part of the talk I will summarize small-angle X-ray scattering (SAXS) investigations that illuminate how the cation structure, presence of cosolvents, and the presence of heteroatoms impact the growth and structure of zeolite precursors/nanoparticles. In the second part of the talk NMR spectroscopy will be used to illustrate the silicate species present in these mixtures and the nature/magnitude of the forces between the organocation and silicate species. The second point is currently being investigated using pulse-field gradient NMR methods. By coupling the NMR with SAXS it is hoped that the growth rates observed via SAXS can be rationalized in the context of the organocation-silicate association strength which can readily be determined using PFG NMR via measurement of organocation self diffusion coefficients.

235f The Formation of Nano Sized Glycine Crystals on Engineered Surfaces

Kitae Kim[1], Insung Lee[2], and Allan S. Myerson[2]. (1) Chemical and Biological Enginnering, Illnois Institute of Technology, 55 west 34th ST,ERB, Chicago, IL 60616, (2) Chemical and Biological Engineering, Illnois Institute of Technology, 55 west 34th ST,ERB, Chicago, IL 60616

The direct formation of organic molecular crystals of below one micron is generally a difficult problem. Methods which employ high supersaturations often produce amorphous solids or oiling out. In addition, high supersaturations make polymorph control difficult often resulting in less stable solid forms. In this study, patterned gold islands of less than one micron are prepared using lithography. Self Assembled Monolayers (SAMs) are used to prepare bifunctional (hydrophilic and hydrophobic) surfaces upon which the droplets of solution are deposited. As shown previously in our work (Lee at al. 2005) the size of the island will influence the final crystal size obtained and the evaporation rate of solvent will determine the crystal form obtained. Supersaturation is controlled by slightly diffusing anti solvent into confined small solution droplets on bifunctional patterned SAMs. This method allows each island to be created single nano-sized crystals. Glycine crystals under 500nm have been made and characterized by Atomic Force Microscopy (AFM).

235g Solvent Topography and Steric Factors: Effects on Crystal Morphology

Charles Acquah, Chemical Engineering, University of Connecticut, Storrs, CT 06269, Arunprakash T. Karunanithi, U.S. Environmental Protection Agency, 26 W. Martin Luther King Drive, Cincinnati, OH 45268, Matthew Cagnetta, Department of Chemical, Materials and Biomolecular Engineering, University of Connecticut, 191 Auditorium Rd, U-3222, Storrs, CT 06269, **Luke E. K. Achenie**, Department of Chemical Engineering, Virginia Polytechnic Institute and State University, Randolph Hall 133, Blacksburg, VA 24061, and Steven Suib, Department of Chemistry, Chemical Engineering and Institute of Material Science, University of Connecticut, Storrs, CT 06269.

Effects of hindrance resulting from solvent topography on the resultant crystal morphology were examined via cooling crystallization of various carboxylic acids in isomeric butyl- and pentyl- alcohols. These isomeric solvents were selected in such a way that the major distinguishing factor was the position of the hydroxyl functional group along the carbon chain. Our experiments show that the magnitude of hindrance is related to the degree of branching at the substituted carbon, with hindrance increasing in the order of 1° < 2° < 3° alcohols. The resulting crystals displayed a trend of low, intermediate, and high aspect ratios corresponding to 1°, 2°, and 3° alcohols, respectively. In particular, 3° alcohols tend to yield significantly different crystal morphologies compared to 1° and 2° alcohols. Hence, the assertion that polar solvents, such as alcohols, always yield low aspect ratio plate-like crystals is incorrect and that the position of hydroxyl functional group plays a major role in enhancing or limiting solute-solvent hydrogen bonding interactions. Infrared spectroscopic analysis revealed that the intermolecular hydrogen-bonding ability of 1° alcohols > 2° alcohols > 3° alcohols, and this helps explain morphological trends observed at the macroscopic level.

238a A Novel Method for Reporting Adsorption Equilibria: Net ADSORPTION

Sasidhar Gumma, Chemical Engineering, Indian Institute of Technology Guwahati, Guwahati, 781039, India and **Orhan Talu**, Cleveland State University, Department of Chemical Engineering, Cleveland, OH 44115.

Usually the experimental data reported in literature for gas adsorption equilibrium is for so-called (Gibbs') excess amount adsorbed. It is obtained by fixing the Gibbs' dividing surface. In practice this is done correcting experimental measurements for buoyancy in case of gravimetric experiments or for void volume in case of volumetric experiments. Both the corrections need a non-adsorbing gas to measure the impenetrable solid volume; helium is a chosen de facto standard for these measurements due to its relatively inert nature compared to any other gas.

In recent years several articles have appeared in literature, which question the validity of non-adsorbing helium assumption. Now, it is established beyond doubt that helium adsorbs at all conditions; at best adsorbed amount may be quite small. In any case, some error and differences result because of such an arbitrary assumption. The magnitude of the error increases when dealing with adsorption of gases whose adsorption is close to that of helium adsorption (for example hydrogen on several adsorbents at room temperature). This error also increases when correction term for experimental measurements becomes significant, usually at very high pressures. These conditions are encountered in most applications of adsorption for energy storage (hydrogen or methane storage).

In this work, we present a new method for reporting adsorption equilibrium data that does not necessitate any arbitrary assumptions or measurement of impenetrable solid volume. We call this as net adsorption. We believe net adsorption is better suited for reporting of adsorption data since correction terms are not needed in its calculation. Moreover, data across different labs at all conditions can be readily compared, since arbitrary assumptions are completely eliminated.

To further validate our claim, we present and compare both excess and net adsorption data for adsorption of methane and argon on a proprietary adsorbent. Amounts adsorbed and heats of immersion are calculated and presented using both excess and net adsorption data. A procedure to calculate mass and energy balances energy balances on an adsorption column is also presented, using this data. It is shown that there will be no difference, when one uses net adsorption data for performing balances on a column.

This work proposes a new method for reporting and using adsorption equilibrium data for designing adsorption processes, without necessity to fix Gibbs' dividing surface through arbitrary assumptions.

238b A Theory of Adsorption/desorption Dynamics for Fluids In Porous Materials

Peter A. Monson, Univ. of Massachusetts, Dept. of Chem. Engineering, 159 Goessmann Lab., Amherst, MA 01003-9303

Recent years have seen dramatic progress in the modeling the thermodynamics of adsorption and desorption for fluids in mesoporous materials based on density functional theories and molecular simulations. The occurrence of adsorption/desorption hysteresis in these systems is an indication of failure by the system to reach thermodynamic equilibrium during condensation and/or evaporation processes and it is important therefore to understand the dynamical behavior that underlies this hysteresis. In this paper we discuss an approach to modeling this dynamics.

We consider mean field kinetic equations describing the relaxation dynamics of a lattice model of a fluid confined in a porous material. The dynamical theory embodied in these equations can be viewed as a dynamical density functional theory, as a theory of diffusion, or as a mean field approximation to a Kawasaki dynamics Monte Carlo simulation of the system. The solutions of the kinetic equations for long times coincide with the solutions of the static mean field (density functional) theory for the inhomogeneous lattice gas. The approach can reveal very useful insights into the nature of dynamical processes in adsorption and desorption.

The approach is applied to a lattice gas model of a fluid confined in a finite length slit pore open at both ends and in contact with the bulk fluid at a temperature where capillary condensation and hysteresis occur. The states emerging dynamically during irreversible changes in the chemical potential are compared with those obtained from the static mean field equations for states associated with a quasistatic progression up and down the adsorption/desorption isotherm. In the capillary transition region the dynamics involves the appearance of undulates (adsorption) and liquid bridges (adsorption and desorption) which are unstable in the static mean field theory in the grand ensemble for the open pore but which are stable in the static mean field theory in the canonical ensemble for an infinite pore.

238c Adsorption on Molecularly Rough Surfaces

Alexander V. Neimark, Department of Chemical and Biochemical Engineering, Rutgers University, 98 Brett Road, Piscataway, NJ 08854-8058

Surface heterogeneity affects significantly the efficiency and selectivity of porous adsorbents and catalysts. However, most of the adsorption theories and simulation techniques assume a standard thermodynamic model of the Gibbs' dividing solid-gas interface, which is molecularly smooth. This assumption gives rise to an artificial layering of the adsorbed phase near the surface that is displayed in oscillating density profiles and leads to step-wise adsorption isotherms and oscillating solvation pressures, which are not observed in experiments. I will present a novel approach to a quantitative description of the effects of surface hererogeneity and microporosity on adsorption in micro- and mesoporous materials based on the quenched solid density functional theory (QSDFT). In QSDFT, the surface roughness is quantified by the roughness parameter, which represents the thickness of the surface "corona" – the region of varying solid density. QSDFT quantitatively describes adsorption/desorption isotherms of simple gases and vapors on various carbons and silica materials in a wide range of relative pressures. The adsorbate density profiles predicted by QSDFT are shown to be in quantitative agreement with the density profiles revealed from the in-situ synchrotron XRD studies of Ar adsorption on MCM-41. Thus, QSDFT reconciles the results of adsorption and XRD measurements. In addition to a better understanding of the fundamentals of adsorption processes on heterogeneous surfaces, an important practical outcome of the proposed approach is envisioned in modeling of selectivity of adsorption separation processes and developing novel methods for calculating pore size distributions, meso- and micropore volumes, surface roughness, and other structural parameters of micro-mesoporous materials employed in separation technologies.

Web Page: sol.rutgers.edu/aneimark.html

238d Deformation of Adsorbents In Gaseous Phase Adsorption – a Computer Simulation Study

Duong D. Do, Division of Chemical Engineering, University of Queensland, School of Engineering, Brisbane, 4072, Australia

GCMC simulations of adsorption of argon in slit-like pores show that adsorbents can swell or shrink with adsorbate loading. It is found that the shrinking or swelling depends on loading, pore size and temperature. The complex functional dependence is illustrated by adsorption in pores of various sizes at a number of temperatures. For pores where the adsorbate layers are not commensurate with pore width; for example when the pore width is 8A, shrinkage is observed under sub-critical conditions; while shrinkage followed by swelling occurs under supercritical conditions. The swelling occurs due to extremely high external pressure. However for commensurate pores that can accommodate an integer number of layers under normal conditions; for example when the pore width is 10A, we observe only moderate swelling at high pressures. For both sub- and supercritical conditions the dependence on loading is quite complex. When these results are compared with the equivalent rigid pores, we find that the difference in adsorption capacity, molar enthalpy of the adsorbed phase and solvation pressure is significant at high pressures under supercritical conditions.

238e Does Water Condense In Hydrophobic Nanopores ?

Alain H. Fuchs, Enscp, Paris Graduate School of Chemical Engineering, Paris, France

We report a molecular simulation investigation of the hydration thermodynamic behavior of purely siliceous, hydrophobic zeolites, in which an increasing amount of hydrophilic surface defects were introduced. An internal surface defect can either be "weak" or "strong", depending on the balance between the defect-water and the water-water interactions. We found that the hydration behavior in presence of weak or strong defects were markedly different. The water hydration process in presence of weak surface defects is basically a homogeneous nucleation process. The effect of the weak hydrophilic defects is to make the framework overall more attractive, and this leads to the progressive shift of the condensation transition, while the framework remains overall hydrophobic. In the case of strong defects, water condensation is characterized by a combination of a strong adsorption of a limited amount of water at very low pressure, followed by a plateau in the intermediate pressure range and by a sudden filling of the nanoporous volume at higher pressure. A progressive change from hydrophobic to hydrophilic behavior is observed, as the defect content is increased. This work provides an insight from adsorption thermodynamics on the issue of hydration of geometrically and chemically heterogeneous pore surfaces. We find that a small increase in hydrophilicity of the porous framework may turn the pore from being dry to being completely filled with water at saturation conditions. We finally discuss the relevance of these results with respect to the issue of hydration of biomolecules and especially the process of high pressure protein denaturation.

238f Modeling Mixture Transport In Nanopores: Departure from Existing Paradigms

Suresh K. Bhatia, Division of Chemical Engineering, The University of Queensland, Brisbane, Australia

A fundamental understanding of the processes affecting fluid behavior in nanoscale confinements is crucial to numerous existing and emerging applications in nanotechnology, materials science, biology, adsorption and membrane transport, as well as a host of other areas. In the last two decades a vast array of new nanoporous materials such as templated periodic MCM-41 silicas, carbon nanotubes as well as various aluminophosphates have been developed, all considered to hold promise for a variety of novel applications. The infiltration of fluid mixtures into the nanopores in these materials is a common feature of most applications being investigated, and this has catalyzed several new developments in the understanding of fluid equilibrium and transport at the nanoscale, most of which have, however, been devoted to pure component systems, with little attention to mixture transport.

For long, models of mixture transport have been based on highly respected approaches such as the dusty gas model, or statistical mechanical treatments such as that of Bearman and Kirkwood. However, they have failed to provide satisfactory solutions, with no definitive treatment even for a classical experiment such as the Stefan tube. We present here a novel theory of mixture transport in nanopores, which represents wall effects via a species-specific friction coefficient determined by its low density diffusion coefficient. This low density diffusion coefficient for each species may be obtained experimentally, or from simulation. Further, the treatment uses a frame of reference based on the individual species velocity, rather than the commonly used mass averaged mixture velocity. Good agreement of the predicted Onsager coefficients with those from molecular dynamics simulations is demonstrated in the mesopore range, while considering inhomogeneity of the density distributions. It is found that the commonly used assumption of a cross-sectionally uniform density in the momentum balance is in serious error, as also the traditional use of a centre of mass based frame of reference. The theory also shows how wall effects can be unambiguously considered through the friction coefficient, avoiding the commonly used arbitrary superposition of viscous and diffusive contributions.

238g Effect of Structural Heterogeneity on Adsorption of Water In Microporous Carbons: Insight from Single-Walled Carbon Nanotubes

Sandeep Agnihotri, Civil and Environmental Engineering, University of Tennessee, 73A Perkins Hall, Knoxville, TN 37996-2010

Single-walled carbon nanotubes (SWNTs) have a structure that exhibits greater order but in essence they are microporous activated carbons. While large-scale applications of carbon nanotubes as adsorbent materials are being debated, the structural simplicity of single-walled carbon nanotubes allows them to be studied with a greater degree of accuracy. Therefore, the knowledge gathered from a rigorous joint experimental and theoretical investigation of adsorption processes in single-walled carbon nanotubes should be transferable to those in microporous activated carbons. We have developed a practical approach for adsorption modeling of heterogeneity of SWNTs. The method integrates experimental analysis with grand canonical Monte Carlo (GCMC) simulation of a small probe molecule, such as nitrogen at 77 K. Using this method it is possible to separately estimate adsorption inside the nanotubes, adsorption on the external surface of the bundles, and adsorptive contributions from the impurities present in samples (Agnihotri et al., J Phys Chem C, 2007). Using this methodology, we recently conducted a study on competitive adsorption of water and organic vapors to elucidate the distinct interactions between the select adsorbates and the nanoporous structure of SWNTs (Agnihotri et al, Langmuir, 2008). Experimental adsorption isotherms on SWNT samples for hexane, methyl ethyl ketone, cyclohexane and toluene individually mixed in carrier gases that were nearly saturated with water vapor were compared with the GCMC simulated isotherms for hexane, as a representative organic, on the external surface of the heterogeneous SWNT bundles. From the nearly perfect overlap between the experimental and simulated isotherms, it is concluded that until near saturation only the internal pore volume of pristine SWNT bundles fills with water. The adsorption of water vapor on the peripheral surface of the bundles remains insignificant, if not negligible, in comparison to the adsorption of water in the internal volume of the bundles. This is in contrast with the adsorption of pure hexane, which exhibits appreciable adsorption both inside the bundles as well as on their external surface. It is also suggested that during competitive adsorption, water molecules take precedence over small non-polar and polar organic molecules for adsorption inside SWNTs and leave unoccupied the hydrophobic external surface of the bundles for other more compatible adsorbates. We notice that the internal volume of nanotubes and the void space between the bundles can be broadly categorized as the microporosity and mesoporosity, respectively, of SWNT samples. Therefore, our study suggests that the structural simplicity of SWNTs may be exploited to develop a broader understanding of water adsorption in microporous carbons.

Figure 1. Competitive adsorption mechanisms for 90% humidity and organic vapor (Agnihotri et al., *Langmuir*, 2008). It suggest water primarily occupying the micropores of SWNTs.

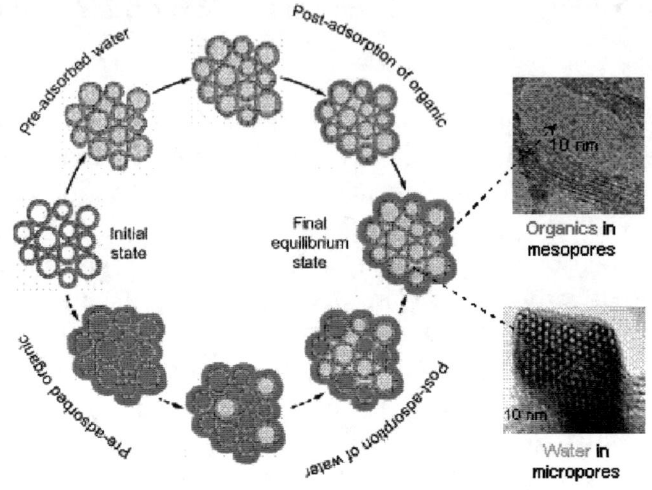

246a Membrane Separation Techniques in the Continuous Fermentation and Separation of Butanol: Possible Applications of a Two-Step Process in a Biorefinery

Jamie A. Hestekin[1], Robert R. Beitle[1], E.C. Clausen[2], D. J. Carrier[3], and Jianjun Du[1]. (1) Ralph E. Martin Department of Chemical Engineering, University of Arkansas, 3202 Bell Engineering Center, Fayetteville, AR 72701, (2) Chemical Engineering, University of Arkansas, 3202 Bell Engineering Center, Fayetteville, AR 72701, (3) Biological and Agricultural Engineering, University of Arkansas, 3202 Bell Engineering Center, Fayetteville, AR 72701

Abstract

Butanol is an excellent substitution of gasoline, which can use in the interior combustion engine without any modification of engine. About 50 years ago, the acetone-butanol-ethanol fermentation process, employing bacterium Clostridium acetobutylicum to convert biomass to butanol, was the most popular way to produce butanol. But this process has some drawbacks in that many bi-products are produced. Besides butanol, the fermentation also produces acetic, butyric and lactic acids when fermentation passes through acidogenesis phase..

In this project, we propose a continuous two stage fermentation/membrane separation system that is based on the work of Ramey (1998) where two organisms, Clostridium acetobutylicum and Clostridium tyrobutyricum, are used in tandem to produce butyric acid (with lactic and acetic) in the first fermentation and butanol (with acetone and ethanol) in the second fermentation. Ramey (1998) proposed liquid-liquid extraction for product removal, but this process has problems with solvent entrainment and selectivity. We are working to demonstrate that selective removal of butyric acid from the first stage by a novel membrane technique that will produce higher yields and productivity than current technology.

During this presentation, we will demonstrate the continuous conversion of sugars to butyric acid using hollow fiber cell recycle and electrodeionization for continuous recovery. We will show how this increases the amount of butyric acid that can be made and will also talk about other applications in the biorefinery as well.

References

Ramey, D.E., "Continuous, Two Stage, Dual Path Anaerobic Fermentation of Butanol and Other Organic Solvents Using Two Different Strains of Bacteria," U.S. Patent 5,753,474, May 19, 1998.

246b Removal and Recovery of Solvents from Fermentation Broth by Pervaporation and Extractive Pervaporation Process Using Liquid Membranes

Atsawin Thongsukmak, John Tang, and Kamalesh K. Sirkar. Chemical Engineering, New Jersey Institute of Technology, Membrane Seperations, 161 Warren st, Newark, NJ 07102

Pervaporation is attractive for separation and recovery of volatile fermentation byproducts present at low concentrations. Although polymeric or ceramic membranes are quite successful in removing most volatile organic compounds (VOCs) from their dilute aqueous solution, their selectivities for removing polar organic compounds are not very high. Previous studies indicated that liquid membranes can achieve much higher selectivity for small molecules such as acetic acid, butanol, etc. Here, we have focused on liquid membranes immobilized in novel porous hollow fiber-based membranes as a support for the pervaporation process. Volatile fermentation byproducts such as acetone, butanol and ethanol present in low concentrations have been separated by novel liquid membranes of tri-n-octylamine (TOA). We have tested the liquid membrane with model solutions (containing acetone-butanol-ethanol (ABE)) as well as a filtered solution from a fermentation broth (ABE fermentation). Very high selectivities achieved for butanol, acetone, and ethanol were 275, 220 and 80 respectively with corresponding mass fluxes of 11.0, 5.0, and 1.2 g/m2-hr at a temperature of 54 °C for a feed solution containing 1.5 wt % butanol, 0.8 wt % acetone, and 0.5 wt % ethanol. High stability of liquid membrane was also observed with virtually no contamination of the broth. This was achieved by having a nanoporous fluorosilicone coating on a porous polypropylene (PP) hollow fiber substrate. The liquid membrane was immobilized in the PP substrate. Thinner liquid membranes have also been developed in different thicknesses. When the thickness of the liquid membrane was reduced to 1/5th of the full PP substrate thickness, the membrane yielded mass fluxes of solvents 5 times higher than those through full thickness liquid membrane; for example, butanol flux was increased to 53 g/m2-hr. Thinner liquid membrane did not lead to a change in the selectivities at various temperatures. A novel extractive pervaporation technique has also been studied to enhance the performance of ethanol separation from ethanol solutions containing 5-10 % ethanol. Ethanol selectivity as high as 100 and a flux of 15.0 g/m2-hr were obtained at 54 °C in this extractive pervaporation process.

246c Adsorption In Zeolites for Detoxification of Lignocellulosic Biomass

Rajiv Ranjan[1], Maybelle Woo[2], Stefan Thust[3], Marc von Keitz[4], Kenneth Valentas[4], and Michael Tsapatsis[1]. (1) Department of Chemical Engg. & Material Science, Univ. of Minnesota, Minneapolis (USA), 421 Washington Ave SE, Minneapolis, MN 55455, (2) (With Millipore at present) Department of Chemical Engg. & Material Science, Univ. of Minnesota, Minneapolis (USA), Minneapolis, MN 55455, (3) Biotechnology Institute, University of Minnesota, 140 Gortner Lab, 1479 Gortner Avenue, St. Paul, MN 55108, (4) BioTechnology Institute, University of Minnesota, 1479 Gortner Avenue, Suite 240, St. Paul, MN 55108

Ethanol produced from lignocellulosic biomass is an alternate energy source to conventional fossil fuel. However, toxic inhibitors produced from hydrolysis of biomass decrease ethanol yield during fermentation. Several methods have been used to pretreat these inhibitors and enhance ethanol productivity. In this work, the use of zeolites for the pretreatment of hydrolyzate in order to remove inhibitors like 5-Hydroxymethylfurfuraldehyde (HMF) and furfural by preferential adsorption from aqueous solution is demonstrated. Ideal Adsorbed Solution Theory (IAST) is used to calculate adsorption isotherm for mixtures using single component adsorption data. The predictions are compared with the experimental adsorption isotherms for aqueous mixtures of HMF, furfural and xylose. Finally, desorption of the industrially valuable HMF from the zeolites after the adsorption process is demonstrated.

246d Energy Efficient Hybrid Vapor Stripping-Vapor Permeation Process for Ethanol Recovery and Dehydration

Leland Vane[1], Franklin Alvarez[1], Yu (Ivy) Huang[2], and Richard Baker[2]. (1) U.S. EPA, 26 W. Martin Luther King Drive, M/S443, Cincinnati, OH 45268, (2) Membrane Technology & Research, Inc., 1360 Willow Road, Suite 103, Menlo Park, CA 94025

Distillation combined with molecular sieve dehydration is the current state of the art for fuel grade ethanol production from fermentation broths. To improve the sustainability of bioethanol production, energy efficient separation alternatives are needed, particularly for lower fermented ethanol concentrations and smaller scale systems. A process, termed Membrane Assisted Vapor Stripping (MAVS), integrating steam stripping with a vapor compression step and a vapor permeation membrane separation step is proposed. The stripping column provides high ethanol recoveries and low effluent concentrations; the vapor compression-vapor permeation membrane component enables the efficient recovery of latent and sensible heat from both the ethanol-enriched retentate and water-enriched permeate streams from the membrane system. Based on computer simulations, recovery of 1 kg of fuel grade ethanol (0.5 wt% water) from a 1 wt% ethanol feed stream can be achieved using 8.9 MJ of fuel-equivalent energy. For a 5 wt% ethanol feed stream, only 2.5 MJ are needed. This represents an energy savings of at least 43% relative to standard distillation producing azeotropic ethanol (7 wt% water). To produce a 99.5 wt% ethanol product from a broth containing 11.5 wt% ethanol, a conventional distillation/mol sieve system is expected to use 5.5 MJ-fuel equiv/kg-ethanol; the MAVS process would require as little as 2.2 MJ-fuel equiv./kg-ethanol. Overall costs are also estimated to be lower for a MAVS system. The energy savings of the MAVS approach were confirmed through experiments. The experimental system consisted of a vapor stripping column with 6 stages of vapor-liquid equilibria, a vapor compressor, and a water-selective vapor permeation membrane module. The membrane was prepared with a solvent resistant perfluoro coating. Energy savings of over 50% were measured for recovery of ethanol from feed streams containing 1.5 and 5.5 wt% ethanol. Experiments with an actual fermentation broth will also be reported.

*This is an abstract of a proposed presentation and does not necessarily reflect USEPA policy.

246e Composite Membranes with Ultra-Thin Polymeric Interpenetration Network and Enhanced Separation Performance Approaching Ceramic Membranes for Biofuel

Lan Ying Jiang, Nanoscience and Nanotechnology Initiative, NUS, BlK E3, 05-29, 2 Engineering Drive 3, Singapore, 117576, Singapore, Hongmin Chen, Department of Chemistry, University of Missouri-Kansas City, Kansas City, MO 64110, Y.C. Jean, Chemistry Dept., University of Missouri - Kansas City, 5009 Rockville Rd., Kansas City, MO 64110, and **Tai-Shung Chung**, Department of Chemical and Biomolecular Engineering, National University of Singapore, 10 Kent Ridge Crescent,, Singapore, 119260, Singapore.

In this study, we report the discovery of novel molecular engineering and membrane fabrication that can synergistically produce polymeric membranes exhibiting separation performance approaching ceramic membranes in biofuel dehydration. Biofuel has emerged as one of the most strategically important sustainable fuel sources. The success of biofuel development is not only dependent on the advances in genetic transformation of biomass into biofuel, but also on the breakthroughs in separation of biofuel from biomass. The "separation" alone currently accounts for 60 to 80% of the biofuel production cost. There mainly exist two kinds of materials applied for biofuel separation by pervaporation, namely, ceramic membranes and polymer membranes. Ceramic membranes made of sophisticated processes have shown separation performance far superior to polymeric membranes. Nevertheless, ceramic membranes seriously suffers fragility and high fabrication cost, and polymeric membrane with excellent flexibility is still attractive. For the polymeric membrane, extensive studies exist on how to "finetune" the membrane's pore structure including it's cross-sectioin morphology by the selection of polymer solvents and non-solvents, additives, residence times and other parameters during non-solvent induced phase separation. The key for the performance is the very thin "skin" layer which enables a high permeability. The newly discovered membranes in current work are fabricated by dual-layer co-extrusion technology in just one step through phase inversion. The best performance obtained was a total flux of ~3.9 kg/m2-hr with a separation factor of ~800 in tert-butanol dehydration. The combined molecular engineering and membrane fabrication approach may revolutionize future membrane research and development for purification and separation in energy, environment, and pharmaceuticals.

260a Assessment of Shear Sensitivity of Breakage and Attrition In Crystal Suspensions

Jan Sefcik[1], Katarzyna Sypek[1], and Miroslav Soos[2]. (1) Chemical and Process Engineering, University of Strathclyde, James Weir Building, 75 Montrose Street, Glasgow, United Kingdom, (2) Institute for Chemical and Bioengineering, Department of Chemistry and Applied Biosciences, ETH Zurich, Wolfgang-Pauli-Str. 10, Zurich, 8093, Switzerland

Flow conditions during crystallization and post-crystallization processing are well known to influence agglomeration and breakage/attrition processes in crystal suspensions. The objective of this work has been to determine a range of flow conditions which can be used for a given crystal suspension in order to avoid significant crystal breakage or attrition. This has been achieved by exposing small volumes of the crystal suspension to a well controlled flow through a small diameter contraction, where the flow rate as well as the contraction geometry can be varied in order to cover a desired range of flow conditions ranging from laminar to turbulent and covering a range of shear rates relevant for industrial operating conditions. Flow fields and corresponding shear rates in contractions were computed with a Computational Fluid Dynamics package Fluent. Experimental diagnostics of crystal breakage has been based on standard in situ analytical technology using Focused Beam Reflectance measurements with Lasentec. We have used model crystal systems such as platelet-like aspirin crystals recrystallized from ethanol. Dry powders were dispersed in saturated solvent solutions and the resulting slurries were exposed to a certain number of passes through specific contraction geometries at a range of flow rates. Effects of flow conditions on breakage of crystals were determined for various initial crystal size distributions, solvents and solid loadings. This allowed us to estimate upper critical shear rate values, below which no significant breakage of crystals was observed in given crystal slurries.

260c Estimating Supersaturation and Nucleation and Growth Kinetics from Chord-Length Measurements

Stephanie Barthe, Martha Grover Gallivan, and **Ronald W. Rousseau**. School of Chemical & Biomolecular Engineering, Georgia Institute of Technology, 311 Ferst Drive, Atlanta, GA 30332-0100

Batch crystallization is widely used in the production of high-value added species. Product properties and downstream processes such as filtration, washing and drying are influenced by crystal morphology and size distribution. Real-time and in-line observation of the evolution of population density, coupled to an ability to compute the kinetics of the system, would constitute major assets in the analysis and control of crystallizer operations and in understanding the phenomena that influence product quality.

Focused-beam reflectance measurement (FBRM) enables monitoring a representation of the crystal population in either batch or continuous crystallization systems. Properly installed, the FBRM records the chord-length density, which is a function of crystal geometry and is related to the population density. A model based on the geometry of the crystal was established to relate both distributions and thus enable computation of the population density from a measured chord-length density. These capabilities have been extended in the present study to include using the FBRM data to determine the prevailing supersaturation and crystallization kinetics in a system of interest. Such capabilities should lead to improved control procedures that, in turn, ensure that downstream processing and product quality meet expectations.

Batch cooling crystallization of paracetamol from ethanol solutions was used to test and illustrate the proposed methodology. Computation of the evolution of the population density and the corresponding solution concentration lead to an estimation of supersaturation, which in turn is used to determine the nucleation and growth kinetics in the system.

260d Crystallization of Refined Sodium Bicarbonate: Influence of Supersaturation, Crystallization Temperature and Sodium Carbonate Concentration

Vanessa Gutierrez[1], Christophe Wylock[2], Benoit Haut[2], Perrine Davoine[3], Thierry Cartage[3], and Marie-Paule Delplancke-Ogletree[1]. (1) Chemicals and Materials, Université Libre de Bruxelles, Av. Franklin D. Roosevelt , 50, cp 165/63, Brussels, 1050, Belgium, (2) Chemical Engineering, Université Libre de Bruxelles, Av. Franklin D. Roosevelt , 50, cp 165/63, Brussels, 1050, Belgium, (3) Direction process engineering, Solvay SA, rue de Ransbeek ,310, Brussels, 1120

Industrial crystallization processes involve a large number of complex physical and chemical phenomena. To control the quality of the product, it is essential to understand the role of each parameter in the nucleation and growth processes. In this study, we are specifically interested in the industrial crystallization of sodium bicarbonate.

Refine sodium bicarbonate ($NaHCO_3$) is produced industrially in bubble columns according to the reaction: $Na_2CO_3 + CO_2 + H_2O \leftrightarrow 2\ NaHCO_3$

The reaction involves three phases and thus transfers between gas and liquid, and liquid and solid. The supersaturation is created by the transfer of CO_2 in the liquid. The presence of crystals in suspension in the liquid phase modifies the bubble behavior and therefore the gas-liquid transfer. The crystal size distribution, the purity and the morphology of the product depend on a wide variety of phenomena and conditions. To elucidate the key parameters, we are developing an in-depth study of the mechanisms controlling the nucleation and growth (liquid-solid transfer) in conditions close to industrial ones.

To be able to study separately the influence of supersaturation, crystallization temperature, sodium carbonate concentration, magma density, agitation and impurities on nucleation and growth, we carried out experiments in a simplified set-up, a 4L - MSMPR (mixed solution mixed product removal) reactor. This type of reactor was chosen for two reasons: first the lowest part of an industrial bubble column can be considered as a MSMPR reactor, second models are available for extracting growth rate and nucleation rate for such a reactor.

The influence of supersaturation level, crystallization temperature, and sodium carbonate concentration on the nucleation and growth kinetics will be presented here.

In all experiments, the supersaturation is obtained by cooling a solution saturated at higher temperature, the residence time is kept constant at 26 min and the product is collected when a steady-state is reached. The sodium carbonate is introduced in the $NaHCO_3$ solution at a continuous rate just before it is fed in the reactor.

The composition of the solution is followed during the crystallization process by titration and densimetry (method developed by Zhu). ICP analyses give us the purity of the solution and crystals. The crystal size distribution (CSD) is obtained by sieving, and the crystal morphology is characterized by scanning electron microscopy (SEM) and optical microscopy.

A new method of data analysis using a fitting square method was developed specially to decrease the error on the data analysis. The growth rate seems to be affected mainly by the crystallization temperature and is barely modified by the two other parameters in the range we investigated in this study. At the contrary, the nucleation rate is strongly affected by the three parameters. Furthermore, an ANOVA analysis of the nucleation rate results shows that parameters are coupled.

260e Effect of Cooling Rate on Oleic and Linoleic Acid Crystal Compositions

Sheena Reeves and Priscilla J. Hill. Dave C. Swalm School of Chemical Engineering, Mississippi State University, Box 9595, Mississippi State, MS 39762

Biodiesel and biodiesel blends are known to cause fuel filter clogging in vehicles at low temperatures. This is often due to waxy fatty acids crystallizing in the fuel at low winter temperatures. The objective of this work is to determine the feasibility of separating a fatty acid from a given stream using low temperature crystallization. Separation by crystallization can reduce the amount of chemicals added to biodiesel that can cause the alternative fuel to lose its environmentally friendly status. Removing the unwanted fatty acids can also lower the fuel viscosity and enhance the cold flow operability of biodiesel.

Crystallization experiments are performed with oleic acid and linoleic acid in acetonitrile. A range of various operating conditions is used to determine the percentages of waxy crystals that can be removed. Since these materials can also be sold as byproducts, the purities of the crystals are also determined. Results from this investigation are presented.

260f Novel Integrated Concepts for Preferential Crystallization

Martin Peter Elsner, Katerina Petrusevska, Grzegorz Ziomek, and Andreas Seidel-Morgenstern. Physical and Chemical Foundations of Process Engineering, Max Planck Institute for Dynamics of Complex Technical Systems, Sandtorstr. 1, Magdeburg, 39106, Germany

Driven by the policy of regulatory authorities such as US Food and Drug Administration and the EU Committee for Proprietary Medical Products the production of pure enantiomers is considerably increasing in several industrial branches, e.g. pharmaceutical, agrochemical, food industries as well as in cosmetic and fragrance industries [1]. An attractive process for gaining pure enantiomers from racemic mixtures is the so-called preferential crystallization [2-4]. The principle of this batch process is quite simple: the tank is filled with a supersaturated solution of the racemate (E_p+E_c as 50%:50% mixture). After addition of homochiral seeds (e.g. E_p) merely E_p is crystallizing within a limited time period. In order to gain this enantiomer as a product of high purity, the process must be stopped before the undesired counter-enantiomer occurs [3]. During this batch crystallization, the concentration of the desired enantiomer in the solution is decreasing, whereas the concentration of the counter-enantiomer remains constant. This phenomenon leads to an arrangement which might provide a better performance where two crystallizers are coupled via the liquid phase, i.e. the crystal free mother-liquor is exchanged between these two vessels. Because of this exchange, the liquid phase shows a higher overall concentration of the preferred enantiomer in that vessel in which the preferred enantiomer was seeded. The supersaturation level which corresponds to the crystallization driving force is higher during the whole process in comparison to the case without an exchange (decoupled simple batch mode). Additionally, the concentration of the counter-enantiomer in the liquid phase for each of the vessels decreases. For the borderline case of infinite exchange flow rate racemic composition is reached in the fluid phase of both vessels. The described effect of decreasing the counter-enantiomer concentration in that crystallizer in which the preferred enantiomer shall be gained makes the probability for primary nucleation lower. This corresponds to higher product purity at the end of the process and enhances also the productivity. This more attractive and effective operation mode using two batch crystallizers coupled via their liquid phases has been investigated theoretically and experimentally for the amino acid threonine in water [5-7]. The influence of specific process parameters, like e.g. the size distribution and the mass of the seeds, and different temperature profiles has been analyzed.

The effect of racemization by exchanging the fluid phase allows the specific manipulation of concentration profiles and seems to be a suitable lever for process intensification on the apparatus level. Similar manipulation of the concentration profiles during the crystallization process can be also realized on molecular level if the racemization is achieved by an enzymatic reaction in which an excess of the counter-enantiomer in the liquid phase is transformed to the preferred one [8, 9]. By coupling crystallization (for conglomerate sys-tems) and racemization (for conversion of the unwanted enantiomer) expected theoretical yield of a pure enantiomer can lead up to 100%. Our goal is to develop a comprehensive study for each operating unit and their subsequent integration in a hybrid process. As a model component for the investigation, the amino acid asparagine in water has been chosen.

The first configuration will be usually applied if there is a need for both enantiomers in pure form (with yields up to 50% for each enantiomer), whereas the second configuration provides just one enantiomer with very high yield (up to 100%). These different configurations for productivity enhancement with regard to preferential crystallization will be presented in this contribution.

[1] CANER, H.; GRONER, E.; LEVY, L.; AGRANAT, I. (2004): *Trends in the development of chiral drugs*. Drug Discovery Today. 9 (3), 105-110

[2] JACQUES, J.; COLLET, A.; WILEN, S.H. (1994): *Enantiomers, racemates and resolutions*. Krieger, Malabar

[3] ELSNER, M.P., FERNÁNDEZ MENÉNDEZ, D., ALONSO MUSLERA, E., SEIDEL-MORGENSTERN, A. (2005): *Experimental study and simplified mathematical description of preferential crystallization*. Chirality 17 (S1), S183-S195

[4] LORENZ, H., PERLBERG, A., SAPOUNDJIEV, D., ELSNER, M.P., SEIDEL-MORGENSTERN, A., (2006): *Crystallization of enantiomers*. Chem. Eng. and Proc. 45(10), 863-873

[5] ELSNER, M.P.; ZIOMEK, G.; SEIDEL-MORGENSTERN, A. (2007): *Simultaneous preferential crystallization in a coupled, batch operation mode. Part I: Theoretical analysis and optimization*. Chem. Eng. Sci. 62 (17), 4760-4769

[6] ELSNER, M.P.; ZIOMEK, G.; SEIDEL-MORGENSTERN, A. (2008): *Efficient separation of enantiomers by preferential crystallization in two coupled vessels*. AIChE Journal (submitted)

[7] ZIOMEK, G.; ELSNER, M.P.; SEIDEL-MORGENSTERN, A. (2008): *Simultaneous preferential crystallization in a coupled, batch operation mode – Part II: Experimental investigations*. Chem. Eng. Sci. (in preparation)

[8] LÜTZ, S.; WANDREY, C.; SEIDEL-MORGENSTERN, A.; ELSNER, M.P. (2006): *Verfahren zur Herstellung chiraler Substanzen durch selektive Kristallisation unterstützt durch eine enzymatische Racemisierungsreaktion*. DE 10 2006 013 725.6 (24.03.2006)

[9] WÜRGES, K.; PETRUSEVSKA, K.; SERCI, S.; WILHELM, S.; WANDREY, C.; SEIDEL-MORGENSTERN, A.; ELSNER, M.P.; LÜTZ, S. (2008): *Enzyme-assisted physicochemical enantioseparation processes – part I: Production and characterization of a recombinant amino acid racemase*. Journal of Molecular Catalysis B: Enzymatic (submitted)

260g Rethinking Industrial Crystallization Processes

Daniel A. Green, DuPont Engineering Research & Technology, DuPont Experimental Station, Building E304-A222, Wilmington, DE 19880-0304 and Changzai Chi, Enginenering Research and Technology, DuPont Engineering Research & Technology, DuPont Experimental Station, Building E304-A222, Wilmington, DE 19880-0304.

Most industrial crystallization process technology is well established, with many years in operation. Consequently, there have been relatively few major improvements recently. Most industrial crystallization is done by creating supersaturation in stirred tanks with simultaneous crystal nucleation and growth. Since both nucleation and growth are strong functions of supersaturation, they can not in general be controlled individually in traditional crystallizers. In addition, new process technology is needed that can commercialize recent advances in the application of molecular and supramolecular control of crystallization, such as template directed crystal nucleation. We will review alternate crystallization process strategies that enable greater control of crystal properties. In particular, we will present research leading to a system that separates crystal nucleation and growth, and that can create templated nuclei.

282a Compressor-Free Heat Integrated Distillation Column System for Dehydration of Fermented Mash In Production of Fuel Bio-Ethanol

Kunio Kataoka[1], Hiroshi Yamaji, Hideo Noda[2], Tadahiro Mukaida[3], Mampei Kaneda[3], and Masaru Nakaiwa[4]. (1) R&D Research Center, Kansai Chemical Engineering Co., Ltd, 2-9-7, Minaminanamatsu-cho, Amagasaki, 660-0053, Japan, (2) Administration, Kansai Chemical Engineering Co., Ltd, 2-9-7, Minaminanamatsu-cho, Amagasaki, 660-0053, Japan, (3) Engineering, Kansai Chemical Engineering Co., Ltd, 2-9-7, Minaminanamatsu-cho, Amagasaki, 660-0053, Japan, (4) National Institute of Advanced Industrial Science and Technology, Central 5, 1-1-1, Higashi, Tsukuba, 305-8565, Japan

This paper reports a simulation analysis of energy consumption about a newly-invented heat integrated distillation column (named "Compressor-free HIDiC") system for production of fuel ethanol from the fermented mash. The cost performance for lowering the fuel ethanol price depends upon the energy-saving technology of dehydration process. The original HIDiC system proposed by Mah et al. has a compressor for raising the boiling point of the rectifying section. This system cannot be applied for the process owing to the fouling problems coming from fermentation residues and dissolved inorganic salts. The new system consists of an ordinary evaporator-like column (1st column, i.e. mash or beer column) and a double-tube heat integrated distillation column (2nd column) without a compressor. The second column has a normal-pressure rectifying section inside the inner tube and a low-pressure stripping section in the annular space outside the inner tube. Both sections are installed with original dual-flow trays named "Lift tray". The overhead vapor issuing from the top of the mash column operated at normal pressure is directly supplied without a compressor into the bottom of the rectifying section operated at a normal or a little bit lower pressure. Another distinctive feature is that the stripping section has an overhead condenser, as different from the original HIDiC system. The condensate from the overhead condenser is partially refluxed to its own top of the stripping section and partially fed back into the top of the mash column as a kind of external reflux.

The target of dehydration from 6 wt% ethanol solution (fermented mash) is that the overhead product of the rectifying section should have an approximately 88 wt% of ethanol suppressed below the azeotropic composition (95 wt%).

The following special problems arise owing to the environmental circumstances of sugar cane island:

(1) The waste water as the final bottom product should not contain above 100 ppm of ethanol,

(2) The residual molasses after sugar production should be employed as the feed for ethanol fermentation. Therefore not only bioprocess residues but also various inorganic salts are contained in the fermented mash even after filtration. Especially much calcium sulfate deposits on the trays of distillation column if the tray temperature is raised and the perforated trays below the feed stage are very often stopped up.

The above problems are based on actual experience in a governmental project carried out in a coral island producing sugar cane.

The mash column should have a very simple inside structure with a very small number of trays for the purpose of maintenance. Therefore the bottom product from the mash column reboiler should be supplied into the top of the stripping section of the second column to lower the bottom product concentration of ethanol below 100 ppm.

The new HIDiC system proposed can solve both problems by supplying the bottom product of the mash column into the top of the stripping section of the second column after filtration and by depressurizing the stripping section to lower the tray temperature in order for calcium sulfate not to deposit.

The simulation analysis has been made by varying the operation pressure of the second column stripping section and the issuing rate of the overhead vapor from the mash column as the control parameters.

It has been indicated that if the HIDiC stripping section is depressurized to 200 mmHg, a very high energy-saving rate above 50% in comparison with an ordinary bio-ethanol dehydration plant can be attained with the new HIDiC system under the same separation conditions.

282b Fifty Years of Fri Contributions to Separation Science

Mike Resetarits and D.W. King. Fractionation Research, Inc., P.O. Box 2108, Stillwater, OK 74076

Abstract: Professor Ohe has served as a consultant with FRI and facilitates FRI communication with the Japanese members of FRI. FRI is a not for profit organization supported by worldwide members. This paper will highlight the influence of people such as Henry Kister that passed through FRI and remain active in FRI planning and testing. A description of FRI is given documenting the testing capabilities of the FRI test columns including Gamma scanning; the move from California to Oklahoma; and the upgrade of the unit to test high capacity devices. A history of the testing emphasis highlights the testing of random and structured packings, trays, distributors and proprietary devices. The release of FRI data to the general public is also described. The paper concludes with comments from Professor Ohe and others as to the value of FRI to Separations Science.

Web Page: fri.org

282c Evaluation of Economical and Environmental Performance for An Internally Heat-Integrated Distillation Column (HIDiC)

Keigo Matsuda, Department of Chemistry and Chemical Engineering, Graduate School of Science and Engineering, Yamagata University, 4-3-16, Jonan, Yonezawa, Yamagata, 992-8510, Japan, Kinpei Horiuchi, Japan Society for Safety Engineering, List Kannai Bldg. 3F, 4-47, Onoecho, Nakaku, Yokohama, Kanagawa, 231-0015, Japan, Koichi Iwakabe, Research Institute for Innovation in Sustainable Chemistry, National Institute of Advanced Industrial Science and Technology (AIST), Energy-Efficient Chemical Systems Group, AIST Tsukuba Central 5, 1-1-1, Higashi, Tsukuba-shi, Ibaraki, 305-8565, Japan, Toshinari Nakanishi, R&D Department, Kimura Chemical Plants Co. Ltd., 2-1-2, Kuiseterajima, Amagasaki,, Hyogo, 660-8567, Japan, and Masaru Nakaiwa, Research Institute for Innovation in Sustainable Chemistry, National Institute of Advanced Industrial Science and Technology, AIST Tsukuba Central 5, 1-1-1, Higashi, Tsukuba-shi, Ibaraki, 305-8565, Japan.

Optimum operation, energy saving, carbon-dioxide emission and cost estimation of a pilot plant of an internally heat-integrated distillation column (HIDiC) separating a twelve-component hydrocarbon mixture to obtain residue-product of crude cyclopentane were studied. From the operation results of the HIDiC at zero external reflux and reboiler free operation, the reduction in CO2 emission by the HIDiC was 56.6 kg/h (453 t/y), which is 63.5% of the total CO2 emission of the conventional distillation column. The HIDiC pilot plant is prospected to be reasonable according to the inflation of the steam cost. The cost estimation results were also compared with those obtained from the conventional column, which showed that the HIDiC pilot plant can indicate a strikingly shorter payback time.

282d Hydrodynamics Modeling of Modern Catalytic Structured Packings

Elisabetta Brunazzi, Dipartimento di Ingegneria Chimica, Chimica Industriale e Scienza dei Materiali, Università di Pisa, Via Diotisalvi 2, Pisa, Italy and **Ricardo Macías-Salinas**, ESIQIE-Departamento de Ingeniería Química, Instituto Politécnico Nacional, UPALM, Av. IPN s/n, Col. Lindavista, México, D. F., 07738, Mexico.

Few modeling efforts have been so far presented in the open literature in an attempt to interpret and reproduce the hydrodynamics behavior of a relatively new family of catalytic structured packings: KATAPAK-SP. The purpose of this work is to present a new hydrodynamic model based on the channel approach for the accurate correlation and/or prediction of pressure drop and liquid holdup under single- and two-phase flow conditions. In this context, a more suitable liquid holdup correlation for the open channels was proposed. Moreover, a new approach has been devised here to reasonably determine the liquid share between the open channels and the catalyst bags.

282e Solvent Recovery by a New Batch or Semi-Batch Rectifying Still Unit

Hideo Noda[1], Kunio Kataoka[2], Hiroshi Yamaji[3], Mampei Kaneda[4], and Tadahiro Mukaida[4]. (1) Administration, Kansai Chemical Engineering Co., Ltd, 2-9-7, Minaminanamatsu-cho, Amagasaki, 660-0053, Japan, (2) R&D Research Center, Kansai Chemical Engineering Co., Ltd, 2-9-7, Minaminanamatsu-cho, Amagasaki, 660-0053, Japan, (3) Kansai Chemical Engineering Co., Ltd, 2-9-7, Minaminanamatsu-cho, Amagasaki, 660-0053, Japan, (4) Engineering, Kansai Chemical Engineering Co., Ltd, 2-9-7, Minaminanamatsu-cho, Amagasaki, 660-0053, Japan

A batch or semi-batch distillation process for rectification of solvent waste mixtures has effectively been intensified by installing a new rotating stirrer into the jacketed tank evaporator. The stirrer named "Wall Wetter" consists of two inclined half-cut pipes (gutters) fixed to the central rotating shaft. The Wall Wetter stirrer can pump up a large volume of liquid through the gutters due to its centrifugal effect from the liquid pool and sprayed over the upper jacket surface above the liquid pool level. The upper heat transfer area is covered with a falling liquid film, so that the entire jacket surface can be utilized effectively over the whole term of batch distillation. This jacketed tank evaporator equipped with the Wall Wetter stirrer (Wall Wetter pot) can effectively reduce the distillation process time providing a very high rate of vaporization utilizing the entire area for heat transfer during the entire distillation term.

Original dual flow trays named "Lift Tray" are installed without a downcomer in the rectifier. This tray is a set of two perforated plates: the lower plate, i.e. a fixed perforated plate is covered from upside with a floating perforated plate which can move up and down depending on the vapor flow rate. Therefore this tray has a great advantage of keeping the pressure drop almost constant over a wide range of F-factor by automatically changing the perforation per cent (max.=36.4%, min.=9.1%) with the height of the floating plate above the fixed plate.

The solvent recovery test was conducted combining a trayed rectifier with the Wall Wetter pot as the new patented rectification still unit. The entire column operates as an enriching section. The experimental rectifier (inside diameter 200 mm) has seven Lift Trays with no downcomers and the stage-to-stage spacing is 400 mm. The solvent mixture feed contains 97.8 wt% of ethyl acetate, 2.1 wt% of n-butyl acetate, and 0.1 wt% of toluene with some non-volatile resin residues.

The primary purpose of distillation is to find an effective operation obtaining pure ethyl acetate as much as possible.

At the beginning, a batch (31.5 kg) of the feed mixture is charged into the Wall Wetter pot, and then the system is brought to and kept at quasi-steady state under total reflux until attaining the desired purity of the overhead product. Then a constant portion of the overhead product equal to the constant feed rate is continuously withdrawn in accordance with the reflux policy (reflux ratio = 1.0) of continuous operation with no discharge of the bottom product. Therefore less-volatile components are accumulated in the liquid pool of the Wall Wetter pot. The second purpose is to make the volume of residuum in the Wall

Wetter pot as small as possible.

After the total reflux operation (85 min), the continuous operation was stably lasted satisfying the specification of the overhead product purity for 115 min.

It has successfully been confirmed that this batch rectification still unit installed with Wall Wetter stirrer has the following advantages:

(1) Stable highly-efficient distillation

(2) Maintenance of very high evaporation rate

(3) Prevention against fouling

(4) Easy washing and saving of solvent for cleaning

(5) Simple inside structure

(6) Easy operation and maintenance

311a Breathing Effects In the Adsorption and Separation of Xylene Isomers on the Metal Organic Framework MIL-53

Joeri Denayer[1], Vincent Finsy[1], Dirk De Vos[2], and Gino V. Baron[1]. (1) Department of Chemical Engineering, Vrije Universiteit Brussel, Pleinlaan 2, Brussels, B-1050, Belgium, (2) Center for Surface Science and Catalysis, Katholieke Universiteit Leuven, Leuven, Belgium

Xylene isomer separation is one of the most important bulk scale adsorptive separation processes. Currently, faujasite based adsorbents are used in this process. Recently, we discovered that MIL-47 is a very effective adsorbent for the liquid phase separation of para-xylene from the other xylene isomers [1]. MIL-47 is a porous terephthalate built from infinite chains of V4O6 octahedra, held together by the dicarboxylate groups of the terephthalate linkers [2]. In this way a 3-dimensional microporous framework with 1-dimensional diamond shaped channels with free internal diameter of about 0.85 nm is formed.

In the present work, adsorption and separation of xylene isomers on a similar structure, MIL-53, was investigated. Low coverage adsorption properties were determined using gas phase pulse chromatography, vapor phase adsorption isotherms were obtained by gravimetry and the separation of C8 alkylaromatics (o-xylene, p-xylene, m-xylene and ethyl benzene) was evaluated by performing breakthrough experiments.

Zero coverage adsorption enthalpies were found to be comparable to those obtained on MIL-47. However, adsorption isotherms showed completely different behavior. All C8 alkylaromatics have 2-step isotherms and show hysteresis. This is attributed to breathing of the MIL-53 framework, as reported before by Llewellyn et al [3]. This breathing effect strongly influences the separation of the C8 alkylaromatics in breakthrough conditions.

[1] Alaerts, L.; Kirschhock, C. E. A.; Maes, M.; van der Veen, M. A.; Finsy, V.; Depla, A.; Martens, J. A.; Baron, G. V.; Jacobs, P. A.; Denayer, J. F. M.; De Vos, D. E. Angewandte Chemie-International Edition 2007, 46, (23), 4293-4297.

[2] Barthelet, K.; Marrot, J.; Riou, D.; Férey, G. Angewandte Chemie - International Edition 2002, 41, (2), 281-284.

[3] Llewellyn PL, Bourrelly S, Serre C, Filinchuk Y, Ferey G, How hydration drastically improves adsorption selectivity for CO2 over CH4 in the flexible chromium terephthalate MIL-53, ANGEWANDTE CHEMIE-INTERNATIONAL EDITION, 45(46), 7751-7754, 2006.

311b Effect of Pore Size, Open Metal Sites, and Functionalized Ligands on Adsorption of Small Molecules In Metal-Organic Frameworks

Krista S. Walton and Jagadeswara R. Karra. Department of Chemical Engineering, Kansas State University, 1005 Durland Hall, Manhattan, KS 66506

Metal-organic frameworks represent a new direction in porous materials research that could lead to the creation of designer-specific multifunctional materials for adsorption applications. The rich field of coordination chemistry provides a versatile platform from which these materials may be assembled from an almost infinite set of building blocks. The key to developing these materials for use in adsorption separations and gas storage is to obtain a fundamental understanding of their adsorption properties.

In this work, we examine the effect of pore size, unsaturated metal sites, and functionalized ligands on the adsorption of carbon dioxide, carbon monoxide, and methane at various temperatures. The specific objective here is to determine the governing structure/property relationships for MOFs that contribute to favorable adsorption properties for polar and nonpolar molecules. Three MOFs were synthesized, and adsorption isotherms were measured. The first material, Cu-BTC, possesses open metal sites and interconnected pores. The other two MOFs were synthesized from a mixed-ligand system of benzene-dicarboxylic acid and triethylenediamine. From this system, we formed a zinc MOF (1) and an isostructural copper MOF (2), which possess two interconnected channels with dimensions of 7.5 x 7.5 angstrom and 4.8 x 3.2 angstrom. In this paper, we will discuss MOF synthesis methods and adsorption equilibrium measurements for CO_2, CO, and CH_4. We will also compare our findings with previously reported MOF results to strengthen the development of structure/property relationships. Recommendations for design of MOFs for trace contaminate applications will be discussed.

Web Page: www.che.ksu.edu/research/walton

311c Hydrogen Adsorption Properties and Structural Stability of Metal-Organic Framework (MOF)-177

Dipendu Saha and Shuguang Deng. Chemical Engineering Department, New Mexico State University, P.O. Box 30001, MSC 3805, Las Cruces, NM 88003

Metal organic framework MOF-177 is probably one of the most promising adsorbents with the highest hydrogen adsorption capacity among all physical adsorbents. MOF-177 was synthesized with a modified approach and characterized with XRD, SEM and nitrogen adsorption for its physical properties. The hydrogen adsorption equilibrium and kinetic data were measured volumetrically at ambient pressure and gravimetrically in a magnetic suspension balance at hydrogen pressure up to 100 bar. The MOF-177 synthesized in this work was found to have a higher specific surface area (Langmuir: 6008 m2/g; BET: 3814 m2/g) and a higher hydrogen adsorption capacity (11.0 wt.% excess adsorption, 19.67 wt.% absolute adsorption) than previously reported values on MOF-177. Freundlich equation fits well the hydrogen adsorption isotherms at low and high pressures. Diffusive time constant and isosteric heat of adsorption were calculated from the hydrogen adsorption kinetics and equilibrium data.

The effects of moisture and oxygen on structural stability of MOF-177 are investigated by X-ray diffraction and thermogravimetric analysis. The MOF-177 sample is exposed to ambient air of 16% RH for five weeks and monitored for its structural changes by XRD. The crystal structure of MOF-177 gradually changes from hexagonal to orthogonal, and then to monoclinic in the five weeks period. The crystal structure of MOF-177 is completely destroyed after it is immersed in water. Weigh loss of MOF-177 is negligible at temperature below 330°C in the presence of oxygen. However, MOF-177 is completely converted to zinc oxide when it is exposed to 420°C. An extension of possible explanation is made for structural shift of MOF-177 compared to previous researchers.

311d Hydrogen Adsorption by the Spillover Mechanism on Carbon-Based Adsorbents: Capacity and Kinetic Studies Using Isotopic Tracer Reactions

Anthony J. Lachawiec Jr. and Ralph T. Yang. Department of Chemical Engineering, University of Michigan, 2300 Hayward Street, 3074 HH Dow Bldg, Ann Arbor, MI 48109-2136

We present results for hydrogen adsorption by spillover of hydrogen atoms from a metallic dissociation source to several carbon-based receptors, namely carbon synthesized from a molecular sieve template and a metal-organic framework. Hydrogen capacity enhancement by adsorption of hydrogen via the spillover phenomenon is reported for room temperature and from ambient pressure to 100 atm, where these materials are expected to find most widespread practical application.

We present kinetic response of such adsorbents for both hydrogen and deuterium. A unique feature of such studies is that desorption is consistently faster than adsorption at room temperature. Temperature programmed desorption studies were performed after sequential dosing of these species. Analysis of product distributions consistently yielded deuterium hydride, a strong indicator of spillover. The sequence of products was dependent on the dosing sequence, with HD always occupying an intermediate position.

We have further explored the difference in adsorption and desorption kinetics with HD in reverse reaction to form molecular hydrogen and deuterium. Based on these studies, we have found evidence that supports the theory of recombination of D atoms to form molecular deuterium, which prematurely desorbs from the carbon receptor and accounts for the apparent increase in reverse spillover kinetics.

311e Adsorptive Membrane and Adsorbent/membrane Hybrid Processes for Hydrogen Separation

Ji-Han Bae, Yoon-Jin Han, and **Chang-Ha Lee**. Department of Chemical and Biomolecular Engineering, Yonsei.university, 262 Sungsanno Seodaemun-Gu, Seoul, 120-749, South Korea

Hydrogen is widely used in various fields of industries, such as chemical, steel, oil refining and petrochemical industries and is also expected as a clean alternative energy source, anticipating enormous demand for hydrogen in the near future. In addition, application of inorganic membranes to practical gas separation has attracted much attention since there is an increasing demand for separation and purification of gases under harsh conditions.

In this study, the transport mechanisms of the MTES (methyltriethoxysilane) templating silica/ƒÑ-alumina composite membrane were evaluated by using CO2, N2, CH4, CO and H2 single gaes and their hydrogen binary systems. And the separation characteristics and dynamics of hydrogen mixture produced from natural gas reformer (H2/CH4/CO/CO2: 69/3/2/26 vol.%) were studied on both experimentally and theoretically. In addition, AMH(Adsorbent/Membrane Hybrid) system, which adsorbents (zeolite 5A) were packed to the MTES membrane, was developed to improve the performance of hydrogen separation in binary gas mixtures(H2/CH4, H2/N2 and H2/CO). Since the permeation flux in the MTES membrane was affected by molecular sieving effects as well as surface diffusion by adsorption, the kinetic and equilibrium separation should be considered simultaneously in the membrane according to molecular properties. The transient permeation/separation behaviours of hydrogen multi-component systems on the MTES membrane were predicted by the Generalized Maxwell-Stefan model incorporating the dusty gas model and Langmuir isotherm model.

311f Diffusion of Hydrocarbons In SSZ42 Materials

Qinglin Huang[1], Zahra Sarshar[2], **Mladen Eic**[1], and Serge Kaliaguine[2]. (1) Department of Chemical Engineering, University of New Brunswick, P.O. Box 4400, Fredericton, NB E3B5A3, Canada, (2) Department of Chemical Engineering, Laval University, Sainte-Foy, Quebec, QC G1K 7P4, Canada

The novel high-silica zeolite SSZ-42 was firstly synthesized in 1997 [1]. It is a representative of an undulating one-dimensional, 12-membered-ring channel system. Although SSZ-42 is one-dimensional material, it has been shown to have remarkably high adsorption capacities for n-hexane, 2,2-dimethylbutane and cyclohexane [2]. The material is also stable up to at least 800 oC under thermal and hydrothermal conditions [2]. These properties suggest that SSZ-42 could be a promising material as catalyst for hydrocarbon processing, and as adsorbent for adsorption separation.

To explore the potential application of SSZ-42 material in adsorption separation and catalysis involving large molecules, it is essential to carry out kinetic characterizations. The present study focuses on a systematic investigation of diffusion properties involving selected organic vapors in SSZ-42 materials, by employing the zero length column (ZLC) technique. Toluene, o-xylene and cumene were used as probe molecules of different (increasing) kinetic diameters, since they are important representatives of aromatics. Two samples of boron SSZ-42 materials (Na- and H-forms designated as SSZ-42-Na+ and SSZ-42-H+, respectively) have been synthesized by using N-benzyl-1,4-diazabicyclo[2.2.2]octane cation as the template molecule following hydrothermal crystallization [2]. The diffusion results reveal that transport processes of all probe molecules in SSZ-42-Na+ and toluene in SSZ-42-H+ were controlled by pore diffusion. On the other hand, transport of both o-xylene and cumene in SSZ-42-H+ was found to be controlled by surface barrier resistance at low temperatures, but pore diffusion resistance became more dominant with the increasing temperature. The transport mechanisms of probe molecules in both samples were verified by the partial saturation method proposed by Brandani and Ruthven [3]. The ion-exchange process may lead to a partial collapse of pore channels of SSZ-42-H+ leading to surface barrier effects, while at relatively higher temperatures pore openings appear to be expanding, rendering negligible barrier resistance. The diffusivities of probe molecules in SSZ-42-Na+ were systematically higher in comparison with SSZ-42-H+, while the activation energies showed the opposite trends. This confirms that the probe molecule interactions with the Bronstead acid sites cause a decrease in the diffusion rate, consequently requiring more energy for the activated transport.

References:

[1] Chen, C.Y., Finger, L.W., Medrud, R.C., Crozier, P.A., Chan, I.Y., Harris, T.V. and Zones, S.I., "SSZ-42: the first high-silica large pore zeolite with an undulating, one-dimensional channel system," Chem. Commun., 1775-1776 (1997).

[2] Chen, C.Y., Finger, L.W., Medrud, R.C., Kibby, C.L., Crozier, P.A., Chan, I.Y., Harris, T.V., Beck, L.W. and Zones, S.I., "Synthesis, structure, and phsicochemical and catalytic characterization of the novel high-silica large-pore zeolite SSZ-42," Chem. Eur. Journal, 4, 1312-1323 (1998).

[3] Brandani, S. and Ruthven, D. M., "Analysis of ZLC desorption curves for gases systems," Adsorption, 2, 133-143 (1996).

311g Imbibition Transition and the "Gate Effect" In Adsorption

Angela D. Lueking, Energy and Mineral Engineering, The Penn State University, 120 Hosler Bldg, University Park, State College, PA 16802 and Milton W. Cole, Physics, The Pennsylvania State University, University Park, PA 16802.

Adsorption in some regular nanoporous materials has been interpreted in terms of a “gate effect” model, in which the low pressure (P) uptake is limited by an initial blocking of access until a sufficiently high pressure opens a “gate”. A simple model (an imbibition transition) is discussed in which a key element in the model is the ability of the pore to expand (or contract, depending on the system) in response to the adsorbate. At low P, the density of the gas is low, so the pore does not respond significantly to the gas's presence. There exists a free energy barrier to pore expansion, since it is coupled to the gas density. At sufficiently high P, this barrier height becomes small; eventually, the system becomes thermodynamically unstable above a critical value of P. At very high P, after the transition occurs, the equilibrium coverage is high because the system's free energy is significantly reduced by the gas-pore coupling. Key parameters in the model are the derivative of the adsorption potential with respect to pore diameter and the force constant for breathing mode-like fluctuations of the pore. Relevant experimental data will be discussed.

312a Investigation of Protein Binding In Ion Exchange and Multi-Modal Chromatographic Systems Using Homologous Protein Libraries, NMR and EPR

Steven M. Cramer, Wai Keen Chung, Alex Freed, Ying Hou, Sarah Dekat, Scott McCallum, K.V. Lakshmi, and George Makhatadze. Chemical and Biological Engineering, Rensselaer Polytechnic Institute, 110 8th Street, Troy, NY 12180

In an effort to better understand the unique selectivities offered by multi-modal chromatographic systems, experiments were carried out with a variety of cation exchange and multi-modal ligands using model protein systems. Nuclear Magnetic Resonance (NMR) was employed to identify and map ligand interaction surfaces between the proteins and anionic and multi-modal ligands in solution. Key chemical features within the ligands that interact with the protein surfaces were identified and the number and locations of interaction sites on the protein were determined and evaluated. Site directed spin-labeling electron paramagnetic resonance (SDSL-EPR) was also performed using a number of cysteine mutants to determine the binding orientation of the wild type protein on both cation exchange and multi-modal resin surfaces. Finally, this information from both free solution and resin binding studies was employed to distinguish "affinity" type multi-modal interactions from multi-site non-specific interactions.

312b Assessment of a Site and Scale Change for the Manufacture of Gardasil

David Wohlpart and Dicky Abraham. Global Vaccine Technology and Engineering, Merck and Company, 700 Sumneytown Pike, WP 60-301, West Point, PA 19486

A small scale purification facility was used to supply Phase III and launch material for GARDASIL®. A larger facility was required to meet the expected demand for the product. A key component of the licensure for this facility was the demonstration of the comparability of the process and the drug substance produced in this in this new facility. We present the framework and strategy used to demonstrate comparability, and the results that clearly show the product made in the new facility is comparable to that in the launch facility.

312c Use of Particle Size Analysis to Characterize Bed Stability Over Multiple Cycles in a Packed Chromatography Column

Brad M. Matanin, John Higgins, and David A. Page. Process Biochemistry, MedImmune, LLC., One MedImmune Way, Gaithersburg, MD 20878

Particle size analysis by laser diffraction was used to correlate the in-situ (batch mode) bead diameter of chromatography media to bed stability of a packed chromatography column. An empirical method was developed to predict the performance of a packed chromatography bed during extended use. For a selected Hydrophobic Charge Induction Chromatography (HCIC) medium (pKa 4.8), the median bead diameter was measured in buffer under a variety of conditions of pH and conductivity. Particle size analysis indicated a significant increase in bead diameter after equilibration in a low pH (<4) buffer. An increased pressure drop across a packed bed at production scale (45 cm diameter) was also observed under this low pH condition. To further investigate this observation, a scale down column (3.2 cm diameter), using a constant diameter-to-height aspect ratio instead of a conventional fixed bed height, was employed to characterize bed stability over multiple cycles. In-situ bead size data was correlated to a measure of breakthrough volume and calculated HETP values when the medium was exposed to repeated cycles of swelling and compression in a packed bed. The analysis indicated that the low pH buffer accelerated bed deterioration. This trend was further supported by qualitative overlays of rate of change of conductivity versus totalized volume. When the low pH buffer was omitted from the process, increased bed stability was observed over multiple cycles. Thus, the trends determined by particle size analysis, that the bead diameter should not change significantly if buffer pH remains above a threshold level (pKa 4.8) throughout the process step, were confirmed. This suggests that changes in bead size could be attributed to induced electrostatic repulsion at low pH and could be modulated to increase usable column lifetime. When correlated with packed-bed performance, particle size analysis can be a valuable tool to better characterize column lifetime at increased scale. This technique can also be used as a predictive tool for screening new chromatography media and developing robust process operations.

312d Innovative Approaches for Removal of Persistent Host Cell Proteins

Sa V. Ho, Global Biologics, Pfizer, 700 Chesterfield Parkway West, Chesterfield, MO 63017

With the acquisition of Esperion by Pfizer in early 2004, the ETC216 project was transferred to Pfizer Global Biologics. ETC216 is an HDL-mimic complex of ApoA-1M and a phospholipid with demonstrated reversed cholesterol transport property. The Esperion process for ApoA-1M manufacture was rather complicated, especially the downstream purification, which consisted of a complex product recovery scheme from fermentation broth, 5 chromatography columns, several ultrafiltration/diafiltration steps, and one reaction step. In its first production campaign to produce additional protein materials at a CMO, the Pfizer Biologics team had to deal with the problem that the first few purified protein batches did not meet the HCP specification, which was also an issue experienced by Esperion in previous campaigns. Through process modification supported by analytical data of key in-process samples, the team was able to finally overcome the HCP hurdle and consistently made materials that met all product specifications. Post campaign development work led to the identification of the main difficult-to-remove host cell proteins. A combination of genetic modification and purification enhancement later resulted in a second generation manufacturing process that not only easily met a more stringent HCP specification but also achieved higher yield and better product quality.

312e Enhancing Affinity Chromatography Resin Efficiency Using 3 Column Periodic Continuous Chromatography

Ekta Mahajan[1], Anupa George[1], Karol M. Lacki[2], Bradley Wolk[1], and Kirk Eppler[1]. (1) Process Development Engineering, Genentech Inc., Mail Box 75B, 1 DNA Way, So San Francisco, CA 94080, (2) GE Healthcare Life Sciences R&D, Bjorkgatan 30, Uppsala, Sweden

Protein A Affinity Chromatography is widely used for purification of monoclonal antibodies from harvested cell culture fluid (HCCF). The HCCF is typically loaded on a single large Protein A affinity chromatography column in cycles until all of the HCCF is processed. Protein A resin costs are significant and can be reduced by operating the process continuously using multiple columns in lieu of one large column. In this study, a series of experiments were performed using three 1ml Mab Select SuRe columns on a modified AKTA system (3C PCC) from GE Healthcare. The columns were loaded individually until the 70% breakthrough point was achieved. The non-bound fluid from the column was then loaded onto the next column to capture the Mab, preventing any protein loss. At any given point, all three columns were in operation; either loading or washing, enabling a reduction in processing time. The product yield and quality were evaluated and compared to current process to determine the affect of using the continuous process. The continuous operation shows the potential to reduce both resin volume and buffer consumption, but the system hardware and the process are more complex than the current process. However, cost savings may make it an attractive option for both new and existing processes.

312f Mathematical Tools for Efficient Purification Process Development

Thomas R. Gervais, Dennis Dong, Nicole Quinlan, Alison Walsh, Marijana Stella, Patricia Alred, and Pedro Alfonso. Purification Process Development, Centocor R&D, Inc., 145 King of Prussia Rd, RA-1-4, Radnor, PA 19087

Efforts are being focused to streamline process development timelines to move products quickly into different stages of clinical trails. Consequently, more efficient methods for process development have become increasingly important for accelerating process optimization and reducing costs. Techniques such as high throughput screening and the leveraging of automation have been used to improve process development efficiency. In addition, mathematical tools such as DOE, mathematical models and MultiSimplex algorithms can also be used to increase efficiency during process development.

Here we present two case studies in which these mathematical tools were utilized to quickly and efficiently establish operating conditions for the purification of antibody-derived fusion proteins. The first involves the development of elution conditions for a Protein-A chromatography operation in which the goal was to achieve a high step yield and reduce product related aggregate. The second concerns the use of a multimodal anion exchange resin for the reduction of product related impurities through the use of an intermediate wash. By using these methods, an optimal elution condition was achieved after only 8 experiments for the Protein-A unit operation, resulting in an 85% process yield with a reduction of product related aggregate from ~20% to less than 2%. These tools also enabled the rapid development of optimal intermediate wash conditions for the multimodal ion exchange process which were found to be robust, scalable and resulted in the removal of the product related impurities while maintaining > 80% step yields.

312g Preparative Separations of Acidic Protein Mixtures and Protein Charge Variants by Chromatofocusing with Unretained Buffers

Timothy M. Pabst, University of Virginia, Charlottesville, VA 22904, **Giorgio Carta**, Department Chemical Engineering, University of Virginia, Charlottesville, VA 22904-4741, Alan K. Hunter, Global Biologics, Pfizer, 700 Chesterfield Parkway West, Chesterfield, MO 63017, and Natarajan Ramasubramanyan, Global Biologics, PGRD, Pfizer Inc, 700 Chesterfield Pkwy West, Chesterfield, MO 63005.

Complex pH transitions occur in exchange columns containing weak acid or weak base during salt steps or gradients if the mobile phase is buffered and the buffering species do not interact with the stationary phase. If not anticipated and planned for within a separation step, they can result in large equilibration volumes and potentially negative effects on pH-sensitive proteins. On the other hand, such pH variations, if controlled and directed, can be beneficial to the separation of proteins. The behavior of chromatographic columns packed with resins containing both weak cation exchange and weak anion exchange groups is investigated in order to obtain protein separations by means of internally generated pH gradients in response to step changes in buffer composition. A local equilibrium model is developed to predict pH transitions using non-adsorbed buffers, i.e. containing neutral or anionic buffering species with cation exchangers and neutral or cationic buffering species with anion exchangers. In agreement with experimental results, the model, based exclusively on the resins' titration curves, predicts practical, fairly linear gradients, which are formed using suitable mixtures of acidic buffers for cation exchange and basic buffers for anion exchange. The separation of mixtures of ovalbumin, albumin, and transferrin is used as a model system with cation exchange columns while the separations of transferrin and of mAb charge variants are used as model systems with anion exchange columns. The peaks obtained with this approach are much sharper than could be obtained isocratically or using externally generated, unretained gradients as a result of the peak compression caused by the axial pH gradient formed along the column. Moreover, separation is obtained at very low ionic strengths (2-3 mS/cm). The effects of flow velocity, mobile phase composition, time of injection, and protein load on retention and elution pH are investigated systematically demonstrating a range of ways in which the separation can be controlled and optimized.

336a Characterization of Pd-Based Membrane Fabricated by Surfactant Induced Electroless Plating (SIEP): Effect of Grain Size In Hydrogen Permeability

Mohammad A. Islam, Mechanical and Chemical Engineering Department, North Carolina A&T State University, Greensboro, NC 27411 and Shamsuddin Ilias, Department of Mechanical & Chemical Engineering, North Carolina A&T State University, 1601 E. Market Street, Greensboro, NC 27411.

Conventional electroless plating method is limited by the non-uniformity in depositions of Pd-grains on porous substrates and results in defective films. Even if the defects are filled, fabricated film becomes comparatively thicker. In order to fabricate defect free, relatively thinner membrane, control of Pd-grain sizes and their subsequent uniform agglomeration are critical. The oxidation-reduction reactions between Pd-complex and hydrazine result in evolution of ammonia and nitrogen gas bubbles. When adhered to the substrate surface and in the pores, these gas bubbles hinder uniform Pd-film deposition. To over these limitations, we investigated the role of surfactants in electroless plating, what we call Surfactant Induced Electroless Plating (SIEP) in fabricating thin and defect-free Pd-film on microporous stainless steel substrate.

Polycrystalline palladium (Pd) was deposited on microporous stainless steel substrate using suitable surfactant in various concentrations in typical electroless plating formulation. To elucidate the role of surfactant in solid-liquid and solid-gas interface during grain coarsening, we used surfactant concentrations as in terms of critical micelle concentration (CMC). The microstructure of the Pd-film was characterized by Field Emission Scanning Electron Microscopy (FESEM), Energy Dispersive Spectroscopy (EDS) and X-ray diffraction (XRD). The type of surfactant (anionic, cationic and nonionic) and their relative CMCs greatly influence the grain size and surface morphology of the deposited Pd-film. We were successful in fabricating defect-free, thin Pd-film on microporous stainless steel substrate using a cationic surfactant. The surfactant molecules play an important role in the plating process in tailoring grain size and the process of agglomeration by removing tiny gas bubbles through adsorption at the gas-liquid interface. We studied the hydrogen permeability and selectivity for each of the membranes fabricated in this study at elevated temperature. Results show that membranes with agglomerated grains (finer grain size) possess relatively higher permeability and selectivity. In this presentation we will present some of these results along with a proposed deposition mechanism for SIEP.

336b The Effect of Complexing Agents on the Electroless Pd-Ag Co-Deposition Bath Chemistry Using Linear Sweep Voltammetry for the Composite Membrane Synthesis

Rajkumar Bhandari, Chemical Engineering, Worcester Polytechnic Institute, 100 Institute Road, Worcester, MA 01609 and Yi Hua Ma, Worcester Polytechnique Institute, Chemical Enginering, Worcester, MA 01609.

Composite Pd-Ag membranes are better suited for H2 separation applications than pure Pd membranes because of their lower critical temperature for the á to â phase transformation and higher H2 permeability (23 Ag wt%).

The electroless co-deposition of Pd and Ag as membrane preparation method offers advantages such as fast synthesis procedure and easy annealing of the deposits in comparison to the sequential plating method. Since the electroless deposition involves redox reactions, the electrochemical technique such as linear sweep voltammetry (LSV) could be very useful to understand the key aspects of electroless plating bath and its components and their effect on the deposit characteristics. For the co-deposition, the deposition potential of co-depositing metals should be close to each other. Complexing agents in the bath influence the metal deposition potential and other deposits characteristics such as morphology and deposits compositional homogeneity. The objective of this study was to investigate the effect of complexing agents (NH3, Na2EDTA and Glycine) in the bath on the Pd-Ag co-deposits characteristics such as morphology and compositional homogeneity using the LSV technique for the synthesis of Pd-Ag composite membrane.

The LSV scans were obtained using the BAS 110B/W electrochemical station. Ag/AgCl (3 M NaCl) was used as a reference electrode. Stainless steel wires plated with Ag or Pd acted as working electrode. For the LSV scans, 3 mM concentration each for Pd and Ag was maintained in the bath. Porous stainless steel coupons were used for the morphology and compositional homogeneity characterization. The deposits were characterized by using SEM, EDX and X-ray differactometer. For the coupon and membrane preparation, the total metal concentration of 4 mM (10 wt% Ag) was maintained in the baths. The coupons with Pd-Ag deposits were annealed at 500 oC for 10 h in H2 to study the annealing characteristics of the co-deposits. Based on the LSV and coupon study, a composite Pd-Ag porous Inconel membrane was synthesized using glycine bath and was characterized for the H2 permeation at successive temperatures of 250 (40 h), 300 (60 h), 350 (70 h) and 400 oC (110 h).

The LSV scans of the NH3 bath showed that the Pd deposition occurred at approximately 650 mV higher cathodic potential than the Ag deposition. The Ag deposition was limited by the mass transfer of Ag ions to the sample surface resulting in inhomogeneous deposits with Ag concentrated more on the tip of the pores of the substrate than inside the pores. The LSV scans obtained from glycine and Na2EDTA baths showed characteristics similar to the NH3 bath. The Pd deposition occurred at 550 (Na2EDTA) and 510 mV (glycine) higher cathodic potential than that for Ag. The deposits obtained from NH3 and

Na2EDTA baths were dendritic and porous in nature in comparison to those obtained from glycine bath which were relatively smooth with compact morphology. Therefore, glycine bath was further used to synthesize a Pd-Ag membrane.

The X-ray patterns of the deposits showed that the deposits were bimetallic in nature i.e. the deposits consisted of Pd and Ag particles and required further annealing for alloy phase formation. The annealing of the coupons with approximate 3.5 μm thick deposits obtained from the glycine bath showed almost complete alloying of Pd-Ag deposits when annealed at 500 oC for 10 h in H2 atmosphere as was evident by the X-ray pattern of the deposits which showed only single phase present after annealing. The annealing time and temperature for the co-deposits was considerably lower than that required for the deposits formed by the sequential coating and diffusion method which requires temperature of 550 oC or more.

No He flow was observed for as synthesized 8 μm thick membrane. The membrane on heating showed increase in the H2 permeance at 250, 300 and 350 oC due to the alloying of the Pd-Ag deposits. The EDS analysis of the cross-section of the membrane (with 6 wt% Ag overall) indicated more Ag concentrated towards the region away from the substrate similar to the coupon study. Also the deposits showed decline in H2/He selectivity (ΔP = 1 atm) on heating between 250-400 oC. The membrane at 400 oC showed H2 permeance of 36.7 m3/m2-h-atm0.5 and selectivity value of 175. The membrane showed slightly less permeance than that of pure Pd foil of the same thickness (46 m3/m2-h-atm0.5) which could be due to the error in the measurement of the thickness of the membrane. Also Ag wt% in the membrane was not significant enough to see the effect of Ag on the membrane permeance. The Ag wt% in the membrane could further be increased by simply increasing the Ag% concentration in the bath.

In conclusion, the large difference in the deposition potential between Pd and Ag resulted in inhomogeous composition of the deposits. Therefore modifying the complexing agents in the bath which provides close deposition potential of Pd and Ag could help to produce homogeneous deposits and requires further investigation. The glycine based bath showed relatively compact morphology and the deposits showed increase in the H2 permeance in the temperature range of 250-400 oC due to the alloying effect. The co-deposits also showed decline in H2/He selectivity on heating in the same temperature range.

Addition-type polynorbornene with trimethylsilyl side groups: study of organic vapors sorption, transport and swelling

M. Galizia[1], M. G. De Angelis[1], G. C. Sarti[1], Y. Yampolskii[2]

[1] *Alma Mater Studiorum* – Università di Bologna, Dipartimento di Ingegneria Chimica, Mineraria e delle Tecnologie Ambientali, via U. Terracini 34, 40131 Bologna.

[2] A.V. Topchiev Institute of Petrochemical Synthesis (TIPS), Russian Academy of Sciences, 29, Leninsky Prospect, 119991, Moscow.

Abstract. High free volume glassy polymers are generally more selective to condensable vapors rather than to light gases. Such a property makes these materials suitable for membrane-based separation processes, as for instance the removal of VOCs. In this work, we focus our attention on poly-(trimethyl-silyl)-norbornene (*PTMSN*). Using both a barometric and an optical apparatus, the sorption and dilation isotherms have been determined at 35°C for n-pentane and n-hexane: the solubility values are very high, like for PTMSP, because of the high free volume trapped into the polymeric matrix. Moreover, the *Lattice Fluid* parameters have been estimated, enabling us to compare the experimental data with the NELF model predictions. Permeability has been calculated as the product of diffusivity and solubility at the same pressure and it results higher for n-C_6 than for n-C_5.

Keywords. *Solubility, Diffusivity, Relaxation, Dilation, Nelf*.

INTRODUCTION

Glassy polymers are typically used in the manufacturing of membranes for gas separation due to high permeability and selectivity. In particular, the selectivity α_{ij} of a membrane towards a component *i* in a mixture of components *i* and *j* can be decomposed into two contributions α_D e α_S, called the diffusivity and solubility selectivity, respectively [1]:

$$\alpha_{ij} = \frac{P_i}{P_j} = \frac{D_i}{D_j}\frac{S_i}{S_j} = \alpha_D \alpha_S \qquad (1)$$

While for many polymers the dominating term is the diffusivity selectivity, in the case of high free volume glassy polymers the solubility selectivity, due to the low size sieving ability, gives the greater contribution, thus these materials are more permeable to condensable vapors than to light gases. The substituted polynorbornenes belong to this family and are potential candidates to be used for separation purposes, in place of poly-(4-methyl 2-pentyne) (PMP) or poly-(trimethylsilyl)-propyne (PTMSP).

EXPERIMENTAL

Poly-(trimethylsilyl)-norbornene

The Poly-trimethylsilyl-norbornene studied here is an addition-type polymer whose monomer unit is obtained by a *Diels-Alder* reaction between cyclopentadiene and vinyltrimethylsilane. It is well known that the polymerization of norbornene can occur, depending on the catalyst used, by ring opening, with the formation of cyclolinear structures or, in the presence of Ni or Pd, by double bond opening, with the formation of bulky monomer units (Figure 1).

The materials obtained by addition reaction are more stable, for the absence of double bonds, and they are characterized by high free volume values, with bulky side groups, stiff backbones and high torsional barrier .

The present polynorbornene presents trimethylsilyl as a side group, its density is equal to 0.883 g/cm^3 (its fractional free volume was estimated to be about 0.275) and shows a glass transition temperature greater than 340 °C, as it has been estimated through TGA analysis: more reliable estimates are not available, because this polymer chemically decomposes before reaching T_g [2]. The PTMSN films have a thickness of about 150μm and they were cast from a dilute toluene solution (2% in weight).

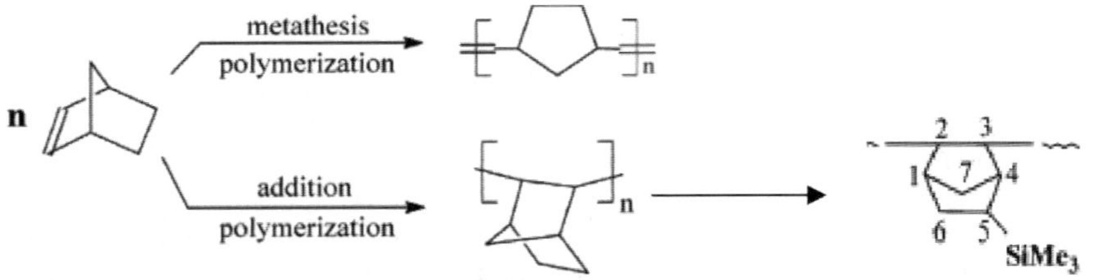

Figure 1: Polymerization of PTMSN and monomer unit obtained by addition.

Solubility and Diffusivity measurement

Sorption experiments of n-pentane and n-hexane were performed at 35 °C using a constant volume/variable pressure apparatus. Initially the film of PTMSN was placed in the sample chamber and exposed to the vacuum overnight to remove the residual air gases. The desired amount of vapor was then introduced into the chamber and allowed to equilibrate. Once the pressure was constant, additional vapor was introduced and again allowed to attain the equilibrium: a capacitance manometer (f.s. 1000 mbar, accuracy 0.1 mbar) monitores the pressure decay of the gaseous phase in the cell, which is related to the amount of mass absorbed by the polymer. The apparatus, whose total volume is around 90 cm^3, is placed in a thermostatic box where the temperature is set to 35 ± 0.1 °C.

Dilation measurement

The dilation of the sample during sorption is measured in a dedicated apparatus, by monitoring the sample elongation in one direction with a CCD camera. It is reasonable to assume, in this case, that the swelling is isotropic and that the volume dilation can be calculated by considering the elongation in one direction only (x), according to the following relationship:

$$\frac{\Delta V}{V_0} = 3\frac{\Delta x}{x_0} \qquad (2)$$

RESULTS AND DISCUSSION

All the sorption experiments shows a wide relaxation stage, in particular during n-hexane sorption. We will not perform a specific analysis of relaxation phenomena in this work, but rather present the equilibrium solubility values at the end of the diffusive and relaxational stages of sorption.

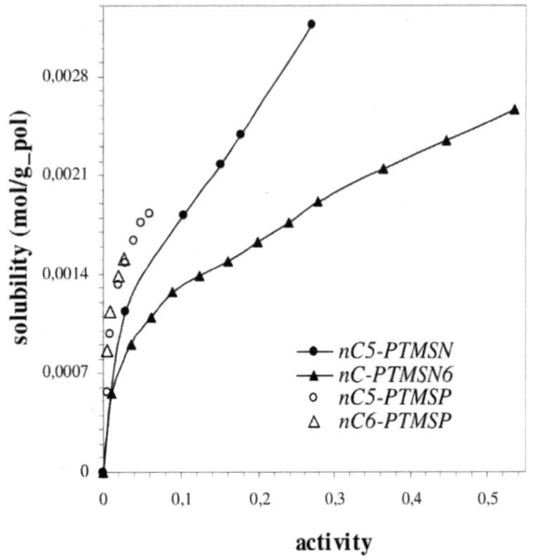

Figure 2a: Solubility isotherms at 35°C for n-C_5 and n-C_6 in PTMSN (closed symbols) as a function of activity and comparison with literature data on PTMSP (open symbols) at 27°C. [3]

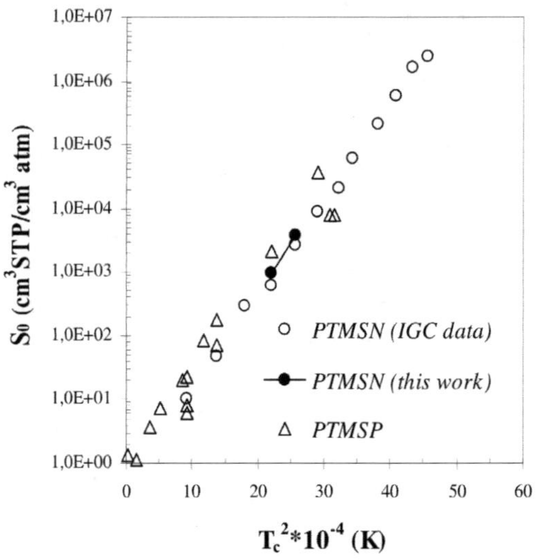

Figure 2b: Infinite dilution solubility coefficients in PTMSP and PTMSN, as obtained with IGC [4] and with pressure decay technique from this work.

In figure 2a we reported the solubility isotherms for n-C_5 and n-C_6, that are concave to the activity axis as it always happens in the case of glassy polymers. The molar solubility of n-C_5 is, at fixed activity, higher than that of hexane, contrary to what happens in the case of PTMSP [3]. The infinite dilution solubility coefficients are reported in Figure 2b and compared to the corresponding values obtained, in the same material, by Starannikova *et al.*,[4] by means of *Inverse Gas Chromatography (IGC)* on a series of gases and vapors. The data are plotted versus the penetrant square critical temperature, and it can be seen that the values estimated in this work are in good agreement with those obtained with a different technique.

In Figure 3a, the diffusivity values are reported as a function of average concentration: the isotherms follow the molecular size of the penetrants, the diffusivity of n-C_6 being lower than that of n-C_5. Moreover, in the case of n-C_6 the plasticization effect, based on which the diffusivity increases rapidly with concentration, is more evident especially in the initial steps. The pure gas permeability, evaluated as the product of diffusivity and solubility, is plotted in Figure 3b, which shows that PTMSN is more permeable, in almost the whole pressure range inspected, to the most condensable component, namely n-C_6, as it is common for this kind of ultra high free volume glassy polymers.

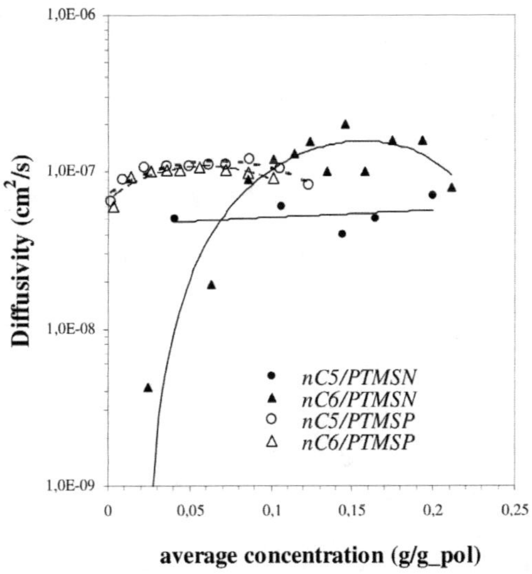

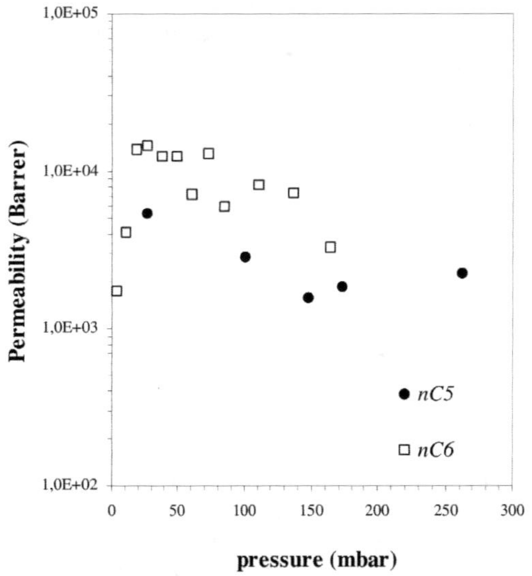

Figure 3a: Diffusivity of n-C_5 and n-C_6 in PTMSN at 35°C (closed symbols) and in PTMSP at 27°C (open symbols).

Figure 3b: Permeability of n-C_5 and n-C_6 in PTMSN at 35°C.

The experimental solubility data were then compared to the predictions of the *Non Equilibrium Lattice Fluid Model (NELF)* [5]. To this aim, it was necessary to determine the characteristic parameters for the *Lattice Fluid* equation of state for the penetrants and for the polymer. For the penetrants, these parameters were taken from the literature, while for the polymer they are usually obtained via a fitting of the pressure-volume-temperature data above T_g. Obviously, in our case this procedure is not suitable because the polymer decomposes before reaching T_g, therefore it has been necessary to use mixture data, in particular infinite dilution solubility data reported in figure 4a [4]. From a regression of these data with the values of S_0 evaluated with NELF model (fig. 4a), one obtains:

$$T^* = 405.98\ K \qquad p^* = 359.82\ MPa \qquad \rho^* = 1.345\ Kg/L$$

The comparison between experimental solubility isotherms and the predictions of the NELF model is rather satisfactory (Figure 4b): in particular, it is interesting notice that the values of k_{ij} are negative in both cases, indicating that there are strong interactions between penetrant and polymer.

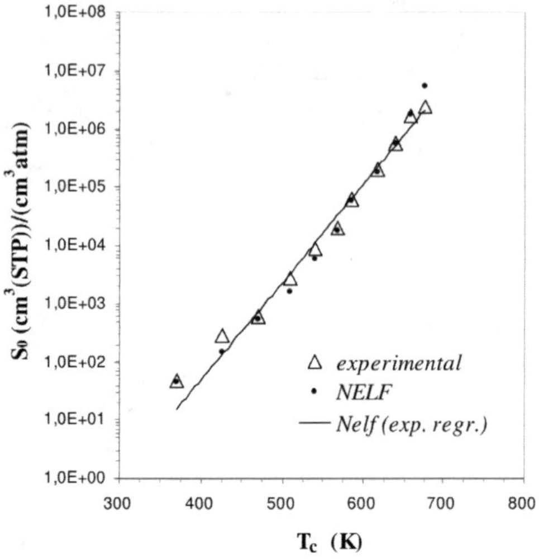

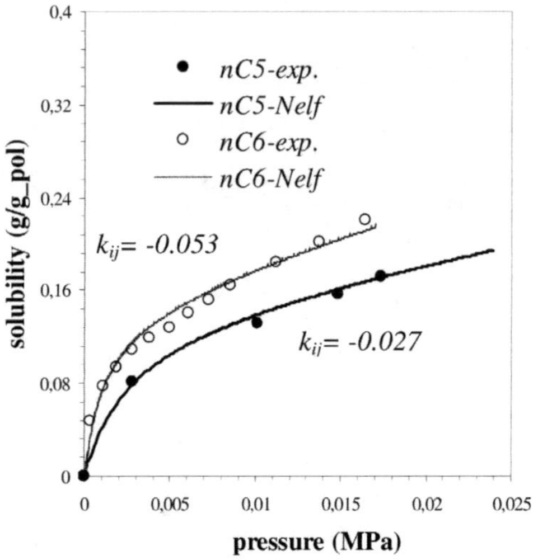

Figure 4a: Experimental [4] and calculated values of S_0 for various penetrants in PTMSN.

Figure 4b: Experimental solubility isotherms and comparison with NELF model predictions.

On the same polymer, we performed also dilation experiments: the experimental curves, plotted in terms of percentage volume dilation versus penetrant concentration are not linear (Figure 5a).

The resulting values of partial molar volume (Figure 5b) show a behaviour that is very typical of glassy polymers [6]: an initial increase with penetrant concentration is observed, corresponding to a "void filling" mechanism, followed by a maximum and a decrease with penetrant concentration, probably due to the fact that the progressive swelling at this high pressure is rather inhibited by the rigidity of the polymer matrix.

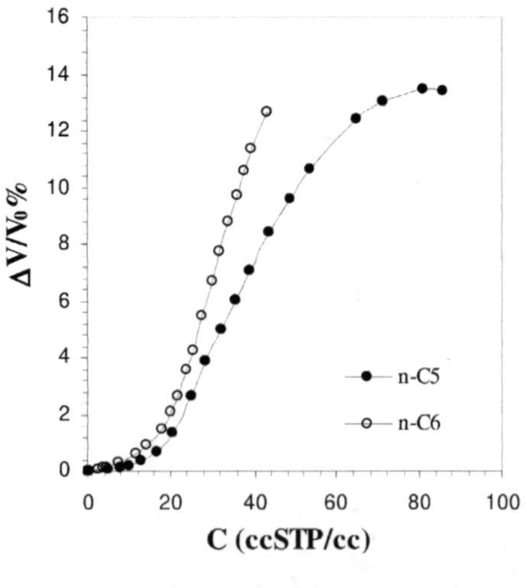

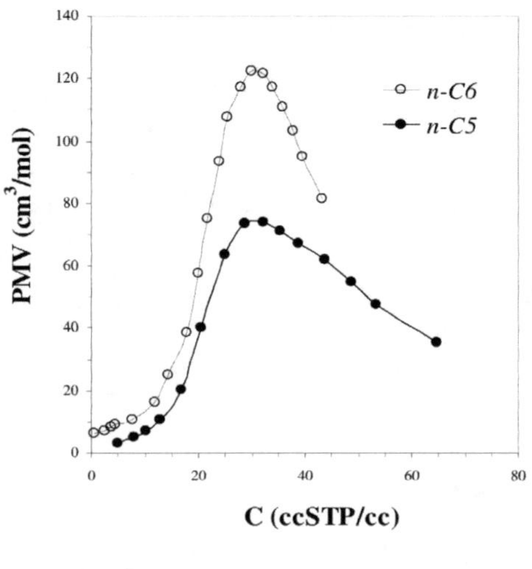

Fig. 5a Swelling isotherm of n-C_5 and n-C_6 in PTMSN at 35°C.

Fig. 5b Partial molar volume of n-C_5 and n-C_6 in PTMSN at 35°C as functions of concentration.

CONCLUSIONS

Sorption and dilation experiments were performed on PTMSN.

The solubilities of organic vapors as n-pentane and n-hexane in the polymer are rather high because of the high fractions of free volume trapped in its glassy matrix, thus the PTMSN is a potential candidate for use in gas separation membranes.

The solubility and diffusivity behaviour is typical of glassy polymers and the values are comparable to those in PTMSP. The permeability is higher for the larger, more condensable hexane than for pentane, as it is the case of vapour-selective polymers.

The solubility isotherms can be modelled satisfactorily with NELF model, even in the absence of experimental volumetric information on the polymer, by fitting the *Lattice Fluid* parameters to the infinite dilution solubility data available in the literature.

The swelling behaviour presents a non monotonous trend that can be explained through the glassy nature of the polymer matrix.

ACKNOWLEDGEMNTS

The financial support of the European project *MultiMatDesign* (NMP3-CT-2005-013644) is gratefully acknowledged.

REFERENCES

[1] Merkel T.C. et al., *Sorption and Transport in Poly(2,2 bis(trifluoromethyl)4,5-difluoro-1,3-dioxole-co-tetrafluoroethylene) containing Nanoscale Fumed Silica*, Macromolecules, **36** (2003), 8406-8414.

[2] Yampolskii Y. et al., *Addition-type Poly-Norbornenes with $Si(CH_3)_3$ side groups : Synthesis, Gas Permeability and Free Volume*, Macromolecules, **39** (2006), 7022-7029.

[3] M. Ghisellini, M. Fehervari, F. Doghieri and G.C. Sarti AIDIC Conferences Series 4 (1999), p. 375.

[4] L. Starannikova, M. Pilipenko, N. Belov, Yu. Yampolskii, M. Gringolts, E. Finkelshtein, *Addition-type polynorbornene with Si(CH3)3 side groups: Detailed study of gas permeation and thermodynamic properties*, Journal of Membrane Science, 323, 2008, 134-143.

[5] Doghieri F., Sarti G.C., *Nonequilibrium Lattice Fluid : a Predictive Model for the Solubility in Glassy Polymers*, Macromolecules, **29** (1996), 7885-7896.

[6] De Angelis M.G., Merkel, T.C., Bondar, V.I., Freeman, B.D., Doghieri, F. Sarti, G.C., *Gas Sorption and Dilation in Poly(2,2-bistrifluoromethyl-4,5-difluoro-1,3-dioxole-co-tetrafluoroethylene): Comparison of Experimental Data with Predictions of the Non Equilibrium Lattice Fluid Model*, Macromolecules, 35, 1276-1288 (2002).

Synthesis and Properties of Tubular Titanate Two-Dimensionally Deposited Carbon Nanosheet Composite

Wen-Qin Peng[1], Mami Aizawa[1], Zheng-Ming Wang[1] and Takahiro Hirotsu[2]
[1]National Institute of Advanced Industrial Science and Technology, Tsukuba, Ibaraki, Japan
[2]National Institute of Advanced Industrial Science and Technology, Takamatsu, Kagawa, Japan

Abstract

Carbon nanosheets were used as template for two-dimensional deposition of tubular titanate, and subsequent calcination at high temperature led to the transformation of titanate nanotubes into titania nanorods. Microscopic observation confirmed that both nanotubes and naorods deposit along the surface of carbon nanosheets, forming a stacking structure. The composites with the novel structure exhibit good adsorbing ability and high photoactivity toward organic molecules due to the synergy effect of adsorption and catalysis.

Introduction

Titanium oxide (TiO_2) have been widely studied as a photocatalytic material because of great potential in environmental purification and pollution abatement.[1-3] A wide variety of organic pollutants can be degraded into CO_2 and H_2O on TiO_2 under UV illumination at ambient temperature and pressure. Low cost, non-toxicity, and chemical stability further prompt its practical application. In recent years, much effort has been devoted to the synthesis of nanosized TiO_2 with larger specific surface area and high pore volume. [4-6]

Adsorption is an important step in the photocatalytic process and effective degradation can be achieved by pre-concentration of target pollutants on the surface of TiO_2 or its vicinity. Porous carbon is widely used as adsorbent because of high permeability to gas and liquid. High photocatalytic activity can be achieved by composing TiO_2 with carbon matrix due to the enhancement of adsorbing affinity toward organic species. Recently, some research has focused on C-TiO_2 composites.[7-9] It was reported that carbon-nanoscaled TiO_2 composites had synergy effect and showed higher photocatalytic activity than pure TiO_2.[7] Compared to the usual TiO_2 nanoparticles, One-dimensional (1D) TiO_2 can achieve higher photocatalytic activity. [5] However, till now, no research has been carried out to produce the composite of 1D TiO_2 and carbon, which is expected to combine excellent adsorptivity of carbon and high photocatalytic activity of 1D TiO_2.

In this paper, we propose the formation of a novel structure which is composed of titanate nanotubes deposited on the surface of carbon nanosheets by a template method. Graphite oxide, which has excellent intercalation properties, was used as template and carbon source.

Experimental

First, titanium tetraisoproxide (Ti(*i*-OC_3H_7)$_4$, TTIP) was interacted with graphite oxide to allow intercalation of Ti species between GO layers, resulting in the formation of a Ti species-pillared GO materials (GOTi). Then, Ti-intercalated GO was mixed with 10 M NaOH solution and treated in a Teflon-lined container at 423 K for 24 h to obtain carbon nanosheet-titanate nanotube composite (C-TNT). Finally, carbon nanosheet-titania nanorod composite (C-TNR) were obtained after calcining the nanotube composite under inert atmosphere at 823 K. XRD, RAMAN, FTIR, FE-SEM, HRTEM, EDX and N_2 adsorption at 77 K were used to characterize structure, morphology and composition of the samples. Their adsorption and photodecomposing activities toward methyl orange (MO) were examined at 303 K.

Results and Discussion

The morphology and size were characterized by FE-SEM and TEM. As shown in Fig. 1 (a), two-dimensional planar structure can be clearly observed in the intercalated sample (GOTi). Though no nanoparticles were observed on carbon platelets, the element analysis confirms the presence of Ti and O components beside C in GOTi.[10] This indicates that Ti species were successfully intercalated in between GO layers. After hydrothermal treatment, tubular morphology was produced (Fig. 1 (b)). These nanotubes deposit two-dimensionally

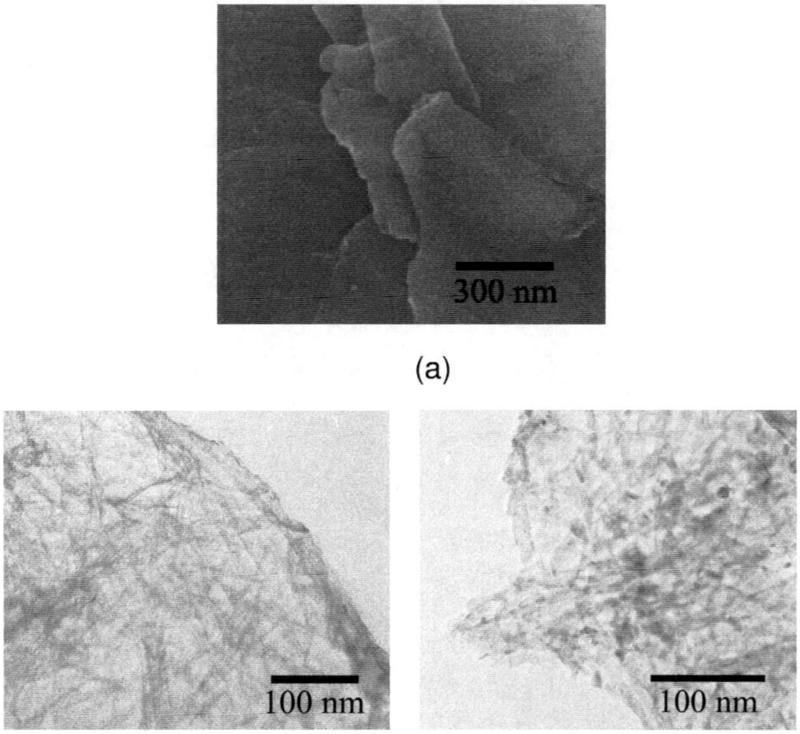

(b) (c)

Fig. 1 (a) FE-SEM image of GOTi, (b) TEM image of C-TNT, and (c) TEM image of C-TNR.

along the surface of carbon nanosheets, forming a stacking composing structure. Further high-resolution TEM observation shows that these tubular materials belongs to hydrogen titanate.

After calcined at high temperature under vacuum, hollow tubes was turned into solid rods. As shown in Fig. 1 (c), these nanorods randomly lie on the surface of carbon nanosheets, with the stacking structure remaining. High-resolution TEM observation reveals that nanorods are single-crystalline anatase titania, obtained by thermal pyrolysis of titanate nanotubes. The anatase structure is proved by Raman spectrum (Fig. 2) from the characteristic peaks of anatase between 140-700 cm^{-1}.[11] In addition, two strong peaks are also found in the high wavelength range, which can be attributed to D and G bands of carbon.

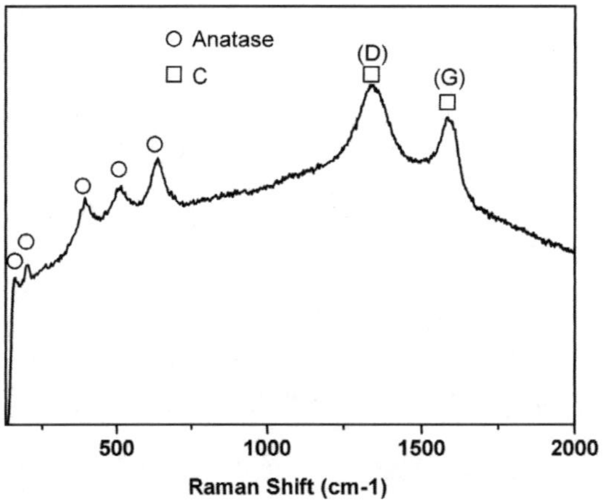

Fig. 2 Raman spectrum of C-TNR.

The adsorbing and catalytic activity of the composites were tested in terms of the degradation of methyl orange in comparison of commercial high-performance anatase ST-01. Before light illumination, the samples were agitated under dark state for 24 h to ensure adsorption equilibrium of MO on the surface of the samples. As presented in the inset

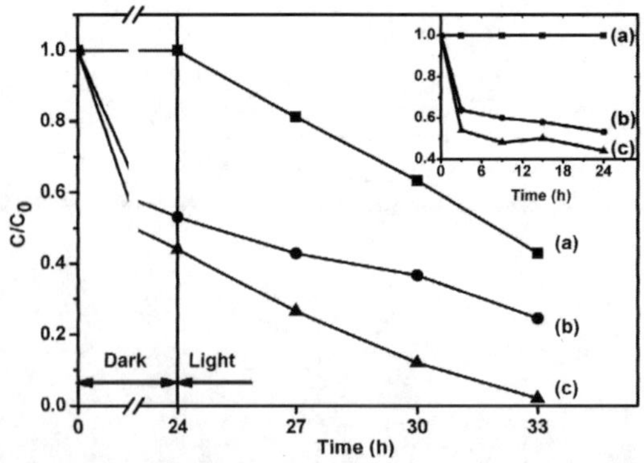

Fig. 3 Adsorption (dark range) and photodegradation (light range) toward methyl orange on (a) ST-01, (b) C-TNT, and (c) C-TNR.

of Fig. 3, under dark state, both the nanotube composite and the nanorod composites exhibit good MO adsorptivity while ST-01 has no adsorption. After dark state, the photocatalytic activity was measured under UV illumination. It can be seen that the photodegradation of MO follows pseu-zero-order equation for all the samples. The calculated degradation rate constants are 3.2, 3.8, and 5.9 x 10^{-3}/h for ST-01, C-TNT, and C-TNR, respectively. The results show that the composites have excellent adsorptivity and photocatalytic activity due to the synergy effect of adsorption and photocatalysis.

Conclusion

In summary, carbon nanosheets have been used as template for the two-dimensional deposition of titanate nanotubes and titania nanorods. The nanotube- and nanorod-based composites exhibit both excellent adsorptivity and photodecomposing ability toward organic pollutants and are promising materials for photocatalytic application.

References

1. Li D, Haneda H, Hishita S, and Ohashi N. (2005), "Visible-light-driven N-F-codoped TiO_2 photocatalysts. 2. Optical characterization, photocatalysis, and potential application to air purification," *Chem. Mater.* , 17 (10), pp. 2596-2602.
2. Goeringer S, Chenthamarakshan CR, Rajeshwar K (2001), "Synergistic photocatalysis mediated by TiO_2: mutual rate enhancement in the photoreduction of Cr(VI) and Cu(II) in aqueous media ," *Electrochem. Commun.*, 3 (6), pp. 290-292.
3. Pichat P, Disdier J, Hoang-Van C,Mas D, Goutailler G, Gaysse C (2000), "Purification/deodorization of indoor air and gaseous effluents by TiO_2 photocatalysis ," *Catal. Today*, 63 (2-4), pp. 363-369.
4. Zhu JF, Chen F, Zhang JL, Chen HJ, Anpo M (2006), "Fe^{3+}-TiO_2 photocatalysts prepared by combining sol-gel method with hydrothermal treatment and their characterization," *J. Photoch. Photobio. A*, 180 (1-2), pp. 196-204.
5. Yu JG, Yu HG, Cheng B, Trapalis C. (2006), "Effects of calcination temperature on the microstructures and photocatalytic activity of titanate nanotubes ,"*J. Mol. Catal. A: Chem.*, 249(1-2), pp. 135-142.
6. Liu YJ, Aizawa M, Peng WQ, Wang ZM, Hatori H, Uekawa N, Kanoh H (2004), "Room-temperature formation of alkoxide-derived anatase nanoparticles by peroxotitanic acid approach ," *Chem. Lett.*, 36 (9), pp. 1094-1095.
7. Liu SX, Chen XY, Chen X (2007), "A TiO_2/AC composite photocatalyst with high activity and easy separation prepared by a hydrothermal method," *J. Hazard. Mater.*, 143 (1-2), pp. 257–263.
8. Lin L, Lin W, Zhu YX, Zhao BY, Xie YC, He Y, Zhu YF (2005), "Uniform carbon-covered titania and its photocatalytic property," *J. Mol. Catal. A*, 236 (1-2), pp. 46-53.

9. Inagaki M, Kojin F, Tryba B, Toyoda M (2005), "Carbon-coated anatase: the role of the carbon layer for photocatalytic performance," *Carbon*, 43(8), pp. 1652-1659.
10. Peng WQ, Wang ZM, Yoshizawa N, Hatori H, Hirotsu T (2008), "Lamellar carbon nanosheets function as templates for two-dimensional deposition of tubular titanate," *Chem. Commun.*, pp. 4348–4350.
11. Balachandran U, Eror NG (1982), "Raman-spectra of titanium-dioxide," *J. Solid State Chem.*, 42 (3), pp. 276-282.

336e Molecular Simulation Study for Chiral Separation of Racemic Mixture

Jianchao Liang[1], Yanping Fan[1], Zhongqiao Hu[2], **Jianwen Jiang**[3], and Raj Rajagopalan[4]. (1) National University of Singapore and The Singapore-MIT Alliance, Singapore, 117576, Singapore, (2) Chemical and Biological Engineering, National University of Singapore, Singapore, Singapore, (3) Chemical land Biological Engineering, National University of Singapore, Singapore, Singapore, (4) Chemical and Biomolecular Engineering, National University of Singapore, Kent Ridge, Singapore, Singapore

Chirality is ubiquitous in living organisms, as nature has evolved to favor one "handedness" over the other. Many pharmaceutically important drugs are racemic mixtures of chiral enantiomers. Clinical use of an improper enantiomer may cause death; therefore, it is important to separate enantiomers before use. Here, a molecular simulation study is presented for the chiral separation of L/D-tryptophan aqueous mixture by â-cyclodextrin film. Consistent with experimental measurements, D-tryptophan transports more rapidly than L-tryptophan. The chiral selectivity depends on the number density of â-cyclodextrin in the film and on the flow rate of the solution through the film. A slow flow rate is found to give relatively larger separation efficiency. The mechanism responsible for the enantio-discrimination is explored at the molecular scale including the preferential binding and the intermolecular forces between the analyte molecule and â-cyclodextrin. The estimated binding energy is stronger between L-tryptophan and â-cyclodextrin, which gives rise to the observed chiral separation.

336f Local Membrane Characterization with Flux Lateral Force Microscopy

Jason P. Killgore and Rene M. Overney. Chemical Engineering, University of Washington, Box 351750, Seattle, WA 98195

Bulk permeability measurements provide an incomplete descriptor of permeability when considering heterogeneous or anisotropic membranes. By positioning a scanning force microscope (SFM) onto a specially mounted membrane, local variations in friction force can be measured directly at the surface, with nanoscale resolution. Analogous to a puck on an air hockey table, permeate flow in membranes contributes to a local reduction in surface friction. Plotting the friction force determined by SFM versus pressure gives a gradient that is nominally proportional to permeance. Here, this new technique coined Flux Lateral Force Microscopy (Flux-LFM) is applied to Zeolites, Nafion, and reverse selective super glassy polymer membranes. In polymer membranes the signal is shown to contain information on permeability, as well as local rheological changes resulting from sorption and phase transitions.

340a Adsorption of Carbon Dioxide on Modified SBA-15

Vladimir Zelenak[1], Maria Badanicova[1], and **Flor R. Siperstein**[2]. (1) Department of Inorganic Chemistry, P.J. Safarik University, Faculty of Science, Moyzesova 11, Kosice, SK-041 54, Slovak Republic, (2) School of Chemical Engineering and Analytical Science, University of Manchester, Po Box 88, Sackville Street, Manchester, M60 1QD, United Kingdom

Adsorption of carbon dioxide and methane up to 20 bar has been measured in modified SBA materials. Measurements were carried out using a gravimetric equipment (IGA-001 from Hiden-Isochema).

We compare adsorption capacity, selectivity and uptake kinetics for materials that have been grafted with diamine ligands with those obtained through cosynthesis of TEOS with alumina, zirconia and titania.

CO_2 adsorption capacity at 20 bar was larger for samples modified with alumina > titania > zirconia > diamine. More importantly, adsorption kinetics was significantly slower for the samples that contained grafted diamines.

Carbon dioxide capture and sequestration will likely become an important technology in the next decades, as concern about climate change drives governments to consider regulating green-house gas emissions. Undoubtedly, one of the largest sources of human-generated CO2 is from coal-fired power plants. Thus, there has been a recent emphasis on assessing current CO2 separation technologies for application to dilute CO2 streams with large water contents, as well as developing new technologies for this separation. One of the most promising is adsorptive separation using amine-modified organic or inorganic polymer sorbents. Unfortunately, a significant drawback with sorbents of this type has been low CO2 sorbent capacities, requiring very large masses of (usually expensive) adsorbent. In this talk, we will describe a new, rationally designed hybrid aminopolymer-silica sorbent, deemed a hyperbranched aminosilica (HAS), with a very large, low temperature adsorption capacity. The material is low cost, owing to the simple method of synthesizing the adsorbent. In addition, the adsorbent can be used with effectively no change in capacity over multiple cycles (10 or more times), demonstrating the potential to have the durability needed for practical application. The sorbent is compared to the leading published supported amine sorbents. The structure of the sorbent and the attributes that endow it with excellent adsorption capacities are described.

340c Development of Sulfur Tolerant High Durability CO2 Sorbents at High Temperatures

Hong Lu and Panagiotis (Peter) Smirniotis. Chemical & Materials Engineering Department, University of Cincinnati, 2600 Clifton Avenue, Cincinnati, OH 45221

Inorganic metal oxides based on reversible reactions have advantages of reacting with CO_2 at high temperature over other sorbents. We presented here that the structure promoted calcium-based sorbent is promising for CO_2 uptake. The sorbents were made by wet chemistry routes, and flame spray pyrolysis in the gas phase. The sorbents were nanostructured particles having specific surface areas as high as 70 m^2/g. The CO_2 uptake at various temperatures was determined by thermogravimetric analysis. To study the reversibility of the sorbents, several dozens of carbonation/decarbonation cyclic operations were performed. The sorbents demonstrated stable and reversible behavior and high CO_2 uptake capacity, sustaining the maximum molar conversion at more than 60% even after 100 cycles of operations. Sulfur tolerant sorbents were also identified.

340d Adsorption of CO2 on Rock Surfaces by Molecular Simulations

Luzheng Zhang[1], Robert Lee[1], and Brian McPherson[2]. (1) Petroleum Recovery Research Center, New Mexico Institute of Mining and Technology, 801 Leroy Place, Socorro, NM 87801, (2) Civil Engineering, University of Utah, 122 South Central Campus Drive, 104 CME, Salt Lake City, UT 84112

It is well knewn now that CO2 emission from the combustion of fossil fuels is a major reseason for the global warming. It is necessary to develop technologies that will allow us to utilize the fossil fuels while reducing the emissions of the green house gas. Existing commercial CO2 capture technology is very expensive and energy intensive. Improved technologies for CO2 capture are necessary to achieve low energy penalties. Preferential adsorption of CO2 on a porous adsorbent is one of key issues for the CO2 remove from gas stream. On the other hand, the CO2 capture from large emission sources is to produce a concentrated stream of CO2 that can be transported and sequestered underground or in deep oceans for different purposes, such as enhanced oil recovery (EOR). Understanding of CO2 adsorption on porous formations under geological conditions plays a key role in evaluating storability and safety in the practice of CO2 sequestration and/or EOR.

In this work, we studied adsorption of CO2, N2 and H2O on difference porous 'rock' surfaces (quartz, calcite, magnesite, etc) by molecular simulations (grand canonical Monte Carlo and molecular dynamics). Adsorption isothermals and dynamic behavior of CO2 on the porous rock surfaces are in detail analyzed. Preliminary results show that the adsorption of CO2 on these porous rock surfaces decreases with temperature, but increases with pressure dramatically and that the adsorption of CO2 is much more than that of N2 or H2O on the porous rock surfaces at certain conditions. In addition, we compared with the CO2 adsorption on well-defined NaY zeolite. This study demonstrates the possibility of CO2 sequestration at difference geological conditions. It also presents a technical approach to evaluating the CO2 storability, and more importantly, a fundamental understanding for further studies relative CO2 sequestration, e.g., gas breakthrough pressure.

340e Metal Organic Frameworks for Carbon Dioxide Adsorption from Flue Gas

Annabelle I. Benin[1], Syed Faheem[1], John J. Low[2], Richard R. Willis[1], Antek G. Wong-Foy[3], Kyoungmoo Koh[3], Adam J. Matzger[3], A. Ozgur Yazaydin[4], Randall Q. Snurr[4], Xiayi Hu[5], Stefano Brandani[5], Jian Liu[6], and **M. Douglas LeVan**[6]. (1) New Materials Research, UOP LLC, 50 East Algonquin Road, Des Plaines, IL 60017, (2) Advanced Characterization, UOP LLC, 50 East Algonquin Road, Des Plaines, IL 60017, (3) Department of Chemistry, University of Michigan, 930 N. University, Ann Arbor, MI 48109, (4) Department of Chemical and Biological Engineering, Northwestern University, 2145 Sheridan Road, Technological Institute E136, Evanston, IL 60208-3120, (5) School of Engineering and Electronics, University of Edinburgh, Sanderson Building, The King's Buildings, EH9 3JL, Edinburgh, United Kingdom, (6) Department of Chemical and Biomolecular Engineering, Vanderbilt University, VU Station B 351604, 2301 Vanderbilt Place, Nashville, TN 37235-1604

Our objective is to incorporate high capacity metal organic framework (MOF) materials into commercially viable carbon dioxide capture applications for coal-fired power plant flue gas. Owing to their very open, accessible and tunable porosity and favorable heats of adsorption, several MOFs show promise in this application. Our approach involves a combination of molecular modeling with synthesis and characterization of novel MOFs. This talk will focus on the synthesis, characterization, and stability and adsorption testing of MOFs for this application. MOFs are a relatively new class of nanoporous materials synthesized in a rational, "building-block" approach by self-assembly of metal or metal oxide vertices interconnected by rigid organic linker molecules. This rational synthesis approach has opened up the possibility of incorporating a wide variety of metal and metal oxide vertices, as well as various functional groups into a MOF structure. Significant progress toward a fundamental understanding of the characteristics required for a MOF suitable for flue gas operation has been achieved. For example, it is now clear that ultra-high surface area MOFs are not suitable for ambient temperature and sub-atmospheric pressure conditions typical for flue gas. A molecular modeling approach on MOF hydrolysis that correlates with experimental data will be provided. The experimental data were generated by subjecting MOF samples to relevant steaming conditions in a combinatorial chemistry heat treatment unit followed by analysis in a high-throughput X-ray diffractometer. Additional results were collected via an in situ high temperature, steam-atmosphere equipped X-ray diffractometer. Many types of MOF materials were prepared and evaluated, ranging from compounds from the open literature to new families of MOF materials prepared via combinatorial and other in-house synthesis strategies. Finally, detailed molecular modeling of adsorption isotherms and MOF hydrolysis show that theory and experiment match closely.

340f Pressure Swing Adsorption Cycles for Carbon Dioxide Capture

Amal Mehrotra[1], Armin D. Ebner[1], and **James A. Ritter**[2]. (1) University of South Carolina, Columbia, SC 29208, (2) Department of Chemical Engineering, University of South Carolina, 301 Main St., Columbia, SC 29208

There are many misconceptions about and misleading reports on the viability of pressure swing adsorption (PSA) cycles for CO2 capture from flue gas. These misconceptions and misleading reports stemmed largely from a 1994 International Energy Agency report [1] that evaluated CO2 separation and capture from flue gas using 13X zeolite in both pressure swing adsorption (PSA) and temperature swing adsorption (TSA) processes. They concluded that PSA and TSA are too energy intensive for use with gas and coal fired power systems. This conclusion has led others to extrapolate their findings and further conclude that adsorption systems, in general, are not applicable for CO2 separation and capture from flue gas. This disturbing misinformation has perpetuated for nearly a decade with essentially the same misleading statements being made in a 2003 IEA report [2]. However, it is strongly contended that those reports might have been factually inaccurate, that some grave errors might have been made in the evaluation of cyclic adsorption technologies for CO2 capture from flue gas, and that PSA might be very applicable to the capture of CO2 from flue gas. These claims are substantiated with evidence from works found in the literature and from some of the author's recent works.

First and foremost, a recent report by Izumi [3] from Mitsubishi Heavy Industry indicated that CO2 was being recovered from flue gas commercially using a layered PSA bed containing X and A type zeolites and activated carbon. Although, not much detail about the PSA cycle was provided, this news was exciting. The Japanese must have discovered something positive about using PSA for CO2 capture from flue gas that has been overlooked by investigators in the USA and apparently everyone else around the world!

Second, in a series of works by the Ritter and co-workers [4-7], novel PSA cycles have been explored systematically for the capture of CO2 from flue gas at high temperature using a K-promoted hydrotalcite like compound (HTlc) as the adsorbent. They revealed that a variety of vacuum swing adsorption (VSA) cycles could be used to concentrate CO2 to over 90 vol% at over 90% recovery from a feed stream containing 15 vol% CO2. The most effective cycles were based on the heavy reflux concept, a concept that has been largely missing from the PSA literature. What is really exciting about these high temperature heavy reflux PSA cycles is that they are completely applicable to ambient temperature adsorbents like zeolites and activated carbons. Moreover, the Ritter and co-workers are still working on the problem and have discovered some better heavy reflux PSA cycles for CO2 capture from flue gas that will be discussed during this presentation. They also feel that the lack of information on and therefore the lack of understanding of the heavy reflux PSA concept for concentrating the heavy component from a feed stream might be the principle reason why the IEA reports were mistaken in their assessments of PSA for CO2 capture from flue gas. To this end, the heavy reflux concept will also be discussed during this presentation

Third, some very recent work by Webley and co-workers [8] substantiated the claims made by the Ritter and co-workers over the past few years on the economic viability of PSA cycles for CO2 capture from flue gas, especially VSA cycles. They proclaimed that the information about the cost of PSA provided in the 2003 IEA report [2] was in error. Zhang et al [8] recalculated the cost of PSA to be around $67/tonne of CO2 produced compared to that in the IEA report of $97/tonne. This new cost for PSA, as verified through the IEA, compared much more favorably with amine scrubbing technology at $60/tonne. What is most exciting about this recent set of events is that the PSA cycles analyzed by the IEA were not designed for heavy component recovery, like CO2; instead, they were designed for light component recovery, like CH4 from a CH4/CO2 mixture.

As alluded to above, there are significant differences between these two kinds of PSA cycles, i.e., between the ones designed for light component enrichment and recovery and those designed for heavy component enrichment and recovery, with understanding of the later lacking, even by some of the best PSA researchers in the field. To make matters worse, the design of a multi-bed PSA process with complex cycle sequencing, like that envisioned for CO2 capture from flue gas, is a non-trivial exercise. PSA beds are always coupled together, usually contain more than one layer of adsorbent, and operate sequentially with each undergoing cycle steps such as pressurization, feed, heavy reflux, equalization, depressurization, light reflux, and repressurization. As an added complication, each cycle step is not necessarily of the same duration. These features make the number of possible cycle configurations very large; and unfortunately, design strategies on how to best configure such a complex PSA cycle is more of an art than a science. To this end, this presentation will also introduce some complex VSA cycle sequences that might be applicable for CO2 capture from flue and stack gases.

References:

1. International Energy Agency, "Carbon Dioxide Capture from Power Stations," 1994.

2. International Energy Agency, "Carbon Dioxide Capture at Power Stations and Other Major Point Sources," 2003.

3. J. Izumi, "Adsorption technology applications to the industries in Japan," Abstract and presentation at the 8th International Conference on Fundamentals of Adsorption, Sedona, AR, May (2004).

4. S. P. Reynolds, A. D. Ebner and J. A. Ritter, New Pressure Swing Adsorption Cycles for Carbon Dioxide Sequestration," Adsorption, 11, 531-536 (2005).

5. S. P. Reynolds, A. D. Ebner and J. A. Ritter, "Carbon Dioxide Capture from Flue Gas by PSA at High Temperature using a K-Promoted HTlc: Effects of Mass Transfer on the Process Performance," Environmental Progress, 25, 334-342 (2006).

6. S. P. Reynolds, A. D. Ebner, and J. A. Ritter, "Stripping PSA Cycles for CO2 Recovery from Flue

Gas at High Temperature Using a Hydrotalcite-Like Adsorbent," Ind. Eng. Chem. Res., 45, 4278-4294 (2006).

7. S. P. Reynolds, A. Mehrotra, A. D. Ebner and J. A. Ritter, "Heavy Reflux PSA Cycles for CO2 Recovery from Flue Gas. Part I. Performance Evaluation," Adsorption, 14 399-413 (2008).

8. J. Zhang, P. A. Webley, and P. Xiao, "Effect of Process Parameters on Power Requirements of Vacuum Swing Adsorption Technology for CO2 Capture from Flue Gas," Energy Conversation and Management, in press on line (2007).

341a Reducing the Energy Demand of Bio-Ethanol through Salt-Extractive Distillation and Electrodialysis

Mohammed A. Hussain and Peter H Pfromm. Chemical Engineering, Kansas State University, 1005 Durland Hall, Manhattan, KS 66506

Bio-ethanol from corn currently consumes about 34,000 BTU in form of natural gas to produce one gallon of ethanol representing 76,000 BTU as lower heating value (U.S. industrial practice data, 2007). Separating ethanol from water consumes about 40% of the natural gas demand cited above. Saline extractive distillation of alcohol-water mixtures and fermentation broth has been considered elsewhere and fairly comprehensive experimental data, thermodynamic data, and simulations are available. Potentially very significant energy savings and process simplifications could be predicted. However, the recovery and recycling of the salt used to facilitate distillation has not been addressed beyond simple evaporation of the bottoms to recover pure dehydrated salt for recycling to the reflux (historic HIAG process) or recovery of a strong brine from the bottoms to feed a separate azeotrope breaking stage (IHI process). Electrodialysis is uniquely suited for salt recovery from the saline extractive distillation column bottoms for the case of ethanol recovery from fermentation broth since salt is selectively removed from the solution (no water is evaporated) and the electrodialysis membranes and overall fluid handling are tolerant to fermentation broth and even to entrained particulate matter. Concepts, modeling, and experimental data for electrodialysis-enabled salt extractive ethanol distillation will be shown. Aqueous/aqueous and aqueous/ethanol electrodialysis will be discussed.

Web Page: www-personal.ksu.edu/~pfromm/index.html

341b Control of the Heterogeneous Azeotropic N-Butanol/water Distillation System

William L. Luyben, Chemical Engineering, Lehigh University, Bethlehem, PA 18015

The interest in n-butanol as a biofuel has increased in recent years because of its superior fuel qualities compared to ethanol: higher octane number, lower heat of vaporization ans lower vapor pressure. However, the concentration of n-butanol coming from the fermenter is lower than that achieved in ethanol fermentation. In addition, acetone and ethanol are also produced. Recent studies to improve yield and increase n-butanol concentration have explored fed-batch systems with stripping, adsorption, liquid-liquid extraction, distillation and/or pervaporation to recover products.

The production of n-butanol involves removing the acetone and ethanol, and separating the n-butanol from the water. This separation is made difficult by the presence of an azeotrope. Fortunately, unlike the homogeneous azeotrope found in the ethanol-water system, the n-butanol-water azeotrope is heterogeneous, so two liquid phases occur in the decanter. Therefore a simple two-column distillation system can be used.

This paper studies the control of the n-butanol-water azeotropic distillation system and develops a simple effective control structure that is capable of handling very large disturbances in throughput and feed composition. The control system is very robust and is demonstrated to be applicable for systems that are designed for a wide range of feed compositions

341c 4-Pass Tray Design Techniques

Daniel R. Summers, Sulzer, 4019 South Jackson Avenue, Tulsa, OK 74107

4-Pass trays have been employed for many decades in the chemical and refining industries. Many papers have been published on this subject, but none give a concise methodology on how to design and balance the V/L ratio on each tray panel. This paper will discuss the methodology of balancing 4-pass trays regardless if the design is based on equal bubbling area or an equal flow path length and whether or not the trays have vapor tunnels. In addition, deficiencies in current pressure drop methods will be discussed and how they relate to and affect 4-Pass tray design.

341d Damage to Column Internals

Daniel Staak, Jens-Uwe Repke, and Günter Wozny. Department of Process Dynamics and Operation, Technical University of Berlin, Straße des 17. Juni 135, Berlin, 10623, Germany

According to Kister [1] damage to tower internals is the number 3 on the top 10 causes of column malfunctions. High costs arise due to repairing costs and production downtime. Most of the destruction is caused by mechanical load, seldom a high temperature is the cause. The design of distillation columns and column internals is directed to the necessary separation performance. In contrast to that the mechanical layout of column internals is of subordinate interest and mostly done according to guidelines applying to stationary operating conditions. For example an industry standard is the differential pressure column trays have to withstand. The minimum differential pressure is in the order of 15 mbar and can be applied from both directions. Due to the enhanced competition most tray manufacturers design for this minimum load. Another industry standard is the vapor pressure drop over a tray which is in the order of 2.5 to 7.5 mbar. However these boundaries are only valid for stationary operation conditions and can be easily exceeded if the column is leaving the field of normal operation. Regarding the fact that the differential pressure is a quadratic function of the vapor flow, an increase of 50% vapor flow (beginning at stationary conditions with ΔP=7.5 mbar) would already lead to a pressure drop which is beyond the standard design limit (ΔP=17 mbar). For a better understanding of the dynamic behavior in this field of operation, experimental investigations were carried out on a pilot scale column. The column has a height of 4 m and a diameter of 0.1 m and is equipped with 16 sieve trays. The column is provided with differential and absolute pressure sensors and automated with Freelance2000 from ABB. A glass section gives an inside view on the events happening in the column. Additionally a measurement technique based on strain gages was developed, tested and implemented inside the column on the tray suspension to measure directly and online the mechanic load on a tray during operation [2]. Two important scenarios were chosen and experimentally investigated: a) Phase separation in reboiler causing first boiling delay and afterwards a violent boiling, producing a rapid upward flow b) Emergency pressure relief with an oversized relief valve causing a rapid upward flow In the presentation a summary of the experimental investigation will be given. Results concerning maximum differential pressure and maximum force on a tray, even lifting of trays after an operation failure will be shown. Measures and methods to prevent comparable incidents in the future will be suggested. [1] H. Z. Kister: Distillation Troubleshooting, New Jersey USA: Wiley-Interscience, ISBN-13 978-0-0471-46744-1, 2006 [2] D. Staak, J.-U. Repke, G. Wozny: Outflow conditions and strains on internals during pressure relief events on distillation columns, Prag, Pres Conference, 2006

Lifted
Downcomer
Tray 1
Tray 2 (lifted)
Glas Section
1cm

341e Molecular Distillation *Modeling and Simulation to Split Heavy Petroleum Fractions*

Lamia Zuñiga Liñan[1], Nádson M. N. Lima[1], Maria Regina Wolf Maciel[1], Rubens Maciel Filho[1], César B. Batistella[1], **Flavio Manenti**[2], Davide Manca[2], and Lílian C. Medina[3]. (1) School of Chemical Engineering, University of Campinas, UNICAMP, P.O. Box 6066, 13083-970, Campinas-SP, Brazil, (2) CMIC Dept. "Giulio Natta", Politecnico di Milano, Piazza Leonardo da Vinci, I-20133, Milano, Italy, (3) Processos Valoração de Petróleo CENPES/PDP/TPAP, PETROBRAS, Rio de Janeiro-RJ, Brazil

Modeling and simulation of molecular distillation to split of heavy petroleum residues (400°C + AET) are proposed. A theoretical study of this problem is illustrated for a heavy petroleum cut (372°C-811°C AET) divided into six pseudo-components.

In the process simulation, a falling film, a heater and an extremely low pressure evaporator are considered.

The mathematical model comprises equations for the evaluation of the physico-chemical properties, in order to characterize the distilled mixture. Physico-chemical properties are formulated through the adjustment of experimental parameters such as molecular mass, specific heat, viscosity, and density.

Heat and material balances on the liquid film are numerically solved by using a central finite-difference method. Specifically, the Crank-Nicolson method is adopted.

The simulation is carried out at the steady-state conditions, where the relevant process variables, such as film thickness, surface evaporation rate, liquid interface temperature, concentration profiles, and amount of distillate flow rate are computed.

In molecular distillation, the concentration of the most volatile components shrinks in both axial and radial directions, especially due to the fast increase of the temperature in the falling film. As a consequence, the less volatile component of the liquid mixture undergoes a fast increase on the liquid interface, by clearly showing the potentialities of the molecular distillation process in separating heavy petroleum residues.

As the results show, the inlet variables such as the temperature and the feed flow rate largely influence the final composition of the condensate flow.

Keywords: Molecular distillation; Simulation; Heavy petroleum residues; Physico-chemical properties.

341f Optimal Design and Operation of Batch Distillation Columns with Regard to Reboiler Liquid Swelling

Levente L. Simon and Konrad Hungerbuehler. Institute for Chemical and Bioengineering, Safety and Environmental Technology Group, ETH Zurich, G-138, HCI, ETH Hoenggerberg, Zurich, 8093, Switzerland

Batch distillation is a widely used unit operation for the separation of liquid mixtures and it is preferred when small quantities of high-cost chemicals have to be separated: fine chemicals, essential perfume oils and pharmaceutical products.

The productivity of the batch distillation columns is affected by the design and operation parameters. While the batch distillation design problem considers the determination of the vapor load, number of stages and column structure (regular or unconventional), the column operation problem deals with the calculation of the reflux ratio as constant optimal value or optimal profile. The batch time is fixed or is an additional optimization variable. Previous works have analyzed the influence of the vapor rate on the batch distillation columns and have concluded that for the unlimited market demand the vapor rate should be as high as possible.

During the operation of the distillation columns at low pressure conditions (depending on the reboiler volume and rectification column diameter) the maximum vapor rate may induce reboiler liquid swelling before causing rectification column flooding. Liquid swelling occurs when the vessel content level rises due to a vapor stream which passes through the liquid. A practical solution to avoid reboiler swelling related problems is to decrease the batch size (filling level) to a safe value; however, the batch size decreases as well. Since the distillation productivity is determined by the batch size and batch time, one should attempt to find an optimum between a large batch size (high filling degree) and fast operation (large vapor rate).

This contribution addresses the optimal design (determination of the batch size) and operation (optimal reflux ratio and vapor load profile) of batch distillation columns under the reboiler liquid swelling and column flooding constraints. To our knowledge the batch distillation design and operation optimization with regard to reboiler liquid swelling has not been tackled previously in the literature.

The batch distillation column is modeled using a tray-by-tray model based on mass balances (differential equations), the true liquid level in the reboiler and the rectification column flooding are calculated using hydrodynamic equations. The design optimization problem is implemented in the outer loop, whereas the optimal control problem which gives the optimal operation is solved in the inner loop. The manipulated variables are the batch size (outer loop), the reflux ratio and vapor flow rate profiles (inner loop). The constraints of the optimization problem are posed by the batch distillation model and the algebraic equations which describe the hydrodynamic conditions. The outer optimization is solved by enumeration since it is one variable and the range of the optimal value is approximately known. The inner optimal control problem, which is computationally much more demanding, is based on the SQP algorithm. The SQP method is implemented in a user-friendly software environment, OptCon, which provides a Matlab interface for easy problem setup. The optimization approach is based on a large-scale nonlinear optimization solver (HQP), which offers a very efficient technique, based on a multiple shooting algorithm. The integrator is the DASPK and the multiple-shooting optimal control algorithm is implemented and compiled as C language library. The model equations are implemented in a Simulink C file.

The OptCon software package solves in about 1-2 seconds the optimal control problem of several hundreds of variables. Therefore, the control profile with large number of discretization intervals can assure a detailed solution and several scenarios can be quickly and efficiently evaluated. The OptCon software is hosted at the Loughborough University, U.K., by Zoltan K. Nagy (http://www-staff.lboro.ac.uk/~cgzkn/).

The optimizations are performed for industrial scale batch distillation reboilers and rectification columns. Furthermore, we analyze the sensitivity of the optimal solution towards the light component ratio in the mixture and towards the start-up time.

341g Optimal Operation and Feedback Control of Reactive Batch Distillation In Middle-Vessel Columns

Harvey Arellano-Garcia, Irisay Carmona, and Guenter Wozny. Chair of Process Dynamics and Operation, Berlin Institute of Technology, Str. des 17. Juni 135, 10623, Berlin, Germany

The transient nature of batch distillation allows for configuring of batch columns in a number of different ways. In particular, when chemical reactions and physical separations have some overlapping operating conditions the combination of these tasks in a single process unit can offer significant benefits.

In this work, we present the optimal operation and feedback control of a new operation mode for reactive batch distillation in middle vessel columns. For this purpose, the feed mixture is loaded into the middle vessel between the two separation sections, and the products are simultaneously obtained from the top and the bottom of the column reducing the way of separation. The special feature of this novel operation mode lies on the fact that depending on the characteristics of the reaction mixture, the reaction will also take place either along the upper column or along the lower column. In the former case, the reaction will be enhanced by adding one of the educts into the top of the column. In the latter, the educt feed will be supplied to the reboiler of the lower column, and the most volatile components will be evaporated out of the lower column towards the upper column through a vapour bypass configuration during the batch. Thus, the efficiency of reactive batch distillation can be improved, e.g., less batch time and energy consumption and more product amount.

The operation and control performance of the new operation mode is studied through simulation of two industrial reactive batch distillation processes. The first case study is a trans-esterification of two esters and two alcohols. To simulate both processes, a dynamic rigorous detailed model is used. Comparisons of the new operation mode with the conventional options clearly indicate the improvement by the new operation mode.

341h Development of Corrosion Control Document (CCD) In Crude Distillation Unit Process

Junghwan Kim[1], Wonsub Lim[2], Sang-Rok Park[3], and Il Moon[2]. (1) Chemical and Biomolecular Engineering, Yonsie University, A446, 134 Shinchondong, Seodaemunku, Seoul, South Korea, (2) Chemical and Biomolecular Engineering, Yonsei University, A446, 134 Shinchondong, Seodaemunku, Seoul, South Korea, (3) Reliability Team, GS Caltex Corporation, 1056, Wollae-dong, Yeosu-si, Jeollanam-do, South Korea

This paper deals with techniques to improve refinery reliability, availability and profitability. Our team developed a CCD for processing of Crude Distillation Unit (CDU). According to recent study the Loss due to corrosion in US is around 276 billion $. It's a big concern for both managers and engineers of refinery industry.

The CCD consist of numerous parts namely damage mechanism (DM), design data, critical reliability variable (CRV), guidelines, etc. The first step in development of CCD is Material Selection Diagram (MSD). We need to choose damage mechanisms affecting equipments and process. These processes are based on API 571. The selected nine DM from API 571 are (1) creep/stress rupture, (2) Fuel Ash Corrosion, (3) Oxidation, (4) High Temperature Sulfidation, (5) Naphthenic Acid Corrosion, (6) Hydrochloric Acid (HCL) Corrosion, (7) Ammonium Chloride (Salt) Corrosion, (8) Wet H2S Corrosion and (9) Ammonia Stress Corrosion Cracking. Each DM related to corrosion of CDU process was selected by design data, P&ID, PFD, corrosion loop, flow of process, equipment's history and experience. Operating variables affecting severity of DM were selected in initial stage of CRV. After the CRV were set, we proposed the guidelines for reliability of equipments. The CCD has been developed on the basis of the corrosion control in refinery industry. It also improves the safety of refinery process and reduces the cost of corrosion by a big factor.

Retrofit of Distillation Columns in a Methanol Plant

Yaşar Demirel
Department of Chemical and Biomolecular Engineering, University of Nebraska Lincoln
ydemirel2@unl.edu

ABSTRACT

Through its 'Column Targeting Tools' the Aspen plus simulator performs the 'Thermal Analysis' for rigorous column (RadFrac) calculations and produce Column Grand Composite Curves (CGCC) and exergy loss profiles. The CGCC displays the net enthalpies for the actual and ideal operations at each stage, and the cold and hot heat utility requirements, while the exergy loss profiles indicate the level of irreversibility at each stage including the condenser and reboiler. Therefore, the thermal analysis can identify the targets for restructuring and modifications, and may be helpful in suggesting retrofits. Some of the retrofits consist of feed conditioning (preheating or precooling), feed splitting, reflux adjustments, and adding side condensers and reboilers. These retrofits target a practical near minimum thermodynamic loss and suggest modifications for existing distillation columns by increasing the efficiency in energy utilization. This study uses the CGGC and the exergy loss profiles to assess the performance of the existing distillation columns, and reduce the costs of operation by appropriate retrofits in a methanol plant. Effectiveness of the retrofits is assessed by means of thermodynamics optimum and economics. The methanol plant is based on steam reforming and utilizes two distillation columns to purify the methanol in its separation section. The first column operates with 51 stages, has a side heat stream to the last stage, a partial condenser at the top and a side condenser at stage 2, and no reboiler. The second column operates with 95 stages, has a side heat stream to stage 95, a total condenser, and high reflux ratio. For the first column, the retrofits consisting of a feed preheating and a second side condenser at stage 4 have reduced the total exergy loss by 21.5%. For the second column, the retrofits of two side reboilers at stages 87 and 92 have reduced the total exergy loss by 41.3%. After the retrofits, the thermodynamic efficiency has increased to 55.4% from 50.6% for the first column, while it has increased to 6.7% from 4.0% for the second. The suggested retrofits have reduced the exergy losses and hence the cost of energy considerably, and proved to be more profitable despite the increases fixed capital costs of the distillation columns of the methanol plant.

Key Words: Thermodynamic analysis, Column grand composite curves, Exergy loss, Retrofitting, Thermodynamic efficiency

INTRODUCTION

Retrofits suggest modifications for existing distillation columns to reduce the costs of operations by increasing the efficiency in energy utilization [1-7]. Thermodynamic Analysis (TA) is one the methods for the retrofits by reducing the thermodynamic losses due to heat and mass transfer, pressure drop, and mixing in an existing design and operation. As a result, for example in a binary distillation, operating curves come closer to the equilibrium curve, and reflux ratio approaches to its minimum value. To analyze the performance of an existing column quantitatively for exploring the energy-saving potential, it is customary to construct the temperature enthalpy, called the Column Grand Composite Curves (CGCC), and the stage exergy loss profiles [1,2,7]. The CGCC displays the net enthalpies for the actual and ideal operations at each stage, and the cold and hot heat utility requirements [1,3], while the exergy loss profiles indicate the level of irreversibility at each stage including the condenser and reboiler [4,8-10]. Therefore, the area between the actual and the ideal operations in a CGCC should be small, and exergy losses should be lower for a thermodynamically efficient operation. The CGCC is constructed by solving the mass and energy balances for a reversible column operation. The stage exergy loss profiles are generated by the stage exergy balance calculations with a reference temperature.

The CGCC and stage exergy loss profiles are becoming readily available even for multicomponent, complex distillation column operations such as crude oil distillation by a suitable simulation package. For design and retrofit purposes, the CGCC and exergy loss profiles can identify the targets for restructuring and modifications, and suggest retrofits. Some of the retrofits consist of feed conditioning (preheating or precooling), feed splitting, reflux adjustments, and adding side condensers and reboilers. These retrofits target a practical near minimum thermodynamic loss [1,3]. This study presents the use of the CGCC and exergy loss profiles generated by the Aspen Plus simulator to assess the existing operations, and suggest retrofits, for the distillation columns in the separation Section of a methanol production plant.

THERMODYNAMIC ANALYSIS (TA)

Distillation columns operate with inevitable thermodynamic losses due to mixing, heat and mass transfer, pressure drops, internal stage design, and configuration of columns, such as the numbers of feeds and side products. TA combines the first and second laws of thermodynamics, and determines the net enthalpy deficits as well as the losses of available energy called the exergy losses due to irreversibilities at each stage of a column. The distributions of the enthalpy deficits and exergy losses can identify the scope and extent of retrofits required [10-12,14,17-19]. The Column-Targeting tool of Aspen Plus based on TA performs the thermal analysis, and produces the CGCC and the exergy loss profiles for rigorous column calculations based on the practical near-minimum thermodynamic condition [1]. This condition targets reversible column operation with negligible entropy production (or loss of exergy). To achieve that, heaters and coolers with appropriate duties would operate at each stage; reflux ratio would be close to its minimum, and hence the operating lines approaches to the equilibrium curve. Whether a retrofit is economical or not would only be known after an overall optimization, which seeks the best solution for a whole plant under specific constraints. Therefore, the relations between the energy efficiency and capital cost must be evaluated [11,12]. In a simpler approach, one may estimate the trade offs between the costs of retrofits and savings due to the reduced exergy loss equivalent of fuel or electricity.

Column Grand Composite Curve (CGCC) - Temperature-enthalpy profiles of CGCC represent the theoretical minimum heating and cooling requirements over a temperature range. Using the equilibrium compositions of light *L* and heavy *H* key components, minimum vapor and liquid flow rates leaving the same stage with the same temperatures can be estimated from the following mass balances

$$V_{\min} = \frac{1}{y_L^*}(D_L + L_{\min} x_L^*) \qquad (1)$$

$$L_{\min} = \frac{1}{x_H^*}(V_{\min} y_H^* - D_H) \qquad (2)$$

The enthalpies for the minimum vapor and liquid flows are obtained from the molar flow ratios

$$H_{\text{Vmin}} = H_V^* \left(\frac{V_{\min}}{V^*} \right) \qquad (3)$$

$$H_{\text{Lmin}} = H_L^* \left(\frac{L_{\min}}{L^*} \right) \qquad (4)$$

where V^* and L^* are the molar flows of equilibrium, and H_V^* and H_L^* are the enthalpies of equilibrium vapor and liquid streams leaving the same stage, respectively. From the enthalpy balances at each stage, the net enthalpy deficits are obtained

$$H_{\text{def}} = H_{\text{Lmin}} - H_{\text{Vmin}} + H_D \quad \text{(before the feed stage)} \qquad (5)$$

$$H_{\text{def}} = H_{\text{Lmin}} - H_{\text{Vmin}} + H_D - H_{\text{feed}} \quad \text{(after the feed stage)} \qquad (6)$$

After adding the individual stage enthalpy deficits to the condenser duty, the enthalpy values are cascaded, and plotted in the CGCC [1,3]. At the feed stage, mass and energy balances differ from a stage

$$H_{\text{def,F}} = Q_C + D[H_D + H_L(x_D - y_F^*)/(y_F^* - x_F^*) - H_V(x_D - x_F^*)/(y_F^* - x_F^*)] \quad (7)$$

The values of y_F^* and x_F^* may be obtained from an adiabatic flash for a single phase feed, or from the constant relative volatility estimated with the converged compositions at the feed stage and feed quality. In a CGCC, a pinch point near the feed stage occurs for nearly binary ideal mixtures. A horizontal distance between the CGCC pinch point and the vertical axis represents the excess heat, and therefore the scope for reduction in reflux ratio [1-3,16]. For smaller reflux ratios, the CGCC will move towards the vertical axis, and hence reduce the reboiler and condenser duties, which may be estimated by [3]

$$Q_R - Q_{R,\min} = Q_C - Q_{C,\min} = D\lambda[R - (x_D - y_F^*)/(y_F^* - x_F^*)] \quad (8)$$

where λ is the heat of vaporization. The horizontal distance of the CGCC from the temperature axis, however, determines the targets for installing a side reboiler or side condenser at suitable temperatures [1,2,16,17]. On the other hand, a sharp change in the enthalpy represents inappropriate feed conditioning, such as feed quality or temperature; a sharp change on the reboiler side may be due to a subcooled feed. Feed conditioning is usually preferred to side condensing or reboiling, since the side heat exchangers are effective at suitable temperature levels or stages only [18].

Exergy Loss Profiles- Exergy Ex ($Ex = H - T_oS$) shows the available energy that can be converted into a useful work in a reversible process based on a reference temperature T_o, which is usually assumed as the environmental temperature of 298.15 K. Exergy balance for a steady state system shows that exergy is not conserved

$$\sum_{\substack{\text{into}\\\text{system}}} \left[\dot{n}Ex + \dot{Q}\left(1 - \frac{T_o}{T_s}\right) + \dot{W}_s\right] - \sum_{\substack{\text{out of}\\\text{system}}} \left[\dot{n}Ex + \dot{Q}\left(1 - \frac{T_o}{T_s}\right) + \dot{W}_s\right] = \dot{X}_{\text{loss}} \quad (9)$$

where $\dot{W}_s$ is the shaft work. The rate of loss exergy $\dot{E}x_{\text{loss}}$ represents the overall thermodynamic imperfections, and directly proportional to the rate of entropy production due to irreversibilities in a column operation. As the exergy loss increases, the net heat duty has to increase to enable the column to achieve a required separation. Consequently, smaller exergy loss means less waste heat or thermodynamic imperfections. For distillation columns, the difference between the exergies of products and feed streams determines the minimum exergy (separation work) necessary for a required separation

$$\dot{E}x_{\min} = \sum_{\text{out}} \dot{n}Ex - \sum_{\text{in}} \dot{n}Ex \quad (10)$$

A conventional column receives heat at a higher temperature level in the reboiler, and discharges about the same amount in the condenser at a lower temperature. Therefore, it resembles a heat engine that produces the separation work. When $\dot{E}x_{\min} > 0$, thermodynamic efficiency becomes

$$\eta = \frac{\dot{E}x_{\min}}{\dot{E}x_{\text{loss}} + \dot{E}x_{\min}} \quad (11)$$

The denominator in Eq. (11) is the total exergy input. The values of efficiencies before and after the retrofits can quantify the improvements, and help assessing the effectiveness of retrofits.

METHANOL PLANT

The methanol plant uses natural gas, carbon dioxide, and water as the basic feed streams, and produces 62000 kg/hr and 99.95% pure methanol. The plant operates with five Sections connected to each other by the material and heat streams, as shown in Figure 1. Section 1 prepares the feeds of 24823 kg/hr carbon dioxide at 1.4 bar and 43 °C, and 29952 kg/hr natural gas containing 95.39 mole% methane at 21.7 bar and 26 °C. Also, there are the circulation water of 410000 kg/hr at 26 bar and 195 °C, and the makeup steam at 26 bar. By adjusting the steam flow rate, the steam to methane ratio of 2.8 is achieved in the reactor. The reactor outlet contains small amounts of dimethylether, n-butanol, and acetone, beside the main product of methanol. The outlet is flashed and the methanol rich liquid stream 407 is fed to the separation Section 4, which consists of two complex columns in series. The columns operate with multicomponent feeds and multiple side products, and use process heats as side heat streams from the other Sections of the plant. The first column operates with a side condenser, and has no reboiler. Separation of the methanol starts with flashing of stream 407. The feed to the first column is the mixture of the liquid outlet of the flash drum and the makeup water at 5 bar and 40°C. Flow rate of the makeup water is adjusted in order to minimize the methanol loss at the bottoms of the second column. The stages are numbered from the top to the bottom. The first column has 51 stages, a partial condenser at the top, and a side condenser at stage 2. It receives the feed at stage 14, a side heat stream of 15.299 MW at stage 51, and operates without rebolier. A pumparound connects the liquid flow between stage 1 and 3. The second column has 95 stages and a total condenser at the top. It receives the feed at stage 60, a side heat stream of 18.9 MW at stage 95, and operates with high reflux ratio. The methanol is a side product of the second column drawn from stage 4. The side heat streams come from Section 2 (Figure 1). Section 5 is the furnace Section, where the offgas from Section 4 is burned.

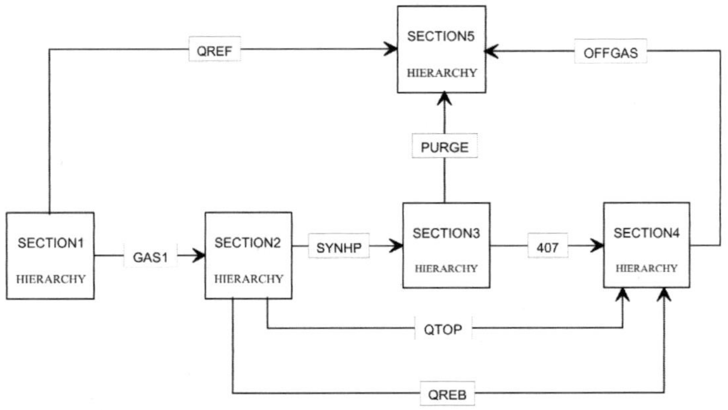

Figure 1. Connection of the Sections of the methanol plant with material and heat streams.

RESULTS AND DISCUSSION

Stream 407 containing 73.45 mole% of methanol and the makeup water are the feed streams to the separation Section. Tables 2 and 3 describe the existing base case columns operations for the columns 1 and 2. The simulations use the thermodynamic method of Redlich-Kwong-Soave (RKW) to estimate the vapor properties, while the activity coefficient model NRTL and Henry components method are used for predicting the equilibrium and liquid properties.

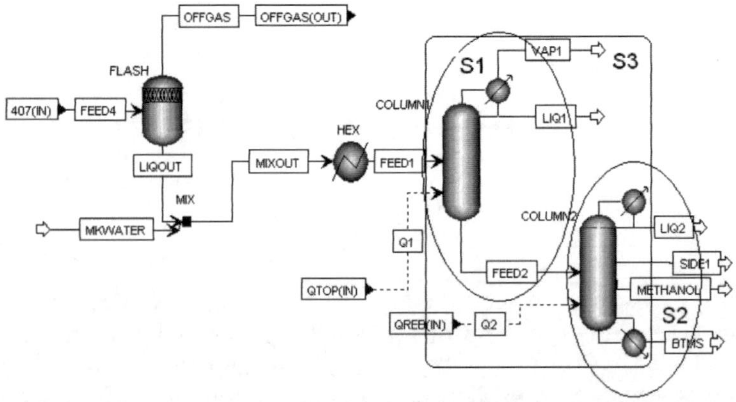

Figure 2. Separation Section of the methanol plant with subsystems: S1-column 1; S2-column 2; S3-column 1 and 2.

Column 1- As the base case design in Table 1 shows, column 1 has 51 stages, and operates with a partial condenser with a duty of 1.371 MW at the top, and a side condenser with a duty of 8.144 MW at stage 2. It has no reboiler, however, it receives a side heat stream with a duty of 15.299 MW to the last stage from Section 2 of the plant (Figure 1).

Table 1 Comparison of operating parameters of base case and retrofitted case for column 1

Parameter	Design 1 (Base case)	Design 2 (Retrofitted case)
No. of stages	51	51
Feed stage	14	14
T (Feed), °C	43.7	65.0
Reflux ratio	3.7	4.5
Condenser duty, MW	1.372	1.691
Distillate rate, kmol hr^{-1}	34.14	34.14
T (Condenser), °C	32.7	32.7
Side condenser 1 stage	2	2
Side condenser 1 duty, MW	8.144	7.700
T (Stage 2), °C	69.4	70.3
Side condenser 2 stage	-	4
Side condenser 2 duty, MW	-	2.100
T (Stage 4) °C	74.4	74.4
Heat stream (Q1) duty, MW	15.299	15.299
Heat stream (Q1) stage	51	51
T(Heat stream) (Q1), °C	104.0	104.0
Boilup rate, kmol hr^{-1}	1551.28	1551.56
Bottom rate, kmol hr^{-1}	2995.14	2995.14
Bottom temperature, °C	85.8	85.8

Within the rectification Section of column 1, there exists a significant area difference between the ideal and actual enthalpy profiles, which identifies the scope for side condensing [21]. As the temperature change after stage 3 is very small, and a side condenser at stage 2 already exists, it has been decided to install second side condenser at stage 4 with a duty of 2.1 MW. Figure 3a compares the base case and retrofitted designs for column 1; the side condenser has reduced the area between the ideal and actual enthalpy profiles. The duty of 2.1 MW is in the range of enthalpy difference between the hot duty of 15.299 MW and the total cold duty of 9.51 MW (side condenser + partial condenser). The existing side condenser duty is reduced to 7.7 MW from 8.144 MW, so that the new total duty of 11.49 MW is close to the previous total of 9.51 MW. The total costs due to retrofits would not change too much, and the need for extra stages would be negligible as the heat changes sharply below the first side condenser.

Figure 3a also displays a sharp change of the enthalpy on the reboiler side. The extent of the change determines the approximate feed preheating duty required [18] as the feed at 43.74°C is highly subcooled. Therefore, a new heat exchanger with a duty of 1.987 MW is used as the second retrofit for the column, and the feed temperature has increased to 65.0°C from 43.74°C. The difference between the hot and cold duties is lower, and the actual and ideal profiles are closer to each other after the retrofits.

The suggested retrofits also aim at reducing the irreversibility due to mixing of the streams at different temperatures on the feed stage, which is at 80.18°C, and throughout the column. The exergy loss profiles (Figure 3b) show that the reduction in exergy loss at the feed stage is about 60% with the values of 0.3865 MW in design 1 and 0.1516 MW in design 2 [21]. However, the exergy loss at the partial condenser increases by 28%, and becomes 0.150 MW in design 2 instead of 0.117 MW in design 1. As Table 3 shows, the reduction in the total exergy loss or the recovered available energy is 21.5 % with the total column exergy losses of 0.837 MW and 0.656 MW in design 1 and 2, respectively.

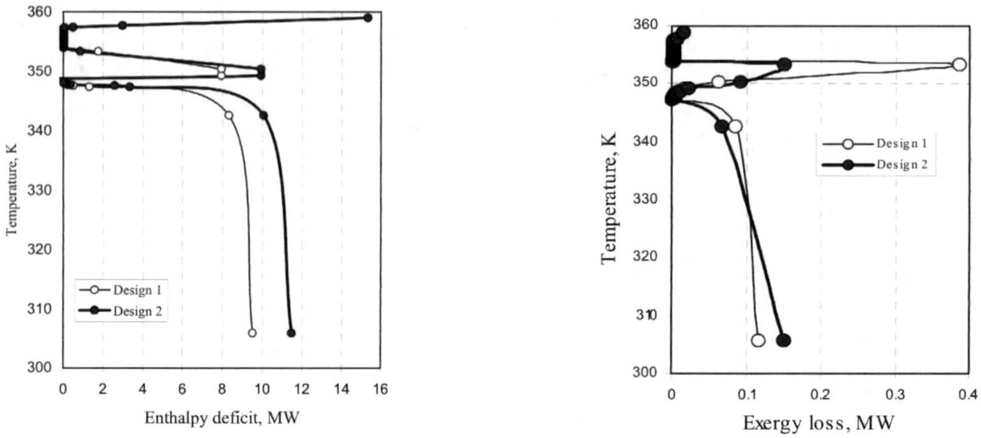

(a) (b)

Figure 3. (a) Temperature-enthalpy deficit curves (CGCC) and (b) exergy loss distribution for column 1.

Column 2- Column 2 has 95 stages, and a total condenser with a duty of 281.832 MW. It operates with a high reflux ratio, and receives a side heat stream of 18.9 MW to the last stage from Section 2 of the plant. One of the side products is the methanol stream and drawn at stage 4 at 348.3 K. The second side product is drawn at stage 86 at 361.2 K. The CGCC of column 2 shows a significant area difference between the ideal and the actual enthalpy profiles above the feed stage representing the pinch, and hence suggests side reboiling at appropriate temperature levels to decrease the difference [19-21]. The existing reboiler duty is 282.28 MW (Table 2). There is a side product at stage 86 and a side heat inlet of 18.9 MW at stage 95. Therefore, it has been decided to install two side reboilers at stages 87 and 92 with the duties of 180 and 50 MW, respectively. Figure 4a compares the base case and retrofitted designs for column 2.

Table 2. Comparison of operating parameters of designs 1 and 2 for column 2.

Parameter	Design 1 (Base case)	Design 2 (Retrofitted case)
No. of stages	95	95
Feed stage	60	60
T (Feed), °C	85.8	85.8
Reflux ratio	188765.0	188765.0
Condenser duty, MW	281.832	281.832
Distillate rate, kmol hr^{-1}	0.15	0.15
Condenser temperature, °C	74.8	74.8
Reboiler duty, MW	282.283	52.292
Boilup rate, kmol hr^{-1}	24890.68	4633.93
Bottoms rate, kmol hr^{-1}	1050.96	1049.66
T (Reboiler), °C	119.7	120.0
Side reboiler 1 stage	-	87
Side reboiler 1 duty, MW	-	180.000
Stage 87 temperature, °C	90.9	93.3
Side reboiler 2 stage		92
Side reboiler 2 duty, MW		50.000
Stage 92 temperature, °C	110.9	110.9
Heat stream (Q2) duty, MW	18.900	18.900
Heat stream (Q2) stage	95	95
Heat stream (Q2) temperature °C	136.0	136.0

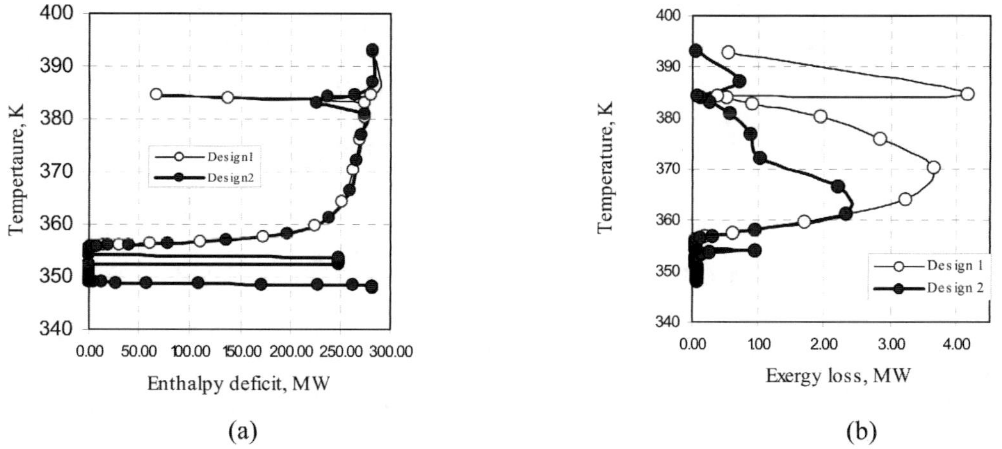

Figure 4. (a) Temperature-enthalpy deficit curves (CGCC) and (b) exergy loss distribution for column 2.

With the two side reboilers, the duty of the reboiler decreases to 52.3 MW from 282.3 MW. Extra stages due to the side reboilers would be minimal since the enthalpy rises sharply at each stage after stage 84. The retrofits reduce the area between the ideal and actual enthalpy profiles [21].

The base case design operates with rather large exergy losses at the feed stage and around the reboiler. The retrofits reduce the total exergy losses by about 41.3 %, and hence save a considerable amount of the available energy (Figure 4b). The minimum values of exergy for the required separation and the thermodynamic efficiencies for designs 1 and 2 are estimated using Eqs. (10) and (11), and compared in Table 3. The estimations are based on the value T_o = 298.15. The reductions in the exergy losses range from 21.5% to 41.35%. The thermodynamic efficiencies have increased considerably, although the low efficiencies are common for industrial column operations [8]. For column 1 the efficiency increases to 55.4% from 50.6%, while it increases to 6.7% from 4.0% in column 2.

An approximate economic analysis has compared the fixed capital costs (FCC) of the retrofits with the savings in electricity due to the retrofits. FCC consists of equipment, materials, construction, and labor cost. Table 4 shows the approximate values of FCC for the heat exchangers needed in the retrofits. The costs are estimated by using the current chemical engineering plant cost index of 420, and the approximate areas of the exchangers are obtained from the individual duties. Estimations of the energy saving are based on the unit cost of electricity of $0.060/kW-hr and a total 8322 hours/year of the plant operation. The costs of retrofits and the saved electricity for each subsystem are compared in Table 4, which shows that the retrofits are effective and save a considerable amount of energy per year.

Table 3. Assessment of the effectiveness of the retrofits: The subsystems S1 to S3 are shown in Fig. 2

	Design 1 (base case)			Design 2 (retrofitted case)						
System	$\dot{E}x_{min}$ (MW)	$\dot{E}x_{loss}$ (MW)	η %	$\dot{E}x_{min}$ (MW)	$\dot{E}x_{loss}$ (MW)	η %	Saved $\dot{E}x_{loss}$ (MW)	Change $\dot{E}x_{loss}$ %	FCC*of retrofits, $	Electricity Saving** ($/year)
S1										
Column 1	0.856	0.837	50.6	0.815	0.656	55.4	0.179	21.5	183,500	89,578
S2										
Column 2	1.136	26.98	4.0	1.135	15.85	6.7	11.133	41.3	409,000	5,558,829
S3										
Column 1+2	1.992	27.82	6.7	1.950	16.51	10.6	11.312	40.7	592,500	5,648,407

$\dot{E}x_{loss}$: Total column exergy loss from the converged simulation by Aspen Plus with SRK, NRTL.

* FCC: Fixed capital cost.

** Electricity equivalent of energy saving is based on a unit cost of electricity of $0.060/kW-hr.

Table 4. Approximate fixed capital cost calculations for the retrofits

Heat exchanger	Type	Duty (MW)	*P* (bar)	Material	Area (m^2)	FCC** ($)
Preheater (HEX) Column 1	S/T* Fixed Tube sheet	1.9	5.0	Carbon Steel	130	90,500
Side condenser Column 1	S/T* Fixed Tube sheet	2.1	1.5	Carbon Steel	130	93,000
Total cost for column 1						183,500
Side reboiler 1 Column 2	Floating head	180.0	2.0	Carbon Steel	600	294,000
Side reboiler 2 Column 2	Floating head	50.0	2.0	Carbon Steel	170	115,000
Total cost for column 2						409,000

* S/T: Shell and tube.

** Approximate fixed capital cost (FCC) with the chemical engineering plant cost index = 420 [20].

SUMMARY and CONCLUSIONS

This study has used thermodynamic analysis for retrofitting the distillation columns within the separation Section of a methanol production plant. The suggested retrofits consist of an additional side condenser at stage 4 and feed preheating for column 1, and two side reboilers at stages 87 and 92, respectively for column 2. Effectiveness of the retrofits has been assessed by the improved column grand composite curves and exergy loss profiles as well as by an approximate economical analysis. The range of reductions in the total exergy losses is 21.5% to 41.3%, which causes a considerable saving in the available energy losses. The thermodynamic efficiencies also increased considerably, and the columns operate with less thermodynamic imperfections. The savings in electricity can payback the initial cost of retrofits in a short time of operation.

Column grand composite curves and exergy loss profiles are becoming readily available through a simulator package. This enables engineers to assess an existing operation and suggest retrofits or design thermodynamically optimum processes. Therefore, Thermodynamic analysis is emerging as retrofitting and optimization tools for existing and new processes.

NOMENCLATURE

D	Distillate, (kmol hr^{-1})
Ex	Exergy (MW)
H	Enthalpy (J mol^{-1})
$\dot{Q}$	Heat flow (W)
L	Liquid flow rate (kmol hr^{-1})
$\dot{m}$	Mass flow rate, (kg hr^{-1})
$\dot{n}$	Molar flow rate, (kmol hr^{-1})
Q_C	Condenser duty (MW)
Q_R	Reboiler duty (MW)
S	Entropy (J $mol^{-1}K^{-1}$)
T	Temperature (K)
x	Liquid mole fraction
y	Vapor mol fraction
V	Vapor flow rate (kmol hr^{-1})
η	Efficiency
λ	Heat of vaporization (J mol^{-1})

subscripts

def	Deficit
D	Distillate
F	Feed
V	Vapor
H	Heavy
L	Light
min	Minimum
R	Reboiler
s	Stream, shaft

REFERENCES

1. Dhole, V.R. and Linnhoff, B. Distillation column targets. Comput. Chem. Eng., 1993, *17*, 549-560.
2. Ognisty, T.P. Analyze distillation columns with thermodynamics. Chem. Eng. Prog., 1995, *91*, 40-46.
3. Bandyopadhyay, S. Malik, R.K. and Shenoy, U.V. Temperature-enthalpy curve for energy targeting of distillation columns. Comput. Chem. Engng., 1998, *22*, 1733-1744.
4. De Koeijer, G.M. and Rivero, R. Entropy production and exergy loss in experimental distillation columns. Chem. Eng. Sci., 2003, *58*, 1587-1597.
5. Al-Muslim. H. and Dincer, I.; Zubair, S.M. Exergy analysis of single-and two-stage crude oil distillation units. J. Energy Resource Tech., 2003, *125*, 199-207.
6. Al-Muslim, H. and I. Dincer, I. Thermodynamic analysis of crude oil distillation systems. Int. J. Energy Res., 2005, *29*, 637-655.
7. Demirel, Y. Thermodynamic analysis of separation systems. Sep. Sci. Technol. 2004, *39*, 3897-3942.
8. Seider, W.D.; Seader, J.D. and Lewin, D.R. *Product & Process Design Principles*; 2nd Ed.; Wiley: New York, 2004.
9. Demirel, Y. *Nonequilibrium Thermodynamics Transport and Rate Processes in Physical, Chemical and Biological Processes*; Elsevier: Amsterdam, 2007.
10. Demirel Y. and Sandler, S.I. Nonequilibrium thermodynamics in engineering and science. J. Phys. Chem. B, 2004, *108*, 31-43.
11. Chang, H. and Li, Jr-W. A new exergy method for process analysis and optimization. Chem. Eng. Sci., 2005, *60*, 2771-2784.
12. Rivero, R.; Garcia, M. and Urquiza, J. Simulation, exergy analysis and application of diabatic distillation to a tertiary amyl methyl ether production unit of a crude oil refinery. Energy, 2004, *29*, 467-489.
13. Demirel, Y. and Sandler, S. I. Linear nonequilibrium thermodynamics theory for coupled heat and mass transport. Int. J. Heat Mass Transfer, 2001, *44,* 2439-2451.
14. Doldersum, A. Exergy analysis proves viability of process modifications. Energy Convers. Mgmt., 1998, *39*, 1781-1789.
15. Liu, Y.A. *Process and Product Design I: Hands-on Training Course on Process Simulation, Retrofit and Optimization Using ASPEN PLUS;* Virginia Tech: Virginia, 2003.
16. Aspen Plus V11.1. Aspen Technology: Cambridge, MA.
17. Agrawal, R. and Herron, D.M. Efficient use of an intermediate reboiler or condenser in a binary distillation. AIChE J., 1998, *44*, 1303-1315.
18. Bandyopadhyay, S. Effect of feed on optimal thermodynamic performance of a distillation column. Chem. Eng. J., 2002, *88*, 175-186.
19. Rivero, R. Exergy simulation and optimization of adiabatic and diabatic binary distillation. Energy, 2001, *26*, 561-593.
20. Turton, R.; Bailie, R.C.; Whiting W.B. and Shaeiwitz, J.A. *Analysis, Synthesis and Design of Chemical Processes*; 2nd Ed.; Prentice Hall: New Jersey, 2003.
21. Demirel, Y., Retrofit of distillation columns using thermodynamic analysis, Sep. Sci. Tech., 2006, *41*, 791-817.

Examining the effect of protein size on transport and adsorption in polymer-modified ion-exchange media

Brian D. Bowes*, Harun Koku, and Abraham M. Lenhoff
Department of Chemical Engineering, University of Delaware, Newark, Delaware, 19716

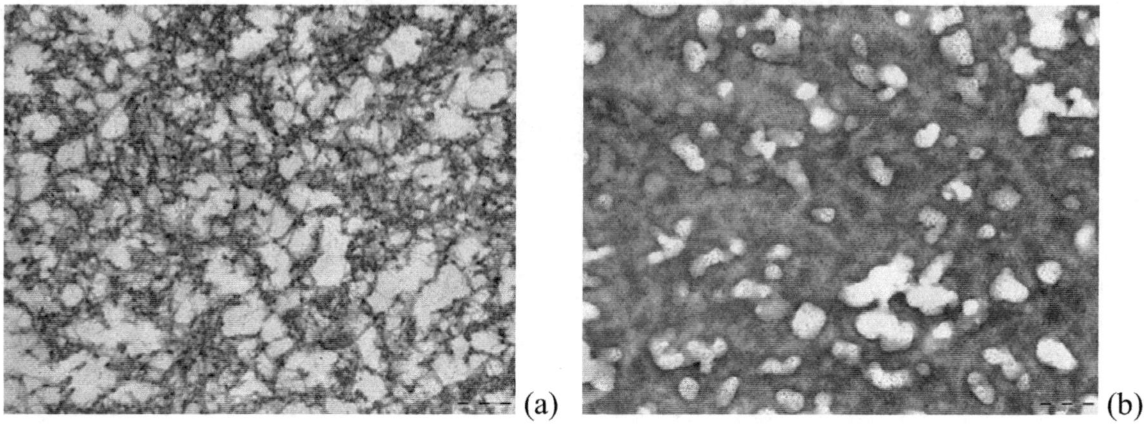

Figure 1: TEM images of (a) Q Sepharose FF and (b) Q Sepharose XL sections at high adsorbed protein concentrations. The dark regions correspond to stained protein and the light regions correspond to void.

Ion-exchange particles in which functionalized secondary polymers have been grafted onto the base matrix have been shown to offer significant performance advantages over traditional particles, especially with respect to increases in dynamic binding capacity (DBC). However, the mechanistic origins of these advantages are not clearly understood. A number of studies have found that, depending on pH and flow rate, a maximum in the DBC can exist as a function of ionic strength, suggesting that the key lies in the balance between improved transport at higher ionic strengths and improved static capacity at lower ionic strengths. The broad goal of this work is to gain a better understanding of this interplay between protein transport and adsorption that accounts for observed column breakthrough performance in both traditional and polymer-modified particles. To this end, the model proteins lysozyme, lactoferrin, and a monoclonal antibody (mAb) are used to examine the role of protein size in column breakthrough performance on the traditional resin SP Sepharose FF and on the dextran-modified resins SP Sepharose XL and Capto S, each having an agarose base matrix. The base matrices for FF and XL are similar, providing an especially good basis for examining the role of the dextran.

In this study, a variety of techniques spanning length scales from the column to local properties within the individual chromatography particle are used to rationalize observed differences in breakthrough for the various protein-resin pairings. Isocratic retention experiments show that the retention mechanism for the three resins is quite similar for the small protein lysozyme. However, the larger proteins can be almost completely excluded from the pores of the polymer-modified resins at higher ionic strengths, which may result from a combination of steric and electrostatic interactions. This exclusion could be important industrially in ion-exchange steps where the ionic strength must be kept relatively high for economic reasons (i.e. minimization of buffer consumption in dilution or dialysis), as the polymer-modified resins may give poor performance, despite their clear advantages at lower ionic strengths. Adsorption isotherms are used to establish the static binding capacities of the resins and knowledge of resin and protein properties is used to interpret observed behavior. The isotherms generally reflect higher static capacities for the dextran-modified resins, except at higher ionic strengths where the exclusion behavior observed in isocratic retention experiments is presumably a limiting factor. Transmission electron microscopy (TEM) images of particles with adsorbed protein, such as those in Figure 1, show that the dextran modification gives a larger available binding volume, providing convincing visual support for the higher static capacities of XL and Capto under most conditions. Manipulation of previously published inverse size exclusion chromatography (ISEC) data allows estimation of the binding volume afforded by the dextran.

The static capacity of a resin sets an upper limit on its dynamic capacity. Comparison of these two capacities nicely demonstrates that, while the dynamic capacity can approach the static capacity at high ionic strengths, transport limitations can be quite significant at lower ionic strengths. In this work, confocal microscopy is used to examine the uptake of protein into individual particles. Observations reveal significant changes with protein size and, as is evident from the representative images in Figure 2, also reflect consistently faster uptake in the dextran-modified resins, which can be understood in terms of the comparative ability of protein to move in a three-dimensional dextran layer and on a two-dimensional agarose surface. The faster transport in XL and Capto shifts the optimum operation to lower ionic strengths where static capacities are higher, resulting in greatly improved dynamic binding capacities.

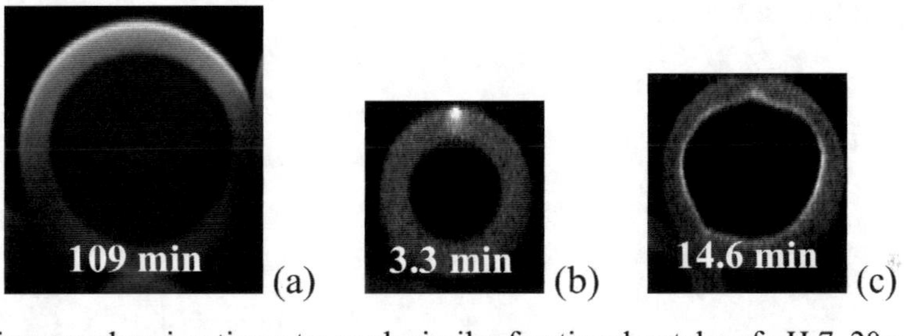

Figure 2: Confocal images showing times to reach similar fractional uptake of pH 7, 20 mM total ionic strength lactoferrin into (a) SP Sepharose FF, (b) Capto S, and (c) SP Sepharose XL. Uptake is clearly much faster in the two dextran-modified resins, even accounting for particle size differences, and is particular quick in Capto S.

The results of this work nicely address some of the broad questions regarding the advantages of the dextran-modified materials in terms of adsorption and transport properties, accounting for improvements in column performance. This work should be helpful in the design of ion-exchange steps and may help in the design of new resins. There are more subtle points that have yet to be adequately addressed to be considered for future work. For example, better understanding of the exclusion of larger proteins from the dextran-modified pores at high ionic strength could help to minimize such exclusion when selecting operating conditions for an ion-exchange step or when developing new particles, or perhaps this understanding could be used to tune the exclusion to actually enhance a separation. Also, improved knowledge of the how the physical and chemical properties of a resin affect the various aspects of its chromatographic performance could help guide the eventual development of a model that provides an appropriate mechanistic description of adsorption and transport in polymer-modified media.

364b Chromatographic Behavior and Protein Adsorption Characteristics of Macroporous Cation Exchangers – Effects of Weak Acid Group Content

Timothy M. Pabst, **Ernie X. Perez-Almodovar**, and Giorgio Carta. Department of Chemical Engineering, University of Virginia, Charlottesville, VA 22904

In this work we analyze the effects of the weak acid content of macroporous cation exchange resins based on acrylamido and vinyl polymers that contain different amounts of weak acid groups in addition to their main sulphopropyl cation exchange functionality. UNOsphere S and Rapid S resins from Bio-Rad Laboratories are used for this comparison. When operated at pH values below 7, the presence of weak acid groups in UNOsphere S results in substantial pH transients, which are induced by changes in salt concentration, even though the mobile phase is buffered. These transients are greatly reduced with UNOsphere Rapid S, which contains a much lower concentration of weak acid groups, permitting very rapid requilibration and avoiding any significant change in pH during protein elution with salt. Protein adsorption equilibrium and mass transfer rates were also studied for both materials using several proteins with molecular masses between 15 and 150 kDa. Despite its much lower concentration of weak acid groups, UNOsphere Rapid S exhibits protein adsorption capacities and effective pore diffusivities that are essentially the same as those observed for UNOsphere S. These results suggest that the weak acid groups in UNOsphere S are primarily associated with the polymer backbone structure and do not interact directly with proteins molecules.

Binding mechanism of proteins and DNAs in electrostatic interaction chromatography –Ion Exchange Chromatography and Hydroxyapatite Chromatography-

Shuichi Yamamoto, Shintaro Higuma, Yuuki Sakaguchi, Tomohiko Yoshitake, Noriko Yoshimoto.

Bioprocess Engineering Laboratory,
School of Engineering and Graduate School of Medicine
Yamaguchi University, Tokiwadai, Ube, 755-8611, Japan
E-mail shuichi@yamaguchi-u.ac.jp

1. Introduction

It is well known that very difficult separation of protein variants can be achieved by electrostatic interaction chromatography (ion-exchange chromatography, IEC) although the mechanism has not yet been fully clarified. DNAs are also efficiently separated not only by IEC but also by hydroxyapatite chromatography (HAC), which is known as a bi-modal or mixed mode interaction system.

In this study we investigated the binding mechanism of proteins and DNAs in IEC and HAC. Especially oligoDNA separation behavior was carefully examined by using different mobile phases for HAC in order to understand its unique property for DNA binding and elution. For this purpose the number of binding sites B values were determined from linear (salt) gradient elution (LGE) experiments as well as the peak salt concentration.

2. Materials and Methods

2.1. Chromatography column

Poly(glycidyl methacrylate-co-ethylene dimethacrylate) disks (12 mm diameter x 3mm thickness) with strong anion exchange group were contained in a specially designed disk holder column from BIA Separations (Ljubljana, Slovenia) This disk is called hereafter QA-CIM.

Two Hydroxyapaptite chromatography columns were used: A pre-packed ceramic hydroxyapatite bead column from Bio-Rad (CHT Type I) and an agarose-hydroxyapatite composite gel (HA-Ultrogel) column packed according to the supplier method (Biosepra).

2.2. Materials

Oligonucelotides (3- to 95-mer poly A and poly T) were purchased from Espec (Tokyo, Japan). Prior to use for chromatography oligonucletotide samples were thermally pretreated according to the following temperature program: from 298K to 368K (20 minutes), 368K (3 minutes), from 368K to 313K (20 minutes) and from 313K to 298K (45 minutes).

Most standard proteins were purchased from Sigma-Aldrich. Other reagents were of analytical grade.

2.3. Chromatography setup

Chromatography experiments were carried out on a fully automated liquid chromatography system ÄKTA explorer 100 (GE Health Care, Uppsala, Sweden).

2.4. Linear gradient elution experiment

The column was equilibrated with a starting buffer (buffer A) containing a specified concentration of NaCl for IEC. Although M (=mol/dm^3) is a non-SI unit symbol, it is used in this paper as it is still common for many chromatography researchers. The same buffer solution containing 0.5-2.0 M NaCl was used as a final elution buffer (buffer B). The linear gradient elution was performed by changing the buffer composition linearly from buffer A to buffer B with time. Namely, the salt concentration was increased with time at a fixed pH and buffer compositions. In this study 14 mM Tris-HCl (pH7.7) was chosen as buffer A for IEC.

For HAC the standard buffer solutions are buffer A 10mM phosphate buffer and buffer B 400mM phosphate buffer.

The linear mobile phase velocity u was calculated with the cross-sectional area A_c and the column bed void fraction ε as $u= F/(A_c\,\varepsilon)$ where F is the volumetric flow rate. The column bed void fraction ε was determined from the peak retention volume of Dextran T 2000 pulses. For the QA-CIM disk, the void faction was assumed to be 0.60. The experiments were performed at 298±1K. Experiments were carried out for at least four different gradient slopes. Each run was repeated twice. The gradient slope g is shown in M/mL.

3. Theoretical

The outline of our model for linear gradient elution (LGE) is briefly explained below. The peak retention volume is a function of gradient slope in linear gradient elution IEC. The peak salt concentration I_R increases with increasing gradient slope, g. The I_R values can be correlated with the following normalized gradient slope,

$$GH = (gVo)[(V_t\text{-}Vo)/Vo] = g(V_t\text{-}Vo) \qquad (1)$$

V_t is the total bed volume, V_o is the column void volume, and g is the gradient slope of the salt. H=(V_t- Vo)/ Vo= (1- ε) /ε is the phase ratio. ε = Vo/ V_t is the bed void fraction (interstitial volume of the bed). The gradient slope g is defined by the following equation.

$$g = (I_F\text{-}I_o)/V_g = (I_f\text{-}I_0)/(F_v t_g) \qquad (2)$$

I_f is the final salt concentration, I_o is the initial salt concentration, V_g is the gradient volume, F_v is the volumetric flow rate, and t_g is the gradient time. Linear gradient elution experiments are performed at different gradient slopes (*GH* values) at a fixed pH. The salt concentration at the peak position I_R is determined as a function of *GH*. The *GH*- I_R curves thus constructed do not depend on the flow velocity, the column dimension, the sample loading (if the overloading condition is not used), or the initial salt concentration I_o. The experimental *GH*- I_R data can commonly be expressed by the following equation.

$$GH = I_R^{(B+1)}/[A\,(B+1)] \qquad (3)$$

From the law of mass action (ion exchange equilibrium), the following relationship can be derived.

$$A = K_e \Lambda^B \qquad (4)$$

Here, *B* is the number of sites (charges) involved in electrostatic interaction, which is basically the same as the *Z* number and the characteristic charge, K_e is the equilibrium association constant, and Λ is the total ion exchange capacity. From the ion-exchange equilibrium model and Eq. (4), the following equation is derived .

$$K - K' = K_e \Lambda^B I^{-B} \qquad (5)$$

K is the distribution coefficient of charged biomolecule, *K'* is the distribution coefficient of salt, and *I* is the ionic strength (salt concentration). The SMA model equation is reduced to Eq.(5) when the sample loading is low (Λ is not influenced by sample adsorption).

4. Results and discussion

The binding site for proteins is a function of pH, which is not as simple as expected from the net charge concept.

There is a good correlation between the number of charges of DNA and the binding site for IEC as shown in Fig.1 Even a small single strand DNA (10-20 mers) is retained with the binding site value greater than 10, and NaCl concentration of 0.6 M or higher is needed for the elution (Fig.2) . Although it is quite difficult to separate large DNAs, we were able to separate DNAs (50 mer poly A-poly T and 95 mer poly A-poly T) by using a monolithic convection-aided IEC disk at high flow velocity and with a shallow gradient.

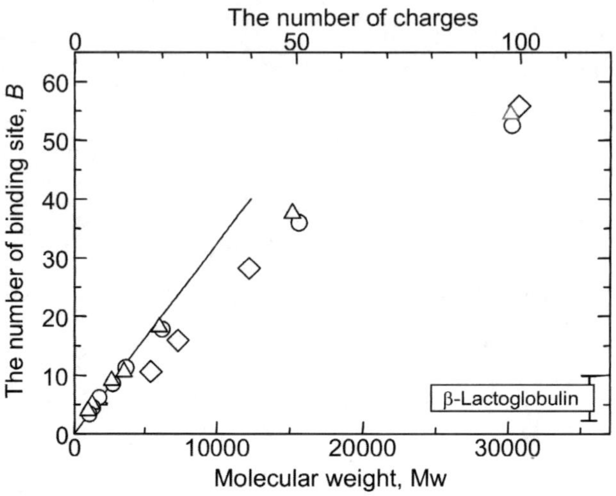

Fig.1 Relationships between the number of binding sites B and the molecular weight or the number of charges of poly A and poly T DNAs. The straight curve in the figure shows B =the number of charges..Note that the range of B values for β-lactoglobulin is shown in the figure for comparison. The B value for the double-stranded 95mer-polyA and poly T was ca.97 although it was not shown in the figure.Circle:poly A, triangle: poly T, diamond: ds-poly A-polyT

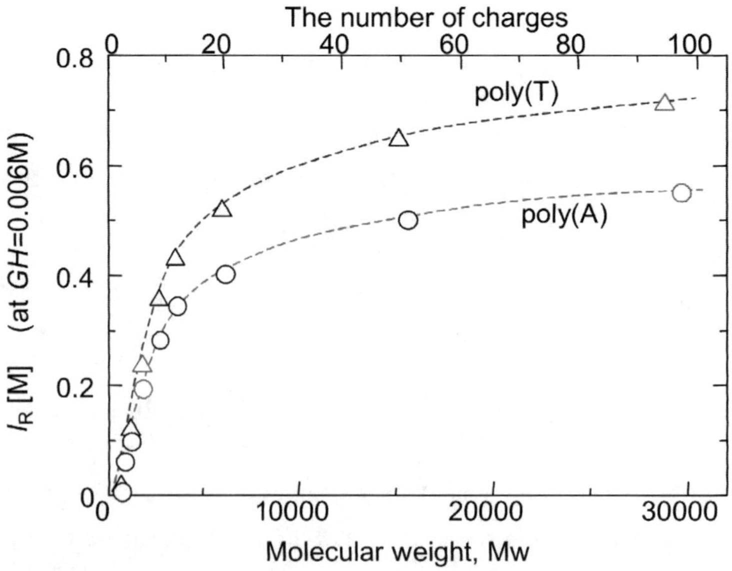

Fig.2 Relationships between the peak salt concentration I_R at GH=0.006 and the molecular weight or the number of charges for single-strand poly A and poly T DNAs.

On the other hand, single strand DNAs were very weakly retained on HAC whereas double strand DNAs were retained. Consequently, separation of single-strand and double strand DNAs are very efficiently performed by HAC(Fig.3). This is due to repulsion of phospate-site of HAC.

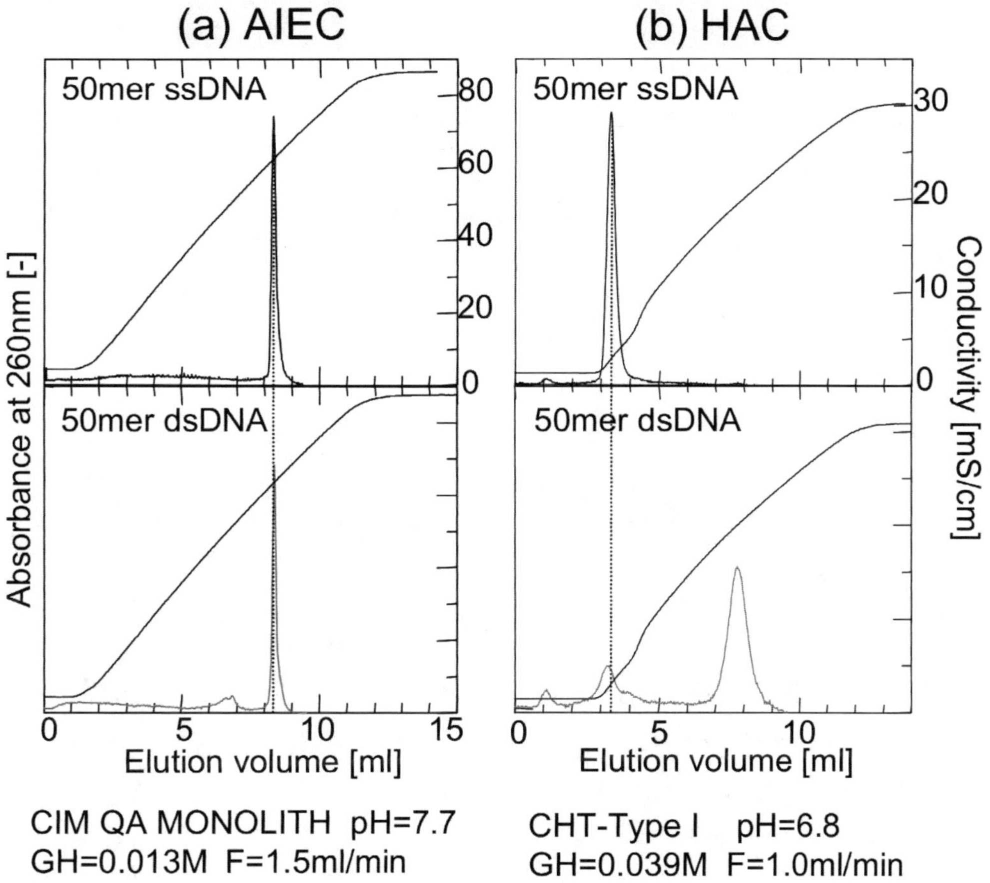

Fig.3 Linear gradient elution curves of ss- and ds-DNA,

Nomenclature

A	$= Ke\Lambda^B$	
B	the number of binding sites	[-]
d_p	particle diameter	[cm]
F_v	volumetric flow rate	[mL/min]
G	gradient slope	[M/mL]
GH	normalized gradient slope = $(gVo)H = g(Vt\text{-}Vo)$	[M]
H	phase ratio $(=(V_t\text{-}Vo)/Vo)$	[-]
I	ionic strength (salt concentration)	[M]
I_f	final salt concentration	[M]
I_0	initial salt concentration	[M]
I_R	peak salt concentration	[M]
K	distribution coefficient	[-]
K'	distribution coefficient of salt	[-]
K_e	equilibrium association constant	
t_g	gradient time	[min]
t_R	retention time	[min]
U	linear mobile phase velocity	[cm/min]
V	elution volume	[mL]
V'	elution volume for the salt	[mL]
V_g	gradient volume	[mL]
V_o	column void volume(interstitial volume)	[mL]
V_R	retention volume	[mL]
V_t	column volume	[mL]
ε	void fraction of column $=V_o/V_t$	[-]
Λ	total ion exchange capacity	[mequiv/mL]

Although M (=mol/dm^3) is not a SI unit, it is used in this paper for the sake of convenience.

References

S.Yamamoto, M.Nakamura, C.Tarmann, A.Jungbauer, J. Chromatogr. A, 1144 (2007)155

S.Yamamoto. K.Nakanishi. R.Matsuno. Ion-Exchange Chromatography of Proteins. Dekker, New York. 1988.

S.Yamamoto, Chem. Eng. Technol., 28(2005)1387

Process engineering approach to large-volume methacrylate monolith synthesis for plasmid purification

The purification of plasmid DNA is hampered by the low performance of conventional chromatographic adsorbents which possess small particle pore diameters. Most of these adsorbents are tailored for the high adsorption capacity of proteins with size less < 10 nm. In columns packed with such adsorbents, large molecules such as pDNA with sizes > 100 nm adsorb predominantly at the particle outer surface, thus resulting in a low binding capacity. A monolith is a continuous phase consisting of a single piece of a highly porous organic or inorganic solid material. The most important feature of this material is that all the mobile phase is forced to flow through its large pores. As a consequence, mass transport is enhanced by convection; dramatically reducing the long diffusion time required by particle based chromatographic supports. Therefore, chromatographic separation process on monoliths is practically not diffusion-limited. The large pores of these monoliths allow penetration of large pDNA molecules to the internal surface area at high flow rate with low pressure drop. Different types of monolithic supports are currently available based on preparation and chemistry. These are silica-based, polyacrylamide-based and gel-based monolithic resins. Polymethacrylate monolithic support is an optimal adsorbent for pDNA separation. These adsorbents have large pore diameters and thus no significant hindrance to convective mass transport. They are resistant to pH, non-toxic, economically favorable to synthesize and can be easily modified by functionalizing with anion-exchange, hydrophobic interaction or affinity ligand. The flexibility and the ease to tailor their pore and surface characteristics to the target pDNA molecule through alteration in synthesis conditions make them more attractive.

The construction of large–volume methacrylate monolithic columns for preparative-scale plasmid purification is obstructed by the enormous release of exotherms, thus introducing structural heterogeneity in the monolith pore system. A remarkable radial temperature gradient develops along the monolith thickness, reaching a terminal temperature that supersedes the maximum temperature required for the preparation of a structurally homogenous monolith. A novel heat expulsion technique is employed to overcome the heat build-up during the synthesis process. The enormous heat build-up is perceived to encompass the heat associated with initiator decomposition and the heat released from free radical-monomer and monomer-monomer interactions. The heat resulting from the initiator decomposition was expelled along with some gaseous fumes before commencing polymerisation in a gradual addition fashion. Characteristics of a 50 mL monolith synthesised using this technique showed an improved uniformity in the pore structure radially along the length on the monolith. Chromatographic characterisation of this adsorbent displayed a persistent binding capacity of 14.5 mg pDNA/mL of the adsorbent. The adsorbent was able to fractionate a clarified bacteria lysate in only 3 min (after loading) into RNA, protein and pDNA respectively. The pDNA fraction obtained was analysed to be a homogeneous supercoiled pDNA.

364e Development of a Realistic 3D Model of Silica Monoliths for Cfd Simulations

Vivek Vasudevan and Kai-Chee Loh. Department of Chemical and Biomolecular Engineering, National University of Singapore, 4 Engineering Drive 4, Singapore, 117576, Singapore

The full potential of therapeutic biomolecules can be realized only if large quantities are available for testing, characterization and quantitative structure–activity analysis. Some of the crucial factors for their successful and fast separation are a reduced mass transfer resistance within large open-ended pores or channels and a small residence time within the chromatographic column to preserve their bio-activity. This has traditionally been carried out using particulate HPLC columns in which separation efficiency increases with decreasing particle diameter. The trade-off between column back pressure and efficiency has resulted in the popular use of monolithic columns for fast and efficient purification. Reduced back pressure, minimal non-specific binding and low product degradation, without loss of resolution, make monoliths ideal candidates as chromatographic matrices for purification of large bio-macromolecules.

Direct visualization, believed to be the only way to capture the true morphology of porous materials, has traditionally been conducted using scanning electron microscopy (SEM) and transmission electron microscopy (TEM). Rapid advances in non–invasive 3D scanning techniques have led to the use of laser scanning confocal microscopy (LSCM) to capture "as is" the inherent morphologies of porous materials. Direct use of reconstructed 3D images as bounding geometries in an environment amenable to solving fundamental transport equations has obviated the need for fitting parameters in transport models, which limit the predictive value of the models. This research is aimed at elucidating the pore structure and transport mechanism characteristics of a silica monolith through CFD simulations by developing a realistic 3D model through image analysis of micro-CT scans of a monolith sample.

Reconstructed images from micro-CT scans of a silica monolith sample were processed to identify a representative pore volume. The images were thresholded and the computed porosity compared with that of the largest scanned image, to identify a representative sized domain. The surface mesh enclosing the flow domain (pore volume) was imported into an environment amenable to using the surfaces as physical boundaries for flow simulations. Simulations of flow hydrodynamics as well as dispersion characteristics in the porous sample were performed using the commercial CFD software – FLUENT. These were verified for grid independence and other geometry-related parameters such as tortuosity and porosity. The effects of flow rate using water as mobile phase were validated against experimental data obtained from commercially available monoliths. Simulation results for dispersion of analyte molecules in non-retained and retained conditions will be discussed. The computing resources utilized in the study will also be highlighted in this presentation.

364f On the Epitope Approach: Oligopeptide Imprinted Polymers Via Conventional and Inverse Suspension Polymerization

Sotiria Chaitidou, Olympia Kotrotsiou, and **Costas Kiparissides**. Department of Chemical Engineering, Aristotle University of Thessaloniki & Chemical Process Engineering Research Institute, P.O. Box 472, Thessaloniki, 54124, Greece

Molecular imprinting is a technique employed in the fabrication of biomimetic polymeric recognition matrices, possessing high substrate site selectivity and specificity. Up to now, MIPs have been prepared as bulk polymer monoliths for the selective recognition of small biomolecules (i.e., amino acids and amino acid derivatives) mainly in organic solvents. However, there is a need for the synthesis of MIPs towards larger biomolecules (i.e., peptides and protein) with well-defined physical characteristics. In the present study, epitope approach was followed for the synthesis of molecularly imprinted microparticles. According to this approach, a short peptide that represents only part of a larger peptide or protein is used as a template, which in turn can be recognized by the synthesized polymer. Conventional and inverse suspension polymerizations were employed for the synthesis of molecularly imprinted microparticles using both hydrophobic and hydrophilic oligo-peptides (i.e., Boc-Trp-Trp-Trp and Ala-Ile-Ser-Tyr-Gly-Asn-Gly-Val-Tyr) as template molecules, aiming to the purification of poly-Trp tagged proteins and class II bacteriocins containing the conserved Tyr-Gly-Asn-Gly-Val motif in their N-terminal region.

The polymerizations were carried out in a laboratory scale reactor, equipped with an impeller, an overhead condenser and a nitrogen purge line. In case of suspension polymerization, the template molecule (i.e, Boc-Trp-Trp-Trp), the functional monomer (i.e., methacrylic acid, MAA), the crosslinker (i.e., ethyleneglycol dimethacrylate, EGDMA) and the initiator (i.e., AIBN) were dissolved into chloroform. The MAA:EGDMA molar ratio was equal to 5:1 and the total amount of the monomers was equal (v/v) to the solvent used as porogen. This organic phase was added into a PVA aqueous solution (1% w/w), under mechanical stirring (600rpm) and nitrogen atmosphere. After the increase of the temperature at 60oC the polymerization was continued for 24hrs at 600rpm. The template molecule was removed from the resulted microparticles by washing with methanol-acetic acid (9:1 v/v). In case of inverse suspension, the heavy oil phase consisting of paraffin oil containing the surfactant Span-80 (2% w/v) was placed into the reactor and the steering speed was set at 500rpm. The reaction mixture was purged with nitrogen while heating to the polymerization temperature (i.e., 30°C or 70°C). The aqueous phase, consisting of the template (i.e., Ala-Ile-Ser-Tyr-Gly-Asn-Gly-Val-Tyr), the functional momomers (i.e., acrylic acid, AA and acrylamide, Am), the crosslinker (i.e., ethylene bisacrylamide, EBAm) and the initiator APS dissolved in distilled water, was added dropwise to the continuous phase. Polymerization was contacted for 16 hrs. When the polymerization contacted at 30°C, 30ë TEMED dissolved in water were added to the medium after the dispersion of the monomer aqueous phase. In all cases, the dispersed aqueous phase was 2,5% v/v and the AA:Am:EBAm molar ratio was equal to 1:3:6. The produced particles were firstly washed with acetone and then with 10% acetic acid, 1% SDS aqueous solution for the template removal. Non-imprinted polymers were also produced.

As far as concerns the synthesis of MIPs towards Boc-Trp-Trp-Trp, a series of experiments were realized using different template concentrations during the polymerization, since the presence of the template at high concentration causes fragility to the synthesized microparticles, while the use of a minimum template concentration results to non specific polymer. The mean particle size of the optimum polymeric microparticles with totally spherical shape was 100ìm. The batch-wise rebinding experiments showed a higher rebinding capacity of MIPs against NIPs. The selectivity of MIPs was also proved since the MIPs rebinding towards Boc-Trp-Trp-Trp peptide was higher than the MIPs rebinding towards other hydrophobic tripeptides (i.e., Boc-Tyr-Tyr-Tyr and Boc-Phe-Phe-Phe). Additionally, the highly crosslinked poly(acrylamide) microparticles prepared by inverse suspension polymerization had particle sizes between 2-25ìm. It was shown that the rebinding capacity of MIPs against NIPs was higher either the polymerization contacted at 70°C or at 30°C using TEMED as catalyst. It was also shown that the thermal-inititated polymers appeared higher rebinding capacity than the polymers prepared at low temperature. Moreover, FTIR analysis revealed very few remaining un-reacted double bonds into the polymer matrix of the microparticles prepared at high temperature and no double bonds in case of the microparticles prepared at low temperature in the presence of TEMED explaining the observed difference in net binding capacity of polymers.

Acknowledgements: EU for supporting this research under the STREP Project, Proposal No. 516981.

364g Modeling Buffer Preparation Operations for Downstream Processing

Charles Siletti and Demetri Petrides. INTELLIGEN, INC., 700 Walton Ave., Mount Laurel, NJ 08054

In most biopharmaceuticals processes, the downstream purification is dominated by chromatographic separations, which require the preparation, holding and distribution of numerous buffer solutions. Buffer preparation is a potential process bottleneck in high-capacity situations. This paper examines the use of macroscopic modeling techniques to evaluate buffer preparation for downstream processing. Cases studies will address the challenges in buffer preparation for multi-line facilities (that share buffer prep tanks), the impact of disposable containers on the demand for utilities and the cost of goods, and the effect of changes in bioreactor titer.

Web Page: www.intelligen.com/literature

364h Development and Scale-up of a Monoclonal Antibody Purification Process

Annika Forss, Carina Engstrand, Gustav Rodrigo, Karol Lacki, and Kristina Nilsson-Välimaa. R&D, GE Healthcare, Bjorkgatan 30, 751 81, Uppsala, Sweden

Time to market is essential for producers of monoclonal antibody therapeutics. When working on process optimization for productivity as well as process economy view points, following a LEAN concept is highly advantageous. Introduction of high throughput formats could significantly reduce the time needed for process development as well as facilitating the development of an optimized production process. Another time-saving approach is to use pre-packed and pre-sanitized equipment, including bioreactors, columns and filters during scale-up.

This presentation describes the development and scale-up of a monoclonal antibody purification process. The initial screening of chromatographic resins and operating conditions are performed in 96-well filter plates. The conditions are verified and further optimized in small columns format. Finally, the purification process is transferred to columns suitable for clinical Phase I/II production.

408a New Insights In Distillation Design Via Column Profile Maps: Sharp Splits

Daniel A. Beneke, **Diane Hildebrandt**, and **David Glasser**. Centre of Materials and Process Synthesis (COMPS), University of the Witwatersrand, Private Bag 3, WITS 2050, Johannesburg, South Africa

In recent years there has been a renewed interest in the design of distillation processes. Holland et al. (2004 a, b) and Tapp et al. (2004) have recently introduced the use of Column Profile Maps (CPMs) to gain insight into complex distillation systems, and have also shown that separations that were previously thought infeasible are in fact feasible at certain operating conditions. They have also shown that CPMs are a simple topological transformation of Residue Curve Maps (RCMs).

CPMs are dependant on two main parameters: $\boldsymbol{R\Delta}$ (reflux) and $\boldsymbol{x\Delta}$ (a pseudo composition vector). This paper focuses on the interaction between these parameters for the special case of sharp splits. With the CPM technique it has been shown that sharp splits behave uniquely to non-sharp splits, so much so that by fixing the aforementioned parameters in a specific manner it may be possible to sample the intermediate boiler in an ideal system (i.e. constant relative volatilities) as a bottoms product. This is possible due to a "node bumping" phenomenon, which could allow for much more efficient designs in distillation. The effect of crossing the discontinuity region for ideal thermodynamics in CPMs has also been investigated. The discontinuity effectively changes the nature of stationary points, which might also be used to design more effective columns.

To summarize, this paper focuses on the use of CPMs to design distillation columns for sharp splits. The interesting behavior displayed by sharp splits can be of great use to the design engineer, in that many new, more efficient columns can be built.

References:

Holland, S.T., Tapp, M.; Hildebrandt, D. and Glasser, D., Column Profile Maps. 2. Singular Points and Phase Diagram Behavior in Ideal and Nonideal Systems, *Ind. Eng. Chem. Res.*,43(14), p3590-3603, (2004 b).

Holland, S.T., Tapp, M.; Hildebrandt, D., Glasser, D. and Hausberger, B., Novel Separation System Design Using "Moving Triangles", *Comp. and Chem. Eng.*,29, p181-189, (2004 a).

Tapp, M., Holland, S.T.; Hildebrandt, D., and Glasser, D., Column Profile Maps. 1. Derivation and Interpretation, *Ind. Eng. Chem. Res.*,43(2), p364-374, (2004).

408b Dynamic Modeling and Optimization of Thermally Coupled Dual-Column Batch Distillation Processes

Girish S. Joglekar[1], Rakesh Agrawal[2], Jong-Ho Moon[3], and Gintaras V. Reklaitis[3]. (1) Batch Process Technologies, Inc., 1291 Cumberland Avenue, West Lafayette, IN 47906, (2) School of Chemical Engineering and Energy Center at Discovery Park, Purdue University, 480 Stadium Mall Drive, West Lafayette, IN 47907, (3) Chemical Engineering, Purdue University, 480 Stadium Mall Drive, West Lafayette, IN 47907

A novel, dual-column batch distillation process has been developed to distill multicomponent mixtures into multiple pure product streams. An important feature of this process is that it retains the flexibility of the original conventional batch distillation column and yet provides large reduction in energy consumption. A simplified model of such a process clearly demonstrated the significant gains in productivity, and reduction in waste stream(1) as compared to the conventional batch distillation process. The dual-column batch distillation processes are expected to find application in pharmaceutical, specialty chemical and agricultural industrial facilities that utilize a wide range of solvents.

In this study, a detailed process dynamics model of the thermally coupled columns was constructed. Distefano's quasi steady-state formulation(2) was used for mass balance on all plates, while the rest of the variables were updated after the specified elapsed time. The important variables that affect the column performance are the reflux ratio, cut durations and fraction of liquid stream within the column that is diverted to the side column. A mixture of Benzene, Toluene and Ortho-Xylene with mass fractions 0.33, 0.33 and 0.34, respectively, was charged to the column.

The following key operating variables associated with each cut affect the overall column performance: reflux ratio, duration and fraction of liquid stream within the main column that is diverted to the side column. To find the optimum values of the operating variables, a two-level approach was used. The Genetic Algorithm function in Matlab was used to create the members in the population. The fitness function value for each member was computed by simulating the column dynamics. The objective function value of each member in turn was used by the GA to create new population. The iterative process was terminated when either the maximum number of iterations were competed or when the best objective function value reached a plateau (change below specified threshold).

For a three component mixture of Benzene, Toluene and Ortho-Xylene, the optimum cycle time for the dual-column process is reduced by 40% compared to the optimum cycle time for a conventional column.

1. AIChE presentation, 2007 Annual Meeting, paper 211a.

2. AIChE Journal, 14(1), 190-199 (1968).

408c A Computationally Efficient Method to Generate a Complete Search Space for Multicomponent Distillation Sequences

Vishesh H. Shah, School of Chemical Engineering, Purdue University, West Lafayette, IN 47906, Arun Giridhar, Chemical Engineering, Purdue University, 480 Stadium Mall Drive, West Lafayette, IN 47907, and Rakesh Agrawal, School of Chemical Engineering and Energy Center at Discovery Park, Purdue University, 480 Stadium Mall Drive, West Lafayette, IN 47907.

In industry, distillation is usually the preferred process to separate multicomponent mixtures. A given multicomponent distillation process can be carried out by various configurations (sequences) that differ significantly in energy consumption. So it is important to be able to generate a complete search space of sequences that are attractive for a given application, and evaluate energy consumption of each. This will enable us to choose a sequence that consumes least energy.

Earlier attempts to generate a search space for zeotropic (ideal to near ideal) distillation were either incomplete, or included a large number of configurations which were never optimal. Our prior work in this area describes a method to generate a search space which is complete in terms of useful sequences and at the same time, does not include the sub-optimal configurations. Because the sub-optimal configurations are not included, the size of the search space is significantly reduced. This method uses binary integer variables associated with edges in a supernetwork. However, though the search space is smaller, it still grows exponentially as the number of components in the feed increases. Hence, a computationally efficient method to generate the search space was desired.

The goal of this work is to describe a novel computationally efficient formulation. The new formulation uses binary integer variables in a node-based network. Representing the decision variables as a matrix easily captures the physical reality of the problem. This matrix representation also helps us to easily automate the method. The new formulation can also be extended to include thermally coupled configurations in the search space almost trivially. This approach is very fast computationally, and the resulting search space can be combined with any desired objective function evaluator to give the truly optimum distillation sequence for the application on hand.

408d Energy-Efficient Complex Column Synthesis

Libin Zhang[1], Jeonghwa Moon[2], Benyamin Grosman[2], and Andreas A. Linninger[3]. (1) University of Illinois at Chicago, 851 S. Morgan St., SEO 218, Chicago, IL 60607, (2) Chemical Engineering, University of Illinois at Chicago, 851 S. Morgan St., Room 218, Chicago, IL 60607, (3) Bioengineering, University of Illinois at Chicago, 851 S. Morgan St., Room 218, Chicago, IL 60607

Complex distillation column configurations are estimated to harness energy savings up to 70%. Distillation synthesis with complex configuration is a meaningful target for energy improvements on an industry-wide scale. In this presentation, we demonstrate a hybrid algorithm combining stochastic search techniques with complex distillation configuration. An evolutionary algorithm will construct automatically structurally different complex separation network with the complex basic distillation configuration. Feasibility of the design will be delegated to an advanced feasibility tests based on temperature collocation of finite elements. Application will demonstrate the algorithms performance in complex column sequencing problems for real industrial application.

In this presentation, a generalized column section approach to draw distillation column configurations to separate an ideal to near ideal n-component mixture into n product streams or less will be presented. The stochastic evolutionary with novel temperature collocation method was incorporated to systematically find a set of optimum distillation schemes for a given application. The structure and length of configuration will not be predefined anymore and will be optimally sought by stochastic evolutionary. After we obtain an array of the optimal configurations and operating conditions, we will identify whether the current separation network is among of those results. The rigorous simulation and energy comparison will be implemented in commercial flowsheet software (HYSYS) with the help of the optimal results from the proposed methodology.

The major innovation of the proposed approach is an evolutionary program combined with a novel temperature collocation algorithm to systematically build and optimize complex column configurations based on column sections. The evolutionary algorithm constructs automatically different separation sequences; feasibility of the synthesized sequences is ascertained by means of the minimum bubble point distance. Finally, realistic solutions will be constructed and determined by the commercial flowsheet simulation (HYSYS) initialized by the proposed approach.

408e Development of Start-up Control Concepts for Dividing-Wall Distillation Columns Based on Simulations and Experiments

Gerit Niggemann, Maria A. Vargas Gomez, and Georg Fieg. Process and Plant Engineering, Hamburg University of Technology, Schwarzenbergstr. 95c, Hamburg, Germany

The separation of multicomponent mixtures in fully thermally coupled distillation columns attracts the chemical engineering world. In that respect, the highest level of process integration is represented by dividing-wall columns. The aspect of getting high value-added products in merely one distillation column with simultaneous savings in both, energy and capital costs, attracts the investors of distillation plants. Furthermore, a growing demand for high-purity products in fine and commodity chemistry within the last years increased the interest of operating dividing-wall columns.

The aim of this contribution is the development of robust and time-optimal process control strategies for the start-up of dividing-wall columns. The strategies allow for significantly shortening the start-up time and for easy application at lab and production-scale columns.

In earlier studies a rigorous process model has been developed, which allows for a detailed analysis of the startup process from ambient conditions. The model considers several features, which are characteristic for a dividing-wall column, such as the self adjusting vapour split and the heat transfer across the dividing wall. An extensive model validation with experimental data of our pilot plant dividing-wall distillation column points out, that the model is capable of predicting the entire startup process. Further comparison with data of a production scale distillation column confirms the significance of the model. This comprehensive validation is indispensable, as a model-based design of startup strategies demands for a reliable process model.

First examinations have revealed, that specific changes in the manipulated variables could significantly reduce the startup time in comparison with conventional strategies. A systematic approach with extensive dynamic optimization studies shows the minimum time requirements and serve as a benchmark for developing the control approach. The various number of manipulated variables and the inherently required inequality constraints for a distillation process result in an elaborate formulation of the dynamic optimization problem. This general formulation of the optimization problem results in an approach which is applicable for a great variety of fully thermally coupled distillation arrangements. Sophisticated time-optimal trajectories of manipulated variables, which are obtained by off-line optimization, are usually not applicable at real plants. Hence, a robust process control strategy is developed, which tracks the time-optimal solution by using a decentralized control scheme and results in a reasonably short startup time.

The presentation starts with a short introduction of the process model and the setup of the pilot plant. Furthermore, simulation results are compared with experiments, whereas the profiles of characteristic process variables are analysed and discussed. The extensive validation addresses the steady state, the dynamic and the startup characteristics of the dividing-wall column. Finally, the results of the dynamic optimization problem and a robust startup strategy are presented, which illustrate the potential for both, a significant reduction in startup time and an easy implementation of the control approach for tracking the time optimal trajectory.

408f On a New Way of Looking at Energy Efficient Distillation Processes with An Emphasis on Dividing Wall (and other complex) Columns

Ross Taylor[1], Lisa Nelson[1], and Angelo Lucia[2]. (1) Chemical Engineering, Clarkson University, Potsdam, NY 13699, (2) University of Rhode Island, Dept of Chem. Engineering, Kingston, RI 02881

Recent work by Lucia and coworkers has shown that the most energy efficient distillation process has the shortest stripping line (this being the length in composition space) of the composition trajectory for the stripping section of a column when measured from the bottoms composition to the pinch point). Lucia and coworkers have demonstrated several advantages of this approach to the determination of minimum energy distillation systems that include the facts that the approach can handle any number of compounds, non-ideal systems, chemical reactions, and mass transfer effects.

In this paper we show how the concept of the shortest stripping line can be applied to the determination of minimum energy design of complex distillation configurations. We place particular emphasis on the modeling of dividing wall columns (DWC). In addition, we demonstrate the influence of mass transfer on the minimum energy design of a DWC.

408g Sharp-Split Kaibel Column Design by Utilising Column Profile Map Techniques

Ronald Abbas[1], Simon. T Holland[2], **Diane Hildebrandt**[1], and David Glasser[1]. (1) School of Chemical and Metallurgical Engineering, Centre of Material and Process Synthesis (COMPS), 1 Jan Smuts Ave, Braamfontein, Johannesburg, South Africa, (2) School of Chemical and Metallurgical Engineering, Centre of Material and Process Synthesis (COMPS), University of the Witwatersrand, 1 Jan Smuts Ave, Braamfontein, Johannesburg, South Africa

This paper considers the implementation of a Kaibel column, (i.e. a fully-thermally coupled column with an adiabatic wall dividing the column into two equal halves for the production of four product streams). The Kaibel Column allows for a feed mixture of four or more components from which it produces a distillate, bottoms and two product side streams. Compared to the conventional 3 column direct split sequence, the Kaibel column can be built in a single shell, making it an attractive alternative in terms of capital cost savings along with its counterpart; the Petlyuk Column. Further, the reduction in the number of reboilers and condensers' required leads to improved operating costs.

In this paper we demonstrate the use of column profile maps (CPMs) for the comprehensive analysis and design of Kaibel columns by applying the CPM technique for a system at sharp-split conditions.

From the results of the topological analysis, it is shown that, for set product composition specifications, when using an ideal system (constant relative volatilities), there is only one set of feasible operating parameters. Future work will extend the assessment of the validity of this result to other zeotropic and azeotropic separation systems as well as the effect of the thermodynamic model.

413a Layer-by-Layer Assembled Proton Exchange Membranes for Fuel Cell Applications

J. Nathan Ashcraft, Avni A. Argun, and Paula T. Hammond. Chemical Engineering, Massachusetts Institute of Technology, 77 Massachusetts Avenue, 66-525, Cambridge, MA 02139

The increasing global focus on alternative energy sources has led to a renewed interest in fuel cells, especially hydrogen and methanol powered fuel cells. Recent research has shown layer-by-layer (LBL) assembly to be a robust technique for the fabrication of solid state electrolytes. Previous LBL electrolyte systems have lacked the high ionic conductivity values which reduce performance losses and are critical for practical applications. We have recently reported the highest ionic conductivity values ever obtained from an LBL system.[1] By sulfonating an aromatic polyether to a high degree and pairing it with an amine-based polycation, we can obtain LBL films with ionic conductivity values up to 3.5×10^{-2} S/cm at 25°C and 98% RH. These multilayer systems also exhibit low liquid methanol permeability and have high chemical and mechanical stability; this provides a direct application as proton-exchange membranes in direct methanol fuel cells (DMFCs). We have demonstrated that simply coating traditional fuel cell membranes with 3 to 5 bilayers of these LBL films improves the power output of DMFCs by up to 50%.

[1] Argun, A. A.; Ashcraft, J. N.; Hammond, P. T., *Highly Conductive, Methanol Resistant Polyelectrolyte Multilayers*. Advanced Materials **2008**, 20, (8), 1539-1543.

413b PEM Fuel Cell Properties of Pre-Stretched Recast Nafion®

Jun Lin[1], Pin-Han Wu[1], **Peter N. Pintauro**[1], Ryszard Wycisk[1], and Zhiqing Shi[2]. (1) Department of Chemical Engineering, Case Western Reserve University, 10900 Euclid Avenue, Cleveland, OH 44106-7217, (2) National Research Council Canada, Institute for Fuel Cell Innovation, 3250 East Mall, Vancouver, BC V6T IW5, Canada

DuPont's Nafion® perfluorosulfonic acid cation-exchange membranes possess many desirable properties for PEM fuel cells which operate at a temperature ≤ 80°C, including high proton conductivity under fully hydrated conditions and good thermal and oxidative stability. Unfortunately, methanol crossover in these membranes is unacceptably high when they are used in a direct methanol fuel cell (DMFC) and the mechanical properties of Nafion under humidity cycling in a hydrogen/oxygen fuel cell are less than satisfactory. A new Nafion membrane fabrication method has been developed, with a new and permanent membrane micromorphology, where a solution cast film is uniaxially stretched and then annealed. Uniaxial stretching of recast Nafion up to a draw ratio of four had no effect on proton conductivity (0.1 S/cm at 25°C for membranes in water), but it did lower methanol permeability by a factor of 1.4-2.0. A pre-stretched recast Nafion film in the 130-180 μm thickness range (draw ratio of 4), when used in a direct methanol fuel cell (DMFC), generated a high power density at 0.4 V (88 mW/cm^2 at 60°C, 1.0 M methanol, and ambient pressure air), as compared to 58 mW/cm^2 for Nafion 117. Membrane-electrode assemblies (MEAs) with pre-stretched recast Nafion consistently out-performed Nafion 117 in a DMFC at methanol feed concentrations in the range of 0.50-10.0 M, with 33-48% higher power densities at 0.4 V, due to a combination of low membrane area-specific resistance (the use of a thinner pre-stretched membrane, where the conductivity was the same as that for commercial Nafion) and low methanol crossover (due to low methanol solubility in the membrane). Equilibrium water sorption, differential scanning calorimetry, electro-osmotic drag coefficient, and pulsed field gradient NMR diffusion coefficient analyses of water in pre-stretched recast Nafion revealed that: (1) the total water content in the pre-stretched films was the same as that in commercial Nafion, (2) there was less free/freezable water in pre-stretched films, and (3) the water self diffusion coefficient was the same in Nafion 117 and pre-stretched recast Nafion, and (4) the electro-osmotic drag coefficient was smaller after Nafion elongation. The experimental data suggested that there were a greater number of small water-filled domains in pre-stretched Nafion, as compared to commercial Nafion (this morphology was confirmed by TEM images of membrane cross sections). Pre-stretched recast Nafion also exhibited properties that are desirable for hydrogen/oxygen fuel cell applications. The proton conductivity of water-equilibrated pre-stretched Nafion films at T=–20°C was 30-50% higher than that of commercial Nafion (conductivity under freezing conditions is important for cold weather fuel cell operation) and the mechanical properties were improved (the films were stronger), as quantified by a high ultimate tensile strength and high storage modulus (the improvement in mechanical properties was associated with an increase in polymer crystallinity upon stretching).

413c Perfluorinated Ionomers Containg the Bis(perfluoroalkyl)Phosphonic Acid Function

Darryl DesMarteau, Li-Mei Jin, and Stephen Creager. Chemistry, Clemson University, AMRL, 91 Technology Drive, Anderson, SC 29625

Perfluorinated ionomers containing the bis(perfluoroalkyl)phosphonic acid

Darryl D. DesMarteau, Li-Mei Jin and Stephen Creager

Department of Chemistry, Clemson University,ARML, 91 Technology Drive, Anderson, SC 29625

Introduction

Bis(perfluoroalkyl)phosphonic acids are known to be very strong protic acids. Incorporation of this acid function into a perfluorinated ionomer might have potential for use in PEM fuel cells. Previous research on related perfluorinated ionomers containing the monoalkylphosphonic acid function had shown some promise for fuel cell applications but the investigations were limited.[1,2] We have developed methods for the synthesis of novel bis(perfluoroalkyl)phosphonic acid vinyl ether monomers and have prepared copolymers with tetrafluoroethylene. Some preliminary evaluation of these polymers will be presented.

.

Experimental

Typical procedure for the polymer synthesis

The copolymerization of the monomer with TFE was initiated by $K_2S_2O_8$ with Na_2HPO_4/NaH_2PO_4 as pH buffer. [3] A 450 ml autoclave was precleaned with nitric acid, rinsed, and thoroughly dried. Approximately 800 ml of distilled water was degassed overnight using nitrogen before use. A sample of monomer $R_FP(O)(ONa)C_4F_9$, **1** or $R_FP(O)(ONa)CF_3$, **2** (R_F: $CF_2{=}CFOCF_2CF(CF_3)OCF_2CF_2$), Na_2HPO_4, and NaH_2PO_4 were dissolved into 210 ml of degassed distilled water and bubbled with N_2 gas overnight. Similarly, the solution of $K_2S_2O_8$ (30ml) was also bubbled with N_2 gas overnight. Then the two solutions were mixed and pulled into the fully evacuated autoclave, and followed by washing with 20 ml of degassed distilled water. The mixture was heated to 65°C, after desired amount of TFE was consumed as determined based on pressure drops which calculated using van der Waals equation, the TFE was vented and then the mixture was acidified with 36% HCl solution (10ml) and a gel-like solid formed. After the polymer was washed with dilute HCl solution, H_2O and Et_2O subsequently, then dried at 120°C under full vacuum for 24h, a white powder polymer was obtained.

Formation of membrane

Membranes were cast on aluminum foil (8x8cm). The polymer (1g) was heated in EtOH(70ml)/DMAc (70ml) at 70°C for 24h and then blended for 2min. After filtered with a nylon cloth, the solution was put onto alumina foil and vacuumed at 100°C for 4h, then heated at 250°C for 2h and a light brown film formed. The film was washed with HCl (36%), refluxed in 70% HNO_3 and then boiled in repetitive deionized water to give a colorless, tough and flexible film.

Characterization Methods. Thermogravimetric Analysis were obtained with a Perkin-Elmer TGA7 thermogravimetric analyzer under N_2 at a heating rate of 20°C/min from 50 to 700°C and Differential Scanning Calorimetry were recorded on a DSC7 at a heating and cooling rate of 10°C/min from 50 to 370°C. Impedance measurements were made using Solartron 1255B gain phase analyzer and Solartron SI 1287 potentionstat. The instrument was used in galvanostatic mode with voltage amplitude of 50mV with over a frequency range from 10 to 20000Hz.

Results and Discussion

Figure 1 shows a thermograph of the bulk copolymer material (from monomer **1**) at different equivalent weight under N_2 atmosphere.

T/°C

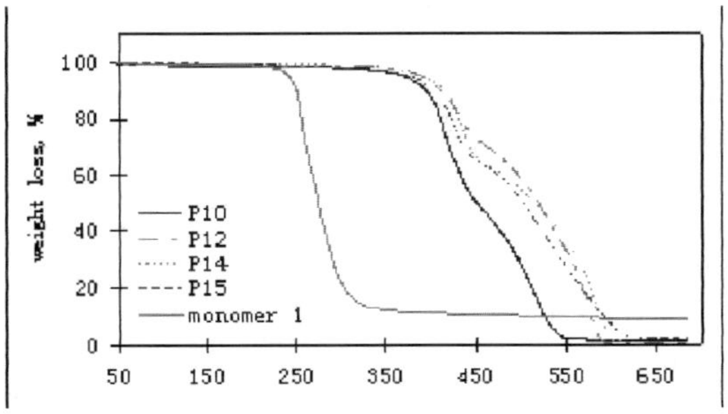

Figure 1. TGA of copolymer **P10, P12, P14, P15** and its monomer **1**.

The first step in the weight (%) loss corresponds to the loss of the perfluoroalkylphosphonic acid function, compared to the decomposition temperature of the monomer (245°C), the typical onset for thermal decomposition of these polymers is around 400°C. The second decomposition temperature (around 500°C) is because of weight loss of the PTFE back-bone. By comparing the second onset temperatures of the polymers, the approximate content of TFE could be estimated. For example, compared to higher onset of **P12** (EW:1250 g/mol), **P14** (EW: 1000 g/mol), and **P15** (EW: 900 g/mol), the lower onset of the **P10** (EW:750) was observed, which means **P10** has lower TFE content. This is consistent with the results from the titration. The DSC curves of the copolymers with various EW also gave similar results. Undoubtedly, the endotherm peak at 320-350°C is due to the melting (Tm) of the PTFE crystalline region in the copolymer. From the DSC measurement, the endotherm (324°C) and exotherm (310°C) peaks were observed for **P12** while no such peaks were observed for **P10**, which indicated the extent of crystallinity increased for **P12** due to more TFE in the polymer. The typical onset for thermal decomposition of the membranes is similar to the bulk polymers.

The water uptake of membranes was measured by soaking the membranes in water at room temperature or 100°C for 24h. The results are summarized in Figure 2.

0.9

1.1

1.0

1.33

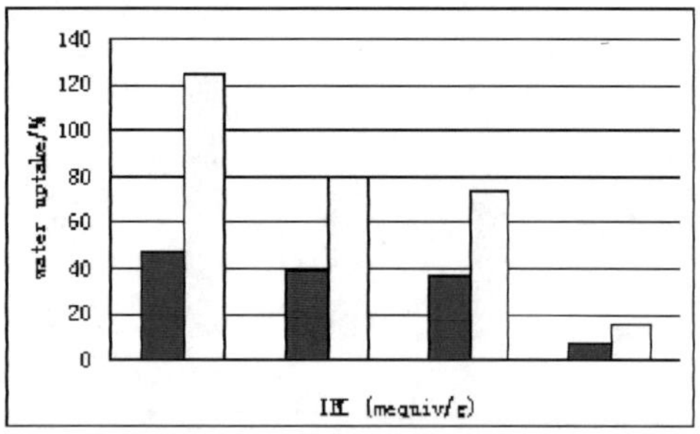

Figure 2. Water uptake at room temperature (left column) and 100°C (right column) *vs* IEC (copolymers

from copolymerization of $R_FP(O)(ONa)C_4F_9$ with TFE).

The water uptake of membranes, treated with refluxing water, with IEC values of 1.33, 1.1, 1, 0.8 mequiv/g was 124, 80, 74 and 16%, respectively. As expected, **P10-mem** (membrane from polymer **P10**) showed the highest water uptake among these membranes, corresponding to absorption of 33 water molecules per phosphonic acid. On the other hand, the **P12-mem** (membrane from polymer **P12**) membrane, with the lowest IEC value, showed the lowest water uptake value, corresponding to 6.6 water molecules per phosphonic acid. The room temperature water uptake of these membranes increased with increasing IEC as well. While increasing temperature, the water uptake value increased. For example, **P10-mem** showed water uptake of 124% at 100°C while 47% at room temperature. It is worthy to mention that the membranes do not show obvious swelling in cool and hot water. Their strength was still retained in a wet state.

Proton conductivity

Proton conductivity data were acquired for samples **P10-mem** (IEC = 1.33 mequiv/g), **P12-mem** (IEC = 0.8 mequiv/g), P14-mem (IEC = 1 mequiv/g) and **P18-mem** of the ionomer at 100% RH at three temperatures between ambient, 80°C and 120°C. The membranes were hydrated by soaking in hot deionized water for 24h prior to study. Representative results are given in Table 1.

Table 1. EW, and conductivity of ionomer membranes[a]

membrane	EW g/mol	Conductivity 25°C	Conductivity 80°C	Conductivity 120°C
P10-mem	750	0.026	0.025	0.046
P12-mem	1250	0.006	0.018	0.013
P14-mem	1000	0.014	0.021	0.035
P18-mem	1100	0.044	0.122	0.124

a. The monomer for **P10-mem**, **P12-mem**, and **P14-mem** is $R_fP(O)(ONa)C_4F_9$; while the while the monomer for **P18-mem** is $R_fP(O)(ONa)CF_3$.

The ionomer **P10-mem** and **P12-mem** showed the proton conductivities 0.025 and 0.019 S/cm at 80°C respectively. Their conductivities increased with increasing IEC. Interestingly, increasing the temperature from 80°C to 120°C, the conductivities of **P10-mem** increased nearly double times, from 0.025 to 0.046 S/cm. The good water holding capability above 100°C might be responsible for high proton conductivity of **P10-mem** membrane. Although the conductivity decreased a little bit (from 0.019 to 0.014 S/cm) for the **P12-mem** (IEC = 0.83), the mechanical strength was unchanged. Interestingly, after adjusting the ionomer structure, to replace the perfluorobutyl group with a trifluoromethyl group, the resulting membrane (**P18-mem**) gave much better conductivity. It showed the conductivity of 0.044 S/cm at room temperature. Importantly, the conductivity of **P18-mem** was 0.122 and 0.124 S/cm at 80°C and 120°C respectively, which is similar to that of the Nafion112.

In conclusion, a series of new bis(perfluoroalkyl)phosphonic acid-based copolymers with TFE were prepared using aqueous emulsion polymerization method. Tough flexible membranes were formed readily by solution casting. The ionomers are stable up to 350°C by TGA (N_2). The ionomer membrane **P10-mem** (from monomer $R_FP(O)(ONa)C_4F_9$, IEC = 1.33 mequiv/g) showed conductivity (100% RH) at 120°C (0.042 S/cm) while the membrane **P18-mem** (from monomer $R_FP(O)(ONa)CF_3$ **,** IEC = 0.9 mequiv/g) gave rather good conductivity at 120°C (0.124 S/cm).

Acknowledgment

This work was supported by U.S. Department of Energy.

References

1. Kotov, S. V.; Pedersen, S. D.; Qiu, W.; Qiu, Z.; Burton, D. J. *J. Fluorine Chem.* **1997**, *82*, 13-19.

2. Yamabe, M.; Akiyama, K.; Akatsuka, Y.; Kato, M. *European Polymer Journal* **2000**, *36*, 1035-1041.

3. Thomas, B. H.; Shafer, G.; Ma, J. J.; Tu, M.; DesMarteau, D. D. *J. Fluorine Chem.* **2004**, *125*, 1231-1240.

413d Evaluation of Resistance Contributions from through-Plane Membrane Proton Transport, Electronic Components and Membrane-Electrode Interface in Fuel Cells

Ruichun Jiang[1], Michael Murphy[1], Cortney Mittelsteadt[2], Timothy Fuller[1], and Craig Gittleman[1]. (1) Fuel Cell Activities, General Motors, 10 Carriage Street, Honeoye Falls, NY 14472, (2) Giner Electrochemical Systems, 89 Rumford Ave, Newton, MA 02466

Proton exchange membrane fuel cells (PEMFCs) are receiving significant research attention for automotive, stationary and portable power applications. A major contributor to PEMFC inefficiency is resistance losses from the membrane, soft-goods (i.e., flow fields, gas diffusion medias), and interfaces between these components. Ascribing these losses to the individual components is a key challenge in evaluating and improving PEMFC performance. Of particular interest is measuring through plane membrane resistance, as unlike the other components this is highly variable with temperature and relative humidity. Though effective techniques have been developed for measuring in-plane membrane resistance through use of separate source and sense leads (four-point-probe measurements) these types of measurements are precluded for through-plane measurements due to membrane thinness (< 50 μm). In-plane membrane resistance measurements may correlate well with through-plane resistance for homogenous isotropic membranes, but have large errors for non-homogeneous membrane structures (e. g. extruded or supported membranes).

In this study, we developed a method to evaluate membrane resistance in the through-plane direction. We also evaluated the soft-good and contact resistances through both ex-situ and in-situ tests. The resistivity/conductivity of membranes with different structures were studied and compared in both through-plane and in-plane directions at various test conditions.

413e The Effect of Polymer Structure and Morphology on Proton Hydration and Diffusion in Perfluorosulfonic Acid Membranes Via Molecular Dynamics Simulations

Junwu Liu[1], Shengting Cui[2], Stephen J. Paddison[2], David J. Keffer[2], and Brian J. Edwards[2]. (1) Department of Chemical Engineering, University of Tennessee, 1512 Middle Drive, Knoxville, TN 37996, (2) Chemical Engieering, University of Tennessee, 1512 Middle Drive, Knoxville, TN 37996

Abstract

Molecular dynamics (MD) simulations have been performed to investigate the effects of polymer architecture on the hydration and diffusion of protons in perfluorosulfonic acid (PFSA) membranes. The structure of the PFSA ionomers were varied in three distinct ways. First, the length of the side chain was varied to correspond to Nafion with its relatively long side-chain, -OCF2CFCF3OCF2CF2SO3H, and to the Short-Side-Chain (SSC) PFSA, -OCF2CF2SO3H. Second, the equivalent weight of the polymer electrolyte was varied in order to probe the effect of polymer equivalent weight. Finally, the molecular weight of the polymer was varied in order to quantify the effects in the molecular simulations on simulated molecules much shorter than those in real systems. We investigate these structure/property relationships as functions of the degree of hydration, including lambda values (ratios of H2O/SO3H) of 3, 6, 9, 15, and 22.

We characterize the effect of these structural changes on the local and global morphology of the aqueous phase, through pair correlation functions and water cluster size distributions, respectively. We also analyze the diffusivities of the hydronium ion and water as a function of polymer architecture. The results from the molecular-level simulations are compared with available experimental data.

413f Molecular Dynamics Simulations of Nanoscale Hydrophilic Domain Structure in Hydrated Nafion

Craig K. Knox, Center for Biophysical Modeling and Simulation, Department of Chemical Engineering, University of Utah, Salt Lake City, UT 84112, Matt K. Petersen, Sandia National Laboratories, Albuquerque, NM, and Gregory A. Voth, Center for Biophysical Modeling & Simulation, Department of Chemistry, University of Utah, Salt Lake City, UT.

Experimental elucidation of the nanoscale structure of hydrated Nafion, the most popular polymer electrolyte membrane (PEM) to date, and its influence on macroscopic proton conductance is particularly challenging. While it is generally agreed that hydrated Nafion is organized into distinct hydrophilic domains within a hydrophobic matrix, the geometry and length scale of these domains continues to be debated. For example, at least half a dozen different domain shapes, ranging from spheres to cylinders, have been proposed based on experimental SAXS and SANS studies. Since the characteristic length scale of these domains is believed to be ~2 to 5 nm, which is approximately equal to the box size of a typical ~5,000 atom PEM system, large atomistic or coarse-grained molecular dynamics (MD) simulations are needed to accurately probe the structure and morphology of these domains, especially their connectivity and percolation phenomena at varying water content. Using classical, all-atom MD with explicit hydronium ions, we have simulated very large hydrated Nafion systems (~2.3 million atoms in a cube with a side length of ~30 nm) to directly observe several hydrophilic domains at the molecular level. The first system was built using a Monte Carlo algorithm without directly assuming any particular geometry or morphological model. For comparison, two more systems were built based on popular morphological models in the literature: the older cluster-network model of Gierke and the newer parallel cylinder model of Schmidt-Rohr and Chen. This talk will explore the results of these simulations, which may help provide guidance for the determination of the morphology of Nafion and the design of new alternative membranes with superior proton conductance that will improve overall PEM fuel cell performance.

421a Surface-Initiated Atom Transfer Radical Polymerization of Poly(ionic liquid)S from Regenerated Cellulose Membranes

Azadeh Samadi and Scott Husson. Chemical and Biomolecular Engineering and Center for Advanced Engineering Fibers and Film, Clemson University, 126 Earle Hall, Clemson, SC 29634

Supported ionic liquid membranes (SILMs) have been used for a variety of separations. Yet, a major weakness of SILMs is that the ionic liquid is held in the membrane pores by capillary forces, such that, when the transmembrane pressure is high enough, the ionic liquid is pushed out of the membrane. We report a solution to this problem. Surface-initiated atom transfer radical polymerization (ATRP) was used to covalently graft poly(ionic liquids) (PILs) from the surfaces of regenerated cellulose (RC) membranes. Because the PIL is attached to the pore surfaces covalently, there is no concern for leakage from the support. In this presentation, we will discuss the methodology in which we first graft a poly[2-(methacryloyloxy)ethyl]trimethylammonium chloride (PMETAC) nanolayer from the support. The thickness of the nanolayer is controlled by changing the polymerization time, and ATR-FTIR spectroscopy is used to follow the modification steps. The kinetics of polymerization was studied in parallel experiments on a model substrate that allows us to measure nanolayer thickness versus time. The chemistry of these charged nanolayers was varied by exchange of the Cl- anion (with, e.g., BF4-, PF6-), as confirmed by XPS.

421b Surface Modification to Enhance the Fouling Resistance of Commercial Reverse Osmosis Membranes

Elizabeth M. Van Wagner[1], Benny D. Freeman[1], and Mukul M. Sharma[2]. (1) Chemical Engineering, The University of Texas at Austin, Center for Energy & Environmental Resources, 10100 Burnet Road Building 133, Austin, TX 78758, (2) Petroleum & Geosystems Engineering, The University of Texas at Austin, Cockrell School of Engineering, 1 University Station C0300, Austin, TX 78712

Due to growing water shortages, reverse osmosis (RO) membranes are increasingly being considered for use in purifying alternative water sources. However, RO membranes are prone to fouling by common contaminants present in these sources, such as oil, surfactants, and particulate matter, making membrane surface modification a current area of significant interest. This study was undertaken to modify the surface of commercial polyamide RO membranes (XLE from Dow FilmTec) to make more fouling-resistant materials.

Short-chain molecules based on poly(ethylene glycol) diglycidyl ethers or fluoroalkyl oxiranes were used to form chemical bonds between their epoxide endgroups and free amines present on the polyamide RO membrane surface. Variables including treatment method (dip coating, spin coating, or stationary top surface isolation), time, temperature, and molecular weight and concentration of grafting molecule were studied for their effect on flux and rejection. The fouling resistance (i.e., flux decline) of modified and unmodified membranes was compared in crossflow filtration studies using simple model foulant solutions containing oil and/or surfactant. Grafting density (ìg/cm2) on the modified membrane surface was measured using a magnetic suspension balance. Surface properties thought to influence fouling (i.e., roughness and charge) were evaluated using atomic force microscopy (AFM, in water and air) and zeta potential analysis.

421c Surface Nano-Structuring by Atmospheric Pressure Plasma Induced Graft Polymerization for the Synthesis of Low Fouling and Scaling Ro/nf Membranes

Nancy H. Lin[1], Gregory T. Lewis[1], Myung-man Kim[1], and Yoram Cohen[2]. (1) Chemical and Biomolecular Engineering, University of California, Los Angeles, 5531 Boelter Hall, Los Angeles, CA 90095, (2) Chemical and Biomolecular Engineering, UCLA, Los Angeles, CA 90095

Rapid population growth and limited potable water resources are generating an increasing demand for water desalination. One of the leading methods of desalination is through the use of reverse osmosis (RO) membranes. However, the major problem associated with desalination through RO is membrane surface scaling by mineral salt, organic and biological fouling. Membrane surface modification has been widely employed to mitigate salt scale formation and biofouling problem. However, traditional RO membrane surface modification methods have been shown to reduce the intrinsic membrane permeability and involve complex chemical steps that are problematic for scale-up commercial production. In the present study, a novel atmospheric pressure plasma-induced graft polymerization method was developed for surface nano-structuring of RO and NF membranes with the goal of synthesizing RO and NF membranes with a highly dense, covalently-bound permselective grafted polymer brush layer. The chemical and physical features of the grafted polymer film were established for the selected monomer by tuning the reaction conditions to achieve unique surface architectures that are optimized for low fouling/scaling RO membrane synthesis. The properties of the graft polymerized surfaces, such as surface topography and surface feature uniformity, were evaluated by Atomic Force Microscopy (AFM), and the surface chemistry was elucidated by Fourier Transform Infrared (FTIR) Spectroscopy. A Quartz Crystal Microbalance (QCM) flow cell was used to investigate the effect of the grafted polymer layer topography on scale formation and the surface scaling/fouling kinetics (e.g., onset of mineral scaling, surface scale coverage) of surface nano-structured RO membranes with typical commercial RO membrane surfaces to determine the impact of the polymer brush layer properties (i.e., polymer chain density, chain length, and monomer chemistry) on surface fouling/scaling. The sensitivity of QCM allowed for the detection of surface scale and protein formation/adsorption in the ng/cm^2 range. The QCM system enabled direct measurements of the kinetics of mineral scaling and organic/biological fouling on the surface nano-structured membranes. The implications of the results from the present study for the development of a new class of RO/NF membranes of low bio-fouling and mineral scaling propensities will be discussed with specific reference to the structural surface properties.

421d PEG Assisted Hydrophilic Enhancement for Fouling Prevention In Ultrafiltration

Tilak Gullinkala, Chemical Engineering, University of Toledo, Toledo, OH 43606 and Isabel Escobar, Department of Chemical and Environmental Engineering, University of Toledo, 3801 W. Bancroft St, Toledo, OH 43606.

Cellulose acetate is widely used in the manufacture of commercial water purification membranes. Although commercially available cellulose acetate membranes have high flux through membrane, they suffer limitations such as susceptibility to microbial attack and low chlorine resistance due to the presence of naturally occurring cellulose acetate backbone structure. Throughout filtration this leads to fouling due to the presence of natural organic matter (NOM) in the feed solution which coordinates with the hydrophobicity of the membrane. Modification of membrane surfaces can lead to improved filtration properties such as amplified permeate flux and high resistance to fouling. In order to reduce the hydrophobic interactions and thereby fouling due to NOM, hydrophilic poly(ethylene glycol) (PEG) monomer chains were attached to membrane via in situ graft polymerization. Different methods of grafting poly (ethylene glycol) (PEG) chains to commercially available cellulose acetate ultrafiltration membrane were considered and compared with respect to permeate flux and fouling prevention. Grafting was attained by forming a reactive radical on the membrane surface by oxidation followed by attaching PEG chains. These chain lengths and density were controlled by using a chain transfer agent.

Graft polymerization was carried out in two different methods named bulk method and drop method. In bulk method membrane samples were completely immersed in the reactive reagents for the modification to take place. Samples were dissolved first in oxidizing agent for 10 minutes then in 10% PEG solution for 5 minutes finally in chain transfer agent foe 2.5 minutes. All the reagents were in vigorous stirring motion during the modification in this process. In another method called Drop method, a membrane sample was placed flat on a glass sample holder and solutions containing the oxidizing agent were added to the membrane sheet drop-wise so that the entire sheet was filled with the oxidizing solution. After ten minutes the oxidizing solution was replaced by PEG solution for polymerization then chain transfer agent was added drop wise for controlling chain length. Energy Dispersive X-ray Spectroscopy, Scanning Electron Microscopy, Atomic Force Microscopy and Fourier Transform Infrared Spectroscopy were performed to assess the receptivity of membrane to the modification via chemical grafting.

Three different feed solutions were used to characterize the modification. Dextran solution was used to determine the effect of modification on uncharged particulate matter, while a modeled sea water solution to determine the influence of graft polymerization on NOM during filtration through cellulose acetate membranes. In these experiments sea water was simulated by forming an aqueous solution composed of 2 mg/l of each Suwannee River Fulvic and Humic Acids, along with 0.1 mM CaCl2 as a representative of divalent cations, 0.1 mM NaHCO3 as buffer system, 1M NaCl as background electrolyte and 1 mg/l of SiO2. Protein filtration was also performed to evaluate the modification influence on ultrafiltartion.

Drop method modification of the membrane resulted in better flux than bulk modified and virgin membranes during ultrafiltration of dextran solution. It also resulted in 10% increase in the rejection capacity than that of virgin membrane. Drop modification of membrane also led to modest decrease in the roughness of the membrane which reduces the membrane susceptibility to fouling. Only bulk modification was used to polymerize the membranes in the case of ultrafiltration of modeled sea water for the ease of use. Different sets of filtration runs were performed such as 1 minute, 5 minutes, 15 minutes and up to 6 hours to determine the fouling patterns due to natural organic matter such as humic and fulvic acids present in the feed solution. In this case modification led to the increase in the permeability of the membrane and finer fouling patterns during filtration.

421e Synthesis of Poly(ethylene glycol)-Grafted Polycaprolactone Porous Membranes by Using Oxygen Plasma to Prevent Fibroblast Adhesion

Chi Yen[1], Hongyan He[2], Zhengzheng Fei[2], W.S. Winston Ho[2], and L. James Lee[1]. (1) Chemical and Biomolecular Engineering, The Ohio State University, 140 WEST 19TH AVE, Columbus, OH 43210, (2) Department of Chemical and Biomolecular Engineering, The Ohio State University, 140 WEST 19TH AVE, Columbus, OH 43210

The adhesion and proliferation of fibroblast onto the surface of an implant biomedical device can lead to fibrous tissue encapsulation. The fibrous encapsulation can be detrimental to various implant applications, such as drug delivery devices. Therefore, the development of a nonfouling surface is necessary for the implantable drug delivery devices.

Poly(ethylene glycol) (PEG) is a nontoxic and nonimmunogenic polymer. PEG has the ability to inhibit cell adhesion on a surface due to its hydrophilicity, steric hindrance, and chain mobility. To produce a nonfouling surface, a new method was proposed for grafting PEG on the polycaprolactone (PCL) porous membrane surface. In this study, PEG(400)-monoacrylate was pre-coated on the PCL membrane surface. By using oxygen plasma, PEG could be successfully bonded on the PCL membrane.

The operating condition of the oxygen plasma was optimized to prevent fibroblast adhesion. When the membrane was treated with soaking in a 0.1M PEG (Mw 400)-monoacrylate solution and the oxygen plasma of 25W power for 5 seconds, there was around 90% less number of fibroblast attached onto the PEG-grafted PCL porous membrane than the untreated membrane surface. Attenuated total reflectance-Fourier transform infrared spectroscopy (ATR-FTIR) showed that the intensity of a PEG characteristic peak (~1100 cm-1) increased significantly for the (Mw 400)-grafted membrane. Moreover, the contact angle for the PEG (Mw 400)-grafted surface was decreased significantly in comparison with the untreated membrane. Therefore, the anti-biofouling surface of the PEG-grafted PCL porous membrane can be obtained for the implant drug delivery devices.

421f Theoretical Prediction of Pka Values for Polymeric Methacrylic Acids Using Combined Ab Initio Molecular Dynamics Simulations and the Continuum Dielectric Model

Haitao Dong[1], Ranil Wickramasinghe[1], Scott Husson[2], and Xianghong Qian[1]. (1) Department of Mechanical Engineering, Colorado State University, Fort Collins, CO 80523-1374, (2) Clemson University, 126 Earle Hall, Clemson, SC 29634

The acidity constants of methacrylic acid polymers are expected to be different from the monomer due to the changes in the dielectric environment surrounding the carboxylic acid groups. This variation provides a great opportunity in designing new bioseparation membranes coated with polymer brushes by tuning the polymer's pKa values to meet the application needs. To understand the pKa dependence on polymer chain length, chain conformation, and the dielectric medium, we used the ab initio molecular dynamics modeling to derive accurate conformations and partial charges for the methacrylic acid oligomers (n = 2 to 20). The reaction field energies of the carboxylic acid groups were calculated by using the quantum-based polarizable continuum model (PCM). For large oligomers, we used the multiconformation continuum electrostatic (MCCE) program that involves solving numerically the Poisson-Boltzmann equation. The advantage of this program is that the variance in conformations is taken into account by utilizing a molecular mechanics force filed in Monte Carlo sampling. The predicted pKa values of the methacrylic acid oligomers show an increase as the chain length grows in agreement with experimental observations

426a Computational Studies of Gas Adsorption In Aerogels: The Effects of Gel Flexibility on Adsorptive Behavior

Lev D. Gelb, Department of Chemistry, Washington University in St. Louis, One Brookings Drive, St. Louis, MO 63130

We present results from simulations of gas adsorption in aerogels in which the flexibility of the gel matrix is treated explicitly. This is accomplished by coupling a coarse-grained model for the aerogel with a mean-field on-lattice density functional model for the adsorbing gas.

Adsorption in aerogel materials has been studied for a number of years both by simulation and experiment. The extremely low density of these materials has made them particularly attractive for studies of the effects of weak confinement on fluid critical properties, but adsorption under conditions nearer to the fluid triple point has also been of considerable interest. Unlike porous materials such as vycor glasses and xerogels, low density aerogels have very low bulk moduli (generally in the range 1-100 bar). As a result they therefore display nontrivial mechanical deformations during adsorption and desorption processes, especially near to the capillary rise and desorption shoulder [1].

Due to the large length scales in aerogels, computational studies of adsorption have generally used lattice-based density functional theory approaches (a la Kierlik, et al. [2,3]) rather than molecular simulation. All such studies to date have kept the gel structure itself (generally obtained by DLCA or similar simulation [4]) frozen, which is a poor approximation for these materials; the purpose of this study is to address this approximation.

We have previously developed a simple coarse-grained model of silica sol-gel materials which was applied to investigate the formation and structural and mechanical properties of low density aerogels [5]. This model is fully flexible and can account for bond breaking and creation upon deformation of the gel structure. This model is structurally reasonable, and for appropriate choice of bond-formation control parameters, quantitative results for bulk modulus and the behavior under mechanical loading can be obtained.

In this work, the commonly used lattice-based fluid adsorption model is coupled to our coarse-grained off-lattice aerogel model using thermodynamic perturbation theory. This provides a complete free energy surface describing the behavior of the gel/fluid system at varying compression and fluid chemical potential. The resulting isotherms and dependence of gel compression on gas pressure are in excellent qualitative agreement with experimental findings, which are intuitively explained as motion on the underlying free energy surface. Perhaps most interestingly, in rigid aerogel models the adsorption isotherms are continuous in the capillary rise region, but when the gel is allowed to relax a sharp jump, corresponding to a first-order transition, is observed.

[1] Reichenauer and Scherer, J. Non-Cryst. Solids 277 (2000) 162.

[2] Kierlik et al., J. Phys.: Cond. Mat. 14 (2002) 9295

[3] Kierlik et al., PRL 87 (2001) 055701

[4] Salazar and Gelb, Mol. Phys. 102 (2004) 1015

[5] Gelb, J. Phys. Chem. C, 111 (2007) 15792

426b A Novel Approach for Comparison of Adsorption Equilibrium Results Between Experiments and Simulations

Pradip Chowdhury, Mayank Singla, and **Sasidhar Gumma**. Chemical Engineering, Indian Institute of Technology Guwahati, Guwahati, 781039, India

The inputs to all molecular simulations involving adsorption are force field / potential parameters (σ, ε/k) between gas and solid, to calculate pair wise potential. Depending on the nature of adsorbent/gas involved and the complexity of the model used in simulations some or all of terms for dispersion, electrostatic interactions, induction, quadrupole etc. are included. In the simplest form, only dispersion interactions are considered (by lumping induced interactions into the dispersion term, if necessary) between pairs of molecules. Pair-wise additivity is assumed to obtain the total potential of the system.

Usually, these parameters are estimated from experimental data of Henry's constants for excess amount adsorbed and isosteric heats.However, this procedure is not straight forward, since the comparison of results between experiments and simulation needs pore volume information of the adsorbent. Moreover, since it is assumed that helium is non-adsorbing to estimate this pore volume, comparison of helium adsorption results between simulation and experiments was ambiguous so far.

In this work, we propose a novel approach to compare results between experiments and simulations. A simple optimization technique to obtain the effective potential parameters is also presented. Using this method helium adsorption results between experiments and simulations can also be compared without ambiguity.

The proposed method was validated by investigating helium adsorption in silicalite. The potential parameters obtained for He-O interactions in silicalite are 3.104 Å and 48.7 K. The helium pore volume for silicalite was found to be 0.155 cm^3/g. We further demonstrate the use of these potential parameters by comparing appropriate simulation results with experimental observations for helium adsorption in silicalite over wide temperature (93-515 K) and pressure (0-3300 kPa) range. The experimental observations used were taken from one of our previous works. The average difference between simulation and experimental results was less 2% using the parameters proposed in this work. All other potential parameter sets used in literature so far yield larger average deviations (greater than 17% in appropriate domain).

426c Molecular Simulation of Selective Paraffin Adsorption from Olefin/ paraffin Mixtures by Aluminum Methylphosphonate-a

Maaike C. Kroon, Section Process Equipment, Process & Energy Department, Faculty of Mechanical, Maritime and Materials Engineering, Delft University of Technology, Leeghwaterstraat 44, Delft, 2628 CA, Netherlands, Carmelo Herdes, Institut de Ciència de Materials de Barcelona, Consejo Superior de Investigaciones Científicas, ICMAB-CSIC. Campus de la UAB, Bellaterra. Barcelona, 08193, Spain, and Lourdes F. Vega, MatGas Research Center and Air Products and Chemicals, Inc, Campus de la UAB., Bellaterra. Barcelona, 08193, Spain.

Due to similar properties, it is hard to separate corresponding olefins and paraffins. Currently, high-pressure distillation is used as purification method, which is energy-intensive and expensive. A cheaper, more efficient, less energy-intensive and cleaner method for separating olefins from paraffins would be very desirable. One of these alternative separation methods is adsorption, which is still in the research-phase. Commonly, it is the olefin that is preferentially adsorbed using this separation method, because of the strong interaction between the unsaturated bond of the olefin and a metal ion on the surface of the adsorbent material, forming a p-complex. Because the olefin is preferably adsorbed, it is the paraffin that is easily obtained in its pure form. Unfortunately, in industrial processes the desired product is the olefin (raw material for polymers) and not the paraffin. Therefore, adsorption is not an option to replace the current distillation columns (yet). However, when it is the paraffin that is selectively adsorbed, the olefin can be easily obtained in its pure form, which would make the adsorption process a very attractive alternative. Recently, for the first time, an adsorption material was found that was able to selectively adsorb the paraffin instead of the olefin. This research involved the selective absorption of ethyl chloride compared to vinyl chloride using the hybrid inorganic-organic material aluminum methylphosphonate polymorph alpha (AlMePO-a)1. The study was performed by a combined experimental-molecular simulation approach. However, it is still unclear why AlMePO-a selectively adsorbs the ethyl chloride compared to the vinyl chloride, and whether this material is also suited for other olefin/paraffin separations. We are using molecular simulations in order to understand the key factors determining the selective paraffin adsorption by AlMePO-a (molecular interactions, steric effects, rigidity) and to investigate the suitability of this material for other olefin/paraffin separations. The molecular model used for the material is the same developed in previous work2, while appropriate force fields are used for the different olefin and paraffins under consideration. Results will be presented and discussed at the conference.

This work is partially funded by the Spanish government (CTQ2005-00296/PPQ) and the Catalan government (SGR2005-00288). Supercomputing time from CESCA is also acknowledged.

References

1. Herdes, C.; Valente, A.; Lin, Z.; Rocha, J.; Coutinho, J. A. P.; Medina, F.; Vega, L. F.; Langmuir 2007, 23 (13), 7299-7305.

2. Herdes, C.; Lin, Z.; Valente, Z.; Coutinho, J.A.P.; Vega, L.F., Langmuir 2007, 22, 3097-3104.

426d A Molecular Dynamics Study of Sodium Dodecyl Sulfate (SDS) at the Silica-WATER Interface: pH Effect

Naga Rajesh Tummala, Dimitrios Argyris, and Alberto Striolo. School of Chemical Biological and Materials Engineering, The University of Oklahoma, Norman, OK 73019

Molecular dynamics simulations will be reported for the adsorption of sodium dodecyl sulfate (SDS) at the silicon dioxide-water interface. The surface hydrophobicity was varied by changing the number of hydroxyl groups on the surface. Three different number densities of hydroxyl groups were chosen that correspond to (1) pH approximately equal to the isoelectric point of the silica surface, (2) pH less than the isoelectric point (acidic solution), and (3) pH greater than isoelectric point. We will discuss how the surface properties affect the molecular adsorption of SDS on the three different surfaces by presenting results for density profiles, planar radial distribution functions, and mean square displacement of both water and SDS molecules. The results will be compared to those of pure water at the SiO2 surface in the absence of SDS. This comparison will allow us to understand whether or not the equilibrium surfactant adsorption is influenced by hydrogen bonding within the water layers and/or by the local dielectric constant. Because pH effects manage to alter the nature of SiO2 surface, spanning from hydrophobic to hydrophilic macroscopic properties, our results will yield a better understanding on the competitive adsorption of water and amphiphilic compounds on charged surfaces, a phenomenon very important for industrial applications ranging from detergency to oil recovery.

Web Page: hotohke.ou.edu/~astriolo/Home2007.htm

426e Hybrid Reverse Monte Carlo Simulations of Activated Carbons Used as Adsorbents for Chemical and Biological Warfare Agents

Jeremy C. Palmer, Chemical and Biomolecular Engineering, North Carolina State University, 911 Partners Way, Engineering Building I, Box 7905, Raleigh, NC 27695-7905, John K. Brennan, Weapons and Materials Research Directorate, U.S. Army Research Laboratory, Building 4600, Aberdeen Proving Ground, MD 21005, Keith E. Gubbins, Chemical and Biomolecular Engineering and Center for High Performance Simulation (CHiPS), North Carolina State University, Raleigh, NC 27695-7905, Margaret Hurley, US Army Research Laboratory, Aberdeen, MD 21005, and Alex Balboa, Research and Technology Directorate, U.S. Army Edgewood Chemical Biological Research Center, 5183 Black Hawk RD, Aberdeen Proving Ground, MD 21010-5424.

Activated carbons are amorphous, microporous materials synthesized by pyrolysis of organic precursors such as nut shells, woods, and charcoals. Their excellent surface activity and low cost of production make them ideal materials for many industrial applications such as gas and liquid filtration, gas storage, and catalytic processes. As a result of their amorphous nature, little is know about how the structural features of these carbons influence their adsorptive characteristics. We have used an atomistic simulation technique known has hybrid reverse Monte Carlo (HRMC) to develop geometrically realistic models form experimental x-ray diffraction data for two activated carbons (BPL and ASZM-TEDA) that have shown promise as efficient adsorbents for chemical and biological warfare agents. Through a detailed structural analysis of our model structures, we have found that our models produce key structural features, such as the carbon-carbon radial distribution function, that are in excellent agreement with experimental measurements. We also performed grand canonical Monte Carlo simulations (GCMC) of argon and nitrogen gas at 77 K in order to calculate adsorption isotherms and isosteric heats of adsorption for our model structures. The results from these simulations indicate that our models for activate carbons are both geometrically realistic and have adsorptive properties that are consistent with those of real activated carbons.

426f Transition from Fickian to Single-File Diffusion for Binary Lennard-Jones Mixtures In Single-Walled Carbon Nanotubes

Joshua D. Moore[1], Qu Chen[2], Ying-Chun Liu[2], Thomas R. Roussel[1], Qi Wang[2], and Keith E. Gubbins[1]. (1) Department of Chemical and Biomolecular Engineering and Center for High Performance Simulation, North Carolina State University, Raleigh, NC 27695-7905, (2) Department of Chemistry, Zhejiang University, Hangzhou, 310027, China

Carbon nanotubes (CNTs) are important building blocks for nanocomposite materials and nanomachinery due to their unique physical properties (electronic, optical, thermal, mechanical, etc.)[1,2] and have received extensive interest in research for materials science. One possible application is the use of CNTs as host structures for gas separation as they yield faster diffusion rates of fluid molecules than other widely used materials (e.g., zeolites). Several diffusion mechanisms are possible in narrow cylindrical pores, but these are still poorly understood[3,4]. Among the most important are ballistic motion, Fickian diffusion, and single-file diffusion. Ballistic motion occurs for very short times, before the molecule has had a chance to collide with anything. For longer times the motion becomes Fickian or single-file. If the pore is large enough, the molecules will diffuse in 3-dimensions as they would in a bulk fluid, and diffusion obeys Fick's Law. If the diameter of the pore becomes small enough, the diffusion will crossover from 3-dimensions to a single-file diffusion mechanism, where the molecules can no longer pass each other. Clearly single-file motion is much slower than Fickian, which is much slower than ballistic.

We report molecular simulation results on the adsorption and the self-diffusion of binary Lennard-Jones mixtures in narrow armchair (n,n), zigzag (n,0), and chiral (n,m) carbon nanotubes using a combination of GCMC (Grand Canonical Monte Carlo) and MD (molecular dynamics) simulations. We investigate the effect of properties such as pore loading, pressure, temperature, intermolecular interactions (fluid-wall and fluid-fluid), and CNT flexibility on the mechanism of diffusion (Fickian or single-file).

1 J. Goldberger, R. Fan, and P. D. Yang, *Accounts Chem. Res.* **39**, 239 (2006).

2 D. M. Guldi, G. M. A. Rahman, V. Sgobba, and C. Ehli, *Chem. Soc. Rev.* **35**, 471 (2006).

3 Dubbeldam, D. and Snurr, R.Q., *Molecular Simulation*, **33**, 305 (2007).

4 Kärger, J., Valiullin, R. and Vasenkov, S., *New Journal of Physics*, **7**, 15 (2005).

426g Glassy Dynamics of Confined Fluids

K. G. Ayappa, Department of Chemical Engineering, Indian Institute of Science, Bangalore, India and S. H. Krishnan, Chemical Engineering, Indian Institute of Science, Bangalore, India.

Molecular dynamic simulations of soft-sphere fluids confined in slit pores reveals that a single component fluid can undergo glass-like dynamics due to confinement. The self intermediate scattering function, of the confined fluid exhibits a two step relaxation at moderate temperatures and follows the time-temperature superposition principle observed in bulk glass formers. The self-diffusivity and relaxation times of the confined fluid are found to follow scalings predicted by mode coupling theories developed for bulk glass formers. The system exhibits a crossover temperature which is above the glass transition temperature. At lower temperatures, signatures of additional relaxation processes are observed in the various dynamical quantities with the self-intermediate scattering function relaxing in three steps. The temperature dependence of the self-diffusivity can be accurately expressed by the Vogel-Fulcher-Tammann equation and the fluid can be classified as a fragile glass former based on the fragility index. Simulations are carried out for pore spacings that can accommodate one to four fluid layers and the influence of the confining surface (structured or smooth) on the glassy dynamics is explored.

426h Structural Changes In Nanoporous Solids Due to Fluid Adsorption: Thermodynamic Analysis and Monte Carlo Simulations

Alain H. Fuchs[1], Anne Boutin[2], and Marie Jeffroy[2]. (1) Enscp, Paris Graduate School of Chemical Engineering, Paris, France, (2) Chemistry, University of Paris-Sud, Batiment 349, Orsay, France

Open framework nanoporous materials are gaining increasing interest on account of their exceptional adsorption properties. Zeolites are now widely used in industry as molecular sieves and catalysts. More recently, metal-organic framework (MOF) materials were shown to be very promising for such applications as carbon dioxide or hydrogen capture and storage.Molecular simulations can help in understanding the adsorption process at the molecular level, and contribute to the design of new materials.

There is ample literature on fluid adsorption in zeolites. In most cases, a rigid framework was assumed for the zeolitic adsorbent. Framework flexibility is generally assumed to play a role in transport properties, but not much on thermodynamics. However, a few examples were reported of "stepped" adsorption isotherms, instead of the usual continuous Langmuir (or type I) isotherm. This was attributed to a structural transition of the host framework taking place upon fluid adsorption. With the advent of "soft" hybrid adsorbents like MOFs, it is likely that framework-adsorbate coupling will become more frequent than in the relatively stiff zeolite solids

The need exists then to devise a method that enables the description and prediction of structural changes in adsorbent upon fluid adsorption. A direct route would of course be to simulate the fully flexible solid in presence of various amounts of fluid adsorbate. Unfortunately the existing forcefields are not always appropriate for such simulations.

We propose here a thermodynamic analysis of the framework-adsorbate coupling, based on the so-called "osmotic" statistical ensemble. This method makes use of standard Grand Canonical Monte Carlo simulations only. It allows to compute the thermodynamic potential of the combined host+fluid system, and thus to predict the occurrence of a structural change upon fluid adsorption. The inputs of this method are the crystallographic structures of the host solid phases and their relative lattice energies.

Our case study is silicalite-1 zeolite which has been observed in three different crystalline structures: MONO (the stable monoclinic structure for the bare zeolite at room temperature) ORTHO (observed at high temperature and also when filled by some guest molecules) and PARA (observed at high loading of aromatic molecules). A stepped adsorption isotherm has been recently reported in the case of tetrachloroethene (4CE). Structural changes of the host framework were reported but no conclusion was drawn on which transition was responsible for this step. Interestingly enough, the adsorption isotherm for a similar molecule: 1,1,2 trichloroethene (3CE), was smooth, although zeolite structural changes were also reported We have thus studied in some detail the 4CE/3CE adsorption process in silicalite-1, and were able to predict structural changes, and to reproduce the jump resulting from one of these transitions. The proposed methodology is very general and can be applied to any type of guest/host systems.

426i A High-Resolution Lattice Model for Adsorption: Weighted Density Functional Theory for Lattice Fluids and Application to Adsorption In Porous Materials

Daniel W. Siderius and Lev D. Gelb. Department of Chemistry, Washington University in St. Louis, One Brookings Dr., Saint Louis, MO 63130

Lattice-gas models have found widespread application in the study of gas adsorption and condensation in porous materials due to their ability to adequately capture the changes in thermodynamic properties brought about by fluid confinement while also being computationally inexpensive. The lattice-gas density functional model of Kierlik et al. [1,2] has found particular success in modeling condensation hysteresis in disordered porous materials [3] and gas adsorption in silica aerogels [4], among other applications. Ultimately, the model is most effective when applied to large length scale problems because the lattice spacing must be at least as large as the molecular dimension. This makes the model incompatible with sub-nanometer scale confinement. Alternatively, one could study such nano-confined fluids via the conventional and well-established density functional theory (DFT) for off-lattice fluids, though at significant computational cost when the confinement is asymmetric [5].

We report progress on a new lattice-gas DFT that retains elements of both the coarse-grained DFT of Kierlik et al. and off-lattice conventional DFT. Our approach is well-suited to the study of fluids in highly-disordered or asymmetric confinements and offers reductions in computational demand compared to off-lattice DFT even in ordered or symmetric confinements. We begin by presenting a new density functional for hard-spheres on a cubic lattice of arbitrary discretization based on the off-lattice 'weighted density approximation' of Tarazona [6], and demonstrate that it satisfactorily reproduces the structural properties obtained by simulation. Accordingly, our lattice DFT is not a 'local' density approximation, but imposes local order on the hard-sphere fluid through an estimation of the direct correlation function. Following, we demonstrate how a mean-field attractive perturbation (e.g., that of Weeks, Chandler, and Anderson [7]) may be applied to our hard-sphere lattice DFT and produce a coarse-grained representation of a fluid with a realistic intermolecular interaction.

As an initial application of this new lattice DFT, we study adsorption hysteresis of a lattice Lennard-Jones fluid in classic systems such as slit pores and cylindrical pores to demonstrate both its flexibility and reduced computational cost. We also investigate its use in the study of hydrogen adsorption in metal-organic frameworks and other complex structures. Finally, we discuss end-use applications such as screening experiments of adsorption or condensation in bicontinuous structures and materials with rough surfaces and three-dimensional reconstruction of mesoporous materials.

[1] Kierlik et al., Phys. Rev. Lett., 74, 4361 (1995)

[2] Kierlik et al., Mol. Phys., 95, 341 (1998)

[3] Kierlik et al., Phys. Rev. Lett., 87, 055701 (2001)

[4] Salazar and Gelb, Mol. Phys., 102, 1015 (2004)

[6] Frink and Salinger, J. Comp. Phys., 159, 407 (2000)

[7] Tarazona, Phys. Rev. A, 31, 2672 (1985)

[8] Weeks, Chandler, and Anderson, J. Chem. Phys., 54, 5237 (1971)

444a Propane/propylene Separation: From Molecular Scale to Process

Alirio E. Rodrigues[1], Miguel Granato[2], Pedro Sa Gomes[2], and Nabil Lamia[2]. (1) Laboratory of Separation and Reaction Engineering, University of Porto - Faculty of Engineering, Rua Dr. Roberto Frias s/n, 4200-465 Porto, Porto, Portugal, (2) LSRE, Dept Chem Eng, UNiv Porto, University of Porto - Faculty of Engineering, Rua Dr. Roberto Frias s/n, 4200-465 Porto, Porto, Portugal

Propane/propylene separation is an energy intensive separation currently done by distillation with consumption of 1.5 MWh/ton propylene in cryogenic distilllation units which can be lowered to 0.3 MWh/ton in well heat integrated units.

Alternative technologies studied in our laboratory include Vacuum Swing Adsorption (VSA) and Simulated Moving Bed in gas phase. In VSA zeolite 4A or 13X have been used in processes controlled by crystal diffusion and macropore diffusion, respectively.

In the case of SMB process the adsorbent chosen was zeolite 13 X. The desorbent was either isobutane or butene-1.

The experimental methodology was the following: measurement of single component adsorption equilibrium isotherms; measurement of binary equilibria; fixed bed dynamics. This was coupled with modeling and simulation of SMB allowing the design and determination of operating conditions.

Molecular simulation studies of propane, propylene and isobutane (or butene-1) were carried out and they are a tool to look for better desorbents.

A recently developed force field for adsorption of linear alkanes in the sodium form of FAU is applied to reproduce experimental data for propylene adsorption in 13X zeolites by fitting LJ parameters between the sp2-hybridized bond of propylene and the other atoms for this system. Similar methodology was extend for the adsorption of isobutane.

References:

A.E. Rodrigues et al. Method of separating propylene in mixture with propane by adsorption in simulated moving bed, WO 2008/012410 A1, 31 January 2008

A.E. Rodrigues et al, Procédé de séparation du propylène en mélange avec du propane par adsorption en lit mobile simulé, French patent application No 06 06827, 24/07/ 2006

M. Granato, T. Vlught and A.E. Rodrigues, "Molecular Simulation of Propane/Propylene Adsorption in Zeolite 13X", Ind Eng Chem Res, 46, 7239-7245 (2007)

N. Lamia, Luc Wolff , P. Sa Gomes , C. A. Grande, and A.E. Rodrigues, "Propane / Propylene Separation by Simulated Moving Bed I. Adsorption of Propane, Propylene and Isobutane in Pellets of 13X Zeolite", Separation Science and Technology, 42, 2539-2566 (2007)

444b Approaches to Achieving High Productivity In Size Exclusion SMB Chromatography Processes

Shawn D. Feist, Inorganic Chemistry and Catalysis, The Dow Chemical Company, 1776 Building, Michigan Operations, Midland, MI 48667, Yogesh Hasabnis, Engineering and Process Sciences, The Dow Chemical Company, Tower C, Panchsil Technical Park, Pune, India, Bruce W. Pynnonen, Dow Water Solutions, The Dow Chemical Company, Larkin Laboratory, 1691 N. Swede Rd., Midland, MI 48674, and Timothy Frank, Engineering & Process Sciences Laboratory, Process Separations Discipline, The Dow Chemical Company, 1319 Building, Midland, MI 48674.

This presentation will describe the generation and modeling of high-productivity SMB miniplant data for a model system involving separation of aqueous bovine serum albumin (BSA, 66 kDa) from equine heart myoglobin (EHM, 17 kDa). The separation was accomplished using a 4-zone, 12-column SMB miniplant packed with cellulose-based porous beads. High purities and recoveries were achieved for each protein, yielding up to 99.5+% recovery of 99.9% pure BSA in the raffinate stream (protein-only basis) and 99.9% recovery of 97.5% pure EHM in the extract. The material balance accountabilities for the proteins were within 100 ± 5%. Productivities as high as 1.7 grams total product (isolated BSA + isolated EHM) produced per hour per liter of media were obtained using total protein feed concentrations up to 30 g/L. This high-productivity performance was achieved at operating pressures within industrially-viable ranges (less than 120 psig or about 8 bar per column) by using relatively large diameter beads (100 – 250 micron) and tall columns. The SMB process was simulated using a rate-based model developed in Athena Visual Studio®. A value for the intra-particle mass transfer coefficient was obtained directly from a previously published study of this system (Houwing et al., AIChE J. 2003, 49 (5), 1158-1167). Linear adsorption isotherms were used to represent the equilibrium, and appropriate values for the isotherm constants were determined by regression of miniplant operating data. The resulting process simulation provided an excellent representation of the miniplant's performance over a wide range of protein concentrations and operating conditions. Strategies used in this work to achieve high-productivity operation will be described.

COMPARISONS OF RECYCLE CHROMATOGRAPHY AND SIMULATED MOVING BED FOR PSEUDO-BINARY SEPARATIONS

Ju Weon Lee, Purdue University, West Lafayette, IN
Phillip C. Wankat, Purdue University, West Lafayette, IN

Abstract

The simulated moving bed (SMB) process has been extensively used in industrial separations for binary and pseudo-binary separations. In pseudo-binary mixtures because the desired component is the most or least retained, the other components do not have to be separated. The UOP *Parex* process for *p*-xylene purification is an example. When non-target components have similar adsorption behaviors, the SMB has higher productivity and requires less desorbent compared to batch chromatography. However, these advantages are dependent on the difference of adsorption behaviors of the non-target components. When the most retained component is the target, productivity and desorbent consumption is related to the retention behaviors of the most and least retained non-target components, respectively.

In this research, the performance of batch chromatography with a single recycle stream is compared to a SMB process for pseudo-binary separations with competitive Langmuir isotherms. A ternary nucleoside separation was used as a model system for the simulation of batch and SMB processes. To determine the optimized operating condition of the SMB process, a multi-component hodograph transformation was numerically calculated to obtain the separation region in the m_{II}-m_{III} plane and the regeneration region in the m_I-m_{IV} plane. To compare the performances of SMB and recycle batch chromatography, each process was simulated with the optimized operating conditions with detailed Aspen Chromatography simulations.

Introduction

SMB processes are commonly used for binary separations, such as chiral separations [1, 2]. It is well known that the SMB process is more efficient than batch chromatographic separations in terms of desorbent consumption and productivity [3]. Single-loop SMB processes have been extensively used for pseudo-binary separations. When non-target products have very similar adsorption behaviors, such as in the *Parex* process [4], it can be handled as a binary mixture. However, the advantages and disadvantages of SMB processes compared to batch chromatographic processes have not been extensively studied for pseudo-binary mixtures with a wide range of adsorption behaviors.

In this work, batch chromatography with a single recycle stream and 4-zone SMB processes are compared for pseudo-binary separations with different selectivities. Four nucleosides, 2'-deoxycytidine (dC), 2'-deoxyguanosine (dG), 2'-deoxythymidine (dT), and 2'-deoxyadenosine (dA), were studied as solutes with feed concentrations in the range of competitive Langmuir isotherms [7].

Recycle Chromatography

In recycle chromatography, a single column can be used to separate pseudo-binary mixtures [5]. Fig. 1 shows a schematic illustration of recycle chromatography with one recycle stream. For separation of the most retained solute (C), pure component (A) mixed in desorbent can be recycled to help elute the other undesired component (B) and thus reduce the desorbent to feed ratio D/F. The recycle is timed so that A and B exit together (A/B). Therefore, there are three different streams leaving the column: *recycle*, *waste*, and *product*.

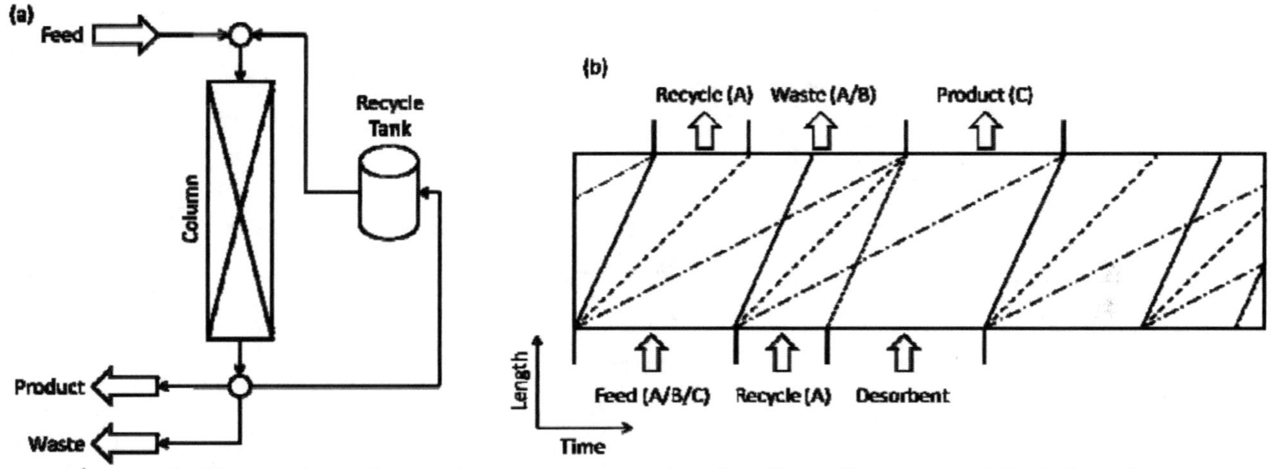

Fig. 1. Schematic illustration of recycle chromatography. (a) Flow diagram and (b) Migration traces of solute in the column.

The operating conditions of recycle chromatography are dependent on the elution bandwidth of each component. Changing the cut point will change the purity and yield. To determine the front and rear ends of elution band, cut points at 0.5% of peak height were used. By changing the flow-rate of the mobile phase, productivity and D/F of batch chromatography were varied. Recycle chromatography and SMB processes were compared at the optimized flow-rate which has maximum productivity and minimum D/F.

SMB Process

Equivalent 4-zone SMB processes composed of 4 and 8 columns, which had the same total bed height as recycle chromatography, were studied. The operating points for m_{II} and m_{III} were chosen based on the even velocity trace which is obtained by the following forms [6]:

$$\frac{m_{\mathrm{II}} - \left(\Delta q_{R,\mathrm{II}} / \Delta c_{R,\mathrm{II}}\right)_{Max}}{\varepsilon_T + (1-\varepsilon_T)\left(\Delta q_{R,\mathrm{II}} / \Delta c_{R,\mathrm{II}}\right)_{Max}} + \frac{m_{\mathrm{III}} - \left(\Delta q_{E,\mathrm{III}} / \Delta c_{E,\mathrm{III}}\right)_{Min}}{\varepsilon_T + (1-\varepsilon_T)\left(\Delta q_{E,\mathrm{III}} / \Delta c_{E,\mathrm{III}}\right)_{Min}} = 0 \tag{1}$$

$$m_j = \frac{Q_j - \varepsilon_p Q_S}{(1-\varepsilon_p) Q_S} \tag{2}$$

where m_{II} and m_{III} are, respectively, the flow-rate ratios of the liquid phase and the solid phase in zone II and III, $c_{i,j}$ and $q_{i,j}$ are, respectively, the concentrations of solute i in the mobile phase and the stationary phase of zone j, ε_T is the total void fraction of the column, ε_p is the interparticle void fraction, Q_j is the volumetric flow-rate of the liquid phase in zone j, and Q_S is the volumetric flow-rate of the solid phase, and solutes E and R represent the least retained component in the extract stream and the most retained component in the raffinate stream, respectively. The purity and yield of extract and raffinate streams are decided by the separation in zone II and III and the regeneration in zone I and IV. The maximum value of $m_{III} - m_{II}$ was investigated with the same or higher purity and yield of target component as in recycle chromatography. With this value of $m_{III} - m_{II}$, the optimum m_I and m_{IV} values were obtained when SMB process had the same purity and yield as recycle chromatography.

Results and Discussion

Three solute mixtures were selected for study of pseudo-binary separations (Table 1). For the most retained solute separation, dA was the target solute and the solute systems were dC/dT/dA, dG/dT/dA, and dC/dG/dA. The selectivities of (intermediate retained/least retained) and (most retained/intermediate retained) are 3.05/2.89, 1.30/2.89, and 2.35/3.74, respectively. The concentration of each component in the feed was fixed as 5.0 g/L.

Table 1. Isotherm parameters and kinetic coefficients of nucleosides [7].

Component	a_i [-]	b_i [L/g]	$a_p k_i$ [1/min]	$D_{e,i}$ [cm^2/min]	$\alpha_{i\text{-}Ref.}$ $=a_i/a_{dC}$
dA	27.7	0.1	6	0.225	8.79
dT	9.60	0.1	30	0.225	3.05
dG	7.40	0.1	30	0.225	2.35
dC	3.15	0.1	60	0.225	1.00

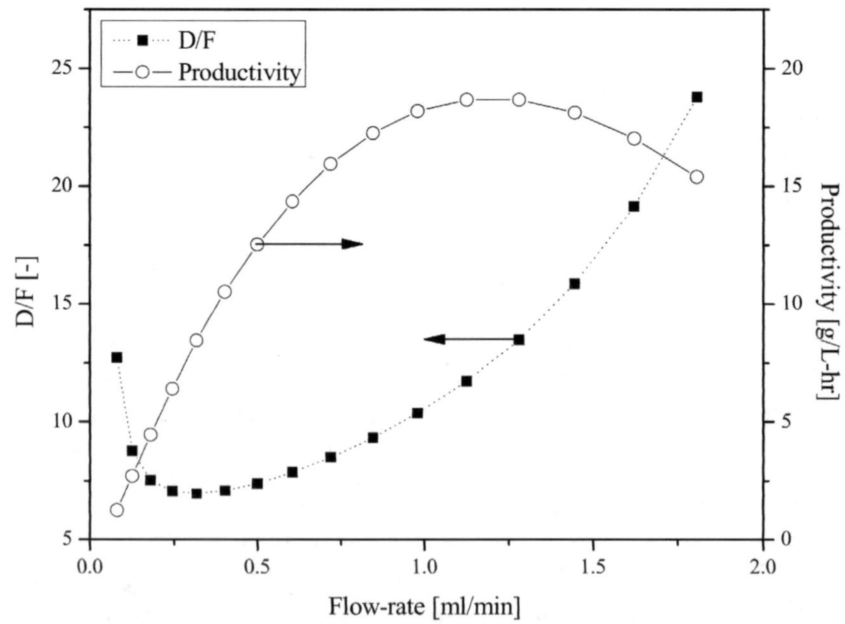

Fig. 2. Changes of D/F and productivity as a function of mobile phase flow-rate in recycle chromatography. Solutes are dC/dT/dA and dA is target. Purity and yield of dA ≥ 0.999

In recycle chromatography, D/F and productivity depend on the mobile phase flow-rate because the efficiency, represented by HETP, depends on the flow-rate of the mobile phase. The apparent dispersion increases when the flow-rate increases. For constant purity, D/F and productivity have minimum and maximum values, respectively (Fig. 2). Table 2 shows the simulation results and the mobile phase flow-rates for the maximum productivity and the minimum D/F in recycle chromatography.

Table 2. Simulation results and mobile phase flow-rates for maximum productivity and minimum D/F in recycle chromatography to separate the most retained solute.

	Solutes	Conc. [g/L]	Flow-rate [ml/min]	D/F [-]	P* [g/L-hr]	ΔP** [bar]	Purity [%]	Yield [%]
Max. P*	dC/dT/dA	5/5/5	1.196	12.493	18.744	4.500	99.896	99.933
	dG/dT/dA	5/5/5	1.232	13.794	18.685	4.633	99.936	99.883
	dC/dG/dA	5/5/5	1.677	11.678	29.035	6.307	99.927	99.915
Min. D/F	dC/dT/dA	5/5/5	0.308	6.942	8.147	1.159	99.936	99.961
	dG/dT/dA	5/5/5	0.308	7.620	8.016	1.158	99.942	99.932
	dC/dG/dA	5/5/5	0.307	5.568	10.032	1.156	99.956	99.967

*Productivity, **Pressure drop

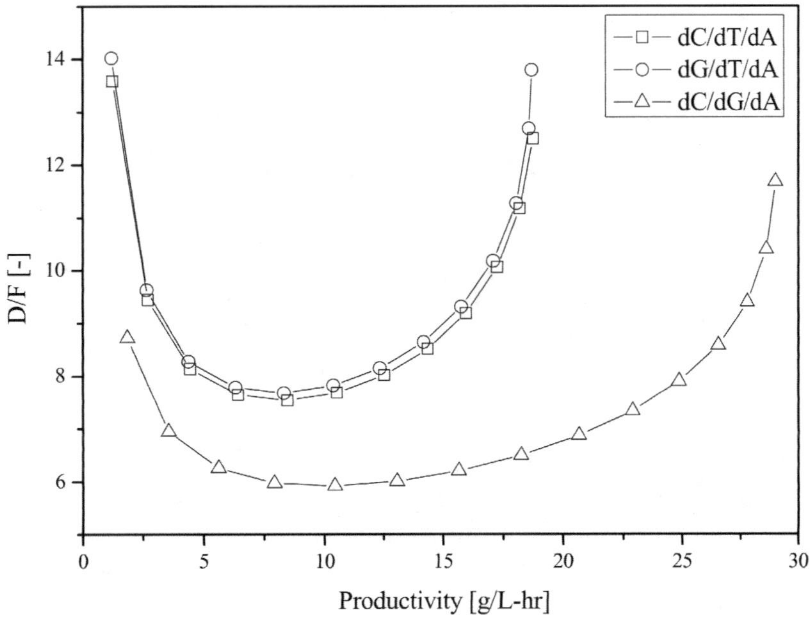

Fig. 3. Relationship between D/F and productivity in recycle chromatography to separate the most retained solute dA.

Figure 3 shows the relationship between D/F and productivity in recycle chromatography. The dC/dT/dA and dG/dT/dA systems show similar tendencies. Partially separated least retained solute (dC or dG) were recycled to be contained in the elution band of intermediate retained solute. There are only two outlet bands; one assigned for *waste* is the elution band which contains the least and intermediate retained solutes. The other assigned for *product* is the elution band of the most retained solute. Therefore, D/F and productivity are strongly related to the separation of the most and intermediate retained solutes which are the same in the dC/dT/dA and dG/dT/dA systems. Figure 4 compares recycle chromatography and a 4-zone, 4-column SMB. For these three ternary mixtures, D/F of the 4-column SMB process is larger than recycle chromatography over the entire range of productivity. The maximum productivities of SMB process are also smaller than recycle chromatography. We will also show results for 8-column SMB systems that have a lower D/F than recycle chromatography for identical purities.

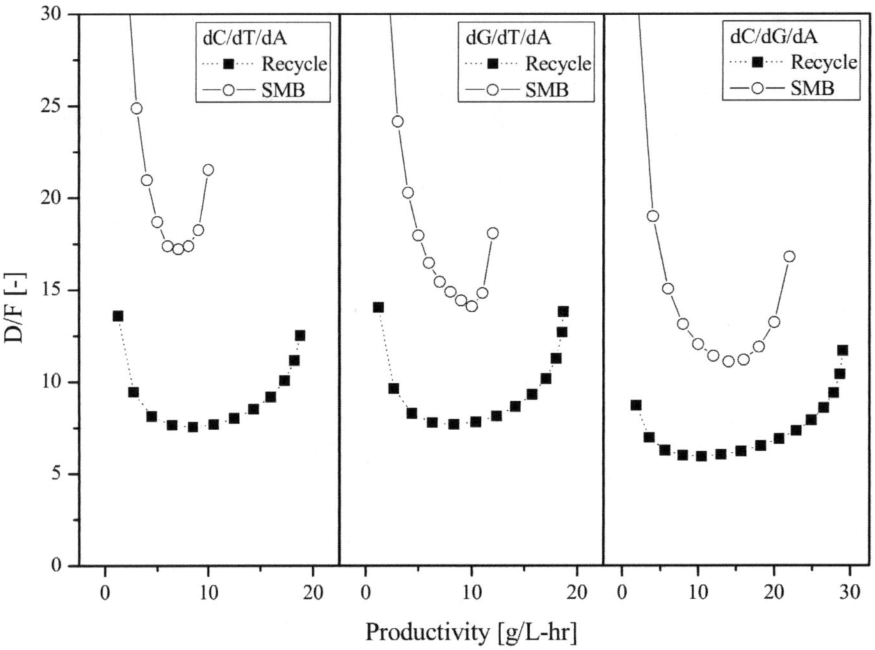

Fig. 4. Comparisons of D/F and productivity for recycle chromatography and 4-zone, 4-column SMB to separate the most retained solute.

Conclusions

Recycle chromatography and 4-column SMB processes were compared for pseudo-binary separations when the most retained solute is desired. While the selectivity of non-target products is varied from 1.30 to 3.05, the performance of recycle chromatography is better than a 4-column SMB in terms of productivity and D/F. With 8 columns the performance of the SMB system was better than the performance of recycle chromatography.

References

1. Pais, S.L., M.J. Loureiro, and A.E. Rodrigues (1997), "Separation of 1,1'-bi-2-naphthol Enantiomers by Continuous Chromatography in Simulated Moving Bed", *Chem. Eng. Sci.*, 52, pp. 245-257.
2. Francotte, E., P. Richert, M. Mazzotti, and M. Morbidelli (1998), "Simulated Moving Bed Chromatographic Resolution of a Chiral Antitussive", *J. Chromatogr. A*, 796, pp. 239-248.
3. Strube, J., S. Haumreisser, H. Schmidt-Traub, M. Schulte, and R. Ditz (1998), "Comparison of Batch Elution and Continuous Simulated Moving Bed Chromatography", *Org. Process Res. Dev.*, 2, pp. 305-319.
4. Broughton, D.B., R.W. Neuzil, J.M. Pharis, and C.S. Brearley (1970), "The Parex Process for Recovering Paraxylene", *Chem. Eng. Prog.*, 66, pp. 70.
5. Lee, J.W and P.C. Wankat (2008), "Optimized Design of Recycle Chromatography for Separation of a Single Component from a Ternary Mixture", *Ind. Eng. Chem. Res.*, *in press*.
6. Migliorini, C., M. Mazzotti, and M. Morbidelli (2000), "Design of Simulated Moving Bed Multicomponent Separations: Langmuir Systems", *Sep. Purif. Technol.*, 20, pp. 79-96.
7. Paredes, G., S. Abel, M. Mazzotti, M. Morbidelli, and J. Stadler (2004), "Analysis of a Simulated Moving Bed Operation for Three-Fraction Separations (3F-SMB)", *Ind. Eng. Chem. Res.*,43, pp. 6157-6167.

444d Modelling and Design Strategies for Lab-Scale SMB Units Accounting for Dead Volumes and Tubing Connections – the Flexsmb-Lsre Unit

Pedro Sá Gomes, Michal Zabka, Mirjana Minceva, and Alírio E. Rodrigues. Laboratory of Separation and Reaction Engineering, University of Porto - Faculty of Engineering, Rua Dr. Roberto Frias s/n, 4200-465 Porto, Porto, Portugal

The practical application of the so-called "non conventional modes of operation" in Simulated Moving Bed (SMB) units is per itself a challenge. Generally an SMB unit, industrial, pilot of laboratory-scale, is limited to one or two modes of operation and to implement a new operating mode is necessary to contact the supplier, do adjustments, even to reformulate the entire unit if not to acquire a new one. Therefore, flexibility (at least at laboratory-scale) is seen as one of the more relevant qualities for this kind of equipment. This aspect was the main reason that led to the design and construction of a 6 columns unit at LSRE, the FlexSMB-LSRE, hereby presented.

Generally, modelling and simulation methodologies for simulated moving bed (SMB) units do not account for pipe transfer lines as the surrounding equipment (valves, degassers, pumps). In the industrial-scale SMB units the relative volume of the connection lines is not so significant as for pilot or lab-scale units, nevertheless, is has been studied in detail and different techniques are applied to overcome the associated negative impacts. For instance, the UOP technique for flushing of the transfer lines before withdrawing the extract, applied in Parex process/(industrial scale SMB unit for p-xylene separation) [1], previously analyzed by Minceva and Rodrigues [2]. In pilot or lab-scale SMB units, the relative volume associated with equipment and transfer lines is indeed an issue and usually these type of units are designed to account for it in its own operation mode. Nevertheless, is still a proxy to the use of more detailed models and some differences are noted [3,4,5]. In this work, a detailed model accounting for dead volumes and tubing connections, as well some compensating strategies and optimization procedures, are presented for pilot (Licosep) and lab-scale (FlexSMB-LSRE) units.

References:

[1] - Stine, L.O., Springs, W., Broughton, D.B., Continuous separation process with emphasis of product purity. US Patent 3,201,491, August 17, 1965.

[2] - Minceva, M., Rodrigues, A.E., Influence of the Transfer Line Dead Volume on the Performance of an Industrial Scale Simulated Moving Bed for p-Xylene Separation, Separation Science and Technology Vol. 38, No. 7, pp. 1463, 2003

[3] - Azevedo, D.C.S., Neves, S.B., Ravagnani, S.P., Cavalcante, C.V., Jr., Rodrigues, A.E. The influence of dead zones of simulated moving bed units. Fundamentals of Adsorption 6; Elsevier: Amsterdam, 1998; 521.

[3] - Migliorini, C., Mazzotti, M., Morbidelli, M., Simulated Moving-Bed Units with Extra-Column Dead Volume, AIChE Journal Vol. 45, No. 7, pp. 1411, 1999.

[4] - Zabka, M, Sa Gomes, P., Minceva, M. and Rodrigues, A.E. Chiral Separation of R,S-a-Tetralol by Simulated Moving Bed, accepted on Separation Science and Technology, 2007.

444e New Developments In Simulated Moving Bed Chromatography

Andreas Seidel-Morgenstern, Max Planck Institute for Dynamics of Complex Technical Systems, Sandtorstr. 1, Magdeburg, 39106, Germany, Christian Kessler, Phys. Chem. Found., Max Planck Institute for Dynamics of Complex Technical Systems, Sandtorstr. 1, Magdeburg, 39106, Germany, and Malte Kaspereit, Proc. Synth. Dynamics, Max Planck Institute for Dynamics of Complex Technical Systems, Sandtorstr. 1, Magdeburg, 39106, Germany.

New Developments in Simulated Moving Bed Chromatography

Andreas Seidel-Morgenstern1,2, Christian Keßler1, Malte Kaspereit1

1Max Planck Institute for Dynamics of Complex Technical Systems, Sandtorstr.1, D-39106 Magdeburg, Germany

2 Otto-von-Guericke-Universitaet, Universitaetsplatz 2, D-39106 Magdeburg, Germany,

seidel-morgenstern@mpi-magdeburg.mpg.de

In the last decade the concept of simulated moving bed (SMB) chromatography has been successfully applied in several fields, e.g. for the separation of enantiomers and the purification of pharmaceutical proteins. Often the separations were realized with the classical implementation of the process using four distinct zones and constant operating conditions (e.g. switching times, liquid phase velocities and solvent strengths). In order to further increase the potential of SMB chromatography, several more sophisticated modes of operation have been developed based on applying gradients and on dynamically varying certain parameters during the separation. Examples for the latter approach are the VariCol [1], PowerFeed [2] and ModiCon [3] processes. There are several interesting attempts to reduce the number of columns needed to realize the basic principle [e.g. 4]. Another promising alternative capable to enhance the process performance is based on the introduction of an enrichment step between zones I and II [5,6]. Besides treating binary mixtures, research increasingly focuses on continuous processes capable to perform multicomponent separations. To achieve separation of a ternary mixture different approaches have been suggested using increased numbers of zones (ranging from 5-Zone open loop up to 14-Zone closed loop setups [e.g. 7]). Recently, a 8-Zone closed loop separation unit was discussed utilizing an external enrichment step [8]. In the presentation will be also discussed the potential of different new strategies based on partial product collection and feed-back of fractions which do not fulfill required specifications [9,10].

The whole presentation intents to give a concise overview to demonstrate the large potential which exists to further improve the classical SMB concept.

[1] M. Bailly, P. Adam, O. Ludemann-Hombourger, R.M. Nicoud, WO 2000025885 (2005).

[2] Z. Zhang, M. Mazzotti, M. Morbidelli, Journal of Chromatography A 1006 (2003) 87.

[3] H. Schramm, A. Kienle, M. Kaspereit, A. Seidel-Morgenstern, WO 2004014511 (2004).

[4] R.C.R. Rodrigues, J. M. M. Araujo, P.J.M. Mota, Journal of Chromatography A 1142(2007) 69.

[5] M. Bailly, R.M. Nicoud, P. Adam, O. Ludemann-Hombourger, US 2006124549 (2006).

[6] G. Paredes, H.K. Rhee, M. Mazzotti, Industrial & Engineering Chemistry Research 45 (2006) 6289.

[7] P.C. Wankat, Ind. Eng. Chem. Res. 40 (2001) 6185.

[8] L.C. Keßler, A. Seidel-Morgenstern, Journal of Chromatography A 1126 (2006) 323.

[9] Y.-S. Bae, C.-H. Lee., Journal of Chromatography A 1122(2006) 161.

[10] L.C. Keßler, A. Seidel-Morgenstern, Journal of Chromatography A, submitted.

444f 'Cycle to Cycle' Optimizing Control of Simulated Moving Beds – Experimental Implementation for a Nonlinear Chiral Separation

Cristian Grossmann[1], Christian Langel[2], Marco Mazzotti[2], Manfred Morari[1], and Massimo Morbidelli[3]. (1) Automatic Control Laboratory, ETH Zurich, Zurich, Switzerland, (2) Institute for Process Engineering, ETH Zurich, Zurich, Switzerland, (3) Institute for Chemical and Bioengineering, Department of Chemistry and Applied Biosciences, ETH Zurich, Wolfgang-Pauli-Str. 10, HCI F 133, Zurich, 8093, Switzerland

Recently, our group has validated experimentally the 'cycle to cycle' optimizing control scheme for simulated moving beds (SMB) for the separation of achiral and chiral components [1, 2]. The feedback information used for control purposes was measured by online optical detectors, i.e. UV and polarized light detectors, positioned at the extract and raffinate streams. The accuracy of such devices is highly sensitive to experimental factors like impurities in the mixture to be separated or pressure fluctuation in the measuring cell; and has a direct impact on the performance of the controller. This can become a limiting factor when extending the control scheme to multicomponent systems and chiral separations.

In the spirit of widening the spectrum of applicability of the controller and improving the accuracy of the measurements and therefore the performance of the controller, we have extended our control scheme to make use of an existing and reliable monitoring technique, namely HPLC.

This approach has, on one hand, the advantage that the HPLC measurements are straightforward, more accurate than online optical detectors, can handle multicomponent systems and are not greatly affected by impurities. On the other hand, these measurements are less frequent and provide only information about the average concentration of the two species at the outlet streams. Moreover, the analysis time may be in the range of the cycle time, which would introduce a significant time delay in the measurements. Hence, the controller relies on less but more accurate information that is available with a lower frequency, i.e. once every cycle, and can be delayed.

This work presents the design, implementation and validation of the SMB controller in combination with the online HPLC monitoring for a nonlinear chiral separation of the guaifenesin enantiomers in Ethanol using Chiralcel OD as stationary phase. Furthermore, it reports the theoretical modifications and extensions that had to be done to the control scheme to cope with the implemented monitoring system.

The performance of the controller is assessed in various scenarios that cover common situations encountered in SMB practice, such as the separation of racemic mixtures with different overall concentrations, in a range from 0.1g/L (linear) to 15g/L (nonlinear) regime; rejection of disturbances due to pump malfunctioning; setpoint tracking in the case of varying purity specifications or if the feed concentration changes in the course of the operation. The results illustrate that the 'cycle to cycle' controller is able to meet the products' purity specifications and operate the process optimally using minimal information about the system regardless of the disturbances that might take place during the operation.

[1] Grossmann, C., Erdem, G., Morari, M., Amanullah, M., Mazzotti, M., Morbidelli, M., 'Cycle to cycle' optimizing control of simulated moving beds. AIChE J., vol. 54, pp. 194-208, 2008

[2] Amanullah M., C. Grossmann, M. Mazzotti, M. Morari, M. Morbidelli. "Experimental implementation of automatic 'cycle to cycle' control of a chiral simulated moving bed separation ", J. of Chromatography A, vol. 1165, no. 1-2, pp. 100-108, 2007.

444g Equilibrium Theory Based Design of Simulated Moving Bed Processes Under Reduced Purity Requirements. Linear Isotherms

Arvind Rajendran, School of Chemical and Biomedical Engineering, Nanyang Technological University, Nanyang Drive 62, Singapore, Singapore

SMB schemes operated under reduced purity requirements can be interesting where complete separation of the binary feed mixture is not essential. Such requirements, which are likely to grow in the coming years, exist in the agrochemical industries, where only an enantiomerically enriched product is often required. This is also useful for hybrid separations such as the SMB-crystallization process where a partial enrichment of the products is performed by the SMB. The design of simulated moving bed processes under reduced purity requirements for systems whose isotherm is linear is considered. Based on the equilibrium theory of chromatography, explicit equations to uniquely identify the separation region that will ensure specified extract and raffinate purities are derived. In the frame of the "triangle theory", this results in a pentagonal region in the (m2, m3) plane, operating within which will ensure an extract and raffinate purity better than the specified one. The identification of the vertices of this region requires only the knowledge of Henry constants of the solutes, the concentration of the solutes in the feed and the purity specifications. The derivation of the design equations is illustrated and these results are validated using numerical simulations.

444h Robust Design and Operation of 2- & 3-Column Streamlined SMB Process for Binary and Ternary Separations

Rui C. R. Rodrigues[1], João M. M. Araújo[1], and **José P. B. Mota**[2]. (1) Chemistry, Requimte/CQFB, Faculdade de Ciências e Tecnologia, Universidade Nova de Lisboa, Quinta Torre, 2829-516 Caparica, Portugal, (2) Chemistry Department, Requimte/CQFB, Faculdade de Ciências e Tecnologia, Universidade Nova de Lisboa, Quinta Torre, 2829-516 Caparica, Portugal

Over the last years there has been intensive research on simulated moving-bed (SMB) chromatography. As a result, several new operating procedures and schemes have been developed to improve the separation efficiency, thus allowing for increased productivity, unit downsizing, or simplification of the process flowsheet.

In this work we report a class of simple SMB schemes, using only two and three columns, to achieve both binary and ternary separations. These schemes have the advantage of employing fewer columns, and requiring only two or three pumps, which renders the system more economic; the set-up is also simpler to build.

However, these compact SMB schemes expand the degrees of freedom for operating the cycle, which renders them more sensitive to disturbances, namely flow-rate stability and parameter uncertainty. Therefore, a robust design of the system is needed.

A robust design of the proposed two- and three-column processes has been developed for protection from flow-rate, dead-volume, and isotherm uncertainties. The optimal cycle parameters are chosen only among candidate solutions that are robust feasible. This gives rise to a robust approach to optimal design in which the nominal problem is replaced by a worst case problem. The nominal optimization problem and its robust counterpart are formulated using a full-discretization approach for steady periodic dynamics. The resulting nonlinear programming problems are solved by an efficient interior-point solver.

The procedure is successfully employed to find robust operating conditions for the linear separation of nucleosides mixtures by reversed-phase and quiral resolution of a racemic antidepressive NRI drug. The robust schemes are validated experimentally. Emphasis is given to the pros and cons of running the processes under nominal and robust operating conditions.

472a Hollow Fiber Membrane Contactor for Characterization of Gas/solvents Thermodynamics and Mass Transfer Properties

Fabien Porcheron and Sophie Drozdz. Separation Department, IFP, BP 3, Vernaison, 69390, France

The control of greenhouse gases (i.e. CO2, CH4, N2O,...) emissions before their release to the atmosphere has became a worldwide concern over the last few years[1]. The principal source of CO2 emissions comes from fossil fuel combustion from refineries and the energy industry. Several technologies are available to reduce CO2 emissions from industrial gas streams but the most popular one involves a liquid solvent which will selectively capture the CO2 through a molecular affinity (physical solvent) or a chemical reaction (chemical solvent). For instance, for low to moderate CO2 pressures, chemical absorption with alkanolamines solutions usually permits the removal of 75 to 90% of the CO2 contained in the gas streams[2]. In order to find the most efficient solvents for CO2 capture, there is therefore an increasing need to understand the mass transfer mechanisms occurring in these systems. These properties are often characterized from lab-scale experiments through the determination of Henry constant (H) and diffusion coefficient (D) of a molecular species within a fluid phase.

However the different experimental techniques used to determine these properties can be time consuming as a set of multiple experiments is required to thoroughly determine the physical properties of the gas/liquid system under study. Moreover these different techniques are often limited by the difficulty to control the interfacial contact area between the gas and the liquid phase due to boundary effects[3].

Different contactors can be used to study gas/liquid systems but hollow fiber membrane contactor (HFMC) geometry has recently arisen as an interesting solution for separation technology[4]. For instance, on a process scale, the use of HFMC has been considered to replace packed columns, to avoid technical problems such as flooding, entrainment and foaming[5-7]. In this geometry, the liquid is usually flowing in the lumen side of solvent repellent hollow fibers while the gas flows in the core shell and diffuses through the pores of a microporous membrane before contacting the immobilized solvent. The contactor geometry ensures a large interfacial contact area between the two phases and prevent any problems induced by the dispersion of one phase into another. Unlike classical membrane separation processes, the membrane does not impact on the separation efficiency but provides an interfacial area for the contact between the two non miscible phases. The thermodynamics of the separation is hence solely controlled by the choice of the solvent flowing in the lumen of the fibers. Besides, hollow fiber membrane contactor displays a fixed and known interfacial contact area which makes it very suitable for gas/liquid characterization.

In this work, we perform transient hollow fiber membrane contactor experiments where a physical solvent is progressively saturated with a gas (CO2 or N2O). We also develop a theoretical model representing the gas/liquid mass transfer occurring during such experiments. The developed methodology allows the determination of gas/liquid thermodynamics and mass transfer properties like the Henry constant and the liquid phase diffusion coefficient. Both parameters are obtained within a single experiment at a constant temperature and the comparison with temperature-dependant correlations yields excellent agreement over the whole range of temperature studied in this work. The plasticizer role of CO2 inside the membrane contactor fibers is also highlighted as simulations show a partial wetting of the membrane pore by the liquid meniscus during a contact between CO2 and H2O. We also discuss the possible extension of this methodology to chemical solvents.

References

[1] Rao, A.B.; Rubin, E.S., 2002. A Technical, Economic, and Environmental Assessment of Amine-Based CO2 Capture Technology for Power Plant Greenhouse Gas Control. Environ. Sci. Technol. 36, 4467-4475.

[2] Bonenfant, D.; Mimeault, M.; Hausler, R., 2007. Estimation of the CO2 Absorption Capacities in Aqueous 2-(2-Aminoethylamino)ethanol and Its Blends with MDEA and TEA in the Presence of SO2. Ind. Eng. Chem. Res., 46, 8968-8971.

[3] Dindore, V.Y.; Brilman, D.W.F.; Versteeg, G.F., 2005. Hollow fiber membrane contactor as a gas-liquid model contactor. Chem. Eng. Sci. 60, 467-479.

[4] Li, J.-L.; Chen, B.-H., 2006. Review of CO2 absorption using chemical solvents in hollow fiber membrane contactors. Sep. Purif. Technol. 279, 301-310.

[5] Gabelman, A.; Hwang, S.T., 1999. Hollow fiber membrane contactors. J. Membr. Sci. 159, 61-106.

[6] De Montigny, D.; Tontiwachwuthikul, P.; Chakma, A., 2005. Comparing the absorption performance of packed columns and membrane contactors. Ind. Eng. Chem. Res. 44, 5726-5732.

[7] Karoor, S.; Sirkar, K.K., 1993. Gas adsorption studies in microporous hollow fiber membrane modules. Ind. Eng. Chem. Res. 32, 674-684.

472b Screening of Metal Organic Framework Based Membranes for Carbon Dioxide Separations Using Atomically Detailed Simulations

Seda Keskin and David S. Sholl. Chemical and Biomolecular Engineering, Georgia Institute of Technology, 311 Ferst Drive NW, Atlanta, GA 30332

Membranes offer a promising approach for energy-efficient separation of gas mixtures. Although large numbers of nanoporous materials exist that are potentially suitable for use as gas separation membranes, the experimental challenge of fabricating defect-free thin films makes screening of these materials a laborious process. Atomically detailed simulations are a useful complement to experimental studies for this problem in order to screen numerous materials prior to fabrication of membranes. These simulations guide experiments to focus on more useful membrane materials by predicting the material properties needed to characterize membrane performance beforehand.

The aim of our work is to use atomically detailed simulations in order to describe macroscopic transport of various gas molecules both as single components and as mixtures in a new group of material, namely metal organic frameworks (MOFs). By using a multi-scale approach in our modeling, we are able to link atomically-detailed descriptions of each material with predictions of membrane performance under practical operating conditions. We focused initially on two well known MOF structures, MOF-5 and CuBTC, to examine the separation capacity of these two materials as membranes. Our results gave the first insight into whether pursuing the fabrication of membranes for gas separations from MOFs would be a useful activity. We will show how insight from these initial calculations can be used to choose candidate materials from the very large number of MOFs that are known for specific membrane applications.

Microporous Silica Coatings on Mesoporous SiO_2: Membranes for High Temperature Gas Separation Applications.

Michael McCann[1], Amsarani Ramamoorthy[1], Mahfujur Rahman[1], Damian A. Mooney[1], Denis P. Dowling[2], J. M. Don MacElroy[1]

[1]School of Chemical and Bioprocess Engineering, University College Dublin, Belfield, Dublin 4, Ireland
[2]School of Electrical, Electronic and Mechanical Engineering, University College Dublin, Belfield, Dublin 4, Ireland

Abstract

In this work, composite asymmetric membranes are prepared by magnetron sputtering deposition (MS), and atmospheric pressure plasma enhanced chemical vapour deposition (APPECVD) of silica films onto flat mesoporous SiO_2. The deposition conditions for both coating types were systematically controlled to determine their effect on the deposited coating architecture (morphology, porosity and thickness) and stoichiometry using scanning electron microcopy (SEM), atomic force microscopy (AFM), X-ray photoelectron spectroscopy (XPS), ellipsometry, X-ray diffraction (XRD) and optical profilometry. These methods were used to provide an accurate depiction of the membrane coating physical and chemical properties. In addition, permeation measurements were made on all membranes in order to assess their perm-selectivity and suitability for membrane applications. APPECVD coatings showed promising permselectivity values with an average pore size of 0.42 nm. The coatings were found to be mechanically stable up to 400^0C and showed good adhesion properties.

Introduction

High atmospheric CO_2 concentrations resulting, in large part, from fossil fuel combustion, are now considered directly responsible for increasing global temperatures [1]. A cost-effective means of separating CO_2 from hot exhaust gases however, represents a major scientific challenge. In addition to the demanding hot exhaust gas environment, there is also the very significant problem of the small differences in molecular size between CO_2 and N_2 (3.3 and 3.7 Å, respectively) [2].

Recent theoretical and experimental studies have demonstrated promising permselectivity results for porous inorganic composite membranes (Fig. 1) fabricated through the chemical vapour deposition (CVD) of ultra-thin layer (≈5–25 nm), microporous, dense films onto carrier mesoporous substrate materials [2, 3]. However, films produced by conventional thermal CVD methods exhibit significant variability in the quality of the controlling deposits, particularly deposition density, which is key to CO_2 separation efficiencies. Plasma-based deposition processes could provide significantly better control of coating properties, including film thickness, density, and, indeed, film chemistry, and therefore offer the potential of fabricating improved membranes.

Experimental methods

Support materials

α-Al2O3 (with average pore size of 3 μm) discs were used as membrane supports. These 1 mm thick discs had a 0.5 um thick top layer of SiO_2 with a pore size of 1 nm. Fig. 1 shows an SEM image of the SiO2/αAl2O3 composite support.

Fig. 1: SEM cross-sectional image of $SiO_2/\alpha Al_2O_3$ composite support.

Magnetron sputtering deposition

SiO_x coatings were deposited onto the composite ceramic supports by magnetron sputtering using a Teer Coatings UDP450 closed field unbalanced magnetron sputtering system. In this deposition process, a high energy argon plasma is controlled and regulated using an Advanced Energy MDX magnetron drive power supply, the output of which is regulated via a Sparc Le V DC pulsing unit. The argon plasma is used to systematically bombard a silicon target in the presence of oxygen to form a silicon oxide coating. The range of deposition parameters are listed in Table 1.

Table 1: Processing parameters - Magnetron Sputtering Deposition

Processing parameter	
Coating thickness (nm)	100
Target Current (amp)	1.0
Flow rate of oxygen (SCCM)	3.0
Working pressure (mbar)	1×10^{-3}
Pulsed DC duty factor (%)	64
Distance between target and substrate	10cm – 17cm

Atmospheric pressure plasma enhanced chemical vapor deposition

SiO_x coatings were deposited onto SiO_2 supports by atmospheric pressure chemical vapor deposition (APPECVD) using a Dow Corning Plasma Stream™ system. In this deposition process, a liquid precursor tetraethylorthosilicate (TEOS) is nebulised inside a helium plasma. The processing parameters used during deposition are listed in Table 2.

Table 2: Processing parameters – Atmospheric Pressure PECVD

Processing parameter	
Precursor	TEOS
Precursor flow rate (SCCM)	0.01
Coating thickness (nm)	100
Height between substrate & nozzle (mm)	2 - 20mm

Results

Gas separation

Gas permeation measurements where carried out at room temperature using a specially constructed membrane test apparatus. Ideal gas selectivity was determined by administering a fixed concentration of permeating gas upstream of the membrane and measuring the differential pressure across the membrane over a period of 1 minute. Prior to permeance measurements, the membranes where heat treated at 400^0C in order to remove all adsorbed material from the membrane micropores. Each of the coatings on the membranes evaluated in this study had a thickness of 100 nm. The thicknesses were determined using both ellipsometry and atomic force microscopy measurements.

Membranes fabricated by magnetron sputtering yielded lower He permeance and He/N_2 selectivity compared with those deposited by APPECVD (Table 3). This may suggest that the coating densities achieved with magnetron sputtering are higher and the low selectivity may suggest non-homogeneous pore size particularly at grain boundaries. The APPECVD technique yielded the highest overall membrane He/N_2 selectivity of 55. The permeances and selectivities listed in Table 3 are referring to those of the actual coating. Calculation of the coating permeance is based on Eqn. 1 where P_{mem}, P_{sup} and P_{coat} are the permeances of the membrane, support and coating respectively. For this equation to be valid it is assumed that the support membrane undergoes no change in thickness, pore size, porosity or tortuosity during the deposition of the selective coating.

$$\frac{1}{P_{mem}} = \frac{1}{P_{\sup}} + \frac{1}{P_{coat}} \quad \text{Eqn. 1}$$

Table 3: Ideal He/N_2 selectivities for silica coatings deposited by APPECVD and magnetron sputtering (1nm SiO_2 support)

Deposition method	He/N2 selectiivity	He Permeance x 10^8 Mol $M^{-2}S^{-1}$ Pa^{-1}	Approx pore size (nm)
APPECVD	83	5.55	0.42
Magnetron Sputtering	5.4	0.85	0.8

Scanning Electron Microscopy images

The SEM images (Fig. 2.) clearly show a more homogenously dense coating exhibited by APPECVD. An issue with these coatings however, is the presence of powder particles which are co-deposited with the coating (Fig. 2b). The presence of powder on the coating surface is undesirable as it leads to lower coating density as well as higher surface roughness.

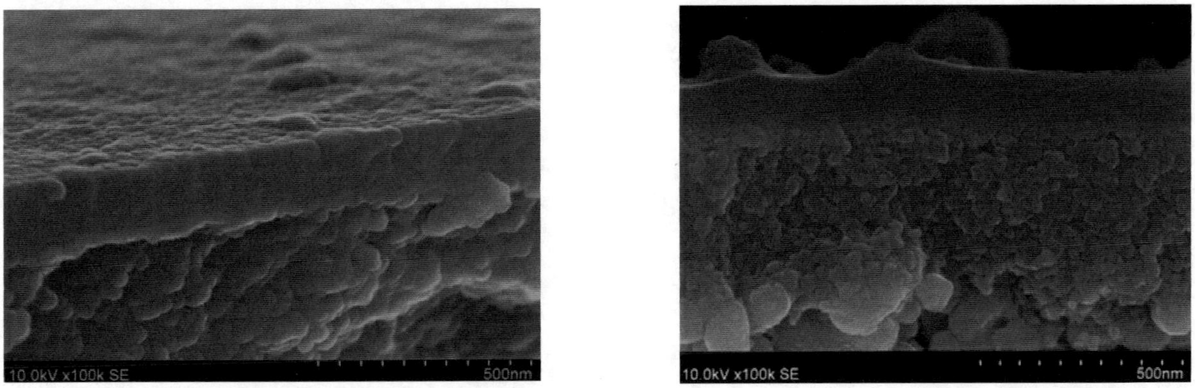

(a) (b)

Fig. 2: Cross-sectional views of selective membrane structure produced by a) magnetron sputtering and b) APPECVD of silica onto microporous SiO_2 support.
Note: the denser structure observed in b.

FTIR spectra

Fourier transform infra red spectroscopy was used to examine coating chemical functionality. A comparison between two types of SiO_x coatings of the same thickness but deposited by APPECVD and magnetron sputtering is given in Fig. 4. Both spectra show the presence of SiO_x in the form of Si-O-Si stretching peak at 1045cm^{-1}. The APPECVD APPLD deposited coating however also exhibits -OH and weak -C peaks at 3400 and 1600 cm^{-1} respectively. For high temperature membrane separation, the incorporation of carbon into the coating may be unfavourable as it could burn out to leave void spaces open for non-selective gas flow.

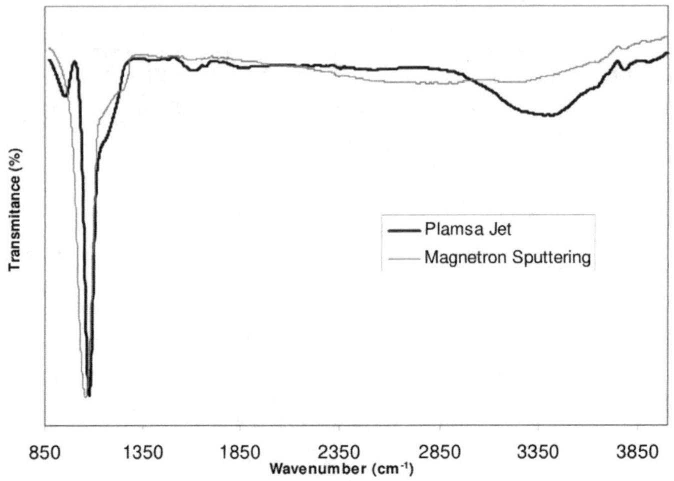

Fig. 4: FTIR spectra of SiO_x coatings deposited by APPECVD and Magnetron sputtering

SiOH is present as powder on the coating surface and is undesirable as it leads to lower coating density as well as high surface roughness. The presence of carbon groups is also not favourable as carbon is removed at high temperature thus leaving void spaces open for non-selective gas flow.

Conclusion

SiO_2 films were deposited onto porous SiO_2 and ZrO_2 support membranes by APPECVD and magnetron sputtering. These methods where optimised to accurately control coating thickness and chemistry. The corresponding permeance of N_2 and He was measured at room temperature for both deposition methods. Magnetron sputtering coatings exhibited higher levels of SiO_x however their columnar structure and grain boundaries led to poor separation efficiency. The denser and more homogenous APPECVD coatings show promising permselectivity values with an average pore size of 0.42 nm. However, the presence of low concentrations of carbon may adversely affect their suitability for high temperature separations. The coatings were found to be mechanically stable up to 400^0C.

Acknowledgements

The financial support of the EU Sixth Framework Programme and Science Foundation Ireland is gratefully acknowledged.

References

1. IPCC, 2007: The Physical Science Basis. Contribution of Working Group I to the Fourth Assessment Report of the Intergovernmental Panel on Climate Change [Solomon, S., D. Qin, M. Manning, Z. Chen, M. Marquis, K.B. Averyt, M.Tignor and H.L. Miller (eds.)]. Cambridge University Press, Cambridge, United Kingdom and NY, USA

2. J.M.D. MacElroy (2002), "Molecular simulation of kinetic selectivity of a model silica system", *Mol. Phys.* Vol.100, pp.2369.
3. Laurence Cuffe, J.M. Don MacElroy, Matthias Tacke, Mykola Kozachok, Damian A. Mooney (2006), "The development of nanoporous membranes for separation of carbon dioxide at high temperatures", *J. Membr Sci.* Vol.272 pp.6–10

Zeolite Membrane Synthesized Employing Crystal Growth Inhibitors for Carbon Dioxide Separation

*Surendar R.Venna and Moises A. Carreon**
University of Louisville, Department of Chemical Engineering, Louisville, KY-40292, USA.

*Corresponding author: macarr15@louisville.edu

Introduction

Silicoaluminophosphate molecular sieves have been the subject of great interest because of their high thermal resistance, mechanical strength, chemical stability, chemical inertness, and uniform-sized pores of molecular dimensions. SAPO-34 with the composition of $Si_xAl_yP_zO_2$ (x = 0.01-0.98, y = 0.01-0.6 & z = 0.01-0.52) is a representative type of silicoaluminophosphate zeolite displaying chabazite structure with uniform microporous channels and cages. SAPO-34 an acid silico-aluminophosphate zeolite with pore size of ~0.38nm, posses unique shape selectivity, which is of great interest in the field of gas separation (in particular in the separation of carbon dioxide from different light gases) and heterogeneous catalysis (in particular especially in methanol-to-olefin reaction in petrochemical industries). Small zeolite crystals are preferred because they display more improved surface properties compared to larger particles. For smaller crystals more number of atoms are exposed to the surface increasing the acidity, which results in higher selectivity when SAPO-34 is used for methanol to olefin reactions (1, 2). Smaller crystals potentially result in thinner membranes which lead to improved separation performance (3, 4).

Lok and co-workers at Union Carbide first reported the synthesis of SAPO-34 in 1984 (5). Later, various routes for the preparation of SAPO-34 have been proposed. SAPO-34 has been synthesized by using TEAOH (6), TEAOH along with DPA (3), different templates (7, 8, 9), via vapor phase transport (10), various sources of silica and alumina (11). Also, the incorporation of different metals in SAPO-34 framework like Mg, Ca, Ti, Ag, Na, Sr, Ce (12) and Cu, Co (13) has been done. SAPO-34 has been used in various functional applications such as separation of carbon dioxide from different gases (4, 14), catalyst for methanol to olefin reactions (1, 2), H_2 purification and storage (15), chloromethane transformation to higher hydrocarbons (16), as a hydrocarbon trap (17).

Here, we present the synthesis of SAPO-34 crystals using various crystal growth inhibitors (polyethylene glycol, polyoxyethylene lauryl ether and methylene blue) and several structure directing agents (tetraethylammonium hydroxide, dipropylamine, morpholine and triethyl amine). The resultant SAPO-34 crystals were characterized by BET, XRD, SEM, FTIR, TGA and CO_2/CH_4 adsorption. The surface area of the SAPO-34 crystals increased as compared to the samples synthesized without crystal growth inhibitors. The crystallinity of SAPO-34 was confirmed by XRD and FTIR. The CO_2 and CH_4 adsorption capacity of SAPO-34 was studied.

Experimental

$Al(i\text{-}C_3H_7O)_3$ (> 99.99%, Aldrich), H_3PO_4 (85 wt% aqueous solution, Aldrich) and Ludox (40 wt%, Aldrich) were used as the inorganic precursors. Tetraethyl ammonium hydroxide and dipropylamine (99wt%, Aldrich) were used as templates. Polyethylene glycol-600 (PEG, Alfa Aesar),pPolyoxyethylene lauryl ether (C12E6, Acros Organics) and methylene blue (MB, Sigma-Aldrich) were used as crystal growth inhibitors (CGI). The composition of the synthesis gel was $1Al_2O_3$:$1H_3PO_4$:$0.3SiO_2$:1TEAOH:1.6DPA:$77H_2O$: xCGI, where x is in the range of 0.037-0.175. $Al(i\text{-}C_3H_7O)_3$ and H_3PO_4 were mixed with distilled H_2O for 2hrs. Then, ludox was added and the resulting solution was stirred for additional 2hrs to form a homogeneous solution. Then the templates TEAOH and DPA were added to the gel with an interval of 30 min. After stirring for 30 min, crystal growth inhibitors were added. The resulting gel was aged for 3 days at 50^0C. It was transferred to a stainless steel autoclave and hydrothermal treatment was done using a conventional oven at 220^0C for 24 hrs. The hydrothermal treatment was also done with controlled temperature ramp (i.e. increasing 1^0C/min till125^0C and hold there for 5hrs and then increasing the temperature to 220^0C at 1^0C/min and hold there for 16hrs). The solution after the hydrothermal treatment was centrifuged at 4000rpm for 20min to separate the seeds from the mother liquor. The resultant seeds were dried overnight at 60^0C. The templates and CGI were removed at 550^0C for 5hrs with heating and cooling rates of 1^0 and 10^0C/min respectively. The calcined SAPO-34 samples were characterized using FE-SEM (FEI Nova 600) to study the size and morphology. Brukar D8 Discover XRD and Spectrum BX FTIR (Perkin Elmer) were used to determine the crystal structure and functional groups at the SAPO-34 surface, respectively. Micromeritics Tristar-3000 model BET porosimeter was used to study the surface area measurements and CO_2/CH_4 adsorption ratios.

Results and Discussion

A typical X-ray diffraction of the SAPO-34 crystals synthesized with MB as CGI is shown in Figure 1. The spectrum shows the chabazite structure confirming the presence of pure SAPO-34 crystals. The position of each peak is identical to those reported for SAPO-34 (6). The synthesized samples were also characterized by the Fourier transform IR to confirm the formation of SAPO-34. A typical FTIR spectrum is shown in Figure 2 for the sample synthesized using PEG as the crystal growth inhibitor. The lattice vibrations in the framework of SAPO-34 crystals are typical for the chabazite framework which is in agreement with previous reported data (18). The most prominent IR band of the SAPO-34 frameworks occurs in the region between 900 and 1300 cm^{-1}, which is asymmetric stretching of AlO_4^{-1}, and PO_4^{-1} tetrahedra. The other peaks corresponding to SiO_4^{-1} bonding, vibrations in the double bond region and -OH bonding are weaker bands in the 400-500, 600-700 and 1600-1700 cm^{-1} region. Therefore, XRD and FTIR confirm the formation of pure SAPO-34 using crystal growth inhibitors.

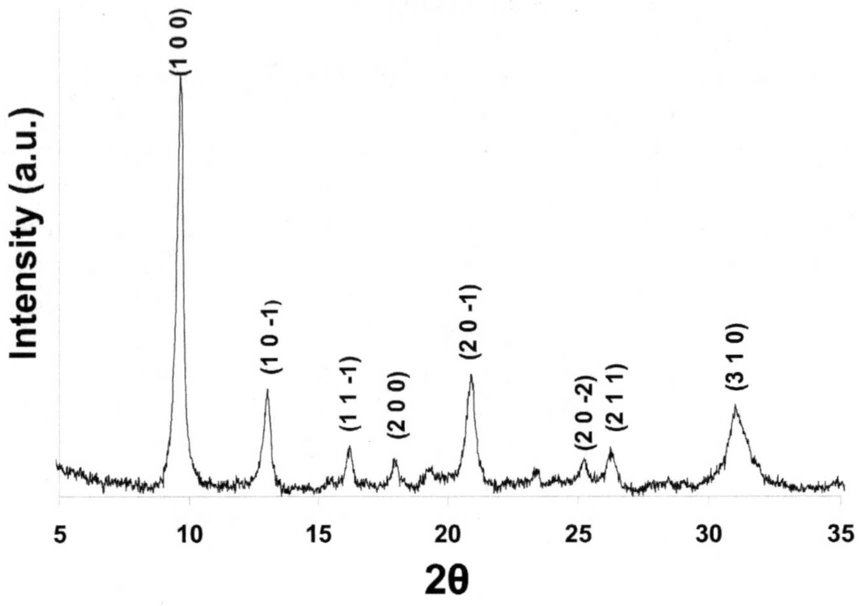

Figure 1: XRD pattern of the chabazite structure synthesized using MB as crystal growth inhibitor

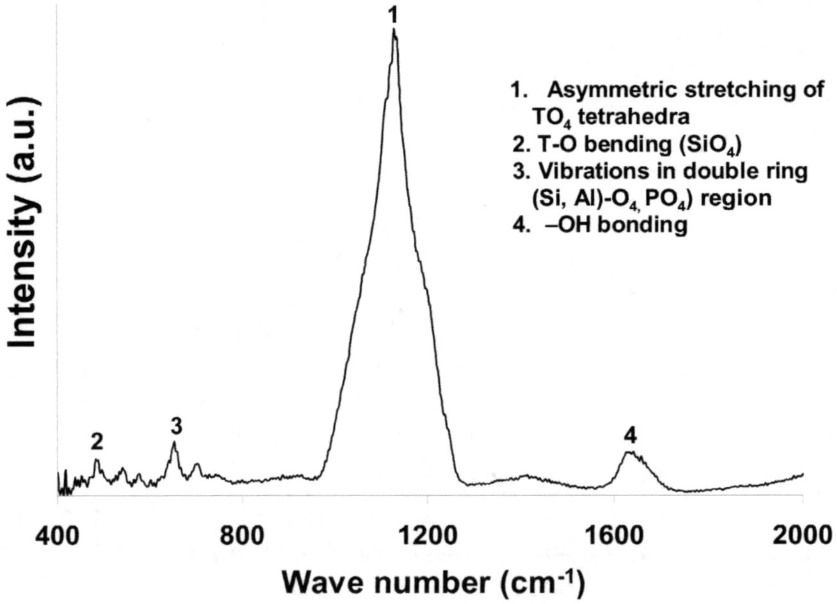

Figure 2: FTIR spectra of SAPO-34 crystals synthesized using PEG as crystal growth inhibitor

SEM images showing the cubic and rhombohedral crystals of SAPO-34 synthesized using the crystal growth inhibitors and ramp heating are shown in Figure 3. The incorporation of crystal growth inhibitors during synthesis led to a remarkable decrease in crystal size from ~2 µm to ~0.5 µm. In general, crystal sizes between ~0.5-1 µm were obtained for samples prepared with crystal growth inhibitors. Crystal growth inhibitors play an important role in changing the alkalinity of the synthesis of gel, which helps to form smaller nuclei and therefore smaller crystals (19). They also interact with reactive sites of the precursors suppressing the crystal growth. They may also adsorb on to the surface of the nuclei to prevent the growth (20).

Figure 3: SEM images of the SAPO-34 crystals synthesized employing a) PEG, b) C12E6, and c) MB as crystal growth inhibitors and d) without crystal growth inhibitors

Nitrogen adsorption isotherms at 77 K were used to determine the BET surface area of SAPO-34 crystals. The surface areas of the different samples synthesized using the crystal growth inhibitors are shown in Figure 4. Surface Area of SAPO-34 crystals significantly increased about 25% - 80%. For example areas up to ~700 m^2/g were obtained when C12E6 was employed as CGI, while surface areas in the range of only ~390 m^2/g to ~500 m^2/g were observed for the samples synthesized without any CGI.

The higher surface areas obtained using crystal growth inhibitors are assumed to be a result of the formation more nuclei during the initial stages of hydrothermal synthesis and controlled the growth of the crystals in later stages.

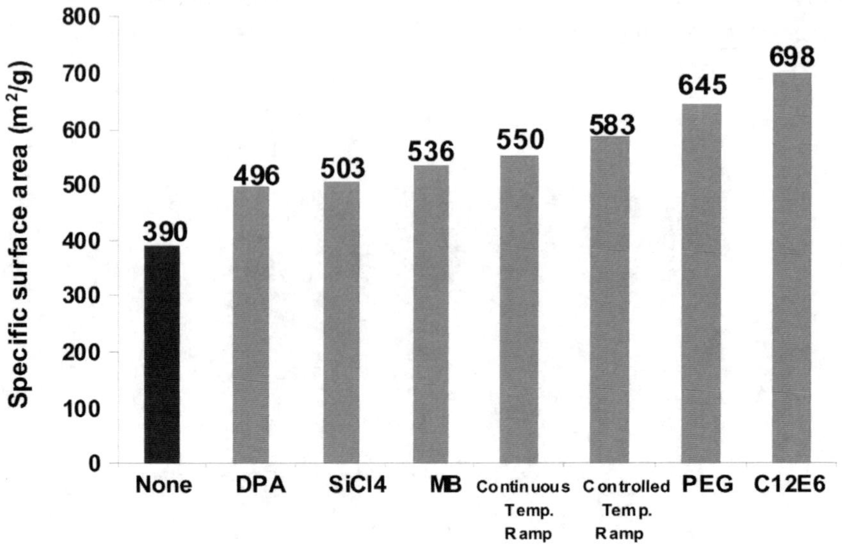

Figure 4: BET surface area of SAPO-34 crystals synthesized using different crystal growth inhibitors and controlled temperature ramp

CO_2 and CH_4 adsorption capacity of SAPO-34 crystals was conducted at room temperature using water the coolant. The CO_2/CH_4 adsorption capacity of the SAPO-34 samples synthesized with different CGI is much higher (~9) than the samples synthesized without growth inhibitors (~5). Typical CO_2 and CH_4 adsorption isotherms are shown in Figure 5 for a SAPO-34 sample synthesized employing C12E6 as the crystal growth inhibitor. The higher adsorption capacity of the samples synthesized using CGI is due to the incorporation of more nitrogen in the framework of SAPO-34, as confirmed by CHN analysis (not shown here).

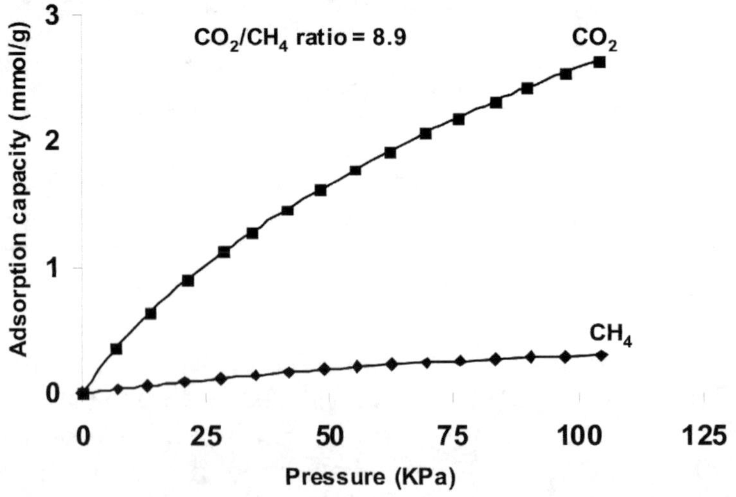

Figure 5: CO_2/CH_4 adsorption capacity of SAPO-34 crystals synthesized using C12E6 as CGI

The synthesized SAPO-34 crystals can be potentially used as *seeds* to prepare thin membranes on porous supports to separate the CO_2 from CH_4. In addition to their chemical and thermal stability, SAPO-34 is attractive as membrane material because of their atomically ordered nanometer-scale pore structures. SAPO-34 also, can be potentially used as catalysts in the methanol to olefins reaction because of its high surface area, and moderate acidity with more active sites exposed to reactants.

Conclusions

- SAPO-34 crystals were synthesized employing crystal growth inhibitors.
- XRD and FTIR confirmed the presence of chabazite structure of typical SAPO-34
- Crystal growth inhibitors (Poly ethylene glycol, polyoxyethylene lauryl ether and Methylene blue) reduced the crystal size and increased the specific surface area of SAPO-34 crystals.
- SAPO-34 crystals (~0.5-1 μm) displaying BET surface areas up to ~700m^2/g were prepared.
- The carbon dioxide adsorption capacity of SAPO-34 increased with the incorporation of crystal growth inhibitors
- Controlled two-stage hydrothermal heating method led to high surface area crystals with narrow size distribution.

References

1. S Y.J. Lee, S.C. Baek, K.W. Jun, (2007), *Appl. Catalysis A: General, 329*, 130–136
2. M. Popova, Ch. Minchev, V. Kanazirev, (1998), *Appl. Cat. A: General, 169*, 227-235
3. S. Li, J.L. Falconer, R.D. Noble, (2008), *Micro. Meso. Mater. 110*, 310-317.
4. M.A. Carreon, S. Li, J.L. Falconer, R.D. Noble, (2008), *J. Am. Chem. Soc. 130*, 5412–5413
5. B. M. Lok, C. A. Messina, R. L. Patton, R. T. Gajek, T. R. Cannan, E. M. Flanigen, (1984), *J. Am. Chem. Soc. 106*, 6092-6093
6. H. V. Heyden, S. Mintova, T. Bein, (2008), *Chem. Mater. 20,* 2956-296.
7. Ø. B. Vistad, D. E. Akporiaye, F. Taulelle, K. P. Lillerud, (2003), *Chem. Mater. 15,* 1639-1649
8. E. Dumitriua, A. Azzouz, V. Hulea, D. Lutic, H. Kessler, (1997), *Micro. Mater. 10*, 1-12.
9. G. Liu, P. Tian, Y. Zhang, J. Li, L. Xu, S. Meng, Z. Liu, (2008), *Micro. Meso. Mater.* 114, 416-423
10. J. Yao, H. Wang, S. P. Ringer, K.Y. Chan, L. Zhang, N. Xu, (2005), *Micro. Meso. Mater. 85*, 267-272
11. M. Mertens, K. G. Strohmaier, (2004), *U.S. Patent 6696032 B2.*
12. M. E. Rivera-Ramos, G. J. Ruiz-Mercado, A. J. Herna´ndez-Maldonado, (2008), *Ind. Eng. Chem. Res. 47,* 5602–5610

13. E. L. Uzunova, H. Mikosch, J. Hafner, (2008), *J. Phys. Chem. C. 112,* 2632-2639
14. J.C. Poshusta, V.A. Tuan, E.A. Pape, R.D. Noble, J.L. Falconer, (2000), *AIChE J. 46*, 779-789
15. M. Yu, S. Li, J.L. Falconer, R.D. Noble, (2008), *Micro. Meso. Mater. 110*, 579-582
16. Y. Wei, D. Zhang , Z. Liu, B. L. Su, (2007), *Chemical Physics Letters*, 444, 197–201
17. S.P. Elangovan, M. Ogura, Y. Zhang, N. Chino, T. Okubo, (2005), *Appl. Cat. B: Environ. 57*, 31–36
18. D. B. Akolekar, S. Bhargava, W. V. Bronswijk, (1999), *Applied spectroscopy*, 53, 931-937
19. H. Hosokawa, K. Oki, (2003), *Chemistry Letters,* 32, 586-587
20. C.M. Lew, Z. Li, S. I. Zones, M. Sun, Y.Yan, (2007), *Micro. Meso. Mater.* 105, 10-14

472e Enhanced Solubility of CO_2 In Aromatic and Alkyl Chain Substituted Imidazolium-Based Room Temperature Ionic Liquids

Alexia Finotello, Jason E. Bara, Suguna Narayan, Douglas L. Gin, and Richard D. Noble. Department of Chemical and Biological Engineering, University of Colorado, UCB 424, Boulder, CO 80309

Room-temperature ionic liquids (RTILs) are molten organic salts that possess a number of desirable properties for gas separations applications. Their high thermal and chemical stability, negligible vapor pressures, and tunable chemistry make them great candidates to replace less environmentally friendly materials such as volatile organic compounds. The versatile functionality of RTILs allows for their potential usage for gas separations as a bulk fluid, a supported liquid membrane, or as a polymerized RTIL solid membrane. Gas separations in these materials are driven by large solubility differences between gases, rather than by diffusion differences. Therefore, it is important to investigate the gas transport properties in RTILs to better understand their gas separation capabilities.

This study focuses on the solubility (S), diffusivity (D), permeability (P), and selectivity behaviors of CO_2, CH_4, and N_2 gases in bulk fluid RTILs at constant temperature (22oC). The bulk fluid RTILs tested are 1-methyl, 3-tolyl imidazolium bis(trifluoromethanesulfonyl)imide ([mtim][Tf2N]), 1-ethyl, 3-paraxylyl imidazolium bis(trifluoromethanesulfonyl)imide ([epxim][Tf2¬N]),1-butyl, 3-paraxylyl imidazolium bis(trifluoromethanesulfonyl)imide ([bpxim][Tf2¬N]), 1-hexyl, 3 paraxylyl imidazolium bis (trifluoromethanesulfonyl)imid ([hpxim][Tf2N]), 1-octa, 3 paraxylyl imidazolium bis (trifluoromethanesulfonyl)imid ([opxim][Tf2N]), and 1-deca, 3 paraxylyl imidazolium bis (trifluoromethanesulfonyl)imid ([dpxim][Tf2N]). Our initial results show that gas solubility increases, diffusivity has a small variance, and permeability slightly increases with increasing alkyl chain length. Solubility and permeability selectivity decreases while diffusion selectivity remains relatively unchanged with increasing chain length. The addition of the aromatic ring substitution shows enhanced solubility over analogous alkyl-chain only substituted RTILs. Also, our work demonstrates that regular solution theory is valid for these RTILs.

472f Molecular Simulation of Pure and Mixture Gases Absorption In Ionic Liquids

Wei Shi and Edward J. Maginn Chemical and Biomolecular Engineering. Chemical and Biomolecular Engineering, Univ. of Notre Dame, Notre Dame, IN 46556

Gases solubility for the mixtures of CO_2/O_2, SO_2/N_2, and CO_2/SO_2, and pure gases of SO_2, O_2, and N_2 absorption in ionic liquid of 1-n-hexyl-3-methylimidazolium

bis(trifluoromethylsulfonyl)imide ([hmim][Tf2N]) were computed using isothermal-isobaric Gibbs continuous fractional component (CFC) Monte Carlo (MC) and osmotic CFC simulations. The predictive capability of regular solution theory (RST) for these systems was also investigated.

Computed isotherms for the pure gases agreed well with available experimental data. Energy analyses showed that the van der Waals (VDW) energy for pure CO_2, SO_2, N_2, and O_2 interacting with the ionic liquid dominates over the electrostatic energy. CO_2 and SO_2 interact more strongly with the anion than with the cation. Typically, CO_2 interacts with anion through VDW and electrostatics forces about 10%-15% more strongly than it does with cation. For SO_2, the difference in interaction energy between the solute-anion and solute-cation is 15%-30%. For N_2 and O_2, the cation and anion interact with solutes comparably.

In the mixed gas simulations, it was observed that the selectivity for CO_2 absorption in [hmim][Tf2N] over O_2 varies between 8 and 20, close to the ideal selectivity of 11.2 computed from the pure gas Henry's law constants. Absorption of CO_2 decreases the solubility of O_2, rather than enhancing the solubility of O_2. This is in contrast to the experimental finding by Hert [1] et al.. The mixture simulation of SO_2/O_2 absorption in [hmim][Tf2N] shows N_2 does not compete significantly for the absorption of SO_2. This is in contrast with the experiment for the same mixture absorption in two ionic liquids of [BMIM][BTA] and [BMIM][BF4] [2] , but in agreement with experimental findings for three other ionic liquids [2]. For the CO_2/SO_2 mixture, both gases compete with each other leading to the decrease of the solubility for both gases. The selectivity of SO_2 absorption over CO_2 varies between 10 and 12, close to the pure Henry's law ratio of 14±2 bar between CO_2 and SO_2.

Overall, calculations show that the RST does not perform well. For pure gases of CO_2,

O_2, and N_2 absorption in [hmim][Tf2N], the mole fractions predicted from RST are about 20% -60% different than those from experiments and simulations. For the mixture of CO_2/O_2 absorption in [hmim][Tf2N], the RST predicts the mole fraction to be about 30% to 40% different from the simulation.

References:

D. G. Hert, J. L. Anderson, S. N. V. K. Aki, J. F. Brennecke, Chem. Commun., 2603-2605, 2005

J. Huang , A. Riisager, P. Wasserscheid, R. Fehrmann, Chem. Commun., 4027-4029, 2006

472g Solubilities and Diffusivities of Carbon Dioxide In Ionic Liquids

Surya Sekhar Moganty, Joshua Close, and Ruth Baltus. Department of Chemical Engineering, Clarkson University, 8 Clarkson Ave, Potsdam, NY 13699

There has been considerable interest in developing technologies for carbon dioxide separations using room temperature ionic liquids. The development of new separation processes requires an understanding of the solubility and diffusivity of target gases in these unique liquids. Recent work in our lab has focused on measurements of carbon dioxide solubility and diffusivity using a gas uptake technique. In this approach, the decrease in carbon dioxide pressure above a thin ionic liquid film is monitored as carbon dioxide permeates into the ionic liquid film. Pressure versus time data are related to the Henry's Law constant and diffusivity using a one-dimensional transport model. Results collected with a number of different ionic liquids at temperatures ranging from 10oC to 40oC will be presented. The role played by the anion and the cation in both solubility and diffusivity will be discussed. A correlation relating carbon dioxide diffusivity to ionic liquid properties will also be presented.

Higher permeability of carbon dioxide is a necessary requirement for ionic liquid selection for carbon dioxide separation. The Permeability of a gas is a function of both diffusivity and solubility. The results from these measurements have also been used to examine choice of ionic liquids for carbon dioxide separation. Results show that ionic liquids with the best carbon dioxide solubility have slower diffusion coefficient. A comparison of selectivity estimations based on these solubility and diffusivity measurements to membrane permeability measurements performed using a supported ionic liquid membrane support this conclusion.

References:

1. Ying Hou and Ruth Baltus, Ind .Eng. Chem. Res, 46, 8166, 2007

474a Design of Reactive Distillation Processes Using Simulated Annealing

J. K. Cheng[1], Hao-Yeh Lee[2], Hsiao-Ping Huang[1], and Cheng-Ching Yu[1]. (1) Chemical Engineering, National Taiwan University, Taipei, 107-16, Taiwan, (2) Department of Chemical Engineering, National Taiwan University, No. 1, Sec. 4, Roosevelt Road, Taipei, Taiwan

Reactive distillation (RD) offers an attractive alternative for the production of many important industrial chemicals. As pointed out by Hung et al. [1], the multifunctional nature leads to input multiplicity which makes the simulation and convergence of the RD flowsheet difficult. Unlike conventional separation units, there are much more design variables, e.g., number of separation trays, number of reactive trays, feed tray location etc., associated with the RD column. The design of the RD column becomes a combinatorial optimization problem where the objective function is the total annual cost (TAC). Conventionally, a sequential design approach is taken where the variable is changed one at a time to minimize the design objective [2-3]. However, as the dimensionality of design variables increases, the sequential method approach imposes heavy computing load, especially for the coupled reactive distillation/separation columns processes [3]. Mathematical optimization algorithms provide an attractive alternative to improve the computational efficiency. As we known, the derivative-based mathematical programming cannot guarantee the global solution for the non-convex problem. Unfortunately, the optimization problems of the RD design are non-convex, and derivative-free algorithms, e.g., Simulated Annealing (SA), Genetic Algorithm (GA), Tabu search (TS) etc, are preferred. Among all of them, SA has the advantage that it is simple to program. The concept of SA was originally proposed by Metropolis et al. [4]. It uses Monte Carlo simulation to calculate the energy distribution of molecules and simulate the cooling of material in a heat bath. Kirkpatrick et al. [5] is among the first to apply this idea to optimization and a traveling salesman problem was solved successfully using SA. After that, many applications of this method were proposed for the optimization of chemical processes.

In recent years, process simulators have been used extensively for the simulation of chemical processes. Because of the "black box" nature of these simulators, derivative-free algorithms can easily be implemented for optimization problems. In this work, we investigate the application of a SA algorithm to obtain the optimal design of two reactive distillation systems. The SA algorithm is implemented in the Visual Basic Application (VBA) which interfaces with the process simulator, Aspen Plus. In performing optimization, parameters associated with the annealing system, such as the initial temperature, temperature decrement factor and quasi-equilibrium detection and the space of design variables, should be defined.

Two reactive distillation systems, methyl acetate and butyl acetate production, are used to illustrated the design using the SA algorithm. The results indicate that improved design can be achieved with relatively efficient computing. The SA approach is recommended for highly nonlinear processes with non-monotonic characteristics such as reactive distillation systems.

Reference

1. Hung, S. B.; Lee, M. J.; Tang, Y. T.; Chen, Y. W.; Lai, I. K.; Huang, H. P.; Yu, C. C. "Control of different reactive distillation configurations" AIChE Journal, 2006, 52, 4, 11423-1440.

2. Tang, Y. T.; Chen, Y. W.; Huang, H. P.; Yu, C. C.; Hung, S. B. ; Lee, M. J., "Design of reactive distillations for acetic acid esterification" AIChE Journal, 2005, 51, 6, 1683-1699.

3. Lin, Y. D., Cheng, J. H.; Cheng, J. K.; Huang, H. P.; Yu, C. C. "Process alternatives for methyl acetate conversion using reactive distillation. 1. Hydrolysis," Chem. Res. Sci., 2008, 63, 1668-1682.

4. Metropolis, N.; Rosenbluth, A. W.; Rosenbluth, M. N.; Teller, A. H.; Teller, E.; "Equation of state calculations by fast computing machines," J. Chem. Phys., 1953, 21, 1087-1092.

5. Kirkpatrick, S., Gelatt, C. D.; Vecchi, M. P.; "Optimization by simulated annealing," Science, 1983, 220, 671-680.

Web Page: www.che.ntu.edu.tw:80/english/people/faculty/YCC.htm

474b Dynamic Control of a Column / Side-Reactor Process

Devrim B. Kaymak, Chemical Engineering, Istanbul Technical University, Istanbul, Turkey and William L. Luyben, Chemical Engineering, Lehigh University, Bethlehem, PA 18015.

If the temperature range suitable for reasonable chemical reaction kinetics does not match the temperature range suitable for vapor-liquid equilibrium, reactive columns are not economically attractive. One way to overcome this temperature mismatch is use a flowsheet that features a distillation column with multiple side reactors. The column operates at a pressure that gives temperatures favorable for separation, while the reactors operate at temperatures (and pressures) favorable for reaction kinetics. This paper discusses the controllability of a column/side-reactor process based on the steady-state design studied by Kaymak and Luyben (Ind. Eng. Chem. Res., 2007). The effectiveness of two different control structures is explored. The first uses two temperature controllers, one with direct action and the other with reverse action. The second control structure uses an internal composition analyzer. One of the main conclusions is that the opposite actions of controllers in the two-temperature control structure result in a breakthrough that moves the column to a different operating condition. The composition and temperature controllers in the second structure have the same action and provide reasonably effective control.

474c Analysis of Absorption with Complex Reversible Reaction

Jerry H. Meldon, Tufts University, Chemical and Biological Engineering Department, Medford, MA 02155

Design of packed scrubbers requires accurate solution to local differential mass balances. The daunting mathematical challenges when there are multiple reversible reactions with nonlinear kinetic expressions are typically addressed using computationally intensive numerical methods. However, as shown here, the kinetic terms in the ordinary differential equations associated with Film Theory may be linearized by either of two different techniques with negligible loss of accuracy. One approach, developed by Van Krevelen and Hoftijzer to analyze absorption with a single irreversible reaction, fixes the concentration of a nonvolatile reactant at its (unknown) value at the gas-liquid interface. The other approach, originally developed by K. A. Smith to analyze permeation with reversible reaction, assumes that departures from local reaction equilibrium are small. Previously, we demonstrated that either approach yields near-exact values for the "Enhancement Factor" by which a single reaction multiplies the local absorption rate. We present results here that demonstrate similar accuracy when either linearization technique is applied to the analysis of absorption and multiple reversible reactions.

Experimental and Modeling Results for Rich Gas Absorption of HCl in Caustic Solutions

Pendergast, John; Lopez-Toledo, Jacinto; Au-Yeung, Patrick; Jewell, Dennis

Developing a modeling capability and providing performance data in the area of absorbing "rich gas" fills a long-standing technical gap in separations technology. The term "rich gas" is used in this context to denote an application in which a substantial portion or a majority of the gas is absorbed from the incoming gas mixture. Closing this gap can lead to more efficient designs of scrubbers, resulting in lower cost and smaller footprints for our processes. Ultimately, the goal is to work toward the more fundamentally predictive methods of calculating mass transfer which rely on the individual mass transfer coefficients, k_g and k_l, along with the effective area for mass transfer across the packing or tray, a. However, since the present state of the predictive methods is not adequate even for the more typical cases encountered, this work must become a part of the total understanding of the larger picture.

As noted above, the term "rich gas" is used in this context to denote an application in which a substantial portion of the gas is absorbed from the incoming gas mixture. It can be easily visualized that under the conditions where a substantial fraction of the incoming gas is absorbed into the liquid mixture, the gas volume diminishes across the tower. Along with the reduction in gas volume is the commensurate reduction in the associated dimensionless numbers, particularly the gas Reynolds number. Where the point "rich gas" applies is a matter of some judgment, but it can certainly be seen that the prediction of the mass transfer coefficient, which is dependent in part on the Reynolds number, is especially problematic in the prediction of the individual mass transfer coefficient. The prediction of "total scrubbers", or rich applications, is especially problematic because of complications from the hydraulic behavior as well as the fact that the controlling mass transfer resistance is not clearly understood.

The system of absorbing HCl in caustic is a classic system that can be used to quantify and test correlations for gas phase mass transfer coefficients because the ionic reactions of HCl_{aq} with excess $NaOH_{aq}$ are extremely rapid. While the statement is true of the HCl / NaOH system as an effective test system, it is not widely reported in literature. The difficulties of handling HCl and NaOH in an academic or university laboratory would make the system uninviting as a "normal" test system. In general, only industrial laboratories or specially equipped laboratories are in a position to handle such reactive and corrosive systems on a routine basis. If we add to this the fact that industrial practitioners are far more likely to have an interest in absorbing rich gas streams that may result from process upsets, it is not surprising to find that there is very little information available in the literature on this type of problem.

An experimental apparatus for rich gas absorption was built, consisting of a 2" (5.08 cm) packed tower with ¼" (0.635 cm) ceramic saddles. This experimental unit was used to measure the absorption performance of hydrogen chloride gas in aqueous sodium hydroxide absorbent. Increasing concentrations of HCl in nitrogen were absorbed across limited packed bed heights to produce meaningful data for absorption performance evaluation. Concentrations of HCl in the incoming feed gas were varied from 5% to 90%. The results of this experimental program are presented in this report.

The results of the experiments on HCl can be viewed below in Table 1. Seven experiments were carried out, varying the concentration of the gas feed from 5 to 90 mole % HCl in the inlet gas. The experiment conducted at 10 mole % concentration was carried out three times. The first experiment seems to be in disagreement with the second two duplicate experiments, but all of the data gathered is presented below. All of the experiments were performed with 10 wt% caustic solution, and arranged such that there was always an excess of caustic over stoichiometric requirements for reaction of HCl and NaOH.

Table 1. Experimental results for several HCL gas concentrations

HCl Concentration Feed Gas Mole %	HCl Concentration Vent Gas Mole %	Gas Feed Rate SCCM	Liquid Feed Rate ml/min	Percent Absorption %
5	0.0429	26463	550	99.1
10	0.0311	26068	550	99.7
10	0.1376	26068	550	98.6
10	0.1775	26068	550	98.2
25	0.3593	25680	550	98.6
50	0.8857	27840	550	98.2
75	0.656	24090	550	99.1
90	1.4088	23640	550	98.4

The experimental conditions were modeled using the rate based modeling tool in Aspen Version 2004 known as RateSep. While it is certainly true that RateSep is a better tool from a modeling standpoint than previous tools and that it constitutes a vast improvement from the user's perspective, one must not confuse ease of use with improvements in the fundamental correlations. At the heart of the rate based modeling effort are still the same fundamental correlations, whether the correlations are those of Onda, Billet and Schultes, or Bravo and Fair. The ultimate goal of this entire work is to improve the correlations that are available by generating new data and regressing existing data in order to provide better, more robust correlations that can be delivered in RateSep.

The experimental configuration was modeled using an electrolyte NRTL model that has been developed for use in Dow applications. The model has been tested a number of times, and gives excellent results across a broad spectrum of HCl concentrations. The tower was modeled with 12 segments (not stages) that represent the 15.2 cm. of packing. Finer discrimination of the grid (more segments) resulted in no numerical changes in the outlet gas result, but did result in decreased stability of the numerical solution. The correlations that were used in this study were the Onda correlations and the Bravo and Fair correlation. The correlations of Billet and Schultes require packing specific parameters for application to a specific type of packing, and so were not utilized in this comparison.

There are several general comments that are appropriate regarding the ability to predict the performance of these types of systems, and lessons that can be learned from the experimental data. The results of the comparison between the simulation and the experimental results can be seen in Figure 1. Note that outlet gas concentration is presented on a logarithmic scale.

- In general, our ability to accurately predict the outlet gas concentration from an absorber is limited. If we were to be tasked with predictive modeling of a system such as this, and the requirement of this is the safety of our facilities and our environment, then these approaches are not yet good enough to employ without resorting to experimental data.

- The older correlation of Onda under predicts the performance of the absorption tower, and the correlation of Bravo and Fair over predicts the performance. Obviously, there are other correlations that can be applied, but none of the available correlations in an unaltered form adequately modeled the system.

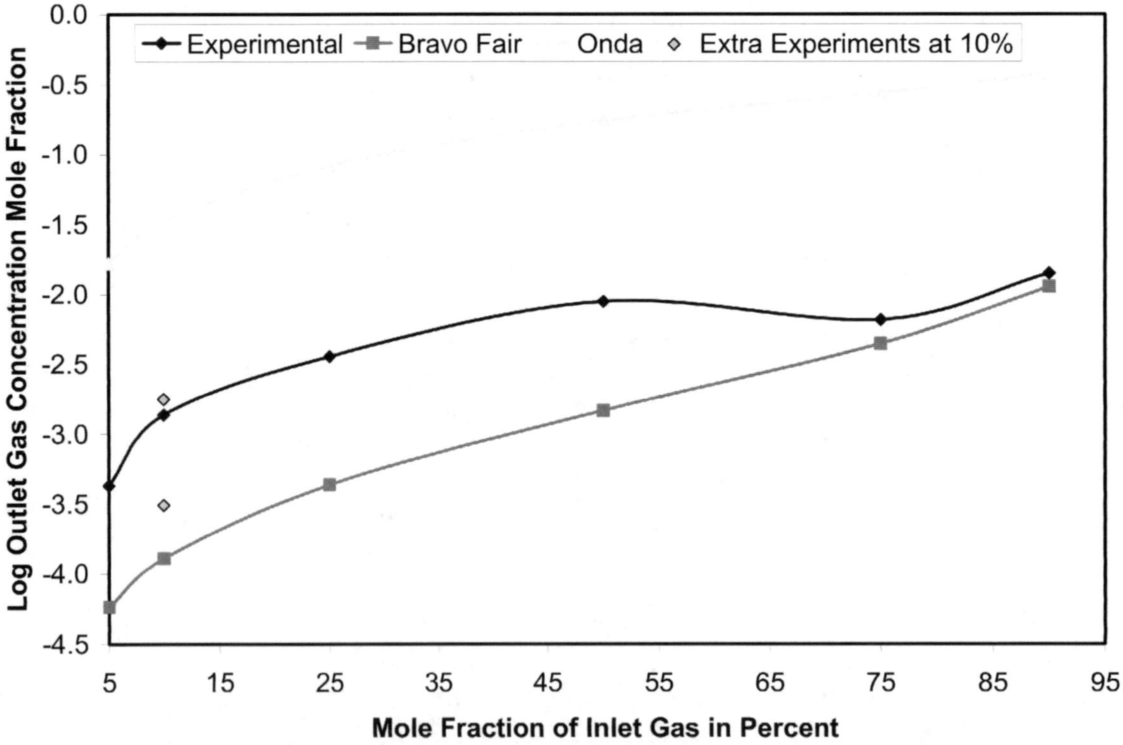

Figure 1. Comparison of experimental outlet concentration of HCl in the gas with the models of Bravo-Fair and Onda.

474e Towards Predicting Catalyst Performance in Reactive Distillation Systems

Heather E. Todd, Katarina Novakovic, Mark J. Willis, and Allen R. Wright. School of Chemical Engineering and Advanced Materials, Newcastle University, Faculty of Science, Agriculture and Engineering, Merz Court, Newcastle upon Tyne, United Kingdom

Some unit operations, such as reactors, are easily miniaturised and incorporated into high throughput equipment such as robotic platforms. However, there are currently no units that can replicate stage-wise unit operations such as distillation or reactive distillation on a small scale. Reactive distillation combines chemical synthesis with separation by distillation, but this leads to a non-trivial system; the hardware selection, the components present in the system, the mode of operation and the operating conditions all affect the performance that can be achieved. An important issue in the design of reactive distillation systems is the identification of new catalysts that perform well under reactive distillation conditions [1]. Catalysts are crucial in increasing reaction rate when the operating temperature range is limited by the evaporation processes taking place. The main goal of this research is to develop a method, utilizing high throughput technology, which can be used to assess potential catalysts for batch or semi-batch reactive distillation systems.

Key to achieving this target will be the ability to identify whether a candidate catalyst could be suitable as early as possible, ideally using information available from knowledge of the catalyst structure. The approach of this research is to correlate catalyst descriptors/properties to performance in a reactive distillation unit. The case study used is batch and semi-batch reactive distillation for the esterification of long-chain fatty acids. Potential catalysts are experimentally studied at small scale in a high throughput Chemspeed SLT 106 synthesizer. The Chemspeed unit cannot replicate distillation, so a dynamic reactive distillation simulation is employed, built in BatchCAD ™ software. Further investigation is performed in a pilot scale batch reactive distillation unit, providing experimental information about the catalyst performance and validation of the strategy.

1. Malone, M. F. and Doherty, M. F. (2000) Reactive Distillation, Ind. Eng. Chem. Res., 39, 3953-3957

474f Optimal Design of Three-Phase Reactive Distillation Columns Using Nonequilibrium/collocation Models

Theodoris Damartzis, Chemical Process Engineering Research Institute, Centre for Research and Technology - Hellas, 6th km Charilaou-Thermi Rd, P.O. Box 60361, Thermi-Thessaloniki, 57001, Greece and **Panos Seferlis**, Mechanical Engineering Department, Aristotle University of Thessaloniki, P. O. Box 424, Thessaloniki, 54124, Greece.

Three-phase reactive distillation offers a challenging case for process modeling and design. It is the inherent complexity of the system arising from the appearance of a second liquid phase that makes process modeling both mathematically strenuous and computationally demanding. Optimal design assisted by mathematical programming techniques of such columns becomes an even more laborious task as alternating three-phase and two-phase columns regions can be identified only at the final solution.

This work combines the predictive power of nonequilibrium (NEQ) models with the approximating properties of orthogonal collocation on finite elements (OCFE) model formulation for the development of an accurate and reliable process model for three-phase reactive distillation columns. Following Higler et al. (2004) the NEQ model is based on mass and energy balances for each phase present while the thin film model is utilized for the description of the transport phenomena and chemical reactions around the three possible interfaces (vapor-liquid I, vapor-liquid II, liquid I-liquid II). Mass and heat transfer in the film region are expressed in terms of the multicomponent Maxwell-Stefan equations.

A key feature in the modeling of three-phase reactive distillation is the accurate identification of the boundaries between three-phase and two-phase column regions. The location of the phase boundary may vary as column conditions change. In particular, when the optimal column configuration and operating points are sought, phase boundaries are not known a priori and may be difficult to estimate. Therefore the model structure should be able to adapt accordingly. The model reduction capabilities of the OCFE model formulation greatly facilitate the tracking of the phase boundary (Swartz and Stewart, 1987) and ultimately the development of a compact in size model that encompasses the rigorous description of the occurring phenomena (Dalaouti and Seferlis, 2006). Phase boundary tracking is achieved with the performance of a thermodynamic phase stability test and the position of a finite element breakpoint at each phase boundary.

Optimal design using a NEQ/OCFE model enables the calculation of the column configuration as described by the number of stages for staged columns or packing height for packed columns, location of feed streams, sizes of reactive and non-reactive column sections, stage holdups, and column operating conditions such as reflux and reboil ratios, catalyst load, and reactant stream ratio using conventional nonlinear programming techniques. Due to the continuous nature of the resulting NEQ/OCFE model, staged columns are transformed into continuous domains thus avoiding the use of integer/binary variables associated with the existence of column stages; a feature that facilitates the design optimization significantly.

The proposed modeling framework has been applied in the optimal design of a staged column for the production of butyl acetate via the esterification reaction of butanol and acetic acid.

Dalaouti N, Seferlis, P., (2006). A unified modeling framework for the optimal design and dynamic simulation of staged reactive separation processes, Comput. Chem. Eng., 30, 1264.

Higler A., Chande R., Taylor R., Baur R, Krishna R. (2004). Nonequilibrium modeling of three-phase distillation. Comput. Chem. Eng., 28, 2021.

Swartz C.L.E., Stewart W.E. (1987). Finite-element steady state simulation of multiphase distillation. AIChE J., 33, 1977.

488a Molecular Simulation of the Transport of Flue Gases through Surface-Modified Inorganic Membranes

Jennifer J. Williams, Tina Düren, and Nigel A. Seaton. Institute for Materials and Processes, School of Engineering and Electronics, University of Edinburgh, Kenneth Denbigh Building, The King's Buildings, Mayfield Road, EH9 3JL, Edinburgh, United Kingdom

Membranes based on porous, inorganic materials such as structured mesoporous silicas (for example MCM materials) are being increasingly considered for industrial separation processes because of their compatibility with high temperature environments and their ability to separate species which are close in molecular size. These materials can be further tailored by introducing functional surface groups. Here, we present the results of our research into the design and optimization of one such mesoporous inorganic silica material, MCM-41, for the removal of carbon dioxide from flue gases.

We have used a number of different molecular modelling techniques including an in-house kinetic Monte Carlo method for the creation and validation of a pore model, grand-canonical Monte-Carlo simulations for the adsorption of pure components and typical flue gas compositions and non-equilibrium molecular dynamics simulations to calculate the flux and transport diffusion coefficients of the components of flue gases in MCM-41 under various conditions. These techniques not only allow for macroscopic predictions of adsorption isotherms and diffusion coefficients but also give a detailed picture of molecular-level phenomena which is not easily accessible using experimental methods.

Here, we present findings and conclusions based on our simulation results. In particular, we describe how the size of the pore wall and pore wall functionalization with organic surface groups can influence diffusant transport and behavior. Our investigations allow us to propose an optimally performing material with a specific pore diameter and surface group functionalization. Importantly, the simulated modifications to the pore wall are experimentally reproducible. We therefore envisage that an optimally performing MCM-41 material designed using these simulation methods can be targeted for experimental synthesis.

488b Pore Network Simulation of Transport and Separation of Gaseous Mixtures through Nanoporous Membranes

Mitra Abdolahi, Rayan Mourhatch, Feng Chen, Theodore Tsotsis, and **Muhammad Sahimi**. Mork Family Department of Chemical Engineering and Materials Science, University of Southern California, 925 Bloom Walk, HED 216, Los Angeles, CA 90089

We develop pore network models of transport and separation of binary and ternary gaseous mixtures, including those involving CO2, CH4, and hydrogen, through silicon carbide and carbon molecular-sieve membranes. A three dimensional network of interconnected pores is used to represent the membrane's pore space, in which the effective pores' radius is distributed according to a pore size distribution (PSD). The connectivity of the pores and the broadness of the PSD are varied in order to study their effect on the transport and separation processes, as are the pressure and temperature in order to understand the effect of the operating conditions. The Maxwell-Stefan equations are used for describing the pore level transport processes, which include Knudsen and hindered diffusion, as well as pore blocking effect and viscous flow. The boundary conditions that are imposed closely correspond to those used in the experimental studies. Good agreement is found between the simulation results and our experimental data for the single and binary gas permeances and separation factors. The results also indicate the fundamental significance to the permselectivity of a membrane of the tail of the PSD, as well as the percolation effect which is manifested through the interconnectivity of the pores that are accessible to the gases.

488c Cfd Modelling of Fluid Dynamics and Mass Transfer in Membrane Modules for Gas Mixture Separation

Mirella Coroneo[1], Jacopo Catalano[2], Marco Giacinti Baschetti[3], Giuseppina Montante[4], and Alessandro Paglianti[4]. (1) Alma Mater Studiorum - University of Bologna, Via Terracini 34, Bologna, Italy, (2) Deapartment of Chemical Engineering, Mining and Environmental Technology, University of Bologna, via Terracini 28, Bologna, Italy, (3) Department of Chemical Engineering, Mining and Environment Technology, Alma Mater Studiorum - University of Bologna, Via Terracini 34, Bologna, Italy, (4) Chemical Engineering, Mining and Environmental Technology, University of Bologna, viale Risorgimento 2, Bologna, Italy

In this work a Computational Fluid Dynamics (CFD) approach for the prediction of the fluid dynamic and the separation performances of complex inorganic membrane modules for gas mixture separation is presented.

The simulations are based on the numerical solution of the Navier-Stokes and local species balance equations on the three dimensional domain representing quite closely the selected module geometry. The membranes are modelled as a selective layer, which allows the permeation of different components as a function of the transport mechanism and the driving force.

The computational approach, previously devised for the simulation of a simple ceramic membrane only, has been further implemented to improve the predictions accuracy and to describe also the case of commercial Pd/Ag membrane modules. Three membranes have been investigated: a ceramic, and two ceramic supported palladium-silver membranes. Since the selected membranes contain a ceramic structure whose behavior is extremely complicated to predict theoretically, the diffusion coefficients of all the membranes have been experimentally evaluated. The experimental data available at different working conditions (pressure and feed flow rate) and different hydrogen concentration in the feed stream have allowed to strictly evaluate the CFD results.

The different membranes have been modelled with simple approach considering four different transport mechanisms: molecular diffusion (self and mutual), Knudsen diffusion, Poiseuille flow and Sieverts' law. The excellent agreement between the experimental and the predicted data suggests that CFD can be considered an useful and reliable tool for designing new modules or optimizing existing ones. In practice, for instance, a CFD simulation allows to evaluate accurately the convective mass transfer resistance thus guiding the module geometry selection. In any case, for industrial applications the CFD simulations may be useful also to identify the influence of some internals on the mass transfer performances, to correctly estimate the driving force across the membrane taking into account the concentration polarization effect, and to identify the effect of different working conditions as the adoption of sweep streams.

488d The Effect of Sweep Uniformity on Gas Dehydration Modules

Pingjiao Hao and **G. Glenn Lipscomb**. Chemical & Environmental Engineering, University of Toledo, Mail Stop 305, Toledo, OH 43606

Air dehydration membranes offer a simple, cost effective solution to humidity control. Membrane units may be installed in-line and require no- auxiliary utilities - only a portion of the feed gas is lost and a small pressure drop is incurred.

To produce desired dew points, a portion of the product gas typically is used as sweep in the module. Sweep lowers the water concentration in the permeate to permit sufficient reduction of the water concentration in the retentate (non- permeate) stream.

The literature reports numerous ways to create the sweep stream including: 1) making the fibers non-selective at the product end, 2) introducing the sweep from an external collar around the module, and 3) inserting tubes through the tubesheet that allow communication between the product header and the shell of the module.

We report simulations of the sweep distribution within the shell and its effect on module performance. Two types of simulations are considered: 1) simulations that explicitly predict flow fields within the shell based on how the sweep gas is introduced and 2) simulations that assume the sweep flow around each fiber is distributed in a Gaussian manner.

The use of fibers that are non-selective at the product end is most efficient based on module capacity and dry gas recovery. Introducing the sweep through an external collar or internal tubes is poorer due to poorer gas distribution in the shell.

Predictions based on explicit calculation of shell flow fields are in good agreement with those based on a Gaussian sweep distribution using a standard deviation in sweep flow equal to ~15% of the average sweep flow rate. We believe the results of this work may be used to evaluate alternative methods providing sweep in a module.

488e Modelling of Ion Transport across Anion Exchange Membranes Under Current Reversal Conditions

Oscar A. Prado, Sten B. Jørgensen, and Gunnar Jonsson. Department of Chemical and Biochemical Engineering, Technical University of Denmark (DTU), DTU-Building 229, Søltoft Plads, DK-2800 Kgs. Lyngby, Denmark

Lactic acid is an important chemical used in industry as pH regulator, emulsifying agent, animal feed supplement, solvent, electrolyte and it is an important potential feedstock for the production of the biodegradable Poly-Lactic Acid (PLA). The PLA has applications in plastic engineering where it can replace hydrocarbon based polymers and thereby reduce the non-biodegradable waste. Besides, PLA will diminish the dependence on fossil feedstock. The production of Lactic acid by Lactic Acid Bacteria is normally impaired by product inhibition like many other fermentation processes at a certain concentration level of the product. Therefore, continuous removal of lactic acid from the fermenter will result in a higher productivity and product yield. Recently, the application of Reversed Electro-Enhanced Dialysis (REED) has shown promising performance for continuous removal of Lactic Acid during fermentation [1].

The REED module combines elements from Electro-Dialysis Reversal (EDR) and Donnan Dialysis (DD) operations. The REED stack is composed only of anion exchange membranes (AEM). Using this unit, the fluxes obtained using Donnan Dialysis can be enhanced by applying electrical current. In addition, the adverse influence of fouling is reduced when the current density is reversed periodically. Another advantage arises if only anion-exchange membranes are utilized, since scaling problems are avoided because only inorganic anions and negatively charged organic acids and amino acids are extracted [1].

A dynamical model is derived to estimate the fluxes and concentration profiles of some ions commonly found in a fermentation broth. The model is based on first engineering principles for diffusion, migration and convection, resulting in a stiff system of partial differential and algebraic equations [2]. The key idea is to reveal the interactions between operation conditions for the REED module, when it is operated with or without applying electrical current. The model is validated against experimental data for dialytic transport of lactate and other carboxylic acids reported in literature [3]. Furthermore, the model can be potentially applied to optimize the design and operation of the REED unit.

References

[1] Rype, J.U. (2003) Separation and recovery of organic acids using Electrodialysis, Ph.D. Thesis. CAPEC, DTU.

[2] Møllerhøj, M. (2006). Modeling the REED process. M.Sc. Thesis. CAPEC, DTU.

[3] Zheleznov, A, et. al. (1998). Dialytic transport of carboxylic acids through an anion exchange membrane. Journal of Membrane Science. Vol 139, pag. 137-143.

488f Separation of Azeotropic Alcohol-Water Mixtures with Fricdiff

Bianca Breure[1], E. a. J. F. Peters[2], and P. J. a. M. Kerkhof[2]. (1) Chemical Engineering and Chemsitry, Eindhoven University of Technology, Den Dolech 2, Eindhoven, 5612 AZ, Netherlands, (2) Chemical Engineering and Chemistry, Eindhoven University of Technology, Den Dolech 2, Eindhoven, 5612 AZ, Netherlands

Separation of azeotropic alcohol-water mixtures with FricDiff

Authors: B. Breure (B.Breure@tue.nl), E.A.J.F. Peters (E.A.J.F. Peters@tue.nl), P.J.A.M. Kerkhof (P.J.A.M.Kerhof@tue.nl)

Address: TU Eindhoven, Laboratory of Separation Processes and Transport Phenomena, Den Dolech 2, 5612 AZ Eindhoven, The Netherlands

Telephone: +31 40 247 3683

Keywords: *FricDiff, kinetic separation, azeotropic alcohol-water mixtures, enhancer, numerical model*

One of the major disadvantages of the separation processes that are currently used in the chemical and petro-chemical industry is their high energy-consumption. This is particularly the case for distillation. Especially when separation processes become more complex, for example in the case of the separation of azeotropic mixtures, energy-demands are high. In addition, hazardous chemicals are often required to facilitate the separation.

Over the past years, the interest for membrane separation processes, like pervaporation and vapor permeation has grown tremendously. The relatively low energy consumption and the fact that no additional chemicals are required, makes these processes interesting candidates to replace, among others, distillative separations. In spite of major efforts in universities and research institutes, membrane processes are still not suited to be applied on a large scale in the chemical and petro-chemical industry. One of the reasons for this is that the fluxes through the membranes are low, implying that large membrane areas are required. Furthermore, the membranes that are used in these processes give rise to problems. Membrane modules equipped with ceramic membranes are very expensive and it is difficult to synthesize defect-free ceramic membranes. Polymeric membranes can also be used, but they have a low chemical and thermal stability.

At our laboratory we are developing a separation process that doesn't have these disadvantages. This separation process, which is named FricDiff ([1] 08D0C9EA79F9BACE118C8200AA004BA90B02000000080000000E0000005F0052006500660031003900310033003700360034003600340000000 , [2] 08D0C9EA79F9BACE118C8200AA004BA90B02000000080000000E0000005F0052006500660031003900310033003700360034003600360000000) (short for Friction-Diffusion), has features in common with methods found in literature such as mass diffusion [3] 08D0C9EA79F9BACE118C8200AA004BA90B02000000080000000E0000005F0052006500660031003900310033003700360034003200300000000 , double diffusion [4] 08D0C9EA79F9BACE118C8200AA004BA90B02000000080000000E0000005F0052006500660031003900310033003700360034003200390000000 , sweep diffusion [5] 08D0C9EA79F9BACE118C8200AA004BA90B02000000080000000E0000005F0052006500660031003900310033003700360034003300380000000 and diffusion distillation [6] 08D0C9EA79F9BACE118C8200AA004BA90B02000000080000000E0000005F0052006500660031003900310033003700360034003400370000000 . In these processes, a binary gas or vapor mixture is partly separated by letting it diffuse through a third auxiliary component ("the enhancer"). This principle is also used in the FricDiff separation process, where we exploit the difference in relative diffusion rates of the components of a feed mixture (gas or vapor) when they counter-diffuse through an auxiliary component. FricDiff can thus be classified as a kinetic separation process. A suitable enhancer for the separation of vapor mixtures is, for example, a permanent gas. For the separation of gas mixtures a condensable vapor can be an option. An important requirement of the enhancer is that it will show a higher interaction with one of the components of the feed mixture than with the others. This component will, while interdiffusing through the enhancer, be more retained and hence will be enriched in the retentate.

A typical FricDiff module consists of two compartments separated by a non-selective macro-porous barrier (macro-porous membrane). Through the compartments the feed mixture and the enhancer are flowing, co-currently or counter-currently. The separation process mainly takes place inside the pores of the barrier. The main function of the barrier is to prevent mixing of the feed mixture and the enhancer gas or vapor. Furthermore, the barrier decreases convective mass transport through the barrier as a result of possible pressure differences over the barrier. The advantage of using a macro porous barrier instead of a selective micro porous membrane is that the fluxes through the barrier are much higher. In addition, the barrier can be made of a robust and relatively cheap material, like stainless steel.

In this project, the applicability of FricDiff for the separation of azeotropic mixtures of alcohols and water is studied. Experiments are performed with a lab scale experimental set-up with various alcohol-water mixtures. The experimental data are compared with results obtained with a detailed numerical model developed in the finite element program Comsol Multiphysics™. The separation process in the FricDiff-module is described by simultaneously solving mass and momentum balances in the porous barrier and in the flow channels where the feed mixture and the enhancer are flowing. The mass transport through the porous barrier is described with an advanced model for transport through porous media: the Binary Friction

Model [7]
08D0C9EA79F9BACE118C8200AA004BA90B02000000080000000E0000005F005200650066003100390031003300370036003500350035000000 .
Dispersion effects, which arise due to the coupling of convection (the flow field in the flow channels is laminar) and diffusion (dominant in the direction perpendicular to the flow field) are included in the numerical model.

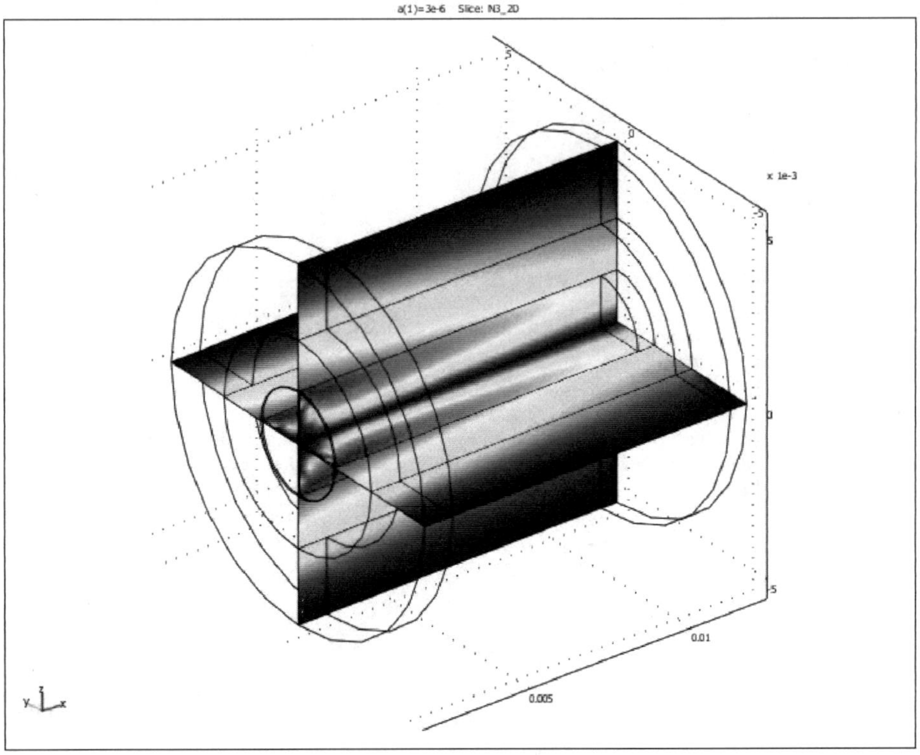

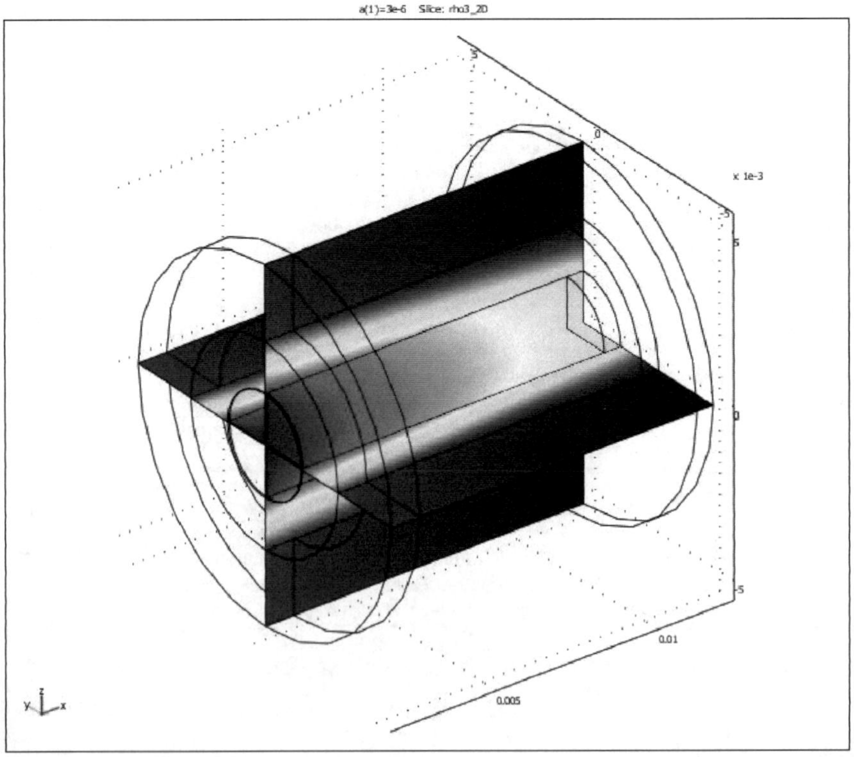

Figure 1. Radial fluxes of IPA (left) and concentration distribution of IPA (right) in a tubular FricDiff-module. Results are obtained in the finite element program Comsol Multiphysics™.

The influence of process conditions (e.g. absolute pressure, ratio of flow rates of feed mixture to enhancer), the geometry of the module (tubular or plate configuration) and characteristics of the porous barrier (e.g. pore size, thickness of the porous barrier) on the separation are studied both experimentally and numerically. Attention is also focused on the negative influences that concentration polarization and concentration boundary layers can have on the separation process in the module. Concentration polarization is a phenomenon that occurs in all membrane separation processes and it adversely affects the separation performance of the module. It reduces the driving force for mass transfer across the membrane and hence is a phenomenon that one wants to prevent from occurring. For this reason the impact of concentration polarization on the process performance of a FricDiff-module under various operating conditions is studied. It will be shown that with decreasing thickness of the porous barrier resistances in the boundary layer adjacent to the barrier start to become important. Therefore, their influence on the separation process has to be taken into account.

By performing the above-mentioned analysis, a set of criteria is defined that can be used to optimize the FricDiff separation process for the separation of a specific mixture.

[1] M. Geboers, P. Kerkhof, P Lipman, F. Peters, FricDiff: A novel separation concept, Sep. Purif. Tech. 56, 47-52 (2007)

[2] B. Breure, E.A.J.F. Peters, P.J.A.M. Kerkhof, Separation of azeotropic mixtures of alcohols and water with FricDiff, Sep. Pur. Techn., accepted for publication (2008)

[3] G.L. Hertz, Method of separating gases from a mixture thereof, patent, 1.486,521 and patent 1,498.097 (1924)

[4] F.A. Schwertz, Separation of gases by single and double diffusion, Amer. J. Phys. 15(1), 31-36 (1947)

[5] M.T. Cichellli, W.D. Weatherford, J.R. Bowman, Sweep diffusion gas separation process, Part I, Chem. Eng. Prog. 47(2) 63-74 (1951)

[6] D. Fullarton, E.U. Schlünder, Diffusion Distillation – A new separation process for azeotropic mixtures. Part I: Selectivity and transfer efficiency, Chem. Eng. Process., 20, 255-263 (1986a)

[7] P.J.A.M. Kerkhof, A modified Maxwell-Stefan model for transport through inert membranes: the binary friction model, Chem. Eng. J., 64, 319-343 (1996)

488g Analysis of Pressure and Sweep Diffusion-Enhanced Membrane Separation

Jerry H. Meldon, Tufts University, Chemical and Biological Engineering Department, Medford, MA 02155

As a rule, concentration polarization in external fluid boundary layers undercuts selectivity in membrane-based two-component separations, because transport of both components in a binary fluid mixture is governed by the same mutual diffusion coefficient. The same is not generally true of multicomponent boundary layer diffusion; e.g., counterdiffusion of gas C retards the transport of heavier gas B more than that of lighter gas A. We have previously reported evidence of this phenomenon in the permeation of hydrogen and ethane in a microporous inorganic membrane. Because pore transport is governed by Knudsen diffusion, the membrane is intrinsically selective to the lighter hydrogen. However, when permeation was sustained by a permeate channel sweep gas, argon, the apparent ratio of H2 and C2H6 permeabilities exceeded the Knudsen value of approximately 4. This was attributed to counterdiffusion of the feed gas mixture and argon in the external boundary layers. Here we present the results of calculations that indicate that when transport of argon is enhanced by a small increase in sweep gas pressure, it can theoretically preferentially retard the flux of heavier ethane to such a great extent that the effective H2:C2H6 permeability ratio approaches infinity.

490a Molecular Modeling of Enantioselective Adsorption In Homochiral Metal-Organic Frameworks

Xiaoying Bao, Linda J. Broadbelt, and Randall Q. Snurr. Department of Chemical and Biological Engineering, Northwestern University, 2145 Sheridan Road, Technological Institute E136, Evanston, IL 60208-3120

Homochiral metal-organic frameworks (MOFs) with a single sense of chirality throughout the framework may be used as chiral stationary phases for preparative enantioselective separations. The highly porous frameworks and high surface areas of MOFs may lead to better production rates compared to conventional non-porous chiral stationary phases. Molecular simulation is employed in this study to predict the enantioselective adsorption of simple chiral hydrocarbons and chiral alcohols in several different homochiral MOFs. The enantioselective sites of adsorption are identified. It is found that both the inherent enantioselectivity and the accessibility of the adsorption sites are important for effective chiral separation. Suggestions on modification of the framework in order to enhance the enantioselectivity/accessibility of the enantioselective sites in the MOFs are made.

490b Predicting the Capability of Metal Organic Frameworks to Selectively Pre-Concentrate Energetic Molecules

Ruichang Xiong[1], Jared T Fern[1], David J. Keffer[1], Miguel A. Fuentes-Cabrera[2], and Don M. Nicholson[3]. (1) Chemical Engineering, University of Tennessee, 419 Dougherty Hall, Knoxville, TN 37996-2200, (2) Joint Institute for Computational Science, University of Tennessee and Oak Ridge National Laboratory, (3) Computer Science and Mathematics Division, Oak Ridge National Laboratory

In order for explosive molecules to be detected, the amount delivered to a sensor must exceed its detection threshold. Pre-concentrators facilitate this process by extracting explosive molecules from the atmosphere and delivering them to sensors in amounts required for reliable detection. Current pre-concentrators are not very selective and hence deliver material that is predominately contaminant. With the ability to tailor Metal Organic Frameworks (MOFs), there is hope that a pre-concentrator can be designed to selectively adsorb explosive molecules in the presence of contaminants. In order to test the feasibility of selectively pre-concentrating explosive molecules for detection, molecular simulations of RDX and TATP within different MOFs were performed with and without contaminants. The simulations give new insight into the competitive nature of the adsorption and the binding of the explosive molecules to the framework. Adsorption isotherms, mobility of the RDX within the framework, the occupancy of each type of cage within the unit cell, and the framework's effect on the configuration of the explosive molecules were investigated by grand canonical Monte Carlo (GCMC) and molecular dynamics (MD).

490c Effect of Water on Binary Adsorption of Nitrogen and Oxygen In Zeolite Na-X: A Molecular Simulation Study

Craig R. Powers and Edward J. Maginn. Department of Chemical and Biomolecular Engineering, University of Notre Dame, 182 Fitzpatrick Hall, Notre Dame, IN 46556

Adsorption isotherms for nitrogen, oxygen, and water, binary nitrogen-oxygen equilibrium, and ternary nitrogen-oxygen-water equilibrium were simulated in Na-X zeolite using the continuous fractional component Monte Carlo method. The effect of three different partial pressures of water on nitrogen-oxygen separation performance is evaluated, and the nature of the inhibitation of separation is examined. Even with very low water partial pressure, significant inhibition of separation and of total gas uptake is observed. The mechanism of this inhibition is shown to be a reduction of nitrogen adsorption near site-III cations.

490d Adsorption of Sulfur Hexafluoride on Mcm-41: Predictions from Molecular Simulations

Aurelio Olivet, Institut de Ciència de Materials de Barcelona, Consejo Superior de Investigaciones Científicas (ICMAB-CSIC) and MatGas Research Center, Campus de la UAB, Bellaterra. Barcelona, 08193, Spain and Lourdes F. Vega, MatGas Research Center and Air Products and Chemicals, Inc, Campus de la UAB., Bellaterra. Barcelona, 08193, Spain.

Sulfur Hexafluoride (SF6) represents the best insulating gas available in the electrical industry, with excellent dielectric and arc-quenching properties. It is also used in several other applications, including thermo-acoustic insulator for windows, as a contract agent in medical applications, etc. In spite of these properties, SF6 is an efficient infrared absorbent, and due to its high chemical stability its atmospheric degradation occurs very slowly, hence, efficient methods are under development for handling and recovering SF6 after industrial uses. One of these methods is related to the optimal adsorption of SF6 in appropriate adsorbent materials. For this reason, several adsorption studies have been carried out on different micro and mesoporous materials. Since SF6 molecules are chemically inert and non-polar, no specific acid-base or pole-pole interactions come out between them and the porous solids commonly used in adsorption systems. Moreover, SF6 molecules are symmetric, large in size and highly polarizable, therefore, their interactions with adsorption sites are dispersive in nature [1]. These features simplify the development of theoretical SF6 adsorption studies, since no complex models have to be used to represent the adsorbate-adsorbate and adsorbate-solid interactions. The goal of the present work is twofold: a) to predict the adsorption capacity of SF6 in the Mobyl Catalytic Material MCM41, and b) to compare the performance of two SF6 force fields in confined media. We have developed a protocol to analyze the Pore Size Distribution (PSD) of structurally different silica-based materials, including the MCM41, by a combination of Grand Canonical Monte Carlo (GCMC) simulations and experimental nitrogen adsorption isotherms [2]. In this work, the PSD obtained for this material is combined with GCMC simulations results of SF6 in cylindrical silica pores, which were modeled by using the structure-less potential proposed by Tjatjopoulos and co-workers [3] in order to predict the overall adsorption behavior of SF6 on this material. For this purpose, SF6 was modeled using the rigid molecular force field proposed by Pawley [4], and also a flexible force field recently proposed by some of us [5]. Discussions regarding the influence of choosing different molecular models for the fluid and their impact on the adsorption behavior are presented. Since the calculated PSD used in this work is in excellent agreement with the one reported in literature for the MCM41 material, we expect a reasonable prediction for the SF6 adsorption.

This work has been partially financed by the Spanish Government (project CTQ2005-00296/PPQ) and the Catalan Government (SGR2005-00288). Additional funding from MATGAS as PhD grants are gratefully acknowledged.

[1] D. V. Cao and S. Sircar. Adsorption 2001, 7, 73-80.

[2] C. Herdes, M. A. Santos, F. Medina and L. F. Vega. Materials Science Forum 2006, 514-516, 1396-1400.; Langmuir 2005, 21, 8733-8742.

[3] G. J. Tjatjopoulos, D. L. Feke and J. A. Mann. Journal of Physical Chemistry 1988, 92, 4006-4007.

[4] G. S. Pawley. Molecular Physics 1981, 43, 1321-1330.

[5] A. Olivet and L. F. Vega. Journal of Chemical Physics 2007, 126, 144502.

490e How Does Nanoscale Roughness Influence Wetting Behavior?

Eric M. Grzelak, Dept. of Chemical and Biological Engineering, University at Buffalo, 303 Furnas Hall, Buffalo, NY 14260 and Jeffrey R. Errington, Chemical and Biological Engineering, University at Buffalo, 303 Furnas Hall, Buffalo, NY 14260.

The way in which a fluid wets a surface it contacts depends greatly on the properties of that surface, including its roughness. This is evidenced by the successes and failures of things such as metal welded ceramics or water repellent fabrics. In this presentation we illustrate our recent computational work designed to better understand the effects on wetting of surface roughness ranging from molecular to nanoscopic length scales. Grand canonical transition matrix Monte Carlo simulations are used to determine the surface free energy of a fluid in contact with an atomistically described surface containing measureable roughness. The free energy is then used to calculate spreading coefficients and prewetting saturation points below and above the wetting temperature respectively. These data are then extrapolated to precisely determine the wetting temperature. In this study we use a Lennard-Jones fluid interacting with a substrate comprised of static particles. These particles are arranged in crystalline configurations and then etched to provide nanoscopic contours. The results of our simulations are compared to predictions from the Wenzel and Cassie wetting models. Lastly, finite-size influences on the technique are explored.

490f Metal–Organic Frameworks for CO2 Storage and CO2/CH4 Separation

Ravichandar Babarao, National University of Singapore, Singapore, 117576, Singapore and Jianwen Jiang, Chemical land Biological Engineering, National University of Singapore, Singapore, Singapore.

By varying the metal center, organic linker, functional group and framework topology, a series of MOFs (IRMOF1, Mg-IRMOF1, Be-IRMOF1, IRMOF3-(NH2)4, IRMOF10, IRMOF13, IRMOF14) are systematically examined for CO2 storage using Monte Carlo simulations. The affinity with CO2 is enhanced by adding functional group, the constricted pore is formed by interpenetrating framework; both lead to a larger isosteric heat and Henry constant and subsequently a stronger adsorption at low pressures. The organic linker plays a critical role in tuning the free volume and accessible surface area, and largely determines CO2 adsorption at high pressures. As a combination of open framework and low density, IRMOF10 and IRMOF14 exhibit higher capacity than other MOFs and even surpass the experimentally reported highest capacity in MOF-177. The gravimetric and volumetric capacities at high pressures correlate well with the framework density, free volume, porosity and accessible surface area of MOFs. These molecular-based structure-function correlations are useful for a priori prediction of CO2 capacity and for rational screening of MOFs toward the high-efficacy CO2 storage. Using Monte Carlo and molecular dynamics simulations, separation of CO2 and CH4 mixture in three different nanomaterials, namely, silicalite, C168 schwarzite and IRMOF-1 is compared. The permselectivity based on adsorption and self-diffusivity in the mixture is marginal in IRMOF-1 and greatest in C168 schwarzite. Although IRMOF-1 has the largest storage capacity for CO2, its selectivity is not satisfactory.

490g Computational Investigation of Window Size Effects on Gas Transport In High Silica 8-Ring Zeolites

Preeti Kamakoti[1], Sang-Eun Jee[2], Ronald R. Chance[1], Sebastian C. Reyes[1], and David S. Sholl[2]. (1) ExxonMobil Research and Engineering, Annandale, NJ 08801, (2) Georgia Institute of Technology, School of Chemical & Biomolecular Engineering, 311 Ferst Drive, N.W., Atlanta, GA 30332-0100

The diffusion of gases and hydrocarbons in microporous materials is of great practical importance in the chemical and petrochemical industry. Small differences in window size and shape can have dramatic effects on molecular transport, an example being methane in 8-ring zeolites, where window size and molecular size are comparable. Pulsed field gradient NMR experiments [1] show that subtle differences in window dimensions of around 0.05-0.35 Å between siliceous LTA, CHA and DDR can cause room temperature methane self diffusivities to span two orders of magnitude, ranging from 1.42×10^{-6} cm^2/s for LTA to 1.6×10^{-8} cm^2/s for DDR. Currently, there is little understanding of the factors affecting window architecture, and no means identified for prediction of the significant structural variations that have been observed experimentally. To provide a better understanding of window architectures, we have performed Density Functional Theory (DFT) calculations on the same three siliceous zeolites LTA, CHA and DDR, wherein a complete relaxation of the cell volume, shape and atomic positions was performed. Results show that DFT is an excellent tool for quantitatively predicting structural information in siliceous zeolites. DFT is also surprisingly successful in capturing the subtle differences between window sizes and puckering in all three structures, those differences being critical to understanding molecular transport. The optimized structures obtained with DFT are used in conjunction with an atomistic model to simulate the adsorption and diffusion of CO2 and CH4. We have used molecular simulations to examine the diffusion of CO2 and CH4 in DDR, CHA, and LTA, both as single components and as mixtures and performed a comparison with experimental rseults. These simulations give insight into the strong dependence of the observed diffusivity on zeolite structure and suggest opportunities for using the unusual properties of diffusing mixtures in some materials for practical applications.

[1] Hedin, N., DeMartin, G.J., Roth, W.J., Strohmaier, K.G., Reyes, S.C. Microporous Mesoporous Mater 109 (2007), 327.

490h Molecular Simulation of CO2 and CO2/CH4 Mixture Adsorption and Diffusion In Metal-Organic Frameworks

Jinchen Liu[1], Seda Keskin[2], Rees B. Rankin[1], J. Karl Johnson[3], and David S. Sholl[2]. (1) National Energy Technology Laboratory, 626 Cochrans Mill Road, Pittsburgh, PA 15236, (2) Chemical and Biomolecular Engineering, Georgia Institute of Technology, 311 Ferst Drive NW, Atlanta, GA 30332, (3) Department of Chemical and Petroleum Engineering, University of Pittsburgh, 1242 Benedum Hall, Pittsburgh, PA 15261

Capture of CO_2 from large point sources such as fossil fuel power plants is important for mitigating global warming. Metal-organic frameworks (MOFs) and Zeolitic imidazolate frameworks (ZIFs) have been proposed as potential candidates for separating CO_2 from gas mixtures. Adsorption and mass transport properties of MOF and ZIF materials for CO_2 and CO_2/gas mixtures are important in designing, screening, and applications of these materials. We have used grand canonical Monte Carlo simulations to calculate the adsorption isotherms for CH_4, CO_2 and their mixtures in MOFs and ZIFs. We have compared our mixture isotherms with predictions from the ideal adsorbed solution theory. We have computed the diffusivities for CH_4, CO_2 and their mixtures from equilibrium molecular dynamics simulations. We have predicted the mixed gas diffusion coefficients from SSK theory, and compared the SSK theory predictions with mixed gas diffusivities computed from equilibrium molecular dynamics. The molecular level information generated from simulations can be used to generate models for the performance of macroscopic membranes. The transport selectivity and permeance can be computed for a given set of conditions, leading to design calculations based on molecular simulations.

490i Molecular Modeling and Adsorption Properties of Silica Templated Mesoporous Carbons

Surendra K. Jain, Department of Chemical Engineering, North Carolina State University, North Carolina State University, Raleigh, NC 27695, Keith E. Gubbins, Chemical and Biomolecular Engineering and Center for High Performance Simulation (CHiPS), North Carolina State University, Raleigh, NC 27695-7905, and Roland J.-M. Pellenq, CNRS, Campus de Luminy, Centre de Recherche en Matière Condensée et Nanosciences, Marseilles, 13288, France.

Mesoporous carbons with highly ordered pore structures can be obtained by templating ordered silica materials. These carbons offer high mechanical and thermal stability and are useful in many applications including adsorption of large molecules, catalyst in fuel cells and capacitor electrodes. Many experimental studies have been reported on obtaining carbon replica using different templating materials like SBA-15, MCM-41, Zeolites etc. The formation of these templated carbons is a two fold process [1]. A carbon rich precursor in vapor or liquid phase is first deposited inside the porosity of the matrix template. The adsorbed precursor is then carbonized by heating at a high temperature in an oxygen free environment to produce a carbon-matrix composite. Finally, the matrix is removed by treating with an acid and the carbon replica is obtained.

In this work, we present molecular models for two types of mesoporous carbons, CMK-3 [2] and CMK-5 [3], using pseudo mimetic methods. CMK-3 is obtained by templating SBA-15 in which carbon is infiltrated in the total porosity of SBA-15. It consists of carbon rods arranged in a hexagonal order and these carbon rods are held together via small interconnections. CMK-5 is obtained by templating SBA-15 in which carbon is introduced as a film on the pore surface of SBA-15. It consists of carbon pipes arranged in a hexagonal order and these carbon pipes are held together via small interconnections. We use these models of mesoporous carbons to study the adsorption behavior of argon using grand canonical Monte Carlo simulations. We investigate the effect of pore size and the role of interconnections between the carbon rods and pipes on the adsorption phenomenon of CMK materials.

[1] Ryoo, R.; Joo, S. H.; Jun, S. J Phys Chem B 1999, 103, 7743.

[2] Jun, S.; Joo, S. H.; Ryoo, R.; Kruk, M.; Jaroneic, M.; Liu, Z.; Ohsuna T.; Terasaki, O. J Am Chem Soc 2000, 122, 10712.

[3] Joo, S. H.; Choi, S. J.; Oh, I.; Kwak, J.; Liu, Z.; Terasaki, O.; Ryoo, R. Nature 2001, 412, 169.

505a Low-Cost H2 Production for Fuel Cells Using Adsorption Enhanced Reforming

Kandaswamy DuraiSwamy[1], Armin D. Ebner[2], James A. Ritter[2], Court Moorefield[3], Donovan Pena[3], and Anand Chellappa[3]. (1) Intelligent Energy, Inc, 2955 Redondo Ave, Long Beach, CA 90806, (2) Department of Chemical Engineering, University of South Carolina, Columbia, SC 29208, (3) Intelligent Energy, 2955 Redondo Ave, Long Beach, CA 90806

Rapid Adsorption Enhanced Cyclic Reformer is being developed at Intelligent Energy for producing low-cost hydrogen at high thermal efficiencies for use in fuel cells for combined heat and power application as a part of a DOE program (DE-FG36-07GO17013). Modeling and simulation is being done at University of South Carolina.

Experimental tests are progressing for ethanol and methane reforming using Rhodium-zirconium stabilized ceria catalyst and potassium-promoted hydrotalcite as sorbent. Results indicate that by adsorptive reforming at ~500 C high thermal efficiencies (~80%) are achievable. The paper will present experimental results and simulations predicting the behavior of the rapid cycle enhanced cyclic reformer.

505b Technology Developments In Psa Ethanol Dehydration

Charles D. Tereck, KATZEN International, Inc., 2300 Wall Street, Suite K, Cincinnati, OH 45212

U.S. production of motor fuel grade ethanol (MFGE) has grown at the rate of 20% per year for the past 12 years outpacing all other chemical and biochemical products. Concurrently, there has been similar growth worldwide. Pressure swing adsorption (PSA) application to the dehydration, using molecular sieve zeolites, of azeotropic ethanol/water mixtures has been key to the economic, environmental, and safety requirements in this expansion. Methods for energy reductions and consequent cost savings will be presented that show dramatic improvements over first generation azeotropic distillation systems as well as resistance to market penetration by various membrane systems. Design parameters and heuristics are discussed that have resulted in more than 1,000,000 cycles on the same initial charge of molecular sieve media. Integration of the adsorption system with the distillation, energy recovery, and vacuum systems are key to the overall operability and viability of the production plant in terms of both a fermentation facility and stand-alone dehydration up-grader unit. Smaller in market but more technically challenging is the production ultra-low moisture product for downstream chemical synthesis and anhydrous high grade ethanol meeting U.S.P. specifications. Developments are presented that allow tailoring the final product quality in light of the catalytic activity of the molecular sieve stationary phase.

505c A Two-Step Pulsed Pressure Swing Adsorption Based Oxygen Concentrator

Vemula Rama Rao, William Bernard Krantz, and **Shamsuzzaman Farooq**. Chemical and Biomolecular Engineering, National University of Singapore, 4 Engineering Drive 4, Singapore, 117576, Singapore

A two-step pulsed pressure-swing adsorption process (PPSA) has been modeled to assess the extent to which an oxygen concentrator might be miniaturized for medical applications. The process consists of a single bed of packed adsorbent particles that is alternately pressurized and depressurized at the feed end. An enriched oxygen product is withdrawn at ambient pressure from the product end when the bed is pressurized at the feed end. The product end remains closed during depressurization. The describing equations were solved using COMSOL® multiphysics software. The effect on the performance of the adsorption time, desorption time, bed length, particle diameter and imposed pressure drop across the bed were investigated. An interesting novel result is that for a chosen particle size, bed length and applied pressure drop, there is an optimum combination of adsorption and desorption times that maximizes the product purity. The results suggest that there is a reasonably wide operating window wherein the product oxygen purity is consistently greater than 80% for both 5A and partially Ag-exchanged Li-substituted 13X zeolite adsorbents. At a given product flow rate within this operating window, the extent of miniaturization is limited by the (maximum) cycling frequency that is practically achievable. Sizing of an oxygen concentrator for personal medical applications will also be discussed.

AUTHOR INDEX

AUTHOR INDEX

AUTHOR INDEX

AUTHOR INDEX

AUTHOR INDEX

AUTHOR INDEX

AUTHOR INDEX

AUTHOR INDEX

AUTHOR INDEX

AUTHOR INDEX

AUTHOR INDEX

AUTHOR INDEX

AUTHOR INDEX

AUTHOR INDEX

AUTHOR INDEX

AUTHOR INDEX

American Institute of Chemical Engineers
3 Park Avenue
New York, New York 10016-5991

CURRAN ASSOCIATES INC
proceedings.com

Separations Division

Presentations at the 2008 AIChE Annual Meeting

Philadelphia, Pennsylvania
16 - 21 November 2008

Volume 2 of 2

AIChE

ISBN: 978-1-61567-217-2

Separations Division

Presentations at the 2008 AIChE Annual Meeting

Philadelphia, Pennsylvania
16 - 21 November 2008

Volume 2 of 2

ISBN: 978-1-61567-217-2

Printed from e-media with permission by:

Curran Associates, Inc.
57 Morehouse Lane
Red Hook, NY 12571

Some format issues inherent in the e-media version may also appear in this print version.

Printed by Curran Associates, Inc. (2009)

For permission requests, please contact AIChE
at the address below.

AIChE
3 Park Avenue
New York, NY 10016-5991

Phone: (203) 702-7660
Fax: (203) 775-5177

www.aiche.org

Additional copies of this publication are available from:

Curran Associates, Inc.
57 Morehouse Lane
Red Hook, NY 12571 USA
Phone: 845-758-0400
Fax: 845-758-2634
Email: curran@proceedings.com
Web: www.proceedings.com

TABLE OF CONTENTS

Volume 1

Volume 2

Feasibility Study of Miniature Oxygen Concentrator via Pressure Swing Adsorption

Siew Wah Chai and Mayuresh V. Kothare

Chemical Engineering, Lehigh University,
111 Research Drive, Iacocca Hall, Bethlehem, PA 18015

Oxygen Generation Process via Pressure Swing Adsorption

The PSA oxygen process uses adsorbents (a synthetic zeolite called molecular sieve), which attracts (adsorbs) nitrogen from the air at high pressure and releases it at atmospheric pressure.

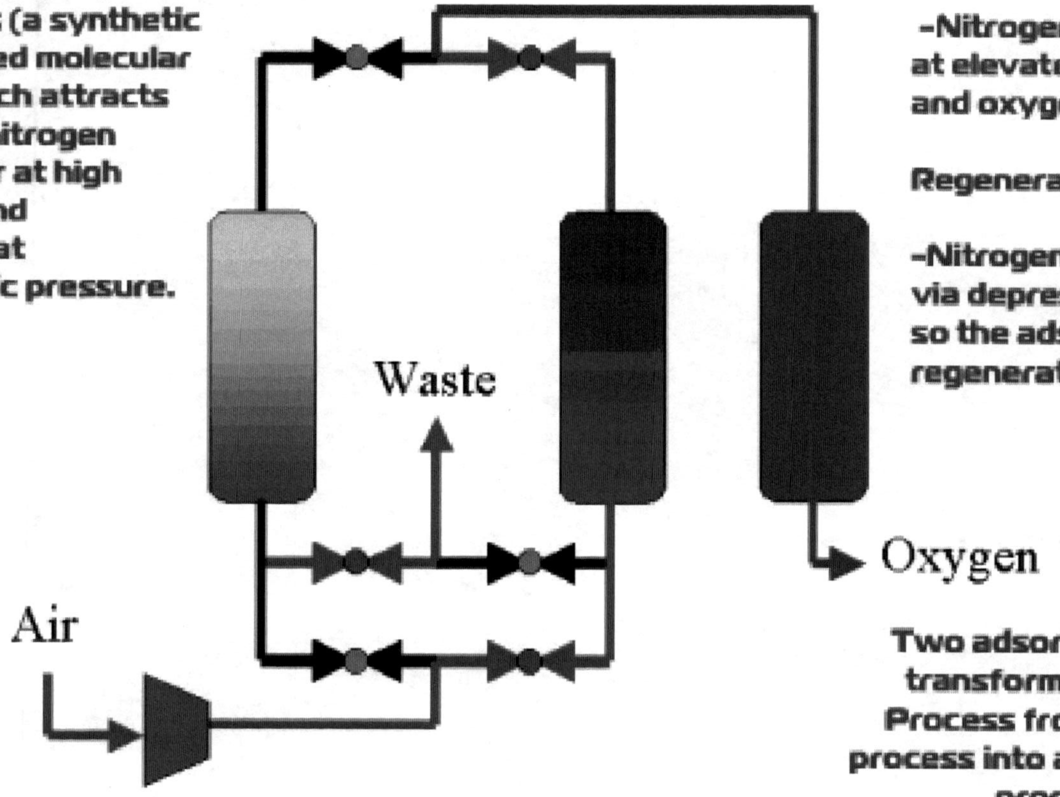

Adsorption Step

-Nitrogen is adsorbed at elevated pressure and oxygen is produced

Regeneration Step

-Nitrogen is desorbed via depressurization so the adsorbent is regenerated

Two adsorbent beds transform the PSA Process from a batch process into a continuous process

Oxygen Generation Process

Why smaller PSA Oxygen Concentrator?

- **N2 adsorbed onto and desorbed from zeolite micro pores easier?**
  (less tortuous gas travel path due to small diameter bed?)
- **Less purging gas required?**
  (more efficient cleaning in small diameter bed?)
- **Faster operation?**
  (shorter time for pressurization in small volume?)
- **More efficient operation?**
  (less dead volume?)
- **Less hazardous?**
  (shorter time for pressurization and less oxygen storage?)
- **More precise system to be controlled?**
  (less dead volume involved?)
- **Less product loss during depressurization?**
  (less product holdup in small bed?)
- **Less energy consumption?**

Application Areas

- Portable respiratory device
- Miniature oxygen provider in electronic device

Objectives

1. To characterize the operating conditions of small PSA systems
2. To compare performances of PSA systems of different sizes
3. To infer feasibility from the performances of current design pieces

Experimental Setup

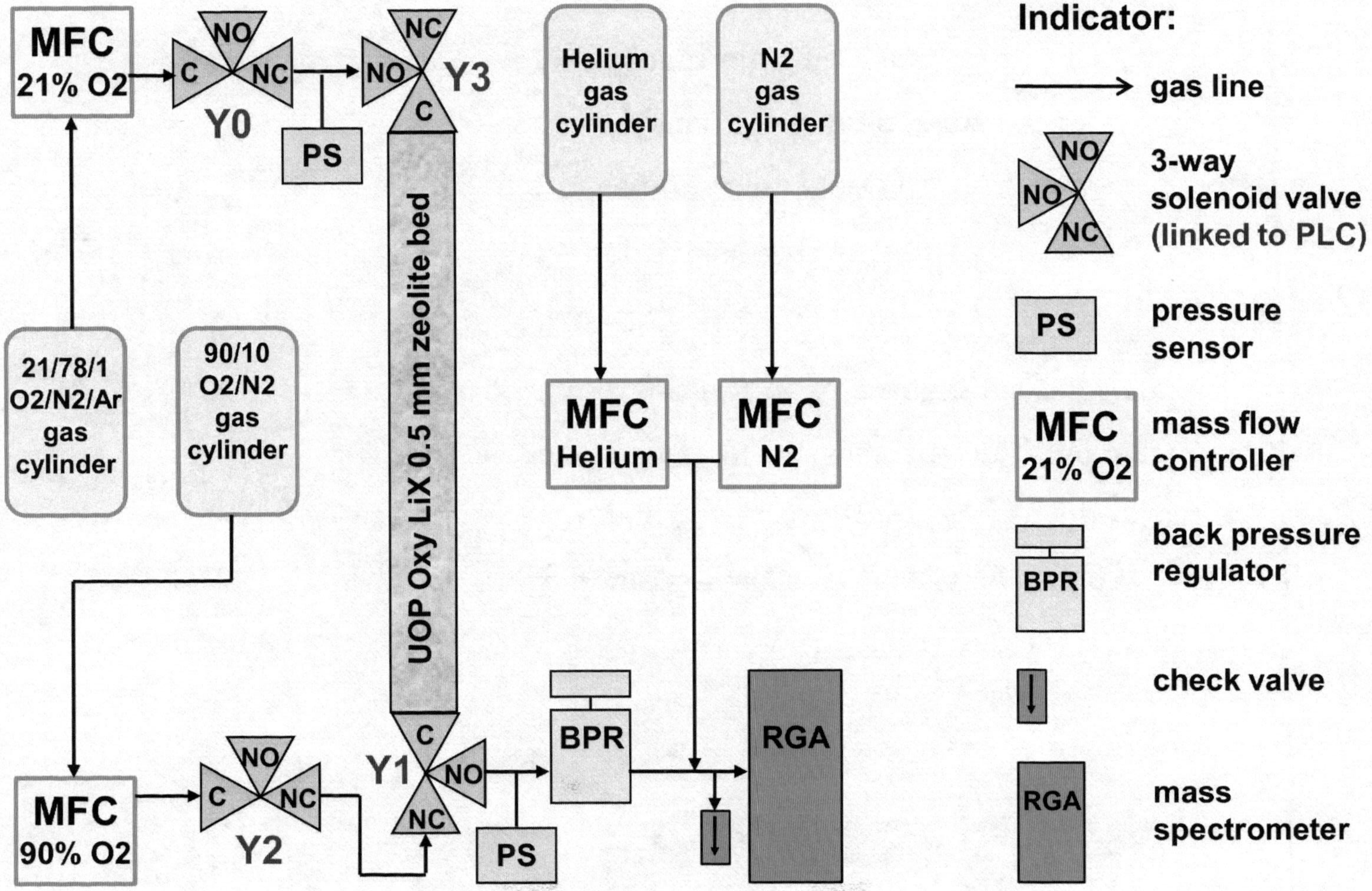

Single Bed Process Cycle

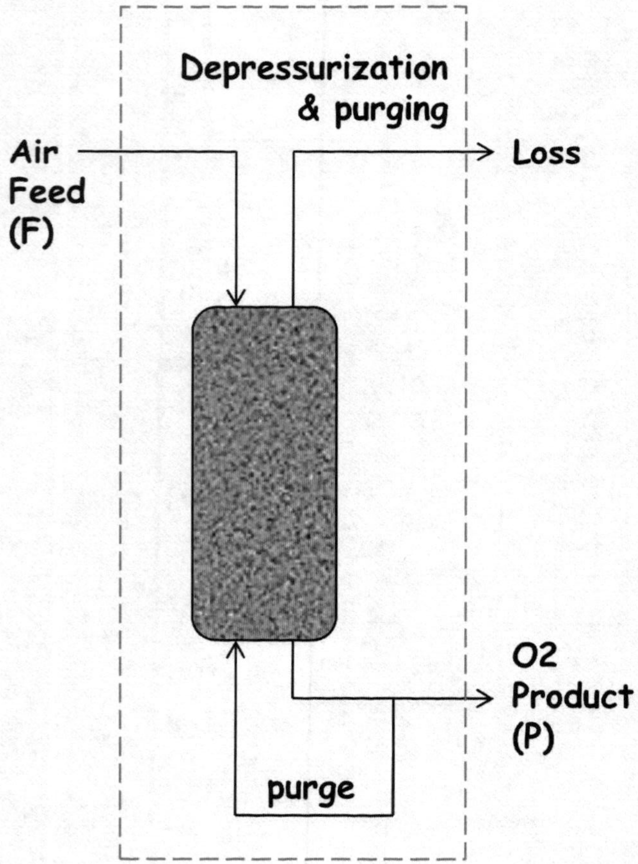

Simplified Skarstrom Cycle	Energize PLC
1. Counter-current depressurization	Y3
2. Counter-current purging using product	Y1, Y2, Y3
3. Co-current pressurization using air	Y0
4. Co-current adsorption & production	Y0

Variables:

- Depressurization time
- Purging time, flow rate
- Pressurization time, flow rate
- Adsorption time, flow rate

Performance Index:

O2 Recovery = $y_p(P)/0.21(F)$

Productivity = $y_p(P)$/g zeolite/hr

Beds of Different Sizes

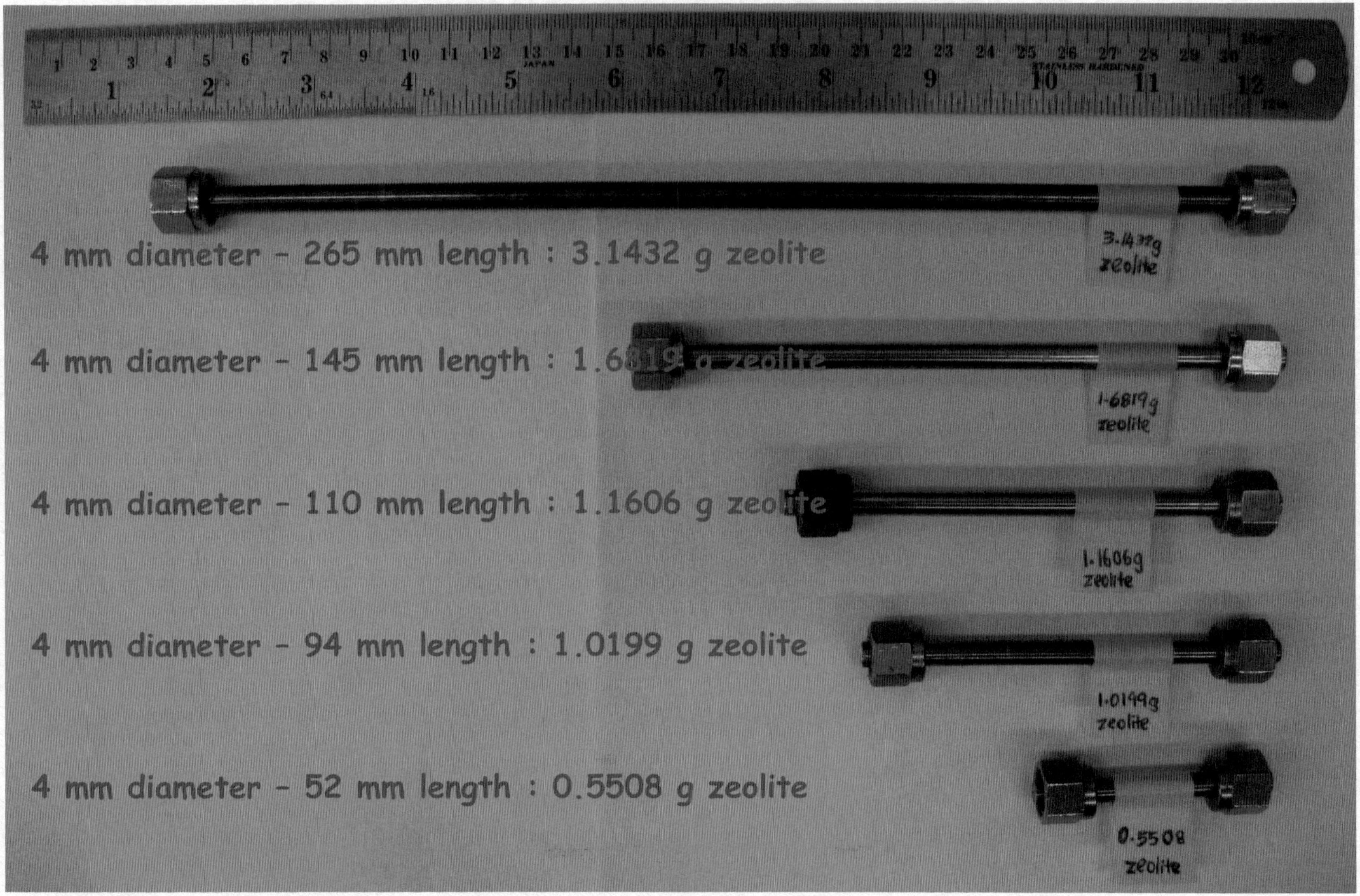

Is PSA possible in long capillary tube?

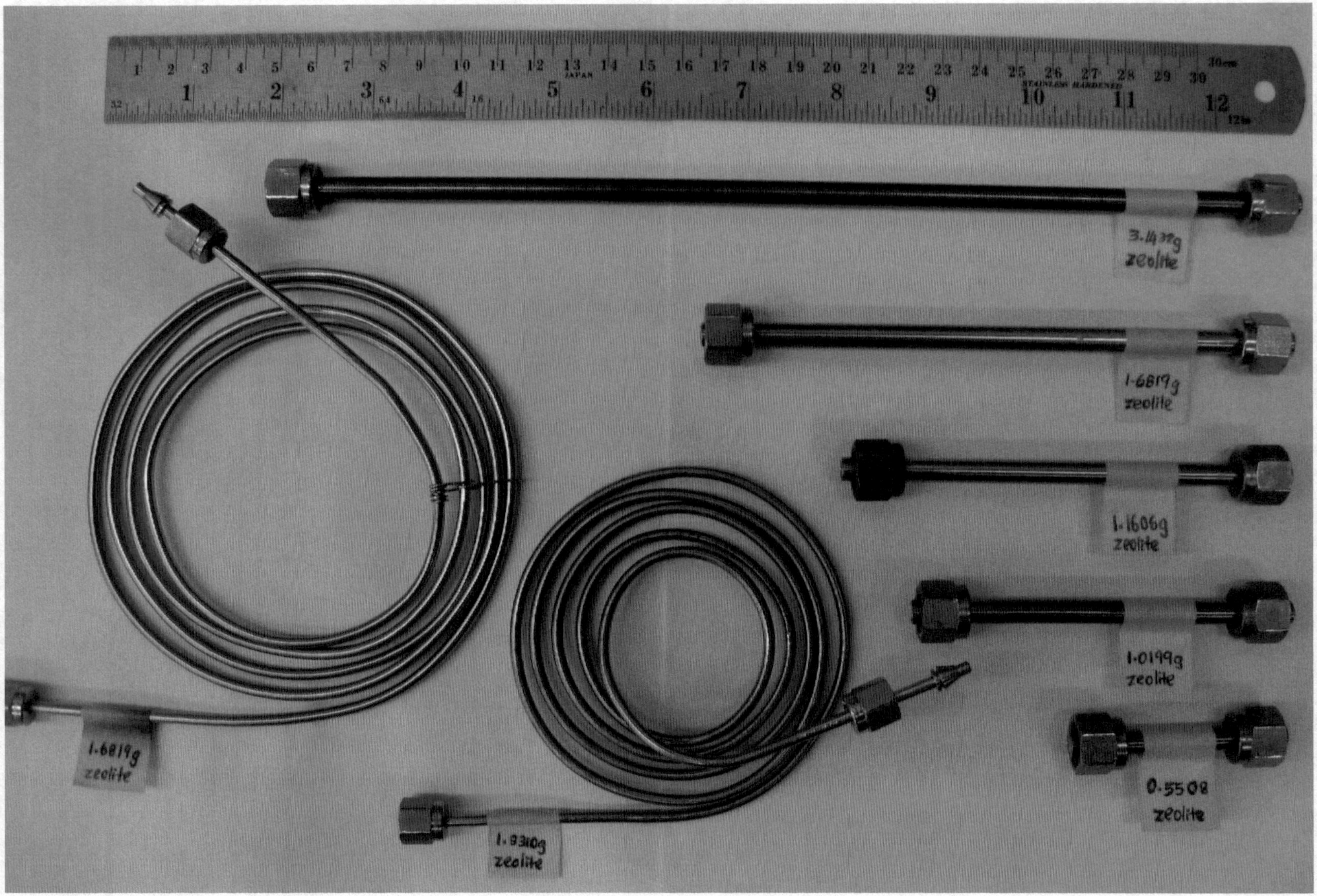

Characterization

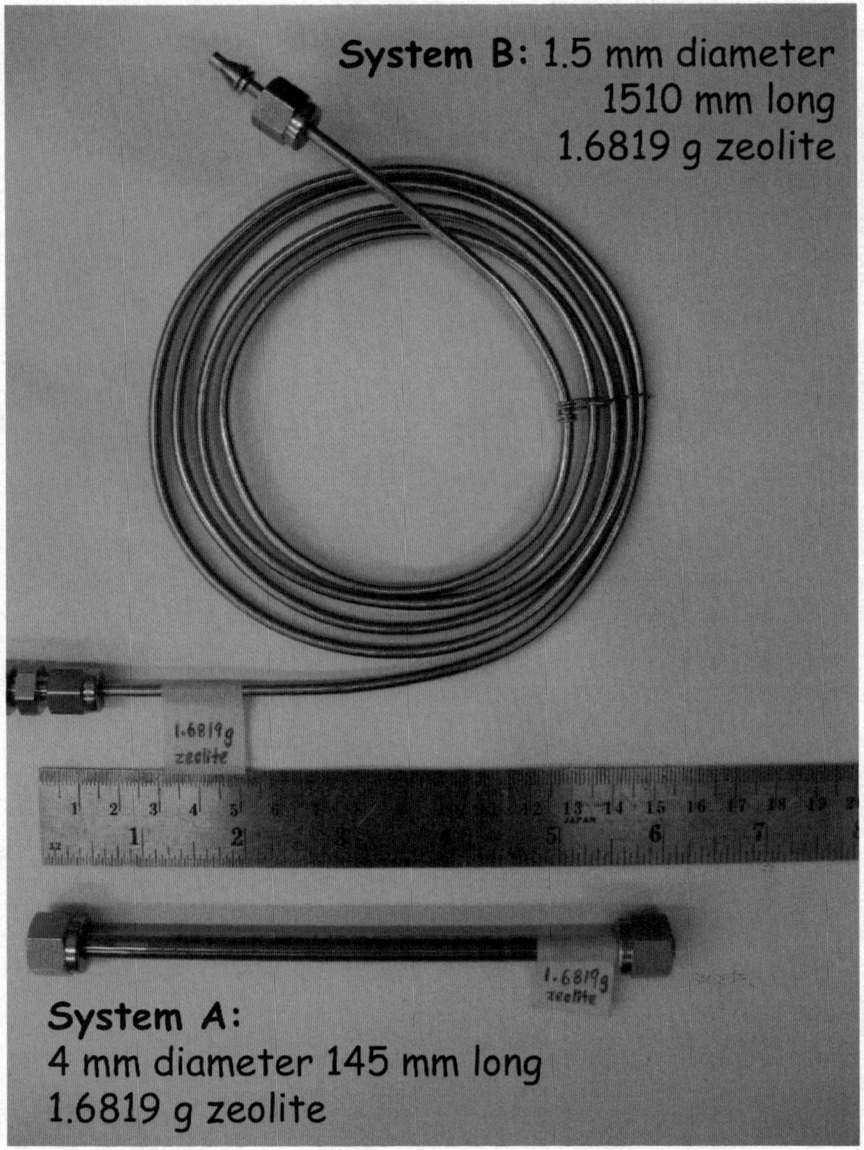

System A and System B contain the same amount of zeolite but in different shapes

Find suitable operational parameters:

1. Depressurization time
2. Purge flow rate
3. Feed flow rate

Compare performances:

1. Product purity
2. O2 recovery
3. Productivity

General Methodology for Characterization

Determine the capacity of the bed

- Purge the bed with product oxygen for long time
- Carry out adsorption step (pressurization and production) for long time
- Observe the breakthrough time of the bed

Set appropriate step times

- Set pressurization and adsorption time shorter than the breakthrough time
- Set a depressurization time long enough to desorb N2 from zeolite (by observing the blown down gas approaching zero flow from the bed)
- Set a purging time a little longer than it needs to purge clean the bed (by gradually reducing purge time but getting the same performance)

Determine the adsorption capability of the bed at the highest pressure

- Tune the back pressure regulator to the highest pressure which is sustainable by all upstream equipment
- To reduce cycle time, use the highest flow rates of air feed and purge gas to carry out PSA cycle
- Slowly reduce purging time to increase oxygen recovery and productivity

Optimize cycle time, compression cost and production

- With the assurance of getting positive adsorption results from the zeolite bed at the highest pressure, now perform optimization
- Carry out PSA cycle at lower adsorption pressure, which shortens the pressurization time and saves compression cost
- Continue to reduce purging time and play with purging flow rate to increase oxygen recovery and productivity

Overall results of System A (16 - 20 September 2008)

Depressurization time (sec)	Purging time (sec)	Pressurization and Adsorption time (sec)	Production time (sec)	Maximum		Average		O2 recovery (%)	Productivity (μmole O2 product/g zeolite/cycle)	Productivity (μmole O2 product/g zeolite/hr)
				mL/min	mole %	mL/min	mole %			
47- 46.6 psi	30 mL		**choosing the appropriate adsorption time**							
4	15	100	24	17.4	94.6	11.1	91.5	-26.02	-75.11	-2272.15
4	15	105	28	17.8	94.1	12.2	92.1	-14.47	-43.85	-1273.09
4	15	108	30	20.3	93.9	13.5	92.1	-5.35	-16.68	-472.82
47- 46.6 psi	30 mL/min	30 mL/min	**4 sec depressurization time yields the best results**							
4	15	102	26	18.3	94.6	11.9	91.4	-19.17	-56.46	-1679.81
4	13	102	26	18.0	94.6	11.6	92.0	-11.45	-33.72	-1020.18
4	11	102	26	17.9	94.2	11.7	91.6	-2.65	-7.81	-240.46
4	10	102	26	17.8	93.8	11.8	91.5	2.01	5.92	183.72
4	**9**	**102**	**26**	**21.1**	**92.3**	**13.1**	**90.1**	**10.61**	**31.24**	**977.88**
4	*8*	*102*	*exhausted*							
47- 46.6 psi	30 mL/min		**6 sec depressurization time causes too much product loss**							
6	15	102	26	17.5	94.5	11.0	91.5	-22.66	-66.74	-1953.31
6	13	102	26	17.9	94.6	11.1	90.5	-13.65	-40.21	-1196.26
6	11	102	26	18.1	94.5	11.0	91.3	-5.11	-15.06	-455.48
6	9	102	26	19.0	93.5	11.6	90.7	5.67	16.71	514.12
6	*8*	*102*	*exhausted*							
47- 46.6 psi		30 mL/min	**2 sec depressurization time causes incomplete N2 desorption**							
2	15	*102*	*24*	*17.9*	*94.6*	11.7	92.3	*-23.06*	*-67.90*	*-2054.04*
2	13	102	24	17.7	94.7	11.5	92.2	-15.17	-44.66	-1374.12
2	11	102	24	18.0	94.4	11.7	92.0	-5.89	-17.33	-542.59
2	9	102	26	21.1	92.6	12.5	90.3	8.73	25.72	819.41
2	*8*	*102*	*exhausted*							
47- 46.6 psi	**30, 20, 10 mL/min**	30 mL/min	**comparing the performances at different purging flow rates**							
4	**9**	**102**	**26**	**21.1**	**92.3**	**13.1**	**90.1**	**10.61**	**31.24**	**977.88**
4	**13.5**	102	26	19.3	93.7	11.6	89.3	6.14	18.09	544.96
4	**27**	102	24	17.6	94.1	10.5	88.2	1.42	4.18	113.11

System A: Results (I)

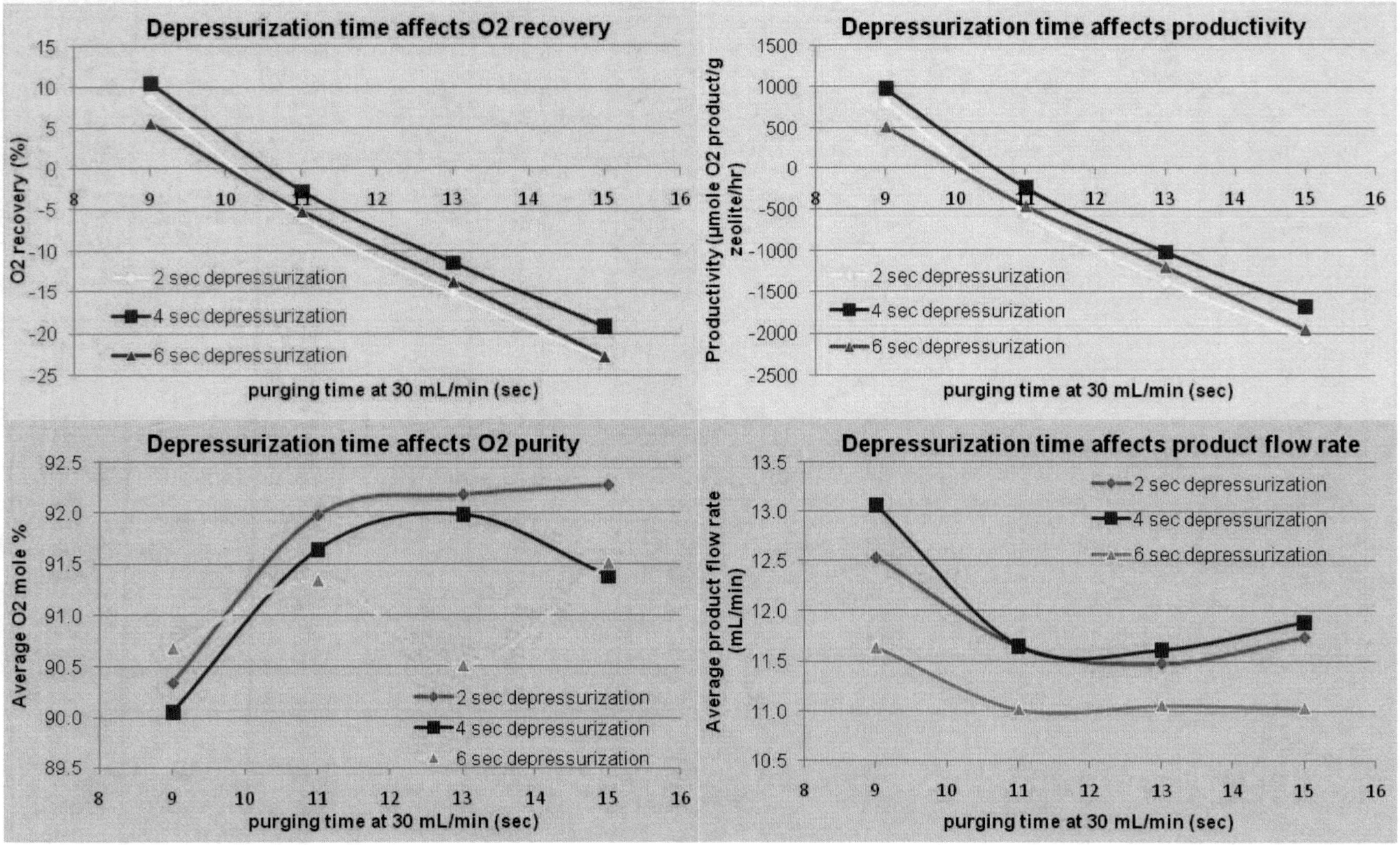

System A: Results (II)

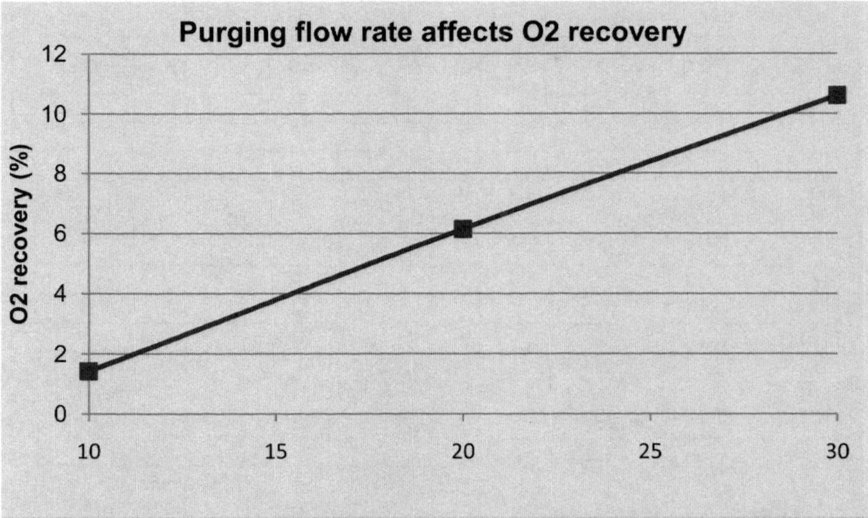

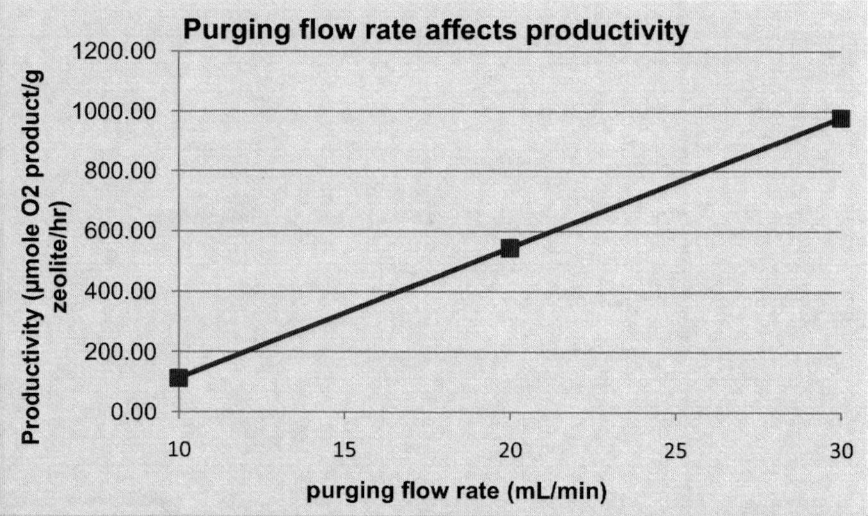

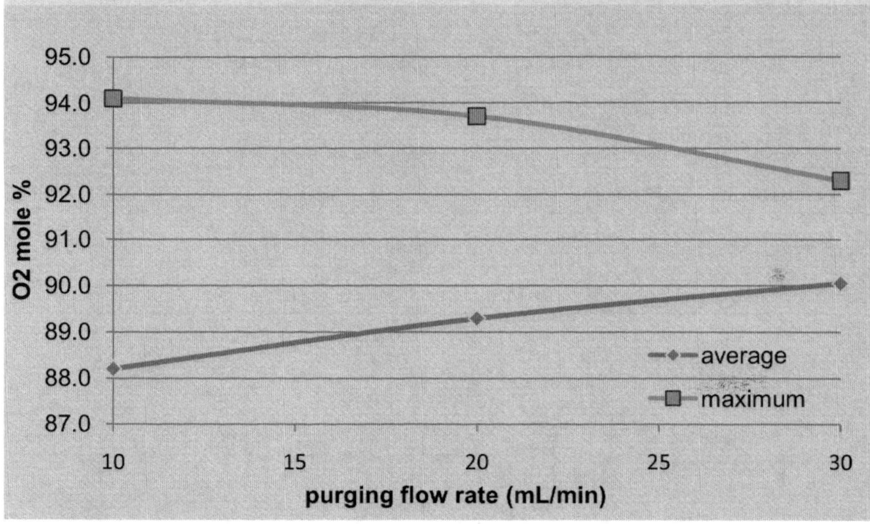

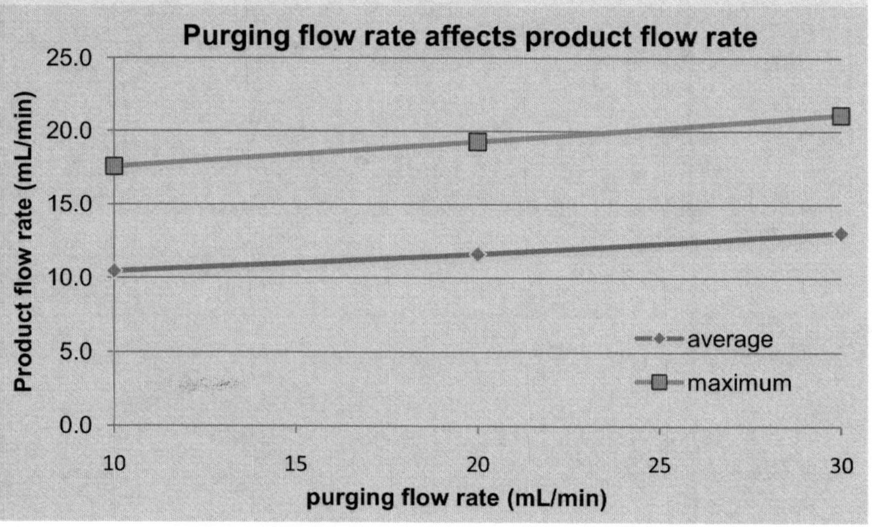

Overall results of System A (27 - 28 September 2008)

Depressurization time (sec)	Purging time (sec)	Pressurization and Adsorption time (sec)	Production time (sec)	Maximum		Average		O2 recovery (%)	Productivity (μmole O2 product/g zeolite/cycle)	Productivity (μmole O2 product/g zeolite/hr)
				mL/min	mole %	mL/min	mole %			
46.2 - 45.8 psi	30 mL/min	**30 mL/min**	**Low feed flow rate yields to slow production rate (low productivity/hr), but experience more complete separation (high O2 recovery and productivity/cycle)**							
4	13	122	36	15.3	94.0	10.2	90.8	2.22	7.49	194.00
4	11	122	36	15.3	93.7	10.7	91.0	11.40	38.44	1010.14
4	*9.5*	*122*	*38*	*21.5*	*90.2*	*13.1*	*88.1*	*28.26*	*95.26*	*2530.88*
4	**9**	**122**	**38**	**17.1**	**91.7**	**11.6**	**89.3**	**23.91**	**80.59**	**2149.03**
4	*8*	*122*	*exhausted*							
46.2 - 45.8 psi	30 mL/min	**40 mL/min**	**this feed flow rate is not too slow to produce product and is not too fast that causing less efficient separation, but the purity is on the verge of exhaustion**							
4	13	91.5	28	19.4	94.3	13.2	91.5	3.36	11.33	375.96
4	11	91.5	30	19.5	93.9	13.3	90.3	13.72	46.33	1565.99
4	*9.5*	*91.5*	*30*	*25.8*	*90.5*	*15.5*	*87.4*	*25.24*	*85.22*	*2921.86*
4	*9*	*91.5*	*exhausted*							
4	13	91.5	30	19	94.2	12.5	91.2	3.78	12.76	423.37
4	11	91.5	30	19.2	94	13.2	91.7	13.53	45.67	1543.77
4	*9.5*	*91.5*	*30*	*27.2*	*90.1*	*15.6*	*88.1*	*25.28*	*85.33*	*2925.75*
4	*9*	*91.5*	*exhausted*							
46.2 - 45.8 psi	30 mL/min	**50 mL/min**	**High feed flow rate yields to fast production rate (high productivity/hr), but experience less complete separation (low O2 recovery and productivity/cycle)**							
4	13	73.2	26	22.6	94.4	13.0	90.7	-0.35	-1.16	-46.49
4	11	73.2	26	22.9	94.2	14.0	91.1	9.71	32.76	1336.95
4	**9.5**	**73.2**	**26**	**25.9**	**92.4**	**15.8**	**89.9**	**20.16**	**67.97**	**2822.16**
4	*9.5*	*73.2*	*26*	*29.5*	*90.4*	*16.3*	*86.5*	*20.57*	*69.35*	*2879.55*
4	*9*	*73.2*	*exhausted*							
46.2 - 45.8 psi	**20 mL/min**	**30,40,50 mL/min**	**Low purging flow rate yields to lower O2 recovery and productivity**							
4	14.3	122	38	16.8	92.5	11.8	90.4	23.59	79.52	2040.41
4	14.3	91.5	30	19.8	93.2	13.3	90.2	18.11	61.13	2004.36
4	14.3	73.2	26	22.9	93.4	14.5	90.5	15.95	53.76	2115.27
46.2 - 45.8 psi	**30 mL/min**	**30,40,50 mL/min**	**High purging flow rate yields to higher O2 recovery and productivity**							
4	*9.5*	*122*	*38*	*21.5*	*90.2*	*13.1*	*88.1*	*28.26*	*95.26*	*2530.88*
4	*9.5*	*91.5*	*30*	*25.8*	*90.5*	*15.5*	*87.4*	*25.24*	*85.22*	*2921.86*
4	*9.5*	*73.2*	*26*	*29.5*	*90.4*	*16.3*	*86.5*	*20.57*	*69.35*	*2879.55*

System A: Results (III)

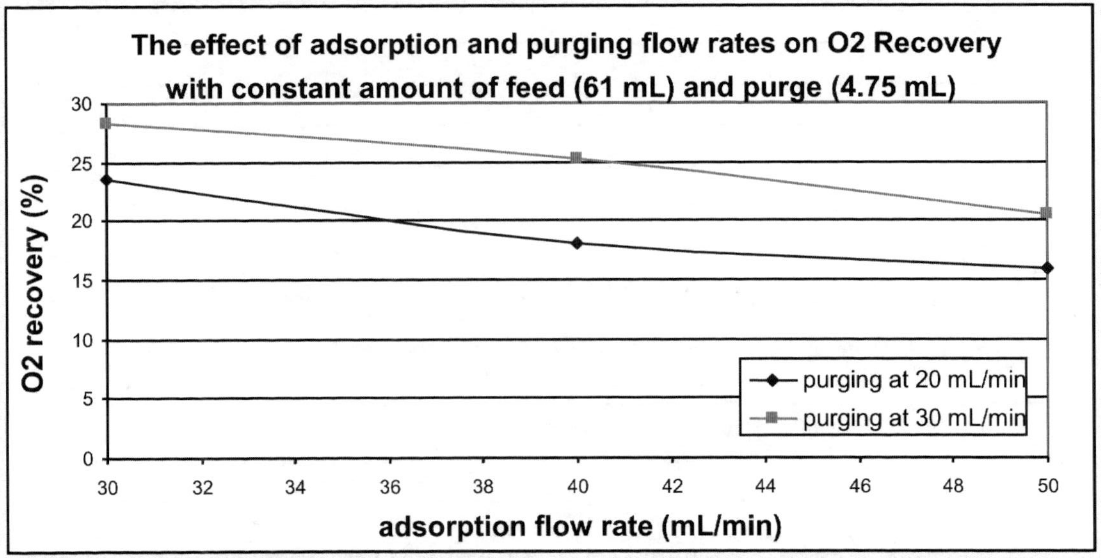

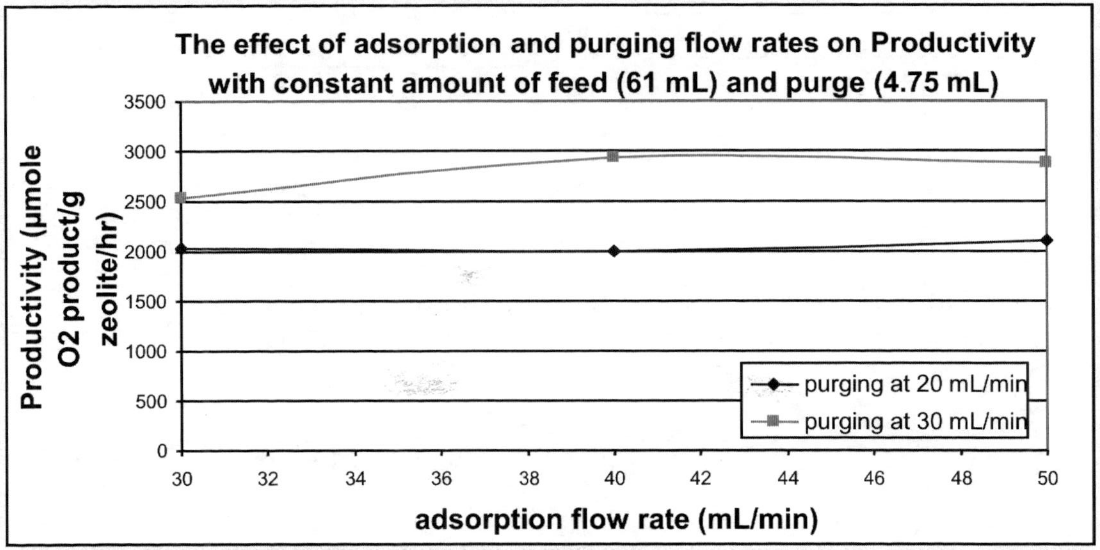

System A: Results (IV)

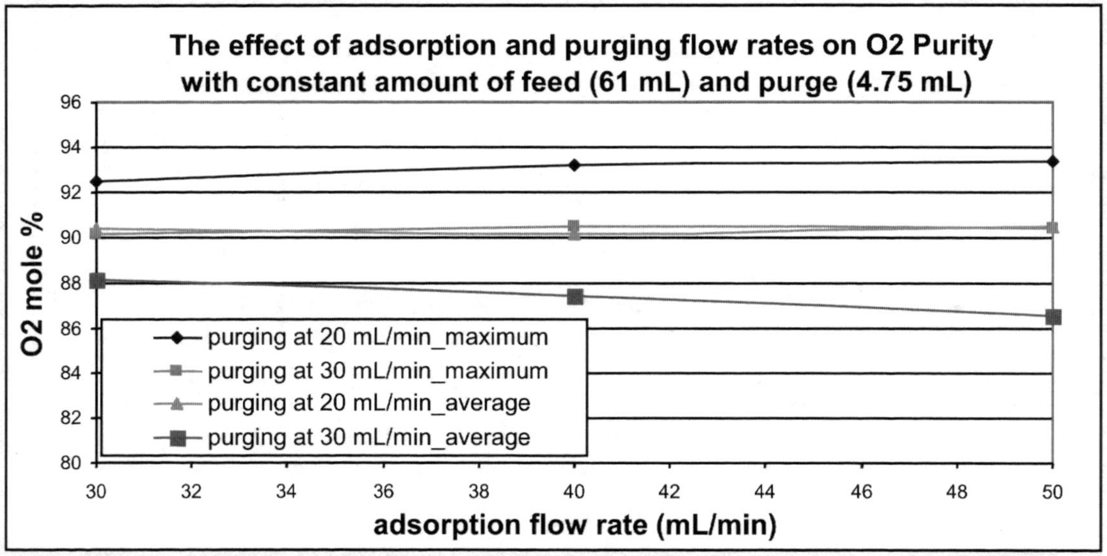

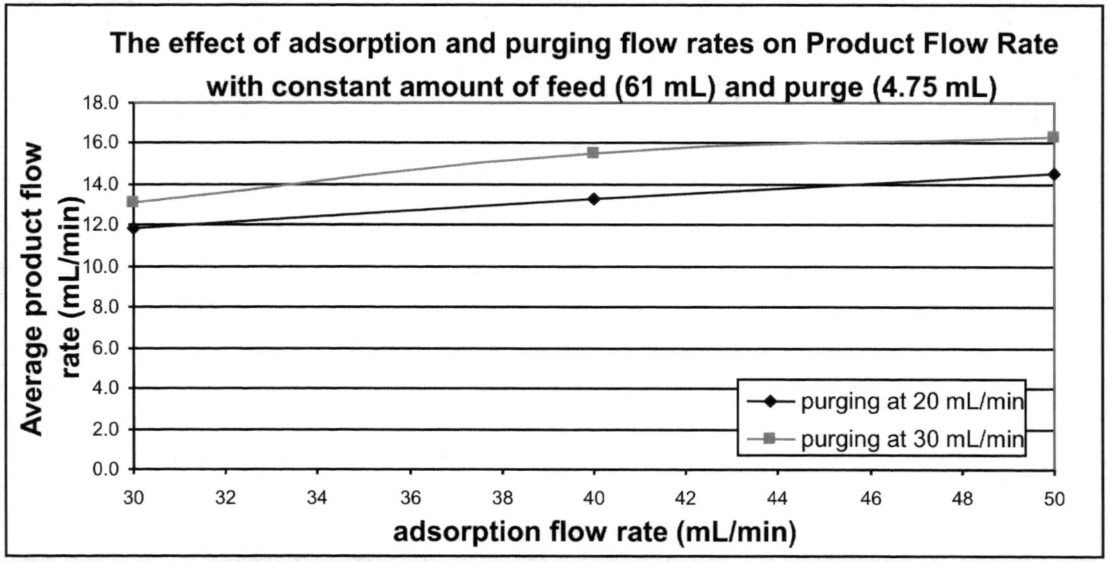

More Stringent Methodology to Characterization System B

1. **Start with the highest pressure, highest feed and purge flow rates**

- 45 psi
- 30 mL/min air feed
- 30 mL/min purge flow
- 15 sec depressurization

Results:

- Cannot produce positive adsorption results
- Gradual narrowing of production band

System B: Results (I)

2. **Reduce feed flow rate**

- 45 psi
- **15 mL/min air feed**
- 30 mL/min purge flow
- 15 sec depressurization

Results:

- The problem of narrowing production band is eliminated

3. Is further reduction of feed flow rate better?

- 45 psi
- **10 mL/min air feed**
- 30 mL/min purge flow
- 15 sec depressurization

Results:

- Improved O2 recovery and productivity
- Considering mass spectrometer sensitivity, 10 mL/min is the minimum feed flow rate suitable for the PSA system

System B: Results (II)

4. Reduce purge flow rate

- 45 psi
- 10 mL/min air feed
- **10 mL/min purge flow**
- 15 sec depressurization

Results:

- Improved O2 recovery and productivity
- Obtain positive adsorption results

System B: Results (III)

5. Is further reduction of purge flow rate better?

- 45 psi
- 10 mL/min air feed
- **5 mL/min purge flow**
- 15 sec depressurization

Results:

- Improved O2 recovery and productivity
- Obtain the best adsorption results for small diameter spiral bed

System B: Results (III)

6. Reduce depressurization time

- 45 psi
- 10 mL/min air feed
- 10 mL/min purge flow
- **5 sec depressurization**

Results:

- Improved O2 recovery and productivity because of lesser product loss
- But cannot give positive adsorption results due to incomplete desorption of nitrogen gas from the bed

System B: Results (IV)

System B: Results (V)

7. Can higher purge flow rate help to clean the bed better?

- 45 psi
- 10 mL/min air feed
- **20 mL/min and 30 mL/min purge flow**
- 5 sec depressurization

Results:

- The bed is not properly cleaned at high purge flow rate and is exhausted

System B: Results (I)

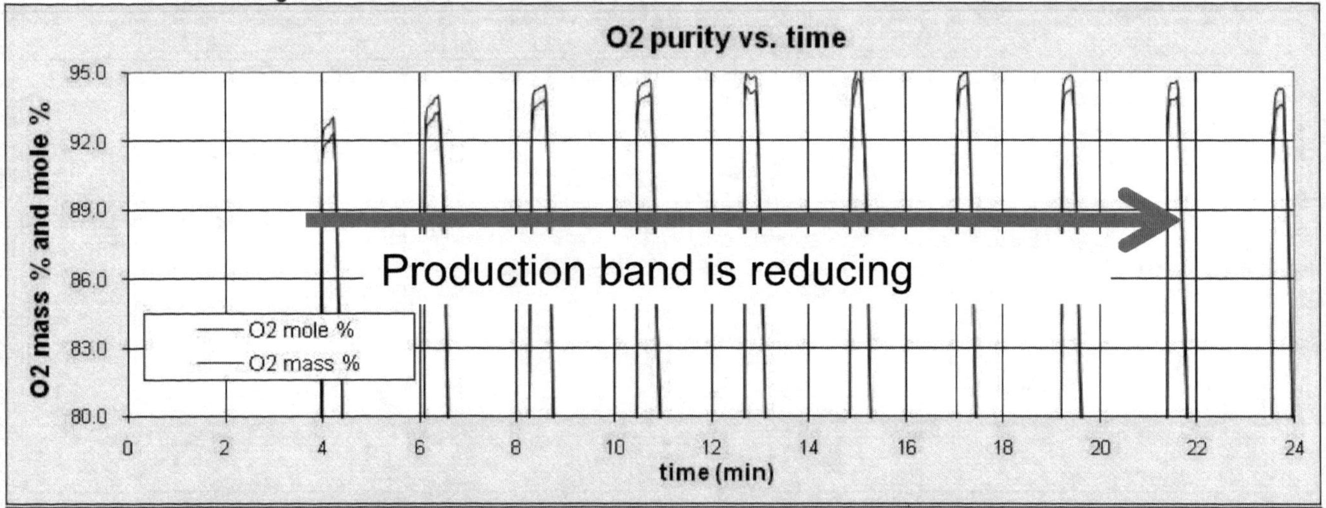

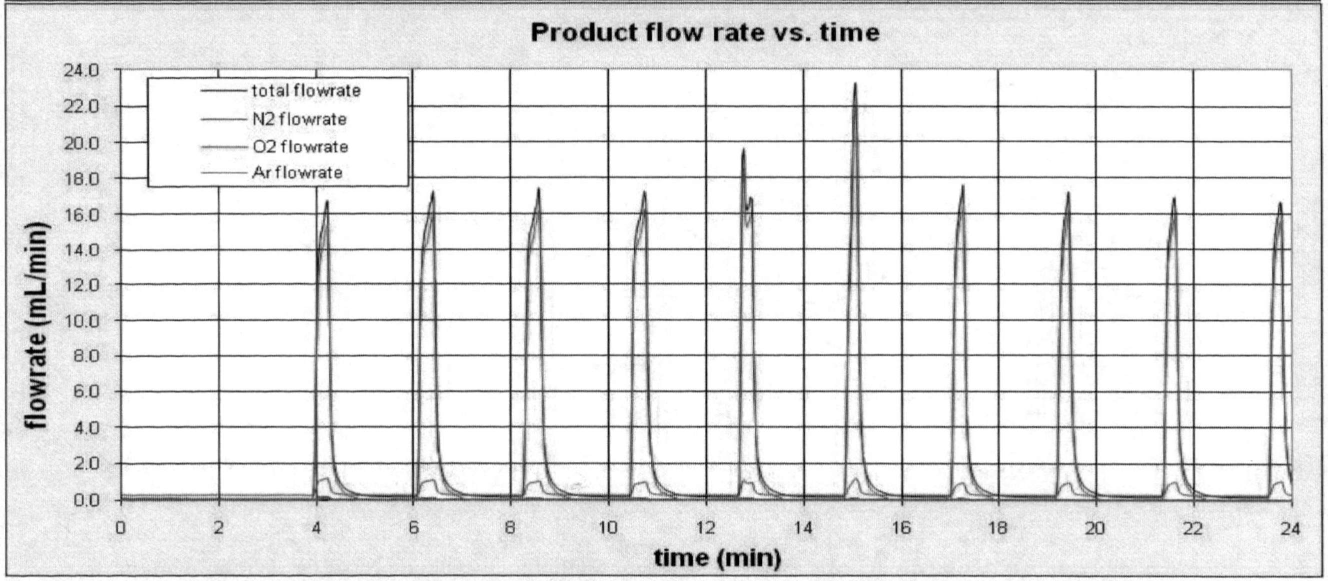

High flow rate of feed air is not encouraged

System B: Results (I)

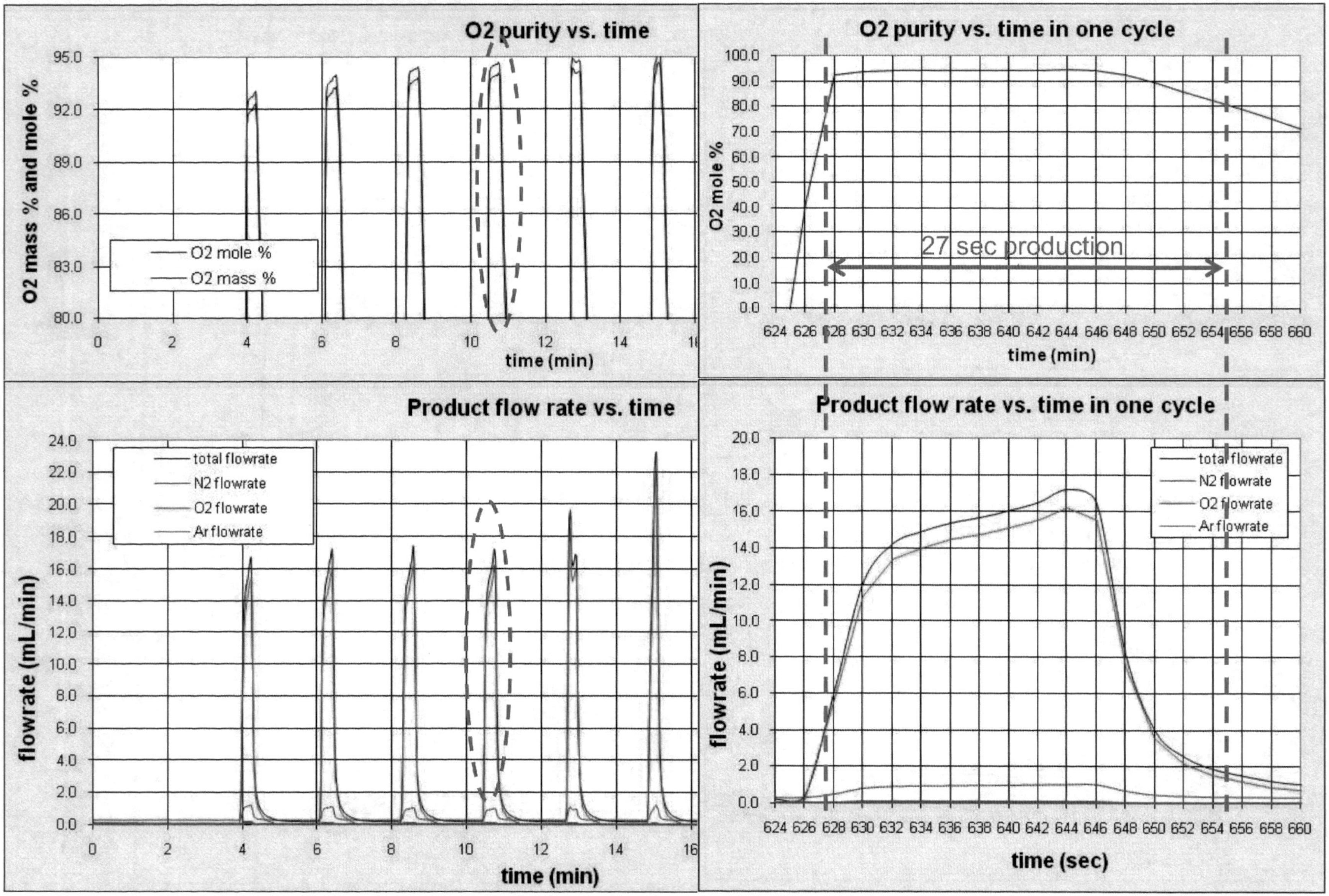

System B: Results (I)

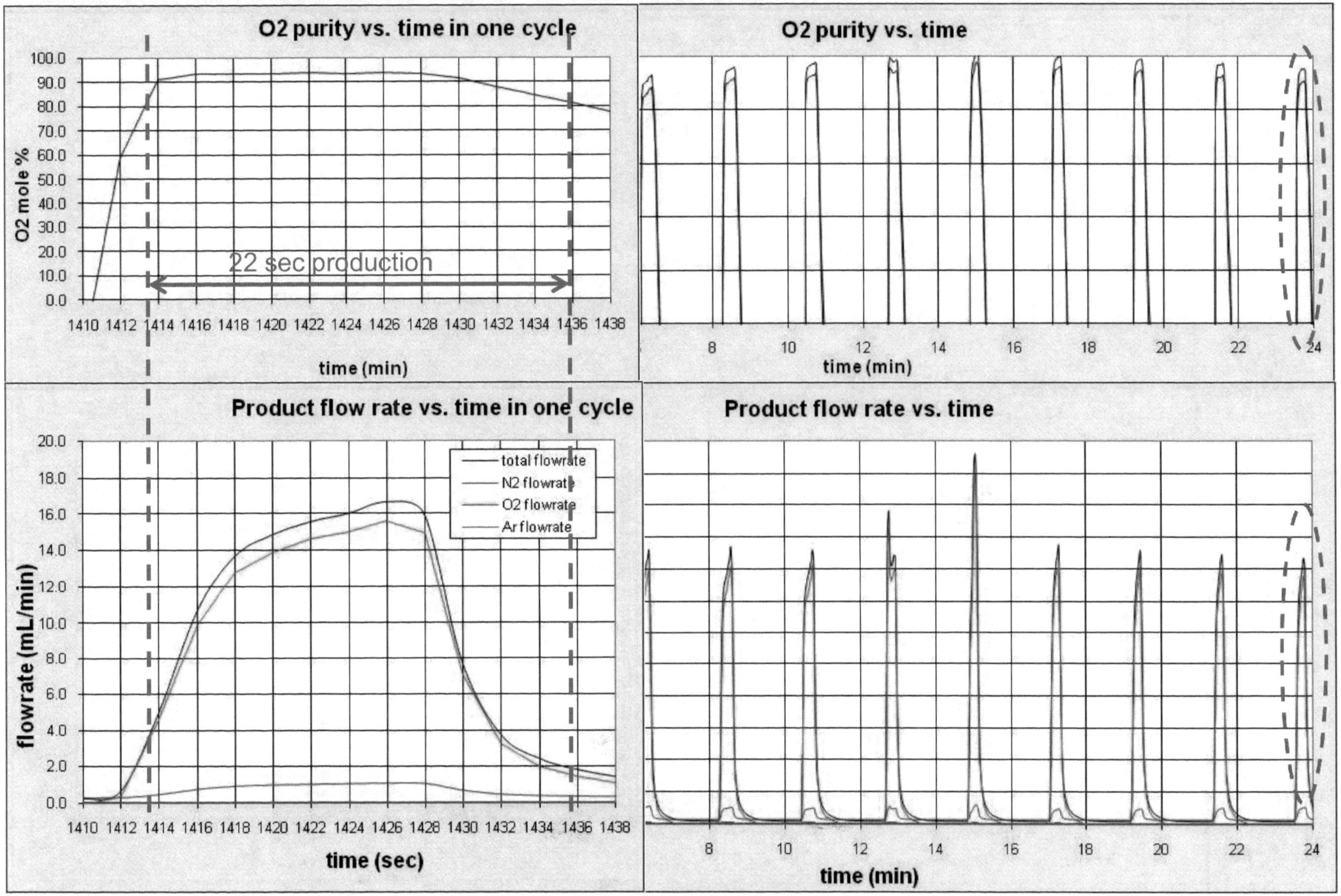

System B: Results (II)

High flow rate of feed air is not encouraged

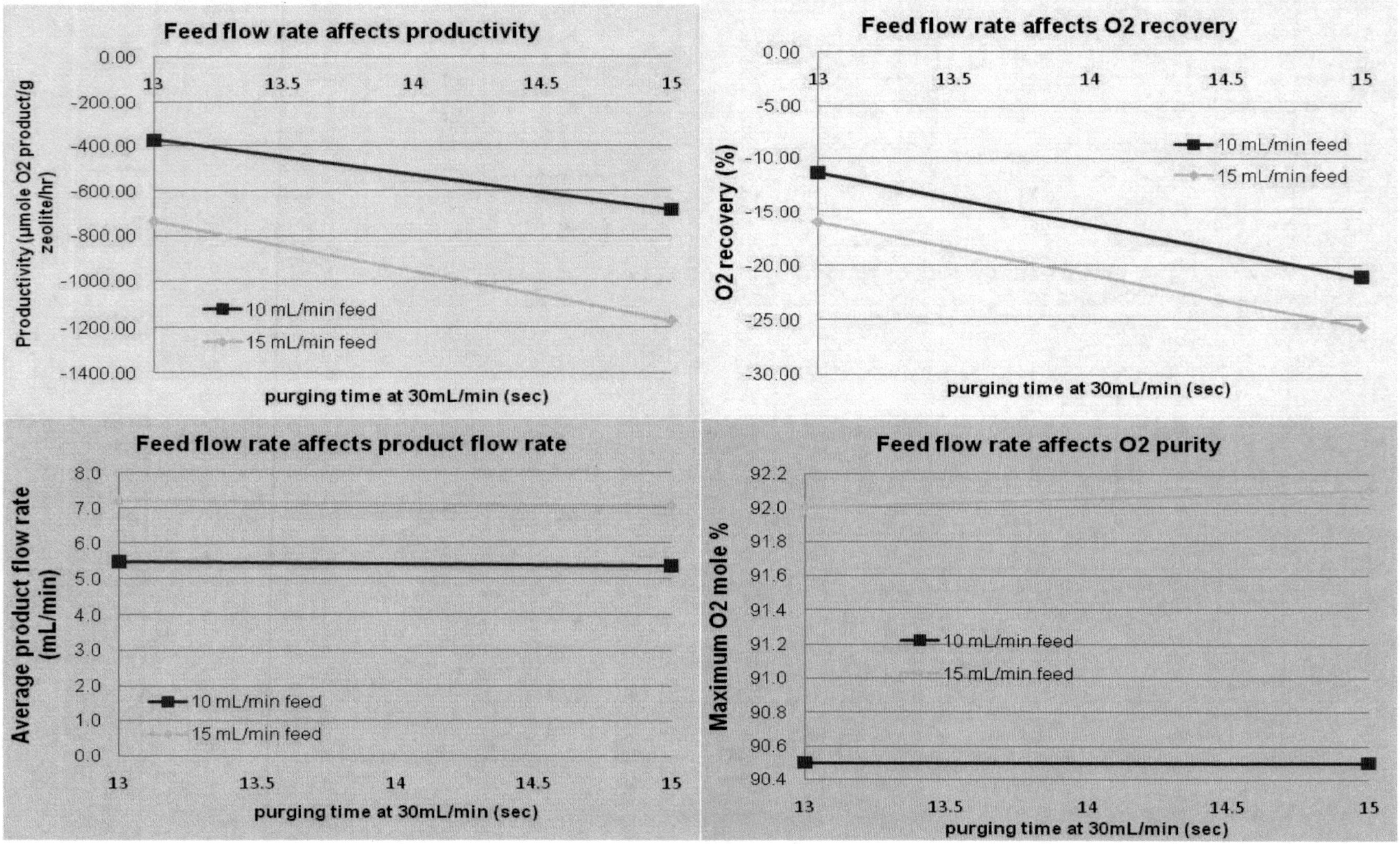

System B: Results (III)

High flow rate of purge gas is not encouraged

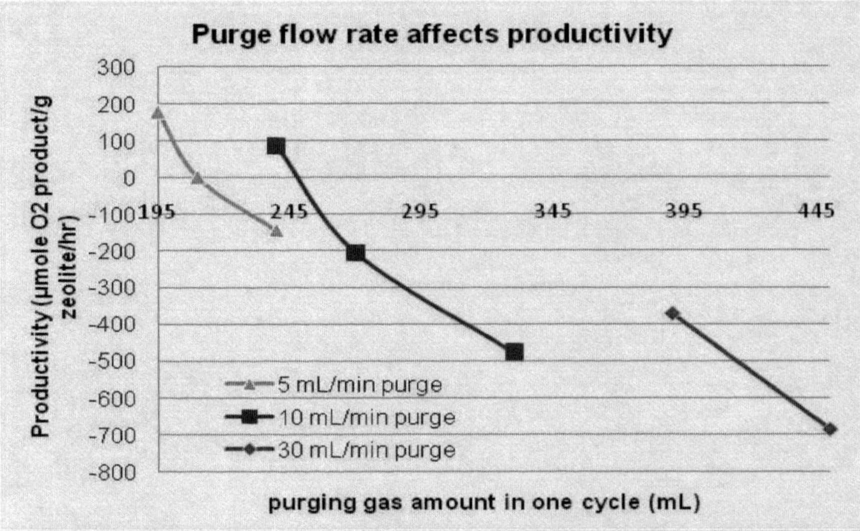

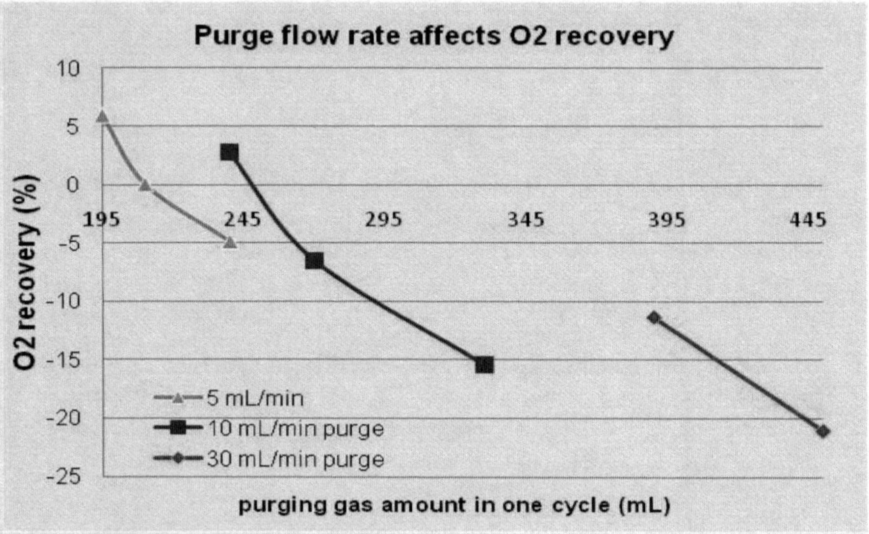

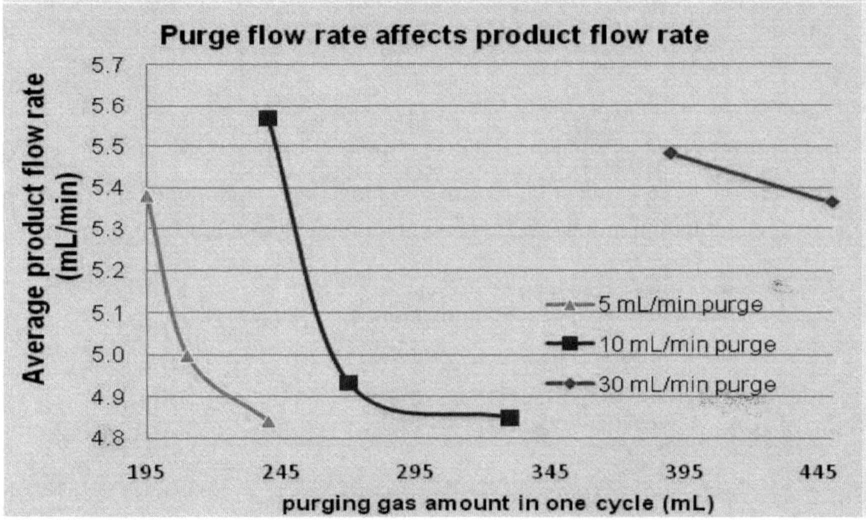

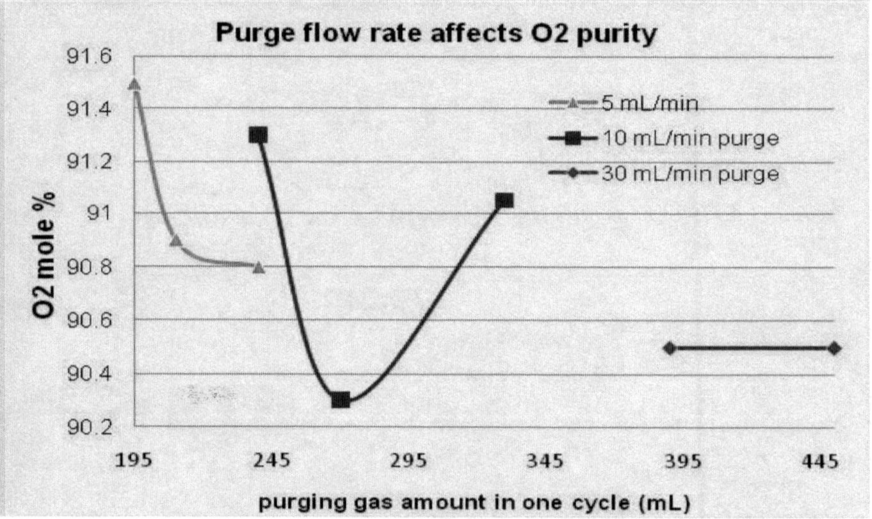

System B: Results (IV)

Short depressurization time is not encouraged

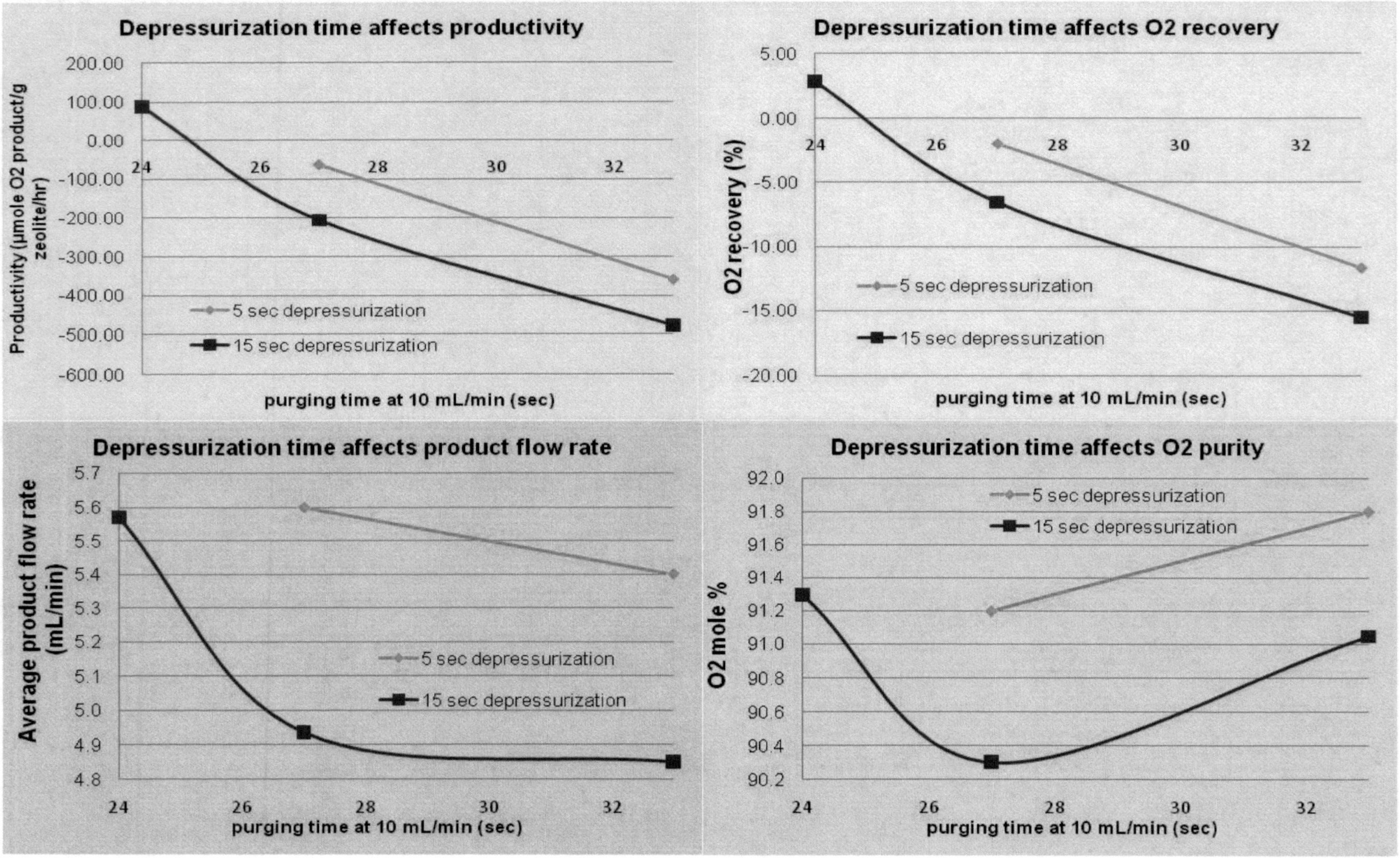

System B: Results (V)

<u>27 sec purging at 10 mL/min:</u>
With low flow rate of purging, the bed is sustained for long time.

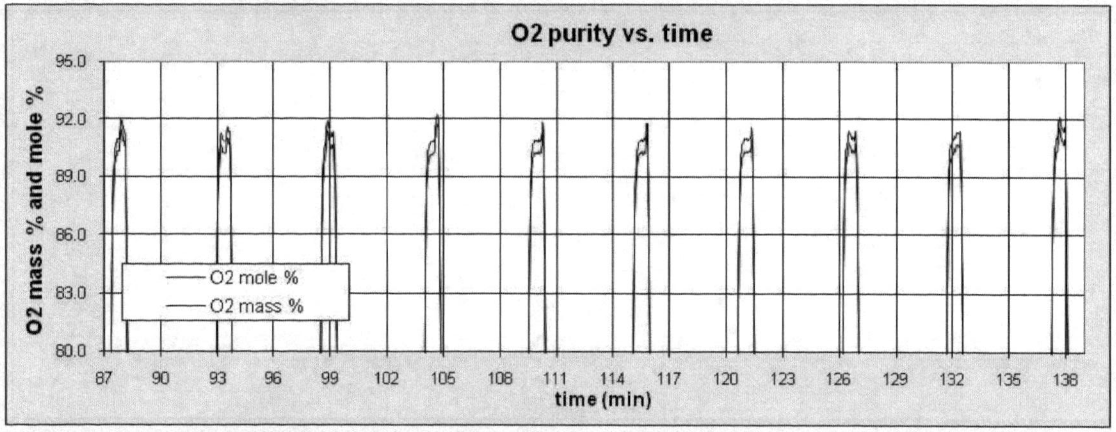

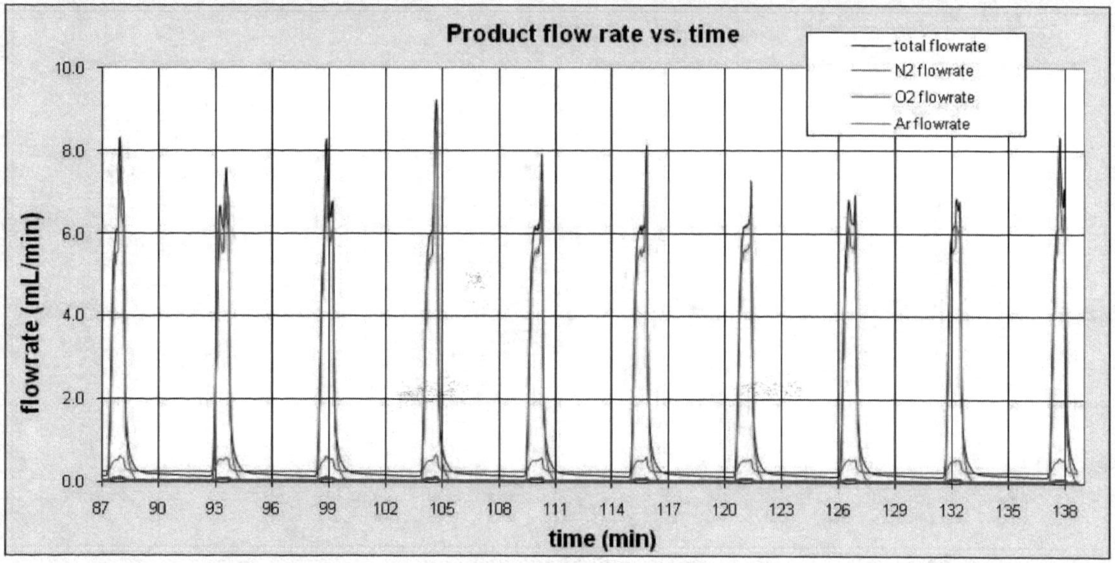

System B: Results (V)

<u>13.5 sec purging at 20 mL/min:</u>

With higher flow rate of purging, the bed is exhausted in 3 cycles.

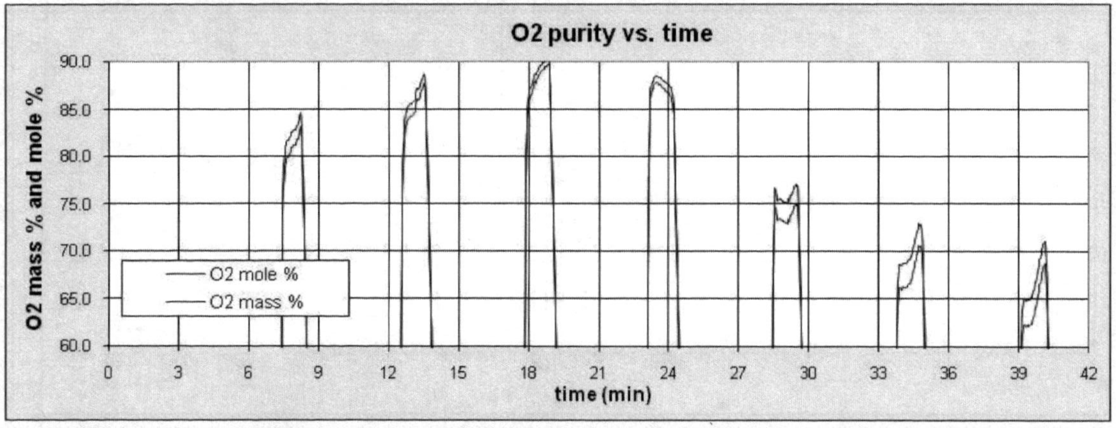

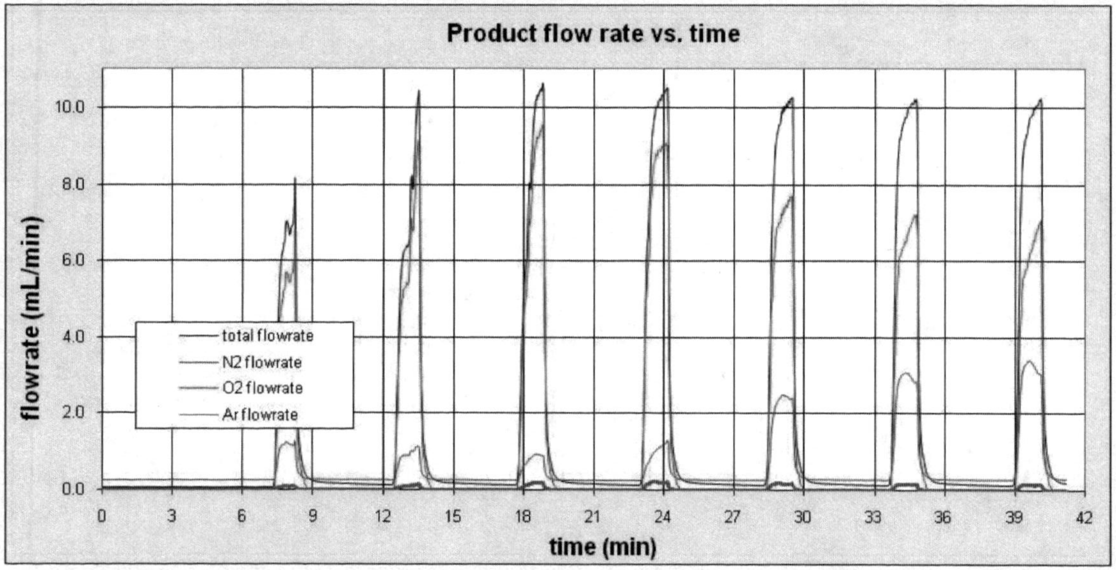

System B: Results (V)

<u>9 sec purging at 30 mL/min:</u>
With highest flow rate of purging, the bed is exhausted in 2 cycles.

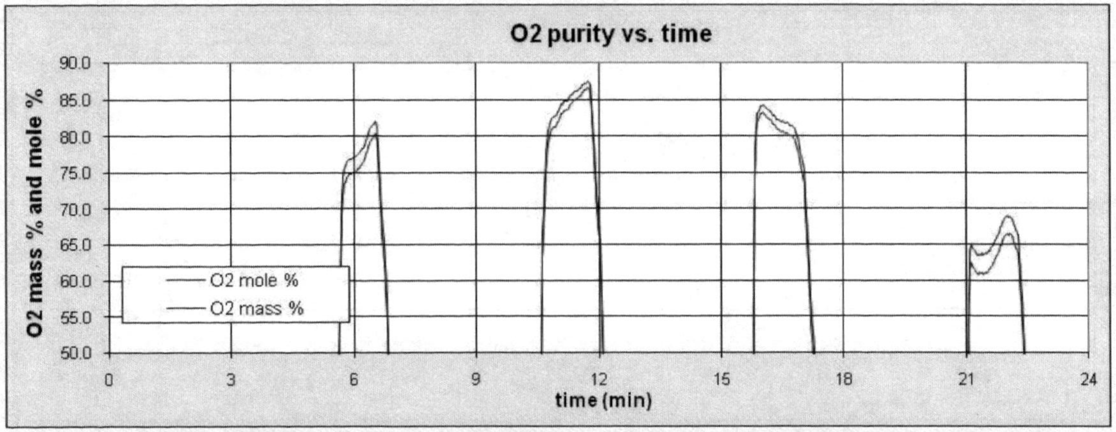

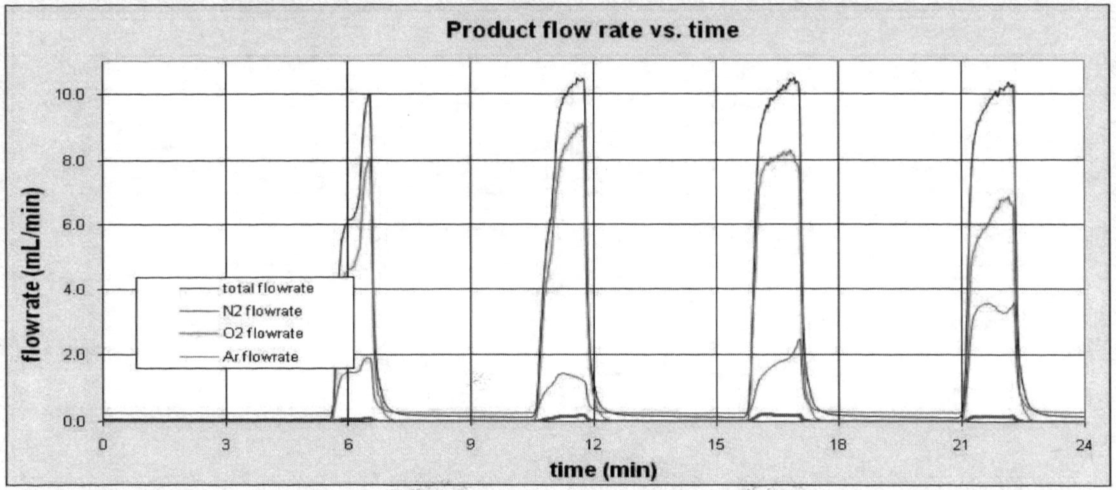

High flow rate of purge gas is not encouraged

Overall results of System B (17 - 22 September 2008)

amount of purging gas (mL)	Depressurization time (sec)	Purging time (sec)	Pressurization and Adsorption time (sec)	Production time (sec)	Maximum mL/min	Maximum mole %	Average mL/min	Average mole %	O2 recovery (%)	Productivity (µmole O2 product/g zeolite/cycle)	Productivity (µmole O2 product/g zeolite/hr)
	46 - 45 psi	30 mL/min	30 mL/min	No matter how high the adsorption pressure is, spiral bed with small diameter cannot produce positive results. Moreover, the bed cannot sustain high feed flow rate (production band is narrowed)							
	15	15	100								
	46 - 45 psi	30 mL/min	**15 mL/min**	**eliminate the narrowed production band problem : lower feed flow rate is better**							
450	15	15	200	40	9.8	92.1	7.1	89.9	-25.64	-74.88	-1172.03
390	15	13	200	40	9.7	92.0	7.2	90.1	-16.03	-46.80	-739.00
330	*15*	*11*	*200*	*exhausted*							
	46 - 45 psi	30 mL/min	**10 mL/min**	**further lower feed flow rate is better**							
450	15	15	300	60	6.6	90.5	5.4	88.7	-21.08	-62.82	-685.33
390	15	13	300	60	6.6	90.5	5.5	88.9	-11.39	-33.93	-372.41
330	*15*	*11*	*300*	*exhausted*							
	46 - 45 psi	**10 mL/min**	10 mL/min	**lower purge flow rate is better**							
330	15	33	300	48	6.4	91.05	4.9	88.4	-15.48	-46.15	-477.33
270	15	27	300	48	6.3	90.3	4.9	88.1	-6.53	-19.45	-204.72
240	*15*	*24*	*300*	*50*	*9.3*	*91.3*	*5.6*	*88.5*	*2.82*	*8.4*	*89.15*
210	*15*	*21*	*300*	*exhausted*							
	46 - 45 psi	**5 mL/min**	10 mL/min	**further lower purge flow rate is better**							
240	15	48 (24)	300	42	6.4	90.8	4.8	88.4	-4.84	-14.72	-142.98
210	**15**	**42 (21)**	**300**	**42**	**7.8**	**90.9**	**5.0**	**88.0**	**0.07**	**0.21**	**2.16**
195	*15*	*39 (19.5)*	*300*	*44*	*9.4*	*91.5*	*5.4*	*88.3*	*5.94*	*17.72*	*180.16*
180	*15*	*36 (18)*	*300*	*exhausted*							
	46 - 45 psi	**10,20,30**	10 mL/min	**high purge flow rate is not good (cannot clean the bed as 10mL/min can do)**							
270	5	27	300	48	6.8	90.7	5.6	89.2	-3.19	-9.50	-100.03
270	*5*	*13.5*	*300*	*exhausted*							
270	*5*	*9*	*300*	*exhausted*							
	46 - 45 psi	10 mL/min	10 mL/min	**shorter depressurization time has lower product loss, thus it is better**							
330	**5**	33	300	48	6.4	91.3	5.4	89.5	-13.97	-41.64	-430.72
270	**5**	27	300	48	6.8	90.7	5.6	89.2	-3.19	-9.50	-100.03
240	*5*	*24*	*300*	*exhausted*							

Overall results of System B (24 - 26 September 2008)

amount of purging gas (mL)	Depressurization time (sec)	Purging time (sec)	Pressurization and Adsorption time (sec)	Production time (sec)	Maximum mL/min	Maximum mole %	Average mL/min	Average mole %	O2 recovery (%)	Productivity (μmole O2 product/g zeolite/cycle)	Productivity (μmole O2 product/g zeolite/hr)
	45.2 - 44.6 psi	10 mL/min	**30 mL/min**	**high feed flow rate**							
240	15	24	100	16	13.6	92.2	7.4	87.9	-14.90	-41.72	-1080.44
180	15	18	100	16	14	92.2	7.5	85.5	-6.26	-17.53	-474.52
150	*15*	*15*	*100*	*exhausted*							
120	*15*	*12*	*100*	*exhausted*							
	45.2 - 44.6 psi	10 mL/min	**20 mL/min**	**lower feed flow rate is better**							
330	15	33	150	22	10.8	92.4	6.5	88.7	-24.17	-68.36	-1242.86
300	15	30	150	22	10.8	92.3	6.7	89.0	-19.18	-54.25	-1001.59
240	15	24	150	24	11	92.4	6.5	88.1	-9.80	-27.71	-527.90
180	**15**	**18**	**150**	**24**	**11.8**	**91.8**	**6.9**	**87.9**	**0.19**	**0.54**	**10.55**
150	*15*	*15*	*150*	*exhausted*							
120	*15*	*12*	*150*	*exhausted*							
	45.2 - 44.6 psi	10 mL/min	**10 mL/min**	**further lower feed flow rate is better**							
240	15	24	300	38	6.00	90.30	4.31	87.70	-9.58	-27.78	-294.98
210	15	21	300	40	6.15	90.25	4.54	87.70	-2.96	-8.58	-91.93
195	**15**	**19.5**	**300**	**42**	**7.20**	**90.33**	**4.89**	**87.70**	**2.35**	**6.82**	**73.37**
180	*15*	*18*	*300*	*exhausted*							
	45.2 - 44.6 psi	**5 mL/min**	10 mL/min	**lower purge flow rate is NOT better**							
195	15	39	300	36	6.45	90.7	4.4	87.5	-4.26	-12.34	-125.50
165	15	33	300	36	6.3	90.1	4.5	87.5	0.51	1.48	15.26
150	**15**	**30**	**300**	**34**	**6.85**	**90.35**	**4.5**	**87.3**	**1.17**	**3.40**	**35.53**
135	*15*	*27*	*300*	*exhausted*							
	45.2 - 44.6 psi	5 mL/min	**5 mL/min**	**further lower feed flow rate is NOT better**							
180	*15*	*36*	*600*	*exhausted*							
150	*15*	*30*	*600*	*exhausted*							
120	*15*	*24*	*600*	*exhausted*							
	45.2 - 44.6 psi	10 mL/min	10 mL/min	**shorter depressurization time is NOT better**							
240	*5*	*24*	*300*	*exhausted*							
210	*5*	*21*	*300*	*exhausted*							

Overall results of System B (28 - 29 September 2008)

Depressurization time (sec)	Purging time (sec)	Pressurization and Adsorption time (sec)	Production time (sec)	Maximum		Average		O2 recovery (%)	Productivity (μmole O2 product/g zeolite/cycle)	Productivity (μmole O2 product/g zeolite/hr)
				mL/min	mole %	mL/min	mole %			
46.2 - 45.8 psi	10 mL/min	**10, 20, 30, 40 mL/min**	**Different feed flow rates: 10 mL/min is the best**							
15	24	360	68	5.9	89.8 + 2.5	5.0	88.2+ 2.5	10.94	38.28	345.41
15	24	360	68	6	89.8 + 2.5	5.0	88.2 + 2.5	10.98	38.40	346.51
15	24	180	34	10.8	91.7	7.4	89.1	2.56	8.64	142.04
15	24	180	34	10.7	91.8	7.5	89.2	2.72	9.18	150.83
15	24	120	26	15	92.4	9.3	89.1	1.97	6.54	148.19
15	24	120	26	15.2	92.3	9.3	89.0	1.81	6.00	135.93
15	24	90	20	18	92.9	10.1	89.2	-2.44	-8.09	-225.90
15	24	90	20	17.6	92.7	10.1	89.0	-2.66	-8.82	-246.03
46.2 - 45.8 psi	**5, 10, 20 mL/min**	10 mL/min	**Different purge flow rates: 10 mL/min is the best**							
15	48	360	50	5.5	89.8+ 2.5	4.4	87.8+ 2.5	-4.63	-16.20	-137.94
15	48	360	52	5.6		4.4	87.7+ 2.5	-3.60	-12.58	-107.07
15	24	360	68	5.9	89.8+ 2.5	5.0	88.2+ 2.5	10.94	38.28	345.41
15	24	360	68	6		5.0	88.2+ 2.5	10.98	38.40	346.51
15	*12*	*360*	*cannot clean the bed throroughly*							
46.2 - 45.8 psi	10 mL/min	10 mL/min	**Optimization by reducing purge amount (> 80 and > 70 mole %)**							
15	22	360	46	5.4		4.2		-3.50	-12.26	-111.14
15	22	360	54	5.4	89.3+ 2.5	3.7	85.0+ 2.5	-2.66	-9.31	-84.46
15	20	360	46	5.4	89.2+ 2.5	4.2	87.0+ 2.5	-1.16	-4.06	-36.99
15	20	360	54	5.4	89.2+ 2.5	3.8	84.9+ 2.5	-0.32	-1.10	-10.05
15	18	360	46	5.4	89.0+ 2.5	4.2	87.1+ 2.5	1.42	4.96	45.42
15	18	360	54	5.4	89.0+ 2.5	3.8	85.0+ 2.5	2.29	8.00	73.31
15	16	360	48	5.3	89.0+ 2.5	4.2	87.1+ 2.5	4.60	16.09	148.18
15	16	360	56	5.3	89.0+ 2.5	3.8	85.1+ 2.5	5.48	19.16	176.43
15	14	360	48	5.3	89.3+ 2.5	4.3	87.6+ 2.5	7.70	26.94	249.30
15	14	360	56	5.3	89.3+ 2.5	3.8	85.1+ 2.5	8.78	30.73	284.41
15	13	360	48	5.3		4.2		7.91	27.67	256.76
15	13	360	56	5.3	89.5+ 2.5	3.7	85.3+ 2.5	8.74	30.59	283.84
15	*12*	*360*	*exhausted*							
46.2 - 45.8 psi	10 mL/min	10 mL/min	**Average the results of optimization (in between 70 - 80 mole %)**							
15	22	360	50	5.4	89.3+ 2.5	4.0	86.1+ 2.5	-3.08	-10.79	-97.80
15	20	360	50	5.4	89.2+ 2.5	4.0	86.0+ 2.5	-0.74	-2.58	-23.52
15	18	360	50	5.4	89.0+ 2.5	4.0		1.86	6.48	59.37
15	16	360	52	5.3	89.0+ 2.5	4.0	86.1+ 2.5	5.04	17.63	162.31
15	14	360	52	5.3	89.3+ 2.5	4.1		8.24	28.84	266.86
15	**13**	**360**	**52**	**5.3**	**89.5+** 2.5	**4.0**		**8.33**	**29.13**	**270.30**
15	*12*	*360*	*exhausted*							

System B: Results (VI)

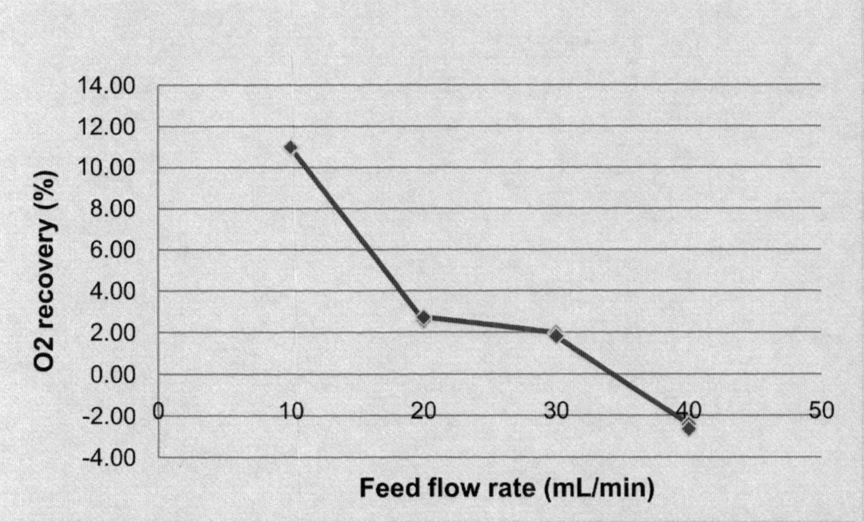

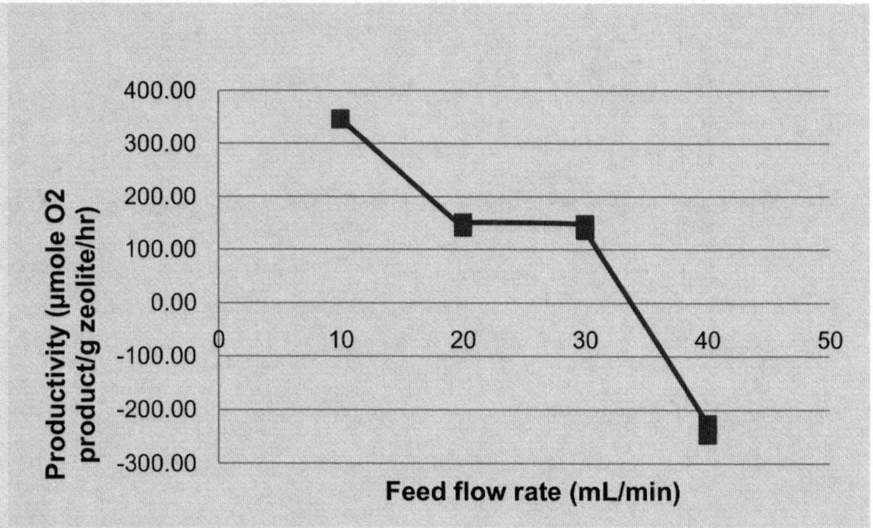

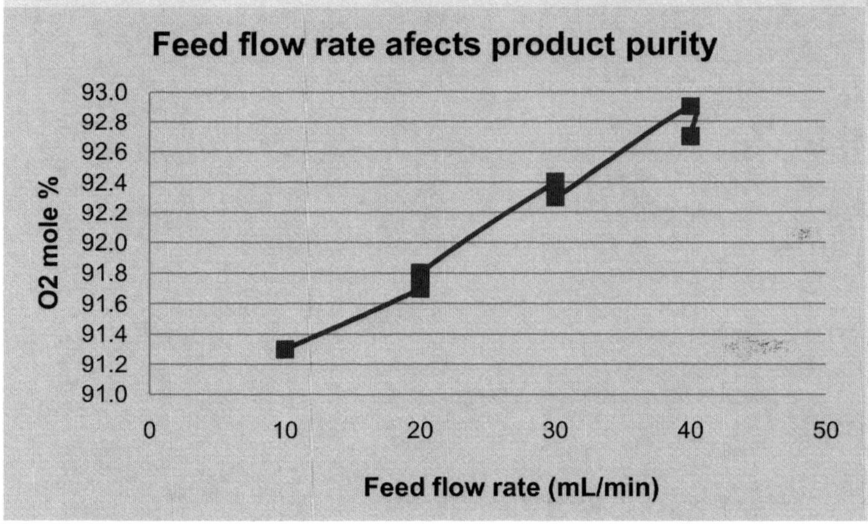

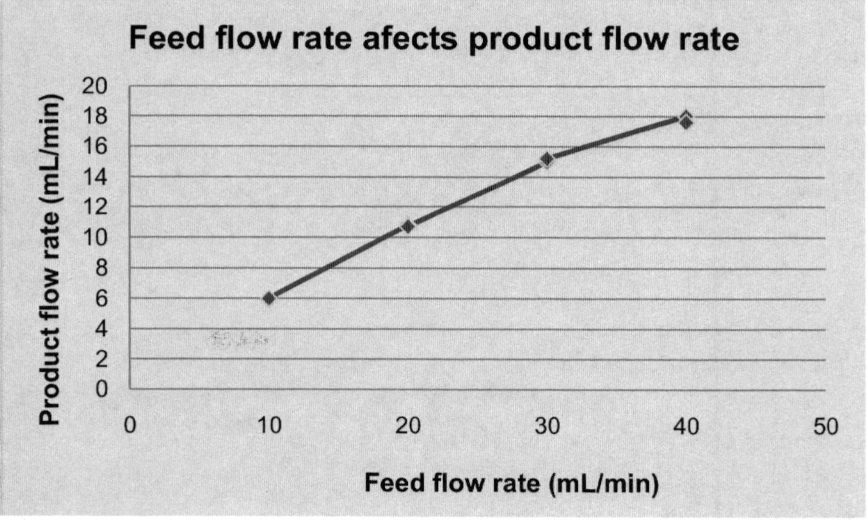

System B: Results (VII)

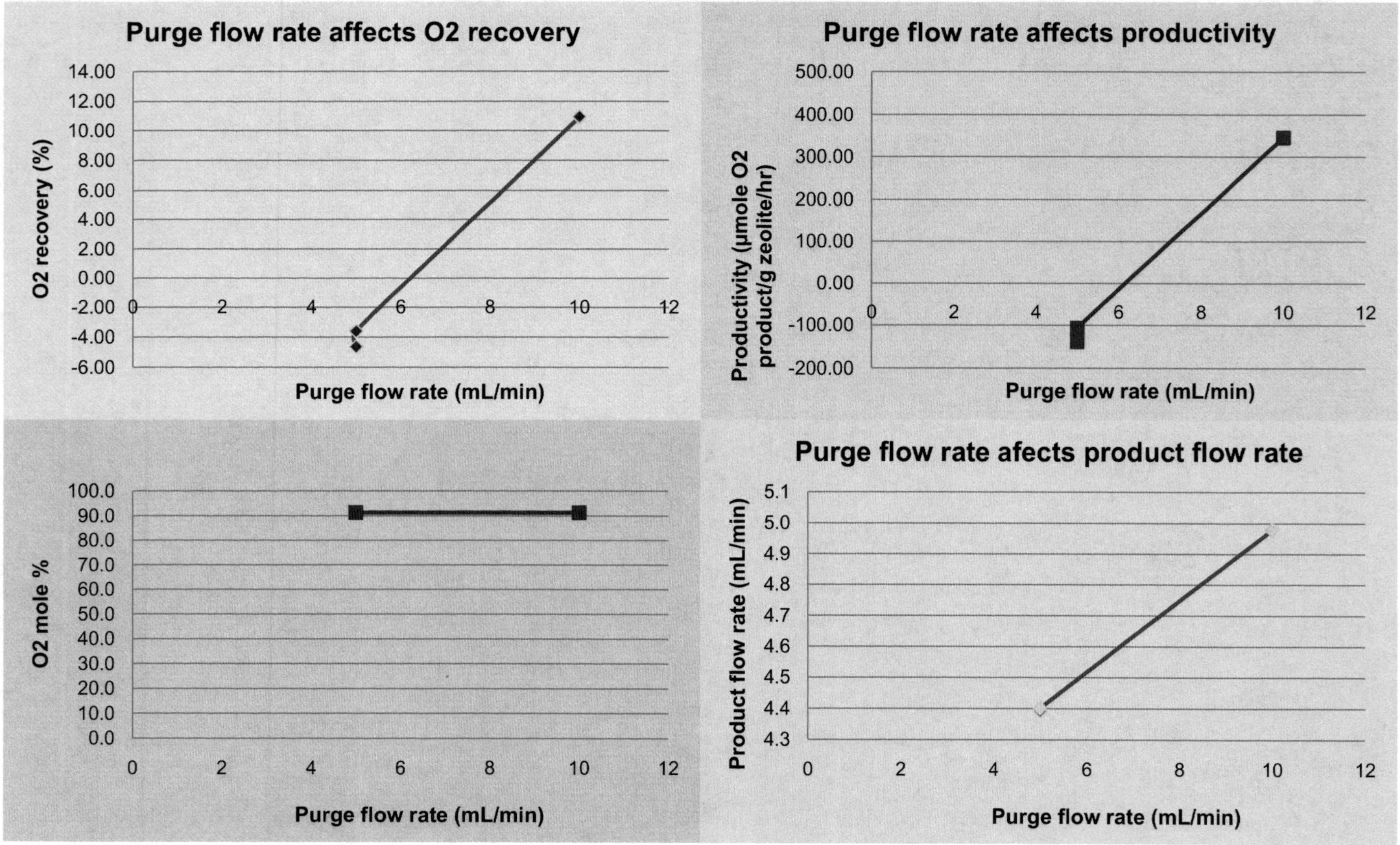

System B: Results (VIII)

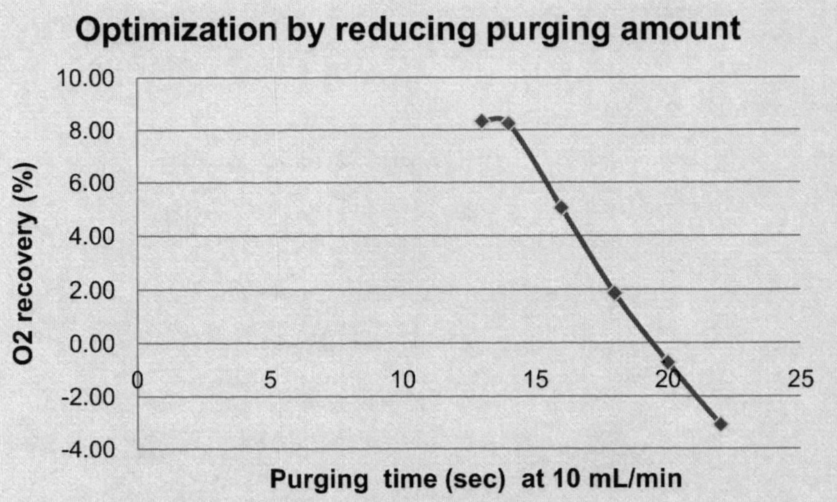

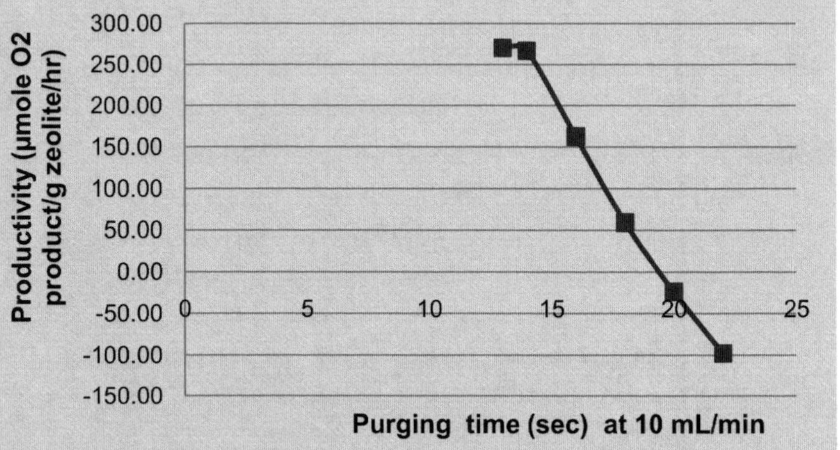

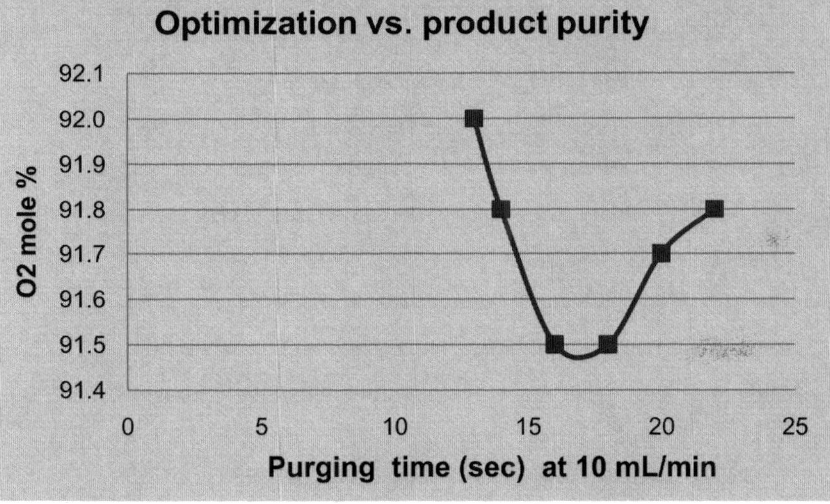

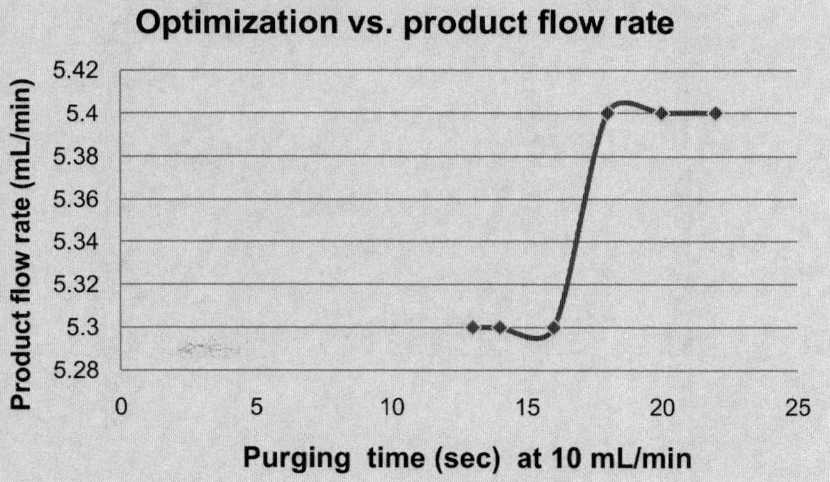

Conclusions

System A

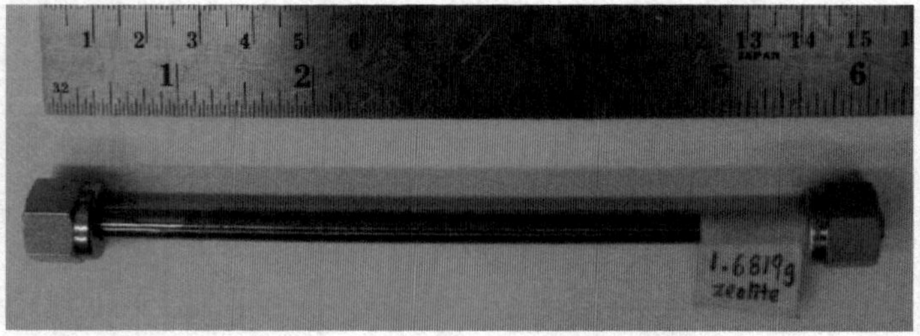

Operating conditions:

- High feed flow rate (30 mL/min)
  can fasten pressurization to increase productivity rate
- High purge flow rate (30 mL/min)
- Short depressurization time (4 sec)

System Outputs:

- 91.7 (max) 89.3 (ave) mole % Oxygen
- Product flow rate: 17.1 (max) 11.6 (ave) mL/min
- 23.91 % Oxygen recovery
- 80.59 μmole Oxygen/g zeolite/cycle
- 2149.03 μmole Oxygen/g zeolite/hr

System B

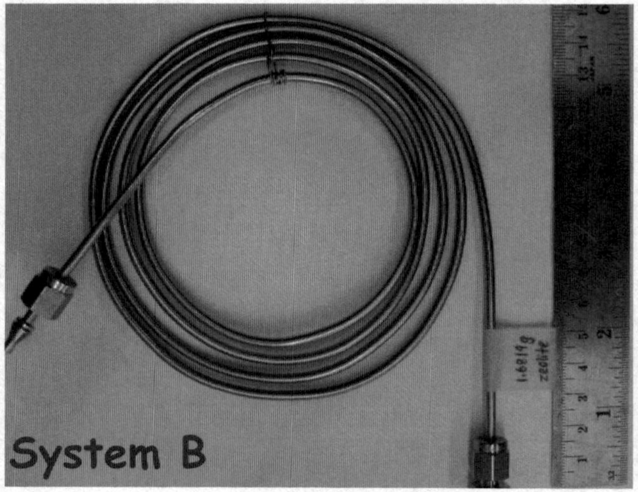

Operating conditions:

- Low feed flow rate (10 mL/min)
  can fasten pressurization to increase productivity rate
- Low purge flow rate (10 mL/min)
- Long depressurization time (15 sec)

System Outputs:

- 92.0 (max) 88.8 (ave) mole % Oxygen
- Product flow rate: 5.3 (max) 4 (ave) mL/min
- 8.33 % Oxygen recovery
- 29.13 μmole Oxygen/g zeolite/cycle
- 270.30 μmole Oxygen/g zeolite/hr

More Areas to be Discussed

- Design Limitations
- Calibration Issue
- Control Issue

Future Work

- Explore ways to improve oxygen recovery and productivity
- Find out suitable bed dimensions and physical ratios that give better productivity
- Multiply bed number to meet production flow rate requirement
- Tailor a shorter cycle time

Challenge:

How to make miniature oxygen concentrator via PSA that meets product purity and flow rate requirements feasible?

505e Optimal Synthesis of a Pressure Swing Adsorption Process for CO2 Capture

Anshul Agarwal[1], Lorenz T. Biegler[1], and Stephen E. Zitney[2]. (1) Department of Chemical Engineering, Carnegie Mellon University, 5000 Forbes Ave., Pittsburgh, PA 15213, (2) Collaboratory for Process and Dynamic Systems Research, National Energy Technology Laboratory, P.O. Box 880, Morgantown, WV 26507-0880

The emission of carbon dioxide from cement industry and power plants that burn fossil fuels is the major cause for the accumulation of CO2 in the atmosphere, which causes long-range environmental problems. One option to mitigate the emission of CO2 is to capture it from the emission sources and store it to the ocean or depleted oil field or use it for enhanced oil recovery. CO2 recovery has been achieved by gas absorption employing solutions of carbonates and alkanolamines. However, this process is energy-intensive for the regeneration of solvent and also faces problems due to corrosion. Recently, the pressure swing adsorption (PSA) process has been considered as an alternative to the absorption process. PSA processes have been widely applied for the removal of CO2 from various feed mixtures, such as CO2 in the steam reformer off gas, landfill gas and natural gas [1]. In all these commercial PSA cycles, the weakly adsorbed component in the mixture is the desired product and enriching the strongly adsorbed CO2 is not a concern. On the other hand, for the capture of CO2 for sequestration, it is necessary to concentrate the CO2 to a high purity to reduce the compression and transportation cost. Thus, it is necessary to develop a PSA cycle by which a high-purity product for the strongly adsorbed component with a high recovery is obtained.

A multitude of PSA cycles and adsorbents have been developed for producing highly pure heavy component (CO2) from feedstock with low CO2 concentration. Kikkinides et al. [2] suggested a 4-bed 4-step process with activated carbon as the sorbent and could recover 68% of CO2 at 99.997% purity. Chue et al. [3] compared activated carbon and zeolite 13X on a 3-bed 7-step process and concluded that the latter is better than the former for CO2 recovery. However, the CO2 recovery was low in their process due to the lack of a countercurrent step in the chosen cycle. Choi et al. [4] reported more than 70% CO2 recovery at more than 90% purity for a modified 3-bed 7-step cycle. However, they solved a very small two variable optimization problem, thus being a specialized case. Zhang et al. [5] have given justifications for using a specific cyclic component step in the adsorption cycle in the context of CO2 capture by using a simplistic mathematical model for the PSA process. Reynolds et al. [6] have suggested a variety of stripping PSA cycles for CO2 recovery at high temperature using a hydrotalcite-like adsorbent.

In this study, a two-bed superstructure of the PSA process has been developed to optimally synthesize an appropriate cycle for CO2 capture. The superstructure considers all the possible operating steps in a PSA cycle with two beds. An optimal control problem with a PDE-based model for PSA system has been formulated in which different steps within a cycle are realized with the help of control variables changing with time. The optimization problem has been solved for three different cases of maximizing

CO_2 recovery (for a given purity), maximizing feed throughput and minimizing specific power (for a given level of CO_2 purity and recovery). Current results indicate the superstructure-based approach as a promising technique for deriving optimal PSA cycles. Different cases with different number of control variables indicate convergence to a particular kind of PSA cycle with over 99% purity and recovery of CO_2. The results obtained from optimization problem will also be compared with the optimal PSA cycle simulated more accurately in a dynamic simulation environment.

References

[1] Sircar, S., Separation of methane and carbon dioxide gas mixtures by pressure swing adsorption. Sep. Sci. Technolog., 1988, 23, 519.

[2] Kikkinides, E. S., Yang, R. T., Cho, S. H., Concentration and recovery of CO2 from flue gas by pressure swing adsorption. Ind. Eng. Chem. Res., 1993, 32, 2714.

[3] Chue, K. T., Kim, J. N., Yoo, Y. J., Cho, S. H., Yang, R. T., Comparison of activated carbon and zeolite 13X for CO2 recovery from flue gas by pressure swing adsorption. Ind. Eng. Chem. Res., 1995, 34, 591.

[4] Choi, W. K., Kwon, T. I., Yeo, Y. K., Lee, H., Song, H. K., Na, B. K., Optimal operation of the pressure swing adsorption process for CO2 recovery. Korean J. Chem. Eng., 2003, 20, 617.

[5] Zhang, J., Webley, P. A., Cycle development and design for CO2 capture from flue gas by vacuum swing adsorption. Environ. Sci. Technol., 2008, 42, 563.

[6] Reynolds, S. P., Ebner, A. D., Ritter, J. A., Stripping PSA cycles for CO2 recovery from flue gas at high temperature using a hydrotalcite-like adsorbent. Ind. Eng. Chem. Res., 2006, 45, 4278.

CO_2 Capture by Moving Bed Temperature Swing Adsorption

Kent S. Knaebel
Adsorption Research, Inc.
6175-D Shamrock Court
Dublin, OH 43016-1200
Tel.: 614-798-9090
Fax: 614-798-9091
e-mail: k.knaebel@adsorption.com

September 27, 2008

EXECUTIVE SUMMARY

An "end-of-pipe" process has been developed to capture carbon dioxide (CO_2) from flue gas from coal-burning power plants, cement plants, and the like. It will treat flue gas immediately after the fly ash is removed. The presence of minor amounts of fly ash, however, should not pose a serious problem. The process employs temperature swing adsorption (TSA). It uses solid adsorbent to take-up CO_2 from flue gas. And it uses the heat contained in the flue gas to drive-off the enriched CO_2 from the adsorbent. The keys to performance are: the flue gas temperature and the CO_2 concentration, which are equivalent to *the available heat* and *the required heat*, respectively. Together, they dictate whether the heat contained in the flue gas is sufficient to drive the separation. When they are properly balanced, the need for parasitic energy is minimal. In contrast, the conventional alternative means for CO_2 capture is scrubbing, which requires about 30% of the gross output of a power plant's steam in order to regenerate the scrubbing solution. Hence, the need for parasitic energy is enormous when scrubbing is used.

The ARI TSA CO_2 capture process has been evaluated via experiments and computer simulations. They revealed important parameters, e.g, the <u>amount of adsorbent</u> required and the <u>size of the equipment</u>, which represent the majority of the capital cost. The ARI TSA process tests showed that both high CO_2 purity (99%) and high recovery (89%) were possible. Additional tests with CO_2, H_2O, and SO_2 in a simulated flue gas showed no signs of adsorbent breakdown. A patent application has been filed (20060230930), and a PCT application (Publication No.=WO 2006/112977) was submitted.

Adsorbent attrition is of minimal concern on account of features that allow less breakage and friction than for other moving bed designs, and a device by which adsorbent fines (if they occur) can be continuously removed.

The design is modular, i.e., the system will be comprised of dozens of identical parallel units. Each one could be amenable to shut-down for maintenance, without adversely affecting the processing capacity of the remainder. If sufficient redundancy is provided for, plant operation and output should not be impacted.

Now that preliminary testing and engineering are complete, we plan to build a pilot-scale TSA process (1 ton of CO_2 per day) to evaluate the process under realistic conditions. The project is being organized by ADA-ES, based in Littleton, Colorado. These tests will yield engineering data and will assess its technical and economic feasibility. The cost for testing will be partly covered by a DOE grant that was recently announced, along with industrial contributions, and will require about 2 years.

PROCESS DESCRIPTION

The process is illustrated below. This diagram was taken from an animated version. As such, it represents a specific instant of its operation, during which the adsorbent, gas, bucket conveyor, and a heat-transfer medium are all moving.

Let's start with the adsorbent beads, being delivered to the top of the adsorption section by the bucket conveyor. At this point the beads (blue) are cool, and relatively free of CO_2. As the beads pass over a screen, any fines that may have been generated (white dots) during adsorption or regeneration are separated. The whole beads fall onto a series of perforated trays. The perforated trays are designed to retard the beads, so that their residence time in the adsorption section is a few minutes. While in the adsorption section, the beads contact flue gas, consisting of adsorbable components such as CO_2 (shown as red squiggles) and non-adsorbable components such as nitrogen (shown as yellow squiggles). Flue gas enters the adsorption section through the perforated distributor pipe at the bottom of that section. The adsorbable components are taken-up the adsorbent as they pass countercurrently, and the non-adsorbable components escape through the stack at the top.

Upon reaching the bottom of the adsorption section, the adsorbent beads are fully loaded, and they proceed to the regeneration section. First, the beads are preheated, as they pass through a countercurrent heat exchanger. In that section, the beads pick-up heat, which is recycled from beads in the heat-recovery section. (The latter beads are hot and relatively free of CO_2.) The beads then contact a heat exchanger section in which the hot flue gas delivers its heat to the beads. Now the beads are red. As can be seen, the hot flue gas enters this heat exchanger section from the power plant, in a duct (at the right-hand side). The flue gas and adsorbent bead remain separated by the heat exchange surface, which is just a metal barrier. The diagram shows the flue gas in a rectangular duct, with triangular sections at the bottom and top, respectively, through which the hot flue gas is delivered and the relatively cool flue gas collected. (The relatively cool flue gas may subsequently be quenched in a separate vessel – not shown.) That relatively cool flue gas is then introduced to the adsorption section through the perforated distributor pipe mentioned earlier. The hot beads readily desorb the CO_2 and any other adsorbable components, and their color changes to yellow. Meanwhile, the enriched CO_2 is collected and withdrawn in a perforated pipe, to sequestration.

As mentioned earlier, the hot adsorbent beads, depleted of CO_2 are cooled by the circulating fluid in the heat recovery section. Afterwards, the cooled beads fall into buckets of the conveyor and are lifted back to the top of the adsorption section. As they are lifted, additional cooling occurs, indicated byt the brown horizontal squiggles, which emanate from the bucket conveyor enclosure, at the left.

A Key identifies the states of the adsorbent beads according to colors at various locations. The block labeled "Transport Systems" refers to animation features, which obviously are all "on" in the snapshot. Those allow one to view the various aspects individually, or in pairs, and to appreciate the relative motion and synchronization, which of course is not possible with a still image.

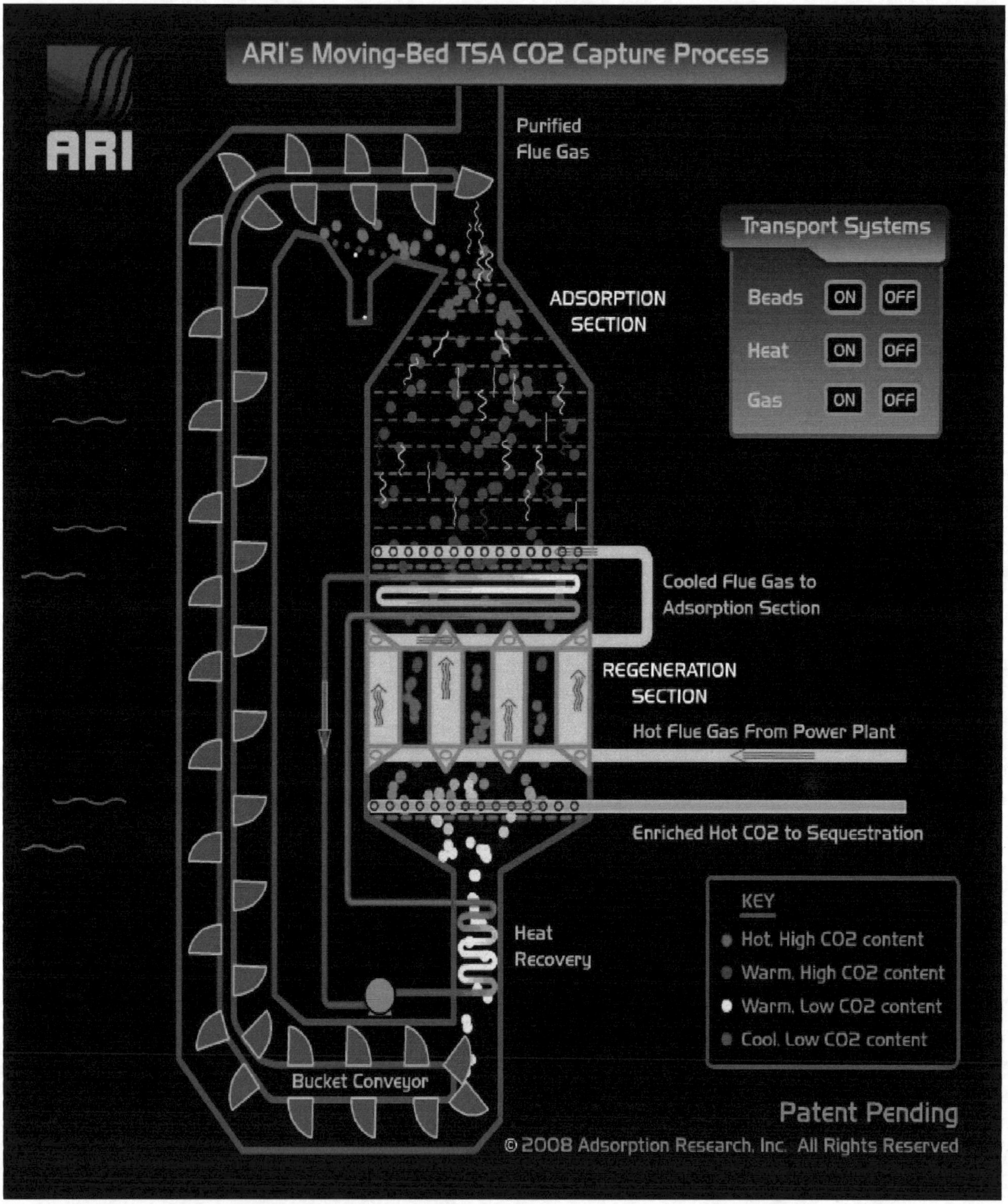

ARI's Moving-Bed TSA CO2 Capture Process
ARI
Purified
Flue Gas
Transport Systems
Beads ON OFF
Heat ON OFF
Gas ON OFF
ADSORPTION
SECTION
Cooled Flue Gas to
Adsorption Section
REGENERATION
SECTION
Hot Flue Gas From Power Plant
Enriched Hot CO2 to Sequestration
KEY
Hot, High CO2 content
Warm, High CO2 content
Warm, Low CO2 content
Cool, Low CO2 content
Heat
Recovery
Bucket Conveyor
Patent Pending
© 2008 Adsorption Research, Inc. All Rights Reserved

Justification

The laboratory data and computer simulations were used to generate approximate capital and operating cost estimates of a commercial CO_2 recovery system. In addition, budgetary quotes were obtained from vendors to arrive at an overall plant cost estimate. The main design parameters and results are shown in Table 1.

Table 1. Operating and Capital Costs for CO_2 Capture by the ARI TSA Process at a 500 MW Coal Fired Power Plant.

Flue Gas Flow Rate:	=	1,500,000 SCFM = 2,200 MMSCFD
CO_2 Emission Rate	=	15,000 tons per day
Flue Gas CO_2 Concentration	=	12% (vol) from the plant
Flue Gas H_2O Concentration	=	12% (vol) from the plant
Outlet Gas CO_2 Concentration	=	1.20% (vol) from the TSA system
Min. CO_2 Recovery Percentage	=	92%
Max. Exhaust Gas Pressure Drop	=	<5 in. H_2O in the TSA system
Adsorbent Reservoir	=	10,000,000 lb
Temperatures		
Flue Gas Inlet Temp.	=	500°F
Outlet Gas Temp.	=	125°F
Adsorbent Regeneration Temp.	=	350°F
Dimensions		
Approximate Footprint and Height	=	300 ft x 200 ft x 50 ft.
Estimated Cost		
Fabricated Steel	=	$147 million
Initial Adsorbent Inventory	=	$ 10 million
Building & Foundation	=	$ 3 million
Installation + Contingency	=	$262 million
Approximate Capital Cost	=	$420 million.
Amortization (6% x 15 yr)	=	$ 43 million per year
Power Cost	=	$ 7 million per year
Total Processing Cost*	=	$9.65 per ton of CO_2

* Assuming the system is on-line 95% of the time, i.e., 347 days per year.

As can be seen, the estimated capital cost includes the structural steel and adsorbent, as well as the building, foundation, installation labor, and contingency. The resulting operating cost per ton of CO_2 captured will depend on the accuracy of these estimates, as well as the amortization rate and term. This is the cost for the captured CO_2, at 1 atm, without drying (which we expect to cost less than an extra $1 per ton).

526a Characterization of Tunable Nanoporous Carbon Films of Capacitance-Storage Function

Katsumi Kaneko[1], Yousheng Tao[2], Rie Oosawa[1], Tomonori Ohba[1], Morinobu Endo[3], and Hirofumi Kanoh[1]. (1) Chemistry, Chiba University, 1-33, Yayoi-cho, Inage-ku, Chiba, 263-8522, Japan, (2) Institute of Carbon Science and Technology, Shinshu University, 4-17-1, Wakasato, Nagano, 380-8553, Japan, (3) Faculty of Engineering, Shinshu University, 4-17-1, Wakasato, Nagano, 380-8553

Nanoporous carbon films have received considerable attention in fundamentals on materials science and commercial applications such as separation and purification, catalysis, electric double layer capacitance, chromatographic separation, biocatalytic sensors, and gas diffusion electrodes in proton exchange membrane fuel cell and anodes in rechargeable lithium ion batteries.1,2 Although numerous methods such as chemical vapor deposition, hydrothermal decomposition of carbide compounds, and polymer coating and pyrolysis have been developed for the fabrication of carbon film, no easily scalable self-support nanoporous carbon films have been obtained with above methods. We report the fabrication of carbon films of tunable nanoporosity. Samples were characterized with field emission scanning electron microscopy, nitrogen adsorption and desorption at 77 K, Raman spectroscopy, thermogravimetric analysis, and the electrical conductivity and cyclic voltammetry measurements. It has been determined with the nitrogen adsorption isotherms that carbon films have surface area of ca. 1000 m2g-1 and total nanopore volume of ca. 1.0 cm3g-1, and bimodal micropores and mesopores with tunable pore sizes. Cyclic voltammetry measurements show that the carbon films have both high mass-specific capacitance and volume-specific capacitance, while Raman spectroscopy and thermogravimetric analysis indicate that the carbon films are chemically, thermally and mechanically stable. Hence, the carbon aerogel films are suggested to have electrochemical applications such as gas sensors, biological sensors and supercapacitors.

[1] B. M. Shiflett and C. H. Foley, Science 285, 1902 (1999).

[2] M. S. Dresselhaus, G. Dresselhaus, J. Electroceram. 1997, 1, 273-286.

526b Enhancing the Capacity of Silver Based Adsorbents for Desulfurization of Liquid Hydrocarbon Fuels

Sachin Nair, Center for Microfibrous Materials Manufacturing, Department of Chemical Engineering, Auburn University, 212 Ross Hall, Auburn, AL 36849 and Bruce Tatarchuk, Chemical Engineering, Auburn University, 230 Ross Hall, Auburn, AL 36849.

Adsorptive desulfurization provides a viable alternative to producing low sulfur fuels from conventional sources such as diesel, gasoline and jet fuel. Such feeds are a prerequisite for the operation of most reformer catalysts as well as fuel cells. Supported silver adsorbents have demonstrated high capacity for sulfur in liquid fuels. For JP5 with a total sulfur concentration of 1172ppmw, this adsorbent demonstrated a 8-10mg/g sulfur capacity. The atomic silver utilization at 2% weight loading for desulfurization was found to be 95% using JP5. Lower atomic utilization was observed at higher silver weight loadings. Thus the sulfur capacity can be significantly improved by increasing the silver utilization at higher loadings. The lowering of silver utilization is attributed to the increase in crystallite size of silver on the support. This study examined the effect of preparative conditions on the dispersion of silver on the TiO_2 support. The factors considered were the support pre-treatment, conditions during impregnation, drying and calcination steps. The desulfurization performance was correlated to adsorbent surface morphology using oxygen chemisorption, XPS and TEM data.

526c Adsorption on Intact Supported Zeolite Membranes: Comparison with Powdered Samples

Karl D. Hammond[1], Geoffrey A. Tompsett[1], Mei Hong[2], Scott M. Auerbach[3], John L. Falconer[2], and W. Curtis Conner[1]. (1) Chemical Engineering, University of Massachusetts, 159 Goessmann Laboratory, Amherst, MA 01003, (2) University of Colorado, 1111 Engineering Drive, Boulder, CO 80309-0424, (3) Chemistry, University of Massachusetts, 701 Lederle Graduate Research Tower, Amherst, MA 01003

Zeolite membranes (layers of zeolite grown on porous supports) have a unique potential for selective, energy-efficient separations based on molecular size and/or shape. Separation experiments using zeolite membranes are typically interpreted with the Maxwell–Stefan model of diffusion, where the Fick diffusion coefficient can be expressed as the product of the Maxwell–Stefan diffusion coefficient and a term involving thermodynamic corrections from the adsorption isotherm. This thermodynamic factor is typically determined from adsorption isotherms collected on a powdered zeolite, as measuring adsorption isotherms on intact membranes is challenging. In this study, we present a non-destructive, reusable apparatus with which to measure adsorption isotherms of intact porous membranes and use it to test the validity of the assumption that using powder adsorption isotherms is equivalent to using membrane adsorption isotherms for the purpose of determining the thermodynamic correction factors. We find that treating a zeolite membrane as a combination of a zeolite powder and a zeolite membrane is a reasonably good approximation to the overall isotherm. The lack of hysteresis in the isotherms indicates that mesopores are not detectable, indicating that the influence of crystal intergrowth defects on membrane transport is limited to their effects on the Maxwell–Stefan diffusion coefficients. The approximation of using zeolite powders to determine the thermodynamic correction factors for zeolite membranes is, in this light, a reasonable one.

526e Discontinuities In the Low Coverage Adsorption of C1-C14 Linear Alkanes on Na-Cha and Sapo-34

Gino V. Baron[1], Sarah Coeck[1], Johan A. Martens[2], Ranjeet Singh[3], Paul A. Webley[3], and Joeri F.M. Denayer[1]. (1) Department of Chemical Engineering, Vrije Universiteit Brussel, Pleinlaan 2, Brussels, B-1050, Belgium, (2) Katholieke Universiteit Leuven, Kasteelpark Arenberg 23, Heverlee, B-3001, Belgium, (3) Chemical Engineering, Monash University, Clayton, VIC 3800, Australia

The isostructural materials Chabazite (CHA) and SAPO-34 are two pertinent examples of small pore or eight membered ring zeolites, offering a highly selective and spatially constrained environment for adsorption, diffusion and reaction. Chabazite has a three dimensional pore system with ellipsoidal shaped cages of 6.7 x 10 Å2 interconnected via 8 membered (8MR) ring windows with pore apertures of 3.8 x 3.8 Å2 for the dehydrated form of CHA. Each cage is connected to 6 neighboring cavities. SAPO-34 is a silicon, aluminum and phosphorous based molecular sieve with the CHA topology. In contrast to the extended literature on adsorption of small gaseous molecules on CHA and SAPO-34, much less attention has been paid to the adsorption of larger molecules, although their pore diameter allows adsorption of e.g. n-alkanes.

In this work, the adsorption of C1-C14 n-alkanes on SAPO-34 and Na-CHA zeolites was studied at low surface coverage using the pulse chromatographic method. On both isostructural materials, containing cages interconnected via small windows, the Henry adsorption constant K' varies in a non-monotonous way with carbon number. K' increases more or less exponentially with carbon number from methane to n-hexane. K' decreases from n-hexane, and reaches a minimum at n-octane or n-decane depending on the material. On SAPO-34, the increase in zero coverage adsorption enthalpy per additional methyl group in the alkane chain becomes smaller for carbon numbers between 5 and 11. A local minimum in adsorption enthalpy occurs at n-undecane. A compensation plot between adsorption entropy and enthalpy shows a pronounced discontinuity at n-undecane. The present observations provide additional proof for the existence of "window effects", in correspondence to recent molecular simulations [1]. C1-C6 alkanes adorb in a linear configuration in the SAPO-34 cages. C7-C10 alkanes adopt a coiled configuration inside the cage, which strongly affects their adsorption enthalpy and entropy. From C11 on, the alkanes are protruding the small windows connecting the cages. The presence of cations on Na-CHA results in stronger energetic interactions compared to SAPO-34, which shifts the window protrusion to lower chain lengths.

[1] Dubbeldam, D., Smit, B., J. Phys. Chem. B., 107, 12138-12152, 2003.

526f Characterization of Cms Adsorbents for Nitrogen Psa Production

R. C. Magalhães[1], D. Ferreira[1], J. Sousa[1], P. Taveira[2], Roger D. Whitley[3], and A. Mendes[1]. (1) LEPAE – Laboratory for Process, Environmental and Energy Engineering, Porto, 4200, Portugal, (2) SysAdvance – Engineering System Lda, Maia, Portugal, (3) Air Products and Chemicals, Inc., 7201 Hamilton Blvd., Allentown, PA 18195-1501

Keywords: Pressure Swing Adsorption, isotherms, breakthroughs

Topic: Characterization of Adsorbents

Nitrogen production from air by PSA (pressure swing adsorption) using CMS (Carbon Molecular Sieve) adsorbents was first proposed by Juntgen et al. in 1981 [1]. Juntgen and co-authors disclosed a process for producing nitrogen at 99.9 % purity with 40 % recovery (though the nitrogen stream also contains about 1 % of argon). Presently a PSA unit can produce nitrogen at purity of 99.9995 % and up to 10000 m3N/h, at 97 % purity. This technology is finding a very significant place in the market for nitrogen production, mainly because the lower costs and higher flexibility compared to cryogenic production [2].

In a PSA adsorption column, nitrogen is separated from air based on the kinetic selectivity of oxygen over nitrogen in absorbents such as CMS. Oxygen is much faster entering the adsorbent micropores while nitrogen, which is much slower is limited to the intraparticle voids and is drawn from the adsorbent column at high pressure. The basic PSA cycle for nitrogen production consists of two adsorbent columns working out of phase and has four basic steps: pressurization, production, blowdown and desorption [3].

The kinetic selectivity of a CMS adsorbent is based on a narrow pore size distribution that can sterically discriminate between oxygen and nitrogen species. Since the expiration of the original patent, several companies began developing and producing new CMS adsorbents that show improved performance. There is then an emerging interest in characterizing these adsorbents to aid in developing optimized PSA cycles for nitrogen production.

This paper reports the characterization of three CMS adsorbent samples from two different suppliers: a CMS D from Carbotech and a KP325 and MSC3K-162 from Japan EnviroChemicals, Ltd. These adsorbents were characterized by measuring the adsorption isotherm and uptake curves of oxygen, nitrogen and argon, from 0 bar to 7 bar and at various temperatures. Afterwards, the effective diffusivity of oxygen, nitrogen and argon was obtained as function of the pressure at the selected temperatures. A model was fitted to the effective diffusivities as function of the pressure and this inserted in a mass balance equation modeling breakthrough experiments. Finally, some breakthrough experiments were performed using a packed bed column and the results used to assess the model proposed.

It was concluded that for all three CMS samples, the effective diffusivity increases with sorbate concentration. Preliminary results show that the oxygen/nitrogen selectivities are similar for all adsorbents, but CMS D has the highest diffusivities followed by MSC3K-162. KP325 shows the highest adsorption capacity. The breakthrough experiments will allow validating the proposed model that will be used to select the best adsorbent and to help the development of the new generations of CMS adsorbents.

References

[1] – Juntgen, et al.; Process for the recovery of nitrogen-rich gases from gases containing at least oxygen as other component. 1981. US 4,264,339

[2] – Agrawal, 2001 R. Agrawal, Separations: perspective of a process developer/ designer, A.I.Ch.E. Journal 47 (2001), p. 967

[3] – Ruthven, D. M., Farooq, S., Knaebel, K. S. (1994). Pressure Swing Adsorption, VCH Publishers, New York.

526g Textural Characterization of Native and N-Alky-Bonded Silica Monoliths by Mercury Intrusion/extrusion, Inverse Size Exclusion Chromatography and Nitrogen Adsorption

Matthias Thommes[1], Romas Skudas[2], Klaus K. Unger[2], and Dieter Lubda[3]. (1) Quantachrome Instruments, 1900 Corporate Drive, Boynton Beach, FL 33426, (2) Institute for Inorganic and Analytical Chemie,, Johannes Gutenberg University, Mainz ,Germany, Duesenbergweg 10-14, Mainz, D-55099, Germany, (3) PLS and Global Production, Merck KGaA, Frankfurter Landstrasse 250, Darmstadt, 64293, Germany

Native and n-alkyl bonded (n-octadecyl) monolithic silica rods with mesopores in the range between 10 and 25 nm and macropores in the range between 1.8 and 6.0 μm were examined by mercury intrusion/ extrusion, inverse size exclusion chromatography (ISEC) and nitrogen sorption. Our results reveal very good agreement for the mesopore size distribution obtained from nitrogen adsorption (in combination with an advanced nonlocal density functional theory (NLDFT) analysis and ISEC. Our studies highlight the importance of mercury porosimetry for the assessment of the macropore size distribution. Mercury porosimetry is further the only method, which allows obtaining a combined and comprehensive structural characterization of macroporous/mesoporous silica monoliths.

Furthermore, our data clearly confirm that mercury porosimetry hysteresis and entrapment have different origin, and indicate the intrinsic nature of mercury porosimetry hysteresis in these silica monoliths. Within this context some silica monoliths show the remarkable result of no entrapment of mercury after extrusion from the mesopore system (i.e for the first intrusion/extrusion cycle). The results of a systematic study of the mercury intrusion/extrusion behavior into native silica monoliths and monoliths with bonded n-alkyl groups reveals that the macro (through) pore structure, which controls the mass transfer to and from the mesopores, here mainly controls the entrapment behavior. Our data suggest that mercury intrusion/extrusion porosimetry does not only allow to obtain a comprehensive pore structure analysis, but can also serve as a tool to estimate the mass transport properties of silica monoliths to be employed in liquid phase separation processes.

543a Designation of a Self-Catalytic Mixed-Conducting Membrane Reactor for Partial Oxidation of Methane to Syngas

Wanqin Jin, Xueliang Dong, **Xiaohua Lu**, and Nanping Xu. State Key Laboratory of Materials-oriented Chemical Engineering, Nanjing University of Technology, Xinmofan Road 5#, Nanjing, China

Increasing concern about world dependence on petroleum oil has generated interest in the more efficient use of nature gas. Methane (CH_4), the main component of natural gas, can be converted to liquid fuels, hydrogen, and other value-added chemicals through a syngas intermediate, a mixture of CO and H_2. Currently, partial oxidation of methane (POM) [1] with pure oxygen in the presence of a catalyst is established to be the most potential process for methane conversion because of its greater selectivity to syngas, its exothermicity, and more desirable CO/H_2 ratio. As for POM, recently, an important advance in directly using air as a feedstock resulted from the application of a mixed-conducting oxide (with high oxygen ionic and electronic conductivities) membrane reactor, which integrated the oxygen separation and POM processes in a single unit. [2] However, one of the formidable problems for this method is that the performance of the membrane reactor is always restricted by the membrane permeability and catalytic activity. For a dense membrane, the oxygen permeability can be improved via decreasing the membrane thickness and/or increasing the surface oxygen exchange rate [3]. In addition, the catalytic performance of the membrane reactor can be improved by increasing the dispersity of catalyst.

According to above ideas to design a high performance membrane reactor, herein we designed a novel self-catalytic membrane reactor based on an asymmetric mixed-conducting membrane for POM reaction. A mixed-conducting oxide La_2NiO_{4+d} (LNO) was used not only for the membrane support but also for the catalyst precursor and $La_2Ni_{0.9}Co_{0.1}O_{4+d}$ (LNCO) oxide was selected as the membrane. At the initial of POM reaction, CH_4 was oxidized by LNO to CO and H_2, and simultaneously parts of LNO on the surface of membrane support were reduced to Ni^0 and La_2O_3, which forms a self-catalytic reaction process because the catalyst (Ni^0) for POM reaction was produced during this process. Moreover, the improved bulk diffusion and surface exchange rate of oxygen ion and high dispersity of catalyst on the surface of membrane support make the membrane reactor possess high catalytic performance for POM reaction. The CO selectivity (S_{CO}) is higher than 95%, the conversion of CH_4 is kept at about 60% and the oxygen permeation flux is about 2.3 ml°¤cm^{-2}°¤min^{-1}, which is nearly 3 times that of the symmetrical LNCO membrane. In addition, the concept for the self-catalytic membrane reactor can be widely used to instruct the design and preparation of high efficiency catalytic membrane reactor for conversion of hydrocarbon.

Keywords: Self-Catalytic; Membrane Reactor; POM

Acknowledgments: This work was supported by Program for Changjiang Scholars and Innovative Research Team in University (PCSIRT 0732), the National Basic Research Program of China (No. 2003CB615702) and National Natural Science Foundation of China (No.20576051, 20436030).

References:

[1] Hickman DA, Schmidt LD. Production of syngas by direct catalytic oxidation of methane. *Science* 1993; 259:343-346.

[2] Jin WQ, Li SG, Huang P, Xu NP, Shi J, Lin YS. Tubular lanthanum cobaltite perovskite-type membrane reactors for partial oxidation of methane to syngas. *J. Membr. Sci.* 2000; 166:13-22.

[3] X. F. Chang, C. Zhang, W. Q. Jin, N. P. Xu, "Match of thermal performances between the membrane and the support for supported dense mixed-conducting membranes", *J. Membr. Sci.* 2006; 285:232-238.

543b Cyclic Hybrid Adsorbent-Membrane Reactor (HAMR) Studies for Hydrogen Production

Aadesh Harale[1], Hyun Tae Hwang[1], Paul K. T. Liu[2], Muhammad Sahimi[3], and **Theodore T. Tsotsis**[4]. (1) Mork Family Department of Chemical Engineering and Materials Science, University of Southern California, 925 Bloom Walk, HED 216, Los Angeles, CA 90089, (2) Media and Process Technology Inc, 1155 William Pitt Way, Pittsburgh, PA 15238, (3) The mork family department of chemical Engineering and Materials Science, University of Southern Californai, 925 bloomwalk HED 216, Los Angeles, CA 90089, (4) Mork Family Department of Chemical Engineering & Materials Science, University of Southern California, Los Angeles, CA 90089

1. Introduction

As a result of stricter environmental regulations worldwide, hydrogen is progressively becoming an important clean energy source. For H2 to replace fossil fuels in mobile applications, it will require the creation of a production and delivery infrastructure equivalent to that currently existing for fossil fuels, which is an immense task. As an alternative, and as an interim step towards the new hydrogen economy, various groups are currently studying steam reforming of methane (SRM) for the on-board generation of hydrogen, or for on site production, in order to alleviate the need for compressed or liquid hydrogen gas storage[1-4]. Conventional technologies are, however, neither convenient nor economical to apply for small-scale (on site or on-board) hydrogen generation. Reactive separation processes have, as a result, been attracting renewed interest for application in H2 production through SRM. One such technology is the hybrid adsorbent-membrane reactor (HAMR) system, which couples reaction and membrane separation steps with adsorption on the reactor and/or membrane permeate side. The HAMR concept was originally proposed by our group[5, 6] for esterification reactions, and it was adapted recently for on-board or on-site hydrogen production applications. Our early studies involved the development of a mathematical model for the HAMR system (applied to hydrogen production through SRM[7]); recently experimental investigations with the water-gas shift reaction[8], using microporous membranes and CO2 hydrotalcite-type adsorbents, were carried out in order to validate the HAMR design models. Experimental data were compared with the model predictions, and found to be consistent. In this paper we focus on the practical process design aspects of the HAMR hydrogen production process. A continuous HAMR process scheme has been investigated, both experimentally and through modeling studies.

2. HAMR Cyclic Process

The steps involved in the proposed cyclic HAMR process for the direct production of pure H2 are described below. It consists of four steps:

1. Adsorption-reaction-membrane-separation step. The reactor is initially pre-saturated with H2 and steam at the desired reaction temperature and pressure. A mixture of steam and CH4 (or CO) at a

prescribed ratio is then fed into the reactor, and an essentially pure H2 product is collected at the permeate side. The reaction step is continued up to the time needed for "breakthrough" (i.e., when the H2 purity and recovery decrease to preset levels) to occur. This time depends upon adsorbent, membrane characteristics, and other reactor parameters. At the "breakthrough" point the feed is diverted into a second identical reactor.

2. Blow-down Step. During this step, the reactor is depressurized to a lower pressure PL, countercurrently to the feed flow direction. The effluent gas stream contains all the components left in the reactor at the end of Step 1, and is either recycled as a feed to another reactor or to be used as fuel.

3. Purge Step. The reactor is counter-currently purged with a weakly adsorbing gas, such as steam or H2, to desorb the CO2. The desorption step operates at PL. The desorbed gas consists of CO, CH4, CO2, H2, and H2O, and is either separated for recycling or used as fuel.

4. Pressurization step. The reactor is countercurrently pressurized to the reaction pressure (for Step 1) with a mixture of steam and H2. At this point, regeneration of the reactor is completed, so that it is ready to undergo a new cycle.

In our studies, a 24 min, 4-bed/4 step cycle was investigated for the water-gas shift reaction. A H2-selective carbon molecular sieve membrane together with a CO2-selective hydrotalcite adsorbent, and a commercial Cu/Zn catalyst was used. Virtually 100% conversion is achieved during the reaction step, while simultaneously 100% of CO2 is being captured during this step. Since the membrane excludes CO, the hydrogen product in the permeate side is highly pure, and ready to use in a fuel cell. A more detailed description of the characteristics of the HAMR cyclic process will be discussed during the conference presentation.

Acknowledgement

The support of the US Department Of Energy and NASA is gratefully acknowledged.

References

1. Y. Choi, H. Stenger, J. Power Source, 124, 432 (2003).

2. N. Darwish,N. Hilal, G. Versteeg, B. Heesink, Fuel, 83, 409 (2003).

3. Z. Liu, H.Roh, S.Park, , J. Power Sources, 111, 83. (2002)

4. T. A. Semelsberger, L. F. Brown, R. L. Borup, M. A. Inbody, Int. J. Hydrogen Energ. 29,1047. (2004)

5. B. Park, Ph.D. Thesis, University of Southern California, Los Angeles, California, (2001)

6. B. Park, T.T. Tsotsis, Chem. Eng. Proc. 43, 1171.(2004)

7. B. Fayyaz, A. Harale, B.G. Park, P.K.T. Liu, M. Sahimi, and T. T. Tsotsis, Ind. Eng. Chem. Res., 44 (25), 9398 -9408, (2005)

8. A. Harale, H. Hwang, P.K. Liu, M. Sahimi, and T.T. Tsotsis, Chemical Engineering Science 62:4126-4137(2007)

543c Aligned Carbon Nanotube Membranes as Chiral Hydrogenation Platforms

Xin Zhan, Chem. & Mater. Eng., University of Kentucky, 177 Anderson Hall, Lexington KY, KY 40506-0046 and **Bruce J. Hinds**, Chemical and Materials Eng., University of Kentucky, 177 Anderson Hall, Lexington KY, KY 40506-0046.

Carbon nanotubes have three key attributes that make them of great interest for novel membrane applications 1) atomically flat graphite surface allows for ideal fluid slip boundary conditions 2) the cutting process to open CNTs inherently places functional chemistry at CNT core entrance and 3) CNT are electrically conductive allowing for electrochemical reactions and application of electric fields at CNT tips. The entrances to CNT's cores were thus functionalized with a series of tethered molecules and simultaneaous permeation experiments showed gatekeeper activity. Changes in the flux and selectivity where applied bias draws tethered charged molecules into the CNT core hindering/gating flux across the membrane. Entrances of CNTs are functionalized with ferrocene based Josiphos ligands and chiral hydrogenation is demonstrated. Enhancement of catalyst activity on CNT surface is observed.

Web Page: www.engr.uky.edu/~bjhinds

543d Hydrogen Membrane Fuel Cells with Nanoengineered Proton-Conducting Layer

Dongxia Liu and **Matt Yates**. University of Rochester, Department of Chemical Engineering, 206 Gavett Hall, Rochester, NY 14627

The hydrogen membrane fuel cell (HMFC) is a new class of fuel cell in which a palladium-based membrane serves simultaneously as the anode and as a hydrogen separation membrane. The cathode consists of a perovskite ceramic and is separated from the anode by a thin layer of proton conducting ceramic. By integrating the fuel cell with a hydrogen membrane, the HMFC can generate electricity directly from hydrogen in gas mixtures produced from steam reforming renewable hydrocarbon fuels, such as ethanol, eliminating the separate hydrogen purification step currently required. Here we report a simple, low cost technique to grow proton conducting ceramic crystals directly onto hydrogen membranes as thin films with nanostructure engineered to enhance performance in HMFCs. The crystals display anisotropic proton conductivity, and are grown so that the crystal axis with the highest conductivity is oriented normal to the surface of the hydrogen membrane. As a result, proton conductivity is significantly enhanced. The novel ceramic structure is characterized by electron microscopy, x-ray diffraction, and electrical impedance spectroscopy. Preliminary fuel cell performance data has been obtained, demonstrating the promise of this novel fuel cell in the temperature range of 300-600°C, temperatures intermediate between state of the art operating conditions of polymeric membranes (<200°C) and ceramic membranes (>600°C). The temperature of operation should allow easy integration of the HMFC into a steam reforming process.

543e Steam Reforming of Ethanol (SRE) In a Palladium Based Membrane Reactor Over Nickel Catalyst

Mohammad A. Islam, Mechanical and Chemical Engineering Department, North Carolina A&T State University, Greensboro, NC 27411 and Shamsuddin Ilias, Department of Mechanical & Chemical Engineering, North Carolina A&T State University, 1601 E. Market Street, Greensboro, NC 27411.

In recent years, there have been renewed interests in bio-ethanol as one of the renewable energy resources for the future energy. It can be easily extracted or blended through biomass fermentation. Direct use of biomass may somewhat reduce our dependence on fossil fuel; its detrimental environmental impact due to CO2 emission is inevitable. Reforming of bio-ethanol is considered as a promising method for hydrogen production from renewable resources. Recent advances in palladium-based thin-film membrane offer an opportunity to develop hydrogen fuel reformer (e.g., steam ethanol reformer) for on-board, one-step generation of hydrogen in a fuel cell vehicle (FCV).

In this work, we used a Pd-thin film membrane in tubular configuration as fuel reformer to study the steam reforming of ethanol (SRE) over nickel based catalyst. The membrane was fabricated in our laboratory using a newly developed surfactant induced electroless plating (SIEP) technique. The Pd-film was deposited on the outside surface of the microporous stainless steel tubular substrate. The effect of steam to ethanol ratio, temperature, and space velocity on ethanol conversion, hydrogen yield, and carbon monoxide suppression was studied for the SRE in the membrane reactor. A two-dimensional, pseudo-homogeneous membrane-reactor model for the SRE reactions was developed to study the membrane reactor performance. Under certain simplifying assumptions, a set of partial differential equations (PDEs) was derived using the continuity equations for the SRE reactions. To account the permeation of hydrogen through the membrane wall in radial direction, radial diffusion was included in the model development. Finite difference method was used to solve the highly coupled PDEs. Experimental ethanol conversions were found to be in good agreement with the model predictions. In this presentation we will discuss some of our experimental and modeling results.

The Interaction Between $SrFeCo_{0.5}O_x$ (SFC) Ceramic Membrane and the Catalyst $Pt/CeZrO_2$ for Syngas Production from Methane

Sedigheh Faraji, Karen J.Nordheden, and Susan M. Stagg-Williams
Department of Chemical & Petroleum Engineering, University of Kansas, Lawrence, KS 66045

There is an increasing interest in the development of a process for producing syngas ($CO+H_2$) from methane using non-porous oxygen-conducting ceramic membranes with mixed ionic and electronic conductivity (MIECs) as an alternative to the conventional expensive process based on air separation plant [1, 2]. Our previous studies [2], which focused on the use of $SrFeCo_{0.5}O_x$ (SFC) dense ceramic membranes for the Carbon dioxide reforming of methane at 800 °C, have shown that the presence of the catalyst $Pt/CeZrO_2$ on the surface of an SFC membrane can give higher methane conversion compared to the methane conversion in the presence of other catalysts like Pt/ZrO_2 [3]. However, the roles of ceramic membrane in the enhancement of this reaction and the interaction between SFC and the catalyst have not yet been investigated. This work presents the results from studies undertaken to clarify the role of SFC ceramic membrane in the enhancement of catalytic conversion of methane to syngas through a series of pulse studies over small particles of SFC (<250 μm) and $Pt/CeZrO_2$ at 800 °C. These pulse studies have been conducted in order to better understand the surface reactions happening during partial oxidation and carbon dioxide reforming of methane.

The SFC powder (from Praxair Specialty Ceramics) was passed through a 60-mesh sieve and coated with 1 wt% ethylcellulose binder prior to pressing. SFC powder was then pressed and sintered in flowing air for 10 hours and finally crushed and passed through a sieve. The catalyst $Pt/CeZrO_2$ was prepared by depositing Pt on commercially available $CeZrO_2$ support material (MEI CHEM) using the incipient wetness impregnation technique (0.5% wt Pt supported on $CeZrO_2$). The catalyst was characterized by conducting ICP and TEM studies.

For some of the reaction tests, a physical mixture of SFC and the unreduced Pt catalyst with a specific mass ratio was fed into the reactor tube, which was made of quartz. For other reaction tests, the catalyst and SFC were loaded into the reactor tube layer by layer. The total material loading was held constant for each experiment (20 mg). The reactor was heated to 800 °C in the presence of argon and then 50 μl pulses of methane or a 1:1 CH_4/CO_2 ratio was injected into the reactor until no changes were observed. The effluent gas was analyzed by mass spectrometer.

It was found out that mixing pattern of SFC and the catalyst and the mass ratio of SFC to the catalyst plays an important role in the methane conversions for both CH_4 injections and CH_4+CO_2 injections. Regardless of the type of reactant, the highest conversion was obtained at mass ratio of one and with the catalyst and SFC well-mixed in the reactor. Adding CO_2 to CH_4 in the inlet stream increased the amount of methane conversion for the case when the SFC and catalysts were well-mixed.

When the $Pt/CeZrO_2$ catalyst and SFC were well mixed, the hydrogen that formed during the decomposition of methane was able to interact with the oxygen in the SFC to form water. In addition, it is believed that the CO can adsorb on the SFC surface and, using oxygen from the SFC structure, react to form CO_2. The water and CO_2 formed can then subsequently react with methane in the presence of the catalyst to form additional hydrogen and CO via steam and carbon dioxide reforming of methane. Since the SFC has a higher capacity to adsorb CO, the H_2/CO ratio for the case of the SFC and catalyst being well mixed is very high, as shown in Figure 1. When the SFC is located below the catalyst in a segregated layer, the CO and H_2 produced on the catalyst still interacts with the SFC to form water and CO_2. However, the water and CO_2 are swept out of the reactor without contacting the catalyst zone and no steam or carbon dioxide reforming occurs. This results in a H_2/CO ratio that is closer to what would be expected for partial oxidation. The final case, when SFC is located as a segregated layer above the catalyst zone results in minimal contact between the SFC and catalyst. The H_2 and CO formed during these studies is removed from the reactor without interacting with the SFC. This results in minimal water and CO_2 formation. In this case, the H_2/CO ratio is less than 1 because some of the hydrogen is consumed during the insitu reduction of the metal. The results of this study clearly demonstrate that a synergistic relationship exists between the catalyst and the membrane material.

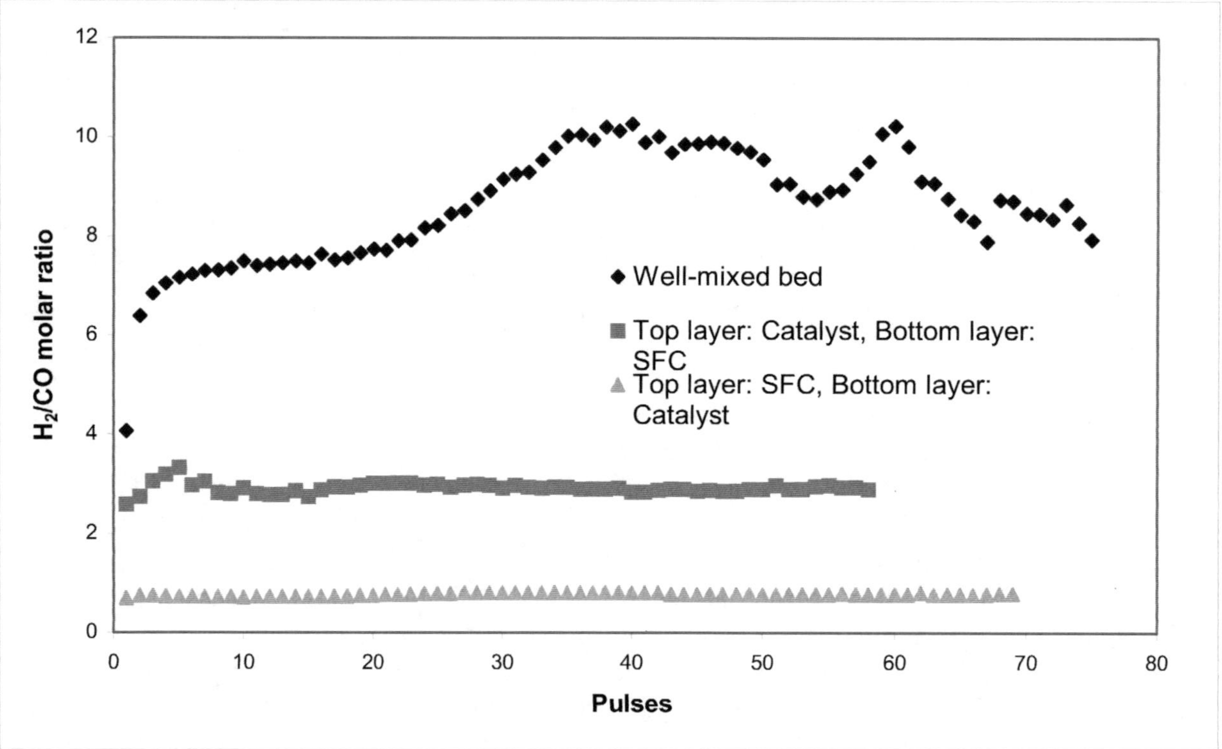

Figure 1: The H_2/CO molar ratio resulted from pulse injections of CH_4 at 800 °C *where the mass ratio of SFC to the catalyst was equal to 1.*

References:

[1] H. J. Bouwmeester, "Dense ceramic membranes for methane conversion," *Catalysis Today*, vol. 82, pp. 141-150, 2003.

[2] D. A. Slade, A. M. Duncan, K. J. Nordheden, and S. M. Stagg-Williams, "Mixed-conducting oxygen permeable ceramic membranes for the carbon dioxide reforming of methane," *Green Chemistry*, vol. 9, pp. 577-581, 2007.

[3] Q. Jiang, S. Faraji, D. A. Slade, K. J. Nordheden, and S. M. Stagg-Williams, "Performance analysis of mixed-conducting ceramic membranes for membrane reactor applications," *AIChE Annual Meeting*, Paper no. 509aO, Salt Lake City, 2007.

543g Microchannel Membrane Reactor Technologies for Artificial Lungs

Jung Kyoo Lee, Lyle F. Mockros, Mayfair C. Kung, and **Harold H. Kung**. Chemical and Biological Engineering, Northwestern University, 2145 Sheridan Road, E136, Evanston, IL 60208-3120

A highly effective blood-oxygen exchange component of an artificial lung would mimic the structure of a natural lung, with blood-size channels about the size of a red blood cell. Calculations show that with 12 μm channels, such a device would require 140 million, 0.8 mm long channels, but the total size of the gas-exchange section would be only 57 mL and a blood prime of only 13 mL. Also attractive are 12 μm high broad open channels with support posts and 40 μm screen-filled rectangular channels. The total size of the former would be 250 mL with a blood prime of 13 mL, and the total size of the latter would be 280 mL with a blood prime of 27 mL.

A major challenge for developing such devices is the requirement that the multitude of channels must be uniform from channel to channel and along each channel in order to avoid preferential blood passages. One strategy is to fill broad rectangular channels with micro scale screens that can provide uniform support and stability. We have explored the effectiveness of 40 μm screen-filled blood-side channels and, as a comparison, 82 μm screen-filled channels. Small concept-devices, consisting of a single 69 mm wide and 3 or 6 mm long channel, were tested using 30% hematocrit blood and oxygen or air on the gas side. The measured oxygen fluxes in the devices were in the range of 4 to 9x10-7 moles/(min.cm2), with the latter close to the theoretical membrane limit. The pressure drop was in the range of 1 to 6 mmHg. Extrapolating the data to a device designed to process 4 L/min suggests a required blood prime of only 35 mL.

Another strategy is to employ lithographic techniques to develop patterned silicone rubber membranes that provide 15 μm high microchannels for artificial lungs. Two types of devices were fabricated: one had a series of parallel, straight, open rectangular channels that are each 300 μm wide, separated by 200 μm walls, and 3 mm long and the other is a wide rectangular channel with support posts, also 3 mm long. Experiments with 30% hematocrit, venous, bovine blood show average oxygen fluxes ranging from 11x10-7 moles/(min•cm2) at a residence time of 0.04 sec to 6.5 x10-7 moles/(min•cm2) at a residence time of 0.20 sec. The average oxygen flux vs residence time, which is due to transverse molecular diffusion, follows the same relation as for all membranes tested. The corresponding increase in hemoglobin saturation ranged from 9% at the residence time of 0.04 sec to 24% at the residence time of 0.20 sec. The support-post channel membranes are attractive for designers because they can be arbitrarily wide and would be less prone blockage.

543h Esterification of Organic Acids with Pervaporation

Satish J. Parulekar, Chemical and Biological Engineering, Illinois Institute of Technology, 127 Perlstein Hall, Chicago, IL 60616

In recent years, membrane technology has emerged as a viable separation process in an effort to combine downstream/upstream separation with reaction to improve process performance. Since membranes allow selective permeation of a component from a multi-component mixture, these can help enhance the conversion of reactants for thermodynamically or kinetically limited reactions via selective removal of one or more product species from the reaction mixture. Pervaporation is used to separate a liquid mixture by partly vaporizing it through a porous or a nonporous membrane. The permeate is enriched in the more rapidly permeating component of the feed mixture, whereas the retentate is depleted in this component. Removal of water from mixtures (dehydration) is the most widely used application of pervaporation. Pervaporation is ideal for enhancing conversion in reversible condensation reactions, such as esterification of carboxylic acids and alcohols, generating water as a product. Substantial acceleration of these reactions can be achieved using appropriate commercially available solid catalysts.

Lactic and succinic acids are important intermediates, which play a significant role in glucose metabolism in most living systems. Diethyl succinate and ethyl lactate are important chemical intermediates. Esterification of each acid with ethanol/methanol is a reversible reaction and hence thermodynamically limited. A detailed analysis of reversible reactions where one of the products is removed by membrane separation is presented here. The reactions of interest are condensation reactions, with one of the products, water, to be removed by pervaporation. The reactor-separator configurations considered are: (1) a semi-batch perfectly mixed reactor with complete recycle of pervaporation retentate, (2) a plug flow reactor (PFR) with in situ pervaporation (tubular membrane reactor), (3) a continuous stirred tank reactor (CSTR) with in situ pervaporation, (4) a CSTR/PFR with external pervaporation and recycle. The analysis is supplemented by appropriate numerical illustrations based on esterification of lactic acid with ethanol. The effects of resistances for reaction and membrane separation on the performance of the reaction-pervaporation systems are studied. Performance of membranes with finite resistance for mass transfer is compared with that of perfect membranes, membranes with negligible mass transfer resistance.

544a A Semi-Empirical Approach for Predicting the Performance of Mixed Matrix Membranes Containing Selective Flakes

Joshua A. Sheffel, Department of Chemical Engineering and Materials Science, Univeristy of Minnesota, 421 Washington Avenue SE, Minneapolis, MN 55455-0132 and Michael Tsapatsis, Department of Chemical Engineering and Materials Science, University of Minnesota, 421 Washington Ave. SE, Minneapolis, MN 55455-0132.

In this presentation we will address a persistent problem in the prediction of mixed matrix membrane permeability and selectivity. On the one hand, analytical models which have been available for some time adopt a view of diffusion that is often too simplistic to accurately apply to zeolites. On the other hand, complex numerical approaches that apply more sophisticated treatments of diffusion to the complex mixed membrane geometry are often inaccessible to the experimentalist for the purpose of membrane design and performance evaluation. We will present an approach for predicting the performance of mixed matrix membranes containing zeolite flakes that combines well-known analytical expressions for mixed matrix membrane performance with Maxwell-Stefan modeling for zeolite diffusion. In so doing we hope to provide a realistic and practical tool for mixed matrix membrane design and performance evaluation. This approach will be illustrated through a case study of normal- and iso-butane separation by a membrane containing silicalite-1 flakes. Predictions will be compared to experimental results for a silicalite-1/PTMSP membrane found in the literature.

544b Molecular Dynamics Simulation of Polyimide-Polyhedral Oligomeric Silsesquioxane Mixed Matrix Nanocomposites

Yin Yani and Monica H. Lamm. Chemical and Biological Engineering, Iowa State University, 2114 Sweeney Hall, Ames, IA 50011

Nanocomposite blends containing polyimide and polyhedral oligomeric silsesquioxanes (POSS) are a promising class of mixed matrix membranes for highly selective transport of small gas molecules such as hydrogen in high temperature applications. In this work, we use atomistic molecular dynamics simulations to model blends of polyimide and either octahydrido silsesquioxane (OHS) or octaaminophenyl silsesquioxane (OAPS). The glass transition temperature of the model polyimide-OAPS blends increases with the incorporation of OAPS, an observation consistent with recent experiments on these systems.[1] We present intermolecular radial distribution functions for POSS-POSS and POSS-PI atoms to show how packing between the two species in the mixed matrix varies as a function of POSS loading and POSS functional group. These results provide molecular insight about the thermal property enhancements afforded by POSS-based additives.

[1] P. Iyer and M. R. Coleman, J. Appl. Pol. Sci., 2008, Vol. 108, 2691-2699.

544c Effect of Fermentation Broth Components on Performance of High Silica Zsm-5/silicone Rubber Mixed Matrix Membranes

Leland Vane, Vasudevan Namboodiri, and Richard Meier. U.S. EPA, 26 W. Martin Luther King Drive, M/S443, Cincinnati, OH 45268

Fermentative organisms produce a range of compounds in addition to the desired product. For example, in addition to ethanol, standard yeast produce longer straight-chained and branched alcohols and organic acids. Additionally, biomass pretreatment process, particularly acid-based treatments, can introduce organic acids into the sugar feed stream to the fermentor. Carboxylic acids, such as acetic acid and succinic acid, are commonly encountered in the conversion of biomass to ethanol and other biofuels. Pervaporation using mixed matrix membranes has proven to be a selective method of removing alcohols from fermentation broths[1]. Such membranes are prepared by dispersing ethanol-selective zeolite particles into a polymer matrix. The carboxylic acids and longer chain alcohols present in the broths, however, compete with ethanol for adsorption sites within the zeolite channels and this affects the permeability and selectivity of the mixed matrix membrane[2].

In this presentation, the effect of actual fermentation broths on the pervaporation performance of mixed matrix membranes will be reported along with methods for mitigating the effect of broth components on membrane performance.

[1] L.M. Vane, V.V. Namboodiri, and T.C. Bowen "Hydrophobic zeolite-silicone rubber mixed matrix membranes for ethanol-water separation: Effect of zeolite and silicone component selection on pervaporation performance," Journal of Membrane Science, 308, 230-241 (2008).

[2] T.C. Bowen, R.G. Meier, and L.M. Vane, "Stability of MFI zeolite-filled PDMS membranes during pervaporative ethanol recovery from aqueous mixtures containing acetic acid," Journal of Membrane Science, 298, 117-125 (2007).

*This is an abstract of a proposed presentation and does not necessarily reflect USEPA policy.

544d Butane Isomers Separation Using Zsm-5 Nanoparticles and 6FDA-Dam Mixed Matrix Membrane

Junqiang Liu and William J. Koros. Chemical and Biomolecular Engineering, Georgia Institute of Technology, 311 Ferst Drive, Atlanta, GA 30332

150nm pure silica ZSM-5 was synthesized and surface modified using a two step Grignard treatment. Mg(OH)2 Whisker was grown on the surface of ZSM-5 particles. Mixed matrix films were cast using a 6FDA-DAM polymer, with upto 35wt% GT-ZSM-5 loading. The permeability of n-Butane more than doubled and selectivity of butane isomers remained the same after adding 35wt% GT-ZSM-5, which means either too permeable ZSM-5 or bad interface. To find the cause, impermeable ZSM-5 particle with template inside pores was synthesized, GT modified and cast into composite membranes with upto 35wt% loading. The permeability of n-Butane decreased 30% at 35wt% loading, which agrees with Maxwell modeling with solid loading, thus proved the defect-free interface between ZSM-5 particles and Polymers. No polymer could match the high permeability of ZSM-5 to make promising composite membrane for butane isomers separation.

544e Gas Separation Nanocomposite Membranes Using a Layered Zeolite Precursor

Sudeep Maheshwari, Jungkyu Choi, Frank S. Bates, and Michael Tsapatsis. Department of Chemical Engineering and Materials Science, University of Minnesota, 421 Washington Ave SE, Minneapolis, MN 55455

Mixed matrix materials comprising of zeolites incorporated in suitable matrix are promising as future membrane material with high permselectivity. However, they suffer from the drawback of low productivity due to increase in membrane thickness by incorporation of zeolites. Nanocomposite membranes consisting of thin zeolite layers (~2 nm) represent a material which combines high selectivity without increasing the membrane thickness due to very small thickness of layers.

In this talk, the fabrication and properties of composite membranes containing ~50-100 nm thick MCM-22 crystals[1] will be presented and compared with nanocomposites containing 2.5 nm exfoliated MCM-22 layers[2,3]. MCM-22 composites were prepared by depositing alternate layers of a matrix phase and oriented MCM-22 crystals and, for a certain appropriately selected matrix, they exhibit high hydrogen/nitrogen and carbon dioxide/nitrogen selectivities. Nanocomposites were prepared by swelling MCM-22 (P) layers by surfactant intercalation and employing melt-processing techniques to separate the layers[3]. A major challenge during nanocomposite processing, layered structure preservation[4], was successfully addressed as revealed by Small Angle X-Ray Scattering and Transmission Electron Miocroscopy[3]. The potential use of the nanocomposite films as easily processable, high-flux, high-selectivity gas separation membranes will be discussed.

[1] Leonowicz, M. E.; Lawton, J. A.; Lawton, S. L.; Rubin, M. K., Science, 1994, 264, 1910-13.

[2] Choi, J.; Lai, Z.; Ghosh, S.; Beving, D. E.; Yan, Y.; Tsapatsis, M., Industrial & Engineering Chemistry Research, 2007, 46, 7096-7106.

[3] Maheshwari, S.; Jordan, E.; Kumar, S.; Bates, F. S.; Penn, R. L.; Shantz, D. F.; Tsapatsis, M., Journal of the American Chemical Society, 2008, 130, 1507-1516.

[4] Choi, S.; Coronas, J.; Jordan, E.; Oh, W.; Nair, S.; Onorato, F.; Shantz, D. F.; Tsapatsis, M., Angewandte Chemie International Edition, 2007, 47, 3, 552-555.

544f Fabrication and Characterization of Sodalite-Polymer Nanocomposite Membranes

Dan Li and Huanting Wang. Chemical Engineering, Monash University, Wellington Road, Clayton, Melbourne, 3800, Australia

Zeolite-polymer nanocomposite membranes have received much attention, because these mixed matrix membrane materials may offer superior performance by taking advantage of the processability of polymers and the selectivity of zeolite crystals. In our research, we are interested in developing sodalite-polymer nanocomposite membranes for gas separation.

Sodalite has a framework consisting of a six-membered ring aperture with a pore size of 2.8 Å, and only allows very small molecules (e.g. H_2) to enter its channels. In order to develop high-selectivity and high-flux sodalite-polymer nanocomposite membranes for hydrogen separation, template-free sodalite nanocrystals with organic functional groups, which would enhance interfacial compatibility of inorganic sodalite materials with the chosen polymer, are required. In our previous research, we have successfully synthesized hydroxy-sodalite nanocrystals with amino organic functional groups (i.e. $=Si-(CH_3)(CH_2)_3NH_2$, denoted Sod-N) by the direct transformation of organic-functionalized silicalite nanocrystals.

This presentation focuses on the fabrication and characterization of sodalite-polymer nanocomposite membranes prepared by using organic functionalized sodalite nanocrystals. In particular, the emphasis will be placed on testing the membranes in hydrogen separation from other gas molecules (e.g. N_2, CO_2 or CO).

557a Concomitant Nucleation of Polymorphs

Aniruddh Singh, In Sung Lee, and Allan S. Myerson. Department of Chemical and Biological Engineering, Illinois Institute of Technology, 10 West 33rd St., Suite 223, Chicago, IL 60616

Polymorphism is the ability of a chemical compound to exist in more than one crystal structure. Different polymorphs of a drug substance can have different physicochemical properties and bioavailability. Thus, polymorph screening is essential in the early stages of the drug development process. The compound 5-methyl-2-[(2-nitrophenyl)amino]-3-thiophenecarbonitrile (known as ROY for its red, orange and yellow crystals) is an often studied system because of the large number of polymorphs which are known to exist. ROY has ten known polymorphs, seven with solved structures and three whose structures have not been solved. Recently, our group has engineered patterned substrates of self-assembled monolayers (SAMs), which consist of lyophilic gold islands surrounded by a lyophobic domain. With this approach, we have demonstrated the concomitant nucleation of multiple polymorphic forms under the same conditions with glycine, mefenamic acid, and sulfathiazole. In this work, crystallization experiments were conducted using patterned substrates of SAMs with solutions of ROY in dimethylsulfoxide (DMSO). The patterned bifunctional surface was immersed and slowly withdrawn from undersaturated solutions. The solution preferentially wetted the metallic islands, and as the solvent evaporated, ROY crystals exclusively nucleated on the lyophilic metallic islands. Raman microscopy was utilized to characterize the crystalline form on each metallic island. We were able to crystallize six of the seven stable polymorphs of ROY using this method, including form YT04, which to the best of our knowledge, has never been obtained from solution crystallization. In one of the experiments, over 10,000 islands were analyzed and three forms of ROY were obtained in very low percentages. This result shows how increasing the number of crystallization trials can help crystallize polymorphs which may not be obtained in a fewer number of trials due to statistical reasons.

557b Modeling and Prediction of the Crystal Structure of Pharmaceutical Co-Crystals

Panagiotis G. Karamertzanis[1], Andrei V. Kazantsev[2], N. Issa[3], G. W. a. Welch[3], Claire S. Adjiman[2], Constantinos C. Pantelides[4], and Sally L. Price[3]. (1) Imperial College London, Centre for Process Systems Engineering, London, SW7 2AZ, United Kingdom, (2) Department of Chemical Engineering, Imperial College London, Centre for Process Systems Engineering, London, SW7 2AZ, United Kingdom, (3) Department of Chemistry, University College London, 20 Gordon Street, London, WC1H 0AJ, United Kingdom, (4) Imperial College London / Process Systems Enterprise Ltd, Centre for Process Systems Engineering, London, SW7 2AZ, UK

Pharmaceutical co-crystals, comprising an active pharmaceutical ingredient (API) and another pharmaceutically acceptable molecule, can be used to address intellectual and physical property issues in pharmaceutical development without changing the chemical composition of the API. The search for suitable co-crystals is hindered by the lack of computational tools to investigate whether the formation of the heteromeric is thermodynamically favoured over the homomeric crystals, and predict their structures.

This paper presents a novel methodology for the prediction of co-crystal structures, based only on the atomic connectivity of the component molecules and assumed stiochiometry. The method uses global crystal energy minimisation, for both the single- and two-component crystals. This allows a comparison of the relative stability of the co-crystal with respect to the energies of the two pure-component crystals, and can therefore predict when co-crystal formation is thermodynamically favoured.

Crystal structure prediction of co-crystals is a challenging problem because of the presence of two, often flexible, crystallographically independent molecules in the asymmetric unit. Hence, a search that is sufficiently broad to reliably locate all energetically competitive minima on the crystal energy surface can typically be performed only with empirical, computationally inexpensive models that may lack sufficient realism for accurate prediction of relative energies. Therefore, the results of such a search need to be refined using more accurate models for both the intra- and inter-molecular contributions to the lattice energy.

In this paper, we propose a 3-stage algorithm:

- **Stage I: Global Search for low-energy minima.**

This is carried out using an approach[1] based on low-discrepancy sequences. The intramolecular energy is approximated using Hermite interpolants[2] on a grid of isolated-molecule energies, and its gradients with respect to all torsion angles that can be affected by intermolecular forces. The grid is pre-calculated using a number of quantum mechanical energy minimisations. The intermolecular interactions are modelled using a computationally inexpensive isotropic model, comprising an empirical repulsion-dispersion potential[3;4] and a set of conformationally invariant atomic charges.

· **Stage II: Re-ranking of low-energy crystal structures based on improved electrostatic interactions.**

The low-energy minima computed at Stage I are used as initial points for local energy minimisations. In this case, intramolecular energy and intermolecular repulsion-dispersion contributions are computed as at Stage I, but the intermolecular electrostatic interactions are modelled[5] by a set of distributed multipole moments on each atom up to hexadecapole.[6] The variation of the charge density with conformation is restricted to the analytic rotation of the atomic multipole moments with the local environment of each atom.

· **Stage III: Re-ranking of low-energy crystal structures based on conformation-dependent electrostatic interactions and accurate intramolecular energies.**

The low-energy minima computed at Stage II are used as initial points for local energy minimisations involving a nested 2-level optimisation approach.[7] The outer level manipulates the molecular conformation (i.e. the flexible torsion angles) while the inner one determines the optimal crystal structure for a molecule of fixed conformation (as determined by the current outer iteration). The intra-molecular energy and the atomic multipole moments are determined at each outer iteration using "on-the-fly" quantum mechanical calculations. Given the cost of this calculation, Stage III is applied only to very few (lowest-energy) crystal structures determined by Stage II.

The above algorithm is applied to the prediction of the structure of the co-crystals of p-aminobenzoic acid with 2,2'-bipyridine. The application of the algorithm to pure p-aminobenzoic acid resulted in its low-temperature polymorph[8] being ranked 1st i.e. being identified as the one of lowest energy. The experimentally known crystal structure[9] of 2,2'-bipyridine was ranked 2nd. The known p-aminobenzoic acid/2,2'-bipyridine crystal structure[10] is also ranked 1st. Since its energy is almost equal to the sum of the global minimum crystal energies of the component molecules, the predicted thermodynamic driving force for forming the cocrystal is small. We discuss the limitations of this methodology, including the possibility that p-aminobenzoic - 2,2'-bipyridine co-crystal formation may be kinetically rather than thermodynamically driven.

Keywords: co-crystals, crystal structure prediction, polymorphism, global optimisation

Reference List

(1) Karamertzanis, P. G.; Pantelides, C. C. *J.Comput.Chem.* **2005**, *26*, 304-324.

(2) Karamertzanis, P. G.; Pantelides, C. C. *Mol.Phys.* **2007**, *105*, 273-291.

(3) Williams, D. E. *J.Comput.Chem.* **2001**, *22*, 1154-1166.

(4) Coombes, D. S.; Price, S. L.; Willock, D. J.; Leslie, M. *J.Phys.Chem.* **1996**, *100*, 7352-7360.

(5) Willock, D. J.; Price, S. L.; Leslie, M.; Catlow, C. R. A. *J.Comput.Chem.* **1995**, *16*, 628-647.

(6) Stone, A. J. *J.Chem.Theory Comp.* **2005**, *1*, 1128-1132.

(7) Karamertzanis, P. G.; Price, S. L. *J.Comp.Theor.Comp.* **2006**, *2*, 1184-1199.

(8) Lai, T. F.; Marsh, R. E. *Acta Crystallogr.* **1967**, *22*, 885-&.

(9) Kuhn, F. E.; Groarke, M.; Bencze, E.; Herdtweck, E.; Prazeres, A.; Santos, A. M.; Calhorda, M. J.; Romao, C. C.; Goncalves, I. S.; Lopes, A. D.; Pillinger, M. *Chemistry-A European Journal* **2002**, *8*, 2370-2383.

(10) Smith, G.; Lynch, D. E.; Byriel, K. A.; Kennard, C. H. L. *J.Chem.Crystallogr.* **1997**, *27*, 307-317.

557c Crystallization and Polymorphism of Chiral Compounds

Qi Gao and Alicia Ng. Analytical R&D, Bristol-Myers Squibb, 1 Squibb Drive, New Brunswick, NJ 08901

It is a commonly shared experience among chemists/engineers working in the field of crystallization - and often a source of frustration - that chiral compounds tend to be more difficult to crystallize than racemic compounds. Indeed, racemic crystals occur more frequently than enantiomorphic crystals. This has been attributed to Wallach's rule, which states that racemic crystals tend to be denser - the left and right hands are arranged around centers of symmetry, and their packing is thus more efficient than their enantiomorphic counterparts, and therefore more stable. In addition, symmetry constraints limit the number of favorable packing arrangements available to chiral molecules. Using our own examples and some from the Cambridge Crystal Structure Database, this paper discusses polymorphism of chiral compounds, structure-stability relationships of polymorphs, and how they are related to their racemates.

557d Pseudopolymorphic Transitions of Sodium Naproxen In Mixed-Solvent Systems

Krystle J. Chavez and Ronald W. Rousseau. School of Chemical & Biomolecular Engineering, Georgia Institute of Technology, 311 Ferst Drive, Atlanta, GA 30332-0100

Solvents are often used in the production of crystalline pharmaceutical species. During processing, these solvents can become incorporated in the lattice of the crystal in a specific stoichiometric ratio as solvates or hydrates (also known as pseudopolymorphs). The properties of these pseudopolymorphs differ from those of the unsolvated species, which should not be surprising because their compositions are different. Accordingly, transformations among pseudopolymorphic forms often produce variations in physical properties such as density, solubility, dissolution rate, and bioavailability that can create difficulties in downstream processing. Thus, predicting the occurrence or transition of crystal pseudopolymorphs is of importance in developing robust, large-scale crystallization processes in the pharmaceutical industry.

Sodium naproxen is a non-steroidal, anti-inflammatory drug which has been shown to exhibit four pseudopolymorphic forms (anhydrous, monohydrated, dihydrated, and tetrahydrated). The purpose of the present study is to examine the transition of pseudopolymorphic forms of sodium naproxen under mixed solvent (methanol/water and ethanol/water) systems with varying temperature. Using batch cooling crystallization, transitions can be seen clearly through a van't Hoff plot of solubility data. Currently, water activity is thought to play an important role in determining when transitions occur. This study explores the role of water activity in pseudopolymorphic transitions for these mixed solvent systems; specifically comparing the water activity at the various transitions of the pseudopolymorphs and how this change in the different alcohol/water systems.

557e In-Situ Controlling and Monitoring of Polymorphic Crystallization and Transformation

Xia Yang[1], Xiujuan Wang[2], and Chi Bun Ching[2]. (1) School of Chemical and Biomedical Engineering, Nanyang Technology University, Singapore 637459, 62 Nanyang Drive, Singapore, Singapore, (2) School of Chemical and Biomedical Engineering, Nanyang Technological University, 62 Nanyang Drive, Singapore, 637459, Singapore

Polymorphism is a widespread phenomenon in organic compounds and is of crucial importance for chemical and pharmaceutical manufacturing. The purpose of this work was to study the influence of various experimental operating factors on the polymorphic crystallization and solution-mediated transformation processes. These factors included initial concentration of solution, temperature, cooling profile, agitation rate, additive, solvent, seeding etc.

The nucleation, metastability and solution-mediated transformation behavior of model compound (Glycine) have been investigated in detail using various off-line and in-situ monitoring techniques. The polymorphs were characterized by Powder XRD, FTIR, Raman microscope, SEM, as well as TGA and DSC. The crystallization experiments under different operating conditions were carried out in a Mettler Toledo LabMax automatic laboratory reactor system, which was equipped with a Lasentec D600L Focused Beam Reflectance Method (FBRM), a ReactIR and a Kaiser RamanRXN1 analyzer. In-situ transformation observation was also carried out with an optical microscope equipped with a Linkam THMS 600 heating/cooling stage.

On the basis of the in-situ measurements and observations of crystallization and transformation behavior with the variation of operating parameters, the nucleation mechanism of the metastable and stable polymorphs was proposed and discussed. The integration of the different off-line and in-situ analytical measurement techniques greatly assisted to accurately and quantitatively investigate the fundamental phenomena that govern the crystallization and transformation process so as to selectively control the polymorphic crystallization.

Title: Achieving Desired Polymorphs in API Process Development

Authors: Chenchi Wang, Qi Gao, Beth Sarsfield, Yuping Qiu, Victor Rosso, Daniel Hsieh, Shih-Ying Chang, and Chiajen Lai

Polymorphism of active pharmaceutical ingredients plays an important role in determining their physical and chemical properties, as well as their suitable use in drug substance application. The search for polymorphs requires thorough and systematic methodologies to gather sufficient understanding of crystal formation and stability in various solvent systems. A compound of interest undergoes multiple phase transformations during material preparations before leading to the desired crystal form. Polymorph control is made possible, from crystallization throughout drying, with knowledge collected in both polymorphism studies and process engineering to ensure consistent generation of the desired product. Our studies will be presented concerning the relation of different polymorphs, crystal form conversions, crystallization techniques and processing needed to tailor crystal morphology, crystallite or particle size, and solid properties.

Adsorption System Based on Microlith® Technology and Its Progress in Fuel Cell, Spacecraft, and Chem-bio Warfare Defense Applications

Christian Junaedi, Subir Roychoudhury, and Dennis Walsh, Microlith® Products Division, Precision Combustion, Inc. (PCI), 410 Sackett Point Rd., North Haven, CT 06473.

James C. Knox, Jay Perry, and David Howard, NASA Marshall Space Flight Center, AL 35812.

Patrick D. Sullivan, Air Force Research Laboratory, AFRL/RXQL, Tyndall AFB, 32403.

Abstract

The development of energy efficient, reduced-weight, regenerable adsorption systems for removal of environmental contaminants, such as CO_2 in spacecraft cabin air, chemical/biological warfare agents (CWA), and toxic industrial chemicals (TICs) is an area of continuing interest to NASA and Department of Defense, respectively. Precision Combustion, Inc. (PCI) has designed and developed a sorption technology using a bed consisting of metal meshes coated with a zeolite or molecular sieve sorbent. The metal mesh elements, trademarked and patented as Microlith® by PCI,[1] can be coated with various zeolite/sorbent materials and can effectively adsorb a number of contaminants of interest. The Microlith® elements also have the potential for direct electrical heating. The ability to directly, resistively heat the Microlith® substrate offers the potential for relatively rapid periodic regenerations instead of the longer thermal cycles typical of packed bed adsorbers. As a result, a regenerable adsorber using zeolite/sorbent coated on Microlith® can reduce the system weight and volume compared to the conventional packed bed configurations.

Another application of this Microlith®-based sorption technology is for efficient hydrogen sulfide (H_2S) removal from the reactor exhaust stream in a fuel reforming-fuel cell system. PCI developed zinc oxide (ZnO) coating process on the Microlith® substrates with a high loading, and demonstrated that ZnO powder deposited on Microlith® is capable to reach its nearly full capacity with high bed utilization. Our test showed that a fresh ZnO-coated Microlith® system can achieve a H_2S sorption capacity of up to ~33 wt.% (i.e., 0.33 gram of H_2S uptake per gram of ZnO). This is a significant improvement from a typical H_2S sorption capacity of 10-20 wt.% obtained using commercial ZnO pellets/extrudates.

This paper describes the design, development, and optimization of the Microlith® sorption technology for each of the above applications. We will also discuss the results obtained from various sorption tests, such as using CO_2, water, and trace chemical contaminants for NASA cabin air cleaning/Environmental Control and Life Support System (ECLSS) application as well as using TICs for chem-bio warfare defense application. Additionally, the development of this technology for sulfur (e.g., H_2S) removal in reforming/fuel cell application will be addressed.

Introduction

Adsorption processes have been of interest in many application fields, such as filtration of environmental contaminants, cabin air revitalization in spacecraft and spacestation, and adsorption of toxic industrial chemicals (TICs). For example, adsorber units have been used for cabin air quality control on all crewed spacecraft, and are expected to continue to remain at the forefront of spacecraft cabin air quality control technologies. As mission durations increase and exploration goals reach beyond Earth orbit, the need for regenerable adsorption processes for continuous removal of CO_2 and trace chemical contaminants becomes paramount. In another application, a regenerable adsorber system can be implemented in fielded military individual and collective protection systems against chemical/biological warfare agents and toxic industrial chemicals.

NASA-supported research at PCI has demonstrated that zeolites deposited on Microlith metal mesh elements can effectively adsorb a number of the contaminants of interest, such as CO_2, ethanol, and ammonia. The inert Microlith substrates and the use of a binder during deposition of the zeolites on them result in volumetric sorbent loadings that are considerably lower than the conventional carbon bed and packed bed systems. However, the unique ability to *directly resistively heat* the Microlith metal mesh support allows for rapid periodic regenerations via direct internal heating.[2,3] Therefore, the weight and volume of the conventional adsorber subassemblies can potentially be reduced by implementing zeolites supported on Microlith and by employing periodic sorbent regeneration.

Furthermore, the ability to coat different sections of Microlith® mesh substrates with different zeolite/sorbent washcoats for removing different chemicals offers a system benefit for combining two or more adsorber assemblies into a single adsorber unit and for tailoring an adsorber system based on the application. This capability allows for weight, volume, and logistic savings. For example, in the NASA program, we have fabricated and delivered an integrated adsorber unit capable of removing both CO_2 and trace chemical contaminants within the same bed by coating two sections of the Microlith with two different zeolites. Both the trace contaminant control system (TCCS) and carbon dioxide removal assembly (CDRA) performance was demonstrated within the volume envelope of the current CDRA unit, thus eliminating the need for the current TCCS unit entirely. Using a similar design concept, we designed, fabricated, and delivered adsorber systems for removing TICs to Air Force Research Laboratory (AFRL).

We recently examined the performance of zinc oxide (ZnO) powder coated on Microlith® substrates for removing sulfur, such as hydrogen sulfide (H_2S), from the product stream of fuel processors. To date, on-board fuel processing of liquid fuels, such as diesel and JP-8, has been investigated by many research groups and of interest to both government and industry as a potential option to supply hydrogen for either polymer electrolyte membrane (PEM) or solid oxide fuel cell stacks for auxiliary power unit (APU) systems.[4-9] The sulfur compounds from the logistic fuels, however, can poison both the fuel reformer catalyst and the fuel cell catalyst. For example, the sulfur level in the H_2-rich stream needs to be less than 100 ppb_v prior to entering a typical PEM fuel cell stack.[4] Therefore, sulfur compounds from the fuels have to be removed before the H_2-rich stream is sent to the fuel cell stacks. Depending on the application and the sulfur concentration in the logistic fuels, the sulfur removal can be performed either in the liquid phase (i.e., prior to the fuel reformer units) by implementing liquid-fuel sulfur sorbent materials or in the gas phase by passing the reformate stream through a ZnO-based unit.[10-15] The results from our sulfur tests showed that the ZnO-coated Microlith substrates were able to reach nearly full ZnO capacity with high bed utilization while maintaining a low pressure drop across the bed.

Microlith® Technology

The Microlith technology consists of a series of ultra-short-channel-length, low-thermal-mass metal meshes as shown in Figure 1. It replaces the long channels of conventional monoliths with a series of short-channel-length substrates. Whereas in a conventional honeycomb monolith a fully developed boundary layer is present over a considerable length of the device, the very short channel length characteristic of the Microlith substrate avoids boundary layer buildup. Since heat and mass transfer coefficients depend on the boundary layer thickness, avoiding boundary layer buildup enhances transport properties. The Microlith can also pack more active surface area into a small volume, providing increased adsorption area for a given pressure drop. The effectiveness of the Microlith technology and the long-term durability of PCI's proprietary zeolite and catalyst coatings have been rigorously demonstrated in applications such as exhaust after-treatment,[16] space station cabin air cleaning,[2,3,17,18] catalytic combustion,[19] and fuel processing.[20-23]

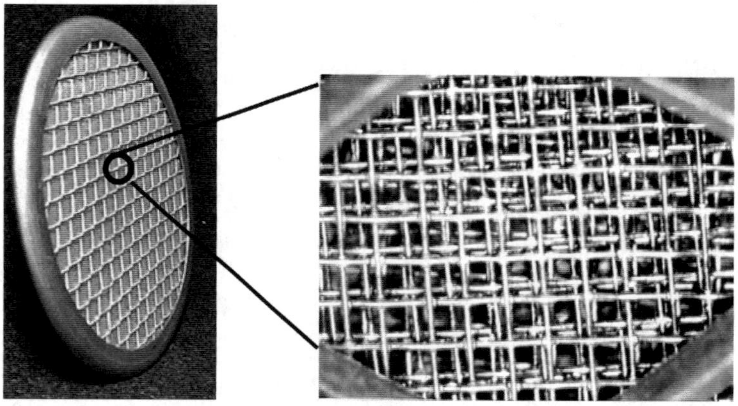

Figure 1. Example of a Microlith substrate assembly.

During the zeolite and other sorbents coating development, several washcoat formulations were evaluated in an effort to produce an adherent coating. The Microlith substrate geometry poses unique challenges in the development of adherent coatings. The washcoats must be easily applied, and the resulting coatings must have a high degree of adhesion and cohesion and must be sufficiently abrasion resistant in order to withstand routine handling and multiple thermal cycles. At the same time, the formulation must retain the desired chemical and physical characteristics of the zeolite and ZnO. The washcoats were applied to the Microlith substrates using methods developed at PCI that allow for rapid application of sorbent coatings to large volumes of substrate. Scanning electron microscopy (SEM) micrographs of the sorbent-coated Microlith substrates are shown in Figure 2 and Figure 3. SEM analyses indicated uniform coatings with complete coverage. Furthermore, the cross-sectional SEM view (Figure 3) indicated that the coating was adherent and that the thickness of the coating was relatively uniform on the surface of the substrate. A few cracks were observed in the coating layer due to cutting of the Microlith substrate prior to SEM cross sectional analysis.

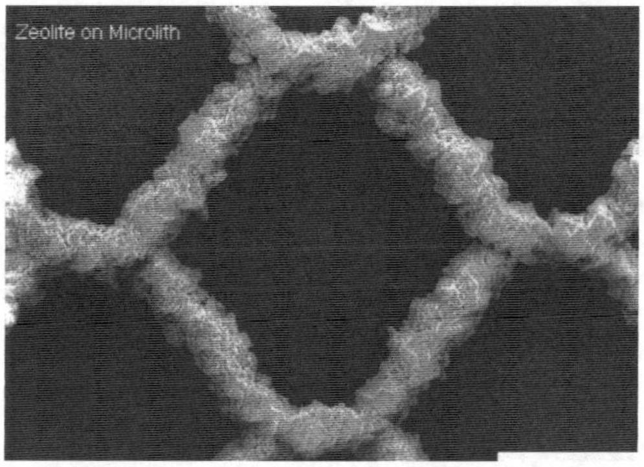

Figure 2. Surface-scan SEM micrograph of sorbent-coated Microlith.

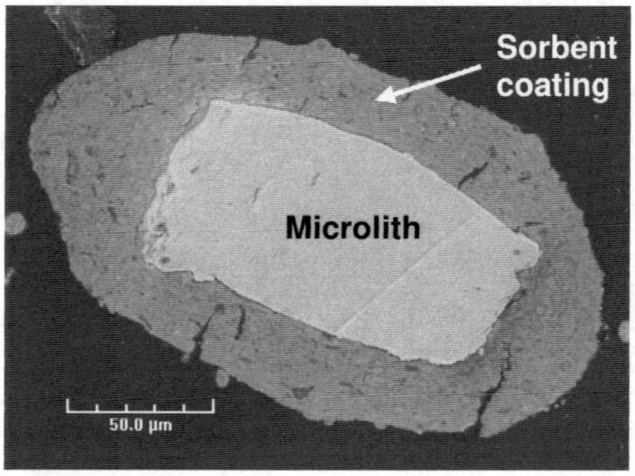

Figure 3. Cross sectional SEM micrograph of the sorbent-coated Microlith substrate.

Regenerable Adsorber Units Design and Fabrication

The original electrothermal swing adsorption (ESA)-based adsorber design consisted of a linear stack of alternate layers of sorbent-coated Microlith screens and sorbent-coated insulating meshes. The mesh serves as an insulating layer to isolate the Microlith screens electrically when applying the resistive heating during the ESA regeneration mode. Implementing this linear-stack design, however, presented practical difficulties.

Subsequently, a unique adsorber design concept consisting of a "jelly roll" coil in a radial flow configuration was implemented.[3,24] The radial design was based on a two-layer "sandwich" system consisting of a *continuous* layer of sorbent-coated Microlith screens and a corresponding insulating layer coated with the same sorbent material. Similar with the previous design, the insulating mesh serves as an insulating layer for isolating the Microlith layer electrically. Upon coiling the two-layer system around a centerline feed tube (i.e., a "jelly-roll" coil), the radial flow configuration can be achieved as illustrated in Figure 4. A preliminary calculation indicated that the radial flow arrangement provided volumetric sorbent loadings that are at least comparable to a linear stack of screen elements. Furthermore, from the electrical and hardware assembly vantage points, a continuous length of coated screens largely mitigated the complicating issues encountered with a stack of screens, such as shorting and reactant channeling. Therefore, direct electrical heating of the metal Microlith substrate to regenerate the sorbent could be implemented more readily in the radial flow arrangement.

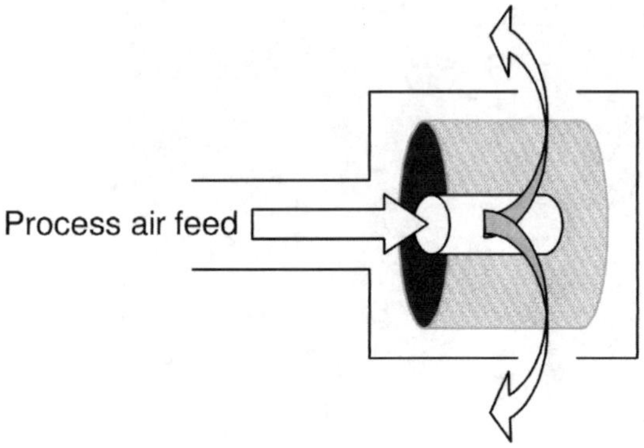

Figure 4. A simplified Microlith®-based radial flow adsorber design consisting of a "jelly-roll" coil of sorbent-coated Microlith screens and sorbent-coated insulating meshes.

Figure 5 shows detailed external and internal cross-section views of a Microlith-based radial flow adsorber unit with internal resistive heating capability. To date, PCI has delivered several adsorber units with this design concept rated for 5-cfm flow rate to NASA and AFRL. A computational fluid dynamic (CFD) simulation software package was used to study the flow pattern of process air within the adsorber unit. The unit design and sizing were optimized based on the modeling results in order to obtain a uniform flow distribution in the radial flow configuration and to avoid the presence of recirculation zones while minimizing the total housing volume (i.e., increasing the overall volumetric efficiency).

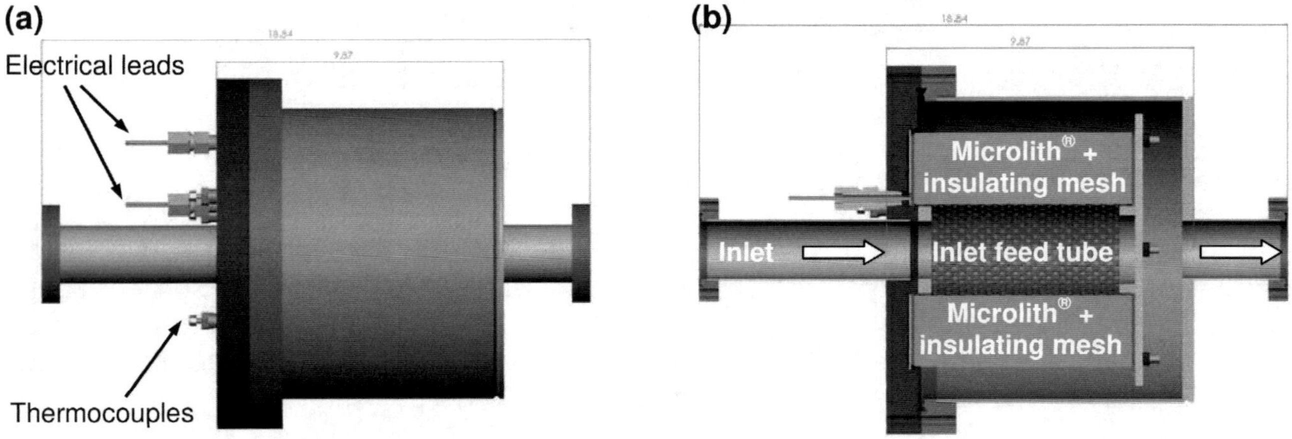

Figure 5. (a) External and (b) Internal cross-section views of the Microlith®-based radial flow adsorber design concept.

Finally, Figure 6 shows the Microlith-based regenerable adsorber unit after integration with the external housing. The adsorber unit was connected both to a vacuum line for the regeneration process and to a process air feed line for the sorption test.

Figure 6. Microlith®-based regenerable adsorber unit rated for 5-cfm flow rate after integration with the external housing.

Results and Discussion

(i) H_2O, CO_2, and trace chemical contaminants sorption for NASA cabin air/ECLSS application

Recent development testing supported by NASA successfully demonstrated "proof of concept" as well as a one-person prototype demonstration of coated adsorbents on the Microlith substrate for a regenerable CO_2 removal and trace contaminants control (TCCS) application. The adsorber unit was exposed to a contaminant-laden airflow containing ethanol, acetone, toluene, ammonia, and CO_2, and the adsorption capacity was evaluated via multiple sorption/regeneration cycles up to at least 500 hours of operation. For a dual function CO_2 removal/TCCS unit, the appropriate lengths of the Microlith and insulating layers in the "jelly roll" coil were coated with the desired amounts of the preferred sorbents as determined from bench-scale tests.

This adsorber unit was targeted to operate at an approximate nominal contact time of ~1.8 sec. This corresponded to a volumetric flow rate of ~5 cfm through the annular sorbent-coated coil volume of 4235 cm^3, which was the total volume of the "jelly roll" coil (CO_2 and trace contaminant removal segments). Part of the coil volume was devoted to CO_2 sorption by coating the appropriate length of the Microlith substrate and the insulating mesh with Molecular Sieve 5A (MS5A), and the remaining part of the coil was designated for the trace contaminants removal. The TCCS sorbents coating was used for the removal of ethanol, acetone, ammonia, toluene, and dichloromethane (DCM). This design was estimated to be sufficient for maintaining the exit concentration of the trace contaminants significantly below their inlet levels during the sorption cycle, whereas the regeneration requirement of the CO_2 removal segment determined the cycle time for the adsorber unit.

Prior to testing, the adsorber unit was heated under vacuum until the lowest registering thermocouple reached ~230-240ºC. The total flow rate of the feed process air was ~5 cfm, and the component concentrations were 3500 ppm CO_2, 500 ppm ethanol, 10 ppm NH_3, 11 ppm acetone, 8 ppm toluene, and 5 ppm DCM. During the regeneration process, the electrical lead terminals (Figure 5a) were connected to a wall AC power supply regulated by a variac. Power requirements were examined for direct electrical heating of the "jelly roll" coil based on estimates of the total energy required to heat the mass of sorbent washcoats and to supply the heat of desorption (dominantly that of CO_2, but also that of the trace contaminants). Assuming a 45-minute sorption/regeneration cycle, ~110–150 watts would be required depending upon the final target regeneration temperature (230–300ºC).

Figure 7 represents the results from the first three prototype testing cycles. The data in the figure indicate poor performance but improving from cycle to cycle, suggesting the need for a better process to remove moisture and/or other sorbed species initially to establish a dry, clean surface effective for sorption. After the fourth cycle, the adsorber unit was regenerated more aggressively by flowing dry nitrogen at ~5 slpm through the unit for ~4.5 hours at the regeneration temperature. After cooling the unit in a dry nitrogen flow to room temperature, sorption cycle 5 was then performed and the results are shown in Figure 8.

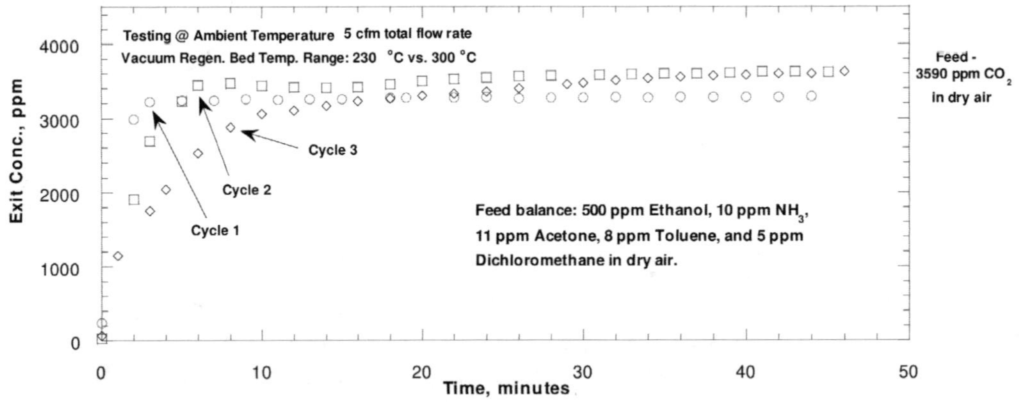

Figure 7. Cycle-to-cycle variation of the CO_2 exit concentration as a function of time observed during the performance test of the CO_2/trace contaminants removal unit.

The results from sorption cycle 5 in Figure 8 indicated 3 wt% CO_2 sorption capacity based on the MS5A washcoat (i.e., 0.03 gram of CO_2 sorbed per gram of MS5A washcoat), and a 45-50 minute cycle time during which ~55% of the delivered CO_2 was adsorbed. The marked improvement in CO_2 sorption capacity compared with the first three cycles underscored the need to establish the correct initial sorbent surface conditions and dryness to achieve proper performance. During this cyclic test, the concentrations of trace contaminants in the exit gas at the end of each cycle were well below their inlet levels. With the exception of ethanol, the exit concentrations of trace contaminants were zero. The ethanol exit level was only 5 ppm or 1% of its inlet concentration. While these end-of-cycle trace contaminant levels might rise over multi-cycle operation of the prototype, our previous durability studies suggest that they will equilibrate at levels appreciably lower than their inlet concentrations. Measurements indicated that the pressure drop of the unit was <0.5 inch of water at 5-cfm process air flow rate. Furthermore, the sorbent coatings on both the Microlith substrate and the insulating mesh showed good adhesion and durability after multiple thermal cycles during the sorption and regeneration cyclic test.

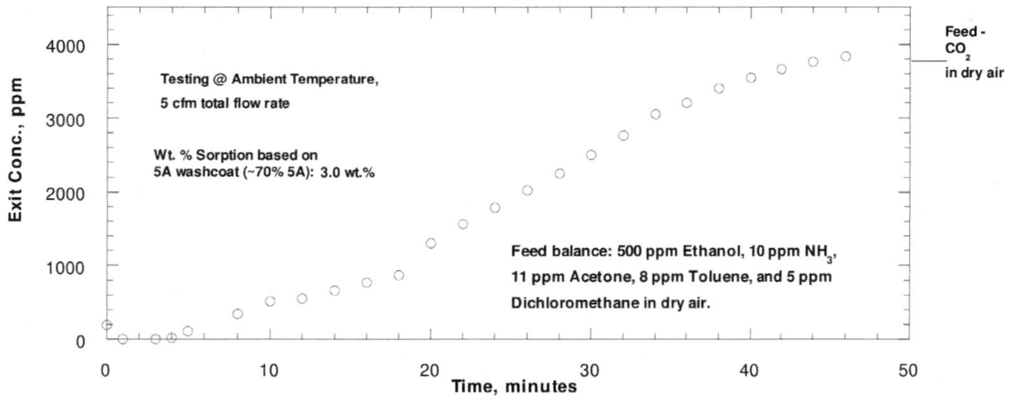

Figure 8. Exit concentration of CO_2 as a function of time obtained from the sorption cycle 5 test of the CO_2/TCCS prototype unit. Cycle 5 of the sorption test was performed following a more aggressive regeneration process to ensure a dry, clean surface effective for sorption.

In the current CO_2 removal assembly located in the ISS, the process air is passed through a water removal unit prior to the CO_2 and trace chemical contaminants removal. Drying of the process air is desirable in order to mitigate the effect of water co-adsorption on the CO_2 and trace chemical contaminants adsorption process and to collect the water to be returned to the cabin for recycling and crew use. PCI has recently developed a Microlith-based regenerable water removal unit by coating both the Microlith substrate and the insulating mesh with zeolite 13X.

Figure 9 shows the results obtained from the cyclic sorption test performed at PCI (Figure 9a) and from a single water sorption test performed at NASA (Figure 9b) using 5-cfm process air feed at room temperature. The water vapor inlet concentration for the cyclic test (Figure 9a) was ~12,100 ppm_v (50% RH, 21ºC, 1 atm). At the end of each cycle, the water removal unit was regenerated for 20 minutes at 200ºC by applying direct resistive heating while flowing dry air at 20 slpm. The subsequent sorption test was performed after the unit was cooled to room temperature by flowing dry air at 5 cfm. The results indicated that the sorption capacity remained unchanged at 4.2 wt.% (based on the zeolite 13X washcoat) after four sorption cycles with 20-minute cycle time. For the water sorption test performed at NASA, the water vapor inlet concentration was ~8300 ppm_v (34% RH, 21ºC, 1 atm). Although the sorption test at NASA was performed prior to a prolonged regeneration process in dry nitrogen, the unit still gave a good performance by removing ~95% of water vapor in the first 17 minutes and was able to maintain the water removal at more than 80% in the remaining 45-minute test.

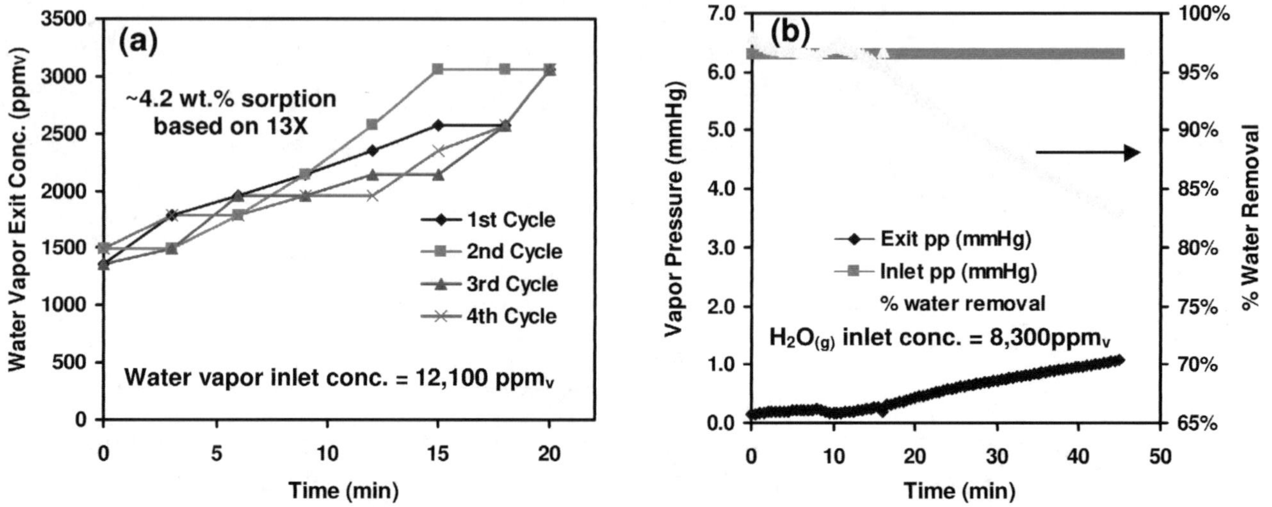

Figure 9. (a) Cycle-to-cycle variation of water vapor exit concentration as a function of time obtained from the cyclic test and (b) Inlet and exit water vapor pressure as well as the % water removal as a function of time from a single water sorption test.

(ii) TICs sorption for chem-bio defense application

Four prototype adsorber units with a design concept similar to that of the NASA adsorber systems were developed for AFRL for testing as filters against toxic industrial chemicals (TICs). The four adsorber units consisted of two water removal units and two units for removing TICs. Two sets of filters working in tandem (i.e., an ESA-based system) are used. Process air will first be passed through the water removal unit in order to reduce the effect of water co-adsorption on the TIC filter and thus maximizing the sorption capacity for TIC components.

Figure 10 shows the preliminary results of TIC adsorption isotherms obtained from various sorbent materials, including (i) activated carbon fiber cloth (ACFC) supplied by American Kynol, (ii) carbon fiber composite molecular sieve (CFCMS) developed by Oak Ridge National Laboratory (ORNL), and (iii) sorbent-coated Microlith substrate and insulating mesh developed by PCI. Based on the % weight change results for each sorbent material shown in Figure 10a, we estimated the volumetric sorption capacity (lb of TIC sorbed per ft^3 of sorbent) for each material (Figure 10b). The results indicated that MS5A coated on insulating meshes gave the highest sorption capacity, much higher than the capacities obtained by the zeolite Y-coated insulating meshes and by ACFC. For a dual-function TICs removal unit, the appropriate lengths of the Microlith and insulating layers in the "jelly roll" coil were coated with the desired amounts of the preferred sorbents. Currently, sorption cyclic tests are being performed at AFRL to evaluate the TICs sorption capacity and to compare the Microlith-based regenerable adsorber performance with the other carbon-based sorbent systems over multiple sorption/regeneration cycles.

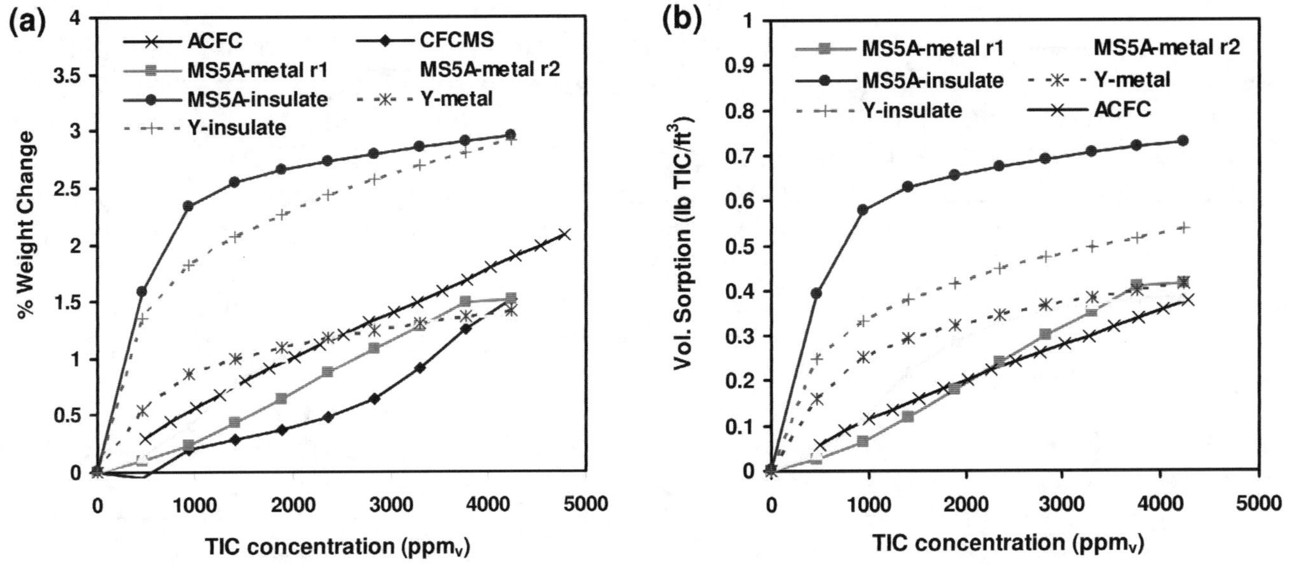

Figure 10. TIC adsorption isotherms: (a) % weight change and (b) volumetric sorption capacity (lb of TIC sorbed per ft^3 of sorbent) as a function of TIC concentration obtained for various sorbent materials.

(iii) H_2S sorption for fuel reforming-fuel cell application

Recently, PCI has developed a formulation and an application method for coating the ZnO powder on PCI's Microlith substrate. We then performed H_2S sorption tests on the ZnO-coated Microlith substrate to evaluate its sulfur sorption capacity, regenerability, and cycle-to-cycle variation. The sulfur sorption testing employed a linear stack of ZnO-coated Microlith screens, which was inserted into a stainless steel tube.

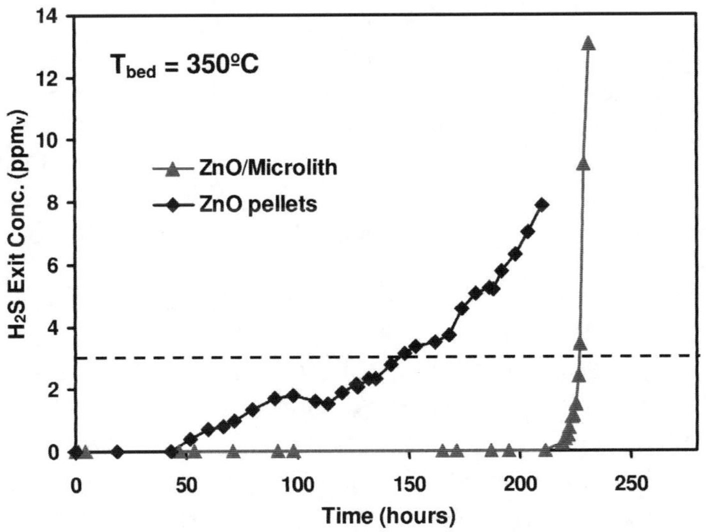

Figure 11. H_2S exit concentration as a function of time obtained from two H_2S sorption tests. The H_2S inlet concentration was 115 ppm$_v$. The H_2S breakthrough level (3 ppm$_v$) is indicated by a dashed line.

Figure 11 shows the H_2S outlet concentration as a function of time obtained from the H_2S sorption tests on: (i) ZnO coated on Microlith substrate and (ii) ZnO pellets bed under the same operating conditions (Table 1). In these tests, a surrogate gas mixture containing 115 ppm_v H_2S was flowed into the ZnO beds. The temperature of the beds was maintained at 350ºC. The surrogate gas was a blend of 23.4 mole % H_2, 16.4% CO, 19.9% H_2O, and 40.3% N_2, which simulates PCI's ATR reformate stream composition when operating with 500 ppm_w sulfur fuel. For the ZnO pellets experiment, the ZnO bed was 1/8 inch long ZnO pellets packed in a stainless steel tube, 1 inch in diameter and 1.01 inch long. In these tests, the concentrations of H_2S at the inlet and outlet streams were measured using a gas chromatograph (GC) equipped with a flame photometric detector (FPD) and were checked periodically using Drager tubes.

Table 1. Detailed test conditions and the resulting H_2S sorption capacities for H_2S sorption tests on: (i) ZnO-coated Microlith and (ii) ZnO pellets.

	ZnO/Microlith	**ZnO pellets**
Gas composition (mole %)	23.4% H_2, 16.4% CO, 40.3% N_2, 19.9% H_2O	23.4% H_2, 16.4% CO, 40.3% N_2, 19.9% H_2O
Total gas flow rate	0.633 slpm	0.880 slpm
V_{eff} or V_{bed}	V_{eff} = 0.57 in^3 = 9.4 cm^3 V_{bed} = 0.90 in^3 = 14.7 cm^3	V_{bed} = 0.78 in^3 = 13.0 cm^3
Bed Temperature	350°C	350°C
ZnO weight	4.011 gm	17.974 gm
Inlet H_2S concentration	Drager analysis: 115 ppm_v	Drager analysis: 115 ppm_v
H_2S uptake	1.405 gm	1.248 gm
Breakthrough time	227 hours	146 hours
Wt.% H_2S sorption (3 ppm_v breakthrough)	**33.0 wt.% (based on ZnO)**	**6.95 wt.% (based on ZnO pellets)**
Vol. sorption capacity (gm H_2S / V_{bed})	**0.096 gm/cm^3**	**0.096 gm/cm^3**

The resulting weight percent H_2S sorption capacity for the ZnO/Microlith was ~33.0 wt.% based on the ZnO (i.e., 0.33 gram of H_2S uptake per gram of ZnO). This is almost five-fold higher capacity than the ZnO pellets capacity obtained at the same operating conditions. The H_2S breakthrough occurred after ~227 hours of continuous exposure to 115 ppm_v H_2S. This high wt.% sorption capacity may be due to the presence of well-dispersed ZnO particles on the surface of Microlith, and thus the ZnO active surfaces are more accessible for the reaction with H_2S. The volumetric sorption capacity for this ZnO/Microlith system based on the bed volume was 0.096 gram H_2S uptake/cm^3, which is expected to be lower than the volumetric sorption capacity for ZnO pellets systems due to the presence of inert Microlith substrate. However, the unique ability to *directly resistively heat* the Microlith metal mesh support allows for rapid periodic regenerations via direct internal heating, which can potentially reduce the weight and volume of the H_2S removal beds.

The most recent H_2S sorption test on ZnO pellets indicated a breakthrough after 146 hours of continuous exposure to 115 ppm_v H_2S, resulting in a H_2S sorption capacity of 6.9 wt.% based on the ZnO pellets (i.e., 0.069 gram of H_2S uptake per gram of ZnO pellets). This corresponds to a volumetric sorption capacity of 0.096 gram H_2S uptake/cm^3, which was surprisingly low and was the same as the capacity for the ZnO/Microlith bed. More experiments still need to be performed on these beds to evaluate and to compare their performance.

Summary

Ongoing efforts with NASA and DoD have shown the potential to implement sorbent-coated Microlith substrates for various sorption applications, such as water removal, CO_2 removal, TIC filtration, and sulfur removal. From these programs, PCI has been able to develop coating formulations and robust application methods for several different sorbents, such as MS5A, zeolite Y, and zeolite 13X, with resulting coatings that are adherent and able to withstand multiple thermal cycles. Additionally, the ability to coat different sections of Microlith® mesh substrates with different zeolite/sorbent washcoats for removing different chemicals offers a system benefit for combining two or more adsorber assemblies into a single adsorber unit and for tailoring an adsorber system based on the requirements. This capability allows for weight, volume, and logistic savings.

Furthermore, the ability to directly, resistively heat the Microlith® substrate offers the potential for relatively rapid periodic regenerations instead of the longer thermal cycles typical of packed bed adsorbers. As a result, a regenerable adsorber using zeolite/sorbent coated on Microlith® can reduce the system weight and volume compared to the conventional packed bed configurations by using a more periodic regeneration. Since these programs were primarily a science- and technology-based discovery effort, there was limited scope for rigorously exploring economic feasibility and long-term performance.

To date, PCI has designed, developed, and delivered several regenerable adsorber units with radial flow concept to NASA and AFRL. These units have been tested for water removal, CO_2/trace contaminant removal, and TICs removal, and the performance results indicated good sorption capacity with the ability to apply direct internal heating. Additionally, we have further developed this Microlith-based sorption technology for efficient sulfur removal (i.e., H_2S) in the reforming/fuel cell application. The results from our sulfur tests showed that the ZnO-coated Microlith substrates were able to reach nearly full ZnO capacity with high bed utilization while maintaining a low pressure drop across the bed.

Acknowledgments

The authors gratefully acknowledge the supports from NASA and DoD (DTRA, AFRL and CERDEC) in conducting a significant portion of this research. Any opinions, findings, and conclusions or recommendations expressed in this publication are those of the authors and do not necessarily reflect the views of NASA or DoD. We are also grateful to the technical support group at PCI.

References

1. U.S. Patent #5,051,241, September 24, 1991.
2. S. Roychoudhury, D. Walsh, and J. Perry, *SAE Int.*, 2004-01-2442 (2004).
3. S. Roychoudhury, D. Walsh, and J. Perry, *SAE Int.*, 2005-01-2866 (2005).
4. C. Song, *Catal. Today*, **77**, 17 (2002).
5. J. R. Rostrup-Nielsen, *Phys. Chem. Chem. Phys.*, **3**, 283 (2001).
6. J. D. Holladay, E. O. Jones, M. Phelps, and J. Hu, *J. Power Sources*, **108**, 21 (2002).
7. E. C. Wanat, B. Suman, and L. D. Schmidt, *J. Catal.*, **235**, 18 (2005).
8. Q. Ming, T. Healey, L. Allen, and P. Irving, *Catal. Today*, **77**, 51 (2002).
9. A. S. Patil, T. G. Dubois, N. Sifer, E. Bostic, K. Gardner, M. Quah, and C. Bolton, *J. Power Sources*, **136**, 220 (2004).
10. A. J. Hernandez-Maldonado and R. T. Yang, *Cat. Rev. - Sci. Eng.*, **46**, 111 (2004).
11. A. Zhou, X. Ma, and C. Song, *J. Phys. Chem.*, **110**, 4699 (2006).
12. H. Y. Yang, Y. Lu, and B. J. Tatarchuk, *J. Power Sources*, **174**, 302 (2007).
13. G. Alptekin, S. DeVoss, M. Dubovik, J. Monroe, R. Amalfitano, and G. Israelson, *J. Mater. Eng. Perform.*, **15**, 433 (2006).
14. Z. Wang and M. Flytzani-Stephanopoulos, *Energy & Fuels*, **19**, 2089 (2005).
15. L. Li and D. L. King, *Catal. Today*, **116**, 537 (2006).
16. S. Roychoudhury, J. Bianchi, G. Muench, and W. C. Pfefferle, *SAE Int*, 971023 (1997).
17. R. Carter, J. Bianchi, W. C. Pfefferle, S. Roychoudhury, and J. L. Perry, *SAE Int.*, 972432 (1997).
18. J. L. Perry, R. N. Carter, and S. Roychoudhury, *SAE Int.*, 1999-01-2112 (1999).
19. G. Kraemer, W. C. Pfefferle, J. Ritter, et al., *Proc. Int. Joint Power Generation Conference*, ASME International (1997).
20. M. Lyubovsky, H. Karim, S. Roychoudhury, et al., *Catal. Today*, **83**, 183 (2003).
21. S. Roychoudhury, M. J. Castaldi, M. Lyubovsky, R. LaPierre, and S. Ahmed, *J. Power Sources*, **152**, 75 (2005).
22. M. Lyubovsky, S. Roychoudhury, and R. LaPierre, *Catal. Lett.*, **99**, 113 (2005).
23. M. J. Castaldi, R. LaPierre, M. Lyubovsky, W. C. Pfefferle, and S. Roychoudhury, *Catal. Today*, **99**, 339 (2005).
24. U.S. Patent #7,141,092 B1, November 28, 2006.

Contact

Author for correspondence: Christian Junaedi, Ph.D. (cjunaedi@precision-combustion.com)

593c Understanding the Binding Mechanism of Mercury on Activated Carbon

Bihter Padak, Energy Resources Engineering, Stanford University, 367 Panama St, Department of Energy Resources Engineering, Stanford, CA 94305-2220 and Jennifer L. Wilcox, Department of Energy Resources Engineering, Stanford University, 367 Panama St., Green Earth Sciences 077A, Stanford, CA 94305-2220.

Coal combustion power plants constitute a significant amount of mercury released into the atmosphere. Oxidized mercury is captured by wet scrubbers, while gaseous elemental mercury passes through the scrubbers readily. Particulate matter such as fly ash, unburned carbon and activated carbon can be used to capture elemental and oxidized mercury through adsorption processes. Activated carbon is currently used in many power plants effectively. Understanding the mechanism by which mercury adsorbs on activated carbon is crucial for developing improved capture technologies for mercury. In this study we examine the possible binding mechanism of mercury on activated carbon, by performing ab initio energetics calculations using Gaussian03. The activated carbon surface is modeled by a single graphene layer in which the edge atoms on the upper side are unsaturated in order to simulate the active sites. In some cases, chlorine atoms are placed at the edge sites to understand the effect of chlorine on mercury binding and to investigate a possible surface reaction between mercury and chlorine.

593d Sulfur Functionalization of Non-Carbon Materials with Applications In Mercury Sorption

Noah D. Meeks, Dept of Chemical and Materials Engineering, University of Kentucky, 177 F. Paul Anderson Tower, Lexington, KY 40506 and D. Bhattacharyya, Chemical and Materials Engineering, University of Kentucky, 177 F. Paul Anderson Tower, Lexington, KY 40506.

Functionalized materials such as silica have been used in a variety of applications such as catalysis, chromatography, and adsorption. Recently there has been increased interest in functionalization of non-carbon materials for the development of effective mercury sorbents, as alternatives to activated carbon sorbents. Previous work has shown that functionalization of both silica and copper doped iron nanoaggregates (1) with tetrasulfur silane compounds leads to particles which have proven effective in lab- and pilot-scale studies for the efficient adsorption of mercury from coal-fired power plant streams.

This study presents the in-depth sulfur functionalization and characterization of functionalized silica and other non-carbon based sorbents, and gives a fundamental understanding of the synthesis and applications of this alternative technology. The functionalization reaction of non-carbon materials is probed using diffuse reflectance infrared spectroscopy (DRIFTS) and other techniques. Preliminary DRIFTS results have given extents of functionalization for various materials. The adsorption of mercury onto these functionalized materials is then presented, including temperature-dependent results. These sorbents undergo various thermal transitions in the temperature ranges of interest, and pyrolysis of the tetrasulfur silane (for example) occurs at about 230 °C. The species involved in these transitions, their impact on mercury adsorption, and the adsorption products, are analyzed using DRIFTS as well as differential scanning calorimetry (DSC), thermogravimetric analysis (TGA), X-ray photoelectron spectroscopy (XPS), scanning and transmission electron microscopy (SEM & TEM) and X-ray diffraction (XRD).

1. Meyer, D. E.; Sikdar, S. K.; Hutson, N. D.; Bhattacharyya, D., Examination of sulfur-functionalized, copper-doped iron nanoparticles for vapor-phase mercury capture in entrained-flow and fixed-bed systems. Energy & Fuels 2007, 21, (5), 2688-2697.

593e A Monte Carlo Study of Molecular Selectivity of Binary Mixtures In Ordered and Disordered Microporous Carbon

Thomas R. Roussel[1], Jeremy Palmer[1], Christophe Bichara[2], Roland Pellenq[2], and Keith Gubbins[1]. (1) North Carolina State University, Engineering Building I, Box 7905, Raleigh, NC 27695-7905, (2) CINAM - University de la Medditerranee, 13009 Marseille, North Carolina State University, Engineering Building I, Box 7905, Raleigh, NC 27695-7905

Carbonaceous microporous materials are useful for storing and/or separating various molecules, such as hydrogen (H2), methane (CH4), and carbon dioxide (CO2). Activated carbons are amorphous materials synthesized from organic precursors pyrolized, such as nut shells, woods, and charcoals. Their excellent surface activity and low cost of production make them ideal materials for many industrial applications such as gas and liquid filtration, gas storage, and catalytic processes. One of the application branches is to capture greenhouse gases, mainly CO2 and CH4 through human activities and industrial process. Since their pore sizes are nanometric, they still represent interesting candidates for fuel cell applications for separation and on-board vehicle storage of hydrogen and methane mixture at room temperature.

The control of their pore size distribution is one of the issues during their synthesis, resulting mainly in size-dispersed and disordered pores. To optimize such materials, we need to identify the structural characteristics of such 3D interconnected microporous carbons as a function of the application. Indeed, the number of obtainable nanostructures is as large as the myriad of precursors. In order to investigate these issues, we performed Monte Carlo simulations to study and compare disordered and ordered microporous carbon properties. Indeed, a new class of well-controlled and 3D-interconnected ordered microporous carbon arises from in-zeolited templating synthesis. We compare ordered carbon templated by faujasite zeolite from Grand Canonical Monte Carlo simulations with several disordered carbon structures derived from an atomistic and geometrically realistic simulation technique, known as Hybrid Reverse Monte Carlo (HRMC), using as input experimental x-ray diffraction data.

593f A New Group-Contribution Model for Prediction of Adsorption Equilibrium of Gases In Porous Adsorbents

Yuqing Ding, Air Purification Engineering, Tyco International Ltd / Scott Health and Safety, 4320 Goldmine Rd, Monroe, NC 28110 and M. Douglas LeVan, Department of Chemical and Biomolecular Engineering, Vanderbilt University, VU Station B 351604, 2301 Vanderbilt Place, Nashville, TN 37235-1604.

A new group-contribution model is developed to predict adsorption equilibrium of non-reactive gases in porous adsorbent media. In this model, we convert the adsorption equilibrium problem into a traditional liquid-vapor equilibrium problem by assuming that the adsorbent can be described by a virtual chemical (a virtual group), and thus the adsorbate-adsorbent system (an adsorbate-adsorbent solution) can be treated like a liquid mixture. We solve the resulting VLE problem with the popular UNIFAC equation, taking advantage of the readily available UNIFAC database for common chemical groups. The UNIFAC parameters for a virtual group are obtained from experimental adsorption equilibrium data on simple molecules. These parameters are then used to predict the adsorption equilibrium of more complex molecules containing the same functional groups. The prediction results for a wide range of pure gases and gas mixtures are satisfactory. The model is particularly useful in predicting the adsorption equilibrium of complex chemicals, on which no data are available, from their molecular structures.

593g Application of a Newly Developed Breakthrough Model for Industrial Air Purifying Cartridges

Yuqing Ding, Michael Parham, and Amy Staubs. Engineering, Tyco International Ltd / Scott Health and Safety, 4320 Goldmine Rd, Monroe, NC 28110

A new breakthrough model recently developed by Ding et al [1] is successfully applied for toxic industrial chemicals (TICs) on adsorptive and reactive industrial air purifying cartridges. The model is derived based on two assumptions: (a) for a constant feed adsorption process, the dimensionless chemical potential changes exponentially along the flow direction; (b) the speed of the concentration wave accelerates with time as it evolves through the bed. Our results show that the model is capable of accurately simulating the breakthrough curve over several orders of magnitude concentration, making it especially useful for describing and predicting breakthrough behavior of toxic chemicals with more favorable adsorption isotherms. All the parameters have well-defined physical significance. Thus, the model can be made predictive if the sensitivities of the model parameters are determined. Simulation results also show that the model can be applied to both adsorptive and reactive adsorbate / adsorbent systems.

603a Filtration Testing with Hanford High Level Waste Samples and Simulants

Reid A. Peterson, Rick Shimskey, Renee Russell, Harry Smith, Justin Billing, and Matthew Edwards. Pacific Northwest National Laboratory, PO Box 999, Richland, WA 99352

The Waste Treatment Plant currently under construction for treatment of High Level Waste at the Hanford Site will rely on ultrafiltration to provide solids liquid separation as a core part of the treatment process. A series of bench scale actual waste and simulant tests have been performed to evaluate the ultrafiltration performance that can be expected as part of this process. This research has focused on correlation of two key parameters to filtration rate: the maximum achievable slurry concentration and the fines concentration present in the slurry feed.

A series of tests were performed using slurries characterized by a ranging of maximum achievable slurry concentration (as defined by centrifuge tests). Similarly, the particle size distribution of these slurries was adjusted by adding small concentrations of fine particulate materials to adjust the particle size distribution of the slurry feeds. These results were then assembled to build a model of filtration behavior that could be used to predict filtration rate based on a limited set of physical characterizations.

603b Assay Development and Filtration Clearance of Dow Corning Medical Antifoam C

William J. Rayfield, Bruno F. Marques, Jennifer M. Pollard, and Kent E. Göklen. BioPurification Development, Merck & Co., Inc., Rahway, NJ 07065

Antifoaming agents are commonly added to mammalian cell culture during production of monoclonal antibodies (MAb), since foaming can have a negative impact on MAb production. The foaming can reduce purification process capacity, increase processing time, contribute to batch contamination (due to foam overflow), and reduce cell productivity due to entrainment and shear , . Even though few regulatory guidelines exist in regard to antifoam compounds (mostly for food production) , these compounds can potentially be carried throughout the process, making it necessary to track them during purification to demonstrate clearance. Several challenges exist in detecting Antifoam C due to poor UV absorbance, low solubility in the solvents screened and its presence at very low predicted concentrations in the process streams. This paper will present an effective detection method for Dow Corning Medical Antifoam C using dynamic light scattering (DLS), total solids analysis (TSA), and inductively coupled plasma atomic emission spectroscopy (ICP-AES). The clearance of Antifoam C through dead-end and tangential flow filtration processes was characterized using both deionized water and cell culture media spiked with Antifoam C at 10x the process concentration. The parameters investigated were Antifoam C feed concentration, flux, filter loading, trans-membrane pressure, and processing temperature. Experimental Antifoam C clearance data were compared to an estimated clearance based on assumptions concerning the theoretical distribution of antifoam in the process streams. At room temperature (20°C), a reduction of 7.1x was observed for bioburden reduction filtration while reductions ranging from 6.6x to 8.0x were observed for depth-filtration.

603c Effect of Protein Charge Distribution on Transport through Semipermeable Ultrafiltration Membranes

Mahsa M. Rohani and Andrew L. Zydney. Chemical Engineering, Pennsylvania State University, 220 Fenske Laboratory, University Park, PA 16802

Several recent studies have clearly demonstrated that protein transport through semipermeable ultrafiltration membranes is strongly affected by electrostatic interactions between the charged membrane and the charged protein. All previous analyses of this phenomenon have implicitly assumed that the key factor is the protein surface charge density (i.e., net charge per unit area), with no effect of the detailed charge distribution over the protein surface. This is in sharp contrast to the behavior seen in ion exchange chromatography where the presence of charge "patches" can have a significant affect on the adsorption behavior. The objective of this study was to examine the possible effects of the protein charge distribution on the magnitude of the electrostatic interactions that occur during protein ultrafiltration.

Experimental studies were performed using cytochrome c and a peracetylated derivative formed by reaction of cytochrome c with acetic anhydride to block all free lysine groups. The peracetylated derivative had almost identical mass as the native protein but with an isoelectric point of pH 4 compared to pH 10 for the native cytochrome c due to the elimination of the positively-charged amino groups on the 19 lysine residues. The net charge of cytochrome c and the peracetylated derivative were determined theoretically, with the results in good agreement with data obtained by capillary electrophoresis. Ultrafiltration experiments were performed using 30 kD Ultracel membranes, both as received and surface-modified by chemical attachment of a quaternary ammonium functionality to generate a positively-charged membrane. The membrane charge was evaluated as a function of pH for both the native and modified membranes using streaming potential measurements.

The transmission of the native cytochrome c was maximum near its isoelectric point when using positively-charged membrane, but was a monotonically decreasing function of increasing pH with the unmodified (slightly negatively-charged) membrane. In contrast, the transmission of the peracetylated derivative increased with increasing pH for the positively-charged membrane but decreased with increasing pH for the unmodified membrane. These differences were a direct result of the different charge-pH profiles for the native and peracetylated proteins, giving rise to different electrostatic interactions. There were also distinct differences in the protein sieving coefficients for the native and peracetylated proteins having the same net electrical charge, which appear to be related to differences in the distribution of charge groups over the protein surface. These results provide important insights into the effects of electrostatic interactions on protein transport through ultrafiltration membranes.

603d Whey Protein Fractionation Using Functionalized Ultrafiltration Membranes

Stephen M. Ritchie, University of Alabama, Department of Chemical Engineering, Box 870203, Tuscaloosa, AL 35487-0203 and Simone Cowan, Chemical and Biological Engineering, University of Alabama, Department of Chemical Engineering, Box 870203, Tuscaloosa, AL 35487-0203.

Ultrafiltration membranes have been functionalized with sulfonated polystyrene grafts to enhance whey protein separation using single-stage ultrafiltration. The mechanism of enhanced selectivity was investigated using dextran, small molecule, and protein permeation studies. Experimental results will be compared to simulations based on pore size and distribution, diffusion, and electrostatic repulsion. The models show excellent correlation with experimental data, and demonstrate that selectivity is enhanced through a combination of size exclusion and electrostatic effects.

603e Separation of Plasma from Blood through Low-Resistance Nanopore Filters

Michael Hill, M. Hill & Associates, PO Box 622, Wyckoff, NJ 07481 and **Edward F. Leonard**, Chemical Engineering, Columbia University, 500 W. 120th St., 801 Mudd Dept of Chemical Engineering, New York, NY 10027.

Nanoporous membranes with precise, short pores and high void fractions are becoming available. Their use in blood filtration strains assumptions inherent in existing theories, introduces design challenges because of the very low hydraulic resistance of the membrane-proper, and provides new insights into several aspects of blood flow in narrow spaces. These aspects include the balance among shear-induced motion away from the filtering surface, convection toward the surface, and shear-enhanced diffusion, as well as the role of erythrocyte motion on the transverse motion of leukocytes and thrombocytes. Different pore sizes in the range 450 to 1200 nm (1.2 um) elicit different behavior of cells adjacent to the filtering surface. Direct observation of cell behavior is accomplished at low hematocrits and is augmented by measurement of transmembrane electrical resistance, allowing extension of measurements to suspensions with higher hematocrits.

604a Desalination Membranes Based on Sulfonated Poly(arylene ether sulfone) Random Copolymers

Wei Xie[1], Ho-Bum Park[2], Benny D. Freeman[1], Mou Paul[3], Abhishek Roy[3], Hae-Seung Lee[4], and James E. McGrath[4]. (1) Chemical Engineering, The University of Texas at Austin, 10100 Burnet Rd., Bldg 133, Center for Energy and Environmental Resources, Austin, TX 78758, (2) Chemical Engineering , Hanyang University, Seoul, 680-749, South Korea, (3) Chemistry, Virginia Tech, Dept. of Chemistry, 107 Davidson Hall, Blacksburg, VA 24061, (4) Macromolecules and Interfaces Institute, Virginia Tech, Blacksburg, VA 24061

The current commercially available membranes for water desalination by reverse osmosis are derived from two basic classes of polymers: cellulose acetate and crosslinked aromatic polyamides. Cellulose acetate membranes are susceptible to microbiological attack, compaction at higher temperatures and pressures, and their operation is limited to a relatively narrow pH range. The polyamide membranes show improved transport properties at given applied pressure and stability over a wider range of pH compared to cellulose acetate membranes. However, these polyamide membranes suffer from poor resistance to continual exposure to oxidizing agents such as chlorine in desalination applications. To overcome these problems, we have synthesized and characterized a new generation of desalination membranes for water purification, based on chemistry that is entirely different from conventional post-polymerization sulfonation technology. As a result, reproducible sulfonated poly(arylene ether sulfone) copolymer membranes have been prepared for desalination. This family of copolymers has been studied for use as proton exchange membranes, while no report has been issued on its water desalination properties. A key technical feature of these new membranes is their high tolerance chlorine in feed water and their excellent reproducibility in fabrication. Our main approach to optimizing polymer design is to use random, cross-linked, block, and blends based on disulfonated copolymers to achieve high water permeability and high salt rejection simultaneously. In this study on random copolymers, we vary the sulfonation level and investigate the solubility, diffusivity and permeability of the desalination membranes towards water and sodium chloride. At a high level of salt rejection, the membranes achieve high water permeability. In addition, the intrinsic properties of polymeric membranes to separate water and salt are being considered in terms of water permeability and salt permeability, and we are using this framework to establish guidelines for material selection and design for desalination membranes.

604b Numericl Modeling of Scale Formation In Ro Memrbane Desalination

Eric Lyster, Chem Eng, UCLA, 5531 Boelter Hall, Los Angeles, CA 90095, Robert Rallo, Universitat Rovira Virgili, Av. dels Paisos Catalans 26, Tarragona, 43007, Spain, Francesc Giralt, Fenòmens de Transport, Departament d'Enginyeria Química, Universitat Rovira i Virgili, Av. dels Països Catalans 26, Tarragona, 43007, Spain, and Yoram Cohen, Chemical & Biomolecular Engineering, University of California, Los Angeles, 5531 Boelter Hall, 420 Westwood Plaza, Los Angeles, CA 90095-1592.

In membrane separation processes, rejection of solutes and solvent permeation lead to increased concentration of dissolved and suspended species in the retentate stream and at the membrane surface. In reverse osmosis (RO) desalination, this effect, known as concentration polarization, can result in salt concentrations exceeding saturation levels, thereby leading to mineral scale build-up on the membrane surface. This scale build-up leads to decreased module performance (e.g., flux decline) and shorter membrane lifetime. To better understand the effects of the integration of concentration polarization with mineral scale formation, 2-D and 3-D numerical models were developed to predict fluid dynamics and solute mass transfer within the feed flow channel of RO modules. This computational model was coupled with a surface crystallization kinetic model to provide a detailed description of the development of mineral scale on the membrane surface. Experimental results from a small RO module with an un-obstructed feed flow channel were used to calibrate and test the mineral scaling simulations, which were then extended to include the impact of channel spacers. The integrated scale formation model provides detailed information on the location and evolution of mineral salt scaling in the RO feed channel. Comparison with experimental scaling results from RO channels with and without spacer filaments will be presented and discussed with respect to the implications for RO module design and optimization of process conditions to reduce the propensity for scale formation.

Web Page: www.watercenter.ucla.edu

604c Direct Contact Membrane Distillation: Studies on Novel Hollow Fiber Membranes, Devices, Countercurrent Cascades and Scaling

Kamalesh K. Sirkar[1], Liming Song[2], Fei He[1], Hanyong Lee[1], Jack Gilron[3], Praveen B. Kosaraju[1], Baoan Li[2], Zidu Ma[4], Xiaohong Liao[4], and James R. Irish[4]. (1) Chemical Engineering, New Jersey Institute of Technology, Membrane Seperations, 161 Warren st, Newark, NJ 07102, (2) Chemical Engineering Department, New Jersey Institute of Technology, Membrane Separations, 161 Warren St, Newark, NJ 07102, (3) Zuckerberg Institute for Water Research, Ben Gurion University, POB 653, Beer Sheva, 84105, Israel, (4) United Technologies Research Center, 411 Silver Lane 129-30, East Hartford, CT 06108

We have recently developed novel hollow fiber membranes and devices for recovering pure water from hot brine via direct contact membrane distillation (DCMD). Hot brine undergoes rectangular crossflow over the outer surface of highly porous hydrophobic polypropylene hollow fibers whose outside surface was coated with porous plasmapolymerized silicone-fluoropolymer coating to mitigate pore wetting and distillate contamination as cold distillate flows through the bores of fibers having a large wall-thickness. The DCMD studies were carried out sequentially with modules having a surface area of ~120 cm^2, with larger modules having a surface area of 0.286 m^2 and recently in a pilot plant at United Technologies Research Center with modules each having 0.61-0.66 m^2 membrane surface area. Extended pilot scale operation demonstrated no salt leakage, stable and repeatable performance with synthetic solutions having salt concentrations of 10 % as well as with sea water. Modeling the DCMD behavior in both small and large-scale modules has been successfully implemented.

Cost-efficient desalination technology was also developed successfully by integrating a countercurrent cascade of these novel crossflow DCMD devices and solid polymeric hollow fiber-based heat exchange devices. Experimental studies were carried out in such a heat-integrated cascade using 2-8 modules representing 2-8 stage achieving a high thermal efficiency (~0.87) and a high GOR (gained output ratio). Modeling results suggest achieving a GOR as high as 12 for ten stages in a countercurrent DCMD cascade.

Laborartory scaling studies were carried out in small DCMD modules using $CaSO_4$ and mixed $CaCO_3$ + $CaSO_4$ systems over a wide range of temperatures ,SI values, two flow patterns as well as different types of fiber surfaces. The results indicate that even though there was significant precipitation, there was no effect on the membrane vapor flux or brine pressure drop. An artificial sea water was concentrated 8 times successfully when a countercurrent cascade composed of 4 stages of the DCMD modules and a heat exchanger was employed during the DCMD process.

Fouling Prevention In Rotating Reactive Membrane Filtration

Shannon Ciston[1], *Yuan Yao*[2], *Richard M. Lueptow*[2], *and Kimberly A. Gray*[3]. (1) Department of Chemical and Biological Engineering, Northwestern University, 2145 Sheridan Rd., Rm A225, Evanston, IL 60208, (2) Mechanical Engineering, Northwestern University, 2145 Sheridan Rd., Evanston, IL 60208, (3) Civil and Environmental Engineering, Northwestern University, 2145 Sheridan Rd., Evanston, IL 60208

Introduction

Current trends of water stress and scarcity worldwide necessitate the development of effective and efficient technologies for water reuse and recycle. Membrane filtration becomes an essential technique for water reuse and recycle of low-quality sources. However, biofouling and biofilm growth are major obstacles for membrane filtration because they cause dramatic losses in efficiency and system flux. By coating ceramic ultrafiltration membranes with titanium dioxide photocatalysts, we have created a multifunctional reactive membrane that resists fouling and biofilm growth by improving hydrophilicity, inactivating microorganisms, and degrading organics. Furthermore, this reactive coating has been incorporated into a rotating cylindrical geometry to take advantage of Taylor-Couette flow for enhanced biofouling prevention.

Background

TiO_2 nanoparticles can be irradiated by near-UV light to excite electrons, leaving behind positively charged holes. In the presence of water, these separated charges may react with hydroxyl groups at the surface of the TiO_2 nanoparticle to produce highly reactive hydroxyl radicals. Titanium dioxide has been demonstrated in a number of environmental applications, including use in aqueous slurries for the degradation of organic chemicals, and immobilization on a substrate for the degradation of organics in gaseous and liquid streams, usually air and water. More recently, its efficacy in the degradation of microorganisms has been shown, both in slurries and in coatings.

Anatase and rutile are the two most commonly discussed crystal phases of titanium dioxide. Anatase requires higher energy light than rutile for excitation (385 nm versus 410 nm respectively), but anatase is regarded as the more catalytically active crystalline phase due in part to its slower rates of photogenerated charge recombination. Nanoclusters containing both anatase and rutile nanoparticles have very high photocatalytic activity. Electrons within rutile are excited and then transfer to trapping sites in defects at the anatase particle surface, allowing extended periods of time for the hole to participate in photooxidation reactions, producing highly reactive oxygen species including hydroxyl radicals.

In previous work we have identified Degussa P25 titanium dioxide as a strong candidate for biofilm prevention based on its excellent bacterial cell attachment resistance and high organic degradation activity, and long-term biofilm growth prevention and reduced flux decline in experiments with Pseudomonas putida on ceramic membrane discs coated with the Degussa P25 photocatalyst. Degussa P25 is a highly characterized commercial catalyst produced by high-temperature flame hydrolysis, with heterogeneous phase composition (15–30% rutile, remainder anatase) and particle size (primary particle size 30 nm, aggregates up to 200 nm). It is considered by many to be the “gold standard” of photocatalysts because of its high reactivity.

Taylor-Couette flow occurs between rotating cylinders when the inner cylinder angular velocity crosses a threshold value, above which instabilities create counter-rotating toroidal vortices all along the annulus (See Figure 1.) Rotating cylindrical membranes are currently used in low-pressure separation systems, especially for the separation of blood and plasma. The instabilities that occur in Taylor-Couette flow between rotating cylinders can be harnessed for the reduction of fouling along the membrane walls. We combine the chemical action of TiO_2 with the physical action of Taylor-Couette flow to create a system with robust biofouling resistance.

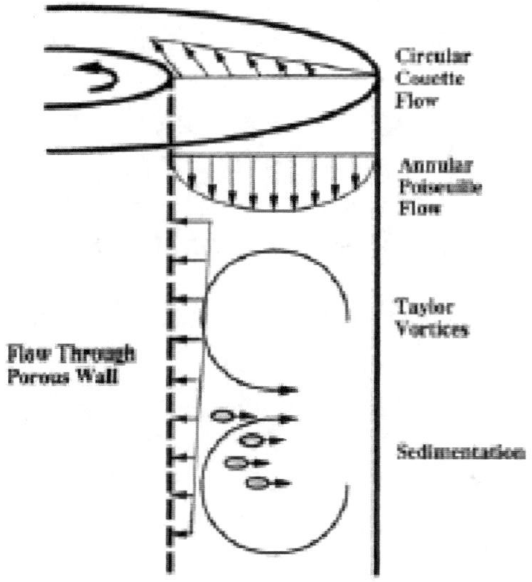

Figure 1-Flow that occurs in a rotating filter separator

Methods

Zirconia ceramic membrane discs were obtained fromSterlitech Corporation. These membranes discs have a layered structure, consisting of an alumina–titania–zirconia support material, and a zirconia pore size-controlling layer with a cut-off of 300 kDa (approximately 0.01_m pore diameter), resulting in awater flowrate of 450–600 L/hm2 at 1 bar trans-membrane pressure. The discs are 47mmin diameter and 2.5mmin thickness. They were rinsed with acetone and then rinsed thoroughly with deionized water to remove any surface impurities. Zirconia discs were coated with Degussa P25 (80% anatase, 20% rutile). To clarify the role of hydrophilicity and the role of photocatalysis at the membrane surface, a set of control discs were coated with non-catalytic rutile as well. A nanocrsytalline pure-phase rutile powder was synthesized by a low-temperature hydrothermal method using a titanium (IV) chloride precursor, as described by Gonghu Li, et al. The high hydrophilicity and low reactivity of the rutile allows for a comparison of the hydrophilic and reactive effects of the biofilm growth prevention observed previously over the Degussa P25 coatings.

Titanium dioxide powders were applied to discs by dip-coating in a slurry of 0.1 g/L catalyst, and 0.05 g/L dioctyl sulfosuccinate surfactant in water [34]. After sonicating the slurry for 1 h, the top surface of each discwas dipped into the slurry and withdrawn by hand at a rate of about 3 cm/s. Discs were dried in a 105 ◦C oven for 2.5 h, and then heat treated in a furnace with a ramp rate of 5 ◦C/min to 450 ◦C, held there for 1 h, and then cooled slowly to room temperature. This process was repeated until 4mg of the catalyst had been deposited on the disc surface (about 5 repetitions).

Pseudomonas putida is a gram-negative, rod-shaped bacterium. It was chosen as the bacterial model because *P. putida* is commonly present throughout the environment, and readily colonizes biofilms. The American Type Culture Collection *P. putida* type strain number 12633 was grown in M9 minimal media at 25° C, maintained at the exponential growth phase in a chemostat supplied with 1 L/min sterile air, and paddle rotating at 100 rpm. It was maintained at a concentration of about 10^8 colony forming units per milliliter (CFU/mL). This solution was diluted in sterile phosphate buffered saline (PBS) to bring the solution cell concentration to approximately 10^6 CFU/mL, a bacterial cell concentration typical of fresh surface waters. This solution was fed over membranes, and confocal microscopy was used to quantify biofilm formation.

Tubular membranes were obtained from Sterlitech corporation. These had a layered structure, with an alumina-titania-zirconia support layer with large pore size on the outside, and a rutile titania inner layer bringing the pore size to 1.4 µm. These were 10 mm in outer diameter, 6 mm in inner diameter, and 250 mm long. Each tube was cut into 3 pieces, 83 mm long. Each tubular membrane was coated by dip-coating into a slurry of 10 g/L zirconia particles (< 5 um, Sigma Aldrich) to create a microfilter layer on the outer surface of the membrane. The membrane was dried at 105 C to drive off water, and then heated in a furnace at a ramp rate of 5 ° C /min up to 450 ° C, where it dwelled for one hour before being slowly cooled to room temperature. The membranes were all dipped in nanoparticle slurries to deposit the final outer layer. Half were dipped in a slurry of Degussa P25 at a concentration of 2 g/L, and half in zirconia <100 nm particles 2 g/L for control samples. These were dried and heat treated similarly, repeating the process as needed until 10 mg of active layer had been deposited on the surface. The pore size of the new membranes was confirmed through bubble point testing according to ASTM F316-03 method.

Each membrane placed into the reactor. There is a bank of 4 reactors, each illuminated by a 4-Watt UV lamp. Each tubular ceramic membrane (OD 10 mm) is surrounded by a 15 mm ID, 17 mm OD quartz tube, creating an annulus for water flow between the two. The tubular membrane is placed inside a stainless steel shaft, which is used to drive the tubular membrane at 100 rpm. A belt and gear system connects the shafts to a motor. Each tubular membrane is sealed on the upper edge to prevent water from bypassing the membrane. There is one inlet tube and two outlet tubes (permeate and retentate).

The water inlet is driven by a set of peristaltic pumps, operating at less than 30 psi transmembrane pressure. Once the membranes are positioned, the entire system is cleaned by pumping ethanol through the system for 30 minutes to sterilize surfaces.

Each day, 120 mL of chemostat outflow *pseudomonas putida* cell culture is added to 12 L of sterile 10% phosphate buffered saline solution in water, to create a synthetic water with environmentally relevant concentrations of cells (10^6 cells/mL) and organic carbon. This solution directly feeds the reactor chambers by pumping at a rate of 2 mL/min.

The pressure drop across the reactor system is measured every 4 hours to monitor fouling. The permeate and retentate flow rates are measured also every 4 hours with a graduated cylinder and stopwatch. Each day, permeate and retentate are collected for standard serial dilution plating for the determination of viable bacterial cell concentration. The test is continued for 5-10 days, until the pressure drop across the membranes becomes a limiting factor.

<u>Results</u>

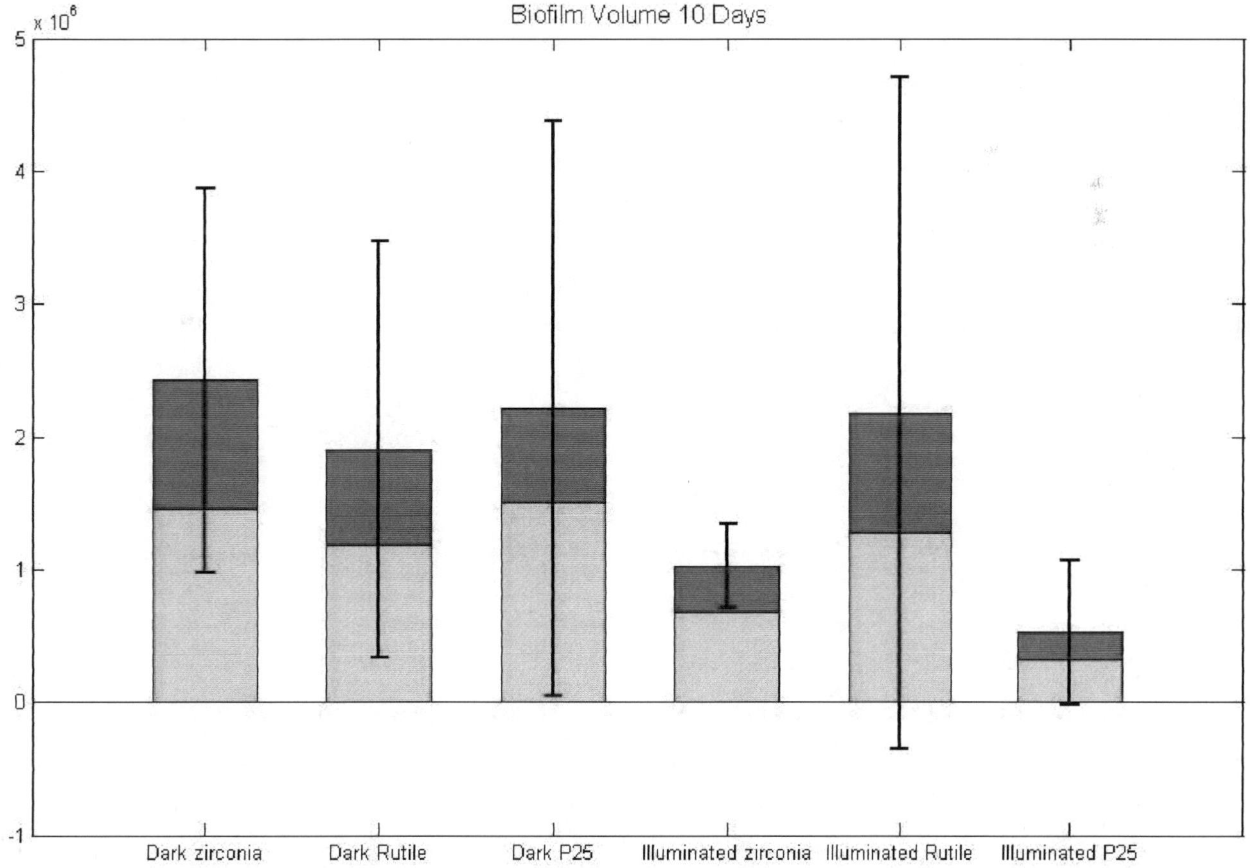

Figure 2

In this study, we confirmed previous results, finding a 90% confidence interval that the illuminated Degussa P25 samples had significantly less biofilm growth than the dark zirconia samples. Furthermore, the results show that there is less biofilm growth over the illuminated Degussa P25 samples than on the illuminated rutile samples (82% confidence interval). This shows that the combination of hydrophilic and photolytic effects over titanium dioxide catalysts are not sufficient to control biofilms, but that the photocatalytic effects observed in P25 are important for biofilm prevention.

Within the rotating reactor system, a control sample with no catalyst, no illumination, and no rotation was found to have significant biofouling within 7 days. I will report the latest results from this experimental setup, including preliminary results on the overall flux decline and biofilm growth reduction.

604e Modeling, Design and Upscaling the Membrane Bioreactor Process for Removing Biodegradable Organic Matter and Disinfection Byproduct Precursors from Water Supplies

Mark D. Williams[1], Varadarajan Ravindran[2], and **Massoud Pirbazari**[2]. (1) Metropolitan Water District of Southern California, 700 Moreno Avenue, La Verne, CA 91750, (2) University of Southern California, Department of Civil and Environmental Engineering, Los Angeles, CA 90089-2531

This research discusses the use of membrane bioreactor (MBR) technology for the removal of biodegradable organic matter (BOM) and disinfection byproduct (DBP) precursors from potable water sources after ozonation. It is indeed well known that ozonation of natural organic matter (NOM) in water supplies results in the cleaving of the organic macromolecules into more easily biodegradable organic matter including aledehydes and ketones. Subsequent to advanced oxidation processes such as ozonation, the BOM essentially consists of smaller molecules that can serve as organic substrates for microbial growth, and can constitute organic molecules that can generate DBPs such as trihalometahnes and haloacetic acids. The purpose of BOM removal after ozonation using the MBR technology was twofold: preventing the growth or regrowth of pathogenic microorganisms in the water distribution system; and reducing the potential for forming DBPs such as trihalometahnes and haloacetic acids.

The removal of BOM was measured in terms of three variables, namely: dissolved organic carbon (DOC), assimilable organic carbon (AOC), and total aledehydes. These parameters were measured after ozonation, and biological treatment in MBR following ozonation. Besides these variables, the DBP formation potentials were also measured in terms of trihalomethane formation potential (THMFP). The AOC was determined using Pseudomonas flourescens P-17 and Aquaspirillum NOX bacterial strains as bioassay organisms. The total aledehydes denoted the total concentration of formaldehyde, acetaldehyde, glyoxal, and methyl glyoxal.

The MBR experiments established the effectiveness of the MBR technology in achieving excellent removals of BOM and DBP precursors. Both AOC and total aldehydes were completely removed under optimal operating conditions, while the DOC removal was 60 percent. The studies further demonstrated showed that PAC application not only enhanced DOC removal but also greatly increased the membrane permeate flux, improving the overall economic viability of the process.

A predictive model was developed for performance forecasting and design of the MBR process. A lumped parameter approach was used in using the three variables for representing BOM (DOC, AOC and total aledehydes). The model involved the phenomenological aspects pertaining to pollutant transport, sorption equilibrium, and biochemical reaction. It considered film transfer from bulk liquid phase to the biofilm-liquid interface, biofilm diffusion, and diffusion into the adsorbent particle. It also considered biochemical reaction in the biofilm immobilized on the PAC adsorbent particle, and in the aqueous suspended phase. These reactions described biofilm and biomass growth and decay within the MBR system, and were governed by Monod kinetics. The adsorption equilibrium and kinetic relationships controlled the sorption of BOM from the aqueous phase, and their release from the adsorbent to the biofilm as well as to the microorganisms in the suspended phase. The BOM removal mechanisms therefore involved a combination of biofilm degradation, suspended phase biodegradation, and adsorption.

The MBR model parameters were determined from independent laboratory-scale experiments and correlation techniques for each of the three BOM variables of interest (DOC, AOC and total aledehydes). The adsorption equilibrium and rate parameters were obtained from batch reactor studies, while the biokinetic parameters were estimated from chemostat studies. The MBR experimental data provided model verification under different process conditions. Simulation studies provided a qualitative evaluation of the parameters influencing the MBR process dynamics under different conditions. These parameters involved liquid film mass transfer, biofilm transport, biological degradation kinetics, influent concentration, and reactor hydraulic retention time.

A new feature presented in the modeling approach is the evaluation of MBR process dynamics under certain real scenarios including process shutdown, variability in influent concentrations of target contaminants, and changes in hydraulic retention time, as well as process upscaling. Process shutdown often upsets biological systems owing to unpredictability in the behavior of microorganisms. Variations in influent concentrations of contaminants arise from fluctuations in the NOM or BOM content of the treated water due to changes in water quality. Changes in hydraulic times are attributed to permeate flux decline arising in membrane systems operated under constant trans-membrane pressures owing to membrane fouling and concentration polarization effects. Besides these aspects, process upscaling by dimensional analysis and similitude are also investigated.

Key words: membrane bioreactor, modeling, dissolved organic carbon (DOC), assimilable organic carbon (AOC), total aledehydes, disinfection byproducts (DBPs), biodegradable organic matter (BOM), process variables and dynamics, upscaling

606a Three Dimensional, Nanoporous and Mobile Containers for Size Selective Sampling

Christina Randall[1], Aubri Gillespie[1], Siddarth Singh[2], Timothy Leong[2], and **David H. Gracias**[2]. (1) Department of Biomedical Engineering, Johns Hopkins University, Baltimore, MD 21218, (2) Department of Chemical and Biomolecular Engineering, Johns Hopkins University, 125 Maryland Hall, 3400 N Charles Street, Baltimore, MD 21218

We describe the fabrication of three dimensional cubic containers with precisely patterned nanoporosity, and their utilization in size selective sampling. We utilized mobile cubic containers to capture objects based on size exclusion.When objects were present in the fluid, those smaller than the size of the pores escaped so that on retrieval the container predominantly captured objects whose sizes were larger than the porosity of the container. Using containers with different pore sizes, we were able to capture sub-micron scale beads and live cells. Additionally, on retrieval, the containers were used to culture and hence multiply entrapped cells.

606b Octanol-Water Partitioning Coefficient Measurements in Picoliter Drops Using a Segmented Flow Microfluidic Device

Nathan A. Marine[1], Steven A. Klein[1], and Jonathan D. Posner[2]. (1) Mechanical Engineering, Arizona State University, 1711 S Rural Rd ECG G346 Mechical and Aerospace Eng, Tempe, AZ 85287-6106, (2) Mechanical & Aerospace Engineering & Chemical Engineering, Arizona State University, 1711 S Rural Rd ECG G346 Mechical and Aerospace Eng, Tempe, AZ 85287-6106

Partitioning of pollutants between water and organic phases is critical in determining their fate in the environment, bioavailability and toxicity. The spatial and temporal mobility of a substance is largely determined by its solubility in water. Currently, the EPA uses partition coefficients to predict the environmental fate, aquatic toxicity and bioaccumulation instead of measurements like molecular weight, conformation, hydration states, ionic charge, etc. Partitioning into an organic phase (typically octanol) is often used as a measure of bioavailability for herbicides, pesticides, pharmaceuticals, and potentially toxic chemicals. The octanol-water partition coefficient is a prerequisite for testing biological degradation and bioaccumulation in water for which the EPA Office of Prevention, Pesticides, Toxic Substances (OPPTS) publishes Harmonized Test Guidelines. Currently, octanol–water partitioning is an important physiochemical property of pharmaceutical drugs and agrochemicals and is required measurement by legislation as part of the profile for high volume production chemicals. The partition coefficient is measured by a variety of methods, with the most common being some variant of the shake-flask method. The shake-flask method typically requires several days to complete and liters of solution.

In this presentation, a novel microfluidic method for measuring octanol-water partitioning coefficients in single picoliter drops is described. Picoliter water droplets are generated in octanol carrier fluid within a T-shaped segmented flow device fabricated in Poly(dimethylsiloxane). The partition coefficient of the fluorescent dyes are measured as a function of pH using epifluorescence microscopy. The results agree well with published measurements based on the batch shake-flask method. The microfluidic partitioning measurements are conducted in minutes across thousands of picoliter droplets. The methods presented here are rapid, provide detailed statistics, and can be run in parallel enabling the simultaneous partitioning of thousands of compounds for combinatorial chemistry applications.

BIOPARTICLE MANIPULATION WITH INSULATOR-BASED DIELECTROPHORESIS

Blanca H. Lapizco-Encinas, Tecnológico de Monterrey, Monterrey, NL, Mexico

Abstract

Dielectrophoresis (DEP) is the motion of particles in due to polarization effects in nonuniform electric fields; this electrokinetic transport mechanism has a great potential for the manipulation of a wide array of bioparticles, ranging from biomolecules to microorganisms. Traditionally, DEP has been carried out employing arrays of microelectrodes, however, employing microelectrodes has some drawbacks such as high cost and loss of functionality due to fouling. Insulator-based dielectrophoresis (iDEP) is an attractive alternative, since it employs arrays of insulating structures, instead of electrodes, to create nonuniform electric fields; resulting in inexpensive and robust devices. This study presents the application of iDEP for the manipulation and concentration of different types of bioparticles (from proteins to microalgae). Glass and polymeric microdevices containing channels with cylindrical insulating posts were employed to trap and concentrate bioparticles. A sample of bioparticles was introduced into the microchannel, then; a direct current (DC) electric field was applied across the post array, creating regions of higher and lower electric field intensity, *i.e.* dielectrophoretic traps. Electroosmotic flow was generated to pump the suspending liquid through the microchannel, while the bioparticles were immobilized and concentrated at the dielectrophoretic traps between the cylindrical posts. The dielectrophoretic response of the bioparticles was recorded in the form of videos and pictures. Experiments were conducted varying the electric field, pH and conductivity of the suspending medium, in order to study the effect of these parameters on the dielectrophoretic response of the bioparticles. Successful trapping of the different types of bioparticles was achieved, demonstrating the great potential of iDEP as a technique for bioparticle manipulation.

Introduction

Miniaturization has brought important advantages to bioseparation technology. Numerous fields, including environmental, clinical, pharmaceutical, etc., have beneficiated from the advances of microanalytical systems. There is a growing interest on the development of separation techniques that can applied in microscale. There is a growing interest on the development of separation techniques that can applied in microscale. During the last decade there has been a significant development of microfluidic devices for analytical applications, quality and performance improvements are making it possible for microanalytical systems to become commons aids to numerous different applications.[1] Dielectrophoresis (DEP) an efficient technique with great potential for miniaturization. It has been applied successfully for the manipulation and concentration of a wide array of particles, including bioparticles such as macromolecules and microorganisms. DEP is an electrokinetic transport mechanism produced by polarization effects when particles are exposed to a nonuniform electric field. This novel technique has the potential for achieving concentration and separation on a single step. The majority of the research studies on DEP have used arrays of microelectrodes and AC electric fields to create no uniform electric fields. Electrode-based DEP allows obtaining high electric field gradients employing low applied voltages. However, there are some drawbacks with this approach: high cost of electrode construction, complex fabrication processes and decrease of functionality due to fouling effects, which is a common effect when handling biological samples.

There is an alternative manner of carrying out DEP, insulator-based DEP (iDEP) is a technique where the voltage is applied employing only two electrodes that straddle an insulating structures array. When an electric field is applied across an insulating structures array, the presence of the structures creates regions of higher and lower field strength, resulting in the nonuniform electric field necessary for DEP to occur.[2] Furthermore, if DC electric fields are utilized, and the microdevice substrate is suitable for electroosmosis, then electroosmotic can be employed to pump liquid through the microfluidic system, eliminating the need of a micropump. Additionally, iDEP systems do not lose their functionally despite of fouling effects, which makes them more suitable for biological applications. Moreover, iDEP systems can be fabricated from a wide variety of materials, including plastics, leading to inexpensive systems, increasing its potential for high throughput applications.

Dielectrophoresis, first discovered by Pohl in 1951,[3] is the movement of particles induced by polarization effects in nonuniform electric fields. In DEP the motion of particles is caused due to the unbalanced force of a nonuniform electric field on the induced dipole moment of the particles: one "side" of the dipole is in a weaker field than the other, causing the particle to be pulled electrostatically along the electric field gradient, generating a net movement of the particles.[4] The dielectrophoretic force acting on a spherical particle can be represented as:

$$F_{DEP} = 2\pi\varepsilon_m r^3 \operatorname{Re}(f_{CM}) \nabla E^2 \quad (1)$$

where ε_m is the permittivity of the suspending medium, r is the radius of the particle, E is the local electric field, and Re(f_{CM}) is the real part of the Clausius-Mossotti (CM) factor:

$$f_{CM} = \left[\frac{\varepsilon_p^* - \varepsilon_m^*}{\varepsilon_p^* + 2\varepsilon_m^*}\right] \quad (2)$$

where ε_p^* and ε_m^* are the complex permittivities of the particle and the medium, respectively. The complex permittivity is related to the real permittivity by $\varepsilon^* = \varepsilon - (j\sigma/\omega)$ where σ and ω are the conductivity and angular frequency of the applied electric field, respectively, and $j = \sqrt{-1}$. Depending on the relative magnitude of ε_p^* and ε_m^*, the dielectrophoretic force acting on a particle can cause it to move toward or away from the high field region, these two effects are called positive and negative DEP behavior, respectively.[3,5] In many practical systems, at frequencies below 100 kHz, the CM factor can be approximated in terms of the real conductivities.[6]

$$f_{CM} = \left[\frac{\sigma_p - \sigma_m}{\sigma_p + 2\sigma_m}\right] \quad (3)$$

DEP is of second order in the applied electric field; at low applied electric fields, low electric fields gradients are produced, which leads to negligible dielectrophoretic force; therefore, the applied electric field has to be high enough for DEP to become significant. The two regimes of iDEP are *streaming* and *trapping* DEP. Streaming DEP occurs when the dielectrophoretic force overcomes diffusion but does not overcome the electrokinetic flow. Trapping DEP is the reversible immobilization of particles in dielectrophoretic traps and it occurs when the DEP force overcomes both diffusion and electrokinetic flow (as well as pressure-driven flow if present).[2,7,8] The electrokinetic flow, which is proportional to the electric field, comprises the effects of electroosmosis and electrophoresis on particle motion relative to a fixed channel. Electroosmotic flow is the movement of liquid when an electric field is applied in a microchannel or a capillary. For electroosmosis to occur it is necessary that the channel or capillary walls posses a net charge in the presence of an electrolyte.

This study presents the application of iDEP for the manipulation and concentration of different types of bioparticles (from proteins to microalgae). Glass and polymeric microdevices containing channels with cylindrical insulating posts were employed to trap and concentrate bioparticles. A sample of bioparticles was introduced into the microchannel, then; a direct current (DC) electric field was applied across the post array, creating regions of higher and lower electric field intensity, *i.e.* dielectrophoretic traps. Electroosmotic flow was generated to pump the suspending liquid through the microchannel, while the bioparticles were immobilized and concentrated at the dielectrophoretic traps between the cylindrical posts. The dielectrophoretic response of the bioparticles was recorded in the form of videos and pictures. Experiments were conducted varying the electric field, pH and conductivity of the suspending medium, in order to study the effect of these parameters on the dielectrophoretic response of the bioparticles. Successful trapping of the different types of bioparticles was achieved, demonstrating the great potential of iDEP as a technique for bioparticle manipulation.

Materials and Methods

A schematic representation of the experimental setup is shown in Figure 1. Experiments were conducted employing different microchannels made from glass and plastic, all microchannels had a similar layout. Each microchannel had an inlet and outlet reservoir and contained an array of cylindrical insulating posts, the post diameter and arrangement varied from channel to channel. In order to prevent particles from crashing against the posts and plugging the system, "dove-tail" geometry was used for the first row of posts on both sides. A 3000 V high voltage sequencer, model HVS448 (LabSmith, Livermore CA) was used to apply electric fields by employing platinum-wire electrodes with a diameter of 0.3048 mm. An inverted epifluorescence video microscope for microfluidics, model SVM340 (LabSmith, Livermore CA) was employed to record the dielectrophoretic response of the bioparticles in the form or videos and pictures. A 4X microscope objective was used for all experiments. Both, the high voltage sequencer and the microfluidics microscope require the use of a personal computer for their operation.

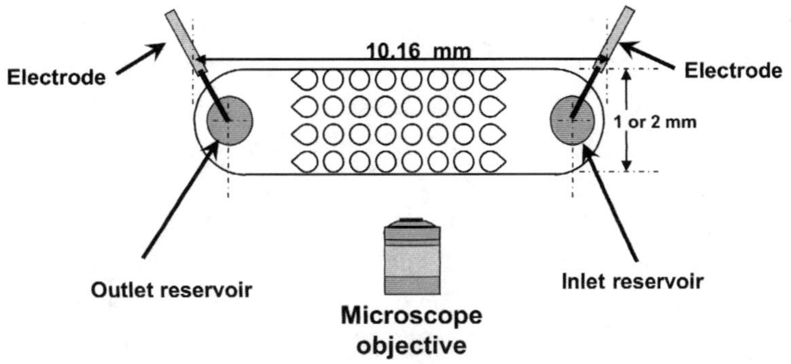

Figure 1. Experimental set-up for iDEP experiments.

Different suspending mediums were prepared, in order to study how the properties of the suspending medium affect the dielectrophoretic response of the bioparticles. The suspending mediums were prepared with the DI water solution with adjusted pH and conductivity. The pH of this solution was adjusted to a value of between 6 and 9 by adding NaOH. The conductivity of this solution was adjusted to a value between 25 and 115 μS/cm by adding K_2HPO_4

A variety of bioparticles were employed for the iDEP experiments: bovine serum albumin (BSA) particles (MW 66 kDa, Sigma, Toluca, Mexico) were dyed with fluorescein isothiocyanate, λex = 492 nm, λem =518 (Sigma, Toluca, Mexico) following a standard protocol provided by the dye manufacturer. Microalgae *Chlorella kessleri (UTEX#*: 397*)* having a diameter of 5 μm was utilized for iDEP experimentation, these particles were dyed with the DNA-intercalator fluorescente dye syto 11, λex = 508 nm, λem =527 (Invitrogen, Carlsbad, CA, USA). *Saccharomyces cerevisiae* (ATCC: 24858) yeast cells having an approximate diameter of 5 μm were also concentrated with iDEP, yeast cells were visualized without the aid of a fluorescent dye. Finally, live and dead suspension-adapted chinese hamster ovary (CHO-S, Invitrogen, Carlsbad, CA, USA) cells were manipulated with iDEP. These cells have an approximate diameter of 20 μm, and were prepared following standard cell growth protocols. In order to differentiate live form dead cells; live cells were not labeled and dead cells were labeled with trypan blue (Sigma, Toluca, Mexico).

Different types microdevices were employed, glass microdevice made from Schott D263 glass were employed with the protein BSA and microalgae, these devices were manufactured following standard wet-etching techniques. Glass microchannels were 10.12 mm long, and 10-□m deep. Plastic microdevices made form a plastic called Zeonor by employing standard photolithography and replication processes including hot embossing and injection molding, were kindly donated by Sandia National Laboratories, these plastic devices were used with yeast and CHO cells. Plastic microchannels were 10.12-mm long and 75-μm deep.

Results and Discussion

The main objective of the present manuscript is to describe the results that our research work has achieved with iDEP, in order to demonstrate some of the potential applications of this novel technique. One field that can be beneficiated by iDEP is downstream processing of valuable biomolecules, such as proteins. Insulator-based DEP was tested as technique for the concentration and manipulation of proteins, Figure 2 shows the results obtained with BSA particles. The dielectrophoretic trapping of protein particles is strongly influenced by the magnitude of the applied DC electric field. Figure 2a shows the dielectrophoretic response exhibited by protein particles when the applied was 900V/cm, by increasing the field to 1200 V/cm the immobilization of BSA particles increases significantly (Figure 2b). These results are in agreement with Eqn. (1), since the magnitude of the dielectrophoretic force is function of the magnitude of the applied electric field. Additionally, the effect of the suspending medium conductivity can be observed by comparing Figures 2a and 2c. From the figures it can be seen that by increasing the conductivity of the suspending medium to 100 µS/cm (employing a pH of 8 and almost the same electric field), protein particles show stronger dielectrophoretic trapping, since protein bands appear brighter. The conductivity of BSA particles has been reported as a maximum value of 25 μS/cm[9], *i.e.* the CM factor of the protein particles is negative under the operating conditions employed here (σ_m from 25 to 100 μS/cm). These results show that increasing the suspending medium conductivity increases the magnitude of the negative CM factor increases (Eqn. 3), leading to an increase in the dielectrophoretic force. The suspending medium pH also affects the dielectrophoretic response of protein particles. By comparing figures 2a and 2d one can see that increasing the ph of the suspended medium produces less dielectrophoretic trapping, this is due to an increase in electroosmotic flow, which makes it difficult for dielectrophoresis to dominate the system.

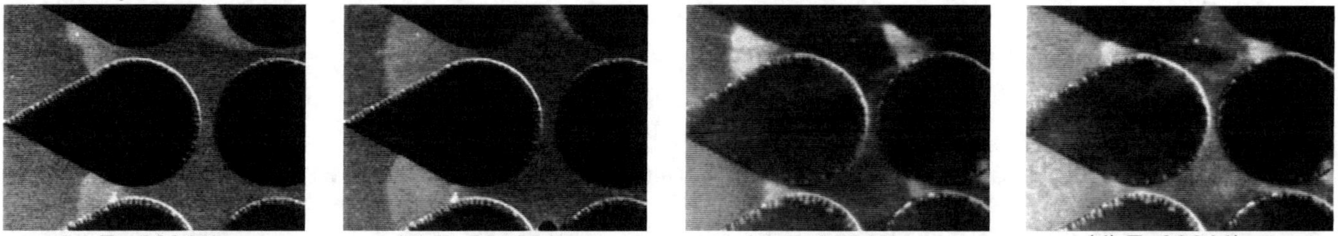

(a) E= 900 V/cm pH=8, σ_m=25 μS/cm (b) E=1200 V/cm pH=8, σ_m=25 μS/cm (c) E=950 V/cm pH=8, σ_m=100 μS/cm (d) E=900 V/cm pH=9, σ_m=25 μS/cm

Figure 2. Dielectrophoretic response of protein of particles (shown green-yellow) inside a glass microchannel with cylindrical insulating structures, flow direction is from left to right, post diameter is 440 μm, arranged 520-μm center-to-center, and microchannel was 2-mm wide.

Another potential application of iDEP is for the concentration of microorganisms, such as microalgae. A sample of water containing a dilute concentration of these algae cells could be processed with an iDEP device, in order to concentrate the microalgae for further laboratory tests. Standard laboratory techniques for microalgae concentration are extremely time consuming. A technique that can concentrate microalgae rapidly has great potential applications. Figure 3 shows the results obtained with microalgae cells, these are relative large particles (5 μm in diameter) compared to BSA particles (5 nm in diameter)[9]. From these results we can observe that iDEP devices have to potential to be used as pre-concentrators of microorganisms.

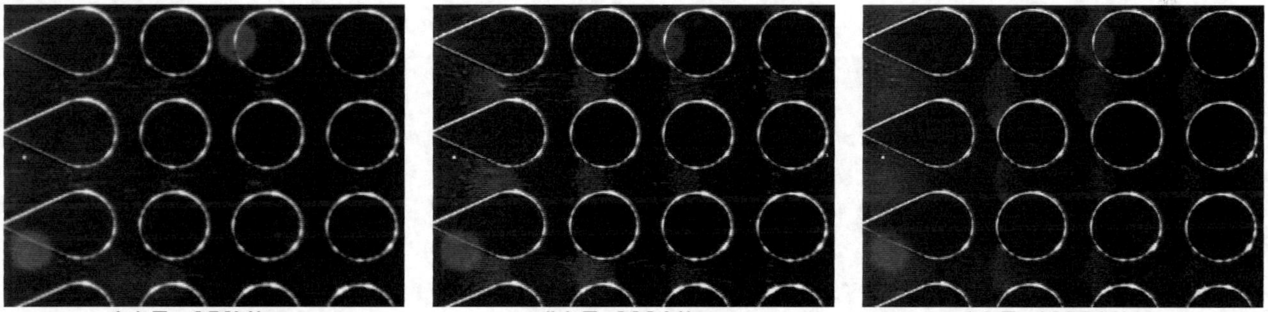

(a) E= 350V/cm pH=6, σ=100 μS/cm (b) E=800 V/cm pH=6, σ=100 μS/cm (c) E=1000 V/cm pH=6, σ=100 μS/cm

Figure 3. Dielectrophoretic response of microalgace cells (shown light gray color) inside a glass microchannel with cylindrical insulating structures, flow direction is from left to right, post diameter is 200 μm, arranged 250-μm center-to-center, and microchannel was 1-mm wide.

Figures 3a shows microalgae cells flowing under electrokinetic flow, when a field of 350 V/cm is applied, the dielectrophoretic force is not strong enough to overcome electrokinetics, and thus, microalgae cells are not trapped. By increasing the field to 800 and 1000 V/cm, Figures 3b and 3c, respectively, microalgae cells are dielectrophoretically trapped. Microalgae cells are exhibiting negative dielectrophoresis, which is caused by the presence of the cell membrane.[10] This means that the cells are repelled form the regions with higher electric field gradient. Comparing Figures 3b and 3c, one can see that when the applied electric field is increased, the band of trapped microalgae is repulsed in a stronger manner from the region of higher electric field gradient. These results are in agreement with Eqn. (1).

In order to extend our work to other types of microorganisms, the potential of iDEP for the concentration of yeast cells was explored. Yeast cells having an approximate diameter of 5 mm were also manipulated with iDEP, but this time, a plastic microdevice was employed. The objective here was to test the functionality of polymer-based devices for iDEP applications. Polymers are inexpensive materials that if used with iDEP open the possibility of this type of particle concentrators for high throughput applications. Figure 4 shows the results obtained with yeast cells.

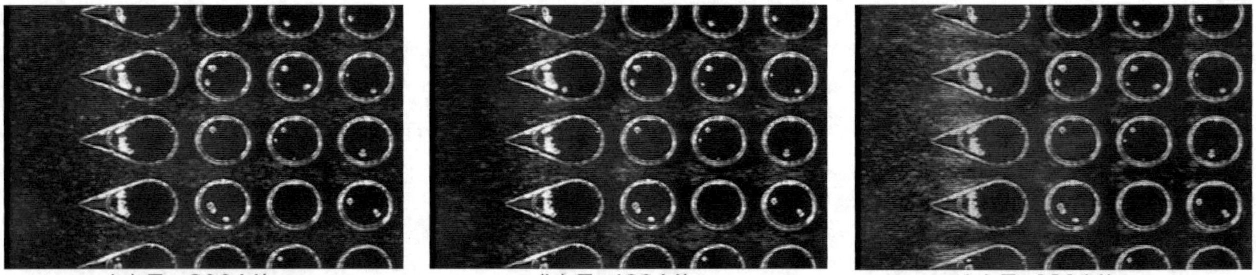

(a) E= 200 V/cm
pH=6, σ=115 μS/cm

(b) E=400 V/cm
pH=6, σ=115 μS/cm

(c) E=600 V/cm
pH=6, σ=115 μS/cm

Figure 4. Dielectrophoretic response of yeas cells (shown white) inside a plastic microchannel with cylindrical insulating structures, flow direction is from left to right, post diameter is 150 μm, arranged 200-μm center-to-center, and microchannel was 1-mm wide

Figure 4a, shows that applying a field of 200 V/cm does not produce dielectrophoretic trapping, increasing the applied field to 400 and 600 V/cm (Figures 4b and 4c, respectevely) produces negative dielectrophoretic immobilization of the yeast cells, where the magnitude of the trapping is a function of the magnitude of the applied field.

Finally, the technique of iDEP was tested for applications in the production of biopharmaceuticals. CHO cells are one of the most employed cells lines in the biopharmaceutical field, where significant effort is devoted to cell growth protocols. The objective was to explore the potential of iDEP as a cell screening method for the separation of live and dead CHO cells. The separation between live and dead cells can be used to optimize cell growth protocols by inoculating only with live cells.

Live and dead CHO cells were manipulated with iDEP, live cells were not dyed, and dead cells were distinguished with trypan blue. Figure 5a shows that applying a field of 400 V/cm does not produce a differentiation between live and dead cells. Increasing the applied field to 600 V/cm (Figure 5b) produces a distinction between live and dead cells. The ongoing goal of this set of experiments is to be able to selectively concentrate the live CHO cells. The results in Figure 5b demonstrate that live and dead cells exhibit different dielectrophoretic behavior, which could be exploited to achieve the desired separation. The differences in dielectrophoretic behavior should be due to changes in cells membrane properties when the cell dies.[10]

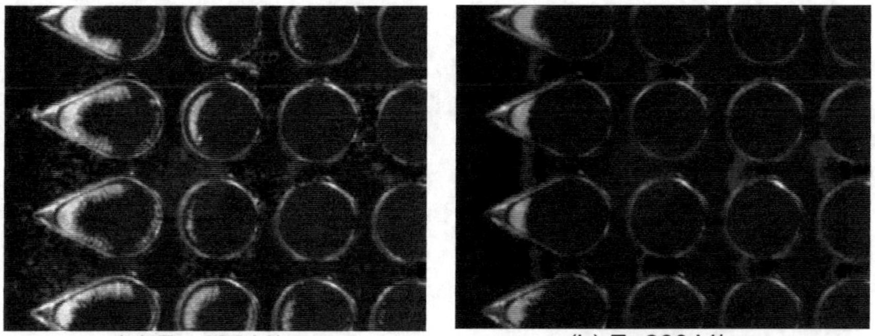

(a) E= 400 V/cm
pH=8, σ=25 μS/cm

(b) E=600 V/cm
pH=8, σ=25 μS/cm

Figure 5. Dielectrophoretic response of live (light gray) and dead (black) CHO cells, inside a plastic microchannel with cylindrical insulating structures, flow direction is from left to right, post diameter is 200 μm, arranged 250-μm center-to-center, and microchannel was 1-mm wide

Conclusions

Manipulation of four different types of bioparticles (BSA protein particles, microalgae, yeast and Chinese hamster ovary cells) employing the technique of insulator-based dielectrophoresis (iDEP) was presented. It was shown that this novel technique has great potential for a wide array of applications, from the concentration of valuable macromolecules to the manipulation and screening of microorganisms. All bioparticles exhibited negative dielectrophoretic behavior when they were immobilized inside a microchannel containing an array of cylindrical insulating structures. Microdevices made from glass and plastic were employed, demonstrating that iDEP can be carried out employing different materials. Additionally, the results showed the effect that operating conditions have over the dielectrophoretic response of the bioparticles. Successful trapping of the different types of bioparticles was achieved, demonstrating the great potential of iDEP as a technique for bioparticle manipulation.

Acknowledgements

The author would like to acknowledge the financial support provided by CONACYT grant No. CONACYT-CB-2006-53603 and *Cátedras de Investigación* (CAT161 and CAT162) of *Tecnológico de Monterrey*. The author is thankful to Sandra Ozuna, Israel Martínez, Roberto Gallo and Héctor Moncada for carrying out the experimental work. The author is thankful to Blake Simmons and Sandia National Laboratories for the kind donation of plastic microdevices.

References

1 Dittrich P.S., Tachikawa K., Manz A., (2006), "Micro Total Analysis Systems. Latest Advancements and Trends," *Anal. Chem.* 78(12), pp.3887-3907.

2 Cummings E.B., Singh A.K., (2003), "Dielectrophoresis in Microchips Containing Arrays of Insulating Posts: Theoretical and Experimental Results," *Anal. Chem.* 75(18), pp.4724-4731.

3 Pohl H.A., (1951), "The Motion and Precipitation of Suspensoids in Divergent Electric Fields," *J. Appl. Phys.* 22(7), pp.869-871.

4 Müller T., Fiedler S., Schnelle T., Ludwig K., Jung H., Fuhr G., (1996), "High frequency electric fields for trapping of viruses," *Biotechnol. Tech.* 10(4), pp.221-226.
5 Jones T.B., (1995), "Electromechanics of Particles," Cambridge University Press, USA.
6 Markx G.H., Dyda P.A., Pethig R., (1996), "Dielectrophoretic separation of bacteria using a conductivity gradient," *J. Biotechnol.* 51(2), pp.175-180.
7 Cummings E.B., Singh A.K., "Dielectrophoretic trapping without embedded electrodes," (2000), in Mastrangelo CH, Becker H (Editors), Proceedings of SPIE: Conference on Microfluidic Devices and Systems III, Santa Clara, CA, vol. 4177, pp. 164-173.
8 Cummings E.B., "A Comparison of Theoretical and Experimental Electrokinetic and Dielectrophoretic Flow Fields," (2002), in 32nd AIAA Fluid Dynamics Conference and Exhibit 2002, American Institute of Aeronautics and Astronautics, St. Louis, Missouri, vol. paper 2002-3193,, pp. 1-17.
9 Zheng L.F., Brody J.P., Burke P.J., (2004), "Electronic manipulation of DNA, proteins, and nanoparticles for potential circuit assembly," *Biosens. Bioelectron.* 20(3), pp.606-619.
10 Lapizco-Encinas B.H., Simmons B.A., Cummings E.B., Fintschenko Y., (2004), "Insulator-Based Dielectrophoresis for the Selective Concentration and Separation of Live Bacteria in Water," *Electrophoresis* 25, pp.1695-1704.

606d Field Effect Transistor Flow Control and Separation of Proteins in Nanofluidic Device

Youn-Jin Oh[1], Danny Bottenus[2], Cornelius F. Ivory[3], and Sang M. Han[1]. (1) Chemical & Nuclear Engineering, University of New Mexico, 1 University of New Mexico, MSC01 1120, Albuquerque, NM 87131, (2) Washington State University, Pullman, WA 99164, (3) Chemical & Bioengineering, Washington State University, Pullman, WA 99164

We have fabricated Si multiple internal reflection infrared waveguides embedded with a parallel array of nanofluidic channels (100 nm W ×500 nm D). This device enable us monitor the flow of proteins visibly using laser-scanning confocal fluorescence microscopy (LS-CFM), and serves *in-situ* analytical tool to probe using multiple internal reflection Fourier transform infrared spectroscopy (MIR-FTIRS). Fluidic field effect transistor (FET) method is used for protein control. A DC potential is applied to highly doped-gate area in mid-section of the channels under longitudinal electric field along the nanochannels. The gate potential controls the surface charge on SiO_2 channel walls and therefore the ξ-potential.

Depending the polarity and magnitude, the gate potential can accelerate, decelerate, or reverse the flow of proteins. In addition, we probed pH variations naturally, or caused by the surface charge modulation and longitudinal electrical field. Our MIR-FTIR analysis demonstrates that Fluorescein dye molecules (pH indicator) are hydrogenated and dehydrogenated in response to the gate bias and subsequent pH shift. This pH shift study enables us to understand FET flow control and abnormal flow characteristics during the FET flow control. For separation of proteins in nanofluidic channels, we conduct isoelectric focusing and separation of proteins, which have different isoelectric point (I_p), by generating

longitudinal pH gradient that is induced by controlled water electrolysis. Separation of proteins using their different valance charges and surface charge of nanochannels will be further disscussed

606e Selective Extraction of Recombinant Proteins by Multiple-Affinity Two-Phase Partitioning in Microchannels

Robert J. Meagher, Yooli K. Light, and Anup K. Singh. Sandia National Laboratories, P.O. Box 969, MS 9292, Livermore, CA 94551

We have demonstrated purification of proteins in a simple aqueous two-phase extraction process in a microfluidic device. The laminar flows inherent to microchannels allows us to perform a binary split of a complex cell lysate sample, in an open channel with no chromatography support and no moving parts. This mild process allows recovery of functional proteins with a modest increase in purity. Aromatic-rich fusion tags are used to drive partitioning of enzymes in a generic PEG-salt two-phase system. Addition of affinity ligands to the PEG phase allows us to exploit other popular fusion tags, such as polyhistidine tags and GST-tags.

627a Adsorption and Diffusion of Gases In ETS-10 and Mof Adsorbents

Hamed Sepehr, Shima Najafi Nobar, Shuli Chen, and **Shamsuzzaman Farooq**. Chemical and Biomolecular Engineering, National University of Singapore, 4 Engineering Drive 4, Singapore, 117576, Singapore

Engelhard titanium silicate ETS-10 is a microporous crystalline solid consisting mainly of an assemblage of titanium oxide (TiO_2) and silicate (SiO_2). The pore size in ETS-10 is uniform and similar in dimension to large-pore classical zeolites. It is also known to have high thermal stability. Metal-organic frameworks (MOFs) are another group of microporous structures made up of organic building blocks (linkers) that are coordinated with metal ions acting as joints. Therefore, compared to conventional zeolites, these metal-organic structures can give various pore properties and sizes by using different metal ions and organic linkers. A variety of MOF known, denoted as Cu-BTC (copper (II) benzene-1,3,5-tricarboxylate), has been synthesized in our laboratory. Synthesis of ETS-10 is also in progress.

Despite the promise of MOF and ETS-10 as potential new adsorbents for gas separation applications, only limited equilibrium studies on these materials are available in the literature. Transport behavior of gases in the micropores of these adsorbents is practically unknown.

We have, therefore, undertaken a program in our laboratory to systematically study adsorption and diffusion of gases in ETS-10, Cu-BTC and a few other varieties of MOF. Gases such as CO_2, O_2, N_2, H_2S, SO_2, C_2H_4, C_3H_6, C_2H_6 and C_3H_8 have been shortlisted for their industrial relevance in CO_2 capture and concentration, natural gas upgrading, air separation and olefin/paraffin separation. Progress made till the time of the conference will be covered in the proposed poster presentation.

627b Temperature and Pressure Dependent Non-Equilibrium Kinetic Model That Describes the Adsorption Behavior of CO2 In K-Promoted Htlc

Hai Du, **Armin D. Ebner**, and James A. Ritter. Department of Chemical Engineering, University of South Carolina, Columbia, SC 29208

Ritter and co-workers [1] recently showed that an initially fast adsorption or desorption phenomenon was followed by a state of extremely slow uptake or release of CO2 in K-promoted HTlc that requires hours perhaps days to reach equilibrium. The CO2 loadings at such an equilibrium state were also considerably different than those attained during the initial stages of adsorption or desorption. Then, they developed a model to describe this behavior in terms of a Langmuir-Hinshelwood type of kinetic mechanism [2].

That model was a based on a non-equilibrium kinetic approach that describes the reversible adsorption and desorption behavior of CO2 on a K-promoted HTlc. It combined adsorption, diffusion and reaction together with a Langmuir-Hinshelwood mechanism to describe the uptake and release processes of CO2 on a K-promoted HTlc. The results from that work convincingly revealed that the adsorption and desorption behavior of CO2 on a K-promoted HTlc is associated with complex, highly coupled, completely reversible adsorption, diffusion and reaction phenomena. It also revealed that a true equilibrium state might never be achieved in this material or that it might take an inordinate amount of time to achieve, with the claimed equilibrium states reported in the literature perhaps only being associated with the initial fast stages of adsorption or desorption observed by Ritter and co-workers [1,2].

The objective here is to report on the development of a temperature and pressure dependent form of the non-equilibrium kinetic model that describes the reversible adsorption and desorption behavior of CO2 on a K-promoted HTlc not only as a function of temperature but also as a function of pressure. This model combines adsorption, diffusion and reaction together with a Langmuir-Hinshelwood approach to describe the uptake and release processes of CO2 on a K-promoted HTlc but now over broad ranges of temperature and pressure. Experimental results and corresponding model predictions will be presented in terms of the CO2 loadings in a K-promoted HTlc during repeated CO2 adsorption and desorption cycles.

References

1. A. D. Ebner, S. P. Reynolds and J. A. Ritter, "Understanding the Adsorption and Desorption Behavior of CO2 on a K-Promoted HTlc through Non-Equilibrium Dynamic Isotherms," Ind. Eng. Chem. Res., 45, 6387-6392 (2006).

2. A. D. Ebner, S. P. Reynolds and J. A. Ritter, "Non-Eqilibrium Kinetic Model that Describes the Reversible Adsorption and Desorption Behavior of CO2 in a K-Promoted HTlc," Ind. Eng. Chem. Res., 46, 1737-1744 (2007).

627c High Temperature Air Separation by Perovksite-Type Oxide Sorbents-Heat Effect Minimization

Jerry Y. S. Lin, Department of Chemical Engineering, Arizona State University, Tempe, AZ 85287 and Qinghua Yin, Department of Chemical and Materials Engineering, Arizona State University, Tempe, AZ 85287.

High temperature air separation for oxygen production can find a number of applications. One application is for oxycombustion where the fuel is burned in oxygen rather than air. This results in pure carbon dioxide exhaust that requires little separation, facilitating carbon dioxide capture. Key barrier for oxycombustion is the high costs of oxygen production with current best cryogenic technology. Significant reduction in the cost of oxygen production is essential to making the oxycombustion power plant a future option when cabon dioxide capture becomes a requirement. An effective high temperature air separation process may offer a significant cost reduction for oxygen production for this application. Oxygen sorption on perovskite-type oxides can be advantageously used for air separation at high temperature. The large heat of oxygen sorption on these oxide sorbents presents a major challenge for the heat management of the high sorption separation process in practical applications. This paper reports a method to minimize the heat effects by taking advantage of an endothermic process of oxygen vacancy order-disorder phase transition accompanying the oxygen sorption process on perovskite-type oxide sorbents. The oxygen sorption isotherms, phase diagram, exothermic heat of oxygen sorption and endothermic heat of order-disorder phase transition for La0.1Sr0.9Co0.9Fe0.1O3-ä were measured by simultaneous TGA/DSC and XRD. The conditions for zero apparent heat of sorption are determined. If the oxygen partial pressure change and adsorption temperature are controlled in such that it gives an oxygen adsorption amount equal to the ratio of heat of phase transition to heat of oxygen sorption, the net heat released from the oxygen sorption step can be minimized or controlled to be negligible. This strategy for heat effect minimization is demonstrated with the results of TGA/DSC measurements at different operating conditions, and air separation by a fixed-bed packed with the perovskite-type oxide sorbent.

627d Direct Cationic and Nonionic Templating Synthesis of Mesoporous Zsm-5 Using Amphiphilic Silylated Surfactant

Rino R. Mukti, Hirotomo Hirahara, Atsushi Shimojima, and Tatsuya Okubo. Department of Chemical System Engineering, The University of Tokyo, 7-3-1 Hongo, Bunkyo-ku, Tokyo 113-8656, Japan

Hierarchical porous structure has been on demand over the past decade. Until now, the most significant progress has been achieved with films while the development of hierarchical porous solids particularly possessing ordered micro- and mesopores is still expecting discoveries and new applications. There have been a number of reports thus far showing that the use of hierarchical zeolite in catalytic reaction has been always superior compared to the one of conventional zeolite. The alkylation of benzene with isopropanol was catalytically favored and selective by using hierarchical Beta zeolite even with undefined mesostructure [1]. Having wormhole-like mesoporous, hierarchical MFI zeolite was reported to be free from deactivation during such reactions as isomerization, cracking, esterification and condensation [2, 3], therefore the synthesis of designed hierarchical zeolite has been very challenging. The achievement of using hard-templating synthesis may result an interesting material with zeolite crystallinity on the pore walls of Al/SBA-15/CMK-5 nanocomposite as elucidated by electron diffraction technique [4]. Moreover, the synthesis resulted from soft-templating method has shown a breakthrough by applying the tunable cationic amphiphilic surfactant with triethoxysilyl group which was self-assembled with zeolite precursors, i.e. assuming the presence of a covalent interaction [5]. However, it should be noted that the resulting composites with inter-connecting particles (physical mixtures) properties are not proposed as the final product, hence intra-connecting particles with new morphology should be designed. In order to demonstrate such properties, the material must show long and short range orders. The relatively intense zeolite crystallinity may bring the result into separated phases while the relatively weak zeolite crystallinity might suggest the interaction with mesopore walls. We are introducing the synthesis of mesoporous ZSM-5 (aluminosilicate zeolite with MFI type) templated by supramolecular nonionic pathway under basic condition. The nonionic surfactant used was silylated oligomeric alkyl(polyethylene oxide). In comparison to cationic route, the nonionic surfactant from block copolymer has more advantages to be used as mesoporous structure-directing agent where this includes larger structural features from solution organization and tunable continuous ordering properties. The new nonionic amphiphilic silylated surfactant can be applied in the synthesis of mesoporous ZSM-5. The nonionic surfactant may feature as nontoxic, biodegradable and low cost raw material if used in large scale.

References

1. Xiao, F. S., Wang, F. L., Yin, C. Y., Lin. K. F., Di, Y., Li, J. X., Xu. R. R., Su, D. S., Schloegl, R., Yokoi, T., Tatsumi, T., Angew. Chemie Int. Ed., 2006, 45, 3090.

2. Shetti, V. N., Kim, J., Srivastava, R., Choi, M., Ryoo, R., J. Catal., 2008, 254, 296.

3. Srivastava, R., Choi, M., Ryoo R., Chem. Commun., 2006, 43, 4489.

4. Fang, Y. M., Hu, H. Q., J. Am. Chem. Soc., 2006, 128, 10636.

5. Choi, M., Cho, H. S., Srivastava, R., Venkatesan, C., Choi, D. H., Ryoo, R., Nature Mater., 2006, 5, 718.

627e Diffusion of CO_2 and N_2 In Zsm-5 Zeolite with Framework Substitutions

David A. Newsome, Delft University of Technology, DelftChemTech, Julianalaan 136, 2628 BL, Delft, Netherlands, Anton J. Dammers, Delft University of Technology, DelftChemTech, Julianalaan 136, Delft, 2628 BL, Netherlands, and Marc-Olivier Coppens, Howard P. Isermann Department of Chemical Engineering, Rensselaer Polytechnic Institute, 110 8th Street, Troy, NY 12180.

Zeolites are inorganic aluminosilicates with an ordered atomistic-scale porous topology that have found usage in membrane separations, catalysis, and sequestration. The zeolite ZSM-5 has a well-defined topology with straight channels in the y-direction, zigzag channels in the x-z plane, and with a typical pore diameter of ~5.5 Å. The zeolite lattice has a nominal composition of SiO_2, but the Si atoms can be replaced by Al atoms, along with extra framework cations, such as Na^+, to maintain charge balance.

One possible application of ZSM-5 is the separation of CO_2 from flue gases. The molecules CO_2 and N_2 have different kinetic diameters and therefore will diffuse through the zeolite pores at different rates, hence separating molecules in a membrane operation.

We have used molecular dynamics (MD) simulations of CO_2 and N_2 diffusing in ZSM-5 with Na^+ and Al present. We performed the simulations with the DL_POLY software package, using the Nose-Hoover thermostat, the Ewald summation for Coulombic forces, a pretabulation of the zeolite potential energy landscape, and for runs over time lengths of 20-26 ns. In the MD runs we collect the trajectories to determine the mean squared displacement over time curve. From this information we calculated the self- and corrected-diffusivities of CO_2 and N_2 at T=200, 300, and 400 K. The diffusivities provide insight into the mobility of molecules as a function of temperature and concentration.

We consider the cases of 0 and 1 Al substitution of Si in the ZSM-5 unit cell to study the role of heterogeneity. In practice the concentration of the cation (that accompanies the Al) is a property that can be controlled via ion exchange to tune the adsorptive and diffusive properties of molecules in the ZSM-5. The CO_2 and N_2 molecules have a strong electrostatic interaction to the Na^+ cation that is tethered near the Al atom. We find that the diffusivities of CO_2 at T=200 and 300 K show a maximum as a function of loading at 1 Al per ZSM-5 unit cell, due to the Na^+ strongly attracting the molecules. The Na^+ ions also cause pore blockage by occupying space in the pore normally available for transport. The activation barrier for CO_2 strongly increases due to the presence of Na^+.

PRE-COMBUSTION CO_2 CAPTURE AT HIGH TEMPERATURE WITH ALKALINE PROMOTED ALUMINAS AND HYDROTALCITES: THE CRUCIAL ROLE OF ALKALINE CATIONS IN CO_2 ADSORPTION[‡].

Stéphane Walspurger *, *Paul D. Cobden, Wim G. Haije, Ruud W. van den Brink, Hydrogen production and carbon capture, Energy research Centre of the Netherlands, Petten, Netherlands*

Abstract

Alkali promotion effects on CO_2 sorption have been characterised by a series of complementary analytical techniques on γ-alumina, magnesia, hydrotalcites and alkaline promoted corresponding materials under conditions representative for WGS conditions. Alkaline carbonates are destabilised when interacting with solid surface as evidenced by thermogravimetric analyses and temperature programmed desorption. Structural rearrangements have been observed in alkali promoted hydrotalcite at the temperature used for capture. Moreover, in the presence of CO_2, the reconstruction of hydrotalcite cooled at room temperature after the experiment could not be observed most probably due to the formation of new carbonate species at high temperature that are stable while temperature decreases and prevent the reconstruction of crystalline hydrotalcite. In addition, in situ vibrational spectroscopies (Raman and DRIFT) showed that carbonates rearrange while heating and that alkali ions strongly interact with carbonate groups coordinated to aluminium oxide centres. Indeed both techniques highlight remarkably the loss of symmetry for carbonates group and comparison between parent material's fingerprints evidenced that basic (KAl_nO_{2n}/CO_3)-like species play a key role in CO_2 fixation at 400°C.

Introduction

Power Generation from coal with reduced CO_2 emissions can be achieved by using Integrated Gasification Combined Cycle (IGCC) with CO_2 capture and storage (CCS) in the near future. This configuration should allow the reduction of the efficiency penalty due to CO_2 capture and be competitive in terms of LHV efficiency with currently widespread used coal-pulverized combustion technologies.[1] Among pre-combustion carbon capture options, the Sorption-Enhanced Reaction Process (SERP) has attracted great interest because of the optimization of reaction yield, hydrogen purity[2] and CO_2 separation from the fuel[3-5]. Process simulations have shown that SERP using the pressure swing adsorption mode (PSA) has a high potential for lower efficiency penalties.[6,7] Gasification products mainly consist of CO which has to be converted into H_2 and CO_2 via the water-gas-shift reaction. In sorption enhanced water-gas-shift (SEWGS) the equilibrium of the reaction is shifted to the product side by removing one of the products using a sorbent.

$$H_2O + CO \leftrightarrows H_2 + CO_2 \qquad (\Delta H_{298}^0 = -41 \text{ kJ/mol})$$

The development of such a process has been started by Air Products in the 1990s.[2] Attention has been focused especially on hydrotalcites because they offer i) relatively fast reversible adsorption/desorption, ii) a stable capacity for CO_2 during long term cyclic experiments, iii) fairly good mechanical strength in high pressure steam and iv) there is no interaction with the physically mixed

‡ Most of these results have been published in ChemSusChem, 2008, 1, 643-650

water-gas-shift catalyst.[8,9] Using hydrotalcites (HT) as sorbent, the process has been operated successfully at temperatures between 300 and 500°C in a PSA unit (pressure between 1 and 30 bars) on a bench scale (Single Column 2 m tall, 38 mm diameter, CACHET project) for more than 6 months at the Energy research Centre of the Netherlands.[10]
In parallel, investigations on HT and derivative materials are ongoing at lab scale. Despite previous studies on CO_2 sorption using HT-based sorbents in water-gas-shift conditions,[8,9,11-13] the true nature of species involved in CO_2 reversible adsorption at such temperatures is still not clearly identified. In particular, the present study is focused on the role of alkali promoters which have been earlier found to significantly increase the reversible CO_2 sorption capacity of HT.[6,14-16] Particularly, HT as acid-base materials[17] may contain sites with an optimal basic strength for CO_2 fixation at high temperature.
Our approach was based on a systematic comparative study of HT "parent" materials such as alumina and magnesium oxide and their potassium carbonate promoted forms. Characterization of these materials has been performed by 1) sorption experiment, 2) in situ XRD to check phase transformation and 3) by infrared spectroscopy to identify rearrangements that occur at high temperatures and the CO_2 adsorption mechanism.

Results and discussions

Reversible capacity and carbonate decomposition

Hydrotalcites used in this study have been synthesized as described earlier[18] and have been first calcined at 400°C, promoted by potassium carbonate by incipient wetness and then dried at 120°C. In the following discussion the material is noted as HT*, alumina is noted as AO and magnesium oxide as MO. These materials have first been investigated for their reversible CO_2 adsorption in the presence of relatively high partial pressure of steam at 400°C. It was found earlier that the presence of steam in the gas stream has a significant effect on the reversible CO_2 capacity, in the case of HT.[13,19,20] Increasing the steam/CO_2 ratio facilitates a quick and efficient desorption and has a positive influence on adsorption capacity as demonstrated recently.[20]

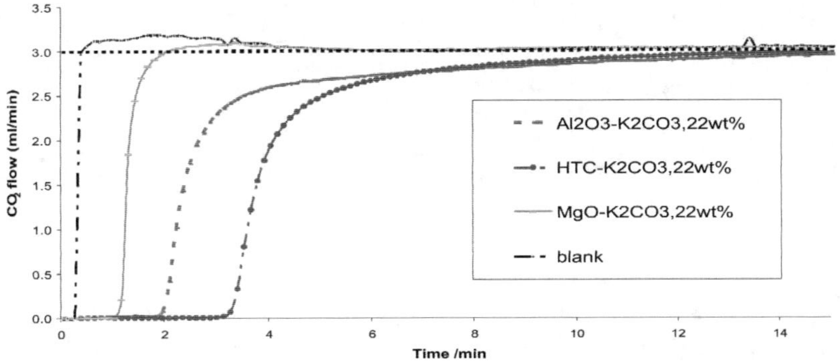

Figure 1: CO_2 breakthrough curves in sorption capacity measurement using 5.8% CO_2, 10.7% H_2O, N_2 balance as gas feed for adsorption step at 400°C, 1 bar. (black) blank, (grey) $K_{22}MO$, (red) $K_{22}AO$, and (green) $K_{22}HT^*$.

Figure 1 shows the typical breakthrough curves for various potassium carbonate loaded materials. Reversible capacities for other materials have been measured using the same method and the results after the third cycle are presented together with nitrogen sorption results in table 1. It is worth to note that generally there was only a very small difference between the 2nd and 3rd cycles and longer tests carried out for some of these materials have confirmed that capacity does not vary significantly after the 3rd cycle under these conditions. Obviously modification by potassium carbonate strongly enhances CO_2 capacity of HT* as demonstrated earlier.[13]

Table 1: CO_2 reversible sorption capacities for hydrotalcite, potassium carbonate promoted hydrotalcites and other selected materials measured after 3 cycles and their specific surface area.

Sample	K_2CO_3 loading (weight%)	BET Specific Surface Area (m^2/g)	CO_2 reversible sorption capacity (mmol/g) at 400°C
K_0HT*	0	48	0.06
K_5HT*	5	41	0.19
K_{11}HT*	11	40	0.30
K_{22}HT*	22	18	0.37
K_0AO	0	256	0
K_5AO	5	-	0.01
K_{11}AO	11	-	0.07
K_{22}AO	22	113	0.27
K_{44}AO	44	-	0.26
K_0MO	0	86	0
K_{22}MO	22	6	0.06

These experiments confirm that there exists an optimal loading of K_2CO_3 on hydrotalcites and that the surface area drops dramatically at high loading.[21] The same conclusion could be drawn from the results obtained with promoted aluminas. Although unpromoted γ-alumina did not show any affinity for CO_2 under the experimental conditions, it clearly appeared that potassium carbonate promotion allows CO_2 fixation despite a pronounced drop in specific surface area. Lower surface area found for high K_2CO_3 loadings are most probably caused by pore or surface blockage due to the formation of bulk K_2CO_3 aggregates as evidenced by the SEM photos (figure 2).

Figure 2: SEM picture of unpromoted HT (A), 22wt% K_2CO_3 promoted HT (B) and 22wt% K_2CO_3 promoted γ-alumina (C)

The typical platelets-like HT particles of the initial material are partially covered by some needle shaped material after potassium carbonate impregnation. In full agreement with the recent contribution of Rodrigues et al. on potassium modified HY,[22] small needle shaped material can also be clearly seen on γ-alumina surface. Although EDX analysis did not allow a definitive answer, it may be assumed that the needles are hydrates of potassium carbonate.[23] It is clear that at least part of the material deposited on the surface is responsible for the increased CO_2 sorption capacities for all those materials. In contrast, it has to be mentioned that commercial bulk potassium carbonate as well as magnesium oxide and other magnesium oxide precursors (Brucite and hydromagnesite) did not show any CO_2 reversible adsorption capacity under the experimental conditions used. Moreover potassium carbonate promoted magnesium oxide showed only a very low reversible CO_2 adsorption capacity compared to the promoted aluminium oxide based materials. Hence it seems that the interaction between γ-alumina

surface or aluminium oxide centres and potassium carbonate does play a key role in the formation of sites[24-26] that do actively and reversibly fix carbon dioxide under the experimental conditions used (1bar- about 60mbar CO_2).
Temperature programmed desorption was carried out to analyse the decomposition of carbonates for various samples. The released CO_2 was analysed by mass spectrometer for HT and aluminas with increasing potassium carbonate loadings and for a potassium carbonate promoted magnesium oxide (Figure 3).

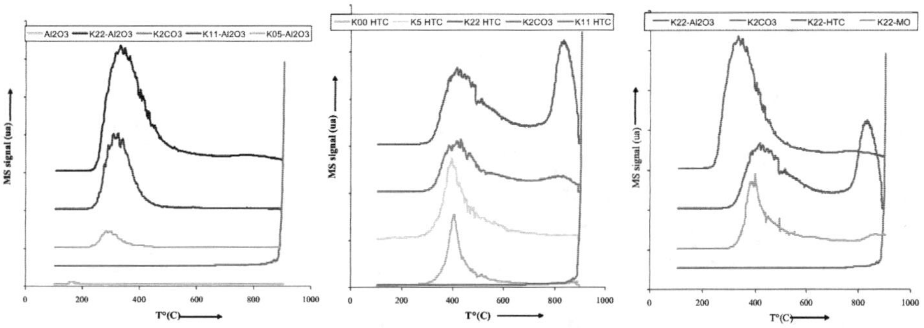

Figure 3: Temperature programmed desorption of CO_2 for K_0AO, K_5AO, $K_{11}AO$, $K_{22}AO$ on the left graph, K_0HT*, K_5HT*, $K_{11}HT$*, $K_{22}HT$* on the center graph and $K_{22}AO$, $K_{22}MO$, $K_{22}HT$* on the right graph, the red line corresponds to the TPD curve of K_2CO_3 bulk.

For unpromoted HT the desorption profile under standard conditions (heating in a flow of nitrogen (50ml/min) to 900°C at a rate of 10°C min^{-1}) showed a maximum at about 400°C with a rather narrow shape although a small tail was present at higher temperature denoting carbonates decomposing at higher temperature. Although the curve was clearly broadened due to the higher concentration of carbonate in the HT promoted with 5wt% potassium carbonate, the decomposition temperature peak was not shifted towards higher temperatures. More interestingly, with higher loadings the maximum of the decomposition peak shifted toward higher temperature while its profile was significantly broadened. Note that with 22 wt% K_2CO_3, a significant second CO_2 peak was observed at 830°C probably due to the presence of bulk carbonate at the surface of the mixed oxide (originally HT). Although commercial bulk potassium carbonate started decomposing above 900°C under these experimental conditions, it can be assumed that bulk potassium carbonate deposited on the surface of the solid (as observed by SEM) is responsible for the CO_2 release at such a high temperature. The corresponding experiments on γ-alumina and promoted γ-alumina are depicted at the left-side of figure 3. As expected unpromoted γ-alumina did not show any carbonate decomposition at high temperature and only a minor amount of residual CO_2 is desorbed between 100 and 200°C. Increasing the amount of potassium carbonate on the surface led to an increasing amount of CO_2 release between 250 and 450°C. Interestingly samples with increased amount of potassium carbonate showed a gradual shift for the maximum of carbonate decomposition from 290 to 335°C with a strong broadening of the temperature range at which CO_2 was released toward high temperatures. Furthermore the shift of the maximum decomposition peak toward higher temperature has been confirmed in extra experiments with various heating rates (not shown here). Most probably, some carbonates more strongly bonded to the surface, which are formed at higher surface coverage, may be responsible for the modification of the TPD profile.
Finally it is of the greatest interest to compare the decomposition profile for potassium carbonate impregnated on various supports, namely alumina, hydrotalcite and magnesium oxide. In comparison

to bulk potassium carbonate it is obvious that potassium carbonate impregnated on these supports undergoes a transformation that decreases the initial decomposition temperature of K_2CO_3. Apparently, γ-alumina can destabilize potassium carbonate to a high extent since some CO_2 was already released at 250°C. However the results from the thermal decomposition curves can hardly be directly correlated to the reversible CO_2 sorption properties shown in table 1. In other words it would be rather difficult to assign a specific group of carbonates responsible for the increased CO_2 reversible capacity on the basis of carbonates decomposition.

It has to be noted that one can try to analyze the results from carbonates decomposition in terms of basic strength. Indeed CO_2 as an acid probe is entrapped in carbonate form on the solid surface. The desorption of CO_2 may restore the initial basic site and as a result the temperature of carbonates decomposition may reflect, to some extent, the bond strength between CO_2 and the basic site, or in other words, it may give insights on the basic strength of the surface sites. Following this concept it appears from TPD profiles that the higher the carbonate loading, the more sites of higher basic strength are formed on the surface, confirming the results found by Wang et al.[25,27,28] It might be that a population of sites with appropriate basic strength is specifically involved in the reversible CO_2 sorption at 400°C. In situ structural analysis and spectroscopic investigations were therefore carried out to identify rearrangements happening in the sorbent under very similar conditions.

Hydrotalcite structural rearrangement during CO_2 capture

In-situ XRD experiments with variable temperature were performed with HT and promoted HT in the presence of steam (35 mbar) with and without carbon dioxide (120 mbar) (Figure 4).

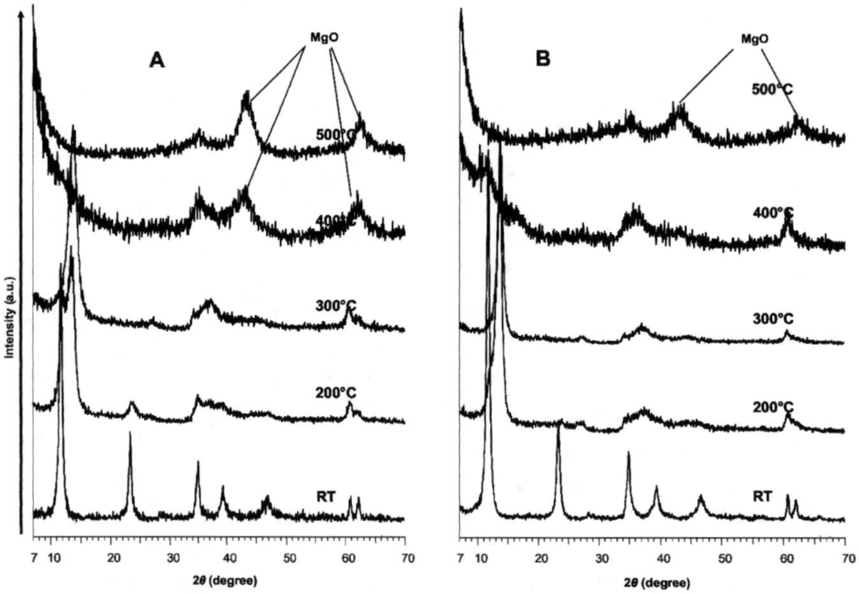

Figure 4: In situ X-Ray powder diffractograms for KoHT* at variable temperature in steam/nitrogen (A) and for KoHT* at variable temperature in steam/nitrogen/CO_2 (B).

Unpromoted hydrotalcite (K_0HT*) and 22 wt% potassium carbonate promoted hydrotalcite (K_{22}HT*) were heated gradually to 500°C in the chosen gas stream and diffraction patterns were recorded at various temperatures. Figure 4A displays the pattern obtained for K_0HT* in a mixture N_2/steam. Note that the diffraction patterns taken at room temperature shows the HT crystalline structure that has been reformed (reconstructed) after calcination at 400°C with subsequent impregnation with water at room temperature and drying at 120°C overnight. The well known decomposition of HT crystalline phase corresponding to the collapse of the layered structure is clearly observed together with the formation of magnesium oxide crystals,[29-32] although neither the $MgAl_2O_4$ spinel crystalline form (at high temperature) nor crystalline magnesium carbonate could be observed under these conditions.[12] The temperature dependence of the crystalline rearrangement was only slightly different for the other experiments. With CO_2 addition in the gas stream crystalline magnesium oxide appeared only at temperature above 400°C (above 300°C without CO_2) for both unpromoted hydrotalcite (Figure 4B) and K_2CO_3 promoted hydrotalcite (not shown here). In the latter case, with CO_2 in the stream, crystalline potassium carbonate was also observed during the experiment.

Remarkably the main differences were observed while cooling the samples to room temperature in a fixed gas stream after the heat treatment at 500°C (Figure 5). Indeed samples in permanent contact with CO_2 (Figures 5 B and 5 D) did not re-crystallise to HT structure contrary to the samples that were flushed with nitrogen/steam where the memory effect of activated HT was confirmed (Figures 5 A and 5 C).

CO_2 might help stabilizing some carbonates species at the surface which are poorly crystalline or non-crystalline and thus it might hinder the HT re-crystallisation. In addition, it seems that potassium promoted HT is reconstructed faster than the unpromoted HT in N_2/steam (figures 5A and 5C).

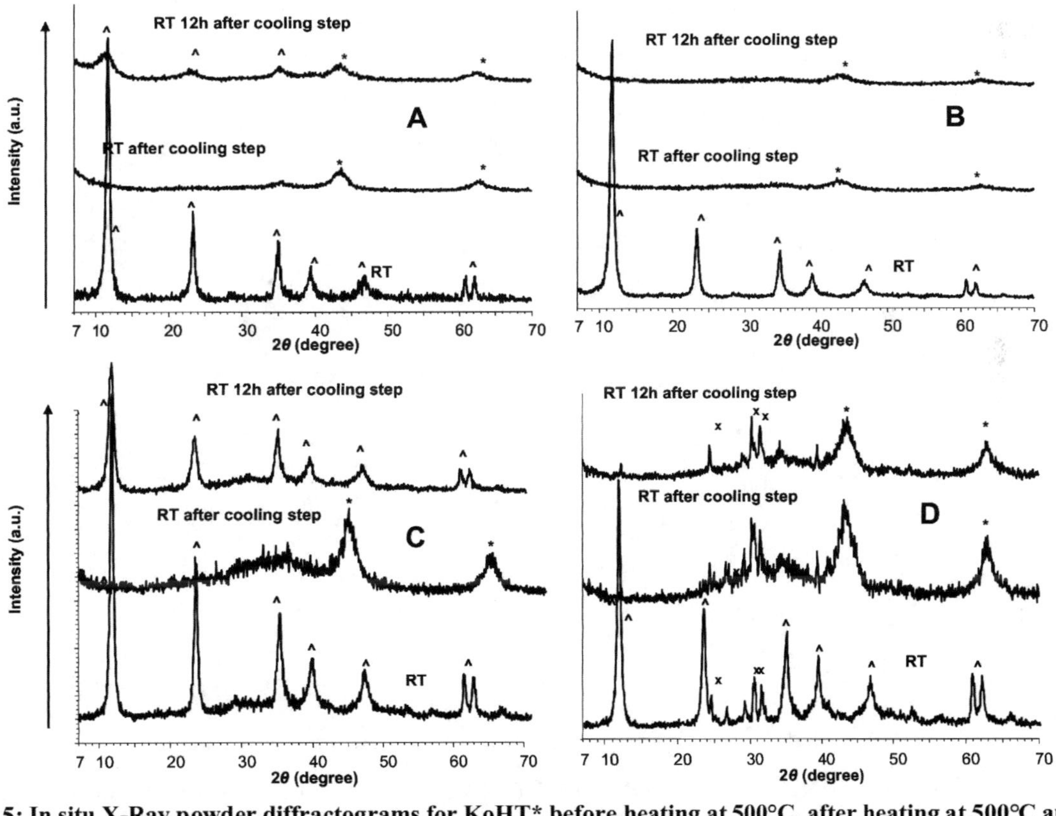

Figure 5: In situ X-Ray powder diffractograms for KoHT* before heating at 500°C, after heating at 500°C and 12h after heating at 500°C flushing with water/nitrogen (A); KoHT* before, after and 12h after experiment flushing with water/nitrogen+$\underline{CO_2}$ (B); for K_{22}HT* before, after and 12h after experiment flushing with water/nitrogen (C); K_{22}HT* before, after and 12h after experiment flushing with water/nitrogen+$\underline{CO_2}$ (D); Crystalline HT (^), MgO (*) and K_2CO_3 (x)

Surface rearrangement during CO_2 capture

In-situ Diffuse Reflectance Infrared Fourier Transformed Spectroscopy (DRIFTS) study was carried out to identify the rearrangements involved in the solids at variable temperature (RT to 400°C) in the presence of N_2/steam. Figure 6 shows the spectra in the 800-2000 cm^{-1} wave numbers range obtained from unpromoted hydrotalcite, 22wt% K_2CO_3 promoted hydrotalcite, γ-alumina and 22wt% K_2CO_3 promoted γ-alumina respectively. Before the heat treatment, a broad absorption band is observed at 1655 cm^{-1} for all samples which is typical of O-H bending mode from water.[34-37] Accordingly, this band disappears after thermal treatment at temperature between 100 and 200°C and reappeared after rehydration of the sample at room temperature (not shown here). It should be noted at this point that this band might also hint at the presence of bicarbonate species at low temperature which are known to decompose at rather low temperature.[21,34]

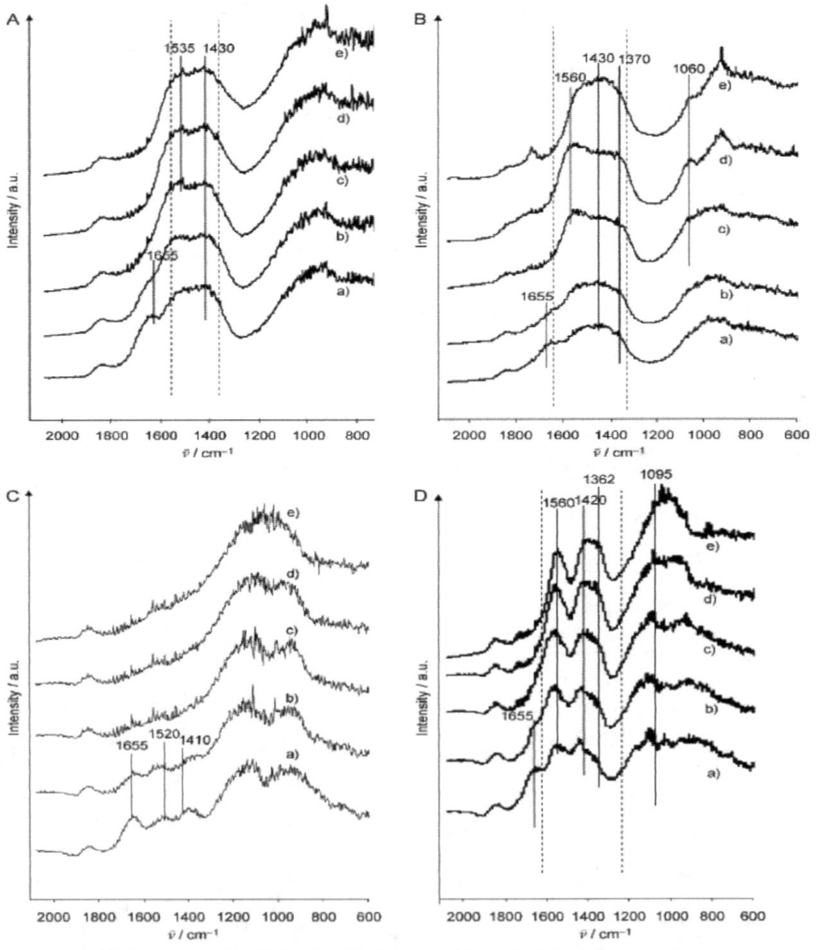

Figure 6: DRIFT spectra of K_0HT^* (A) $K_{22}HT^*$ (B) K_0AO (C) $K_{22}AO$ (D) in the carbonate region at RT in N_2/steam stream at room temperature (a), 100°C (b), 200°C (c), 300°C (d), 400°C (e)

The unpromoted HT sample exhibited a strong and broad absorption band between 1400 and 1600 cm^{-1}. When the temperature is increased there are clearly two maxima appearing at 1535 cm^{-1} and 1430 cm^{-1} (Figure 6A). These bands are generally assigned to ν_3 asymmetric stretching of carbonate groups.[34,36-40] The splitting is most probably caused by the rearrangement of surface carbonates that

involves a lowering of their symmetry. According to Busca et al.[41] the splitting Δv 105cm^{-1} can be assigned to bidentate carbonates in this case, because the polarisation might be rather low for Mg-Al atoms. For the 22 wt% K_2CO_3 promoted HT (Figure 6B), band splitting also occurs when the temperature is increased. Between 200 and 400°C there are two broad bands peaking at 1560 and 1370 cm^{-1} corresponding to the v_3 vibrational mode of carbonate groups. Strikingly, the splitting is increased to Δv 190cm^{-1} denoting a stronger perturbation of the symmetry in carbonate groups which is confirmed by the appearance of the band at 1060 cm^{-1}, infrared inactive for symmetric carbonates. The main reason for this significant change may be the close interaction of surface carbonates with potassium ions provided by potassium carbonates. The band at 1430 cm^{-1} and the weak absorption band at 1740 cm^{-1} maximised at 400°C can be unambiguously assigned to the vibration frequencies v_3 and $2v_2$ (harmonic signal of v_2, usually found at 870-880 cm^{-1}) respectively of bulk potassium carbonate present in large amount, in agreement with the observation made by TPD and SEM [42].
In contrast, experiment on non-promoted γ-alumina (Figure 6C) showed that no carbonates were present on the surface at high temperature. At room temperature in the presence of steam and residual CO_2 from the atmosphere bands were observed at 1655, 1520 and 1410 cm^{-1} corresponding most probably to surface bicarbonates formed during the storage of the material in ambient atmosphere.[43]
More interestingly 22 wt% K_2CO_3 promoted γ-alumina showed a very clear picture on rearrangement of surface carbonates occurring at elevated temperature (Figure 6D). Two bands peaking at 1560 and 1362 cm^{-1} (Δv 198cm^{-1}) corresponds to v_3 stretching frequencies for bidentate carbonates. The second band is rather large probably because of the presence of some bulk potassium carbonate with its typical single v_3 frequency around 1430 cm^{-1}. Furthermore the bands broadening may be explained by the presence of K-O-(CO)-O-Al species which usually exhibit a split v_3 frequency and v_1 at 1550, 1420 and 1095 cm^{-1} respectively.[44,45] These species have previously been invoked to rationalize the enhanced basic strength of alkali promoted aluminium oxides.[24,46,47]
Hence the observed similarities between hydrotalcites and K_2CO_3 promoted γ-alumina may be of the utmost importance for the comprehension of the promoting mechanism. Carbonates provided by the addition K_2CO_3 are rearranged at the surface as proved by the splitting of v_3 frequency for both materials. Moreover by comparison of the splitting values observed for promoted hydrotalcite and promoted γ-alumina it clearly appeared that the carbonates symmetry may be affected in a similar way. Thus it is obvious that aluminium centres play a major role in the destabilisation of potassium carbonate and may be at the origin of the increased CO_2 capacity observed in our experiments at 400°C in agreement with previous reports.
Although Raman spectroscopy was found to be of limited help concerning HT carbonates analysis because of a pronounced fluorescence which prevents a proper analysis, the comparison between potassium bicarbonate and $K_{22}AO$ has confirmed that γ-alumina destabilises potassium carbonate (figure 7).
Very interestingly, Raman spectroscopy allows to observe very clearly v_1 stretching frequency of carbonate groups which absorb intensively in the region 1000-1150cm^{-1} contrary to DRIFTS. As a reference, potassium bicarbonate shows a single absorption band at 1027cm^{-1} at 100°C, whereas its decomposition product at 400°C (potassium carbonate) exhibits a band at 1061cm^{-1}. Although $K_{22}AO$ suffered from some interference due to fluorescence, two bands could be clearly detected at 100°C. The one around 1060cm^{-1} can probably be assigned to residual bulk potassium carbonate, whereas the one around 1094cm^{-1} is to be assigned to a carbonate group with a completely different symmetry. At 400°C it seems that both types of carbonate are still present although the relative intensity i.e. presence, obviously changed. Raman spectroscopy investigations are ongoing.

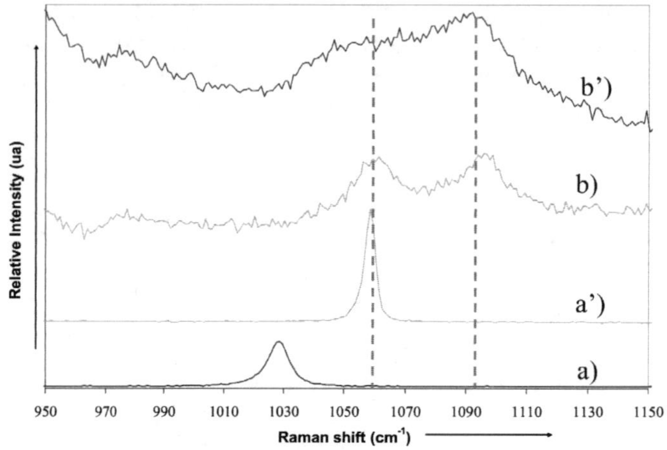

Figure 7: in situ Raman spectra -within 950-1150 cm^{-1}- of potassium bicarbonate at 100°C (a) 400°C (a') and 22wt% K_2CO_3 on γ-alumina at 100°C (b) and 400°C (b')

Conclusion

Hydrotalcites, potassium carbonate promoted hydrotalcite and potassium carbonate promoted γ-alumina have been proved to be excellent materials for in situ CO_2 capture during water gas shift process (SEWGS). Significant enhancement of CO_2 capacity can be achieved by alkaline promotion of the sorbents. We have unambiguously demonstrated that potassium carbonate is destabilized or transformed at the surface of metal oxides and particularly γ-alumina and hydrotalcites. Although no new carbonates structure or species could be identified by *in situ* XRD studies we showed that fixed CO_2 at 300-500°C hinders the reconstruction of hydrotalcites at room temperature most probably because of the formation of carbonate species. Moreover the *in situ* DRIFTS study confirmed that surface carbonates rearrangement occurs at high temperature and that the interaction between aluminium oxides centres and potassium carbonate plays a crucial role in the formation of active sites (strongly basic) for CO_2 capture at 300-500°C. These findings are furthermore corroborated by the very low sensitivity to potassium promotion of magnesium oxide.

References

1. F. Li, L.S. Fan, *Energy Environ. Sci.*, **2008**, *1*, 248.
2. S. Sircar, C.A.M Golden, PSA process for removal of bulk carbon dioxide from a wet high-temperature gas, US Patent 6322612, **2001**
3. IEA Prospects for CO_2 Capture and Storage. International Energy Agency, Paris, **2004**.
4. B. Metz, H.C. Coninck, "CO_2 capture and storage: a future for coal?" ECN policy study available at http://www.ecn.nl/publicaties/default.aspx?nr=ECN-B--07-020
5. C. Gough, *Int. J.Greenhouse Gas Control* **2008**, *2*, 155.
6. J.R. Hufton, S. Mayorga, S. Sircar, *AIChE J.* **1999**, *45*, 248.
7. M.C. Carbo, D. Jansen, J.W. Dijkstra, R.W. van den Brink, A.H.M. Verkooijen, "Pre-combustion decarbonisation in IGCC: Abatement of both steam requirement and CO_2 emissions" *ECN publication* **2007** http://www.ecn.nl/publications/default.aspx?nr=ECN-M--07-055
8. K.B. Lee, A. Verdooren, H.S. Caram, S. Sircar, *J. Colloid. Interf. Sci.* **2007**, *308*, 30.
9. K.B. Lee, M.G. Beaver, H.S. Caram, S. Sircar, *AIChE J.* **2007**, *53*, 2824.
10. Results to be published. Preliminary presentation at CACHET workshop available at http://www.cachetco2.eu/publications/conf_pubs.html
11. Z. Yong, V. Mata, A.E. Rodrigues, *Ind. Eng. Chem. Res.* **2001**, *204*, 40.
12. N.D. Hutson, S.A. Speakman, E.A. Payzant, *Chem. Mater.* **2004**, *16*, 4135.

13. H.T.J. Reijers, S.E.A. Valster-Schiermeier, P.D. Cobden, R.W. van den Brink, *Ind. Eng. Chem. Res.* **2006**, *45*, 2522.
14. Y.Ding, E.Alpay, *Chem. Eng. Sci.* **2000**, *55*, 3461.
15. S. Mayorga, S. Weigel, T.R. Gaffney, J.R. Brzozowski, US Patent No 6,280,503B1 **2001**.
16. A.D. Ebner, S.P. Reynolds, J.A. Ritter, *Ind. Eng. Chem. Res.* **2006**, *45*, 6387.
17. K. Tanabe, H. Hattori, T. Yamaguchi, T. Tanaka, Acid-Base Catalysis, Proceedings of the international symposium on Acid base catalysis, Sapporo, VCH Publishers, New York, **1989**
18. F. Cavani, F. Trifirò, A. Vaccari, *Catal. Today* **1991**, *11*, 173.
19. S. Abello, J. Perez-Ramirez, *Micropor. Mesopor. Mater.* **2006**, *96*, 102.
20. M.K.R. Reddy, Z.P. Xu, G.Q. Lu, J.C. Diniz da Costa, *Ind. Eng. Chem. Res.* **2008**, *47*, 2630
21. J.I. Yang, J.N. Kim, *Korean J. Chem. Eng.* **2006**, *23*, 77.
22. E.L.G. Oliveira, C.A. Grande, A.E. Rodrigues, *J. Sep. Purif. Tech.* **2008**, *62*, 137.
23. A. Pabst, *Am. Mineral.* **1930**, *15*, 69
24. T. Horiuchi, H. Hidaka, T. Fukui, Y. Kubo, M. Horio, K. Suzuki, T. Mori, *Appl. Catal. A* **1998**, *167*, 195.
25. Y. Wang, J. H. Zhu, W. Y. Huang, *Phys. Chem. Chem. Phys.* **2001**, *3*, 2537.
26. A.G. Okunev, V.E. Sharonov, A.V. Gubar, I.G. Danilova, E.A. Paukshtis, E.M. Moroz, T.A. Kriger, V.V. Malakhov, Y.I. Aristov, *Russ. Chem. Bull. Int. Ed.* **2003**, *52*, 359.
27. Y. Wang, X.W. Han, A. Ji, L.Y. Shi, S. Hayashi, *Micropor. Mesopor. Mater.* **2005**, *77*, 139.
28. T. Yamaguchi, Y. Wang, M. Komatsu, M. Ookawa, *Catal. Surv. Japan* **2002**, *5*, 81.
29. F. Rey, V. Fornes, J.M. Rojo, *J. Chem. Soc. Faraday Trans.* **1992**, *88*, 2233.
30. M.L. Valcheva-Traykova, N.P. Davidova, A.H. Weiss, *J. Mater. Sci.* **1993**, *28*, 2157.
31. J.C.A.A. Roelofs, J.A. Bokhoven, A.J. Dillen, J.W. Geus, K.P. de Jong, *Chem. Eur. J.* **2002**, *8*, 5571.
32. T. Stanimirova,; T. Hibino, V. Balek, *J. Therm. Anal. Cal.* **2006**, *84*, 473.
33. J. Perez-Ramirez, S. Abello, N.M. van der Pers, *Chem. Eur. J.* **2007**, *13*, 870.
34. M. Del Arco, C. Martin, I. Martin, V. Rives, R. Trujillano, *Spectrochim. Acta* **1993,** *49*, 1575.
35. Y. Kim, W.Yang, P.K.T. Liu, M. Sahimi, T.T. Tsotsis, *Ind. Eng. Chem. Res.* **2004**, *43*, 4559.
36. J.T. Kloprogge, The application of vibrational spectroscopy to clay minerals and layered double hydroxides, CMS workshop lectures, vol. 13, Kloprogge J.T. Ed., The clay mineral society, Aurora, CO, **2005**, p.203-238.
37. R.L. Frost, B.J. Reddy, *Spectrochim Acta A* **2006**, *65*, 553.
38. T. Hibino, Y. Yamashita, K. Kosuge, A. Tsunashima, *Clays Clay Miner.* **1995**, *43*, 427.
39. F. Prinetto, G. Ghiotti, R. Durand, D. Tichit, *J. Phys. Chem. B* **2000**, *104*, 11117.
40. F. Millange, R.I. Walton, D. O'Hare, *J.* Mater. *Chem.* **2000**, *10*, 1713.
41. G. Busca, V. Lorenzelli, *Mater. Chem.* **1982**, *7*, 89.
42. R.A. Nyquist, R.O. Kagel, Handbook of infrared and Raman spectra of inorganic compounds and organic salts, vol. 4, Academic Press, San Diego, **1997**.
43. C. Morterra, A. Zecchina, S. Coluccia, A. Chiorino, *J. Chem. Soc. Faraday Trans.* **1977**, *73*, 1544.
44. A.S. Berger, N.P. Tomilov, I.A. Vorsina, *Russ. J. Inorg. Chem.* **1971**, *16*, 42.
45. A. Iordan, M.I. Zaki, C. Kappenstein, *J. Chem. Soc. Faraday Trans.* **1993**, *89*, 2527.
46. S.C. Lee, B.Y. Choi, T.J. Lee, C.K. Ryu, Y.S. Ahn, J.C. Kim, *Catal. Today* **2006**, *111*, 385.
47. S.C. Lee, J.C. Kim, *Catal. Surv. Asia* **2007**, *11*, 171.

Synthesis of a Zeolite Column with a Monolithic Microhoneycomb Structure Using the Ice Template Method

Shin R. Mukai, Shinya Murata, Kazufusa Onodera and Izumi Yamada
[*1] *Graduate School of Engineering, Hokkaido University, Sapporo, Japan*

Introduction

Various porous materials are widely used as adsorbents and catalysts and are supporting our daily lives in many different ways. In such materials, active sites are usually dispersed throughout the whole volume of the material and provide the basic function(s) of the material.

Generally, the performance of such materials can be maximized by keeping the lengths of the diffusion paths, which lead substances from the outer surface of the material to the active sites within them, as short as possible. As porous materials are generally synthesized in the form of particles, the lengths of the diffusion paths in them can be shortened by simply reducing the size of the particle itself. However, small particles cause a severe resistance when fluids are passed through them, therefore this method does not always lead to the improvement of the total performance of the material. In fact, particles with fairly large sizes are usually used in industrial applications just only to avoid significant resistance to flows. This indicates that there are many cases that the performance of such materials can be further improved, if such materials can be synthesized to have a structure in which short diffusion paths and low resistance to fluid flows are compatible.

A monolithic microhoneycomb which has straight and aligned micrometer-sized channels within it may be an ideal structure, if the walls which form the channels are thin enough. It was thought to be difficult to obtain porous materials with such a structure through conventional synthesis methods, but recently we found that porous materials attainable through the sol-gel method can be molded into a monolithic microhoneycomb just by freezing their parent hydrosol or hydrogel unidirectionally. As ice crystals which grow within the precursor during freezing act as the template, we named this method the "Ice Template Method."

We also found that this method can also be applied to hydrogels including fine particles. The materials obtained through this process have a monolithic microhoneycomb structure and can be practically regarded as a monolithic column of the fine particles, in which the gel acts as the binder.

In this work we synthesized columns with a monolithic microhoneycomb structure from zeolites, a typical catalyst used in various petrochemical processes, using the ice template method, and checked their characteristics.

Experimental

Commercial sodium silicate solutions (Wako Pure Chemical Industries) were diluted with ion-exchanged and distilled water, and the SiO_2 concentration (C_s) in them were adjusted to a desired value set in the range 1.0 mol L^{-1} to 1.9 mol L^{-1}. Then the pH of the solutions was adjusted using an ion-exchange resin. To these solutions, high silica Y-type zeolite particles (HSZ, Tosoh, average particle size: 300 nm) were added. First, the particles were dispersed in the solutions using an ultrasonic homogenizer, and the resulting slurries were poured into polypropylene tubes (i.d.: 10 mm, L: 100 mm) and were aged for different periods. Next the tubes were dipped at a constant rate v, set in the range 6 cm h^{-1} to 20 cm h^{-1}, into a cold bath maintained at 77 K. After the tubes were completely frozen, they were taken out from the bath and thawed at 323 K. Then the water in the tubes was exchanged with *t*-butanol, and the samples were freeze-dried.

The morphology of the samples was directly observed using a scanning electron microscope (SEM). Next the samples were analyzed using an X-ray Diffractometer (XRD) to confirm whether the included zeolite particles maintained their structures after monolith synthesis. The porous properties of the samples were evaluated through nitrogen adsorption-desorption experiments. First, nitrogen adsorption-desorption isotherms of the samples were measured at 77 K. Next, BET surface areas of the samples were calculated using the BET method. The compression strengths (axial direction) of the monoliths were measured using a load cell. Finally, stainless steel columns in which the monoliths were fitted in were synthesized, and the pressure drop which occurs when air (293 K) was passed through the columns was measured. The dimension of the monoliths used in measurements was $\phi 10$ x 20.

Results and Discussion

Sample Morphology

In this work, samples including zeolite particles in the range of 0.2 g-zeolite $(g\text{-}SiO_2)^{-1}$ to 1.0 g-zeolite $(g\text{-}SiO_2)^{-1}$ were obtained. Figure 1 shows SEM images of cross sections of typical samples obtained in this work. The channels of the microhoneycombs were not as straight and as regularly ordered as those in the microhoneycombs obtained through the original ice template method. However, it was confirmed that fairly straight channels run from one side to the other side of the honeycomb through SEM observation and ink penetration tests.

Figure 2 shows the XRD patterns of typical samples. Peaks corresponding to the crystal structure of the included zeolite particles can be clearly identified, which indicates that the included particles maintained their nanostructure during freezing. Therefore, it is assumed that the particles maintain their unique functions.

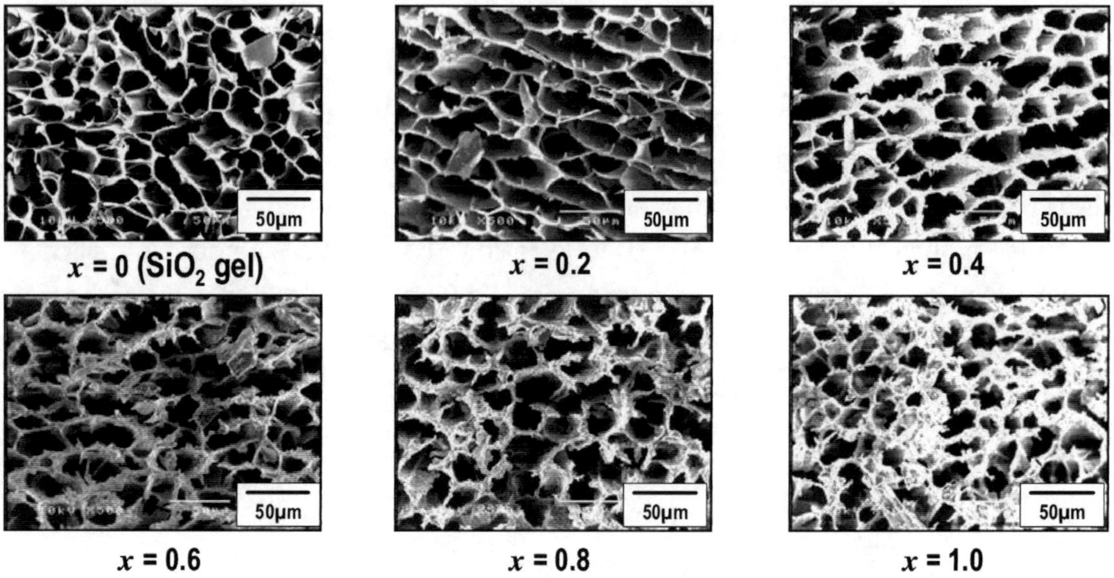

Fig. 1 Cross sectional SEM images of typical samples obtained in this work

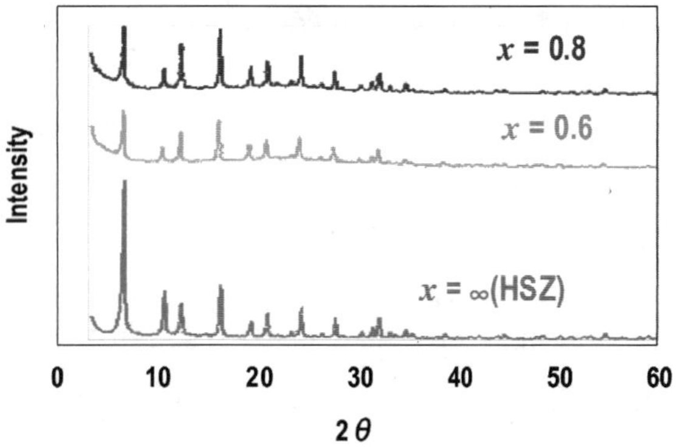

Fig. 2 XRD patterns of typical samples obtained in this work

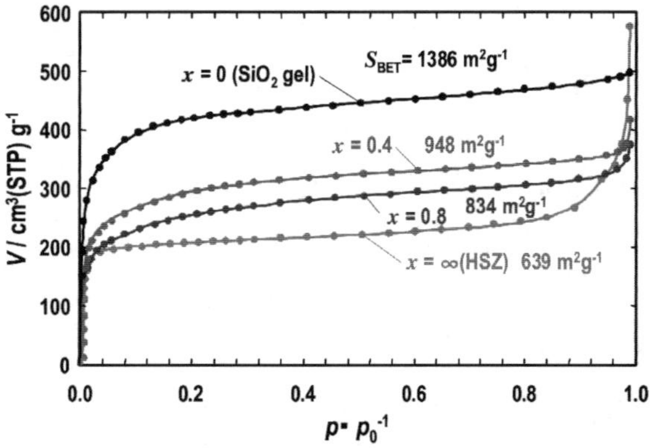

Fig. 3 N2 adsorption isotherms of typical samples obtained in this work

The porous properties of the samples were evaluated through nitrogen adsorption experiments. Figure 3 shows nitrogen adsorption isotherms of typical samples measured at 77 K. The BET surface areas (S_{BET}) are also listed in the figure. Silica gels prepared using the ice template method has developed micropores and possess fairly high BET surface areas. Zeolites are also microporous, but their isotherms indicate that they also have a slight amount of mesopores, with are thought to exist within or between the particles. Silica including zeolite samples show isotherms which correspond to the amounts of silica gel and zeolite included within them. Therefore, it is assumed that the pore inlets of the zeolite are not blocked by the silica gel matrix.

Next a typical sample was divided into 5 equal parts in the axial direction, and structural analysis of each part was conducted. Figure 4 shows SEM images and nitrogen adsorption isotherms measured at 77 K of the divided parts. A significant difference among the samples cannot be recognized, which indicates that the ice template method gives a fairly uniform sample.

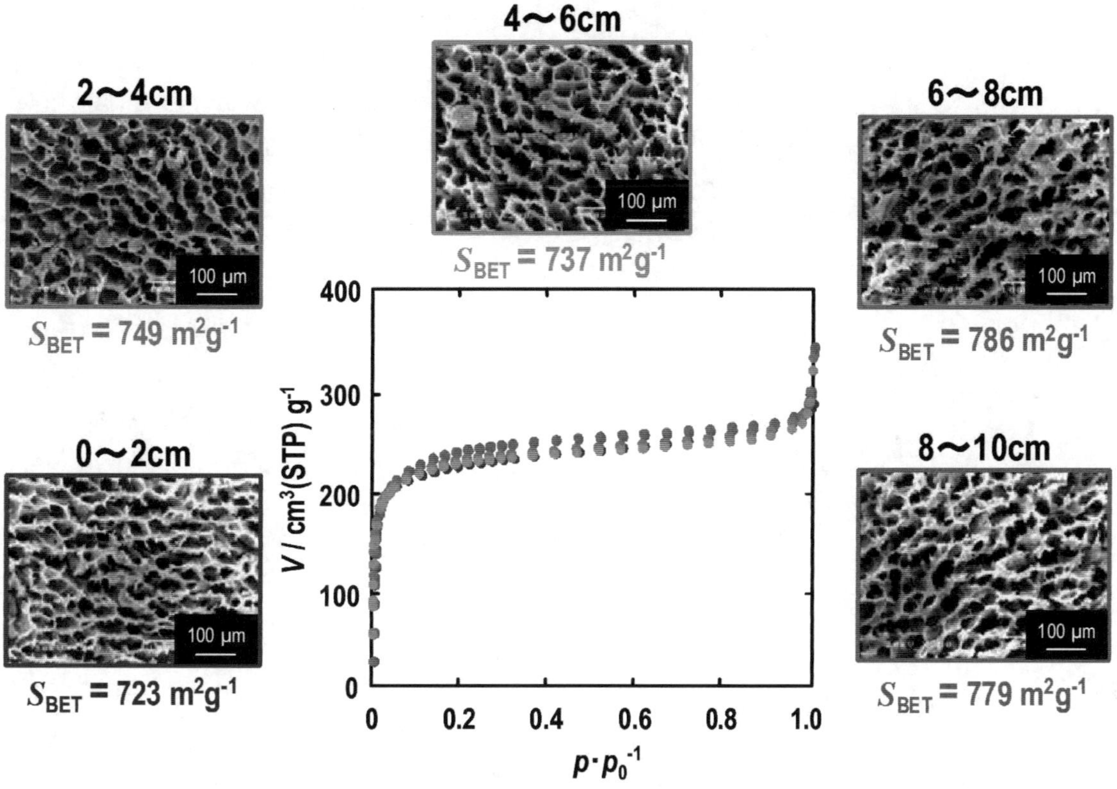

Fig. 4 SEM images and N2 adsorption isotherms of divided sections of a typical sample in the axial direction

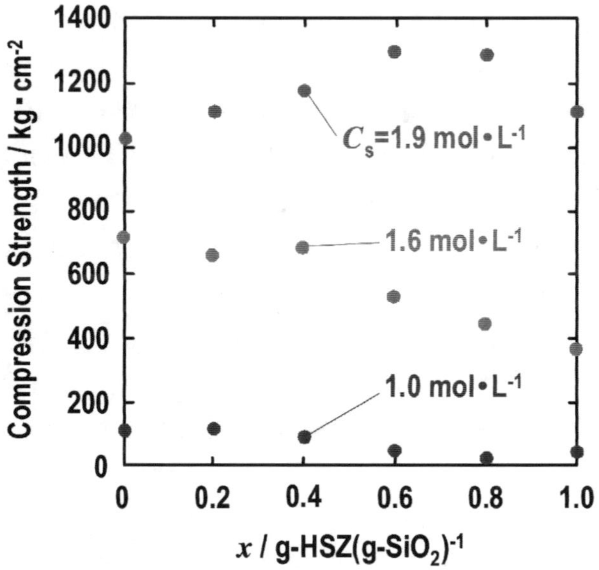

Fig. 5 Compression strengths of typical samples obtained in this work

Figure 5 shows the compression strengths of typical samples obtained in this work. As expected, the strength of samples synthesized using sodium silicate solutions with low C_s values was low, and the strength tended to decrease as the amount of zeolite particles was increased. However, moderate strengths can be achieved when C_s is increased, and the strength is hardly affected even when the amount of zeolite particles is significantly increased.

As it was confirmed that uniform zeolite monoliths with a microhoneycomb structure can be obtained through the ice template method, next we attempted to synthesize such monoliths directly in a stainless steel column. Figure 6 shows a

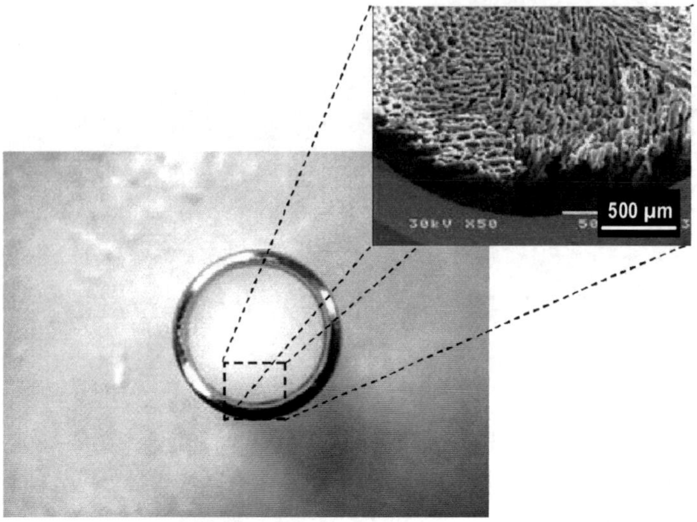

Fig. 6 Photograph and a SEM image of a typical column obtained in this work

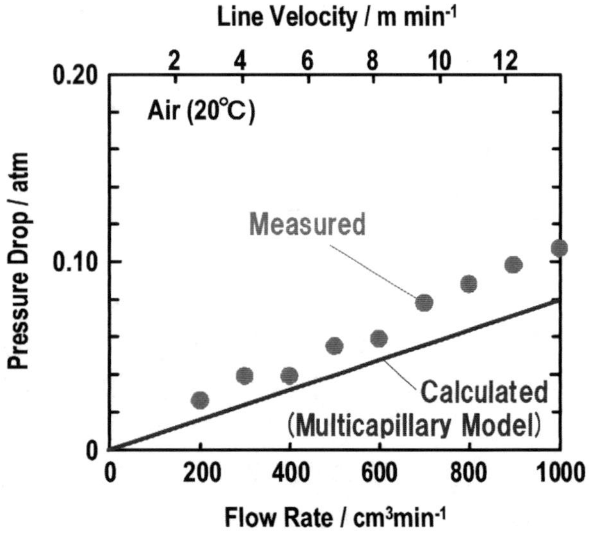

Fig. 7 Pressure drop which occurs in a typical column
(x = 0.6, i.d. 10 mm, L 30 mm)

photograph and an SEM micrograph of the cross section of a typical column. The existence of a slight gap between the monolith and the stainless steel column can be confirmed. However, the size of this gap is comparable to the sizes of the channels within the monolith, therefore this gap will probably not become a serious problem in most of the applications this monolith is expected to be used for.

Using these stainless steel columns, the pressure drop which occurs when air was passed through them was measured. All measurements were conducted at room temperature. Figure 7 shows the results, where the pressure drop is plotted against the flow rate of air. The pressure drop linearly increased with the increase in flow rate, but was extremely low even at high flow rates. The line in the same graph shows the predicted pressure drop of the column which was calculated using a multi-capillary model to represent the structure of the monolith. The measured values are slightly higher than the predicted values, which indicates that the channels in the monolith are basically straight and are fairly aligned, but slight irregularity which disturbs the flow also exists within them.

Conclusion

In this work, a zeolite monolith with a monolithic microhoneycomb structure was synthesized by applying the ice template method to a silica hydrogel including zeolite nanoparticles. The monolith was found to have a microhoneycomb structure and the included zeolite nanoparticles were found to be evenly distributed within it. The zeolite nanoparticles maintained their crystal structure, and there was no sign of pore blocking due to the inclusion in silica. As expected, the resistance to flows of the monoliths was extremely low. These monoliths can be used to provide a zeolite column in which short diffusion paths and low pressure drops are compatible.

This work was supported by the Industrial Technology Research Grant Program in 2006, 06B44702a from New Energy and Industrial Technology Development Organization (NEDO) of Japan and the Japan Society for the Promotion of Science (JSPS), Grant-in-Aid for Scientific Research (B), 19360355 (2007)

References

1. Mukai, S. R., Nishihara, H. and Tamon, H. (2003), " Porous Properties of Silica Gels with Controlled Morphology Synthesized by Unidirectional Freeze-Gelation," *Microporous and Mesoporous Materials, 63*, pp. 43-51

2. Mukai, S. R., Nishihara, H. and Tamon, H. (2004), "Formation of Monolithic Silica Gel Microhoneycombs (SMHs) Using Pseudosteady State Growth of Microstructural Ice Crystals," *Chemical Communications*, pp. 874-875

3. Nishihara, H., Mukai, S. R., Yamashita, D. and Tamon, H. (2005), " Ordered Macroporous Silica by Ice-Templating," *Chemistry of Materials*, 17, pp. 683-689

Comparison of Processes for Capture, Separation, and Sequestration of CO_2 from the Combustion of Coal

Michelle Czarniak, Elisa Lau, Candice McLeod, Stuart W. Churchill
Dept.of Chemical and Biomolecular Engineering
University of Pennsylvania
William B. Retallick, Retallick Associates
West Chester, PA

mvczarni@seas.upenn.edu
elisalau@wharton.upenn.edu
mcleodc@seas.upenn.edu
churchil@seas.upenn.edu
wbr@catacel.com

Introduction

CO_2 is a major "greenhouse" gas and combustion of carbonaceous fuels is one of the primary sources of its addition to the atmosphere. The sequestration of CO_2 from combustion or other sources is not currently prescribed by law and has no direct economic advantages but a recent study by the National Academy of Engineering has listed it among the principal technological challenges of the 21st century. Since the time that this work was completed, hardly a day has gone by without the appearance of a technical report or a news item concerning the sequestration of CO_2. However, none of this new information and opinion negates or renders our study obsolete, because it had the narrow objective of examining the technical feasibility and relative economic costs of several processes for the recovery, separation, and sequestration of CO_2 from the combustion of a particular coal in a particular location to generate electricity at a representative rate. The optimality of the choices made in our work can be questioned but the results are nevertheless presumed to remain valid as a point of reference.

Location

Our analysis is based on the combustion of coal, which, despite widespread opposition, is generally expected to be a major source of energy for the generation of electricity in the next several decades. The Illinois Basin, which underlies Illinois, western Indiana, and western Kentucky was arbitrarily chosen for this analysis because (1) it has both large and accessible holdings of bituminous coal, (2) includes a number of geological formations suitable for sequestration, and (3) produces 255 million metric tons of CO_2 per year (11.7% of total US emissions), 92% of which are from coal-fired power plants. (The other major sources of emission in this region include petroleum refineries, cement plants, and miscellaneous heavy industries.)

Venues for Capture and Separation

The possible venues for sequestration include algae, the ocean depths, depleted oil and gas reservoirs, coal seams that are not suitable for mining, and deep saline formations. Algae are one of the most productive users of CO_2 in the plant world, and after saturation the can be burned, which produces thermal energy with no net production of CO_2. However, before it can be burned the mass of algae must be dried, which is very energy-consumptive. The rate, potential capacity, and economic feasibility of this form of sequestration are difficult to estimate at the present time because of a lack of the necessary technical information or industrial experience. Accordingly, this alternative was not examined in detail.

Venues for Sequestration

The choice of the Illinois Basin as a site eliminated the possibility of sequestration in the ocean. The oil and gas reservoirs, the coal seams, and the saline formations of the region have an estimated capacity of 282 million metric tons, 282 million metric tons, and 4.14 billion metric tons of CO_2, respectively. Insertion of CO_2 in the oil and gas reservoirs may enhance their recovery and in coal seams may replace and release methane that can be recovered, but the fractional offsets of the cost of sequestration appear to be negligibly small. In saline formations, the CO_2 dissolves in part and may react in part to form carbonates, but the latter do not appear to have any economic potential. If all the electricity produced in the Illinois Basin were produced by the combustion of coal, and if 90% of the corresponding production of CO_2 were sequestered in the oil and gas reserves there they would become saturated in only 1.12 years. The coal seams have an almost equal capacity, while the saline formations have a capacity of 16.2 years. Because of its greater capacity, attention was focused on sequestration in the saline formations.

This total of only 18.4 years is discouraging at first sight, but it is an unrealistic extreme case. Over this period of time sequestration, as a new technology, would be expected to be applied to only a very small fraction of the coal burned to produce electricity in the basin. This time period provides a window of opportunity in time to seek solutions other than sequestration and other than the particular processes for sequestration examined herein.

Representative Conditions

For the quantitative calculations of the combustion of coal and of the separation and sequestration of CO_2, "Old Ben 26" coal and a conventional furnace and turbine capable of producing 500MW of electricity at an overall efficiency of 36% and operated at 65% capacity on an annual basis, were chosen as representative. The corresponding annual requirement of coal and the annual production of CO_2 were calculated to be 1.0 and 2.47 million metric tons, respectively. These quantities amount to approximately 1% of those for the whole basin. The flue gas produced by the combustion of this low-sulfur bituminous coal contains a significant amount of ash in the form of particulate matter, CO, and N_2, and traces of NO_x and SO_2. The ash is postulated to be removed by electrostatic precipitation, and the NO_x and SO_2 catalytically. However, these costs were not assigned to that of sequestration because they are required in its absence. The concentration of CO was based on the use of 25% excess air, which results in a mass concentration of 58% N_2 in the flue gas. Gaseous N_2 is not ordinarily considered a pollutant but it is in terms of the sequestration of CO_2. It can be almost totally avoided in the stack gas by burning the coal with oxygen rather than air, but at the price of a whole set of new problems, including the high cost of oxygen and the necessity of a massive recirculation of flue gas, or of some other expediency, to reduce the flame temperature. This alternative was not examined. The separation of CO_2 from the flue gas was thereby considered to be a necessary part of the process of sequestration.

Processes for Separation

Three processes for separation of the CO_2 from the resulting "clean" flue gas were examined, namely absorption in an aqueous solution of monoethylamine (*MEA*), adsorption by and reaction with calcium oxide (CaO) , and adsorption by and reaction with magnesium oxide (MgO). Because the use of aqueous *MEA* as an absorbent for various chemicals is a well-developed technology, it was adopted as the base case for fiscal comparisons. The use of CaO as an adsorbent and reactant is also well developed but it has a serious shortcoming when used for CO_2, namely the temperature of 1175K required for the decomposition of the calcium carbonate (Ca_2CO_3) that is formed. Magnesium oxide (MgO) was considered as a possible alternative to CaO because the decomposition of magnesium carbonate (Mg_2CO_3) occurs at the much lower temperature of 700K, but the physical handling of powdered MgO was immediately recognized to be a possible serious impediment. The discovery of a Department of Energy report by Breault and Reasbeck that described the impregnation of MgO on solid granules of alumina (Al_2O_3) resulted in our reconsideration of MgO, and the subsequent detailed calculations revealed the resulting process to be more economical than those utilizing *MEA* or CaO. These calculations were based on the recovery of 90% of the CO_2 in 245 kg/s of cleansed stack gas at 311K and slightly above 1bar.

***MEA*.** The *MEA* system was conventional except for its size, consisting of 12 identical 18-tray columns, 4.27m in diameter and 15.85m high. The estimated utility costs, which included 1.7kg of steam per kg of CO_2 totalled \$21 per metric ton of CO_2, which corresponds to the value of 21% of the output of the generator. Liquid absorption might be more favorable if a solvent could be found that degrades more slowly, is less corrosive, has a higher capacity for CO_2, and requires less energy for regeneration than *MEA*.

***MgO*.** As contrasted with the conventionality of the *MEA* system, the MgO system has three unique characteristics: 1) the preparation of the impregnated tablets; 2) the utilization of six identical packed-bed adsorber/reactor/desorbers 5.0m in diameter, and 6.3m high; and 3) their operatin in parallel in batch cycles at different stages. To produce the combined MgO $/Al_2O_3$ granules, alumina tablets 5mm in diameter are soaked for up to 3 days in an aqueous solution of 20-80 gm/l of magnesium hydroxide. The water of hydration is then driven off by heating, resulting in a mass ratio of MgO to Al_2O_3 of 1.2. The total requirement for MgO supported on alumina in the six adsorbers was estimated to be 990Mg with an annual replacement of 10% due to attrition. Clean stack gas at 384K and 1.5bar is compressed to 14.6bar before entering a packed bed in the first step of the cycle. The adsorption and reaction are very exothermic and recirculation and cooling of a portion of the exiting gas by heat exchange with water is necessary to avoid an excessive temperature. The recirculation to feed ratio is 12/5. The stream exits the adsorber/reactor at 475K with 10% of the CO_2 , 90% having been converted to Mg_2CO_3. The desorption is very endothermic and the same of rate of recycling, in this case with heating by exchange with flue gas, is necessary to maintain a sufficient temperature.

Storage and Compression

The captured CO_2 at 600K, after being cooled to 311K, passes at a flow rate of 254 Mg/hr into two 11,300m^3 holding tanks, which are maintained at slightly above atmospheric pressure. Before its sequestration the CO_2 must be converted to a condition within the supercritical phase, which requires a pressure of at least 73.3bars at 311K. A still higher pressure, namely 95.2bar is required to compensate for the pressure drop over a representative distance of 167 km in a 10-in (0.254-m) pipeline leading to the site of sequestration.

Because the this compression accounts for over half of the total cost, several alternative schemes were considered. The final choice was a centrifugal blower and three-stage centrifugal compression with intercooling. Two-stage compression would require a slightly lower fixed capital investment (by 0.7%) but a slightly higher operating cost (by 4%) as well as more electricity and cooling water. The use of holding tanks for the compressed gas in the event of a temporary shutdown of the compressors was examined but concluded to be impractical.

Estimated Overall Costs

The estimated operating and fixed costs of separation and sequestration of CO_2 using the *MEA* and *MgO/Al*$_2$*O*$_3$ processes for the indicated conditions (a 500MW combustor/generator in the Illinois Basin) are summarized in Table 1.

Table 1. Comparative Costs

Process	Operating cost			Capital investment
	10^6\$/yr	\$ per metric ton of CO_2	¢ per KWH	10^6\$
Capture				
MEA	139	56.70	8.9	463
MgO	44	17.80	1.5	178
Sequestration	34	14.00	1.2	108

Conclusions

Based on the preceding analyses, it is concluded that separation using *MgO* is superior to that using MEA in every respect except operational experience. The fixed and variable costs are both much less and the plant equipment is more compact. Even for this new process, the overall costs for capture, separation and sequestration are huge and no easy means of accomplishing their reduction is apparent. Because it produces no income, this process of sequestration can be justified only for environmental reasons or legal requirements. Our estimate of a total cost of 3.7 ¢ per KWH provides some guidance in those respects.

631b Modeling CO2 Capture In Amine Solvents with An Advanced Association Model: Process Optimisation and a Platform for Solvent Design

Niall Mac Dowell[1], Claire S. Adjiman[2], Amparo Galindo[2], and George Jackson[2]. (1) Centre for Process Systems Engineering, Chemical Engineering Department, Imperial College London, Roderic Hill Building, South Kensington Campus, London, SW7 2AZ, United Kingdom, (2) Department of Chemical Engineering, Imperial College London, Centre for Process Systems Engineering, South Kensington Campus, London, SW7 2AZ, United Kingdom

The reduction in CO_2 emissions from anthropogenic sources has become a topic of widespread interest over the past number of years. As the power generation sector is by far the largest stationary-point-source of CO_2, being responsible for approximately 35% of total global CO_2 emissions[1] this question has special relevance for this industry. As the inclusion of carbon capture facilities incurs a significant energy penalty on the efficiency coal-fired power-stations, there is a strong requirement for the improvement of these systems in terms of the minimisation of operation and maintenance costs, capital costs and the maximisation of efficiency and flexibility. This last issue has relevance for start-up times and ramp-rates. Post-combustion capture methods based on the chemisorption of CO_2 in aqueous amine solutions are among the most mature and accepted technologies for CO_2 capture from power plants[2]. The objective of this work is to provide a model-based platform for assessing alternative designs and solvents for post-combustion CO_2 capture. In this talk, we report on the modelling of an absorption/desorption process using an aqueous solution of ammonia. The mechanism for the absorption of CO_2 is complex and special emphasis is placed on capturing the thermodynamic and kinetic behaviour of the solvent/CO_2/N_2 system. The models thus derived are incorporated within a process model, which is used to investigate the performance of the process. To account for the hydrogen-bonding interactions that are typical of ammonia and water, the statistical associating fluid theory (SAFT)[3] is used. This is a molecular approach, specifically suited to associating fluids. In this contribution we use the SAFT approach for potentials of variable range (SAFT-VR[4]). The SAFT formalism is used to represent some of the equilibrium reactions present in the system, thereby simplifying the kinetic model. The molecules are represented by homonuclear chains of tangentially bonded square-well segments of variable range, and a number of short-ranged off-centre attractive square-well sites are used to mediate the anisotropic effects due to association in the fluids. We determine values of the pure component and parameters and of the binary interaction parameters for all the pairs of the compounds in the system. These are obtained by fitting to experimental data and are used in a predictive manner to obtain ternary and quarternary phase behaviour. We also estimate reaction rate constants for the kinetic model. We also present a model of an absorption process, implemented in the gPROMS[5] software package. The overall process model is used to assess the performance of the system under different operating scenarios. The sensitivity of the process to several design variables is studied and provides some insight into the key decisions in developing such processes.

1 Steeneveldt, R., Berger, B. & Torp, T.A., ChERD, 84(A9): 739-763, 2006

2 Rao, A.B.; Rubin, E.S., 2002. A Technical, Economic, and Environmental Assessment of Amine-Based CO2 Capture Technology for Power Plant Greenhouse Gas Control. Environ. Sci. Technol. 36, 4467-4475

3 Chapman, W.G., Gubbins, K.E., Jackson, G. & Radosz, M., Ind. Eng. Chem. Res., 1990. 29, 1709-1721

4 Gil-Villegas, A., Galindo, A., Whitehead, P. J., Mills, S. J. & Jackson, G., J. Chem. Phys. 106 (10), 8 March 1997

5 Process Systems Enterprise (PSE) Ltd.

631c Modeling of Carbon Dioxide Absorption/desorption In Amine Solutions

Jerry H. Meldon, Tufts University, Chemical and Biological Engineering Department, Medford, MA 02155

Design of packed scrubbers for CO2 capture by amine solutions requires accurate solution to local differential mass balances. Because the analysis is complicated by complex reaction kinetics, solutions are typically obtained via computationally intensive numerical methods. However, the kinetic terms in the ordinary differential equations associated with Film Theory may be linearized by either of two different techniques with negligible loss of accuracy. One approach fixes the concentration of nonvolatile reactants at their (unknown) values at the gas-liquid interface. The other approach assumes that departures from local reaction equilibrium are small. We present results here that demonstrate high accuracy when either linearization technique is applied to the analysis of CO2 absorption in and desorption from amine solutions.

631d Modeling Gas Hydrates Containing Carbon Dioxide, Hydrogen and Methane Using Simplified Pc-Saft

Saravanan Swaminathan[1], Michael L. Michelsen[2], Georgios M. Kontogeorgis[2], and Nicolas von Solms[2]. (1) IVC-SEP, Department of Chemical and Biochemical Engineering, Technical University of Denmark, Søltofts Plads, Building 229, room 254, Kgs. Lyngby, 2800, Denmark, (2) Department of Chemical and Biochemical Engineering, Technical University of Denmark, Kgs. Lyngby, 2800, Denmark

Gas Hydrates are ice-like crystalline solids formed with a water host-lattice that encages guest-gas atoms or molecules. Under the right conditions of temperature (low) and pressure (high), water molecules form lattice structures on hydrogen bonding, with cavities large enough to incorporate small gas molecules, thereby, providing a potential method for CO_2 capture and sequestration. The other interest to studying these gas hydrates is to provide more insights in using the vast source of methane (natural gas) hydrates available in the subsurface. The added motivation to understand hydrates is the oil and gas industry, as a lot of money is spent in order to prevent the formation of natural gas hydrates as they clog and block the flowlines leading to disastrous economic ramifications.

The basis of modeling hydrate phase behavior is to use the van der Waals Platteeuw (vdW-P) model[1] for the hydrate phase and to use an equation of state (EOS) (and/or activity coefficient model) to model the fluid phases (vapor and liquid). The vdW-P model was first published almost five decades ago, and generalized by Parrish and Prausnitz[2] in the early 1970s. Various thermodynamic models have been used to describe the liquid and vapor phases depending on the authors. One of the important parts of modeling the hydrate phase behavior is describing the fluid phases accurately; the predictive ability of the over-all model is constrained by EOS used. Due to the nature of compounds present in these systems such as water, CO_2, alcohols etc., it is essential to choose thermodynamic models that can describe the mixture very well and have good predictive nature. The statistical associating fluid theory (SAFT) developed based on the Wertheim's perturbation theory has been successful in modeling various associating systems with water, alcohols etc. Hence, it should be able to provide excellent representation of the thermodynamics of the guest gases. However, there have been very few attempts to use molecular-based equations of state such as the Statistical Associating Fluid theory (and its variants)[3-5]. One of the variants of SAFT, the Perturbed-Chain SAFT has been very successful in modeling various highly asymmetric and associating mixtures. A modification to PC-SAFT was proposed by von Solms et.al6 that reduced the computational times markedly for associating systems (which would be the case in hydrate systems). Hence, in this work, we use the Simplified PC-SAFT[6] with the vdW-P to model the hydrate phase behavior with a better degree of predictability.

1. J.H. van der Waals and J.C. Platteeuw, Adv. Chem. Phys. 22-57 (1959)

2. W.R. Parrish and J.M. Prausnitz, Ind. Eng. Chem. Process. Des. Develop. 11 26-35 (1972)

3. X. Li, H. Wu, Y. Li, Z. Feng, L. Tang, and S. Fan, J. Chem. Therm., 39, 417- 425 (2007)

4. X. Li, H. Wu, and P. Englezos, Ind. Eng. Chem. Res., 45, 2131-2137 (2006).

5. A. Haslam, A. Galindo and G. Jackson, 'Modelling gas clathrate hydrates: incorporating an advanced thermodynamic description of the fluid' presented at the Eleventh PPEPPD Conference in Crete, Greece (2007)

6. N. von Solms, M.L. Michelsen and G.M. Kontogeorgis, Ind. Eng. Chem. Res. 42, 1098-1105 (2003)

Demonstration of Carbon Dioxide Capture from Coal-fired Flue Gas Using Calcium-based Sorbents

Songgeng Li[a], William Wang[a], Shwetha Ramkumar[a], Siddharth Rao Gumuluru[a], Danny Wong,[a] Mahesh Iyer[a], Robert M. Statnick[b], Liang-shih Fan[a]
[a]Department of Bio-molecular and Chemical Engineering,
The Ohio State University, Columbus, OH 43210
[b]Clear Skies Consulting, Cornelius, NC 28031

Abstract

Carbon dioxide (CO_2) is one of the most important greenhouse gases causing global climate change. To minimize the undesirable impact of carbon dioxide emissions on the environment, many new technologies are being developed to reduce its emissions from the combustion of fossil fuels. A promising technology is the multi-cyclic Carbonation and Calcination Reaction (CCR) process. The CCR process utilizes a calcium-based sorbent in a high temperature reaction (carbonation) to capture the CO_2 from the flue gas stream. The subsequent calcination reaction releases a pure, dry stream of CO_2, which can be sequestered. A 20 pph-scale facility firing coal has been constructed at The Ohio State University to demonstrate the CCR process. A number of calcium-based sorbents were tested at this facility. The effect of residence time and Ca/C ratio were evaluated and over 90% CO_2 capture was achieved on a once-through basis. One of the calcium based sorbents has exhibited stable reactivity for CO_2 capture over multiple cycles.

Introduction

CO_2 is one of the major greenhouse gases contributing to global climate change. In the United States, coal-combustion for electricity generation is responsible for approximately one-third of anthropogenic carbon dioxide emissions.

There are three main approaches to capturing CO_2 generated from fossil fuels: pre-combustion capture system, post-combustion capture system and oxy-combustion [1]. Pre-combustion capture systems process the primary fuel in a reactor to produce separate streams of CO_2 for storage and H_2, which is used as a fuel. Post-combustion system separates CO_2 from flue gas generated by combustion of fossil fuels. Oxy-combustion uses oxygen instead of air for combustion, producing a flue gas that is mainly H2O and CO_2, which is readily captured. Both pre-combustion and oxy-combustion produce a CO_2 stream by making significant design changes to a conventional boiler, which is a good strategy for new power plants. For existing electric power plants, post-combustion capture might be the best option.

A number of technologies are currently being developed to separate CO_2 from coal combustion flue gas. These include sorbents/solvents separation, membrane separation and cryogenic distillation. The calcium based solid sorbents technology has received increasing attention [2-5]. This technology utilizes a calcium-based sorbent in a carbonation reaction at high temperature to capture the CO_2 from the flue gas stream and releases a pure, dry stream of CO_2 in the subsequent calcination reaction for sequestration. One of the advantages of this technology over low temperature process (such as amine scrubbing) is that the heat released at the temperatures of 600~700 C during the carbonation process is utilized.

In this work, a 20 pph-pilot scale facility was constructed to demonstrate the calcium-based carbonation/calcination reactions (CCR) process. The objective of this work includes (1) testing the feasibility of this process; (2) determining CO2 capture capacity of different calcium-based sorbents over multi-cycles; and (3) determining the effect of residence time and calcium/carbon ratios on CO_2 removal.

Test facility

The experiments were conducted in the facility shown in Figure 1. The system consists of a 20lb/hr coal-fired stoker furnace, a rotary calciner, sorbent and ash injection systems, upstream and downstream gas sampling systems, fans, particulate control devices and data acquisition, control systems and associated instrumentation. The sorbent is injected at temperatures between 600-700 C, which is the optimal temperature window for the carbonation reaction. The carbonated sorbent is then separated from flue gas and sent to the rotary calciner for regeneration. The calciner regenerates the spent sorbent for a subsequent CO_2 capture cycle. The calciner, which is 20 ft. long and 2 ft in diameter, is electrically-heated. The maximum calciner operating temperature is 980 C.

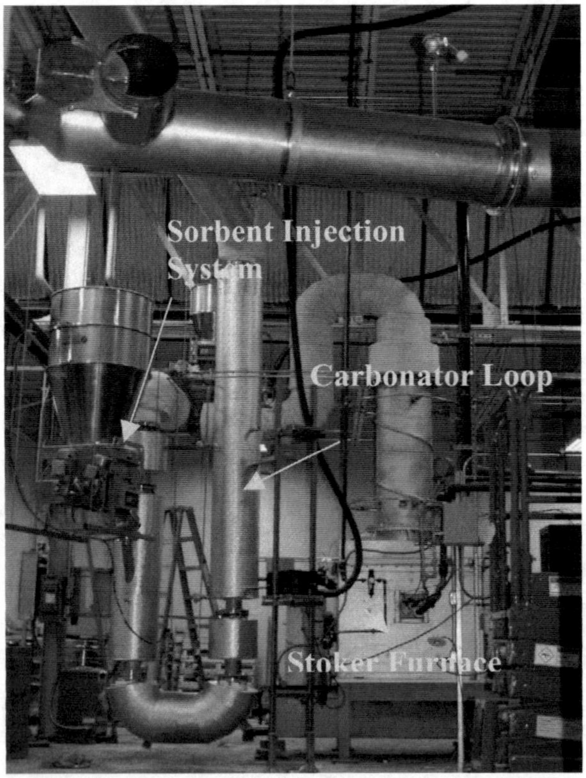

(a)

(b)

Figure 1 Demonstration facility for CO_2 capture

Test results

Figure 2 shows the effect of increasing Ca/C ratio on CO_2 removal on a once-through basis. It can be seen that CO_2 removal increases with an increase in Ca/C ratio. Among the three tested calcium based sorbents, sorbent 1 gave the highest CO_2 capture. CO_2 removal efficiency significantly increases from 46% to nearly 100% as Ca/C ratio increases from 0.6 to 1.6. However, CO_2 removals by sorbents 2 and 3 seem less sensitive to the variation of Ca/C ratio in comparison to sorbent 1.

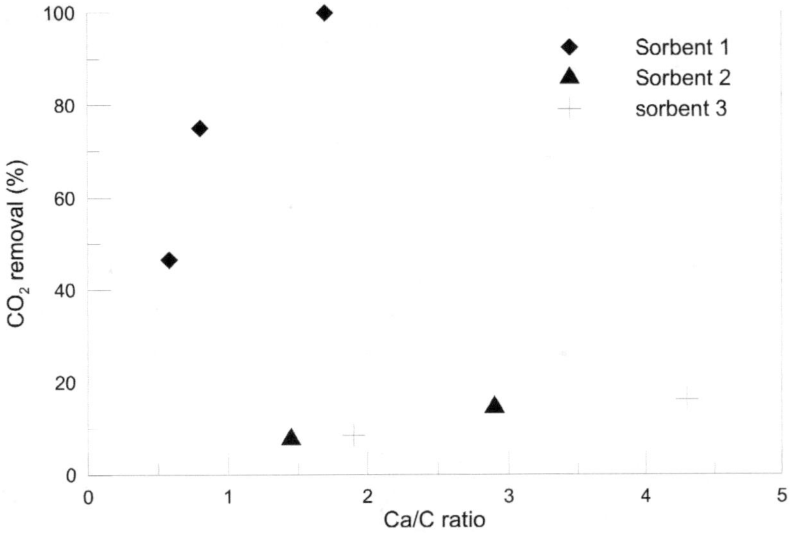

Figure 2. Effect of Ca/C ratio on CO_2 removal

Figure 3 presents the effect of the residence time on CO_2 removal on a once-through basis. It is obvious that the residence time has a significant impact on the CO_2 removal. CO_2 removal increases from 38% to 55% as the residence time is doubled.

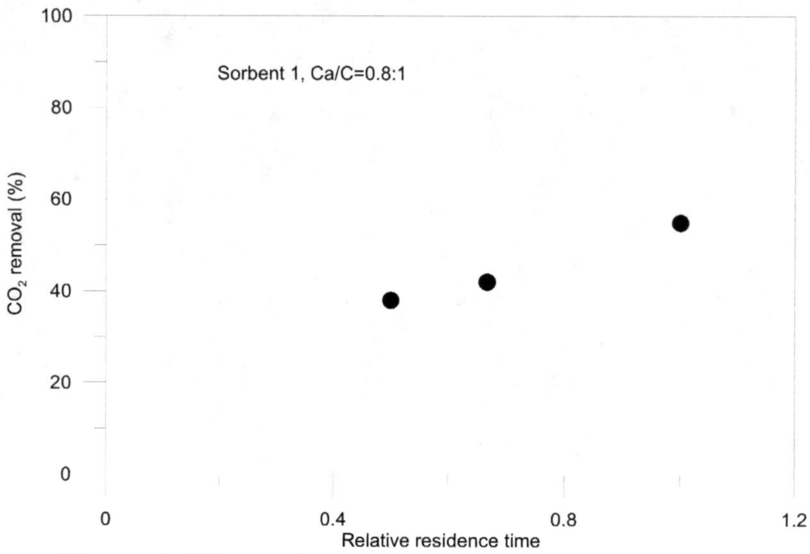

Figure 3 Effect of residence time on CO_2 removal

A critical issue for this process is the stability of sorbent reactivity after multiple cycles. No obvious decay was observed for sorbent 1 after several cycles. We are continuing to conduct recycle tests to confirm sorbent 1 recycle performance.

Summary

A 20lb/hr combustion facility was constructed at The Ohio State University for the purpose of demonstrating the CCR process. Three types of calcium based sorbents were tested on a once-through basis. Sorbent 1 achieved greater than 90% CO_2 capture at Ca/C ratio of 1.6. Multi-cyclic CO_2 capture test show that there is no obvious decay of sorbent 1 after three recycles.

Acknowledgements

The authors are grateful to the Ohio Coal Development Office of the Ohio Air Quality Development Authority, Babcock & Wilcox, American electric Power (AEP), CONSOL Energy Inc., Duke Energy Corp. and Specialty Minerals Inc., and Carmeuse North America for providing financial and technical support for this project.

References

[1] Working group III of the Intergovernmental panel on climate change (2005), "IPCC special report on carbon dioxide capture and storage".

[2] Iyer, Mahesh V.; Gupta, Himanshu; Sakadjian, Bartev B.; Fan, Liang-Shih. (2004), "Multicyclic Study on the Simultaneous Carbonation and Sulfationof High-Reactivity CaO"*Ind. Eng. Chem. Res.*, pp.: 3939-3947.
[3] Sun, Ping; Grace, John R.; Lim, C. Jim; and Anthony, Edward J. (2007), "Removal of CO_2 by Calcium-Based Sorbents in the Presence of SO_2", *Energy & Fuels*. pp:
[4] Alvarez, Diego and Abanades, J. Carlos. (2005), "Pore-Size and Shape Efects on the Recarbonation Performance of Calcium Oxide Submitted to Repeated Calcination/Recarbonation Cycles", *Energy & Fuels.* pp:.
[5] Gupta, Himanshu and Fan, Liang-Shih. (2002), "Carbonation-Calcination Cycle Using High Reactivity Calcium Oxide for Carbon Dioxide Separation from Flue Gas" *Ind. Eng Chem Res*. pp:

631f Effect of Carbonic Anhydrases on Carbonation of Mg-Bearing Minerals for Environmentally-Benign Carbon Sequestration

Ah-Hyung Alissa Park, Earth and Environmental Engineering and Chemical Engineering, Columbia University, 918 S.W. Mudd Hall, MC 4711, 500 W. 120th Street, New York, NY 10027

Carbon dioxide is a greenhouse gas that causes global warming. Historically, the atmospheric concentration of carbon dioxide fluctuated naturally on the timescales of ice ages. Concerns, however, stem from the recent dramatic increase in the CO2 concentration, which coincides with global industrial development. In order to meet the ever-increasing global energy demands while stabilizing the CO2 level in the atmosphere, it is widely believed that current carbon emissions must be reduced by at least a factor of three.

Mineral carbonation is a new and consequently less studied method of sequestering CO2. This method mimics the natural mineral weathering process involving the exothermic reaction of CO2 with non-carbonate minerals to form geologically and thermodynamically stable mineral carbonates. Hence this sequestration process truly presents a safe and permanent method of CO2 containment that is based on chemical fixation of CO2 in a solid mineral matrix.

Earlier studies by the PI on mineral carbonation indicated promise with regard to enhanced dissolution rates of serpentine in a mineral slurry system using a mixture of weak acids and chelating agents. This previous research resulted in the development of the pH swing process that fixes CO2 into a solid matrix while producing value-added solid products such as high surface area silica and iron oxide. The next challenge for carbon mineral sequestration is how to recycle or conserve acid and base required for the pH swing. Thus, this study is performed to search for chemical additives that can promote the carbonation reaction near neutral pH.

The proposed research focuses on the effect of carbonic anhydrases on the carbonation of Mg-bearing minerals. Carbonic anhydrases are found in many organisms, including human, and can catalyze CO2 hydration. Compared to water, a zinc-bound hydroxide ion which is sufficiently nucleophilic is much more reactive to CO2, and thus, the concentration of HCO3- and CO32- in an aqueous solution would increase as carbonic anhydrase is introduced. In this study, the reaction mechanisms and the reaction kinetics are investigated while determining the optimum pH conditions.

631g Use of Ionic Liquids as Physical Solvents for Selective Capture of CO2 from Fuel Gas Streams

Yannick J. Heintz, Chemical and Petroleum Engineering, University of Pittsburgh, 1249 Benedum Hall, 3700 O'Hara Street, Pittsburgh, PA 15261

This study is to investigate the potential use of ionic liquids (ILs) as physical solvents for selective CO_2 capture from post water-gas-shift reactor streams at elevated pressures and temperatures. The equilibrium gas solubility (x*) and the volumetric mass transfer coefficients (k_La) for CO_2, H_2 and a gas mixture (CO_2, H_2, CH_4, CO, Ar) in three ILs (TEGO IL K5, TEGO IL P51P and TEGO IL P9) were determined. The data were obtained in an agitated reactor, equipped with sight-windows, in wide ranges of pressure, temperature, mixing speed, and liquid height.

Under the operating conditions investigated, the CO_2 solubilities in the three ionic liquids increased with pressure at constant temperature and decreased with temperature at constant pressure. Also, the volumetric liquid-side mass transfer coefficients of CO_2 increased with mixing speed, pressure, and temperature and decreased with liquid height. At 500 K, the CO_2 solubilities in the TEGO IL K5 were greater than those in the other two ILs. Under similar operating conditions, the CO_2 solubilities in the three ILs were greater than those of H_2, which reflects the selective nature of ILs to capture CO_2. In addition, the ILs appeared to have negligible vapor pressure up to 500 K, which presents an advantage over conventional physical solvents currently employed for CO_2 capture from post water-gas-shift reactor streams.

This study demonstrated the thermal stability of the ILs and highlighted their ability to selectively capture CO_2 at temperatures up to 500 K and pressures as high as 30 bars.

ACKNOWLEDGMENT

The technical effort was performed in support of ongoing Carbon Dioxide research at the National Energy Technology Laboratory of the U.S. Department of Energy under RDS contract DE-AC26-04NT41817.

631h Temperature Effects on the Fuel Reactor Behavior of a Chemical Looping Combustion System

Jonghwun Jung, Technical Research Laboratory, POSCO, 1, Goedong-dong, Nam-gu, Gyeongbuk 790-785, Pohang, South Korea and Isaac K. Gamwo, National Energy Technology Laboratory, U.S. Department of Energy, Pittsburgh, PA 15236-0940.

CO2 emissions from fossil fuels combustion in power plants have increased the concentration of atmospheric greenhouse gases, raising concerns on global climate change. The foresights show that fossil fuels will be the dominating energy source worldwide at short or medium term, which encourages the development of new technologies for CO2 capture and sequestration. Chemical looping combustion (CLC) is a good candidate for fossil fuel combustion with negligible atmospheric CO2 emissions. CLC produces a relatively pure stream of CO2 ready for compression and storage in geological settings. Furthermore, CLC also minimizes NOx emissions since the fuel combustion is flameless in the absence of nitrogen.

We previously simulated the fuel reactor behavior of a chemical looping combustion process. Here, we report on the effects of temperature on the fuel reactor behavior. We found that unburned fuel weight fraction was reduced by nearly half when the temperature increased from 690 0C to 950 0C indicating a substantial increase in the fuel conversion rate. The simulations also showed that the bubble size in the fuel reactor increases with temperature. This study clearly indicates the need to optimize the temperature in the fuel reactor with respect to the fuel conversion rate.

643a Activated Carbon Derived from Fast Pyrolysis Liquids Production of Agricultural Residues and Energy Crops

Isabel M. Lima[1], Akwasi Boateng[1], Thomas Klasson[2], and Kevin B. Hicks[3]. (1) U.S. Department of Agriculture, ARS, Agricultural Research Service, 1100 Robert E. Lee Blvd, New Orleans, LA 70124, (2) USDA-ARS, 1100 Robert E Lee Blvd, New Orleans, LA 70124, (3) Eastern Regional Research Center, Agricultural Research Service, 600 Mermaid Lane, Wyndmoor, PA 19038

Fast pyrolysis is a thermochemical method that can be used for processing energy crops such as switchgrass, alfalfa, soybean straw, corn stover as well as agricultural residuals (broiler litter) for bio-oil production. Researchers with the Agriculture Research Service (ARS) of the USDA developed a 2.5 kg/h biomass fast pyrolyzer that generates bio-oil and non-condensable gases as the main products as well as a char fraction as a by-product. This study characterizes these chars for their potential for further conversion into value-added activated carbons to be used as adsorbents for environmental remediation. The final physical, chemical and adsorptive properties of the resulting activated carbons depend on the the biomass precursor. Char and activated carbon yields, surface areas, elemental analysis, electron scanning microscopy, microporosity and adsorption of both metal ions and trihalomethanes for the chars and after further activation (ex-situ), will be compared and contrasted with available commercial samples.

643b Sustainable Arsenic Removal Using Hybrid Anion Exchange Resins (ArsenXnp) In Remote Villages of West Bengal, India

Sudipta Sarkar[1], Lee M. Blaney[1], Debabrata Ghosh[2], Anirban Gupta[2], and Arup K. SenGupta[1]. (1) Department of Civil and Environmental Engineering, Lehigh University, 13 E. Packer Avenue, Bethlehem, PA 18015, (2) Department of Civil Engineering, Bengal Engineering and Science University, Botanical Garden, Shibpur, Howrah, India

Community level arsenic removal devices in remote villages of West Bengal, India, now use hybrid anion exchange resins (HAIX) which are essentially spherical anion exchange resin beads containing dispersed nanoparticles of hydrated ferric oxide (HFO). HAIX, now commercially available in the US as ArsenX^{np}, is a highly selective media for removal of oxyanions of arsenic. The high selectivity of HAIX towards arsenic oxyanions is attributed to the Donnan membrane effect. The sorption columns used in the field for removal of arsenic are either single column or split column design. The sorption columns are designed to allow flow of atmospheric oxygen dissolution of which in arsenic-contaminated raw water promotes oxidation of dissolved Fe(II) species to insoluble Fe(III) oxides or HFO particulates. Apart from the usual role played by the sorbents like ArsenX^{np} or activated alumina towards arsenic removal, the HFO particulates formed inside the column also aid in the treatment process. Each unit is attached to a hand-pump driven well and capable of providing arsenic-safe water to three hundred (300) households or approximately one thousand villagers. No chemical addition, pH adjustment or electricity is required to run these units. On the average, every unit runs for more than 20,000 bed volumes before a breakthrough of 50 mg/L of arsenic, the maximum contaminant level in drinking water in India, is reached. In addition to arsenic removal, significant iron removal is also achieved throughout the run. Upon exhaustion, the media is withdrawn and taken to a central regeneration facility where 2% NaCl and 2% NaOH solution is used for regeneration. Subsequently, the regenerated resin is reloaded into the well-head sorption column. Laboratory investigations confirmed that the regenerated ArsenX^{np} is amenable to reuse for multiple cycles without any significant loss in capacity. Field studies are under way to supplement the laboratory observation. The process of regeneration and subsequent sludge formation reduce the volume of disposable arsenic-laden solids by nearly two orders of magnitude. Finally, the arsenic-laden solids are contained on well-aerated coarse-sand filters with minimum arsenic leaching. This disposal technique is scientifically more appropriate than dumping arsenic-loaded adsorbents in the reducing environment of landfills as currently practiced in developed countries.

SYNTHESIS AND CHARACTERIZATION OF NOVEL SORBENT FOR BORON REMOVAL

*Yu-Ting Wei; Yu-Ming Zheng; J. Paul Chen**
National University of Singapore, Singapore
**Corresponding author, email: esecjp@nus.edu.sg*

Abstract

Excess intake of boron containing water can pose serious problems on human beings. Traditional technologies are limited by insufficient efficiency especially for trace boron removal. In this study, a novel sorbent was developed. A series of batch sorption experiments was conducted to investigate the boron sorption behaviors of the synthesized chitosan based sorbent. Our results show, comparing to raw chitosan, the boron uptake capacity of modified chitosan is significantly enhanced and the sorption capacity is determined to be 11.82 mg B/g sorbent. The equilibrium time of boron sorption can be achieved in less than 10 h. At a dose of 1.3 g/L, boron in seawater can be lowered to 0.5 mg/L to meet the WHO standard for drinking water.

Key words: Atom transfer radical polymerization; Boron; Chitosan; Sorption

1. Introduction

Boron is widely distributed in our environment, existing naturally in water, soils, plants and animals, or coming from human activities. Generally, boron is found in the form of boric acid or borate salts. Long-term consumption of water with increased boron content can cause nausea, headache, diarrhea, kidney damage, reproductive and nervous system disease, and even death [1, 2]. Consequently, the World Health Organization (WHO) has declared that the water quality for drinking water must meet the 0.5 ppm boron guideline [3]. The strict regulations bring about a great challenge to traditional water treatment technologies such as chemical precipitation, electrocoagulation, membrane process, adsorption on activated carbon (AC) and solvent extraction. However, these technologies are insufficient efficiency under normal conditions, especially for trace boron removal [4-6]. Polymers with high content of OH functional groups are able to bind boron in aqueous solutions. The aim of the study was to design and synthesize an innovative sorbent with good affinity for boron. Moreover, boron sorption behavior on the synthesized sorbent was investigated by batch experiments including isotherm, kinetics etc. The results from this work will be used for the preliminary understanding of the process, which is helpful for future optimization and modeling simulation.

2. Experimental

An innovative sorbent was prepared. Sorption kinetic experiments were conducted for determination of equilibrium time. Solution pH was initially controlled at 7. 0.5 gram of sorbents was added to boron solution with initial concentration at 5 mg/L. Samples were taken at appropriate time intervals and boron concentrations were determined by ICP-ES.

In isotherm experiments, 40 ml of boron solution with various initial concentrations was prepared. A solution of pH 7 was adjusted initially. The airtight bottles were rotated on a shaker for 20 h to reach equilibrium. At the end of the experiments, 20 ml samples were extracted and filtered for analysis of boron concentrations.

3. Results and discussion

3.1. Sorption Kinetics

The rate of boron uptake on the synthesized sorbent is shown in Fig. 1. Most of boron adsorption rapidly takes place in the first 2 h, followed by a slightly slower process. The equilibrium can be attained within 10 h.

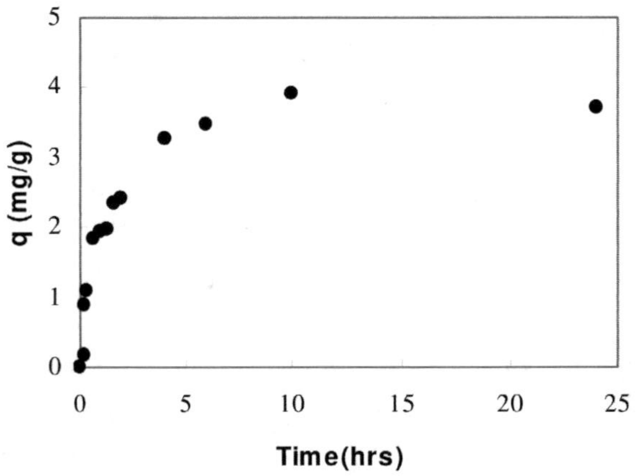

Fig. 1: Boron sorption kinetics by modified chitosan. m = 0.5 g/L.

3.2. Sorption Isotherm

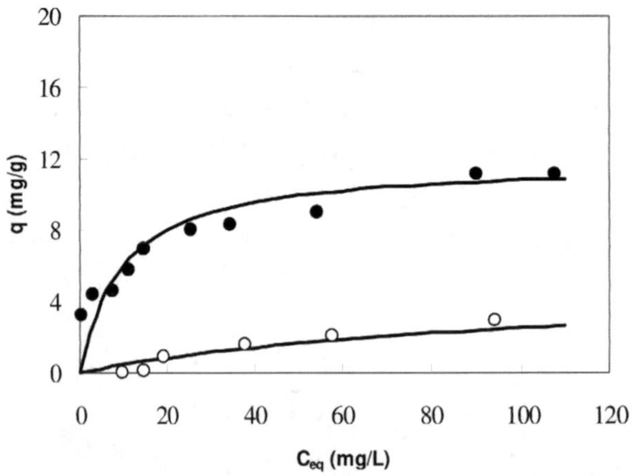

Fig. 2: Boron sorption isotherms by raw chitosan and modified chitosan. ○ Raw chitosan, •Modified chitosan, — Langmuir fitting, m = 0.5 g/L.

The sorption isotherm on boron was investigated over a range of 2 to 120 ppm using 0.5 g/L dose of the sorbent (see Fig. 2). Obviously, the boron sorption of raw chitosan is very low in the range of 0 to 3 mg/g; and boron sorption capacity is definitely enhanced after modification, which is around 11.82 mg/g.

3.3. Boron Sorption from Seawater

In order to understand the sorption efficiency of newly synthesized sorbent to the actual water samples, seawater locally collected was used in sorption experiments. As shown in Fig. 3, boron initial concentration in West Coast seawater is about 4 mg/L. At 1.3 g /L of sorbent dose, boron concentration is below 0.5 mg/L to meet WHO standard.

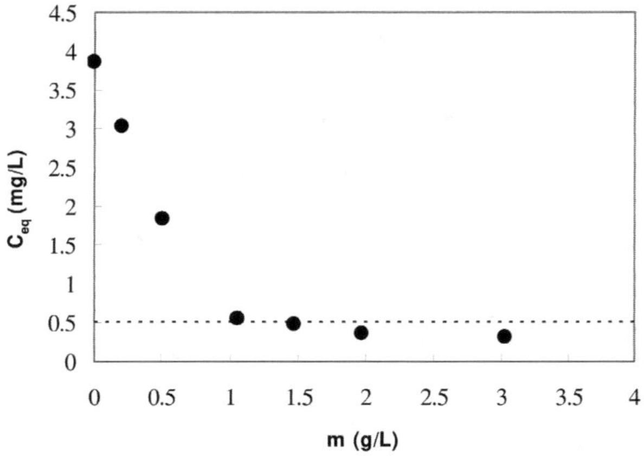

Fig. 3: Boron sorption from the coastal seawater

4. Conclusions

In this study, a new functional polymer sorbent is successfully prepared. Sorption of boron can reach equilibrium within 10 h. Isotherm study indicates the sorption can be well fitted by Langmuir isotherm model, as well as the sorption capacity of boron is around 11.82 mg/g, much higher than that of unmodified chitosan. The synthesized sorbent showed a potential applicability for boron removal from seawater.

References

1. Parks, J. L.; Edwards, M., Boron in the environment. Critical Reviews in Environmental Science and Technology 2005, 35, (2), 81-114.
2. Xu, Y.; Jiang, J. Q., Technologies for boron removal. Industrial & Engineering Chemistry Research 2008, 47, (1), 16-24.
3. Guidelines for Drinking-Water Quality, 3rd Ed; Vol. 1. Recommendations; World Health Organization: Geneva, 2004.
4. Kluczka, J.; Trojanowska, J.; Zolotajkin, M.; Ciba, J.; Turek, M.; Dydo, P., Boron removal from wastewater using adsorbents. Environmental Technology 2007, 28, (1), 105-113.

5. Bektas, N.; Oncel, S.; Akbulut, H. Y.; Dimoglo, A., Removal of boron by electrocoagulation. Environmental Chemistry Letters 2004, 2, (2), 51-54.
6. Prats, D.; Chillon-Arias, M. F.; Rodriguez-Pastor, M., Analysis of the influence of pH and pressure on the elimination of boron in reverse osmosis. Desalination 2000, 128, (3), 269-273.

643d Fluoride Removal from Water by Adsorption Onto Chitin-Based Biocomposites

Jose Luis Davila-Rodriguez[1], Jose Rene Rangel-Mendez[1], Vladimir A. Escobar-Barrios[1], and Keiko Shirai[2]. (1) Division of Environmental Sciences, IPICyT, Camino a la Presa San José 2055. Col. Lomas 4 sección, San Luis Potosí, 78216, Mexico, (2) Biotechnology Department, Universidad Autónoma Metropolitana, Av. San Rafael Atlixco No. 186, Mexico City, 09340, Mexico

Fluoride is present in high concentration in drinking water of many places in our planet, which causes negative effects on human health.

Adsorption has been the most used method to remove fluoride from water, and in recent years researches have studied the use of new adsorbents for this purpose. Some of these new materials are biosorbents.

Chitin and chitosan are biosorbents that have showed an acceptable adsorption capacity for cations and recently, their anions (like fluoride) adsorption capacity has been demonstrated. Nevertheless, characteristics of chitin and chitosan as low mechanical resistance and high solubility in acid medium limit their application in real conditions found in water treatment. However, such disadvantages can be overcome by obtaining chitin and chitosan based biocomposites (mixtures of the biopolymer and synthetic polymers).

In this study, chitin-based biocomposites were physicochemically characterized and their fluoride removal capacity from water was evaluated. The ionic exchange capacity, solubility (pH 2-12), surface charge distribution, specific surface area and pore volume of the studied materials were determined. The morphology and chemical composition of both biosorbent and biocomposites were analized. Finally, equilibrium and kinetic studies of fluoride adsorption onto biocompoistes were carried out between pH 4 and 9 at 25°C.

Results showed an improvement in both mechanical properties and particle morphology of biocomposites with regard to the original biopolymer. This contributes to obtain good hydraulic flow and low pressure drop in packed column with biocomposites. Furthermore, the biosorbents were chemically and mechanically stabilized when these were mixed with a polymeric matrix, which caused that the biocomposites were less soluble and brittle than the biopolymer.

On the other hand, the fluoride adsorption results showed a higher adsorption capacity of biocomposites as the pH decreased. In adition, at optimum conditions, biocomposites had competitive adsorption capacities compared to a commercial activated alumina (DeseCamen®). The adsorption kinetic of biocomposites showed a rapid adsorption process, which indicates the possibility of a practicable operation in packed columns with these materials.

These findings encourage the use of chitin to remove anions from water, and when these are mixed with synthetic polymers, the resultant biocomposites offer a viable option to remove fluoride in packed columns.

Keywords: chitin, fluoride, adsorption, polymer, biocomposite

Sorption and Desorption of Chromium (III) from Aqueous Solution by Agro-Waste Materials

Refugio Bernardo García-Reyes and José René Rangel-Méndez
Instituto Potosino de Investigación Científica y Tecnológica, A.C., San Luís Potosí, S.L.P, México.

Abstract

Agro-waste materials such as sorghum straw (SS), oats straw (OS) and agave baggasse (AB) contain several functional groups (carboxyl, carbonyl, acetamido, amino, hydroxyl, etc.) which are able to bind heavy metal ions present in aqueous solution. The major advantages of these materials are: low-cost, high efficiency and regeneration. The agro-waste materials studied in this research (water washed and acid washed biosorbents) were chemically characterized and their chromium sorption capacity was obtained. Also, the regeneration of partially saturated biosorbents was carried out by using HNO_3, NaOH and EDTA at different temperatures. Finally, the sorption mechanism of chromium (III) was discussed.

The results show that the main functional groups can be associated to carboxyl, nitrogen, and hydroxyl groups. In addition, acid treatment shifted the pH_{PZC} to acid pH values. The maximum chromium (III) sorption capacity, for acid washed OS, was 12.9 and 8.3 mg/g at pH 4 and 3, respectively: the sorption capacity decreased at pH 3 since H^+ ions competed with chromium (III) ions for the biosorbents sorption sites. On the other hand, the biosorbents regeneration increased with rising the temperature. Based on the ATR-FTIR spectra and previously results reported in literature, it was suggested that chromium (III) has special affinity for the biosorbents oxygen-containing sites.

In conclusion, the agro-waste materials studied herein could be used to remove chromium (III) and they can be efficiently regenerated by an EDTA solution at 55°C.

Keywords: Chromium (III), agro-waste materials, sorption, regeneration.

Introduction

Recently Suwalsky *et al.* (2008) demonstrated that trivalent chromium ions exert stronger structural effects than Cr (VI) on the human erythrocyte membrane and molecular models. This perturbation could affect the functions of ion channels, receptors and enzymes immersed in the erythrocyte membranes.

Biosorption process is a cost-effective alternative for the removal of chromium (III) from water. Some biosorbents that have been tested for the removal of Cr (III) from aqueous solution are marine seaweed (Yun *et al.*, 2001; Cossich *et al.*, 2004), agricultural by-products (Romero-Gonzalez *et al.*, 2006; Sawalha *et al.,* 2007), etc.

Agro-waste materials are mainly formed by natural polymers such as cellulose, hemi-cellulose, pectin and lignin. These components contain carboxyl and hydroxyl functional groups which can be involved in metal binding (Reddad *et al.*, 2002). However, desorption studies are not often reported when agro-waste materials are used as adsorbents.

The objectives of this research are to chemically characterize and to obtain the chromium (III) sorption/desorption capacity by agro-waste materials (sorghum *bicolor* straw (SS), oats *sativa* straw (OS) and agave *salmiana* baggasse (AB)) available in Mexico.

Methodology

Agricultural residues such as sorghum *bicolor* straw (SS), oats *sativa* straw (OS) and agave *salmiana* baggasse (AB) were used as adsorbents after being de-ionized water washed (W) and hydrochloric acid (0.01 N) washed (A). These capital letters were used throughout the document to identify each biosorbent; for example acid washed agave *salmiana* baggasse is represented as AAB.

To obtain the agro-waste materials surface charge distribution, potentiometric titrations were carried out by adding 125 mg of biosorbent in 25 mL of solution at different pH. The ionic strength solution was fixed by NaCl solution (0.01 N). To estimate the functional groups and equilibrium constants, the experimental data obtained from potentiometric titration were processed according to the method proposed by Yun *et al.* (2001). Attenuated Total Reflection Fourier Transform Infrared (ATR-FTIR) was used to identify the biosorbents functional groups.

Batch sorption experiments were carried out by adding 100 mg of biosorbent in 100 mL of chromium (III) solutions (ranging from 5 to 100 mg/L) at 25°C. These materials were continuously stirred. The solution pH was adjusted daily to pH 3 or 4 by adding 0.1 N NaOH and/or HNO_3 until the equilibrium was reached.

Partially saturated biosorbents were obtained by adding 100 mg of biosorbent in 100 mL of chromium (III) solutions (20 mg/L) at pH 4 and 25°C. Regeneration of partially saturated biosorbents was carried out by using 100 mL of HNO_3 (0.1 and 1.0 N), NaOH (0.1 and 1.0 N) or EDTA (0.05 and 0.1 M) at 25, 35 and, in some cases, 55°C. The initial and equilibrium chromium concentration were determined by an atomic absorption spectrophotometer.

Results and Discussion

In general, the biosorbents functional groups were similar, although these exist in different amounts. The first functional group can be associated to carboxyl (pk_a value 3.8-4.1), these were confirmed by ATR-FTIR analyses at 1745, 1650, 1260, 1160, and 1060 cm^{-1} (Wade, 1993; Socrates, 1998). The second functional group cannot be associated to a specific group by the great amount of functional groups (amine, phosphoryl, amide, imidazol, etc.) associated to the pk_a values between 6.3 and 7.1. On the other hand, the acid treatment applied to the agro-waste materials reduced the amount of functional groups that have pk_a values around 7.0. Finally, the third functional group can be associated to hydroxyl groups (pk_a 9.1-12.6), these were confirmed by ATR-FTIR analysis in a broad band at 3350 cm^{-1}.

From the surface charge distribution curves, it was observed that the acid treatment, applied to the agro-waste materials, shifted the pH_{PZC} to acidic values (from pH 5.5 to 3.5 for WAA and AAB). In addition, acid treatment removed soluble material attached to the biosorbents surface such as carbonates, silicates, calcium, magnesium, etc.

The maximum chromium sorption capacity, at pH 4 and 25°C, obtained by the Langmuir model was 6.96, 12.97 and 11.44 mg/g for ASS, AOS and AAB, respectively. No significantly differences were observed between the chromium sorption capacity of water and acid treated adsorbents.

On the other hand, although the agro-waste materials presented less chromium sorption capacity than seaweed biomass, activated carbons, or ion exchange resins, the agro-waste adsorbents are an available and cost-effective option for the removal of chromium (III) from aqueous solution. However it is necessary to find a way to regenerate them and reuse them to make their application more attractive in real systems. In that sense, batch desorption experiments were carried out by using biosorbents partially saturated with chromium. These chromium-loaded biosorbents were obtained at pH 4, 25°C and 20 mg/L of initial chromium concentration. Desorption results from acid washed oats straw (AOS) by HNO_3, NaOH and EDTA are given in Figure 1. The eluents were selected according to the possible chromium-biosorbents interaction.

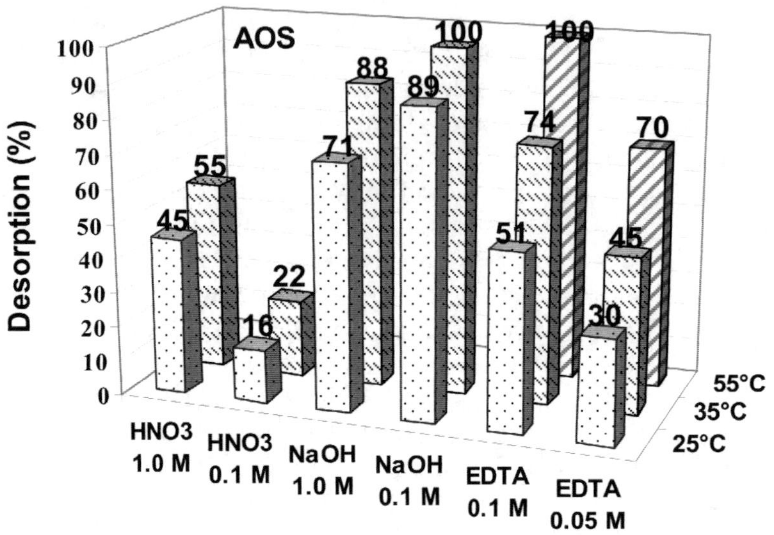

Figure 1. Chromium (III) desorption from oats straw acid washed at several eluents and temperatures. The chromium sorption capacity was 2.5±1.0 mg/g.

From Figure 1, it is clearly observed that the agro-waste regeneration was affected by temperature, for instance the regeneration by EDTA 1.0 M, at 25 and 35°C the chromium (III) recovered was 51 and 74%, respectively, but at 55°C the biosorbent was completely regenerated. In addition, regeneration by EDTA did not cause initial weight lost as HNO_3 or NaOH 1.0 N. The regeneration by HNO_3 (1.0 N) at 35°C was 55% which implied that the chromium ions were adsorbed on different functional groups including carboxyl groups. NaOH regeneration was better than HNO_3 (1.0 N) but produced more changes in the agro-waste materials.

Chromium (III) adsorption mechanism is not well understood due to the complex nature of the biosorbents. Nevertheless, some insights have been established. For example, Yun *et al.* (2001) demonstrated by potentiometric titration and FTIR analyses of seaweed biomass that trivalent chromium is adsorbed on carboxyl groups. In the same way, Sawalha *et al.* (2007), showed more evidences that the trivalent chromium adsorption was mainly due to carboxyl groups of chemically modified saltbush biomass (*Atriplex canescens*) by means of FTIR analyses and chemical modification. Finally, Ibánez and Umetsu proposed the following chromium sorption mechanism based on H^+ release when calcium-alginate beads were used as adsorbent:

$$3\{R\text{-}COOH\} + Cr^{3+} \leftrightarrow \{R\text{-}COO\}_3\text{–}Cr + 3H^+ \quad (1)$$

$$2\{R\text{-}COOH\} + CrOH^{2+} \leftrightarrow \{R\text{-}COO\}_2\text{–}CrOH + 2H^+ \quad (2)$$

Moreover, poly-nuclear chromium species are presented at pH 4, and then reaction (3) has to be considered as follows:

$$n\{R\text{-}COOH\} + Cr_2(OH)_2^{4+} \leftrightarrow \{R\text{-}COO\}_n\text{–}[Cr_2(OH)_2]^{(4-n)} + nH^+ \quad (3)$$

However, the low percentage of chromium desorbed by HNO_3 (1.0 N) at 35°C (see Figure 1) indicated that chromium (III) ions first bind to carboxyl groups but form covalent bonds with their surrounding which contains oxygen-binding sites. For that reason, regeneration by EDTA (a chelating agent) at 55°C allowed to release the chromium from agro-waste materials.

Conclusion

Agro-waste materials contain carboxyl and hydroxyl functional groups which are able to bind chromium (III) from aqueous solution. In addition, agro-waste materials can be efficiently regenerated by an EDTA solution at 55°C without apparent changes in the structure. The possible chromium (III) sorption mechanism involved complexation on oxygen-containing groups.

Acknowledgements

R.B. García-Reyes thanks CONACyT through the scholarship (204219) received. Also, the authors thank the Environmental Science Division, IPICyT, for the infrastructure and Dulce Partida for technical support to carry out the project.

References

Cossich ES, Da Silva EA, Tavares CRG, Cardozo L, Ravagnani TMK (2004), Biosorption of chromium(III) by biomass of seaweed S*argassum sp.* in a fixed-bed column, Adsorption, **10(2)**, 129–138.

Ibanez JP, Umetsu Y (2004), Uptake of trivalent chromium from aqueous solutions using protonated dry alginate beads, Hydrometallurgy, **72(3-4)**, 327–334.

Reddad Z, Gerente C, Andres Y, Le Cloirec P (2002), Adsorption of several metal ions onto a low-cost biosorbent: kinetic and equilibrium studies, Environmental Science Technology, **36(9)**, 2067-2073.

Romero-Gonzalez J., Peralta-Videa J.R., Rodrıguez E., Delgado M., Gardea-Torresdey J.L. (2006), Potential of agave *lechuguilla* biomass for Cr(III) removal from aqueous solutions: thermodynamic studies, Bioresource Technology, **97(1)**, 178–182.

Sawalha MF, Peralta-Videa JR, Saupe GB, Dokken KM, Gardea-Torresdey JL (2007), Using FTIR to corroborate the identity of functional groups involved in the binding of Cd and Cr to saltbush (*Atriplex canescens*) biomass, Chemosphere, **66(8)**, 1424–1430.

Socrates G. (1998) Infrared Characteristic Group Frequencies, 2nd ed., John Wiley & Sons, New York, 249 pp.

Suwalsky M, Castro R, Villena F, Sotomayor CP (2008), Cr(III) exerts stronger structural effects than Cr(VI) on the human erythrocyte membrane and molecular models, Journal of Inorganic Biochemistry, **102(4)**, 842–849.

Wade, L.G. Jr. (1993), QUIMICA ORGANICA, 2nd. Ed., Prentice Hall, 1312 pp.

Yun YS, Park D, Park JM, Volesky B (2001), Biosorption of trivalent chromium on the brown seaweed biomass, Environmental Science Technology, **35(21)**, 4353-4358.

643f Recovery of Zinc (II) Ions from Aqueous and Industrial Effluent Using a Continuous Foam Fractionator

Muthukumarasamy Arulmozhi[1], Dr.K.M. Meera S.Begum[2], and **Dr.N. Anantharaman**[2]. (1) Department of Chemical Technology, Anna University,, Tiruchirappali-24, India, (2) Department of Chemical Engineering, National Institute of Technology,Trichy, Tiruchirappali-15, India

Many methods are available for the separation of liquid mixtures and metal ions. Among them, foam separation technique holds great promise especially when the concentrations involved is very low. It is a simple and low cost method with less maintenance as there are no moving parts. It comes under adsorptive bubble separation techniques. In foam separation, the more surface active compounds are attracted by adsorption to gas bubbles, which then rise to the top of the liquid at the surface, the less surface active molecules remained entrapped in the interstitial liquid. When these bubbles rise out of the solution, foam is formed. Only a small fraction of the liquid is carried with the bubbles into the foam phase due to gravity drainage. The foam phase can be collapsed into a new liquid foamate by releasing the gas bubbles.

The concentration of the surface active solute in the new liquid is usually several times of that in the initial liquid solution. The success of this technique depends on the stability and characteristics of foam. In this study we designed and fabricated a single continuous foam fractionator and investigated a continuous operation in the recovery of Zinc (II) anions from synthetic and industrial effluent using sodium lauryl sulphate (SLS) as foaming agent. The effects of conditions of the feed solution and surfactant (pH and concentration) and operational parameters of the column (air and feed flow rate, liquid pool height) on the separation characteristics were investigated and enrichment ratio (E) was calculated.

'Enrichment ratio' or 'Separation factor' E is defined as the ratio of the concentration of Zinc (II) ions in the foamate (CP) to the concentration of zinc (II) ions in the bulk liquid (Cb) from which the foam has been generated.From these studies it has been observed that as the flow rate of air increases from 1.0 to 10 LPM, the separation factor decreases which is due to the fact that more solution drainage occurs at low flow rates which results in dry foam. The enrichment ratio is observed to be high whenever the foam is dry. The increasing concentrations of the surfactant and zinc (II) ions from 0.1 to 0.5% and 50 to 100ppm respectively resulted in the reduction in enrichment ratio. It is caused by lower surface tension and higher surface liquid viscosity which causes decrease in the rate of film drainage. As the liquid pool height increases from 10 cm to 30 cm, the separation factor increases which is due to the fact that the residence time of bubbles in the liquid pool is more. It takes longer time for bubbles to rise through the solution and adsorption of Zn (II) ions at the gas-liquid interface could approach an equilibrium level more closely, and raise the enrichment ratio. As the pH of the solution determines the sign and the magnitude of the charge on a variety of inorganic particulates, the extent of removal of the particulates by foam separation technique is controlled by pH of the solution. The optimum pH obtained from the experiment is around 7.3 which reveals that Zn (II) reverses its charge at pH 7.3(isoelectric point).

Key Words: Foam Fractionation, Zinc (II), Enrichment ratio, Foaming agent, Sodium lauryl sulphate.

643h The Binding of Hg on Pd Binary Alloys and Monolayers

Erdem Sasmaz, Shela J. Aboud, and Jennifer L. Wilcox. Department of Energy Resources Engineering, Stanford University, Green Earth Sciences, 367 Panama Street, Stanford, CA 94305

Many sorbent materials such as activated carbon, metal oxides, metal sulfides and pure metals can capture mercury under specific conditions. Among these materials activated carbon is one of the most-used sorbents due to its high removal capacity. Unfortunately, it has been reported that activated carbon can prevent concrete from meeting the freeze-thaw requirements in the recycling step involving the addition of particulate matter for concrete manufacturing. It would be much more efficient to use a material that is proficient at removing mercury and also concrete-friendly.

Density Functional Theory (DFT) calculations have been performed to predict the binding mechanism of element mercury on the binary alloys PdAu(111), PdAg(111), PdCu(111). The Pd surface has been found to be more reactive than the Au, Ag and Cu surfaces, although small additions of Au, Ag and Cu to bulk Pd surface increases the overall mercury binding energy. Specifically, it has been found that Au atoms in the PdAu alloys are most beneficial to Hg adsorption when they remain below the surface layer. This prompted studies using a monolayer of Pd overlayed on an Au substrate to further enhance binding compared to the PdAu alloys. The use of monolayers not only removes the dependence on the random atomic arrangement, but it may also lead to a higher capacity because of the surface composition uniformity and subsequent increased number of binding sites.

647a Sulfonated Polyimide-Silica Proton Exchange Nanocomposites

Mitchell Anthamatten and Lijun Zou. Chemical Engineering, University of Rochester, 206 Gavett Hall, Rochester, NY 14627

Sulfonated polyimides (SPIs) are promising materials for polymer electrolyte membrane fuel cells (PEMFCs). Aromatic polyimides have excellent thermal stability, high mechanical strength, and superior chemical resistance. However, current SPIs show insufficient proton conductivity at low humidities and high temperatures, and they suffer from hydrolytic degradation. Influencing the nano-scale ordering of SPIs is expected to affect ionic ordering, proton conductivity, and the rate of hydrolytic degradation.[1]

We are developing materials to learn about the interrelationships between morphology and proton conductivity in SPIs. Targeted materials exhibit phase segregated nanostructure on the same length scale as ionic ordering. In our earlier study, hydrophobic polysiloxane segments were attached to SPIs to form segregated copolymers.[2] The presence of hydrophobic siloxane segments did not interfere with water swelling. However, ionomer morphology was found to be significantly different than parent SPIs.

Here we report a second approach to introducing nanostructure into SPIs. Inorganic silica nanoparticles were co-synthesized during polymerization and curing to form SPI-silica nanocomposites. Sol gel processing, using amine-terminated silanes, was employed to ensure that particles are uniformly dispersed and are connected to polyimide host. TEM and XPS studies confirmed that particles were successfully incorporated into the material. Water sorption studies show that silica particles encourage membrane hydration at lower humidity. On the other hand, silica particles also act as covalent crosslinks and influence solvent resistance. By preventing excessive swelling of the membrane, the inorganic nanoparticles are expected to improve hydrolytic stability and durability under aggressive conditions.[3] The silica content is limited to below 20 wt% to prevent loss of mechanical properties.[4-6] Impedance and water uptake measurements are reported as the first step in evaluating their potential as fuel cell membranes.

1. Asano, N.; Aoki, M.; Suzuki, S.; Miyatake, K.; Uchida, H.; Watanabe, M. J. Am. Chem. Soc. 2006, 128, 1762-1769.

2. Zou, L. J.; Anthamatten, M. Journal of Polymer Science: Part A: Polymer Chemistry 2007, 45, 3747-3758.

3. Lee, C. H.; Hwang, S. Y.; Sohn, J. Y.; Park, H. B.; Kim, J. Y.; Lee, Y. M. Journal of Power Sources 2006, 163, (1), 339-348.

4. Chang, C. C.; Chen, W. C. Chemistry of Materials 2002, 14, (10), 4242-4248.

5. Chen, Y.; Iroh, J. O. Chemistry of Materials 1999, 11, (5), 1218-1222.

6. Yen, C. T.; Chen, W. C.; Liaw, D. J.; Lu, H. Y. Polymer 2003, 44, (23), 7079-7087.

Web Page: www.che.rochester.edu/users/anthamat/

647b Development of a Novel Technique for Functionalization of Nanoporous Silicon Membranes

Saeed Moghaddam, Mechanical Science and Engineering & Chemical and Biomolecular Engieering, University of Illinois at Urbana-Champaign (UIUC), 1206 West Green Street, Urbana, IL 61801, Rich Masel, University of Illinois, Urbana-Champaign, 600 South Mathews, Room 204, Urbana, IL 61801, and Mark Shannon, Mechanical Engineering, University of Illinois, 1206 W Green St, Urbana, IL 61801.

A novel technique has been developed to assemble molecules with functional end groups inside pores of a nanoporous silicon membrane. A 20-um thick p-doped silicon membrane was anodized in HF electrolyte to produce 5-7 nm in diameter pores through the entire membrane thickness, with the exception of a 50-nm thick layer at the backside of the membrane that was later etched using reactive ion etching process. A recipe was developed to hydroxylate the surface of the pores. It was shown that a simple process of soaking the membrane in self-assembly solution, as done in prior studies, only results in self-assembly in a very shallow depth inside the pores. Thus, a new method was developed to conduct self-assembly down to the bottom of the pores (the aspect ratio of the pores is approximately four thousand). The molecule assembled on the pores' wall is (3-mercaptopropyl)trimethoxysilane (MPTS). The thiol end group of this molecule was later oxidized to sulfonate group in nitric acid. Penetration depth of the functional group inside the membrane was verified using Time of Flight-Secondary Ion Mass Spectroscopy (TOF-SIMS) and Auger Electron Spectroscopy (AES). In TOF-SIMS, a spot on the membrane was carved down gradually and its elemental composition was measured. The sulfur concentration was found to stay relatively constant through the membrane thickness. Analysis of the membrane cross section using AES also showed uniform coverage. The developed membrane is the first of its kind, with uniformly covered covalently bonded sulfonate groups, that can be used as proton exchange membrane (PEM). This development has a significant impact on silicon-based micro fuel cell technology where a silicon-based MEMS/CMOS compatible PEM membrane has been sought.

647c Nanofiber Network Membranes for Fuel Cell Applications

Jonghyun Choi[1], **Peter N. Pintauro**[1], Kyung Min Lee[1], Ryszard Wycisk[1], and Patrick Mather[2]. (1) Department of Chemical Engineering, Case Western Reserve University, 10900 Euclid Avenue, Cleveland, OH 44106-7217, (2) Biomedical and Chemical Engineering Department, Syracuse University, Syracuse, NY 13244

New generation of proton conducting membranes for PEM fuel cell applications will combine the latest developments in both materials chemistry and membrane nanomorphology control. A critical challenge is to achieve high proton conductivity under low humidity fuel cell operating conditions. The most obvious way to enhance proton transport in sulfonic acid containing membrane polymers is to increase the sulfonation degree so as to maximize the number of protogenic carriers and the membrane's water retention capability, which are especially important for applications in hydrogen fuel cells. Unfortunately, this approach leads to problems with membrane dimensional/mechanical stability. Recent studies on multiblock sulfonic acid copolymers open up an interesting avenue of approach, but the use of block copolymers has limits imposed by the monomer/oligomer reactivity, block stoichiometry and the identification/availability of a good casting solvent.

An entirely new approach for fabricating fuel cell membranes has been developed by the present authors. The method can be universally applied to a wide range of ion/proton conducting materials. Briefly, a three-dimensional, interconnected network of proton-conducting polymer nanofibers is fabricated via electrospinning and then embedded in an inert/impermeable polymer matrix. The resulting nanomorphology is similar to that desired in a block copolymer material. The nanofiber network, occupying about 70% of the dry membrane volume, is composed of a high ion-exchange capacity (low equivalent weight) sulfonic acid polymer to ensure high water affinity and a high concentration of fixed charge acidic sites. The inert (hydrophobic) polymer matrix controls water swelling of the nanofibers and provides mechanical strength to the membrane. Unlike other fuel cell membranes, the role of the mechanical support is decoupled from that of the proton-conductor.

The talk will review the experimental details for nanofiber network membranes fabrication, where the nanofibers were composed of sulfonated poly(arylene ether sulfone) with/without molecular silica (sulfonated octaphenyl polyhedral oligomeric silsesquioxanes). The measured water swelling, proton conductivity, and thermal/mechanical properties of the resulting membranes will be discussed.

NANOCOMPOSITE FUEL CELL MEMBRANES

A. Fuchs[1], A.Anasuya[1], J. Sutrisno[1], C. R. Henderson [2], B. Wallis [2], C. Evrensel [2], F. Gordaninejad[2]
1) Chemical Engineering Department, University of Nevada, Reno – Nevada, USA
2) Mechanical Engineering Department, University of Nevada, Reno – Nevada, USA

Introduction

Nanostructured polymers are synthesized from macromolecules demonstrating supramolecular interactions. Examples are: hydrogen bonding, hydrophobic interactions, organometallics, electrostatic interactions, liquid crystal polymers and van der Waals [1]. Composite membranes are prepared using polyimide (PI) and polybenzimidazole (PBI) with heteropolyacids (HPAs) [2]. Nanostructured PEMs may be prepared from supramolecular polymers and HPAs.

Nanocomposite membranes have been investigated by several groups [2-9]. Nanoporous tungstosilicate based on 12- phosphotungtic acid has been prepared using crosslinked linear polyethyleneimine as a polymer matrix [2]. Nanocomposites of Nafion and silica are being made using sol-gel process [3]. Heteropolyacids (HPAs) are used as inorganic proton conductors because they exhibit high proton conductivity but also are stable at high temperatures [4].

Surface Polymerization – covalently bonded polymers and supramolecular polymers

Most work on surface polymerization has involved covalent bonding of polymers to a particle surface. It is also of interest to investigate the use supramolecular polymers for surface polymerization. Surface polymerization has been widely done in with covalent bonding for silicon dioxide and Si- H groups. Germanium has also been surface polymerized by Xu et al [10] by covalently immobilized polymer brushes from surface-initiated ATRP. Well-defined polymer-Ge hybrids, consisting of covalently tethered polymer brushes of pentafluorostyrene (PFS), (2-dimethylamino)ethyl methacrylate (DMAEMA), and PFS-DMAEMA diblock copolymers, were prepared. This is a process for covalent bonding of an ATRP initiator monolayer on the Ge-H surface. Well-defined polymer-Ge hybrids could be readily prepared via surface-initiated ATRP on the substrates.

Membrane blends

Blended membranes are made so that the properties of both the polymers can be useful to the ultimate purpose of the membranes. Polyether sulfone (PES) have been blended with polymers such as polyvinylidenefluoride (PVDF), polyacrylonitrile, polyurethane, chitosan, cellulose acetate, sulfonated poly(ether ether ketone) (SPEEK) and poly (amide- imide) (PAI). The PAI blends with PES have been made to be used in ultrafiltration membranes [11]. Blended membranes were formed from SPEEK and PES and thermal degradation studies, glass transition temperature, proton conductivity and water uptake were studied for these blended membranes [12]. Blends have also been made with polybenzimidazole (PBI) and sulfonated partially fluorinated poly (arylene ether sulfone) [13]. Blends of PES with PBI have been made as a part of this project.

Surface polymerization

Polymer coating of heteropolyacids (HPA) is done in order to prevent the HPAs from being washed out of the membrane. In addition, functional groups attached to the polymer can provide enhancement of proton conductivity because they can be reacted with sulfonic acid.

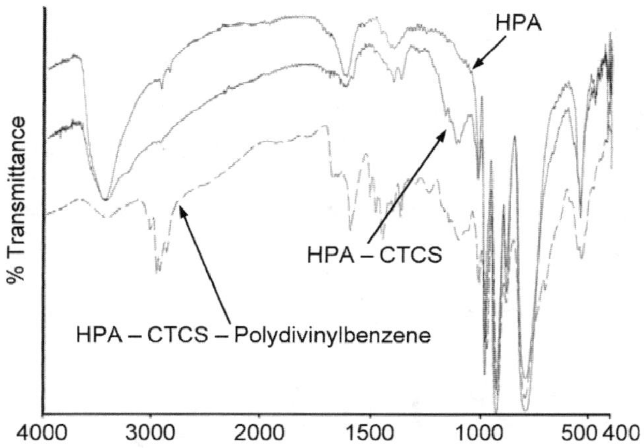

Figure 1 – FTIR of surface polymerized HPA.

Atom transfer radical polymerization (ATRP) was used for as a surface polymerization technique. The surface initiator is grafted onto the HPA surface and is initiated by electrons from the redox reaction of metal halide (CuBr) [14 –16]. Then, the monomer is initiated and followed by propagation and termination.

Polymer Sulfonation

The sulfonation of polyether can be done in two ways, in addition to making a sulfonated polymer from sulfonic groups attached to a monomer. The sulfonation can be achieved by either using chlorosulfonic acid in dichloroethane or using sulfur trioxide in dichloromethane [17]. Sulfonation can also be carried out by the reaction of the monomers with concentrated sulfuric acid [18]. The process with chlorosulfonic acid is carried out at room temperature. Sulfonation of vulcanized ethylene-propylene- diene terpolymer has been sulfonated by acetyl sulfate [19]. Acetyl sulfate was prepared by reacting acetic anhydride and sulfuric acid. Polyarylene ether sulfone was sulfonated by acetyl sulfate as well [20]. SPEEK membranes have also been prepared by sulfonation by chlorosulfonic acid in dichloroethane. Sulfur trioxide, a gas at room temperature, has been used in dichloromethane to sulfonate polyether sulfone pellets [21]. The polymer was dissolved in dichloromethane and stabilized sulfur trioxide was then added in dichloromethane under nitrogen atmosphere.

Results

Composite membranes and composite blends were prepared with PES and HPA. Three types of HPAs were used for the study. Phosphomolybdic acid, phosphotungstic acid and silicotungstic acid (SiW) were used to study the proton conductivity of the membranes. PES and PBI was dissolved in dimethylacetamide (DMAc) and followed by adding and mixing of HPA at 50-60^{0}C with specified weight ratio. The solution was poured in the mold and solvent was then evaporated in a vacuum oven at 80-100^{0}C. The blended membranes were acid doped in 85% phosphoric acid for 7 hours.

More permanent ways of sulfonation are being explored and the most appropriate methods to sulfonate the polymer are by reacting with acetyl sulfate or chlorosulfonic acid. The SiW could not be added to PES at greater than 50% by weight. The membrane, at even 50 %, was extremely brittle and had reached the saturation concentration of SiW.

Surface polymerization of SiW was done to investigate the effect of concentration of SiW on the membrane characteristics.

The surface polymerization procedure was as follows: SiW (16 g) was pulverized, sieved, and dried in a vacuum oven at 100 °C. Dried SiW particles (12 g) were added and reacted at 85 °C with 2-4(-chlorosulfonylphenyl)-ethytrichlorosilane (CTCS, 4 g)) for 36 hours in anhydrous toluene (110 g) as a solvent in inert gas (nitrogen). The mixture was then filtered and washed with anhydrous toluene in order to remove excess CTCS. The residue (SiW-CTCS) was dried in a vacuum oven at low temperature (50 °C) for 24 hours. Functionalized SiW-CTCS (6 g) was reacted with.

CuBr (0.06 g), $CuBr_2$ (0.03 g), Spartein (0.06 g), and monomer (divinylbenzene) in anhydrous toluene (60 g) at 85 °C for 24 hours under nitrogen.

Finally, the mixture was filtered, washed several times with anhydrous toluene and dried in a

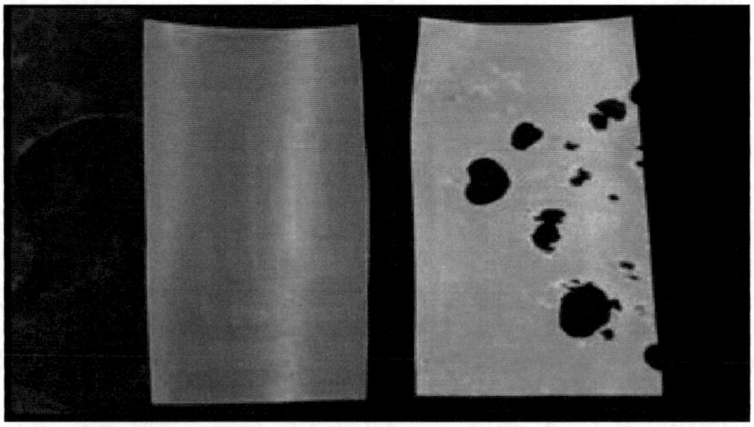

Figure 2 – Surface-coated and non-surface-coated SiW composite (50 wt.%) PES membrane.

vacuum oven at low temperature (50 °C) prior to use.

Membrane Characterization

The grafted surface initiator and polymer on the SiW was characterized using FTIR. The SiW sample was scanned from 400 to 4000 cm^{-1}. SiW, SiW – CTCS, and SiW – CTCS – polydivinylbenzene transmission data are shown in the **Figure 1** from the top to the bottom respectively. Comparing the SiW, SiW – CTCS curves, the peak is shifted at 1100 cm^{-1} which can be associated with the presence of surface initiator (CTCS) on the SiW. In addition, the peak is shifted at 2900 cm^{-1} on the SiW – CTCS – polydivinylbenzene if compared with SiW – CTCS which means the polymer has been attached onto the SiW surface through surface initiator. The HPA content within the composite membrane could be increased when the HPA was coated with polydivinylbenzene. The comparison of surface-coated (left) and non-surface-coated (right) SiW composite membrane pictures are shown in **Figure 2**. The higher quality of the surface coated SiW composite membrane indicates that the interface compatibility of polymer matrix and HPA may have been increased by the surface coating. The characterization of the composite membranes was done using an electrochemical impedance spectroscopy. The EIS plot of composite blended membrane with 60 wt. % phosphotungstic acid has been shown in **Figure 3**. This is a plot of real and imaginary impedance. The real impedance is the resistance while the Y- axis represents the imaginary impedance. Blended composite membrane with 40 % by wt. content of silicotungstic acid has conductivity of 1.2 x 10 $^{-2}$ S/cm with acid doping. It is expected to increase further after acid doping. The conductivities were measured from 1Hz to 1MHz at 700mV potential. The four point probe method was used to check for the conductivity of the membranes. The summary of the composite membranes conductivity is listed in Table 1. These values were found to be fairly less than Nafion®, whose conductivity is 0.1 S/cm [22]. The process of sulfonation is expected to increase the conductivity of these membranes.

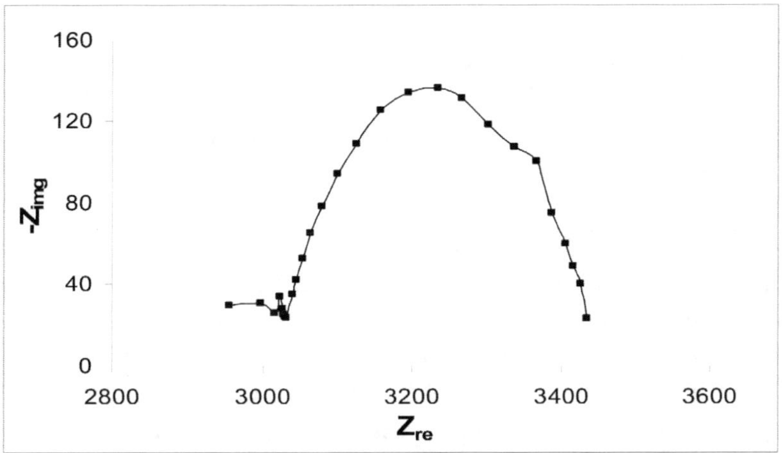

Figure 3 – Nyquist plot of 60 wt. % PWA – PES – 10%PBI (of PES) acid doped.

Future work

The SiW will be surface polymerized using ATRP with styrene as monomer. The surface polymerized SiW can be sulfonated through the reaction with acetyl sulfate in dichloromethane at 40 °C through the styrene functional in order to increase the conductivity [23]. The synthesized composite membrane will be tested using the single test cell for generating current – voltage polarization curve. This test will be done at 20% - 80 % RH and 25°C – 120 °C.

A single membrane PEM fuel cell has been designed. This fuel cell has an active area of 25cm^2, so commercially available membrane electrode assembles can be used for testing. The flow paths in the fuel cell were designed to allow flexibility with the monopolar plate flow patterns. They can accommodate any flow pattern without changing the location of the ports or sizes. The bipolar plates can also be stacked without changing anything to analyze a stacked system. Testing will be done with this prototype on a commercially available membrane electrode assembly (MEA) to achieve a standardized system. From that point we will create our own MEA using a new membrane. If that test is a success we are then going to move onto the monopolar plates by changing the flow patterns, coating, and material type. The goal of these experiments is to find the highest power density possible. While looking at the plate designs and flow patterns we will look at the gas diffusion layer (GDL) and catalyst layer to find

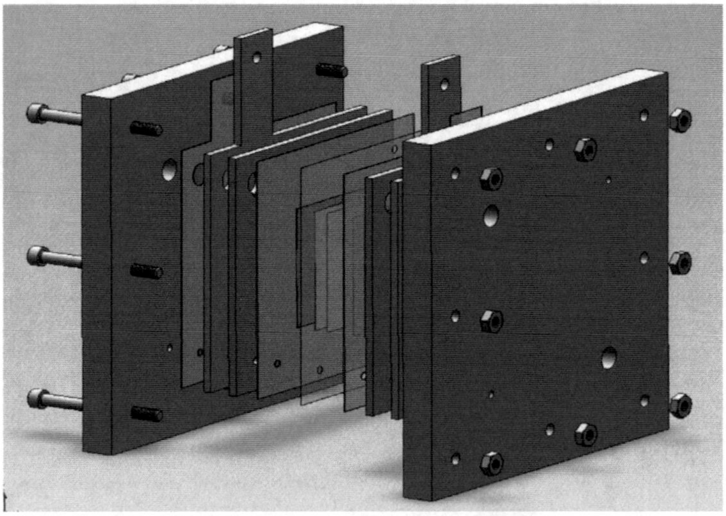

Figure 4 – Representation of MEA.

the correct amounts to use with the new membrane.

The Computational Fluid Dynamics (CFD) modeling of the existing fuel cell is being developed using commercial code FLUENT. The two-dimensional geometry of the flow field for a homogeneous single phase fluid, either hydrogen or oxygen is created, meshing and its optimization for the flow conditions is currently in progress. Meshing is one of the most important steps of CFD modeling. It should capture the details of the flow field at locations with high gradient such as boundary layer and corners using small enough elements while minimizing the number of elements used in order to keep computational time at a reasonable level. GAMBIT, the preprocessor to FLUENT, is used for construction of main flow field geometry, mesh generation and optimization. Once flow through this main 2-D geometry is successfully simulated it will be expanded to 3-D model and generalized to multi phase flow.

Overall operation of the fuel cell requires simulation of electrochemical reactions, modeling MEA and prediction of the power generation. The representation of the MEA is shown in Figure 4. For this purpose FLUENT add-on module, PEMC is being evaluated. The predictions of the completed model will be compared with the experimental measurements using the existing fuel in our laboratory. Initial model and testing will use a well studied and characterized commercially available Nafion®

References

1. H. Jiang et al., 'Infinite chains constructed from flexible bis(benzimidazole)-based ligands', Polyhedron (2008), in press.
2. He et al, 'Proton conductivity of phosphoric acid doped polybenzimidazole and its composites with inorganic proton conductors' , Journal of membrane sciences, 226,169-184, 2003.
3. Haolin Tang , Zhaohui Wan, Mu Pa, San Ping Jiang, Electrochemistry Communications, 9(8), 2003-2008, 2007.
4. A. Herring, "Inorganic–Polymer Composite Membranes for Proton Exchange Membrane Fuel Cells", Journal of Macromolecular Science, Part C: Polymer Reviews, 46:245–296, 2006.
5. R. He, Q. Li, A. Bach, N. J. Bjerrum, Journal of Membrane Science, 2005.
6. S. W. Chuang, S.L. C. Hsu, Y. H. Lui, Journal of Membrane Science, 305, 353-363, 2007.
7. Meng.F, A. Herring, et al., Electrochemica Acta, 53, 1372-1378, 2007.
8. Honma. I, Nomura. S, Journal of Membrane Science, 185, 83-94, 2001.
9. Licoccia.S, Traversa, E, Journal of Power Sources, 159, 12-20, 2006.
10. Xu et al, 'Well-Defined Polymer-Germanium Hybrids via Surface-Initiated Atom Transfer Radical Polymerization on Hydrogen-Terminated Ge(100) Substrates', Organometallics, 24, 1768, 2005.
11. A. Rahimpour et al, Journal of membrane science, 311(2008) 349-359.
12. Han- Lang Wu et al, European Polymer Journal, 42, (2006), 1688-1695.
13. F. Schonberger et al, , Solid State Ionics, 178, 2007, 547-554
14. G. Odian, John Willey and Sons, Inc., 2004.
15. A. Fuchs, et. al., Polymer, Vol. 47, pp. 7653 – 7663, 2006.
16. J. Sutrisno, Chemical Eng. Master Thesis UNR, 2008.
17. A. Noshay et al, Journal of Applied Polymer Science, 20 (1976) 1885-1903.
18. C.A. Linkous et al, Intl. J. Hydrogen Energy, 23 (1998), 525-529.
19. F. Barroso-Bujans et al, Acta Materialia, (2008) (In Press).
20. X. Zhang et al, Polymer, 46 (2005), 1719-1723.
21. R. Nolte et al, Journal of Membrane Science, 83 (1993), 211-220.
22. http://www.espec.co.jp/english/tech-info/tech_info/pdf/05_10/contents2.pdf
23. C. P. L. Rubinger, et. al., Sensor and Actuator B, Vol. 123, pp. 42 – 49, 2007.

647e Improved Performance of Composite Proton Conducting Membranes in Fuel Cells through Electric Field Control of Membrane Structure

Xue Wei, Dongxia Liu, and **Matt Yates**. University of Rochester, Department of Chemical Engineering, 206 Gavett Hall, Rochester, NY 14627

Electric fields were used to control the structure of polymer composite proton conducting membranes. Under appropriate conditions, functional particles dispersed in the polymer can be aligned into chains in the direction of the applied field. The resulting composite membranes are anisotropically structured with proton conduction greatly enhanced in the alignment direction. Compared to randomly structured composites, the electric field aligned membranes have dramatically improved performance in direct methanol fuel cells. The field-induced structure results in anisotropic water swelling behavior of the membranes, with greater swelling in the direction of alignment. By tailoring the membrane structure with the electric field, the mechanical and transport properties of the membrane can be optimized for fuel cell application.

647f Polyelectrolyte Multilayers for Direct-Methanol Fuel Cells

Avni A. Argun, J. Nathan Ashcraft, and Paula T. Hammond. Chemical Engineering, Massachusetts Institute of Technology, 77 Massachusetts Avenue, 66-525, Cambridge, MA 02139

The global need for clean and sustainable energy is ever increasing and electrochemical devices such as fuel cells, batteries, and solar cells show great potential. A major component of these devices is an ion-conducting electrolyte which enables fast charge transport between electrodes. The alternating adsorption of oppositely charged molecular species, known as the electrostatic layer-by-layer (LBL) technique, is a simple and elegant method of constructing highly tailored polymer electrolytes and organic-inorganic composites. We have utilized this method to develop proton-exchange membranes for direct-methanol fuel cells (DMFCs). This approach presents strong advantages as it allows the incorporation of many different functional materials within a single film at a full range of compositions with exceptional homogeneity. By pairing a sulfonated poly(2,6-dimethyl 1,4-phenylene oxide) (sPPO) with various amine-based polymers, we obtain ionic conductivity values of up to ***35 mS/cm*** *at fully humidified conditions.*[(1)] *Furthermore, these multilayer systems exhibit low liquid methanol permeability and have high chemical and mechanical stability due to the use of aromatic polymers. Coating a traditional fuel cell membrane with these highly conductive LBL films reduces the methanol crossover and improves the power output of DMFCs by up to 50%.*

(1) Argun, A.A.; Ashcraft, J.N.; Hammond, P.T., "Highly Conductive, Methanol Resistant Polyelectrolyte Multilayers." Adv. Mater. ***2008****, 20*, 1539-1543.

661a Propylene/propane Separation through 6FDA-Based Polyimide Membranes

Arnaud Baudot[1], Hélène Rodeschini[1], and Gonzalez Serge[2]. (1) Catalysis and Separation Division, IFP Lyon, PO Box 3, 69360 Solaize, France, (2) Physical Chemistry, IFP Lyon, PO Box 3, 69360 Solaize, France

Propylene/propane mixtures separation experiments by gas permeation through dense films composed of 6FDA-DAPF (a fluorinated polyimide resulting from the polycondensation of 2,2'-Bis (3,4-dicarboxyphenyl) hexafluoropropane dianhydride and di-9,9'(4 aminophenyl fluorene) have been carried out.

The influence of various synthetis parameters of 6FDA-DAPF such as polymerization routes (chemically or thermally induced), polymerization solvents, casting solvents as well as heating post-treatements have been studied. This polymer offers high propylene/propane selectivities, even at high hydrocarbon partial pressures, as well as high propylene permeabilities.

661b a Comprehensive Study on the Diamino Modification of Polyimides

Lu Shao, Dept of Chemical and biomolecular Eng,, National U of Singapore, 10 Kent Ridge Crescent, Singapore, 117576, Singapore, Tai-Shung Chung, Department of Chemical and Biomolecular Engineering, National university of Singapore, 10 Kent Ridge Crescent, Singapore 119260, Singapore, Singapore, Bee-Ting Low, Chemical and Biomolecular Engineering, National University of Sigapore, Singapore, 117576, Singapore, and You-Chang Xiao, Chemical and Biomolecular Eng, National Univ of Singapore, 10 Kent Ridge Crescent, Singapore, 117576, Singapore.

Diamino modification represents a potential approach for enhancing the gas transport properties of polyimide membranes. For effective membrane modification, several critical intervening factors are in play, namely electrophilicity and free volume of the polyimide, and nucleophilicity and molecular dimensions of the diamines. The extent of the crosslinking reactions between various polyimides and diamines were investigated both in solution and membrane forms. The reaction of a specific diamine with Matrimid/CH2C12 and 6FDA-durene/CH2C12 solutions revealed that the appearance of a cross-linked network was more rapid in the former. This is attributed to the higher electrophilicity of Matrimid compared to 6FDA-durene. A series of aliphatic and cyclic diamines were used to effect the cross-linking reaction in Matrimid and 6FDA-durene solutions. The cross-linking rate was highly dependent on the nucleophilicity of the diamines.

Solid state diamino modification of polyimide membranes was performed. In addition to the reactivity of the diamines and polyimides, the degree of cross-linking was related to the molecular dimensions of the diamines which in turn influence the diffusivity of the diamines within the polymeric matrix. In our study, polyimides with considerably different free volume were employed i.e. free volume of 6FDA-durene > 6FDA-ODA/NDA > Matrimid. Polyimides with high free volume favor the access of diamine molecules to the reactive sites, thereby giving better crosslinking effects. Hence, for the same modification reagent, the greatest decline in gas permeability was observed in 6FDA-durene membranes. Similar extent of cross-linking can be achieved at shorter modification duration by using polyimides with greater free volume. Two other side reactions occur during diamino modification, namely chemical grafting and etching (polyimide main chain scission). The varying extent of each reaction occurring is dependent on the nucleophilicity and molecular dimensions of the diamines. It has been proven that the diamino modification approach is useful for enhancing the H2/CO2 separation performance of polyimide membranes. For 6FDA-durene membranes, H2/CO2 selectivity increased from an intrinsic value of ~1 to 100 after modification with 1,3-diaminopropane (PDA) for 5 mins. Similarly, an enhancement in H2/CO2 selectivity from 2 to 64 was obtained by modifying 6FDA-ODA/NDA membranes with PDA for 90 mins. Therefore, the appropriate selections of polyimide-diamine pairs and modification condition can successfully alter the intrinsic gas separation performance of polyimide membranes.

661c Gas Permeability and Physical Aging In Ultra-Thin Glassy Polymer Films

Brandon W. Rowe[1], Anita J. Hill[2], Steven J. Pas[2], Ryoichi Suzuki[3], Benny D. Freeman[1], and Donald R. Paul[1]. (1) Chemical Engineering, The University of Texas at Austin, 1 University Station CO400, Austin, TX 78712, (2) CSIRO Manufacturing Science and Technology, Clayton, VIC 319, Australia, (3) National Institute of Advanced Inductrial Science and Technology (AIST), Tsukuba, Japan

Most gas separation membranes are formed from glassy polymers because of their exceptional permeability-selectivity properties. However, glassy polymers are non-equilibrium materials that will spontaneously, but usually slowly, change over time towards an equilibrium state by a process known as physical aging. The physical aging rate becomes orders of magnitude more rapid when film thickness is decreased below one micron, indicating a strong influence of the free surface on aging behavior.1 This phenomenon is an intrinsically fascinating scientific issue, and understanding physical aging has broad impacts in several technologies including gas separation materials, sensors, and microelectronics.

Previous studies have investigated the aging behavior of free-standing films with thicknesses down to approximately 400 nm. These studies showed accelerated aging rates, as compared to bulk-like films, for samples with thicknesses below one micron, and the aging rates increased with decreasing film thickness. 1 Creating defect-free films below this limit for gas permeability measurement is extremely difficult. A coating technique has been developed in this work to enable the study of gas transport properties in glassy films with thicknesses down to 20 nm. This thickness is similar to, if not below, the thickness of the selective skin in state-of-the-art hollow fiber membranes. A thin coating of a highly permeable, rubbery polymer, poly(dimethylsiloxane) (PDMS), effectively blocks the convective flow through the selectivity-destroying defects in the glassy film. The resistance to mass transfer of the PDMS layer is constant, due to its equilibrium nature at the testing conditions, and accounting for this resistance enables calculation of the glassy layer properties. This is the first report, to our knowledge, on the aging behavior of ultra-thin (<100 nm) glassy polymer films monitored by gas permeability.

The gas transport properties of films with thicknesses down to 20 nm were monitored during physical aging at 35 °C. All films exhibited decreasing permeability and increasing pure gas selectivity with time, as expected from the densification caused by physical aging. As film thickness was decreased, the initial permeability decreased while the pure gas selectivity increased. These trends are consistent with the idea of a rapidly aging region near the surface, whereby the thinner films would exhibit a greater aging response before the first measurement could be completed, ~ 1 hour. The aging rate increases with decreasing film thickness, down to ~75 nm. Below 75 nm, the aging rate depends less on thickness and even slightly slows as film thickness decreases. It is suggested that this behavior is caused by the very rapid initial aging in the near-surface region of the film, i.e. the apparent aging rate of the ultra-thin films is reduced because the film is initially closer to the equilibrium state than the thicker films. These results are consistent with the findings of a recent variable energy positron annihilation spectroscopy study where the size of free volume elements was found to be smaller near the surface of these glassy films.

(1) Huang, Y.; Paul, D. R. Polymer 2004, 45, 8377-8393.

661d Effect of Porous Nanoconstraints on Ptmsp Membrane Performance

Jason P. Killgore and Rene M. Overney. Chemical Engineering, University of Washington, Box 351750, Seattle, WA 98195

The development of large polymer membranes with sub micron thickness is a significant challenge. Here, porous anodisc membranes, with extremely high permeability, are used to support thin reverse selective Poly(trimethylsilyl propyne) (PTMSP) films. The permeance of the PTMSP is characterized by a series resistance model, eliminating variance in the support membranes. Supports with 20 nm, 100 nm, and 200 nm diameter pores are investigated. By using dilute polymer concentrations, it is also possible for the polymer to impregnate the pores of the anodisc, providing a permeability enhancing interfacial confinement effect similar to that seen with particle nanocomposites, but without selectivity reducing agglomeration.

Synthesis and Characterization of Cross-linked Polyvinylalcohol/Polyethyleneglycol Blend Membranes for CO_2/CH_4 Separation

Rong Xing, W.S. Winston Ho
Department of Chemical and Biomolecular Engineering
Department of Materials Science and Engineering
The Ohio State University, Columbus, OH, 43210

Abstract:

Polymeric membrane technology has received increasing attention in the field of gas separation, e.g., natural gas sweetening [1], hydrogen recovery and purification [2], and flue gas separation [3], due to its low cost, energy efficiency and ease of operation. Natural gas sweetening aims to separate acid gas, mainly CO_2, from CH_4 and other mixed gas components, to meet pipeline specifications (2% CO_2 in the U.S.A.). Currently, the polymer membranes, either commercially or reported for CO_2/CH_4 separation in the literature, show either a high permeability with a low selectivity [4] or a high selectivity with a low permeability [5,6], which potentially restrict their scale-up application. Recently, polymer membranes containing polar ether segments have been investigated to separate CO_2 from CH_4 due to the strong affinity between CO_2 and polar ether segments. For example, pure acrylated PEG has been reported to fabricate cross-linked polymer membranes via UV-irradiation, and the resulting membranes showed good performance in CO_2/CH_4 separation [1,7,8]. However, it was difficult to control the formation of large, homogeneous and defect-free ultra thin films. Also, the membranes have relatively low mechanical strength and high cost.

The objective of this research, therefore, is to develop an economically and technically viable polymer membrane for CO_2/CH_4 separation. The target membrane should have desirable separation performance, good mechanical strength and low cost. A cross-linked polyvinyl alcohol (PVA)/polyethylene glycol (PEG) blend membrane was developed to meet the above requirements. The ethylene groups in PEG enhances CO_2 transport through membrane while PVA provides mechanical strength of polymer matrix. The cross-linking and entanglements of PVA and PEG create robust and homogeneous network. In order to improve separation performance, zeolite 5A was chosen as the inorganic filler due to its stronger adsorption of CO_2 than CH_4 [9].

The cross-linking time, molecular weight of PEG and PEG (MW 200) content were found to significantly affect the transport properties of the zeolite unfilled PVA/PEG membrane. The optimal cross-linking time was found to be 84 minutes in order to obtain simultaneously good CO_2 permeability and CO_2/CH_4 selectivity. PEG (MW 200) showed the maximum CO_2 permeability among the PEGs with different average molecular weight investigated. Figure 1 shows the effect of PEG (MW 200) content on the separation performance of CO_2/CH_4, and it was found that, at 64 wt.% PEG (MW 200), the membranes showed good CO_2 permeability of 80.2 Barrers (1Barrer = 10^{-10} cm^3 (STP)•cm/cm2•s•cmHg) and CO_2/CH_4 selectivity of 33.

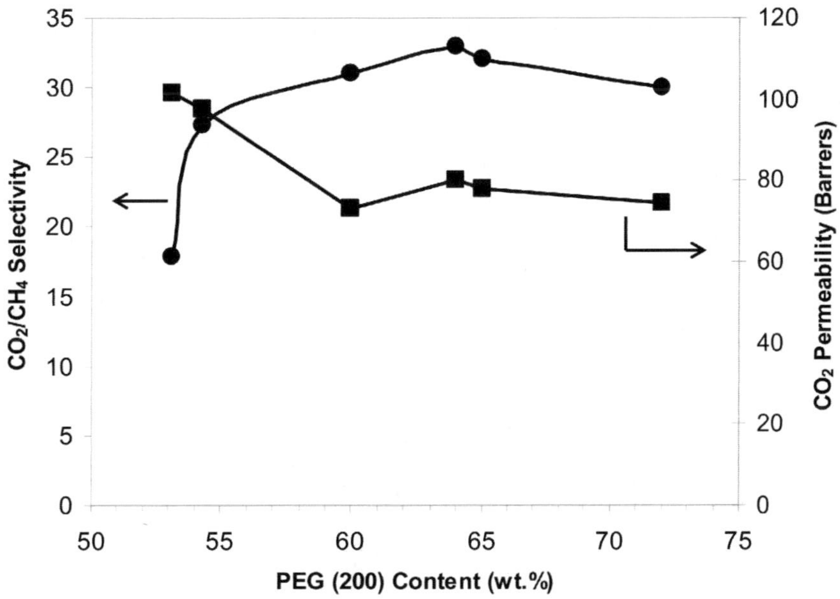

Figure 1. Effect of PEG(MW 200) content on the separation performance of CO_2/CH_4 for the PVA/PEG membrane at 30°C and 150 psi.

The effects of zeolite 5A loading, temperature and feed pressure on the gas separation performance of both zeolite-filled and unfilled PVA/PEG membranes were investigated. Figure 2 illustrates the effect of zeolite 5A loading on CO_2 permeability and CO_2/CH_4 selectivity for the PVA/PEG membrane. As can be seen that the CO_2/CH_4 selectivity decreased as the amount of zeolite 5A loading increased; while the CO_2 permeability first decreased, then drastically increased. The drop in CO_2/CH_4 selectivity could be ascribed to the formation of interfacial cracking at the polymer-zeolite 5A interfaces, which have been evidenced by SEM pictures. The initial decrease of CO_2 permeability was mostly likely caused by the partial pore blockage of the zeolite by polymer chains; while the drastic increase of CO_2 permeability suggested that, with the increase of zeolite 5A loading, CO_2 molecules permeated not only through the polymer matrix, but an additional network created by interconnected zeolite 5A particles. The effect of temperature on the separation performance of both zeolite 5A filled and unfilled PVA/PEG membranes was investigated in the range of 16.4 - 40°C at the feed pressure of 150 psi. On the one hand, the CO_2/CH_4 selectivities of all the membranes decreased linearly as temperature increased due to an increase in polymer chain mobility and thereby, diffusivity at the increased temperatures. However, compared to the zeolite-unfilled membranes, the CO_2/CH_4 selectivity could be influenced to a lesser extent from the incorporation of zeolite particles into the polymer matrix. On the other hand, the permeability of both CO_2 and CH_4 increased as temperature increased for all the membranes. The trade-off relationship exhibited by the CO_2/CH_4 selectivity and CO_2 permeability suggested that overall gas permeation in the PVA/PEG membranes follows the solution-diffusion mechanism. Thus, the gas permeability could be approximately described using the Van't Hoff-Arrhenius approach. Based on the calculation results, we could conclude that (1) the zeolite-filled PVA/PEG membrane had higher activation energy for both CO_2 and CH_4 than that of the unfilled PVA/PEG membrane, and (2) CO_2 always showed a lower pre-exponential value and activation energy than CH_4. These results indicated that the

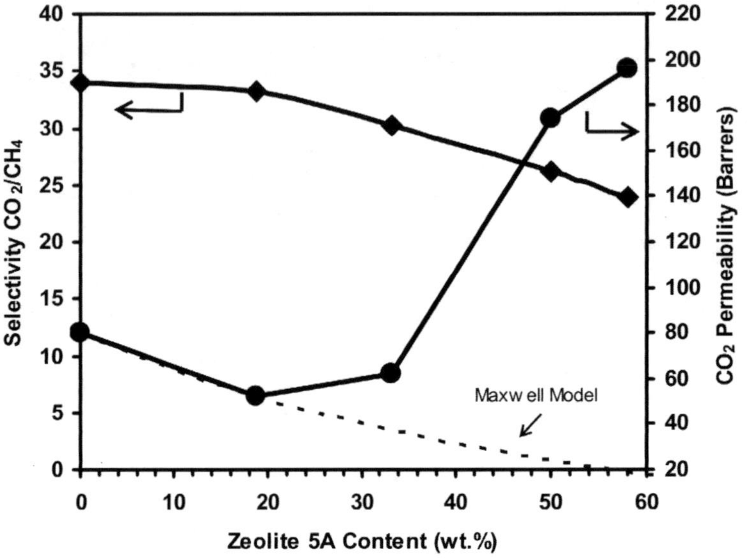

Figure 2. Effect of zeolite 5A loading content on the separation performance of CO_2/CH_4 for the PVA/PEG membrane at 30°C and 150 psi.

temperature effect on the zeolite-filled PVA/PEG membrane was more significant than the unfilled ones, and that CO_2 had a lower enthalpy of sorption in the PVA/PEG membrane than CH_4, possibly due to the strong interaction of CO_2 with ethylene oxide segments. It is noteworthy that the unfilled PVA/PEG membrane could perform beyond the Robeson's upper bound as temperature decreased. As the feed pressure increased, the CO_2/CH_4 selectivity and CO_2 permeability decreased for the unfilled PVA/PEG membranes, while the filled ones were not affected significantly, indicating that the incorporation of zeolitic particles could significantly improve the performance stability of membranes at high pressures.

The separation performance results of some zeolite-filled and unfilled PVA/PEG membranes were summarized with reference to the Robeson's upper bound line for the CO_2/CH_4 separation as shown in Figure 3. The performance of the zeolite-filled membranes approached the upper bound line with the increase of zeolite 5A loading. The performance of unfilled membranes gradually approached the bound line, and finally exceeded the boundary as the temperature decreased from 40 to 16.4°C. In comparison with the commercialized CA and Matrimid®, both the zeolite-filled and unfilled PVA/PEG membranes showed potentially to be industrially attractive.

Future work is required to eliminate or at least minimize the interfacial voids in order to improve the CO_2/CH_4 selectivity, and reduce membrane thickness to obtain high CO_2 permeance at low cost.

Acknowledgement: This report is based on work funded by Shell Oil Company, the National Science Foundation, and The Ohio State University. Part of this material is based upon work

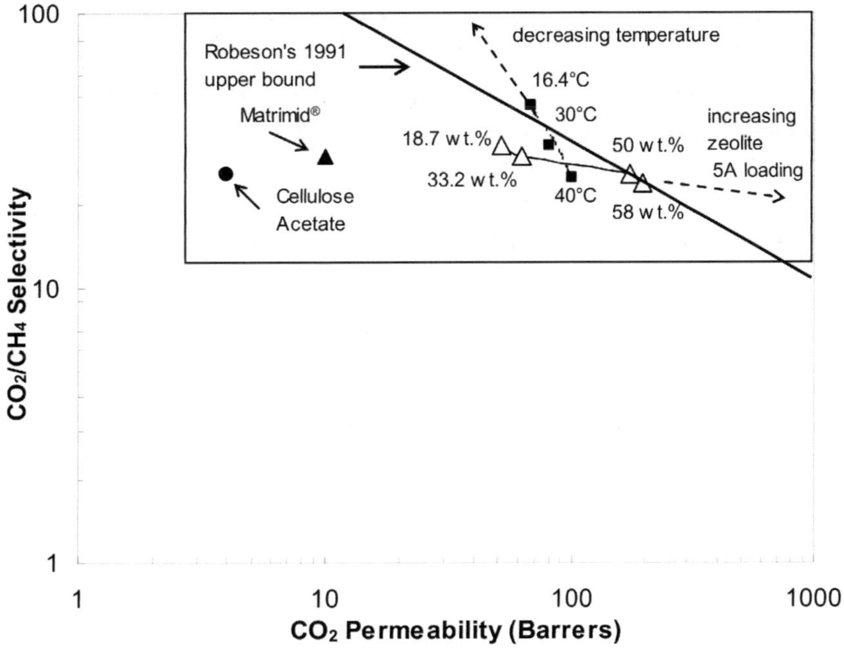

Figure 3. CO_2/CH_4 selectivity versus CO_2 permeability for zeolite-filled and unfilled cross-linked PVA/PEG blend membranes. The solid line represents Robeson's 1991 "upper bound". Solid squares refer to zeolite-unfilled membranes and open triangles represent zeolite-filled membranes. The right-top shaded area is commercially attractive.

supported by the National Science Foundation under Grant No. 0625758. We would like to thank Chi Yen for taking the SEM pictures.

Reference:

[1] H.Lin, E.V. Wagner, R. Raharjo, B.D. Freeman and I. Roman, High performance polymer membranes for natural-gas sweetening, Adv. Mater. 18 (2006) 39-44.

[2] J. Zou, J. Huang, W.S.W. Ho, CO_2-selective water gas shift membrane reactor for fuel cell hydrogen processing, Ind. Eng. Chem. Res. 46 (2007) 2272-2279.

[3] A. Chakma, Separation of CO_2 and SO_2 from flue gas streams by liquid membranes, Energy Convers. Mgmt. 36 (1995) 405-410.

[4] M.P. Chenar, M. Soltanieh, T. Matsuura, A. Tabe-Mohammadi, K.C. Khulbe, The effect of water vapor on the performance of commercial polyphenylene oxide and Cardo-type polyimide hollow fiber membranes in CO_2/CH_4 separation applications, J. Membr. Sci. 285 (2006) 265-271.

[5] Y. Li, T.S. Chung, Highly selective sulfonated polyethersulfone (SPES)-based membranes with transition metal counterions for hydrogen recovery and natural gas separation, J. Membr. Sci. 308 (2008) 128-135.

[6] C. Staudt-Bickel, W.J. Koros, Improvements of CO_2/CH_4 separation characteristics of polyimides by chemical crosslinking. J. Membr. Sci. 155 (1999) 145-154.

[7] H. Lin, E.V. Wagner, J.S. Swinnea, B.D. Freeman, S.J. Pas, A.J. Hill, S. Kalakkunnath, D.S. Kalika, J. Membr. Sci. 276 (2006) 145-161.

[8] N.P. Patel, A.C. Miller, R.J. Spontak, Highly CO_2-permeable and selective membrane derived from crosslinked poly(ethylene glycol) and its nanocomposites, Adv. Funct. Mater. 14 (2004) 699-707.

[9] S. Pakseresht, M. Kazemeini, M.M. Akbarnejad, Equilibrium isotherms for CO, CO_2, CH_4, and C_2H_4 on the 5A molecular sieve by a simple volumetric apparatus, Separation and Purification Technology, 28 (2002), 53-60.

661f Synthesis and Characterization of Modified Polysulfones for Gas Separation Applications

Claudia Camacho-Zuñiga and Francisco A. Ruiz-Treviño. Departamento de Ingenierías, Universidad Iberoamericana, Prolongación Paseo de la Reforma 880. Lomas de Santa Fe., México, D.F., 01219, Mexico

Structure/property relationships are exhaustively studied as a way to design materials with better permeability-selectivity properties and therefore to increase the applications of gas separation technologies. Many works focused in this address have lead to new materials with exceptional gas separation properties. In fact, a previous job has shown that an alternative to increase selectivity of a bisphenol-A polysulfone, BPA, with very small losses in permeability is achieved by introducing compact and rigid connectors as 1,1 bi-2-naphtol, BN[2]. Whereas it has also been shown that its permeability increases by incorporating polar and bulky connectors like tetrametilhexafluoro bisphenol A, TMHFBPA[2], at expenses of barely significant losses in selectivity. The aim of this job is to analyze the effect on thermal, volumetric and gas transport properties of the simultaneous presence of two different substituents with this characteristics. It presents a series of modified polysulfones copolymers that were synthesized by polycondensation in solution. The structures resulted from the polymerization of two diols with an acid group (4,4'-difluoro biphenyl sulfone). Such diols were added in equimolar proportion and they included either hexafluro bisphenol A (HFBPA), tetrametil bisphenol A (TMBPA), or tetrametilhexafluoro bisphenol A (TMHFBPA); while the second diol being dihidroxinaphtalen (N). The resultant polysulfones formed clear films from solution cast, and they were characterized by FTIR, intrinsic viscosity, DSC, TGA, specific volume and gas transport properties to H_2, O_2, N_2, CH_4 and CO_2.

(1) C. Lagunas-Fuentes and F. A. Ruiz-Treviño, J. of Polym. Sci.: Part B. Polym. Phys., 42, 226 (2004).

(2) J. S. McHattie, W. J. Koros, and D. R. Paul, Polymer, 32, 2618 (1991).

661g Oxygen/Nitrogen Enriched Air for Energy Applications

Stuart Nemser[1], Donald Stookey[1], and Eugene L. Keating[2]. (1) Compact Membrane Systems, Inc, 335 Water Street, Wilmington, DE 19804, (2) Environmental Kinetics, Ltd., 1687 Camden Court, Arnold, MD 21012

The benefits of increased oxygen partial pressure over air fed processes for power generation from combustion have been well known and documented for at least 50 years. As typically practiced, oxygen enrichment involves dilution of high purity oxygen with air to the desired oxygen concentration with the unavoidable associated loss in the separation energy required for isolation of the high purity oxygen.

Membranes offer potential for a lower total energy approach by supplying oxygen enriched air (OEA), within limits of separation, at the required oxygen concentration. The portable nature of OEA membrane systems also affords their application in remote locations and mobile uses. CMS is presently investigating the combination of these features in a variety of applications, including the increased power density of diesel engines.

Parallel efforts to control diesel engine NOx Emissions with nitrogen enriched air (NEA) will be discussed. By broadly reducing diesel NOx emissions (e.g. 50%) the use of diesel engines is enhanced. Diesel engines are over 30% more energy efficient than gasoline powered spark engines.

This presentation will outline recent feasibility studies for increasing the power and energy efficiency on a portable diesel engine driven electric power generator with membrane supplied OEA.

675a Computational Design of Metal-Organic Frameworks for the Adsorption of CO2

A. Ozgur Yazaydin, Chemical and Biological Engineering, Northwestern University, 2145 Sheridan Road E136, Evanston, IL 60208, John J. Low, Advanced Characterization, UOP LLC, 50 East Algonquin Road, Des Plaines, IL 60017, and Randall Q. Snurr, Department of Chemical and Biological Engineering, Northwestern University, 2145 Sheridan Road, Technological Institute E136, Evanston, IL 60208-3120.

Metal-organic frameworks (MOFs) are nanoporous materials with potential applications in adsorption separations, catalysis, and gas storage. They are synthesized in a "building-block" approach by self-assembly of metal or metal-oxide vertices interconnected by rigid organic linker molecules. Several MOFs have been shown to exhibit interesting CO2 adsorption behavior and have high adsorption capacity. Using computational modeling tools we have screened a wide variety of MOFs for CO2 adsorption, including MOFs of various topologies, metal corners and organic linkers. Adsorption isotherms were calculated from atomistic grand canonical Monte Carlo simulations. Mixture simulations or single-component isotherms plus ideal adsorbed solution theory were used to investigate the selective adsorption of CO2 over other gases such as N2, CH4 and H2 to explore separations such as the removal of CO2 from flue gas. Comparisons of predicted single-component isotherms and heats of adsorption with experiment were made where possible. In addition, we are designing new hypothetical MOFs by modifying existing MOFs, for example by proposing new linkers, and simulating their adsorption performance.

675b Exploring the Distribution and Mobility of Nonframework Ions in Charged Metal-Organic Frameworks

Ravichandar Babarao[1], Anjaiah Nalaparaju[1], and Jianwen Jiang[2]. (1) National University of Singapore, Singapore, 117576, Singapore, (2) Chemical land Biological Engineering, National University of Singapore, Singapore, Singapore

With the emerging of metal-organic frameworks over the past few years, the quest for developing zeolite-like metal-organic frameworks (ZMOFs) has attracted much interest recently. Analogue to cation-exchanged zeolite structures, the presence of nonframework ions in metal-organic frameworks can increase the strength of interactions between the host and the guest molecules and in turn enhances the storage, separation and ion-exchange capability of the framework. A wealth of studies on zeolitic materials have showed that the static and dynamic properties of the nonframework ions are crucial to understanding the fundamental chemistry of these materials and also in tailoring their properties. In this work, using the atomistic simulations we explore the distribution and mobility of the nonframework ions in ZMOFs with both anionic and cationic frameworks. The nonframework ions are found to be essentially immobile and show an oscillating motion around their favorable sites. Hydrogen adsorption in the ZMOFs is also investigated.

675c Impact of Synthesis Conditions on the Nanoporosity of Carbon-Silica Composites

T. Grant Glover and M. Douglas LeVan. Department of Chemical and Biomolecular Engineering, Vanderbilt University, VU Station B 351604, 2301 Vanderbilt Place, Nashville, TN 37235-1604

Several carbon-silica composite materials have been synthesized from MCM-41 and furfuryl alcohol. The materials are completely nanoporous, have high surface areas, and pore size distributions centered near 12 Angstroms. In order to determine the sensitivity of the resulting carbon to synthesis conditions, materials were prepared using different furfuryl alcohol solvents and were polymerized and carbonized at either atmospheric pressure, 5 atm, or under vacuum. The surface area of the materials is affected by both solvent selection and the pressure at which carbonization takes place. The results are consistent with work done by others, which documents furfuryl alcohol decomposing, liberating gaseous products, and spontaneously generating nanoporosity when carbonized. The high surface area and narrow pore size distribution of these materials provides evidence that carbon-silica composites may be suitable candidates for novel separations.

675d Zinc-Based Metal-Organic Frameworks as High-Efficiency Adsorbents for Organic Vapor Separations

Mathew Luebbers, Chemical and Biomolecular Engineering, University of Illinois, 600 S. Mathews St., Urbana, IL 61801, Adarsh Radadia, Department of Chemistry, University of Illinois at Urbana-Champaign, 600 South Mathews, 294 Roger Adams Laboratory, 42-6 CLSL, MC 712, Urbana, IL 61820, Tianjiao Wu, Department of Chemistry, University of Illinois Urbana-Champaign, 600 S. Mathews St., Urbana, IL 61801, and Richard Masel, Chemical and Biomolecular Engineering, University of Illinois at Urbana-Champaign, 294 RAL, MC-712, 600 S. Mathews Ave., Urbana, IL 61801.

For industrial applications as well as microdevices such as gas analyzers, there is a need for adsorbents with improved separation efficiencies. The work reported here regards the usage of Zn-based metal-organic framework (MOF) materials for the separation of hydrocarbon mixtures. An example of such a separation with these materials is the chromatographic separation of hexane isomers. The highly ordered nature of MOF materials as well as the ability to tailor the framework's chemical functionality by modifying the organic linker ligands gives the materials great opportunities as high efficiency adsorbents. X-ray diffraction and gas sorption studies of the MOF materials such as isotherms, breakthrough studies, and thermogravimetric analysis will be used to study the material properties and explore the feasibility of separations.

675e Adsorption of Gases on Cu-Btc and Cr-Bdc Frameworks

Pradip Chowdhury[1], Chaitanya Bikkina[1], Frieder Dreisbach[2], Dirk Meister[2], and Sasidhar Gumma[1]. (1) Chemical Engineering, Indian Institute of Technology Guwahati, Guwahati, 781039, India, (2) Rubotherm GmbH, Universitätsstr. 142, Bochum, 44799, Germany

Adsorption characteristics of metal organic frameworks are strongly affected by their physical properties like pore volume and surface area. This creates discrepancy when one tries to compare the isotherms on adsorbent material obtained from different laboratories and/or synthesis procedure. The discrepancy makes it difficult to use the existing experimental data for applications like adsorptive separations or energy storage. The objective of this work is to relate physical properties of the adsorbent to its adsorption characteristics, for several gases.

Cu-BTC was synthesized using two different procedures, to obtain materials with widely different surface areas (1482 and 857 m^2/g). A completely different frame work viz. Cr-BDC with a surface area of about 2900 m^2/g was also synthesized. Isotherms for seven permanent gases viz. N_2, O_2, Ar, CH_4, CO_2, C_3H_8 and SF_6 were measured on these three materials at two different temperatures and for pressures up to 8 bar. The selectivity between N_2, O_2 and Ar was close to unity in all cases. Henry constants increase in the order $C_3H_8 > SF_6 > CO_2 > N_2$. C_3H_8 and SF_6 seem to approach their monolayer coverage values. While CO_2 had the highest capacity on Cu-BTC, capacity for C_3H_8 was highest on Cr-BDC framework.

The isotherms on the two Cu-BTC samples can be readily related to one another through use of a scaling factor. This scaling factor is not unique but is a function of loading; however, it is independent of nature of the gas, temperature or pressure.

Ordered Mesoporous Carbon as an Efficient and Reversible Adsorbent for the Adsorption of Fullerenes

Huan Wang, Frank L.Y. Lam, Xijun Hu and Ka Ming Ng*

Department of Chemical Engineering, Hong Kong University of Science and Technology

Clear Water Bay, Hong Kong

ABSTRACT. An ordered mesoporous carbon, CMK-3, was synthesized using mesoporous siliceous material, SBA-15, as the template. The CMK-3 was characterized and used for the adsorption of fullerenes, C_{60} and C_{70}. It was found that the adsorption capacity of CMK-3 is four times higher than that of activated carbon (AC), while the other two adsorbents being investigated, alumina and silica gel, have little adsorption capacity for fullerenes. A detailed study was undertaken to determine the adsorption equilibrium isotherms of C_{60} and C_{70} on CMK-3 for both single and binary systems. The reversibility of fullerene adsorption on CMK-3 was also studied. It was found that nearly 24 % adsorbed C_{60} and 8 % adsorbed C_{70} can be desorbed from CMK-3 within 10 minutes at 25 □ when a volume of 200 ml pure toluene is used to desorb 25 mg of pre-saturated adsorbent in a batch desorption process. The effects of time, volume of toluene, and temperature on the desorption of fullerenes from CMK-3 were examined. It was shown that the volume of toluene and the temperature are two important factors for fullerenes, especially for C_{70}, to be desorbed from CMK-3. Increasing the volume and the temperature will help C_{70} to be desorbed from CMK-3. The above results indicate that CMK-3 is an efficient and reversible adsorbent for the separation of fullerenes by adsorption.

KEYWORDS: mesoporous carbon; fullerenes C_{60} and C_{70}; adsorption; desorption.

- **Introduction**

The discovery of fullerene series, C_{60} and C_{70}, was first reported by Kroto, Curl and Samlley in 1985.[1] Fullerenes are described as the third form of ordered carbon having a structure very different from diamond and graphite. Over the last decade many scientists have been intrigued by the exploitation of the physical and chemical properties of these novel molecules.[2] It is found that fullerenes has installed in scientists a variety of application hopes, such as superconductors, large-scale industrial catalysts, and perhaps even antiviral agents.[3] Because the preparation of pure fullerene compounds on a large scale is very difficult, the number of their successful applications is still limited.[4] This academic interest and enthusiasm in fullerene science has prompted us to develop a safe, fast, effective, and inexpensive method for the separation and purification of fullerenes, which can produce large quantities of pure C_{60} and C_{70}, the most popular members of the fullerene family.[5]

The main methods for the separation and purification of C_{60} and C_{70} are selective complexation,[6-10] fractional crystallization,[11-14] and liquid chromatography.[15-19] Fractional crystallization is a very useful method to separate C_{60} and C_{70}, but it requires stirring for a long time at elevated temperature, and repeating the precipitation-filtration sequence a few times.[20] Liquid chromatography seems to be so far the most convenient method for the laboratory-scale separation of C_{60} and C_{70}, but it also has fundamental limitations in its application.[4] The most serious problem limits the application of liquid chromatography is the unavailability of an effective and reversible adsorbent for fullerene adsorption. Many researchers have carried out studies on the adsorption of fullerenes on different adsorbents. Taylor[16] firstly used neutral alumina as the stationary phase to separate C_{60} and C_{70} and found that large amounts of expensive alumina and intensive labor are needed. Then, activated carbon, for its huge specific surface area and high adsorption capacity, becomes the most popular adsorbent for fullerene separation during the following years.

The high capacity of activated carbon to adsorb C_{60} and C_{70} allows the quantity of the stationary phase to be reduced.[19] However, the drawback of this method is the microporous structure of activated carbon (AC), leading to irreversible adsorption of some C_{60} and nearly all of the C_{70}. The discovery of ordered mesoporous materials[21-25] has opened new possibilities in the field of adsorption and catalysis. The template synthesis of ordered mesoporous materials has received widespread attention recently since this technique allows the preparation of materials with a controllable architecture.[26] In 2000, in order to solve the irreversibility problem in fullerene adsorption, Piwonski et al.[27] first reported the adsorption of pure C_{60} from toluene solution on ordered mesoporous silica material, MCM-41. Although the amount of C_{60} adsorbed on MCM-41 was much smaller than that obtained with AC,[4] the whole process appeared to be completely reversible. This quality gave MCM-41 certain advantages over other stationary phase materials. Thus, if a new adsorbent can be found, which not only have a good adsorption ability for fullerene, but also is reversible for this process, the overall cost of fullerene production can be significantly reduced.[4] Ordered mesoporous carbon, which has a well-defined mesoporous pore structure, a large surface area, and a high adsorption capacity, is an ideal adsorbent to solve the problems in fullerene adsorption.

In this study, a new type of adsorbent, ordered mesoporous carbon (CMK-3) is synthesized in our group and firstly used for the adsorption of fullerenes for both single and binary systems.

• Materials and Methods

Synthesis of CMK-3. Siliceous material SBA-15 was synthesized by the sol-gel method. The tri-block copolymer (Pluronic 123, pure, Aldrich) and the organic silica material, tetraethylorthosilicate (TEOS, 98 %, Aldrich), were employed as sources of template and silica, respectively. In a typical synthesis, 4 g of Pluronic 123 was first dissolved in 150 g of 1.6 M HCl to form a template solution. It was then stirred vigorously until a milky solution was observed, followed by adding 8.5 g of TEOS into the milky solution at 35 °C. The resulting solution was aged at 35 °C for 24 h and subsequently at 110 °C for another 24 h. Afterwards, the SBA-15 was obtained by calcination at 550 °C for 9 h in air.

The template-free SBA-15 material was then used as a mold for the synthesis of mesoporous carbon. Sucrose and sulfuric acid were used as the carbon source and the dehydrating agent, respectively. Sucrose solution is obtained by dissolving 1.25 g of solid sucrose into 5 ml of water. Then, 0.14 g of sulfuric acid was added into the sucrose solution under stirring. Finally, 1 g of the template-free SBA-15 was impregnated in the acid-containing sucrose solution and the mixture was put into an oven at 100 °C for 6 h and subsequently at 160 °C for another 6 h. The impregnation step was repeated once again but with different amounts of the substances (0.8 g sucrose, 0.09 g H_2SO_4, and 5 ml H_2O). After the impregnation step, the dark brown solid products, which were the composites of carbon and SBA-15, was carbonized at 1050 °C under vacuum for 6 h. The generated black powders (composites) were leached with a mixture of ethanol and 1 M sodium hydroxide solution to remove the silica, and further recovered by filtration, washed with distilled water and dried at 60 °C overnight to produce the ordered mesoporous carbon (CMK-3).

Fullerenes. C_{60} (99.5 %) and mixture of fullerenes (approx. 79 % C_{60}, 13 % C_{70}) were supplied by Frontier Carbon Corp., Japan. C_{70} (98 % +) was obtained from Nano-C, USA.

Characterization. Low angle X-ray diffraction patterns of samples were recorded with a powder x-ray diffraction system (Model X'pert, Panalytical). The step width was 0.02°at an acquisition time of 8 s per step in the range of $2\theta = 0.85° – 5.0°$. TEM image was obtained using a JEOL JEM-2010F transmission electron microscope. The TEM image was taken from the thin edges of carbon particles mounted on a porous carbon grid.

Nitrogen adsorption isotherms were measured with an ASAP2010 adsorption analyzer (Micromeritics) at liquid nitrogen temperature. Prior to the measurements, all the samples were degassed at a temperature of 200 °C for 6 h. The adsorption data in the relative pressure range from 0.04 to 0.2 were used to determine the BET surface area. The pore size distribution (PSD) was calculated by the BJH method from the desorption branch.

Measurement of Concentrations of C_{60} and C_{70}. High Performance Liquid Chromatography (HPLC) was used to determine the concentrations of C_{60} and C_{70} in the solution. Different known concentrations of pure C_{60} and pure C_{70} solutions in toluene were prepared to calibrate the relationship between the peak area and the fullerene concentration, which was found to be linear. The HPLC analysis was performed on a Hewlett Packard I090 liquid chromatography equipped with a YMC-Pack ODS-A column (4.6×150 mm). The detection was performed with UV-visible light at 330 nm. Toluene/2-propanol (1:1 v/v) was used as the mobile phase and the flow rate was set to 1 ml/ min.

Determination of Fullerene Adsorption Equilibrium Isotherm on Different Adsorbents. The time required for the adsorption of fullerene to reach equilibrium on different adsorbents was studied in a batch system. Because the adsorption equilibrium time is much longer for a binary system than a single component system, a mixture of C_{60} and C_{70} was used to determine the required equilibrium time. Once this equilibrium time is obtained, it is used to measure the adsorption equilibrium isotherms of pure C_{60} and pure C_{70} as well as their mixture. The adsorption process of mixed C_{60} and C_{70} from its solution in toluene (initial concentration of fullerene was 100 mg/L, weight ratio of C_{60}:C_{70} = 79:13) by different adsorbents at 25 °C was studied in a batch system. The four adsorbents were alumina, activated carbon mixed with silica gel (weight ratio is 1:1), activated carbon, and ordered mesoporous carbon (CMK-3). The alumina was Aluminium Oxide 90 Active Neutral for column chromatography, 0.063-0.2 mm (70-230 mesh ASTM) supplied by Merck. Activated carbon was Norit Row 0.8 Supra supplied as cylindrical extrudate but cruched into around 0.1 mm particles before usage, and silica gel was Silica gel 60, (0.04~0.063 mm) for column chromatography, (230-400 mesh ASTM), also supplied by Merck. The weight of each adsorbent used was 25 mg. The adsorption solution volume was 20 ml. During the adsorption process, samples were drawn out (0.3 ml per sample) at different time intervals and the fullerene concentrations in the samples are analyzed by the HPLC.

Adsorption isotherms of Binary C_{60} and C_{70} on Different Adsorbents. Once the time required to reach equilibrium was determined (24 hours), it was used to measure the adsorption equilibrium isotherms of fullerenes on CMK-3 and AC at 25 ℃. A volume of 20 ml of different initial concentrations of fullerene in the toluene solutions was mixed with 25 mg adsorbent and stirred for 24 hours for the equilibrium to be established.

Adsorption of pure C_{60} and pure C_{70} on CMK-3. The adsorption equilibrium isotherms of pure C_{60} and pure C_{70} on CMK-3 at 25 ℃ were measured by changing the initial concentrations of pure C_{60} or pure C_{70} solutions. The weight of CMK-3 was 25 mg. The adsorption volume was 20 ml. After 24 h, the adsorption equilibrium was reached in the system.

Desorption of pure C_{60} and pure C_{70} from CMK-3 and AC.

25 mg of CMK-3 or AC was placed in 200 ml solution of 100 mg/L pure C_{60} or C_{70} at 25 ℃ for 24 h for the equilibrium to be established. Then the solution was filtered and the solid was dried under vacuum at 60 ℃ for 10 h to remove toluene. The recovered solid was then placed in 200 ml fresh toluene to start the desorption of pure C_{60} or pure C_{70}.

Effect of solvent volume on the desorption of pure C_{60} and pure C_{70} from CMK-3

25 mg of dried CMK-3 pre-saturated with pure C_{60} or C_{70} was placed in different toluene volumes of 50 ml, 150 ml, 250 ml, 450 ml, 650ml, 1000ml for the desorption to proceed. The desorption time for each desorption volume was 12 h.

Effect of temperature on the desorption of pure C_{60} and pure C_{70} from CMK-3

25 mg of dried CMK-3 pre-saturated with pure C_{60} or C_{70} was placed in 250 ml fresh toluene. Again the desorption time was 12 h. The desorption was done for different temperatures. The recovered CMK-3 was placed into another 250 ml fresh toluene solution to start the second desorption cycle. This was repeated for 4 desorption cycles.

• Results and Discussion

The structures of SBA-15 template and the corresponding CMK-3 were studied using low angle XRD. Their XRD patterns are shown in Figure. 1. The well resolved XRD peaks of SBA-15 which can be assigned to (100), (110), and (200) diffractions, indicate that such a material has an ordered structure with 2-D hexagonal symmetry. Therefore, CMK-3 should have a hexagonal structure, which corresponds to the negative replication of silica template. The XRD pattern of CMK-3, characteristic of the hexagonal structure, was similar to that of SBA-15 as shown in Figure. 1. The d-spacing values of the peak (100) of SBA-15 and CMK-3 are 9.7 nm and 8.3 nm, respectively.

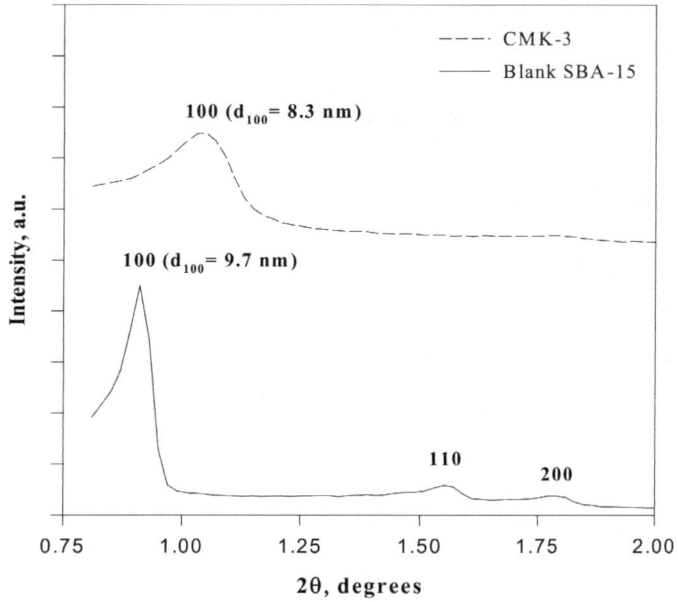

Figure 1. XRD spectrums of SBA-15 and CMK-3

The transmission electron microscopy (TEM) image of the CMK-3 mesoporous carbon was shown in Figure 2, which illustrates the hexagonal arrays of carbon rods, 6.5 nm diameter, similar to the structure of CMK-3 reported in the literature.

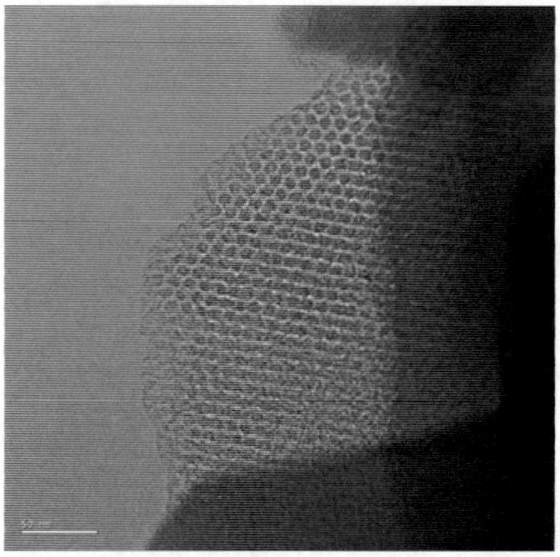

Figure 2. TEM image of CMK-3

The porosity of the SBA-15 and CMK-3 were investigated by nitrogen sorption measurements. The isotherms are presented in Figure. 3. The corresponding pore structure parameters, including BET surface area, mean pore diameter and particle size, are listed in Table 1 together with those of Norit activated carbon for comparison. The nitrogen sorption isotherm (Fig. 3) is of essentially type IV with a small hysteresis loop, indicating that the pore structure of CMK-3 is reasonably regular. Compared with SBA-15, CMK-3 has a smaller pore size and a higher surface area.

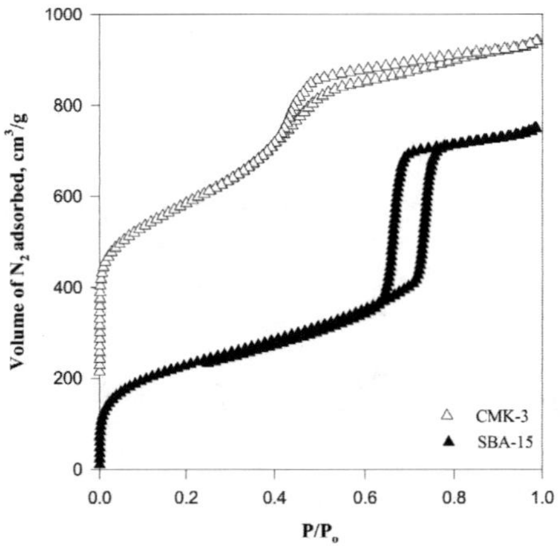

Figure 3. Nitrogen sorption isotherms of SBA-15 and CMK-3

Table 1. Texture parameters of SBA-15, the corresponding CMK-3 and AC

Sample	S_{BET} (m^2 g^{-1})	V_t (cm^3 g^{-1})	V_{im} (cm^3 g^{-1})	Pore Size (nm)
SBA-15	815	1.02	0.06	6.5
Ordered Mesoporous Carbon (CMK-3)	1366	1.14	0.12	3.5
Norit Row 0.8 Supra (AC)	911	1.18	0.58	1.16

V_t, total pore volume calculated at p/p_0= 0.99. V_{im}, microporous volumes (diameter < 2.0 nm) calculated by (Haba and Kleinert) HK method.

The pore size distributions of SBA-15 and CMK-3 are shown in Figure. 4, which possess mode values of. 6.5 nm and 3.5 nm, respectively. For comparison, the pore size distribution of Norit activated carbon is also plotted in Figure 4, which has a peak at around 1.1 nm, much smaller than those of SBA-15 and CMK-3. The specific pore size of the CMK-3 depends on both the wall thickness of SBA-15 and also the structural shrinkage effect during the carbonization process. It is noted that such a pore size (3.5 nm) is large enough for the adsorption of large organic molecules such as benzene, toluene, dyes, and especially in this study, fullerenes.

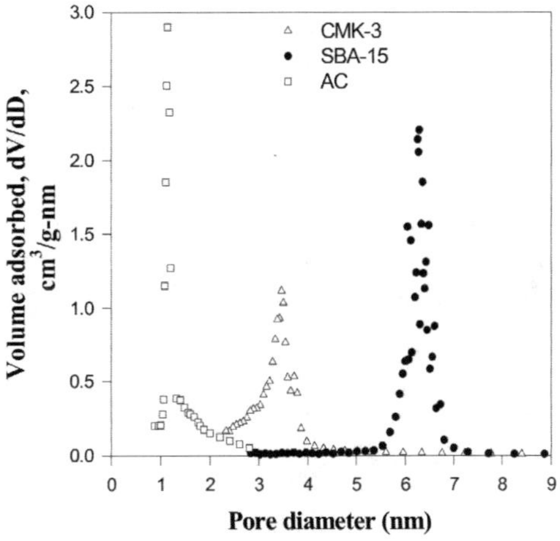

Figure 4. Pore size distributions of SBA-15, CMK-3 and AC

Fullerene Adsorption Equilibrium Time on Different Adsorbents. Figure 5. shows the binary adsorption process of C_{60} and C_{70} at 25 □ on four different adsorbents versus time. The four adsorbents were silica gel, neutral alumina, Norit activated carbon Row 0.8 Supra (AC) and CMK-3. It is seen that the adsorbed phase concentration becomes stable after 10 h for all CMK-3, AC and silica gel. Alumina has nearly no adsorption capacity for fullerenes and the process is very slow. Because the adsorption process is quicker for a single component system than a binary system, 24 h was considered long enough to reach adsorption equilibrium and was chosen as the equilibrium time in the following experiments for both single and binary systems.

It is also seen that alumina and silica gel have almost no adsorption capacity for both C_{60} and C_{70}. The two carbon based adsorbents, AC and CMK-3, showed significant adsorption capacity for the fullerenes. The difference in the adsorption capacity of different adsorbents may be explained by the interaction between fullerenes and the adsorbents. CMK-3 and AC are carbon based materials. Fullerenes are also made up of carbon atoms. Therefore, the adsorption of fullerenes on AC and CMK-3 is caused by the π-π interaction between the fused aromatic rings in the pores of carbons and the surface of fullerenes. This interaction is much stronger than the Van der Waals forces between the pores of alumina/silica gel and the surface of fullerenes.

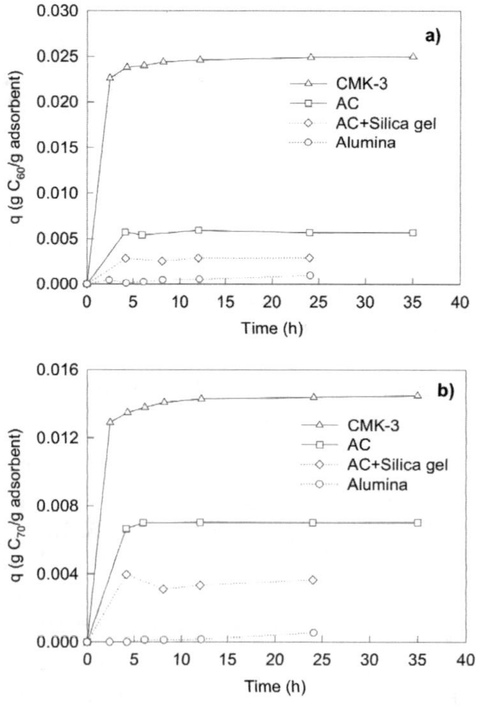

Figure 5. Binary adsorption of C_{60} and C_{70} versus time on different adsorbents: (a) C_{60}; (b) C_{70}.

Binary Adsorption Isotherms of C_{60} and C_{70} on Different Adsorbents. The binary adsorption equilibrium isotherms of fullerenes on CMK-3 and AC are shown in Figures 6. It is seen that the adsorption capacity of fullerenes on CMK-3 is nearly four times higher than that on activated carbon. This indicates that ordered mesoporous carbon is a more effective adsorbent than activated carbon for the adsorption of fullerenes. This can be explained by the pore size distribution and pore structure of these two adsorbents. Because the size of C_{60} and C_{70} molecules is around 1.0 nm to 2.0 nm, CMK-3, having a well-defined mesoporous structure and a pore size distribution centered at 3.5 nm, can accommodate the molecules of fullerenes efficiently. However, activated carbon is a microporous material and its pore size distribution is peaked at 1.1 nm, as shown in Figure 4. This pore size is too small to adsorb fullerenes.

Since CMK-3 is the best adsorbent for binary C_{60} and C_{70} adsorption, further study on the adsorption of fullerenes on CMK-3 will be carried out as below.

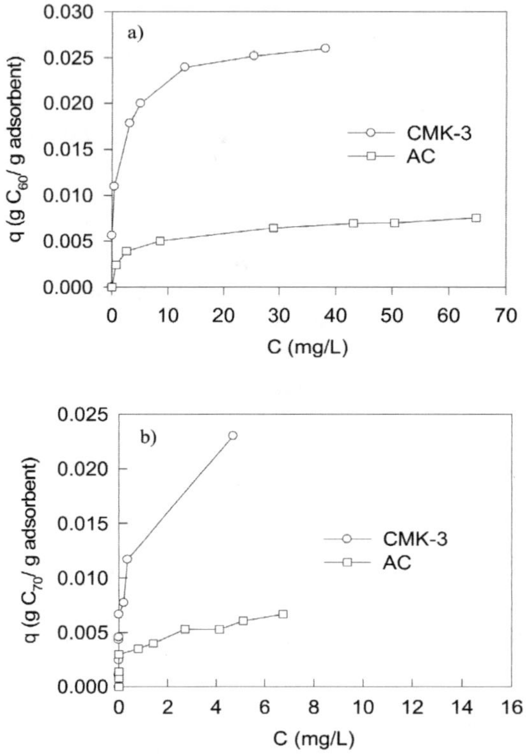

Figure 6. Binary adsorption isotherms of fullerenes on CMK-3 and AC at 25□. (a) C_{60}; (b) C_{70}.

Adsorption Equilibrium Isotherms of Pure C_{60} and Pure C_{70} on CMK-3. The adsorption equilibrium isotherms of pure C_{60} and pure C_{70} on CMK-3 from its toluene solutions at 25 □ are shown in Figure 7. It can be seen that C_{70} is adsorbed more strongly than C_{60} on CMK-3 at the same liquid phase concentration. The interaction between C_{70} and the pores of CMK-3 is stronger than that between C_{60} and the pores of CMK-3. This is due to the shape of the fullerene molecules. C_{60} molecules are spherical whereas C_{70} molecules are elliptical. Hence, the interaction area between C_{70} and CMK-3 is relatively larger due to the C_{70} molecular shape being flatter. It is easy for C_{70} to be adsorbed on CMK-3 and difficult to be desorbed. The difficulty of desorption of C_{70} from the adsorbent has been a problem for a long time which remains unsolved.

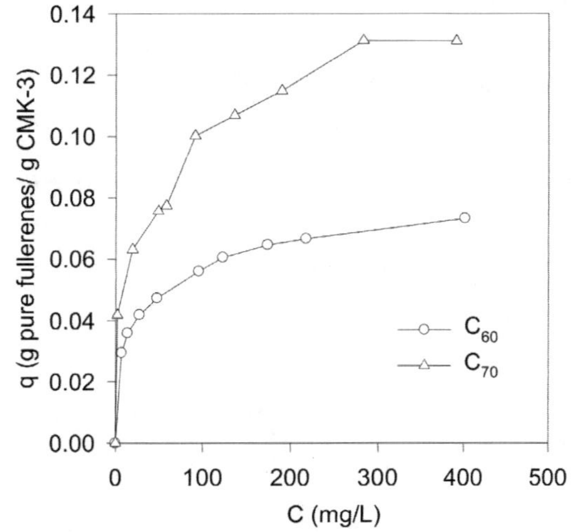

Figure 7. Adsorption equilibrium isotherms of pure fullerenes on CMK-3 at 25□

Adsorption Equilibrium Isotherms of Binary C_{60} and C_{70} on CMK-3. The equilibrium isotherms of binary C_{60} and C_{70} adsorption on CMK-3 at 25 °C are shown in Figure 8. The adsorption equilibria of C_{60} and C_{70} mixture were also measured and the results showed that the temperature effect is not significant for the adsorption of fullerenes on CMK-3.

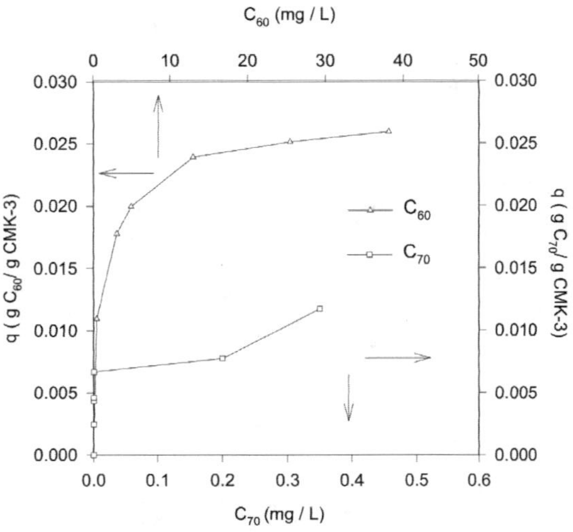

Figure 8. Adsorption equilibrium isotherms of binary C_{60} and C_{70} on CMK-3 at 25□.

Desorption of Pure C_{60} and Pure C_{70} from CMK-3 and AC.

The desorption process of pure C_{60} and pure C_{70} from CMK-3 and AC at 25 □ is shown in Figure 9. For both cases, 25 mg adsorbents of pre-adsorbed pure C_{60} or pure C_{70} at 25 □ were added to 200 ml pure toluene. It is seen that for both adsorbents the desorption process can quickly reach the new equilibrium in five minutes. Nearly 24% of adsorbed pure C_{60} and 8% of adsorbed pure C_{70} can be desorbed from CMK-3. For comparison, only 9.5% of adsorbed pure C_{60} and 6% of adsorbed pure C_{70} can be desorbed from activated carbon. Thus, the ordered mesoporous carbon, CMK-3, not only has a higher adsorption capacity, but also is a more reversible adsorbent than activated carbon.

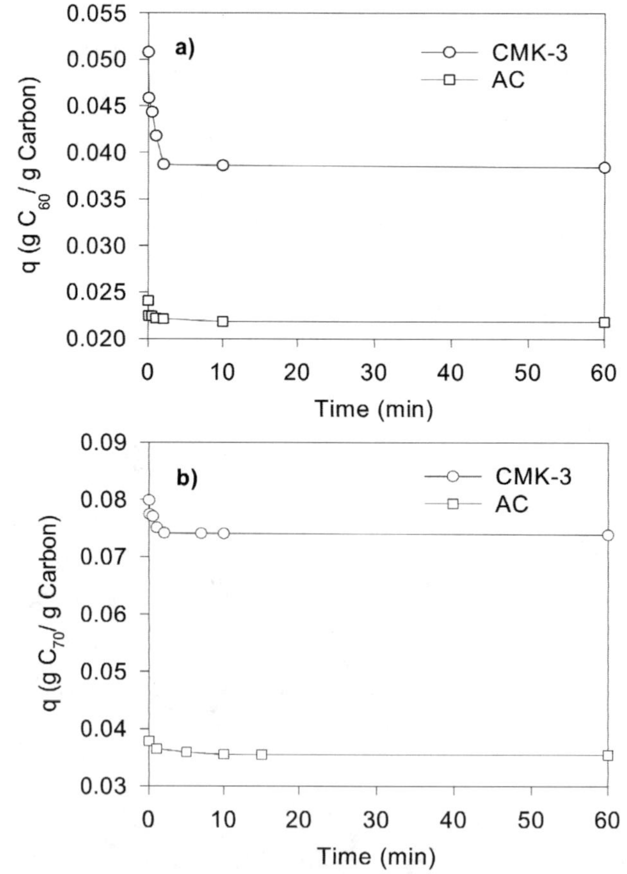

Figure 9. Desorption evolution of pure fullerene from CMK-3 and AC at 25°C: (a) pure C_{60}; (b) pure C_{70}

Effect of solvent volume on the desorption of pure C_{60} and pure C_{70} from CMK-3

The effect of solvent volume on the desorption from 25 mg CMK-3 of pre-adsorbed pure C_{60} or pure C_{70} at 25 □ is shown in Figure 10. As expected, the amount of fullerenes desorbed increases with more solvent volume used. About 25 % pure C_{60} and 13 % pure C_{70} can be desorbed when a solvent volume of 1000 ml is used to desorb 25 mg of pre-saturated CMK-3 at 25 □. When only 200 ml toluene is used, only 8 % pure C_{70} can be desorbed. The desorption of C_{70} can be significantly improved by using more solvent. The desorption of C_{60} can also be improved by using more solvent but not as significantly as C_{70}.

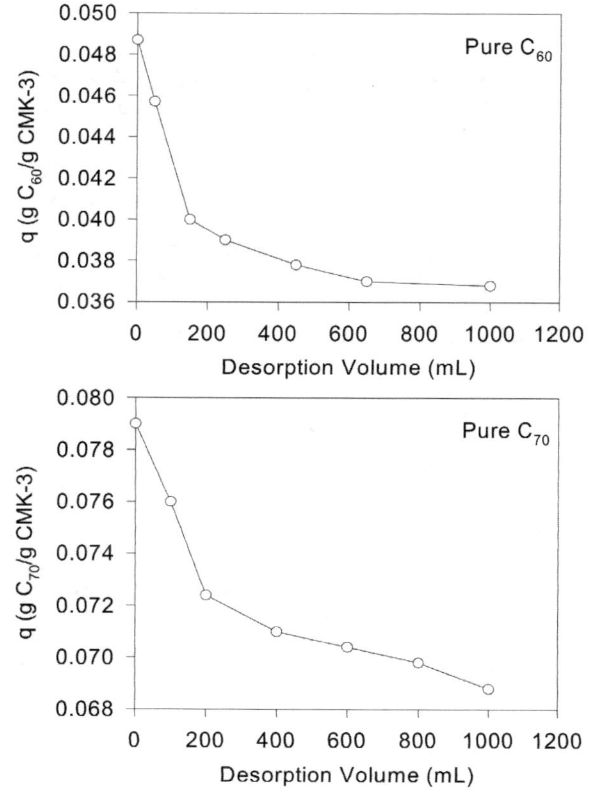

Figure 10. Effect of solvent volume on the desorption of pure fullerene from CMK-3 at 25°C.

Effect of temperature on the desorption of pure C_{60} and pure C_{70} from CMK-3

Because adsorption is an exothermic process, increasing the temperature will be helpful for the desorption. The effect of temperature on the desorption from 25 mg CMK-3 pre-adsorbed pure C_{60} or pure C_{70} at 25 □ is shown in Figure 11. The solvent volume used is 250 ml. It is seen that the amount of pure C_{60} desorbed from CMK-3 changes very little by increasing the temperature from 25 □ to 55 □. However, the amount of pure C_{70} desorbed from CMK-3 increases with the temperature increasing. At the highest temperature of 55 □, the amount of pure C_{70} desorbed from CMK-3 is the highest. Therefore, temperature is another important factor for C_{70} desorption. High temperature will help to improve the reversibility of CMK-3 for C_{70}. Figure 11 also shows the effect of desorption cycles on the desorbed quantity of pure fullerenes. In the subsequent desorption cycles, the desorption is repeated by drying the desorbed adsorbent from the previous cycle and then placing it into another 250 ml fresh toluene. Nearly 47 % pure C_{60} can be desorbed from CMK-3 after four cycles of desorption process. About 16 % pure C_{70} can be desorbed from CMK-3 after three cycles of desorption process.

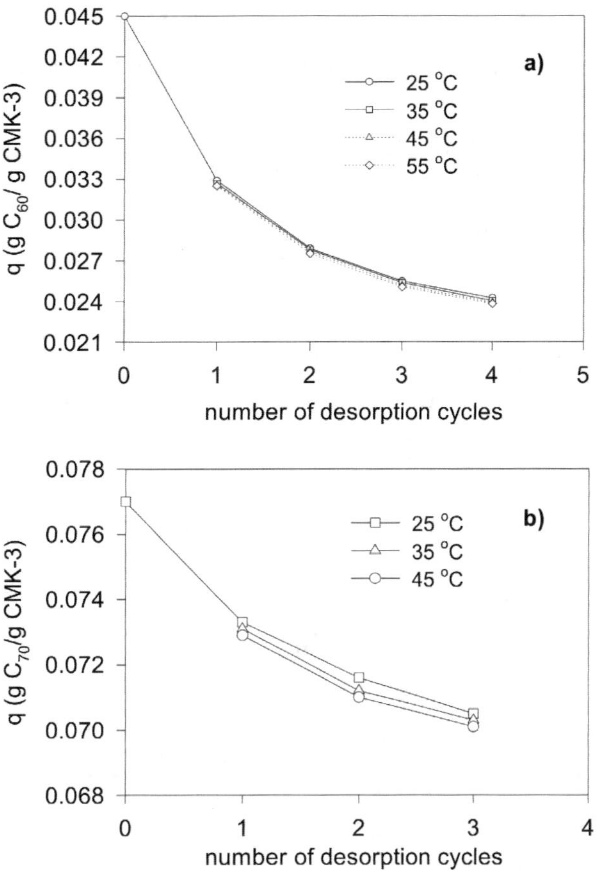

Figure 11. Effect of temperature on the desorption of pure fullerene from CMK-3:
(a) pure C_{60}; (b) pure C_{70}

• Conclusions

The adsorption of fullerenes, C_{60} and C_{70}, on four adsorbents including alumina, silica gel, activated carbon (AC), and ordered mesoporous carbon (CMK-3) was investigated using a batch system. Alumina and silica gel have little adsorption capacity for fullerenes. Both CMK-3 and activated carbon can adsorb fullerenes but the adsorption capacity of CMK-3 is four times higher than that of AC. The detailed study shows that C_{70} is more strongly adsorbed on CMK-3 than C_{60}. The effect of temperature on the adsorption equilibrium of fullerenes on CMK-3 is marginal.

The reversibility of fullerene adsorption on CMK-3 was also studied. Nearly 24 % adsorbed pure C_{60} and 8 % adsorbed pure C_{70} can be desorbed from 25 mg pre-saturated CMK-3 within 10 minutes at 25 □ using 200 ml pure toluene. More fullerenes can be desorbed by using more solvent. For a toluene volume of 1000 ml, about 25 % pure C_{60} and 13 % pure C_{70} can be desorbed at 25 □. The desorption can also be improved by using more cycles of desorption process. Nearly 47 % pure C_{60} can be desorbed after four cycles of desorption and around 16 % pure C_{70} can be desorbed after three cycles of desorption. The amount of pure C_{70} desorbed increases with increasing temperature.

- **Acknowledgment**

Financial support from the innovation and technology commission (ITS/069/02) is gratefully acknowledged.

• References

(1) Kroto, H. W.; Heath, J. R.; OBrien, S. C.; Curl, R. F.; Smalley, R. E. *Nature* **1985**, *318,* 162-163.

(2) Crowley, C.; Taylor, R.; Kroto, H. W.; Walton, R. M.; Cheng, P. C.; Scott, T. *Synthetic Metal.* **1996**, *77*, 17-22.

(3) Baum, R. *Chem. Eng. News*. **1993**, *71*, 3-5.

(4) Zajac, J.; Groszek, A. J. *Carbon* **1997**, *35*, 1053-1060.

(5) Steven, P. West.; Thomas, Poon.; Jamey, L. Anderson.; Michele, A.; Christopher, S. Foote. *J. Chem. Educ.* **1997**, *74*, 311-313.

(6) Komatsu, N. *Org. Biomol. Chem.* **2003**, *1*, 204-209.

(7) Komatsu, N. *Tetrahedron. Lett.* **2001**, *42*, 1733-1736.

(8) Atwood, J. L.; Koutsantonis, G.A.; Raston, C. L. *Nature* **1994**, *368*, 229-231.

(9) Ryuichi, S.; Kazuyoshi, T.; Yukiko, K.; Shigeo, I. *Chem. Lett.* **1994**, *4*, 699-704.

(10) Bucsi, I.; Aniszfeld, R.; Shamma, T.; Prakash, G. K. S.; Olah, G. A. *Proc. Natl. Acad. Sci. USA*. **1994**, *91*, 9019-9021.

(11) Doome, R. J.; Fonseca, A.; Richter, H.; Nagy, J. B.; Thiry, P. A.; Lucas, A. A. *J. Phys. Chem. Solids.* **1997**, *58*, 1839-1843.

(12) Zhou, X.; Gu, Z.; Wu, Y.; Sun, Y.; Jin, Z.; Xiong, Y. *Carbon*. **1994**, *32*, 935-937.

(13) Wu, Y.; Sun, Y.; Gu, Z.; Zhou, X.; Xiong, Y.; Sun, B. *Carbon* **1994**, *32*, 1180-1182.

(14) Coustel, N.; Bernier, P.; Aznar, R.; Zahab, A.; Lambert, J. M. *J. Chem. Soc., Chem. Commun.* **1992**, *19*, 1402-1403.

(15) Ajie, H.; Alvarez, M. M.; Anz, S. J.; Beck, R. D.; Diederich, F.; Fostiropoulos, K.; Huffman, D. R.; Kratschmer, W.; Rubin, Y.; Schriver, K. E.; Sensharma, D.; Whetten, R. L. *J. Phys. Chem.* **1990**, *94*, 8630-8633.

(16) Taylor, R.; Hare, J. P.; Abdul-Sada, A. K.; Kroto, H. W. *J. Chem. Soc., Chem. Commun.* **1992**, *19*, 1423-1425.

(17) Vassalo, A. M.; Palmisano, A. J.; Pang, L. S. K.; Wilson, M. A. *J. Chem. Soc., Chem. Commun.* **1992**, *1*, 60-61.

(18) Scrivens, W. A.; Bedworth, P. V.; Tour, J. M. *J. Am. Chem. Soc.* **1992**, *114*, 7917-7919.

(19) Manolova, N.; Rashkov, I.; Legras, D.; Delpeux, S.; Beguin, F. *Carbon* **1995**, *33*, 209-213.

(20) Naoki, Komatsu.; Toshiyuki, Ohe.; Kazumi, Matsushige. *Carbon* **2004**, *42*, 163-167.

(21) Beck, J. S.; Vartuli, J. C.; Roth, W. J.; Leonowicz, M. E.; Kresge, C. T.; Schmitt, K. D.; Chu, C. T. W.; Olson, D. H.; Sheppard, E. W.; McCullen, S. B.; Higgins, J. B.; Schlenker, J. C. *J. Am. Chem. Soc.* **1992**, *114*, 10834-10843.

(22) Monnier, A.; Schu¨ th, F.; Huo, Q.; Kumar, D.; Margolese, D.; Maxwell, R. S.; Stucky, G. D.; Krishnamurty, M.; Petroff, P.; Firouzi, A.; Janicke, M.; Chmelka, B. F. *Science* **1993**, *261*, 1299-1303.

(23) Inagaki, S.; Fukushima, Y.; Kuroda, K. *J. Chem. Soc., Chem. Commun.* **1993**, *8*, 680-682.

(24) Tanev, P. T.; Pinnavaia, T. J. *Chem. Mater.* **1996**, *8*, 2068-2079.

(25) Jones, D. J.; Jime´nez-Jime´nez, J.; Jime´nez-Lo´pez, A.; Maireles-Torres, P.; Olivera-Pastor, P.; Rodrý´guez-Castello´n, E.; Rozie`re. *Chem. Commun.* **1997**, *5*, 431-432. Jones, D. J.; Aptel, G.; Brandhorst, M.; Jacquin, M.; Jime´nez-Jime´nez, J.; Jime´nez-Lo´pez, A.; Maireles-Torres, P.; Piwonski, I.; Rodrý´guez-Castello´n, E.; Rozie`re, J.; Zajac, J. *J. Mater. Chem.* **2000**, *10*, 1957-1963.

(26) Branton, P. J.; Hall, P. G.; Sing, K. S. W.; Reichert, H.; Schuth, F.; Unger, K. K. *J. Chem. Soc., Faraday Trans.* **1994**, *90*, 2965-2969.

(27) Piwonski, I.; Zajac, J.; Jones, D. J.; Roziere, J.; Partyka, S.; Plaza, S. *Langmuir.* **2000**, *16*, 9488-9492.

675g Metal-Organic Aerogels-a New Kind of Porous Material

Qiang Wei[1], Dali Yang[2], Zhijun Lin[1], Hongwu Xu[3], and Yusheng Zhao[4]. (1) LANSCE-LC, Los Alamos National Laboratory, los Alamos, NM 87545, (2) Mst - 7, Los Alamos National Laboratory, Los Alamos, NM 87545, (3) Ees-6, Los Alamos National Laboratory, Los Alamos, 87545, (4) Lansce-12, Los Alamos National Laboratory, Los Alamos, NM 87545

New highly porous aerogel materials based on the coordination of metal ions and organic ligands are developed and characterized by PXRD, BET and TEM analysis. The new metal-organic aerogels (MOAs) have typical specific BET surface area above 1000m2/g with the thermal stability up to 250oC, which is generally larger than the normal silica based or carbon aerogels. The intrinsic chemical compositions of meal-organic aerogel provide itself the advantages of easy synthesis, feasible modification and functionalization on the properties and structures. One example of using such aerogel as reaction matrix to synthesis metal-organic frameworks, such as HKUST-1, is discussed. It generates size-homogenized crystals of HKUST-1 around 20ìm. This new kind of porous materials gives great potentials in the catalysis, adsorption/separation and sensors applications.

High Density, Vertically-Aligned Carbon Nanotube Membranes

Miao Yu, Hans H. Funke, John L. Falconer*, Richard D. Noble
Department of Chemical and Biological Engineering, University of Colorado
Boulder, CO 80309-0424

August 21, 2008

*Corresponding author: john.falconer@colorado.edu Ph: (303) 492-8005

Abstract

A method is presented to prepare high-density, vertically-aligned carbon nanotube (VA-CNT) membranes. The CNT arrays were prepared by chemical vapor deposition (CVD), and the arrays were collapsed into dense membranes by capillary-forces due to solvent evaporation. The average space between the CNTs after shrinkage was ~ 3 nm, which is comparable to the pore size of the CNTs. Thus, the interstitial pores between CNTs were not sealed, and gas permeated through both CNTs and interstitial pores. Nano-filtration of gold nanoparticles and N_2 adsorption indicated the pore diameters were approximately 3 nm. Gas permeances, based on total membrane area, were 1 to 4 orders of magnitude higher than VA-CNT membranes in the literature, and gas permeabilities were 4 to 7 orders of magnitude higher than literature values. Gas permeances were approximately 450 times those predicted for Knudsen diffusion, and ideal selectivities were similar to or higher than Knudsen selectivities. These membranes separated a larger molecule (triisopropyl orthoformate (TIPO)) from a smaller molecule (n-hexane) during pervaporation, possibly due to the preferential adsorption, which indicates separation potential for liquid mixtures.

Several studies have reported vertically-aligned, carbon nanotube (VA-CNT) membranes with high fluxes [1-3]. The nanotubes were sealed in a polymer or inorganic matrix, however, so that CNTs comprised less than 3% of the membrane permeation area; most of the membrane area was not available for permeation. Hinds *et al.* [1] prepared composite membranes with 6-7 nm (inner diameter) multi-walled CNTs by using polystyrene to fill the spaces between the CNTs. The flow velocities of molecules in their CNT membranes in pressure-driven permeation were 4-5 orders of magnitude higher than those calculated from fluid flow theory due to the smoothness of the CNT walls [4]. This is consistent with previous simulation results[5]. Holt *et al.* [2] grew vertically-aligned, dual-wall CNT membranes with much smaller pores (1.6 nm average) using a pre-deposited nano-catalyst (10-nm Al_2O_3, 0.3-nm Mo, 0.5-nm Fe) on a silicon wafer. They sealed the spaces between the CNTs with silicon nitride that was deposited by chemical vapor deposition (CVD). They reported gas fluxes that were one to two orders of magnitude higher than those predicted by the Knudsen model, and ideal gas selectivities were similar to Knudsen selectivities for non-hydrocarbon gases and slightly higher than Knudsen selectivities for hydrocarbons. They also reported water fluxes that were 3 orders of magnitude higher than those calculated by the Hagen-Poiseuille equation. Kim *et al.* [3] used a different approach to prepare CNT membranes. They filtrated aligned, single-walled CNTs (1.2-nm average pore size) with a PTFE filter, and then sealed the space between the CNTs with polysulfone polymer by spin coating. Their membranes had ideal gas selectivities similar to Knudsen selectivities. Their membranes also had a CO_2/CH_4 separation selectivity equal to 1.6, whereas the single component selectivity was 0.6.

These composite membranes had low CNT porosity, and thus the fractions of the permeation areas that were CNTs were 2.7%, 0.5% and 0.079% in these three studies [1-3]. Therefore, although the fluxes through individual nanotubes were high, the fluxes per cm^2 of membrane area were limited because of the low porosity. In addition, the fabrication processes of the composite CNT membranes were complicated and thus expensive. In contrast to the above studies, which used CNT pores as transport pathways, Srivastava *et al.* [6] made a CNT filter from vertically-aligned CNT forests without a filler. Because their nanotubes were mostly blocked by catalyst particles, transport was in the interstitial spaces, which were approximately 20-30 nm across. In the current study, free-standing VA-CNT membranes with high CNT densities (~ 20% CNT porosity) were fabricated by shrinking free-standing VA-CNT arrays that were grown by CVD [7]. Since no filler was used, gases also permeated through the interstitial spaces between the CNTs. The diameters of both the CNT and the interstitial pores were approximately 3 nm (see below).

The membrane fabrication process is illustrated Fig. 1a. Vertically-aligned CNT forests were grown by water-assisted, CVD synthesis at 1020 K, as reported by Hata *et al.* [8] (see supporting information). Carbon nanotubes grew approximately linearly with growth time, and reached approximately 2 mm in length in 120 min (see supporting information Fig. S-1). The arrays used to prepare membranes were grown for 40 min and were 750 μm thick. Water etching at 1020 K, as described by Ci *et al.* [9] (see supporting information), was used after CVD to detach the CNT arrays and obtain free-standing layers. Figure 1b shows a SEM image of a detached, CNT forest after water etching at 1025 K for 15 min The procedure reported by Futaba *et al.* [7] was then used to make densely-packed, CNT forests by collapsing the CNT forests with capillary forces. They found that evaporation of solvents from between the CNTs shrank the layer to appropriately 5% of its original size. In our study, the free-standing arrays were soaked in n-hexane, and the layers were dried at room temperature to collapse them. As indicated in Fig. 1c, the as-grown VA-CNT forest shrank to approximately 5% of its original area after n-hexane evaporation. The dense membranes were then mounted on silver-plated filters (5-μm pore size) that were embedded in nickel VCR gaskets (Swagelok), as shown in Fig. 1d. Gas and liquid permeations were measured by placing the membranes in standard VCR fittings.

The catalyst most frequently used for water-assisted CVD growth of VA-CNTs is 1~3 nm iron deposited on a 10~30 nm alumina layer [8,10-12]. The resulting pore sizes and the number of nanotube walls depend strongly on the iron thickness. For example, for 1-nm Fe/10-nm Al_2O_3 catalyst films, the CNT layers that formed have been reported to vary from 1-3 nm SWCNTs [8] to 5-nm DWCNTS [12]. A TEM of CNTs prepared in the current study (Fig. S-2) showed that that the inner pore diameters ranged from 2.2 to 4.7 nm, with an average of 3.6 ± 0.9 nm. This is consistent with Hata *et al.*'s results [8].

As shown by SEM imaging in Fig. 2a, the as-grown CNTs were aligned. After the CNT layer was collapsed by n-hexane evaporation, the CNTs were still aligned, but packed more densely (Fig. 2b). No cracks were seen, and the surfaces were smooth (Fig. 2c). Pore size distributions were calculated from N_2 adsorption and desorption at 77 K and the Barrett–Joyner–Halenda (BJH) method. As shown in Fig. 2d, two peaks, at approximately 3 and 30 nm, were seen for the as-grown CNT forest. These correspond to the CNT pores, which were estimated to be 3.6 nm from TEM measurements (see supporting information Fig. S-2), and the interstitial pores. The distance between the centers of CNTs in the as-grown CNT forest was estimated to be 28 nm (Table 1 & Fig. 1a) from the CNT porosity (~ 1%) and the average size of CNTs (~ 3 nm from N_2 adsorption/desorption). After shrinkage by solvent evaporation, the interstitial pores reduced to an average size of 2.8 ± 0.7 (std dev) nm with a range from 1.4 to 7 nm. This spacing is consistent with the average space between CNT centers (6 nm) calculated from the dense membrane density and the average CNT size. A comparison of the properties of the as-grown arrays and the dense membranes with CNT membranes in the literature is presented in Table 1.

Two advantages of these dense membranes are their CNT density is 8 to 270 times that of previous CNT membranes, and they do not contain a binder. Two studies [9,10] showed that water opened the CNT caps and etched the interface between CNTs and the Fe catalysts during water-assisted CVD growth. Therefore, the tops of our VA-CNT are expected to be opened by water etching. The substrates could be reused several times after detaching the arrays by water etching.

This indicates that most of the iron catalysts remained on the substrates and did not block the bottoms of the VA-CNTs. Reactive ion etching was used to slightly etch the top of a dense CNT membrane, but this did not change the gas flux, indicating the CNTs were already open on the top. Thus, etching was not necessary for this membrane preparation.

Hydrogen permeances were measured as a function of feed pressure for two dense membranes; the permeate pressure was held at 84 kPa. As shown in Fig. 3a, the H_2 permeance decreased with pressure drop. Permeance for viscous flow increases with feed pressure and is independent of feed pressure for Knudsen flow. Thus, neither viscous nor Knudsen flow dominates the transport across these membranes. The H_2 permeances were 450 times the values calculated for Knudsen diffusion (Fig. 3a). Simulations by Newsome and Sholl [13] predict significant enhancement of gas permeance through CNTs over Knudsen diffusion.

The single gas permeances decreased with pressure drop for six gases with molecular weights from 2 to 146 (Fig. 3b). This behavior is different from that expected for Knudsen flow, and Chen et al. [14] and Skoulidas et al. [5] reported that their simulations indicated that Knudsen behavior is not expected for CNT membranes. The ideal selectivities (H_2 relative to other gases) were similar to Knudsen for membranes 1 and 3, but higher than Knudsen for membrane 2 (Fig. 3c). Because membrane shrinkage is not completely controlled, membrane quality varied. Membrane 2 may have had a larger number of smaller pores that favor H_2 permeation. The H_2 and N_2 permeances decreased as the temperature increased for membrane 2 (Fig. 3d), and exhibited $T^{-1/2}$ dependence on temperature, which is the same as seen for Knudsen diffusion. Even though these dense CNT membranes exhibited single-gas selectivities higher than Knudsen values, no separation was seen at room temperature for several 50/50 mixtures (CH_4/CO_2, CO_2/H_2, CH_4/N_2) for pressure drops from 10 to 250 kPa. The high permeances (relative to Knudsen diffusion), decreasing permeance with pressure drop, and no separation selectivity indicate that transport though these dense CNT membranes is not Knudsen, even though the ideal selectivities and temperature dependence of permeance were similar to Knudsen values

Nitrogen permeabilities at room temperature through membrane 1 and CNT composite membranes from the literature are compared in Table 1. Nitrogen permeability ((flux × membrane thickness)/pressure drop) for the dense membrane, based on membrane permeation area (CNT pore area + interstitial area), is 1 to 4 orders of magnitude higher than N_2 permeabilities through CNTs in CNT composite membranes. The N_2 permeability is 3 to 7 orders of magnitude higher than those for CNT composite membranes based on total membrane area. Newsome and Sholl[13] found, using molecular simulations, that surface resistance (resistance of entering and leaving pores) reduced CH_4 flux at room temperature for SWNTs (diameter 1.56 nm) that are shorter than 5 μm. Thus, surface resistance may lower the permeability (compared to thicker membranes) for the composite membranes in refs *(2)* and *(3)*.

The N_2 permeability for dense membranes (based on open permeation area) is 4 orders of magnitude higher than that for Kim *et al.*'s membrane[13]. Because they spin-coated a polymer to seal the space between their CNTs, some of their CNTs may not have been open, and thus not available for permeation; they did not use a procedure to open the CNTs. Hinds *et al.*'s membranes[13] had N_2 permeances similar to that predicted by Knudsen diffusion. Because most of their CNT pores were blocked by catalyst, their N_2 permeation area was probably smaller than the 2.7% of the total permeation area that they calculated. If a smaller permeation area were used to calculate N_2 permeability for the membranes prepared by Hinds *et al.*[1], the ratio of permeability of dense membrane to Hinds *et al.*'s membrane would be less than 160, which is the value calculated from Table 1. The dense CNT membranes have greater than 95% of their area (CNT pore area + interstitial pore area) available for permeation, whereas the CNT composite membranes used less than 3% of their membrane area for permeation. This leads to significantly higher permeabilities for the dense membranes (Table 1).

The maximum size of the pores in the dense membranes was estimated by filtering 3.2-nm gold nanoparticles in DI water (Purest Colloids Co.). Water with 12,800 ppb gold was introduced to the feed side of the membrane, which was pressurized to 184 kPa. The liquid filtrate was analyzed by inductively coupled plasma mass spectrometry. The filtrates for membranes M1 and M2 contained 72 and 35 ppb gold, respectively (Table 2). That is, more than 99% of the gold particles were rejected by dense CNT membranes. Because the gold particle size is comparable to the estimated pore size (from N_2 adsorption), the water flux decreased to 2 to 3% of the pure water flux (Table 2). The membranes were cleaned so that the water flux recovered after 30 min sonication in DI water.

Triisopropyl orthoformate (TIPO, 1.0 nm) was separated from n-hexane (0.43 nm) by a dense CNT membrane by pervaporation, which uses a liquid feed and vacuum on the permeate side. The permeate vapor was condensed in a liquid nitrogen trap and analyzed by GC. A dense CNT membrane separated TIPO from n-hexane with a separation factor (TIPO wt%/n-hexane wt%)$_{permeate}$/(TIPO wt%/n-hexane wt%)$_{feed}$) of approximately 3 for a 50wt%/50wt% feed mixture. Because both molecules are smaller than the membrane pore size, and TIPO has a much higher molecular weight, this separation may result from preferential adsorption of TIPO.

In summary, dense VA-CNT membranes have advantages over CNT composite membranes. Eliminating the spaces between CNTs in CVD grown CNT forests by collapsing the structure by solvent evaporation results in much higher CNT porosity (~ 20%). Because interstitial pores are comparable to CNT pore sizes (3 nm), no sealing materials were used so membrane fabrication was simplified, and additional permeation area was available. Gas diffusion through these membranes is not Knudsen; permeances are 2 orders of magnitude higher.

Supporting information available: experimental set-up and procedures, additional characterization, and membrane preparation.

Acknowledgment

We gratefully acknowledge support by the National Science Foundation Grant CBET-0730047.

References

1. Hinds, B.J. et al. Aligned multiwalled carbon nanotube membranes. *Science* **303**, 62-65 (2004).
2. Holt, J.K. et al. Fast mass transport through sub-2-nanometer carbon nanotubes. *Science* **312**, 1034-1037 (2006).
3. Kim, S., Jinschek, J.R., Chen, H., Sholl, D.S. & Marand, E. Scalable fabrication of carbon nanotube/polymer nanocomposite membranes for high flux gas transport. *Nano Lett.* **7**, 2806-2811 (2007).
4. Majumder, M., Chopra, N., Andrews, R. & Hinds, B.J. Nanoscale hydrodynamics - Enhanced flow in carbon nanotubes. *Nature* **438**, 44-44 (2005).
5. Skoulidas, A.I., Ackerman, D.M., Johnson, J.K. & Sholl, D.S. Rapid transport of gases in carbon nanotubes. *Phy. Rev. Lett.* **89**, - (2002).
6. Srivastava, A., Srivastava, O.N., Talapatra, S., Vajtai, R. & Ajayan, P.M. Carbon nanotube filters. *Nat. Mater.* **3**, 610-614 (2004).
7. Futaba, D.N. et al. Shape-engineerable and highly densely packed single-walled carbon nanotubes and their application as super-capacitor electrodes. *Nat. Mater.* **5**, 987-994 (2006).
8. Hata, K. et al. Water-assisted highly efficient synthesis of impurity-free single-walled carbon nanotubes. *Science* **306**, 1362-1364 (2004).
9. Ci, L.J., Manikoth, S.M., Li, X.S., Vajtai, R. & Ajayan, P.M. Ultrathick freestanding aligned carbon nanotube films. *Adv. Mater.* **19**, 3300-+ (2007).
10. Zhu, L.B., Xiu, Y.H., Hess, D.W. & Wong, C.P. Aligned carbon nanotube stacks by water-assisted selective etching. *Nano Lett.* **5**, 2641-2645 (2005).

11. Chakrabarti, S., Nagasaka, T., Yoshikawa, Y., Pan, L.J. & Nakayama, Y. Growth of super long aligned brush-like carbon nanotubes. *Jpn. J. Appl. Phys. 2* **45**, L720-L722 (2006).

12. Chakrabarti, S., Kume, H., Pan, L.J., Nagasaka, T. & Nakayama, Y. Number of walls controlled synthesis of millimeter-long vertically aligned brushlike carbon nanotubes. *J. Phys. Chem. C* **111**, 1929-1934 (2007).

13. Newsome, D.A. & Sholl, D.S. Influences of interfacial resistances on gas transport through carbon nanotube membranes. *Nano Lett.* **6**, 2150-2153 (2006).

14. Chen, H.B. & Sholl, D.S. Predictions of selectivity and flux for CH_4/H_2 separations using single walled carbon nanotubes as membranes. *J. Membr. Sci.* **269**, 152-160 (2006).

Table 1: Properties of CNT membranes

Membrane Properties	This study: As grown	This study: Dense membrane	Reference (1)	Reference (2)	Reference (3)
Average CNT inner diameter, nm	3.0	3.0	7.5	1.6	1.2
Thickness, μm	750	750	5	2	0.6
Morphology	—	CNT and interstitial pores	CNT/polymer composite	CNT/Si_3N_4 composite	CNT/polymer composite
Density, g/cm^3	0.010	0.21	—	—	—
CNT density, 10^{10} CNT/cm^2	14 *	290 *	6.0	25	7.0
Average distance between CNTs, nm	28	6	41	20	38
CNT volume occupancy, %	1	21	2.7	0.50	0.079
N_2 permeability based on open permeation area, 10^{-4} cm^3 (STP)·cm/(cm^2·s·atm)	—	17,800**	110	2,000	1.3
N_2 permeability based on total membrane area, 10^{-4} cm^3 (STP)·cm/(cm^2·s·atm)		17,000**	3.0	10	0.001

* Estimated from the N_2 adsorption/desorption measurements at 77 K

** Membrane 1 at a pressure drop of 200 kPa

Table 2: Removal of nano-gold particles from DI water

Membrane	Water flux, $kg/m^2 \cdot h \cdot bar$	Water flux with 12,800 ppb Au particles, $kg/m^2 \cdot h \cdot bar$	Au concentration in filtrate, ppb
M1	2,130	44	72
M2	2,740	80	35 ± 2

Figure Captions

Figure 1: Preparation of high density, vertically aligned carbon nanotube membranes. a) schematic representation 1) chemical vapor deposition of vertically-aligned CNT forest; 2) in-situ water etching to detach the CNT forest; 3) soaking and evaporating to densify CNT forest. b) SEM image of a VA-CNT forest detached by water etching. c) Photograph of a dense CNT membrane; white dotted lines indicate the original size of as-grown CNT forest. d) Photograph of a porous filter and a dense CNT membrane glued onto it.

Figure 2: a) SEM image of the cross-section of an as-grown CNT forest. b) SEM image of the cross-section of a dense CNT membrane (scale bar 1 μm). c) SEM image of the top surface of a dense CNT membrane (scale bar 100 μm) and magnified top surface in the inset SEM image (scale bar 5 μm). d) Pore size distribution of as-grown CNT forests and dense CNT membranes calculated from N_2 adsorption/desorption isotherms at 77 K.

Figure 3: Gas permeation through dense VA-CNT membranes: a) hydrogen permeance versus pressure drop for membranes M1 and M2; b) single gas permeances through membrane M1 versus pressure drop; c) ideal selectivity for three membranes (M1: open triangles; M2: open squares; M3: solid circles) at 50 kPa pressure drop. Line is Knudsen selectivity; d) hydrogen and nitrogen permeances through membrane M3 versus $T^{-1/2}$.

.

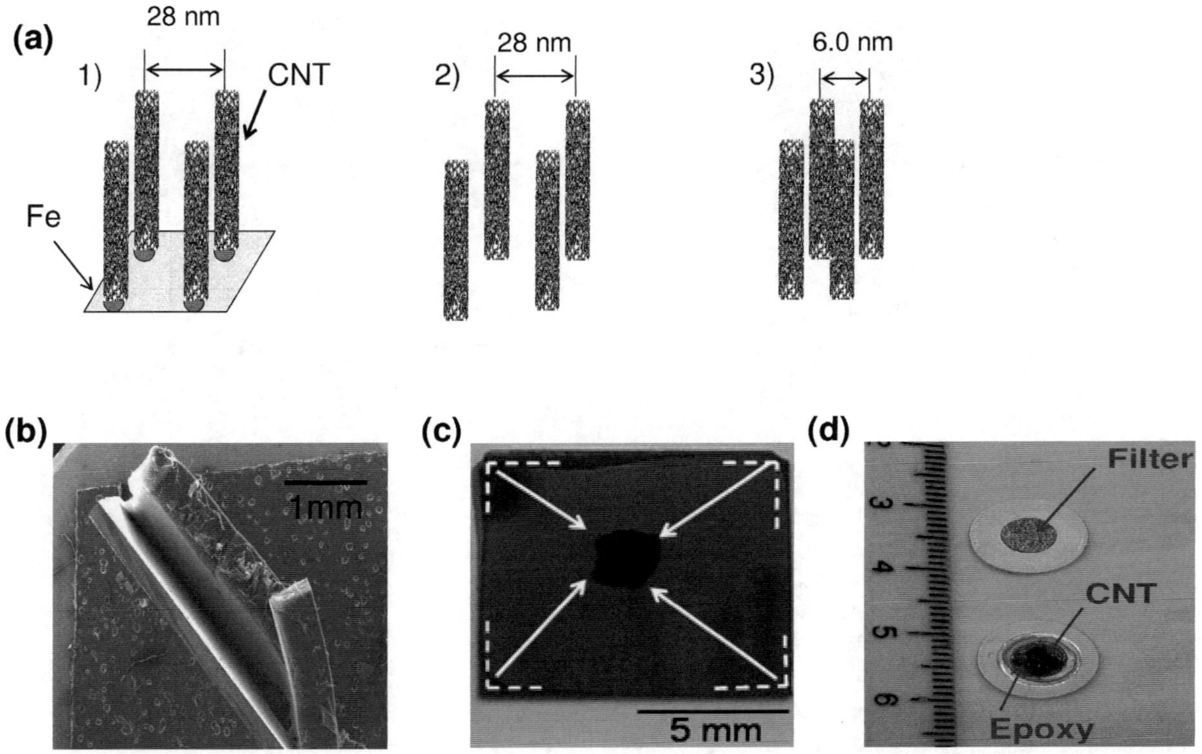

Figure 1: Preparation of high density, vertically aligned carbon nanotube membranes. a) schematic representation 1) chemical vapor deposition of vertically-aligned CNT forest; 2) in-situ water etching to detach the CNT forest; 3) soaking and evaporating to densify CNT forest. b) SEM image of a VA-CNT forest detached by water etching. c) Photograph of a dense CNT membrane; white dotted lines indicate the original size of as-grown CNT forest. d) Photograph of a porous filter and a dense CNT membrane glued onto it.

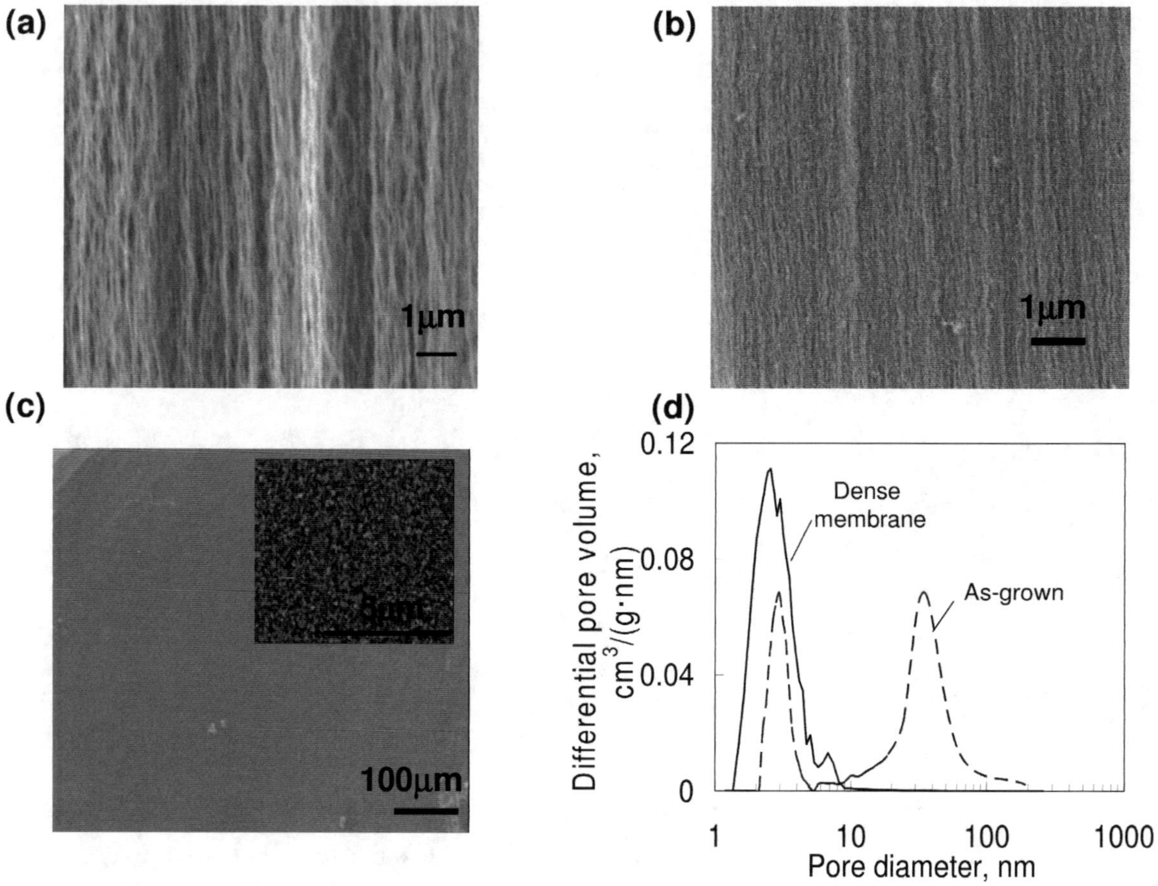

Figure 2: a) SEM image of the cross-section of an as-grown CNT forest. b) SEM image of the cross-section of a dense CNT membrane (scale bar 1 μm). c) SEM image of the top surface of a dense CNT membrane (scale bar 100 μm) and magnified top surface in the inset SEM image (scale bar 5 μm). d) Pore size distribution of as-grown CNT forests and dense CNT membranes calculated from N_2 adsorption/desorption isotherms at 77 K.

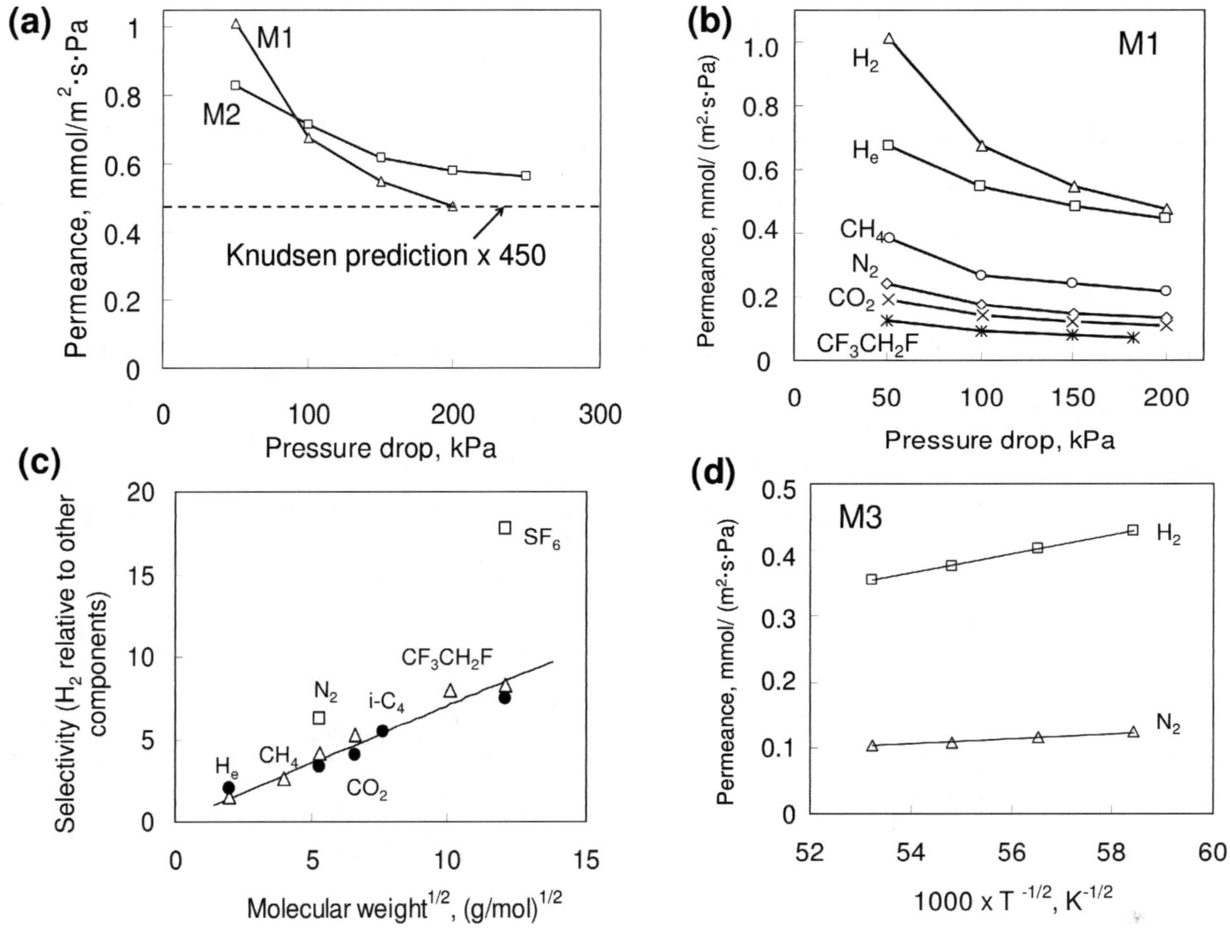

Figure 3: Gas permeation through dense VA-CNT membranes: a) hydrogen permeation at different pressure drop for 2 dense CNT membranes; b) single gas permeation through membrane M1; c) ideal selectivity for three membranes, membrane M1 (open triangle), membrane M2 (open square) and membrane M3 (solid circle) at 50 kPa pressure drop; solid line is Knudsen selectivity; d) permeation of hydrogen and nitrogen through membrane M3 at different temperature.

SUPPORTING INFORMATION

1. Growth of vertically-aligned carbon nanotube array

The growth procedure of vertically-aligned carbon nanotube arrays follows water-assisted highly efficient chemical vapor deposition (CVD) synthesis suggested by Hata *et al.* (*1*). Chemical vapor deposition (CVD) was conducted in a horizontal quartz tube (5.1 cm diameter and 90 cm long) housed in a Lindberg furnace. The catalyst thin films (1 nm Fe/ 10 nm Al_2O_3) were deposited on silicon wafer with 500 nm oxidation layer by sequential e-beam evaporation. At room temperature substrates were placed into the oven, and argon was introduced at 600 sccm to purge out the air. At the same time, start to heat the oven to 750 °C. When the oven temperature reached 350 °C, hydrogen (400 sccm) was introduced, and after 20 min (oven temperature, 660 °C) an argon stream (30 sccm) bubbled through a water bubbler to introduce small amount of water (~ 730 ppm) to the reactor. After another 10 min, oven temperature reached 750 °C and stabilized, ethylene (100 sccm) was introduced for a designated time for growth. In order to detach the as grown VA-CNT array easily, water etching process was applied after CVD growth, as suggested by Ci *et al.* (*2*). Specifically, after CVD the ethylene flow was stopped and argon flow rate decreased to 400 sccm and hydrogen to 270 sccm, and argon flow rate through water bubbler increased to 50 sccm (~2000 ppm). During the water etching process, the oven temperature kept at 750 °C, and the etching process lasted for 15 min. Figure S-1 shows growth kinetics of the VA-CNT arrays. Thickness of VA-CNT layers increases appropriately linearly with the growth time, and for all the membranes for shrinkage the growth time is 40 min, and thus a thickness of 750 μm.

2. Preparation of dense VA-CNT membranes

We used a similar VA-CNT array shrinking procedure as Futaba et al. (*3*). After detaching an as-grown VA-CNT array from the silicon substrate, it was soaked in organic solvents, such as n-hexane, for 30 min. The wet CNT array was dried at room temperature naturally and shrank on a silicon wafer. The shrunk membrane was then stored in a vacuum oven at 70 °C for 24 h for later N_2 adsorption/desorption measurements or making a membrane. After storing in vacuum oven for 24 h, the membrane was glued on a porous filter (5 μm pores) using epoxy, which was ready to use for gas permeation measurements.

3. Nitrogen adsorption/desorption

Appropriately 20 mg as-grown or shrunk VA-CNT arrays were used for adsorption in an Autosorb-1 (Quantachrome Corp. Model AS1-C-VP-RGA) system. Prior to each adsorption experiment, the sample was outgassed in vacuum at 473 K for about 2 h. Nitrogen adsorption/desorption measurements were performed at 77 K, and BJH desorption and HK methods were used for mesopore size distribution and pore volume calculation and micropore size distribution and pore volume calculation respectively.

4. Single gas permeation measurements

A dead end system was used for single gas permeation measurements. A gas pressure regulator was connected to the gas cylinder before the membrane to adjust the gas pressure in the system and a pressure gauge was used to read the pressure. Gas flux through membranes was measured by a bubble flow meter.

5. Nano-gold particle filtration experiments

Figure S-3a shows the experimental set-up for nano-gold particle filtration measurements. Dense CNT membranes were connected to a liquid reservoir by standard VCR fitting, and the liquid was pressurized by N_2. The filtrated liquid at atmosphere pressure was collected by a small vial and analyzed by inductively coupled plasma mass spectrometry (ICP-MS). Figure S-3b and c show that after filtration the solution became clear in contrast with the feed red solution.

5. Liquid mixture separation by pervaporation

Liquid mixture pervaporation (liquid feed, vacuum on permeate side) was carried out at 300 K in a liquid recirculation system similar to that described previously *(4)*. The dense CNT membrane was connected to a liquid pump by VCR fitting, and liquid circulated at a flow rate of 1 L/s. The permeate was collected in a liquid nitrogen trap, and a permeate sample was weighed every 1-2 h to determine the flux, and its composition was measured by gas chromatography (GC). The pressure was below 0.25 kPa on the permeate side.

References

1. K. Hata *et al.*, *Science* **306**, 1362 (2004).
2. L. J. Ci, S. M. Manikoth, X. S. Li, R. Vajtai, P. M. Ajayan, *Advanced Materials* **19**, 3300 (2007).
3. D. N. Futaba *et al.*, *Nature Materials* **5**, 987 (2006).
4. T. C. Bowen *et al.,* Microporous and Mesoporous Materials 71, 199 (2004).

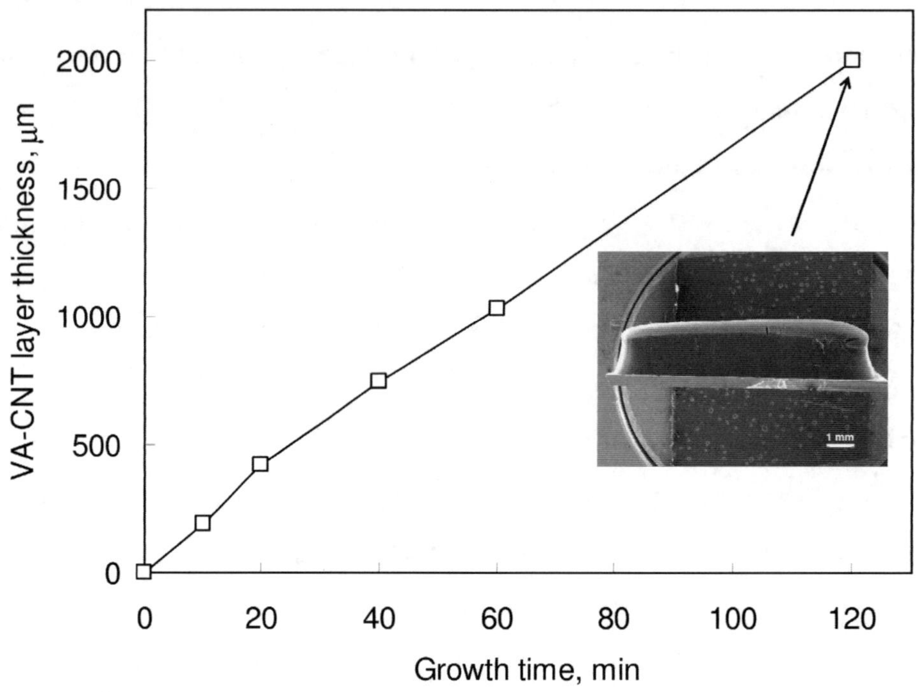

Figure S-1 Time evolution of VA-CNT array growth. Plot of the VA-CNT array thickness as a function of the growth time. Inset picture shows a sample of 2 mm thickness VA-CNT array.

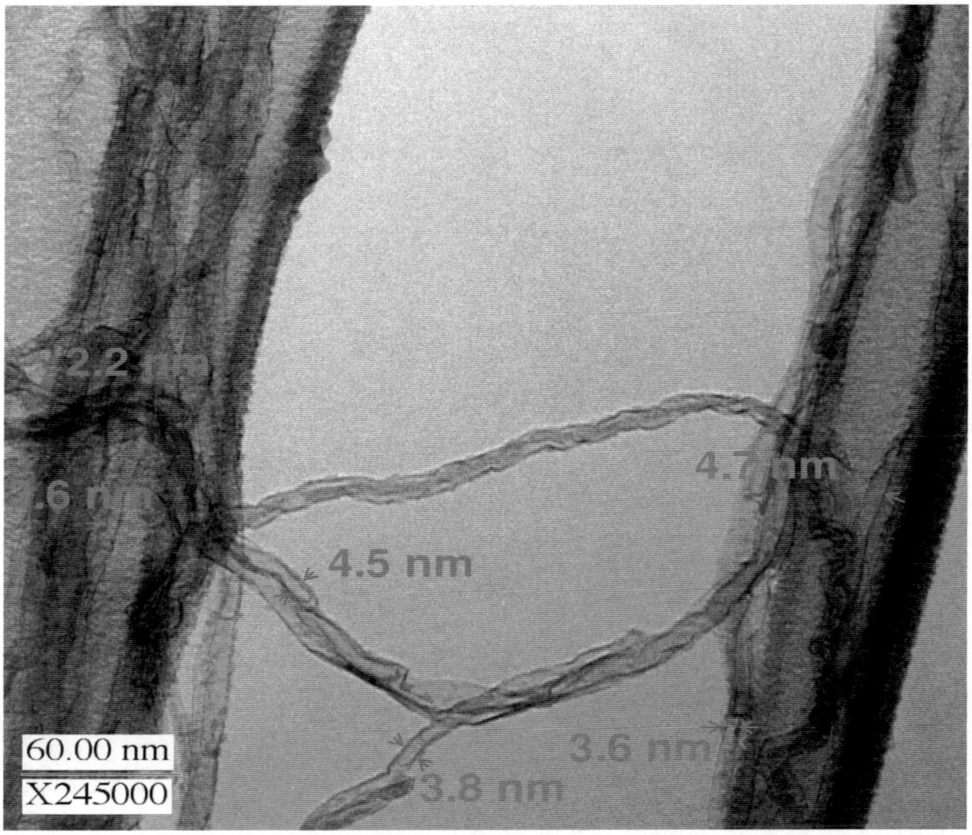

Figure S-2 Transmission electron microscopy (TEM) of carbon nanotubes grown from 1.0 nm Fe thin film.

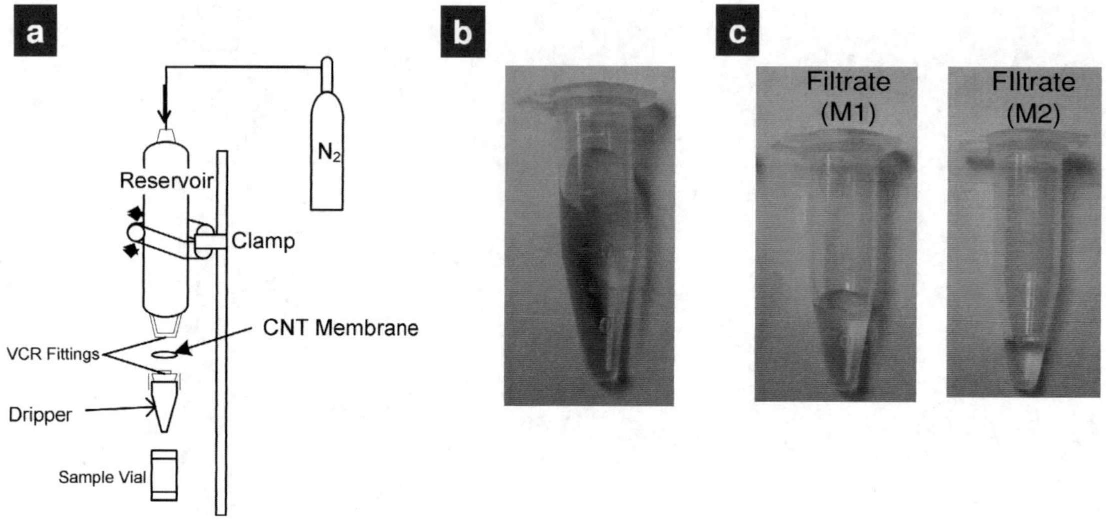

Figure S-3: Nano-gold particle removal measurements by pressurized liquid filtration. a) Experimental set-up for measuring filtration. b) Digital picture of feed solution of 12800 ppb nano-gold particles in DI water. c) Digital pictures of filtrates of two dense CNT membranes after membrane rejection.

695b Preparation of Silicon Carbide Membranes Using Nanoporous Fillers

Bahman Elyassi, Muhammad Sahimi, and Theodore T. Tsotsis. Mork Family Department of Chemical Engineering & Materials Science, University of Southern California, Los Angeles, CA 90089

The increased interest in the efficient production of hydrogen, and in environmentally-benign power generation, is driving current development efforts on the preparation of nanoporous, hydrogen-permselective, inorganic membranes which can be used in processes that take place at high temperatures and pressures. Due to many unique properties, such as high oxidation resistance and hydrothermal stability, silicone carbide (SiC) is a promising membrane material for such applications in harsh environments.

Previously, we used slip casting of SiC powders on tubular SiC supports to reduce defect propagation from support into the membrane layer [1,2]. However, the SiC particles used were nonpermeable. The idea of the present work is to replace the nonpermeable SiC particles with permeable structures, and to investigate their effect on the membrane performance.

To do so, using a template technique, nanoporous SiC nanofibers with high surface area were prepared and used in the slip casting solution. The supports were slip casted in a solution of nanofibers and the pre-ceramice polymer, allyl-hydridopolycarbosilane (AHPCS). Using a sacrificial interlayer based method that we recently developed [1], additional layers of the AHPCS and sacrificial polystyrene were coated on the slip-casted support and, then, the entire structure was pyrolyzed at 750 ^{o}C for 2 h. The membrane results obtained with these materials and technique are compared with those obtained the previous method in which we used nonpermeable SiC powders for slip casting.

1. B. Elyassi, M. Sahimi, T.T. Tsotsis, A novel sacrificial interlayer-based method for the preparation of silicon carbide membranes, Journal of Membrane Science, 316 (2008) 73-79.

2. B. Elyassi, M. Sahimi, T.T. Tsotsis, Silicon carbide membranes for gas separation applications, Journal of Membrane Science, 288 (2007) 290-297.

695c Surface Functionalized Composite Membranes for Solubility Based Separations

Achyuta Teella[1], David M. Ford[1], Sukjoon Yoo[2], Seunguk Yeu[3], and Daniel F. Shantz[4]. (1) Chemical Engineering, University of Massachusetts, Amherst, 159 Goessmann Laboratory, North Pleasant Street, Amherst, MA 01003, (2) School of Chemical and Biomolecular Engineering, Georgia Institute of Technology, 311 Ferst Drive, N.W., Atlanta, GA 30332, (3) Department of Chemical Engineeing, Texas A&M University, TAMU 3122, College Station, TX 77843-3122, (4) Department of Chemical Engineering, Texas A&M University, TAMU 3122, College Station, TX 77843-3122

Surface-functionalized composite membranes were synthesized by modifying commercially available mesoporous and macroporous supports with selective materials. The membranes were evaluated for performance in separating light gases and vapors based on solubility differences. Three different types of membranes were studied.

Membralox gamma-alumina membranes with 5 nm nominal pore size were decorated with linear alkyl trichlorosilanes or hyper-branched triazine based dendrimers. These organic molecules were delivered in an organic solvent and anchored to porous inorganic surfaces as filling agents to give hybrid materials. Pure gas permeance of the alumina membranes decreased by two to three orders of magnitude after altering with organic groups. Our membranes showed a wide range of propane/nitrogen selectivity. The effect of residual solvent on composite membrane performance is also correlated with conditions employed in cleaning and drying the composite membranes.

Alumina-ordered mesoporous silica (OMS) hybrid membranes were fabricated using a variation of the evaporative-induced self-assembly process by coating of a Membralox 200 nm membrane with solution containing Brij surfactant and a silica source. The effective pore size of the membrane can be tuned based on the number of coating cycles used. Single gas permeation measurements were performed using hydrogen, helium, and nitrogen. The observed permselectivity of helium over nitrogen is slightly higher than that of commercial 5-nm Membralox membranes, which might be ascribed to the lack of pinhole defects. Also, from the permeation measurements and other materials characterization experiments we conclude that the final composite membrane has a pore size around 5 nm.

Finally, we are also exploring modification and functoinalization of nanoporous track-etched polycarbonate membranes for the separation of gases based on solubility.

695d Synthesis and Modification of Sodalite (SOD) and Deca-Dodecasil 3R (DDR) Membranes Grown on Alpha-Alumina Supports

Zhenkun Zheng, **Gangzhi Mei**, and Vadim V. Guliants. Chemical and Materials Engineering, University of Cincinnati, Cincinnati, OH 45221

Sodalite (SOD) [1-3] and Deca-dodecasil 3R (DDR) [4-6] are promising membrane materials for hydrogen separation from larger gas molecules due to their 6-membered ring (0.28 nm) and 8-membered ring pore (0.36×0.44 nm), respectively. The SOD and DDR membranes were hydrothermally synthesized on porous alpha-alumina supports by in-situ crystallization and seeded secondary growth methods. The X-ray diffraction (XRD) and scanning electron microscopy (SEM) techniques were applied to investigate the phase composition and microstructure of the membranes and corresponding bulk zeolite phases. Defective sodalite membranes were successfully modified by chemical vapor deposition (CVD) technique using tetraethylorthosilicate (TEOS) as the silica source. The CVD-modified sodalite membranes exhibited significantly enhanced H2/CO2 permselectivity of 41 and 45 at 550 oC for low- and high-silica sodalite membranes, respectively. H2/CO2 permselectivity of defective DDR membranes obtained by in-situ crystallization increased from 2.6 to ~33 at 550 oC after CVD modification. For largely defect-free DDR membranes obtained by seeded secondary growth, H2/N2 permselectivity increased from ~10 to 14 by the atomic layer deposition (ALD) of thin Al2O3 layer on the surface of the DDR membrane at 25 oC. These results indicated that ultramicroporous SOD and DDR membranes are highly promising for H2 separation from various H2-containing process streams, such as syngas from the water-gas-shift reaction.

Acknowledgements

The financial support from Ohio Coal Development Office (OCDO) is gratefully acknowledged.

References

[1] J. Motuzas, et al., Microporous and Mesoporous Materials, 2006. 92(1-3): p. 259-269.

[2] X.C. Xu, et al., Chemical Communications, 2000(7): p. 603-604.

[3] Z. Zheng, V. V. Guliants, and S. Misture, Journal of Porous Materials, 2008, accepted

[4] T. Tomita, K. Nakayama, and H. Sakai, Microporous and Mesoporous Materials, 2004. 68(1-3): p. 71-75.

[5] S. Himeno, et al., Microporous and Mesoporous Materials, 2007. 98(1-3): p. 62-69.

[6] Z. Zheng, A. S. Hall, and V. V. Guliants, J Mater Sci (2008) 43: p. 2499–2502.

Oxygen Permeation Studies of $Ba_{0.5}Sr_{0.5}Co_{0.8}Fe_{0.2}O_x$ Asymmetric Membranes

Qiying Jiang, Susan M. Stagg-Williams, Karen J. Nordheden
Chemical and Petroleum Engineering Department, University of Kansas, 1530 W. 15th Street, 4132 Learned Hall, University of Kansas, Lawrence, KS 66045, USA

Due to their infinite theoretical oxygen separation factor and staged addition of oxygen at higher temperature, mixed ionic-electronic conductive (O-MIEC) membranes not only show great potential applications in the pure oxygen production, but also be expected to be desirable catalytic reactor materials for many chemical reactions of the industry interest. Preparing asymmetric O-MIEC membranes, which contain one thin film and one porous substrate, is considered to be practically efficient means to improve the oxygen permeation performance of O-MIEC membranes due to two facts:(i) it can reduce the diffusion resistance of oxygen permeation through membranes;(ii) it also can maintain the necessary mechanical strength of such membranes. $Ba_{0.5}Sr_{0.5}Co_{0.8}Fe_{0.2}O_x$ (BSCF) is the new O-MIEC material, which has been demonstrated to have the promising potentials in oxygen permeation reactors [1]. Our previous studies showed for the dense thick BSCF, the bulk diffusion is dominating factor for the oxygen permeation through it [2]. Therefore, the objective of this study is to investigate the performance of BSCF asymmetric membranes fabricated by the dry-pressing method. Furthermore, the important factors affecting the oxygen permeation performance of those membranes including the thickness of thin film and the porosity of the substrates were examined carefully.

BSCF powders were prepared by Citrate-EDTA method [3,4]. The asymmetric BSCF membranes were prepared by the dry-pressing method [5]. The weighed amounts of BSCF green powders were first added into a stainless die with 2mm diameter. The surface of BSCF powders was softly leveled off by the isostatical press. Then, the mixture of BSCF green powders with ethyl-cellulose (EC) and carbon fiber (CF) was added into the die. The die was pressed under 250 MPa for 5-7 minutes to form the green BSCF asymmetric membranes. Finally, the membranes were sintered at 1150°C for 5 hours with ramping and cooling rate of 1°Cmin^{-1} in the muffle furnace. The burning-out of ethyl-cellulose and carbon fiber resulted in the formation of the porous support. No binder was added during the preparation process to avoid the possible effect on the performance of membranes. Oxygen flux tests were performed in the two-sided concentric quartz tube reactor. To avoid the gas leakage from the ambient air atmosphere, the reactor was sealed at 1073K using gold ring gaskets between the outer quartz tubes and the membrane surfaces. Pressure against the gold seal was maintained by an external nitrogen pneumatic press. The whole reactor was installed in a tubular furnace. The K-type thermocouple was used to control the temperature of the furnace. Air was introduced into the bottom side of reactor (oxygen supply side of the membrane), and 20 mlmin^{-1} Ar was used to sweep the topside of reactor (oxygen permeate side of the membrane). The effluent gases from the oxygen-lean side of the membrane were analyzed simultaneously with the Balzers Omnistar mass spectrometer and oxygen sensor.

The typical morphology of the asymmetric membranes is shown in Figure 1. A clear interface between the thin film and porous support was observed. To ensure the mechanical integrity of the membranes, helium was used on the air side of the membrane and the permeate side was monitored for helium leaks. No helium was observed on the permeate side which indicates that no defects or leaks are present at 1073K.

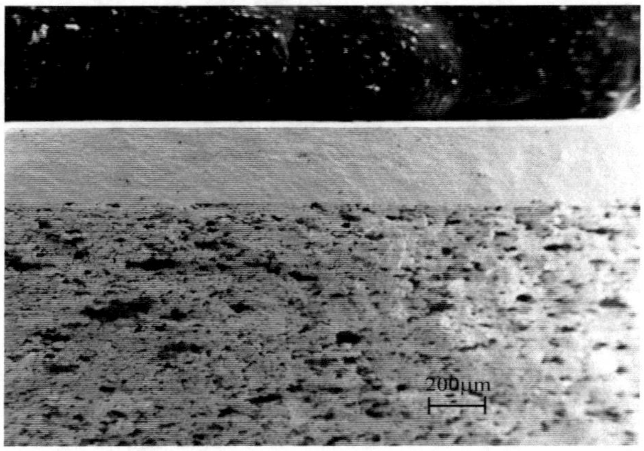

Figure.1 SEM micrographs of BSCF asymmetric membranes (cross-section view)

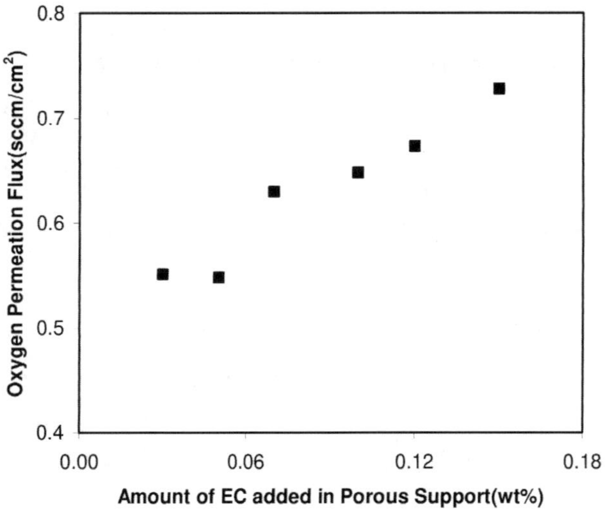

Figure.2 Effect of EC on oxygen fluxes of BSCF asymmetric membranes (1073K)

To investigate the contribution of EC to oxygen fluxes of asymmetric BSCF membranes, different amounts of EC were mixed during the porous support preparation, while keeping the thickness of thin film same. When the EC is burned off, the resultant porous supports will have different void fractions. The results presented in Figure 2 show the oxygen fluxes increase with the increase of the amount of added EC. However, the flux increase is not linear. We believe although the removal of EC results in the formation of void, however parts of those voids (pores or holes) are discrete and not connected, especially when only small amount of EC was added. Those unconnected pores or holes cannot facilitate the diffusion of oxygen, therefore cannot lead to the increase of the oxygen flux. To demonstrate this explanation, small amounts of CF were added into BSCF powder mixture during porous supports preparation while keeping the EC content at 10wt% and the thickness of thin film same. The corresponding fluxes show that the addition of CF to the porous support increases the oxygen flux dramatically. This increase is attributed to the ability of the CF to connect some of the discrete voids that were isolated in the supports without CF.

To understand the oxygen permeation through the asymmetric membrane, the oxygen fluxes of asymmetric membranes with different thin film thicknesses were investigated and the results are shown in Figure 3. Our previous studies showed for the thick symmetric BSCF membranes, the bulk diffusion is the dominating step for the oxygen permeation through it [2]. According to the Wagner equation under the bulk diffusion control, the oxygen flux is inversely proportional to the thickness of the membrane [6]. This inversely proportional relation is also presented in Figure 3 (shown as the solid line). The oxygen flux of asymmetric membranes increases with the decrease of the thin-film

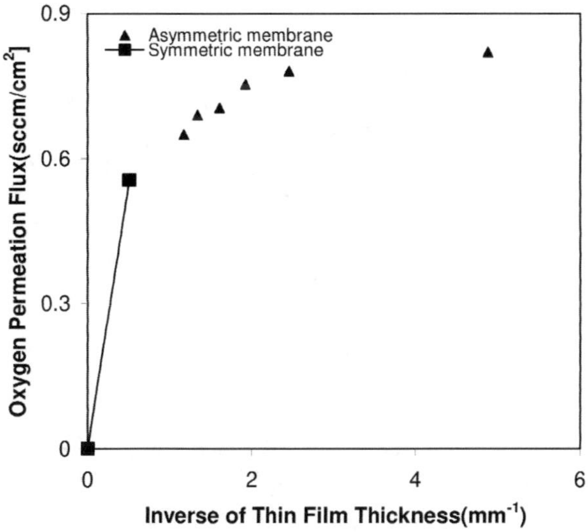

Figure.3 Dependence of oxygen fluxes of BSCF asymmetric membranes on the thickness of thin film (1073K)

thickness as expected. That indicates the bulk diffusion plays the role in the oxygen permeation through the membranes. However, those oxygen fluxes are lower than the corresponding theoretical values, and the increasing trend diminishes. The first explanation is the porous support still exerts the resistance to the oxygen permeation. Though the addition of CF connected parts of originally close-ended pores, but it isn't guaranteed that the fully connected pores network is formed in the whole porous support. Therefore, the resistance exerted by the porous support is inevitable. Another possible explanation is that as the thin-film thickness decreases, the influence of the surface exchange increases, and the oxygen permeation shifts to the transitional stage, in which the bulk diffusion and surface exchange all play the roles. As the thickness then approaches the characteristic thickness, surface exchange becomes dominating. During the transfer of the rate-limiting step, the relation of the oxygen flux to thickness is not inversely proportional, which also results in the deviation of the oxygen flux from the theoretical values calculated according to Wagner's equation.

Reference

[1] Q. Jiang, S.Faraji, D.A.Slade, K. J. Nordheden and S. M. Stagg-Williams, *Performance Analysis of Mixed-Conducting Ceramic Membranes for Membrane Reactor*, AICHE Annual Meeting, paper no.509ao, Saltlake City, 2007.

[2] Q. Jiang, K. J. Nordheden, S. M. Stagg-Williams and D.A.Slade, *Oxygen permeation studies of* $Ba_{0.5}Sr_{0.5}Co_{0.8}Fe_{0.2}O_x$, in preparation.

[3] Z. Shao, W. Yang, Y. Cong, H. Dong, J. Tong, G. Xiong, *Investigation of the permeation behavior and stability of a* $Ba_{0.5}Sr_{0.5}Co_{0.8}Fe_{0.2}O_{3-\delta}$ *oxygen membrane,* Journal of Membrane Science 172 (2000) 177-188.

[4] L. Tan, X. Gu, L. Yang, W. Jin, L. Zhong, N. Xu, *Influence of powder synthesis methods on microstructure and oxygen permeation of* $Ba_{0.5}Sr_{0.5}Co_{0.8}Fe_{0.2}O_{3-\delta}$ *perovskite-type membranes*, Journal of Membrane Science 212 (2003) 157-165.

[5] M. Ikeguchi, K. Ishii, Y. Sekine, E. Kikuchi, M. Matsukata, *Improving oxygen permeability in* $SrFeCo_{0.5}O_x$ *asymmetric membranes by modifying support-layer porous structure,* Materials Letters 59 (2005) 1356-1360.

[6] C. Wagner, *Equations for transport in solid oxides and sulfides of transition metals*, Prog. Solid State. Chem. 10 (1975) 3.

Effect of Heating Sources and Template-Free Secondary Growth Synthesis on MFI-type Zeolite Microstructure

*Seok Jhin Kim, Zhong Tang, Junhang Dong**

Department of Chemical & Materials Engineering, University of Cincinnati, Cincinnati, OH 45220, USA

ABSTRACT

MFI-type zeolite membranes have great potential application in the separation of gas and liquids, which are difficult to achieve by traditional techniques [1]. Thus, research on high-quality MFI zeolite membrane and zeolite microstructure is increasing rapidly. In general, MFI-type zeolite membranes are synthesized by conventional heating (CH), which needs a long crystallization time ranging from several hours to a few days and usually results in the formation of impure materials because of the low heating rate and the inhomogeneous heating [2]. Additionally, MFI-type zeolite membranes are usually synthesized in the presence of organic templates. The organic agent, however, needs to be removed by calcination in order to open up the zeolite channels; then intercrystalline gaps tend to enlarge in the membranes, which affect the separation performance of the membranes [3]. It was found that in order to obtain a high quality MFI-type zeolite membrane, a homogeneous heating source and template removal are important factors. Therefore, it is very important to examine the influence of heating sources and template-free synthesis on zeolite microstructure.

This paper reports synthesis of MFI-type zeolite powders by a template-free method using microwave heating (MH). The influence of heat sources, the amount of silicalite nanoparticle seeds, treatment with an NaOH solution, etc., on the formation of MFI zeolite microstructure were studied. In the present work, MFI zeolite powder was prepared according to both heating sources (CH or MH) and the methods (with template or without template). In case of using template, MFI zeolite powders were synthesized by in situ crystallization with a composition of 18.7g TEOS/20ml TPAOH/0.27g NaOH/75g H_2O/Xg Al2(SO4)318 H_2O. The X values varied were 0 and 0.25 for silicalite-1 (Al-free) and ZSM-5 (with Al), respectively. In preparing template-free samples, silicalite nanoparticle seeds were prepared by hydrothermal treatment (at 80□×84h) from a molar ratio of 0.33 SiO_2/0.1 TPAPH/0.035 NaOH/5.56 H_2O. After hydrothermal synthesis, the resultant silicalite nanoparticle slurry was washed in DI water with the assistance of centrifuge. MFI zeolite powders were synthesized by in situ crystallization with a composition of 2.0g silicalite nanoparticle seed/5.0g SiO_2/1.9g NaOH/100g H_2O/Xg $Al_2(SO_4)_3 18H_2O$. The X values varied were 0 and 0.25 for silicalite-1 and ZSM-5, respectively. After the hydrothermal synthesis, the zeolite powder was washed and dried overnight. The powder was then fired at 450□ for 8h with a heating and cooling rate of 1.0□/min. The names and synthesis conditions for all samples are presented in Table 1.

---

*Corresponding author: Phone:(513)556-3992; email: dongj@UCMAIL.UC.EDU

Table 1. Preparation conditions for MFI zeolite powder samples

Sample	Type	Template	Heating source	Si/Al	Synthesis temperature	Silicalite seed (g)	Alikali-treatment (mol l^{-1})
1	ZSM-5	TPAOH	CH	111	180□×10h	-	-
2	ZSM-5	TPAOH	MH	111	180□×4h	-	-
3	Silicalite-1	TPAOH	CH	∞	180°C×10h	-	-
4	Silicalite-1	TPAOH	MH	∞	180□×4h	-	-
5	ZSM-5	-	CH	111	180□×10h	0.5	-
6	ZSM-5	-	MH	111	180□×4h	0.5	-
7	Silicalite-1	-	CH	∞	180°C×10h	0.5	-
8	Silicalite-1	-	MH	∞	180□×4h	0.5	-
9	Silicalite-1	-	CH	∞	180°C×10h	1.0	-
10	Silicalite-1	-	CH	∞	180°C×10h	1.5	-
11	Silicalite-1	-	CH	∞	180°C×10h	2.0	-
12	ZSM-5	-	MH	111	180□×4h	0.5	0.025

The results of pore-structural parameters derived from the nitrogen adsorption data are shown in Table 2. The pore structures of the samples were investigated by physical adsorption of nitrogen at 77K on an automatic volumetric analyzer (Micromeritics ASAP 2020). Prior to measurements, the samples were degassed at 350□ for 10h under vacuum. The specific surface areas (S_{BET}) were determined according to BET method. The total volumes (V_t), the micropore volumes (V_{mi}), the mesopore volumes (V_{me}) and the median pore width were also obtained. Samples between 1 and 4 possess similar BET surface areas and pore volumes. MH synthesis drastically reduced the crystallization time of zeolites, which resulted from the simultaneous nucleation and homogeneous heating by microwave. Samples between 5 and 8 display lower pore volumes and smaller BET surface areas than the former samples. Since there is no template in the synthesis solution, zeolite growth is limited to silicalite seed crystals [4] and the amorphous materials can be made. BET surface areas of silicalite-1 are lower than that of ZSM-5, which means that the particle obtained from the aluminum free precursor did not have enough intergrown polycrystalline structures.

Table 2. Characterizations of MFI zeolite powder samples

Sample	S_{BET} (m^2/g)	V_t (cm^3/g)	V_{mi} (cm^3/g)	V_{me} (cm^3/g)	Median pore width (Å)
1	421	0.18	0.13	0.05	5.51
2	410	0.18	0.13	0.05	5.47
3	430	0.16	0.14	0.03	5.53
4	427	0.18	0.13	0.05	5.58
5	354	0.15	0.12	0.04	5.35
6	371	0.15	0.11	0.04	5.35
7	106	0.05	0.03	0.02	5.21
8	56	0.03	0.01	0.02	-

To further investigate the template-free zeolite microstructure, the amount of silicalite nanoparticle seeds was increased. Table 3 also shows that the amount of silicalite nanoparticle seeds has obvious influence on both BET surface areas and pore volumes of template-free silicalite-1 powders. After 1.5g or more of seeds is added, BET surface areas were found to be slightly larger than those of powders with template. These observations are in accordance with the XRD characterization. It means that the amount of crystals increases with the amount of silicalite nanoparticle seeds, on which zeolite can grow.

Table 3. Characterizations of MFI zeolite powder samples

Sample	S_{BET} (m^2/g)	V_t (cm^3/g)	V_{mi} (cm^3/g)	V_{me} (cm^3/g)	Median pore width (Å)
7	106	0.05	0.03	0.02	5.21
9	140	0.06	0.04	0.02	5.21
10	435	0.17	0.14	0.03	5.51
11	465	0.19	0.15	0.04	5.50

Alkali-treatment is another way of modifying zeolite. The results of porous structural parameters of alkali-treated zeolite samples and the morphology of them are shown in Table 4. When MFI type zeolite powder is treated with NaOH aqueous (0.025mol l^{-1}) solution, the surface area increased slightly in the micorpore region, and pore volume also increased in the mesopore region. It means that the large pores were newly generated at the boundary of the microcrystallites by alkaline dissolution of Si and Al. SEM and XRD measurements revealed that amorphous phases present at the boundary of the crystallities were dissolved. By NaOH treatment, the aggregates containing of the primary microcrystallines and the amorphous SiO_2 as cement changed to those of the primary microcrystallines through the removal of the amorphous phase [5].

Table 4. Characterizations of MFI zeolite powder samples after alkali-treatment

Sample	S_{BET} (m^2/g)	V_t (cm^3/g)	V_{mi} (cm^3/g)	V_{me} (cm^3/g)	Median pore width (Å)
6	371	0.15	0.11	0.04	5.35
12	399	0.18	0.12	0.06	5.81

MFI zeolite powders synthesized by a template-free secondary growth method display lower pore volumes and smaller BET surface areas than MFI zeolite powders synthesized with template. However, both the amount of silicalite nanoparticle seeds and the alkali-treatment have obvious influence on both BET surface areas and pore volumes of template-free MFI zeolite powders, which were found to be similar to those of powders with template.

Reference

[1] J. Dong, Y. S. Lin. Peascoe, E. A. Payzant, *Micropor. Mesopor. Mat.* 34 (2000) 241-253
[2] X.C. Xu, Y. Bao, C.S. Song, W.S. Yang, J. Liu, *Micropor. Mesopor. Mat.* 75 (2004) 173-181
[3] M. Masakoto, J. O'Brien, Y.S. Lin, *J. Membr. Sci.* 286 (2006) 213-222
[4] M. Pan, Y.S.Lin, *Micropor. Mesopor. Mat.* 43 (2001) 319-327.
[5] Tetsuo Suzuki, Toshio Okuhara, *Micropor. Mesopor. Mat.* 43 (2001) 83-89

695g Synthesis, Characterization and Separation Property of Zeolite Membrane for Steam Separation

Ken-ichi Sawamura, Yasushi Sekine, Eiichi Kikuchi, and **Masahiko Matsukata**. Department of Applied Chemistry, Waseda University, 3-4-1 Okubo, Shinjuku-ku, Tokyo, 169-8555, Japan

Membrane technologies will offer us great opportunities to reduce energy demands revolutionary in future chemical and petroleum industries. Among emerging membrane technologies, great attention has been paid to zeolites as membrane materials. Zeolites, microporous aluminosilicates crystals, have been widely used as adsorbents, catalysts, and ion exchangers. In addition to their high thermal, chemical, and structural stability, zeolite micropore systems can uniquely recognize molecules by molecular sieving and/or preferential adsorption.

In commercial applications, a membrane to remove steam selectivelyfrom syngas above 200°C seems to be promising for a membrane extractor inreactions such as Fischer-Tropsch synthesis and methanol synthesis, where water is a by-product. These syngas reactions (C1 chemistry) require further technical development considering future growth in alternate fuel sectors, driven by both strategic and environmental considerations.

In the 2006 AIChE annual meeting, we reported that a ZSM-5 type zeolite membrane highly permeated and separated steam from hydrogen above 200°C. This study aimed at clarifying correlation among synthesis condition, membrane structure, and separation property of ZSM-5 membranes.

ZSM-5 membranes were prepared on the outer surface of porous alumina tubular support by a secondary-growth method. The molar composition of a synthesis mixturewas $10Na_2O$: $0.15Al_2O_3$: $36SiO_2$: $1200H_2O$. Hydrothermal treatment was carried out at 180°C with prolonged synthesis time from 6 to 18 h. Sort of cations in the zeolite membranes were controlled by ion exchange. Membrane structures were characterized by FE-SEM and permporometry. Separation properties of membranes were evaluated by water/hydrogen separation in the temperature range of 150-250°C.

Permporometry measurements suggest that voids among crystal boundaries were filled after the crystallization for 12 h and that remaining inter-crystal pathways readily lead to a deterioration of water/ hydrogen separation performance. The sizes of pores and voids were tentatively estimated using the Kelvin equation. Gas permeation via pores with Kelvin diameter of larger than 0.55 nm was not detected in permporometry measurements through a Na-exchanged ZSM-5 (Na-ZSM-5) membrane with a separation factor of water/hydrogen = 74 (at 150°C and partial pressures of water/hydrogen = 10/91 kPa). On the other hand, 3-7 % of gas permeation was via pores with the Kelvin diameter of 0.55-1.2 nm through a Na-ZSM-5 membrane with separation factor of water/hydrogen = 26, while the voids of larger than 1.2 nm was not detected.

Separation performance of water/hydrogen mixture was influenced by Na cation occluded in the micropores of zeolite. With increasing partial pressure of water, hydrogen permeation through Na-ZSM-5 membrane was effectively blocked, whereas such hindrance of hydrogen permeation with water was difficult to observe with H^+-exchanged ZSM-5 (H-ZSM-5) membrane at higher temperature. Adsorption experiments revealed that the amount of water adsorbed on Na-ZSM-5 was much larger than that on H-ZSM-5. When comparing hydrogen permeation through Na-ZSM-5 and H-ZSM-5 membranes at the same loading of water, the Na-ZSM-5 membrane showed more remarkable inhibition of hydrogen permeation than the H-ZSM-5 membrane. Differences in permeation and separation properties between Na-ZSM-5 and H-ZSM-5 membranes were more significant at higher temperature.

699a Palladium Membranes on Mesoporous Ceramic Supports: Effect of Metal-Support Interface on Hydrogen Permeation

Jerry Y. S. Lin, Department of Chemical and Materials Engineering, Arizona State University, Tempe, AZ 85287 and Ke Zhang, Dept of Chemical Engineering, Arizona State University, Tempe, AZ 85260.

Thin palladium membranes are synthesized on mesoporous ã-alumina/á-alumina and yttria stabilized zirconia (YSZ)/á-alumina composite supports by sputter deposition and electroless plating method. The effect of metal-substrate interaction on H2 permeation was investigated by comparing H2 fluxes through the Pd/γ-alumina/α-alumina and Pd/YSZ/α-alumina supprots in 633-713K. H2 fluxes through Pd/γ-alumina/α-alumina are lower than those through Pd/YSZ/α-alumina membrane of the same thickness under the same conditions. The permeation data are analyzed by a model considering hydrogen diffusion in the bulk of the Pd layer and surface reaction at the metal-support interface. The Pd-alumina interface provides much higher surface reaction resistance for hydrogen permeation than the Pd-zirconia interface due to formation of Pd-Al alloy at the Pd-alumina interface. The results show the importance of the selection of a chemically inert material as the support for thin metal membranes.

699b A New Method to Minimize Brittle Fracture Effects on Zeolite Membrane Performance: Demonstration In Columnar C-Oriented Zeolite Mfi Films

Jungkyu Choi[1], Hae-kwon Jeong[2], and Michael Tsapatsis[1]. (1) Chemical Engineering and Materials Science Dept., University of Minnesota, Department of Chemical Engineering and Materials Science, 421 Washington Avenue SE, Minneapolis, MN 55455-0132, (2) Chemical Engineering, Texas A&M University, 200 Jack E. Brown 3122 TAMU, College station, TX 77843

It is well-known that MFI type zeolites are selective for p-xylene over the other xylene isomers. There are many reports on the xylene separation performance of MFI membranes.[1-6] Among the available microstructures, 15~30 μm thick, columnar, c-oriented MFI membranes made by secondary growth of randomly oriented seed layers, are attractive for large scale industrial use, because they can be easily fabricated with high reproducibility compared to other MFI membranes made by in-situ and secondary growth methods.[7, 8] However, although single component permselectivity (or ideal selectivity) of p-xylene over o-xylene through c-oriented MFI membranes is high (up to ~ 100), the separation factor for the corresponding binary mixture is significantly lower, down to 2 ~ 4.[3] The significant reduction is attributed to the presence of non-zeolitic paths (defects) in the membranes. In particular, cracks are often generated during the synthesis, calcination or use due to the mismatch of thermal expansion behaviors of MFI films and porous supports.

We will report a new preparation method that minimizes defect formation and results in c-oriented columnar MFI membranes with unprecedented selectivities (e.g., xylene and butane isomer selectivities as high as 120 and 50, respectively). Implications in membrane manufacturing and scale up will be discussed.

References

1.Gump, C.J., et al., Industrial & Engineering Chemistry Research, 2001. 40(2), 565-577.

2.Sakai, H., T. Tomita, and T. Takahashi, Separation and Purification Technology, 2001. 25(1-3), 297-306.

3.Xomeritakis, G., Z. Lai, and M. Tsapatsis, Ind. Eng. Chem. Res., 2001. 40(2), 544-552.

4.Hedlund, J., et al., Microporous Mesoporous Mater., 2002. 52(3), 179-189.

5.Lai, Z., et al., Science (Washington, DC, U. S.), 2003. 300(5618), 456-460.

6.Yuan, W., Y.S. Lin, and W. Yang, Journal of the American Chemical Society, 2004. 126(15), 4776-4777.

7.Gouzinis, A. and M. Tsapatsis, Chem. Mater., 1998. 10(9), 2497-2504.

8.Xomeritakis, G., et al., Chem. Eng. Sci., 1999. 54(15-16), 3521-3531.

699d Preparation and Modeling of Nanoporous Silicon Carbide Membranes by Chemical-Vapor Infiltration/Chemical-Vapor Deposition Techniques

Rayan Mourhatch, Feng Chen, Theodore T. Tsotsis, and **Muhammad Sahimi**. Mork Family Department of Chemical Engineering and Material Science, University of Southern California, 925 Bloom Walk HED 216, Mork Family Department of Chemical Engineering and Material Science, Los Angeles, CA 90089-1211

Silicon carbide (SiC) has many attractive chemical and physical properties, which make it an excellent candidate for preparation of nanoporous membranes, especially those that are used for gas separation and hydrogen production under harsh conditions. Preparation of asymmetric nanoporous SiC membranes using chemical-vapor infiltration/chemical-vapor deposition (CVI/CVD) is described. We use macroporous SiC disks and tubes as supports, and tri-isopropylsilane as the precursor. Experimental data for the permeation and selectivity of the membranes are presented. We also develop two dynamic models to describe the process of preparation of the membranes. The models are shown to provide accurate predictions for the experimental data for the permeation characteristics of the membranes, as a function of the preparation conditions. Moreover, we have used an optimization technique in order to specify the best operating conditions for preparation of high quality SiC membranes. Preparing SiC membrane using the CVD technique operating under the optimal conditions results in significant improvement in the transport properties of the membranes. The SiC membranes that we have been preparing are to be utilized in reactive separation applications with the water-gas shift and methane steam reforming reactions, where the membrane must function in the presence of high-temperature steam.

699e Asymmetric Membrane Formation by Hollow Fiber Spinning - Theory and Experiment

Yang Su[1], Holly Balasubramanian[2], **Glenn Lipscomb**[1], and Doug Lloyd[2]. (1) Chemical and Environmental Engineering Department, The University of Toledo, MS 305, 2801 West Bancroft Street, Toledo, OH 43606-3390, (2) Chemical Engineering Department, The University of Texas at Austin, 1 University Station, C0400, Austin, TX 78712-0231

We present a theoretical study of the momentum, heat and mass transfer that occur in the draw zone of dry and dry- wet hollow fiber spinning processes. The governing two-dimensional conservation equations are solved for the axisymmetric geometry of the moving filament.

A numerical approximation to the solution of the conservation equations is obtained using the FIDAP computational fluid dynamics software package. FIDAP allows for determination of the location of the two free surfaces of the filament simultaneously with the velocity, concentration, and temperature fields within it.

The simulations reveal the presence of a recirculation region in the bore of the filament under typical fiber spinning conditions. This recirculation region leads to larger than expected fiber dimensions initially in the draw zone and a more rapid decrease to the final fiber size along the draw zone than predicted by the one-dimensional approximation to the conversation equations (the thin filament analysis).

The simulations also indicate that mass transfer is confined to a thin region near the external surface of the filament. The extent of this region and the concentration profile within it are predicted well by a mass transfer boundary layer analysis. Additionally, the predicted extent of the region agrees well with experimental data obtained for a polyethylene-dodecanol filament.

699f Spinneret Design in Fabricating Ultra-Thin Defect-Free and Macrovoid-Free Hollow Fiber Membranes for Gas Separation

Na Peng, Chemical and Biomolecular Engineering, National university of Singapore, 10 Kent Ridge Crescent, Singapore 119260, Singapore, Singapore, Natalia Widjojo, Biomolecular and Chemical Engineering, National University of Singapore, 10 Kent Ridge Crescent, Singapore, 117602, Singapore, and Tai-Shung Chung, Department of Chemical and Biomolecular Engineering, National University of Singapore, 10 Kent Ridge Crescent,, Singapore, 119260, Singapore.

Since the phase inversion process was developed to fabricate membranes in late 1950s, how to fabricate macrovoid-free and defect-free hollow fiber membranes with an ultra-thin dense selective layer for gas separation has always been a scientifically interesting topic. To the present, most literatures on fabricating hollow fiber membranes mainly focus on the effects of process parameters such as dope formulation, coagulant chemistry, dope extrusion rate, elongation stress, as well as air gap distance on membrane morphology and separation performance. However, membrane scientists must not ignore the importance of a spinneret's internal design and annulus dimension, because they play an utmost important role in determining not only the molecular orientation, chain packing, macro- and micro-pore formation of the resultant membranes, but also the membranes' separation performance and long term stability. Therefore, the main objective of this work is to investigate how a spinneret's internal design and annulus gap dimension affect 1) the macrovoid formation and 2) gas separation performance of the resultant hollow fiber membranes. Several spinnerets with different dimensions were designed and several interesting phenomena have been observed.

As for the morphological aspect, four types of macrovoids (i.e. inward- and outward-pointed, elliptical, and tear-drop shapes) can be observed simultaneously in the P84 copolyimide hollow fiber membranes. Hollow fibers spun from thinner spinneret annulus gaps show only inward-pointed long, tear drop, and elliptical shape macrovoids, while those spun from thicker annulus gaps have both inward and outward-pointed macrovoids. In addition, the number of inward-pointed macrovoids increases, while the number of outward-pointed macrovoids decreases with an increase in air gap distance. In terms of the effect of spinneret dimension in the formation of hollow fiber membranes for gas separation, defect-free Torlon® hollow fiber membranes with ultra-thin dense selective layers have been successfully fabricated without any post treatments. It is observed that narrower annular gap of the spinneret facilitates the formation of macrovoid-free hollow fibers, while broader one produces membranes with higher permselectivity and thinner dense selective layer. The highest ideal O_2/N_2 permselectivity is 9.06, with a dense layer of 540 •. The average selectivity of CO_2/CH_4 mixed gas is around 40.

This is a pioneer work in studying the effect of spinneret dimension in the formation mechanism of hollow fiber membranes. Besides providing an integrated understanding of elongation induced molecular orientation and shear induced molecular orientation in different spinnerets for hollow fiber formation, it is also believed our idea in spinneret design may start a new step in improving hollow fiber performance, not only for gas separation but also for all industrial applications.

699g Polymeric Nanofiltration Membranes with Engineered Molecular Weight Cutoff for Molecular Separations In Aggressive Solvents

Ines R. Baptista[1], Yoong H. See Toh[1], Issara Sereewatthanawut[2], Fui W. Lim[2], and Andrew G. Livingston[1]. (1) Chemical Engineering and Chemical Technology, Imperial College London, Prince Consort Road, SW7 2AZ, London, United Kingdom, (2) Membrane Extraction Technology Ltd., Unit 25 Talina Centre, Bagley's Lane, Fulham, London, SW6 2BW, United Kingdom

In recent years, there has been a growing interest in the application of Organic Solvent Nanofiltration (OSN) to many areas including the fine chemical, pharmaceutical and petrochemical industries [1]. OSN applications include the purification and fractionation of natural products, homogeneous catalyst recovery, colour removal from pharmaceutical products, and monomer separation from oligomers. Although there is no shortage of polymeric materials for membrane fabrication, current challenges facing the further industrial application of OSN include: i) the lack of commercially viable membranes which are stable in a broad range of organic solvents including polar aprotic solvents, and ii) the ability to control the permeation pathways to give tunable molecular weight cutoff (MWCO) profiles in the molecular weight range 200 – 1000 g mol-1 [2]. This paper will describe a range of new membranes prepared by phase inversion which meet these challenges, and have been scaled up as far as small spiral wound elements capable of performing OSN in DMF and other aggressive solvents.

Available OSN membranes fall into two types (i) composites made from polydemethylsiloxane films on polyacrylonitrile supports, or integrally skinned asymmetric membranes made from polyimides (PI). PIs offer several attractive mechanical and physicochemical properties, including high glass transition temperatures, good thermal stability and chemical stability in many solvents and weak acids. Commercial PI OSN membranes have been shown to give good performances in several organic solvents (toluene, methanol and ethyl acetate), but have generally poor stability and performance in polar aprotic solvents such as dichloromethane (DCM), tetrahydrofuran (THF), dimethyl formamide (DMF) and n-methyl pyrrolidone (NMP), in which most of these membranes are soluble. Inorganic membranes have been developed which offer good stability in organic solvents, but they are often more expensive and difficult to handle [3]. To date, there are few reports of OSN in aggressive solvents such as THF, DMF and NMP, which are of particular industrial interest.

Control over the separation performance of OSN membranes is desirable to achieve improved species separation. Ideally, a single membrane formation platform might be used to make a range of membranes with varying, controlled separation performances. At present, although other properties such as solute-polymer interactions can affect the membrane separation performance, the selection of OSN membranes is still based upon the MWCO. This is defined as the molecular weight at which 90% rejection is obtained by interpolation from a plot of solute rejection against molecular weight [4]. Although some relatively small changes to the MWCO were reported through changes in some membrane formation parameters (evaporation time prior to phase inversion, dope solution polymer concentration, thermal annealing of the membrane and use of additives in dope solution), successful control over the membrane separation performance was by and large still not achieved [5].

This paper will present new, integrally skinned asymmetric polyimide membranes made from the PI polymer Lenzing P84. These membranes were prepared by immersion precipitation from a casting solution containing DMF as the solvent and 1,4-dioxane as the co-solvent. The variation of the solvent compositions during membrane formation allowed for control over the molecular weight cut off membranes for the use in organic solvent nanofiltration. The chemical stability of these membranes was improved through chemical crosslinking using aliphatic diamines. The resultant membranes had a spongy structure and were stable in many organic solvents including toluene, methanol, DCM, THF, DMF and NMP. Operation in DMF and THF for 120 h showed that the membranes had stable fluxes and good separation performances, with DMF permeability in the range of 1–8×10-5 Lm-2 h-1 Pa-1 (1–8 Lm-2 h-1 bar-1) and a MWCO between 250 and 400 g mol-1. Possible re-imidization and loss of crosslinking at elevated temperatures limited their range of application to temperatures <100 °C. Their ease of preparation makes the membranes easily scalable and opens up future possibilities for further applications in harsh solvent environments.

In conclusion, this presentation will show that chemical crosslinking of PI membranes improves the chemical stability of these membranes allowing extended operation in harsh solvents such as DMF; also, that different dope compositions allows for good control over the MWCO in the NF range.

References:

1. U. Razdan, S.V. Joshi, and V.J. Shah, Novel membrane processes for separation of organics, Curr. Sci. Ind., 85 (2003) 761.

2. P. Vandezande, L.E.M. Gevers, and I.F.J. Vankelekom, Solvent resistant nanofiltration: separating on a molecular level, Chem. Soc. Rev., 37 (2008) 365.

3. Y.H. See-Toh, F.W. Lim, and A. Livingston, Polymeric Membranes for Nanofiltration in Polar Aprotic Solvents, J. Membrane Sci., 301 (2007) 3.

4. Y.H. See Toh, X.X. Loh, K. Li, A. Bismarck, and A.G. Livingston, In search of a standard method for the characterisation of organic solvent nanofiltration membranes, J. Membrane Sci., 291 (2007) 120.

5. Y.H. See-Toh, F.C. Ferreira, and A.G. Livingston, The influence of membrane formation parameters on the functional performance of organic solvent nanofiltration membranes, J. Membrane Sci., 299 (2007) 236.

700a Nano-Structured Ro/nf Membrane with the Resistance for Scaling In Desalination and Fouling In Bio-Separation

Myung-man Kim[1], Nancy H. Lin[1], Gregory T. Lewis[1], and Yoram Cohen[2]. (1) Chemical and Biomolecular Engineering, University of California, Los Angeles, Los Angeles, CA 90095, (2) Chemical and Biomolecular Engineering, UCLA, Los Angeles, CA 90095

Reverse Osmosis (RO) and Nanofiltration (NF) membranes used for surface and groundwater desalination are susceptible to bio-organic fouling (i.e., proteins, humic acid, fulvic acid), colloidal fouling and mineral salt scaling. Membrane fouling and/or scaling not only results in a decreased membrane permeate flux but also protein adhesion and mineral salt scale formation that may permanently alter the physical features of the surface and lead to irreparable membrane damage. Previous strategies for mitigating membrane fouling/scaling (i.e., polymer surface adsorption and UV, gamma irradiation, and low-pressure plasma graft polymerization) have relied on alteration of the membrane surface chemistry and topography by addition of a permselective polymer thin film that would act both as a separation layer and a physical boundary to prevent adsorption of organic and mineral salt species. In the present study, a novel atmospheric pressure plasma-induced graft polymerization method was developed to enable the generation of a high surface density of active surface sites for subsequent graft polymerization using a suitable monomer. The properties of the grafted polymer on the RO and NF membranes, specifically the surface density, polymer chain length, and monomer chemistry, were evaluated with respect to the membrane performance (i.e., onset of mineral scaling, water permeate flux decline and surface scale coverage) to determine the optimal surface structuring conditions required reduce surface fouling/scaling. Bio-organic fouling studies were conducted in a dilute aqueous feed stream of model proteins. Surface scaling was also evaluated by subjecting the surface structured membranes to a dilute aqueous mineral salt solution with the onset of mineral scaling detected by a novel scale-observation imaging system. The results suggest that surface modification of both RO and NF membranes by plasma-induced graft polymerization can be an effective tool for increasing membrane performance by decreasing the propensity for scaling and fouling.

700b Pretreatment Strategies for Nanofiltration of Natural Organic Matter in Groundwater

Wonho Song, Varadarajan Ravindran, and Massoud Pirbazari. Civil and Environmental Engineering, University of Southern California, Los Angeles, CA 90089-2531

Pretreatment strategies could significantly affect the performance of nanofiltration membranes used in water treatment scenarios. Pretreatment techniques including coagulation and advanced oxidation processes (AOPs) could alter the surface properties of membranes that are critically important in establishing their fouling potential and performance with reference to permeate flux and solute rejection. This research evaluates surface characteristics and fouling potential of membranes subjected to pretreatment by coagulation and AOPs, as well as cleaning agents containing acid, base, or surfactant. The study investigates the permeate flux of nanofiltration membranes used in water treatment for the removal of NOM from groundwater. The investigations compare the permeate fluxes and flux decline patterns for nanofiltration membranes subjected to pretreatment by coagulation and by AOP using ultraviolet radiation and hydrogen peroxide.

In this investigation, nanofiltration membranes at various stages of filtration operation were subjected to chemical coagulation pretreatment and AOP pretreatment followed different cleaning solutions. The permeate flux decline patterns were compared for raw water, raw groundwater subjected to coagulation pretreatment, and AOP pretreatment prior to nanofiltration. The study also investigated the influence of cross-flow filtration rates on the fouling and flux reduction patterns. The permeate flux data provided sufficient information to determine which pretreatment strategy produced less membrane fouling and minimized permeate flux decline.

Additionally, the study evaluates morphological and/or structural damage occurring on membranes due to residual oxidant species and hydroxyl, peroxyl or other free-radical species generated during water treatment. This is an important consideration in membrane processes such as nanofiltration and reverse osmosis, because membranes fabricated from polymeric materials by sol-gel or other processes are prone to polymer decomposition and structural alteration. The techniques employed for evaluating fouling potential and morphological details of membrane surfaces include atomic force microscopy (AFM), Fourier transform infrared spectroscopy (FTIR), and x-ray photoelectron spectroscopy (XPS). AFM studies provided information on membrane surface topography under different operating conditions, and XPS and FTIR provided details on numbers and types of organic functional groups on membrane surfaces attributable to different pretreatment techniques. The results provided useful information on several aspects: amelioration of membrane fouling (and permeate flux recovery) due to the application of cleaning agents, alteration of membrane surface characteristics attributed to membrane cleaning, and damage to membrane material caused by oxidative reactions associated with AOPs.

Keywords: Nanofiltration, pretreatment strategies, coagulation, advanced oxidation processes, natural organic matter, membrane fouling, surface characterization

700c Cellulose Acetate Ultrafiltration Membranes Modified with Temperature-Sensitive Polymers for Fouling Resistance

Colleen Gorey[1], Isabel Escobar[1], Amr Zaky[2], Guang Cai[2], and Cyndee L. Gruden[2]. (1) Department of Chemical and Environmental Engineering, University of Toledo, 3801 W. Bancroft St, Toledo, OH 43606, (2) Department of Civil Engineering, University of Toledo, 3801 W. Bancroft St, Toledo, OH 43606

The presence of microorganisms in feed water can exacerbate fouling due to the accumulation of microorganisms onto the membrane surface and on the feed spacer between the envelopes, or biofouling. Microorganisms transported to the membrane element can attach to the feed side of the membrane and the spacer. Attachment depends on Van der Waals forces, hydrophobic interactions and electrostatic interactions between the microorganisms and the surface. Biofouling control has been attempted via biocide additions; however, while a biocide may kill the biofilm organisms, it usually does not remove the biofouling layer, and may cause bacteria that survive disinfection to potentially become more resistant.

The goal of this study was to produce a fouling-resistant membrane by attaching a stimuli-responsive polymer on the membrane surface, which collapsed or expanded as a response to the stimulus. The phase change arises from the existence of a lower critical solution temperature (LCST) such that the polymer precipitates from solution as the temperature is increased. The polymers studied for this application were hydroxypropyl cellulose and N-isopropylacrylamide, which have LCSTs of 46 C and 32 C, respectively.

Roughness measurements, using a wet atomic force microscopy (AFM) cell, and filtration experiments (to monitor flux declines) were performed at cold temperatures (25oC), at hot temperatures (60oC) and with temperature oscillations. The unmodified membrane roughness values were unaffected by temperature changes, and it displayed flux declines under all temperature conditions. When the modified membranes were tested, both roughness values and filtration experiments supported temperature activation.

Developing a membrane with high flux and selectivity along with low fouling is one of the goals of membrane research. There are physical and chemical limitations, however, that cannot be disregarded. For microfiltration/ultrafiltration systems higher flux operation requires energy input, in the form of back flushes, air scouring, physical agitation, etc, to reduce fouling. The work presented here developed a membrane with desired inherent properties along with fouling control.

700d Fouling Resistant Coating Materials for Oil/water Separation

Hao Ju, Bryan McCloskey, Alyson Sagle, Yuan-Hsuan Wu, and **Benny D. Freeman**. Chemical Engineering, The University of Texas at Austin, 10100 Burnet Rd., Bldg 133, Center for Energy and Environmental Resources, Austin, TX 78758

Potential fouling reducing coating materials were synthesized via free-radical photopolymerization of aqueous solutions of poly(ethylene glycol) diacrylate (PEGDA) and other related classed of hydrogel polymers. Crosslinked PEGDA (XLPEGDA) exhibited high water permeability and good fouling resistance to oil/water mixtures. Water permeability increased strongly with increasing the water content in the prepolymerization water mixture, going from 10 to 150 L μ m/(m2 h bar) as prepolymerization water content increased from 60 to 80 wt.%. However, molecular weight cutoff decreased as water content increased. These materials were applied to polysulfone (PSF) UF membranes to form coatings on the surface of the PSF membranes. Oil/water crossflow filtration experiments showed that the coated PSF membranes had water flux values 400% higher than that of an uncoated PSF membrane after 24 hours of operation, and the coated membranes had higher organic rejection than the uncoated membranes. Results for other polymer systems and their use as fouling-reducing coatings will also be presented.

700e Performance Characteristics of Virus Filtration Membranes

Meisam Bakhshayeshi and Andrew Zydney. Chemical Engineering, The Pennsylvania State University, 219 Fenske Laboratory, University Park, PA 16802

In recent years, virus filtration has been introduced as a robust, size-based operation capable of providing the viral clearance needed for production of therapeutic proteins. Several different companies make virus filtration membranes for bioprocessing applications. Although these membranes all provide at least 3-logs of viral clearance, the membranes have very different capacity and filtrate flux due to large differences in clean membrane permeability and protein fouling characteristics. The objective of this study was to develop a fundamental understanding of the mechanisms governing protein fouling and capacity of virus filtration membranes and the relationship between these performance characteristics and the underlying membrane structure.

Experiments were conducted using commercially available Ultipor DV20 (Pall Corp.) and Viresolve 180 (Millipore Corp.) virus filters, both made from polyvinylidene fluoride. The Viresolve 180 is a composite structure with the pore size increasing from less than 35 nm at the upper skin to micron size in the substructure. In contrast, the DV20 has a fairly uniform pore size throughout the structure, resulting in a hydraulic permeability that is more than 25 times smaller than that of the Viresolve 180. Bovine serum albumin (BSA) was used as a model protein. Data were obtained in a stirred ultrafiltration cell at both constant pressure and constant filtrate flux. The effects of fouling on the membrane were examined from both buffer permeability and dextran sieving measurements obtained with the clean and fouled membranes.

The Viresolve 180 membrane displayed a very sharp flux decline during constant pressure operation, with large differences in performance depending upon the orientation of the membrane (skin-side facing up or down). The dominant contribution to the flux decline in the skin-up orientation was concentration polarization, with membrane fouling being most important with the skin-side down. In contrast, the rate of flux decline was quite slow with the DV20 membrane even over 24 hr, which was directly related to the very low flux arising from the low hydraulic permeability of these membranes. The flux decline with the DV20 membrane was due almost entirely to fouling, with no measurable effect of concentration polarization. The volumetric capacity (L/m^2) of the DV20 membrane was an order of magnitude larger than that of the Viresolve 180 for constant pressure operation, although the DV20 required much longer filtration times. In addition, the volumetric capacity of the DV20 was nearly independent of the bulk protein concentration, while the volumetric capacity of the Viresolve 180 membrane decreased almost linearly with increasing protein concentration. The very different behavior of these membranes is discussed in terms of the differences in underlying pore structures. These results provide important insights into the design and operation of virus filtration membranes.

701a Prediction of H2 Permeance through Pd-Based Binary and Pdcu-Based Ternary Alloys Via Ab-Initio Calculations

Lymarie Semidey-Flecha, Chen Ling, and David S. Sholl. Georgia Institute of Technology, School of Chemical & Biomolecular Engineering, 311 Ferst Drive, N.W., Atlanta, GA 30332-0100

First-principles calculations offer a useful complement to experimental approaches for characterizing hydrogen permeance through dense metal membranes. A challenge in applying these methods to disordered alloys is to make quantitative predictions for the net solubility and diffusivity of interstitial H based on the spatially local information that can be obtained from first-principles calculations. In this study we used a combination of Density Functional Theory calculations and a cluster expansion method to describe interstitial H in alloys of composition Pd96M4 where M included Ag, Au, Cu, Ni, and Pt. We also applied these methods to alloys of composition Pd70Cu30 and Pd70Cu26M4 where M = Ag, Au, Ni and Pt. The cluster expansion approach highlights the shortcomings of simple lattice models that have been used in the past to study similar binary systems. We use Sieverts' law to calculate H solubility and a Kinetic Monte Carlo (KMC) scheme to find the diffusion of H in PdAg, PdAu, PdCu, PdNi, PdPt, PdCuAg, PdCuAu, PdCuNi, and PdCuPt alloys at a temperature range of $400 \leq T \leq 1200$ K. From these results we are able to predict the permeability of hydrogen through thick membranes made from these Pd-based binary alloys and PdCu-based ternary alloys.

THE EFFECT OF H_2S ON THE LONG-TERM STABILITY OF Pd/Cu MEMBRANES AND THE CHARACTERISITICS OF H_2S POISONING OF ELECTROLESS DEPOSITED Pd

Natalie L. Pomerantz, Worcester Polytechnic Institute, Worcester, MA, USA
Elisabeth Shaw, Massachusetts Institute of Technology, Cambridge, MA, USA
Yi Hua Ma, Worcester Polytechnic Institute, Worcester, MA, USA

Introduction

Hydrogen selective Pd and Pd alloy membranes can improve the efficiency and cost effectiveness of the coal gasification process by recovering high purity H_2 from the syngas. Also, the high pressure CO_2 in the retentate reduces the cost of carbon capture required for carbon sequestration. However, small quantities of H_2S present in the gas stream poison the Pd membranes and reduce the hydrogen permeance by either adsorbing on the surface or reducing the selectivity of the membrane by forming a sulfide scale. Pd/Cu membranes have been tested and shown to be more tolerant to H_2S than pure Pd, but bulk analysis methods such as EDS (Energy Dispersive Spectroscopy) and XRD (X-ray Diffraction) were not sufficiently surface sensitive to detect the adsorbed sulfur on poisoned membrane surfaces. Therefore, the objective of this study was to examine the interactions of H_2S with Pd that led to membrane poisoning with the more surface sensitive technique of XPS (X-Ray Photoelectron Spectroscopy), and to test the long-term stability of Pd/Cu membranes in the presence of H_2S with the purpose of designing sulfur tolerant membranes.

Experimental

All of the supports used in this study were provided by Mott Metallurgical, Inc. Coupons (1 cm x 1 cm x 0.1 cm) consisted of 316L PSS (porous stainless steel) with a 0.5 μm media grade. The tubular membrane supports (length - 6 cm, OD - 1.25 cm, thickness - 0.16 cm). consisted of a porous Inconel alloy with a 0.1 μm media grade. All supports were cleaned with an alkaline solution to remove impurities as described elsewhere [1] and oxidized at 700°C for 12 hours in order to form an oxide intermetallic diffusion barrier [2, 3] between the support metals and the H_2 selective Pd layer.

To block the larger pores of the support before plating the dense Pd layer, the support was attached to a vacuum and immersed in an acidic Al_2O_3 slurry for 10 seconds and dried at 140°C for 12 hours. To seal the slurry within the pores of the support, the support was activated once under vacuum and plated with Pd for 10 mintues. A Pd/Ag barrier was deposited which further bridged the larger pores of the support and also functioned as an intermetallic diffusion barrier between the support metals and the hydrogen selective layer, further details of the procedure are described elsewhere [4, 5]. The supports were activated and plated until dense, according to the procedure outlined in Mardilovich, et al. [1]. The Cu plating bath was adapted from Ma, et al. [6] and the bath composition is shown with the Pd and Ag plating bath compositions in Table 1. Cu was plated for 30 – 60 minutes, depending on the desired Cu content. After the electroless deposition of Cu, the membranes were immediately immersed in 0.01 M HCl to neutralize any residual plating solution, rinsed with DI water and then rinsed with ethanol to facilitate drying in order to prevent oxidation of the Cu layer.

The coupons were annealed in a tubular furnace and the membranes were tested in a shell and tube apparatus [1]. The He leaks of the membranes were measured at a pressure difference of 1 atm and the ideal H_2/He separation factors were calculated according to the ratio of H_2 to He fluxes through the membrane at a pressure difference of 1 atm (P_{feed} = 2 atma, $P_{permeate}$ = 1 atma).

Table 1. Plating bath compositions and conditions.

Component	Pd bath	Cu bath	Ag bath
$Pd(NH_3)_4Cl_2{\cdot}H_2O$ (g/l)	4		
$CuSO_4{\cdot}5H_2O$ (g/l)		25	
$AgNO_3$ (g/l)			0.519
$Na_2EDTA{\cdot}2H_2O$ (g/l)	40.1	47.5	40.1
NH_4OH (28%) (ml/l)	198		198
H_2NNH_2 (1 M) (ml/l)	5.6		5.8
H_2CO (37%) (ml/l)		25	
EDA (ppm)		112	
$K_4Fe(CN)_6{\cdot}3H_2O$ (ppm)		35	
$(C_2H_5)_2NCS_2Na{\cdot}3H_2O$ (ppm)		5	
pH	10 - 11	12.0	10 - 11
Temperature (°C)	60	20 - 25	60

Surface characterizations were carried out with the Amray 1610 Turbo scanning electron microscope (SEM) equipped with a Princeton Gamma-Tech Avalon energy dispersive X-ray light element detector and a RBA-1610 5MC type Robinson retractable backscattered electron detector for qualitative and quantitative analysis. The spatial resolution for SEM – EDS lies between 0.8 – 1.2 μm and the penetration depth for EDS was 1 μm for the samples investigated. Surface elemental analysis was performed with a Kratos AXIS Ultra X-ray Photoelectron Spectrometer using a monochromatized aluminum source operated at 150 W, with an analysis area of approximately 300 x 700 um and a maximum information depth of approximately 10 nm. The pass energy used for high energy resolution spectra was 20 eV. CasaXPS software [7], equipped with a Scofield-cross-section-based sensitivity factor library, was used for the peak-fitting and quantitative analysis. All spectra were calibrated using the adventitious carbon C 1s peak at 285.0 eV and a Shirley-type background was subtracted from the spectra.

Results and Discussion

Pd coupon study

In order to observe the Pd – H_2S interactions at membrane operating conditions, and to assess the possible effects on membrane performance, electroless deposited Pd samples were exposed to 54.8 ppm H_2S/H_2 and pure H_2 for 24 hours over a range of 350 - 500°C. Figure 1a shows the XPS survey spectra of three of the annealed samples. The top spectrum was a sample annealed in pure H_2 and exhibited no peaks characteristic of sulfur. In the case of the Pd samples annealed at 350 and 500°C in 54.8 ppm H_2S/H_2, the peaks identifying the 2s and 2p orbitals of sulfur were present, showing that sulfur had adsorbed on the Pd surface at both temperatures. Similar results were seen with the Pd samples annealed at 400 and 450°C in 54.8 ppm H_2S/H_2.

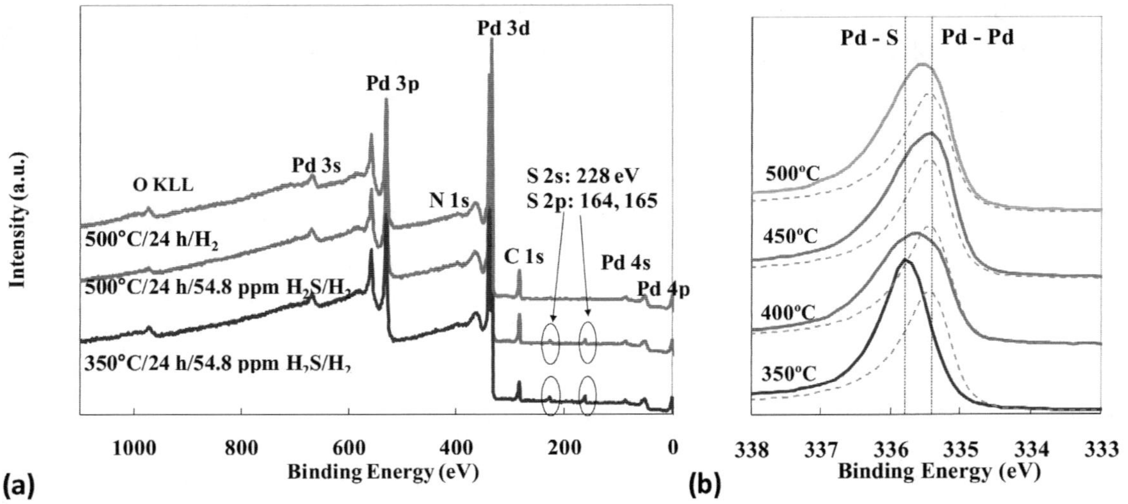

Figure 1. (a) XPS survey spectra of the annealed Pd samples and (b) High resolution XPS scan of the $3d_{5/2}$ peaks of the annealed Pd samples. The solid and dotted lines correspond to the samples annealed in 54.8 ppm H_2S/H_2 and pure H_2, respectively.

Figure 1b shows the high resolution XPS scans of the $3d_{5/2}$ Pd orbital. The peak broadening and shift to higher binding energies than that of the Pd – Pd peak maximum (dotted lines) at 400, 450 and 500°C showed that Pd – S bonds had formed on the surface, indicating that H_2S had undergone dissociative chemisorption and formed a surface sulfide with the Pd. Bulk sulfides had not formed as evidenced by the broad peaks between 400 - 500°C that indicated the presence of the Pd – Pd chemistry near the surface as well as the Pd – S chemistry. However, at 350°C the $3d_{5/2}$ peak of Pd was not as broad and the maximum was at the corresponding binding energy of the Pd – S peak maximum. The Pd – Pd chemistry was not detected at this temperature, indicating that bulk Pd sulfide had formed. The values of the Pd – Pd and Pd – S chemistries were confirmed with the literature [8 – 10].

Figure 2 shows the amount of adsorbed sulfur on the Pd surface as a function of temperature. The S/Pd atomic ratio was calculated from the survey scans shown in Figure 1a and the amount of the Pd – S and Pd – Pd chemistries were calculated from the high-resolution scans shown in Figure 1b. The amount of sulfur adsorbed and consequently, the amount of sulfur which bonded with Pd, increased with decreasing temperature. Sulfur adsorption is an exothermic reaction and higher temperatures would cause less sulfur to adsorb [11]. A similar trend was seen with bulk Pd sulfides in previous works [12, 13] indicating that bulk sulfides would only form at lower temperatures. Indeed, at 350°C, the S/Pd atomic ratio was 0.25, which corresponded to the atomic ratio of Pd_4S. The XRD analysis further confirmed that the bulk sulfide Pd_4S phase was indeed present, in contrast to the samples annealed at 400, 450 and 500°C which only resulted in surface sulfide formation.

Figure 3a shows the uniform granular morphology of Pd clusters annealed in H_2. Figure 3b shows that the surface sulfide formation caused a change to a more chunky morphology (encircled on the left side of the micrograph) and the formation of pores (encircled in the center of the micrograph) at the cluster boundaries. The pores and morphology change were both probably caused by the restructuring of the Pd surface atoms as they bonded with the sulfur. Figure 3c shows the Pd_4S scale which had a drastically different morphology than the annealed Pd. While Pd has a FCC lattice structure, Pd_4S has a tetragonal lattice structure and a lattice parameter almost twice that of pure Pd. The lattice reordering caused by the formation of the bulk Pd sulfide caused a coral-like scale to form on top of the Pd clusters. While the performance of a Pd membrane with adsorbed sulfur would deteriorate due to a decrease in the permeance caused by sulfur preventing the H_2 adsorption, and a decrease in the selectivity caused by the pore formation, a Pd membrane forming a Pd_4S scale during operation would suffer such an irreparable damage to the selectivity that the membrane would only have Knudsen selectivity [14].

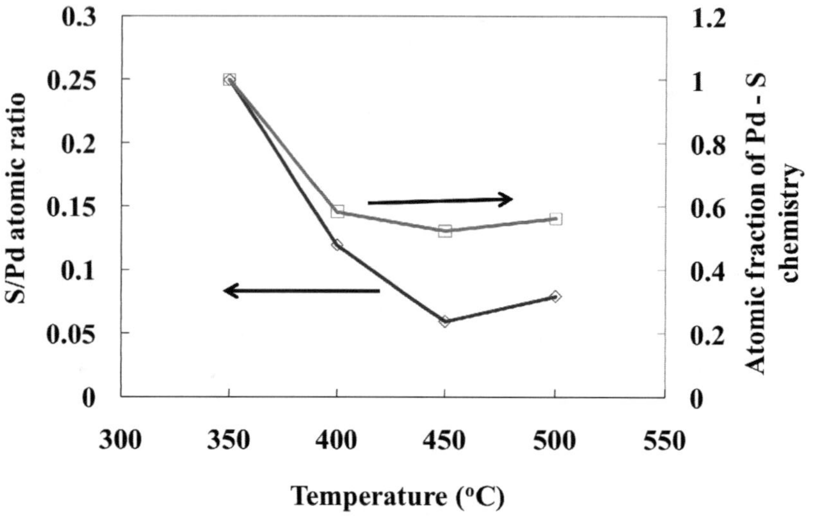

Figure 2. Dependence of the amount of adsorbed sulfur on temperature.

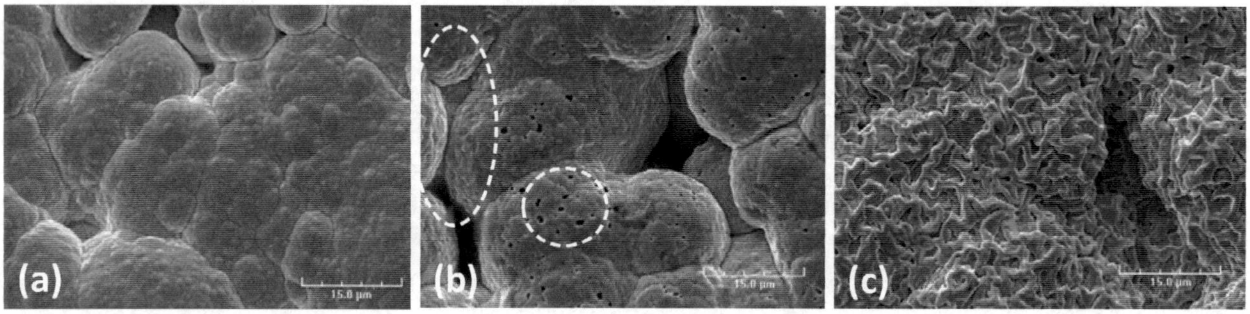

Figure 3. Scanning electron micrographs of (a) deposited Pd annealed at 500°C in H_2 for 24 hours, (b) deposited Pd annealed at 450°C in 54.8 ppm H_2S/H_2 for 24 hours, (c) deposited Pd annealed at 350°C in 54.8 ppm H_2S/H_2 for 24 hours.

Pd/Cu membrane study

The Pd/Cu membrane consisted of a 5.1 μm porous Pd/Ag barrier, a 7.6 μm Pd layer and a 3.8 μm Cu layer for a total Cu composition of 18 wt%. The membrane was characterized between the temperatures of 250 - 450°C for over 800 hours before the sulfur testing. The permeance at 450°C was 19.9 $m^3/m^2*bar^{0.5}*h$ and the activation energy of permeation was 15 kJ/mol. The ideal H_2/He separation factor decreased from 1800 to 220 during the characterization. At the end of the pure H_2 permeationg characterization, the Pd/Cu membrane was exposed to a 54.8 ppm H_2S/H_2 mixture at 450°C.

As seen in Figure 4a, the permeance of the membrane dropped instantaneously to 21% of the original permeance at 450°C upon the introduction of 54.8 ppm H_2S/H_2. During the exposure to H_2S, the permeance remained stable for 125 hours, signifying that the cause of the permeance drop was from the surface sulfide formation since it was observed that bulk sulfide formation caused a continuous decrease in the permeance over time as the sulfide scale formed and thickened [15]. The permeance was recovered by reintroducing pure H_2 to the feed stream, which further substantiated the notion of the surface sulfide formation. During the recovery, the hydrogen permeance increased at a rate of 0.017 $m^3/(m^2*bar^{0.5}*h)*h$ for roughly 300 hours until the permeance stabilized at 64% of the original permeance. The inability to completely recover the permeance was an indication that part of the sulfide poisoning was irreversible at these testing conditions. The sulfur adsorption consisting of both irreversible and reversible surface sulfides has also been observed at 500°C with poisoned and recovered Pt/Al_2O_3 catalysts [16].

The activation energy of permeation following the permeance recovery was 13.8 kJ/mol. The fact that the activation energy did not significantly change before and after the H_2S testing indicated that the irreversible sulfur poisoning resulted in a decrease in the effective membrane area for the H_2 adsorption, and did not change the hydrogen transport mechanism. In accordance with the XPS studies, operating the Pd/Cu membrane at a higher temperature could result in less of a permeance drop and more permeance recovered in pure H_2, due to less of the surface sulfide formation.

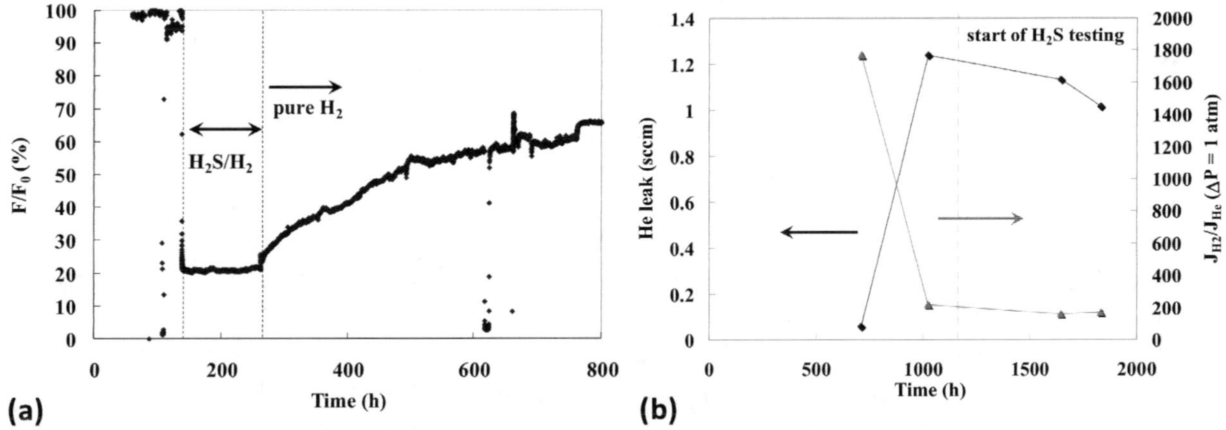

Figure 4. (a) The relative permeance of the Pd/Cu membrane as a function of time during poisoning and during recovery in pure H_2 at 450°C. (b) The leak and ideal H_2/He separation factor as a function of time for the Pd/Cu membrane at 450°C.

Figure 4b shows the change in leak for the duration of the Pd/Cu membrane testing. At temperatures lower than 450°C, non-permeable gases in the permeate were undetectable, but at 450°C a very small He leak of 0.06 sccm was detected which increased to 1.24 sccm after 300 hours, resulting in a decrease in the ideal H_2/He separation factor. The development of leaks at high temperatures over long periods of time has been documented previously [17, 18] and the cause was attributed to pinhole formation due to grain coarsening. Contrary to these results, the He leak of the Pd/Cu membrane decreased following the H_2S testing to 1.15 and then 1.01 sccm over a period of 700 hours at 450°C. The decrease in the He leak could have been caused by sulfur segregating to the grain boundaries of the deposits, as has been previously observed [19, 20], which would reduce or block pathways for non-permeable gases. In contrast to the implications of the Pd coupons study, the Pd/Cu membrane showed leak stability in the presence of H_2S.

Conclusions

It was found that the bulk sulfide Pd_4S phase transformation did not occur at temperatures higher than 350°C. At 400 - 500°C, H_2S formed Pd surface sulfides, which caused slight morphology changes to the Pd surface and pore formation at the cluster boundaries. In addition, the sulfur adsorption on the Pd decreased at higher temperatures due to the exothermic nature of H_2S adsorption. While the surface sulfide formation would reduce both the permeance and selectivity of a Pd membrane, the change in lattice parameter caused by bulk Pd_4S sulfide formation would render a Pd membrane useless.

During exposure to the 54.8 ppm H_2S/H_2 mixture, the Pd/Cu membrane had a reduced permeance due to the surface sulfide formation, but did not show an increase in leak that a pure Pd membrane would exhibit. Even more so, the He leak decreased during the sulfur testing, most likely due to sulfur segregating in the grain boundaries. The poisoning of the Pd/Cu membrane was found to be partly irreversible in that the full H_2 permeance could not be recovered. In accordance with the Pd coupon study, increasing the operating temperature should increase the tolerance of the membrane.

Acknowledgements

The authors gratefully acknowledge the financial support provided by the University Coal Research Program, U.S. Department of Energy, Award No. DE-FG26-04NT42194.
This work made use of the Shared Experimental Facilities supported by the MRSEC Program of the National Science Foundation under award number DMR 02-13282.

References

1. Mardilovich, P.P.; She, Y.; Ma, Y.H. and Rei, M. "Defect free Palladium membranes on porous stainless steel support," *AIChE Journal* **44**(2) (1998) 310 - 322.
2. Ma, Y.H.; Mardilovich, P.P. and She, Y. "Hydrogen gas-extraction module and method of fabrication." US Patent 6,152,987, 2000.
3. Ma, Y.H.; Akis, B.C.; Ayturk, M.E.; Guazzone, F.; Engwall, E.E. and Mardilovich, I.P. "Characterization of intermetallic diffusion barrier and alloy formation for Pd/Cu and Pd/Ag porous stainless steel composite membranes." *Industrial Engineering and Chemistry Research* **43** (2004) 2936 – 2945.
4. Ayturk, M.E.; Mardilovich, I.P.; Engwall, E.E. and Ma, Y.H. "Synthesis of composite Pd-porous stainless steel (PSS) membranes with a Pd/Ag intermetallic diffusion barrier." *Journal of Membrane Science* **285** (2006) 385 – 394.
5. Ma, Y.H.; Mardilovich, I.P. and Engwall, E.E. "Composite gas separation modules having intermediate porous metal layers." US Patent 7,175,694, 2007.
6. Ma, Y.H.; Mardilovich, I.P. and Engwall, E.E. "Thin composite Palladium and Palladium/alloy membranes for Hydrogen separation." *Annals of the New York Academy of Sciences* **984** (2003) 346 – 360.
7. Fairbanks, N. and Carrick, A. The Casa Cookbook. Acolyte Science, 2005.
8. Matsumoto, Y.; Soma, M.; Onishi, T. and Tamaru, K. "State of sulfur on the Palladium surface studied by Auger electron spectroscopy, electron energy loss spectroscopy, ultraviolet photoelectron spectroscopy and X-ray photoelectron spectroscopy." *Journal of the Chemical Society Faraday I* **76** (1980) 1122 – 1130.
9. Endo, K.; Ohno, H.; Matsuda, K.; and Asakura, S. "Electrochemical and surface studies on the passivity of a dental Pd-based casting alloy in alkaline sulphide solution." *Corrosion Science* **45** (2003) 1491 – 1504.
10. Chaplin, B.P.; Shapley, J.R.; and Werth, C.J. "Regeneration of sulfur-fouled bimetallic Pd-based catalysts." *Enviornmental Science Technology* **41** (2007) 5491 – 5497.
11. Bartholomew, C.H.; Agrawal, P.K. and Katzer, J.R. "Sulfur poisoning of metals." *Advances in Catalysis* **31** (1982) 135 – 242.
12. Taylor, J.R. "Phase relationships and thermodynamic properties of the Pd-S system." *Metallurgical Transactions B* **16B** (1985) 143 – 148.

13. Iyoha, O.; Enick, R.; Killmeyer, R. and Morreale, B. "The influence of hydrogen sulfide-to-hydrogen partial pressure ratio on the sulfidization of Pd and 70 mol% Pd-Cu membranes." *Journal of Membrane Science* **305** (2007) 77 – 92.
14. Kulprathipanja, A.; Alptekin, G.O.; Falconer, J.L. and Way, J.D. "Pd and Pd-Cu membranes: inhibition of H_2 permeation by H_2S." *Journal of Membrane Science* **254** (2005) 49 – 62.
15. Morreale, B.D.; Howard, B.H.; Iyoha, O.; Enick, R.M.; Ling, C. and Sholl, D.S. "Experimental and computational prediction of the hydrogen transport properties of Pd_4S." *Industrial Engineering and Chemical Research* **46** (2007) 6313 – 6319.
16. Barbier, J.; Lamy-Pitara, E.; Marecot, P.; Boitiaux, J.P.; Cosyns, J. and Verna, F. "Role of sulfur in catalytic hydrogenation reactions." *Advances in Catalysis* **39** (1990) 279 – 318.
17. Guazzone, F. and Ma, Y.H. "Leak growth mechanism in composite Pd membranes prepared by the electroless deposition method." *AIChE Journal* **54**(2) (2008) 487 – 494.
18. Guazzone, F.; Payzant, E.A.; Speakman, S.A. and Ma, Y.H. "Microstrains and stresses analysis in electroless deposited thin Pd films." *Industrial Engineering and Chemical Research* **45** (2006) 8145 – 8153.
19. Ainslie, N.G.; Hoffman, R.E. and Seybolt, A. U. "Sulfur segregation at α-iron grain boundaries." *Acta Metalligraphica* **8** (1960) 523-527.
20. Iyoha, O.; Enick, R.; Killmeyer, R.; Howard, B.; Ciocco, M. and Morreale, B. "H_2 production from simulated coal syngas containing H_2S in multi-tubular Pd and 80 wt% Pd-20 wt% Cu membrane reactors at 1173 K." *Journal of Membrane Science* **306** (2007) 103 – 115.

Characterization of sulfur resistance of Pd/Au hydrogen separation membrane prepared by galvanic displacement

Chao-Huang Chen and Yi Hua Ma, Worcester Polytechnic Institute, Worcester, MA, USA

Introduction

Sulfur containing impurities are known to cause severe and irreversible poisoning of Pd membranes during the hydrogen separation application (Especially in the coal gasification) resulting in the reduction of hydrogen permeability, selectivity and structure deterioration of Pd membranes[1,2,3]. Pd/Au alloys are found to show high sulfur resistance[1,4] in comparison to pure Pd and other Pd based alloys (i.e., Pd/Ag, Pd/Cu, etc.) in addition to its high hydrogen permeability[5,6]. As a result, Pd/Au alloys have recently regained great interests for membrane separation process. The preparation of Pd/Au alloys can be carried out by the galvanic displacement method[7,8,9], which does not require external sources of current, reducing agent and complexing agent (such as cyanide) and provides simplified plating equipment and environmentally preferable procedure. The main objectives of this study were to investigate the effect of temperature on sulfur poisoning of pure Pd and Pd/Au alloy morphologically. In addition, the hydrogen permeation characteristics of the Pd/Au membrane in the presence of H_2S and the permeance recovery after the sulfur poisoning at different temperatures were investigated.

Experimental

Porous 316 stainless steel (PSS) plates were used for the coupon preparation, while porous Inconel tubes were used for the membrane preparation. Both were media grade of 0.1 μm and purchased from Mott Metallurgical Corp. Pure Pd coupons were prepared by the electroless plating[10] on the oxidized (700°C, 12 hours) PSS plates (1 cm x 1 cm), and Pd/Au alloy coupons were prepared by the Au displacement plating on the Pd plated PSS coupons (6-9 μm thick Pd) followed by annealing at 550°C for 24 hours in H_2. The prepared pure Pd and Pd/Au alloy coupons were then exposed to a 54.8 ppm (±1ppm) H_2S/ H_2 mixture for 24 hours at the temperature from 350°C to 500°C with the interval of 50°C, and characterized by SEM (including EDX) and XRD. A Pd/Au membrane, designated as C04, was prepared on the porous Inconel support (1.27 cm in O.D. and 5.08 cm in length) by the electroless plating of Pd followed by the displacement plating of Au. Prior to the plating, the support was oxidized at 700°C for 12 hours followed by the successive Pd and Ag deposition to form the inter-metallic diffusion barrier[11,12], and between the oxides and Pd/Ag layer, the support was graded with the Al_2O_3 slurry. The final thickness of C04 was estimated gravimetrically as 18 μm with approximately 10wt% Au. The prepared membrane was then characterized for its gas permeating characteristics in a presence of a 54.8 ppm (±1ppm) H_2S/ H_2 mixture at the temperature from 350°C to 500°C.

Results and Discussion

In order to investigate the sulfur poisoning effect morphologically, poisoning tests of pure Pd and Pd/Au alloy coupons with a 54.8 ppm (±1 ppm) H_2S/ H_2 mixture for 24 hours at the temperature from 350°C to 500°C with interval of 50 °C were performed. XRD patterns and surface morphology of pure Pd coupons after poisoning were shown in Figure 1. From the X-ray patterns, a number of new peaks appeared in addition to the pure Pd peaks in all plane indicating the formation of new phases after the poisoning. The new peaks were identified as the palladium sulfide (Pd_4S) phase. The formation of the Pd_4S indicated that the chemical reaction between Pd and H_2S and the diffusion of adsorbed sulfur into the interior of the Pd layer forming bulk sulfide compounds. Further, the Pd_4S peak intensity was found to decrease as the poisoning temperature increased and disappeared at 500°C, indicating that the formation of palladium sulfide and the sulfur poisoning effects were temperature dependent. The less poisoning effect at higher temperatures might be resulted from the higher desorption rate of H_2S at higher temperatures. Further, the surface morphology showed the "net-like" surface structure with numerous pinholes after poisonings, which might be the result of the incorporation of sulfur into the Pd lattice, causing stress in the Pd lattice. The number of pin holes and the sulfur content detected by EDX decreased as the temperature was increased, indicating the less sulfur adsorption and poisoning at higher temperatures, which agreed with the results from the XRD analysis.

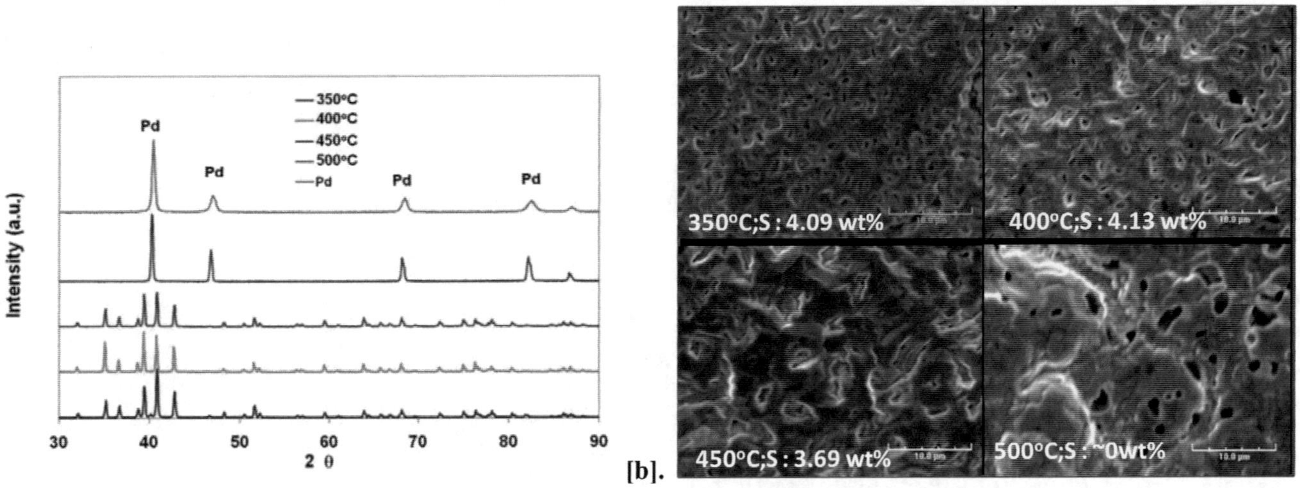

[a]. [b].

Figure 1. Temperature effect of sulfur poisoning of Pd on [a] crystalline phase, and [b] morphology.

Similar to Pd coupons, Pd/Au alloy coupons were also poisoned under the same conditions and characterized, and the resultant XRD patterns and morphology were shown in Figure 2. Unlike the pure Pd samples, the Pd/Au alloy samples showed no new peaks formation after the poisoning in the entire temperature range, and the XRD patterns of all tested temperatures remained almost identical in comparison to the one before poisoning. This indicated that the adsorption of H_2S on Pd/Au alloy was extremely small or even negligible with the same poisoning conditions as that of pure Pd coupons, suggesting the better sulfur resistance of Pd/Au alloy than pure Pd. Further, the surface morphologies of Pd/Au after poisoning at all tested temperatures showed no significant morphology changes and pin holes formation, remaining nearly identical in comparison to the one before poisoning with undetectable sulfur content by EDX. This was in agreement with the XRD results, indicating the small amount of the H_2S adsorption on the Pd/Au alloy surface. Since EDX and XRD were not able to detect the small amount of S on the very surface due to the long penetration depth of the X-ray emitted from EDX and XRD, the characterization via X-ray Photoemission Spectroscopy (XPS) (In cooperation with University of Connecticut) was further performed on the poisoned Pd/Au coupons. An average 3.4 at% S was detected on the Pd/Au sample that was poisoned at 350°C in the 54.8 ppm (±1 ppm) H_2S/ H_2 mixture for 96 hours, indicating the small amount of S adsorbed on the Pd/Au alloy surface. More comprehensive XPS studies will continue to be performed.

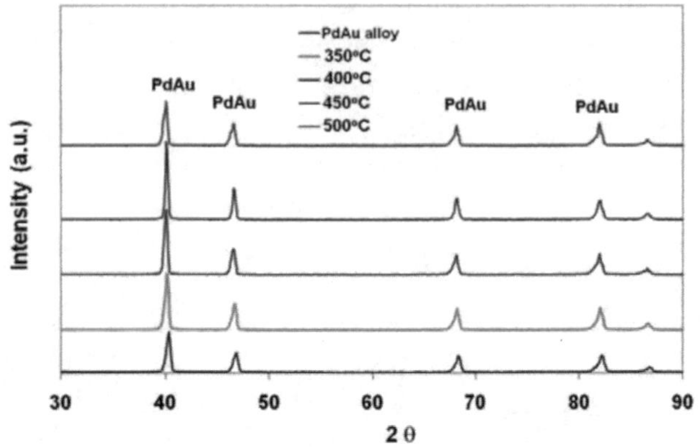

[a].

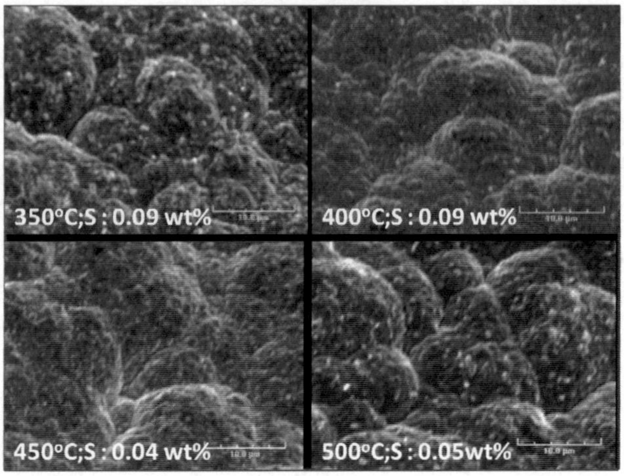

[b].

Figure 2. Temperature effect of sulfur poisoning of Pd/Au alloy on [a] crystalline phase, and [b] morphology.

In order to examine the sulfur poisoning effect on hydrogen permeating characteristics, a pure Pd membrane and the Pd/Au alloy membrane, C04, were characterized for their hydrogen permeance in the presence of a 54.8 ppm (±1ppm) H_2S. Firstly, the pure Pd membrane with a stable permeance of 30 $m^3/(m^2*h*atm^{0.5})$ at 400°C was characterized for its sulfur tolerance by switching the gas feed to the 54.8 ppm (±1ppm) H_2S/H_2 mixture, and the result is shown in Figure 3. It showed that the permeance dropped instantaneously after the 54.8 ppm (±1ppm) H_2S/H_2 mixture introduction at 400°C, and continued to decline for over 24 hours before it reached a steady state value. This indicated two poisoning mechanism, the instantaneous permeance drop at the beginning might be resulted from the surface adsorption which was rapid, while the slow and long decline of the permeance afterward might be resulted from the reaction between the Pd and the adsorbed sulfur which was a long process. Further, when the pure H_2 was re-introduced, the permeance remained the same value, showing no recovery, indicating the irreversible reaction occurred between Pd and H_2S during the H_2S exposure.

The same poisoning test was performed on C04 and the result is also shown in Figure 3. Prior to the sulfur poisoning tests, C04 was characterized in H_2 and He at the temperature from 250°C to 550°C for over 2000 hours and showed a steady state hydrogen permeance of 1.2 $m^3/(m^2*h*atm^{0.5})$ at 250°C and increased with increasing temperature reaching 6.5 $m^3/(m^2*h*atm^{0.5})$ at 500°C with a stable H_2/He selectivity of 100. The activation energy for the H_2 permeation was 13.1kJ/mol. For the sulfur poisoning result, it showed that the permeance also dropped sharply right after the H_2S introduction (As shown in Figure 3.). However, it reached a steady state value immediately without continuing decline. Further, the permeance was found to be recovered to some extent when re-introducing the pure H_2 at the same temperature, indicating the reversible sulfur poisoning effect on Pd/Au alloy. Moreover, when increasing the temperature to 500°C during the recovery in H_2, and permeance was found to be near fully recovered (As shown in Figure 4). The results suggested that the Pd/Au alloy membrane did show better sulfur resistance than the pure Pd membrane, and the minor and reversible sulfur poisoning effect on the Pd/Au alloy membrane was possibly due to the adsorption of H_2S on the Pd/Au surface only (i.e. No further reaction between adsorbed S and Pd/Au alloy).

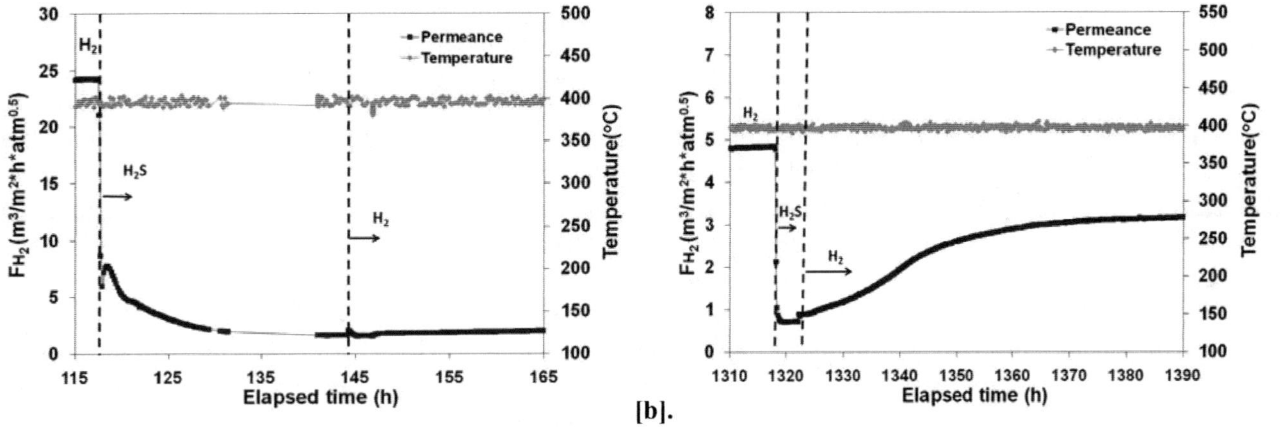

[a]. [b].

Figure 3. Effect of H_2S poisoning on hydrogen permeance at 400°C: [a] pure Pd membrane, and [b] Pd/au alloy membrane.

Similar poisoning/recovering tests have been performed on C04 at other temperatures, and the percentage of permeance decline during the H_2S exposure was found to decrease as the temperature was increased, from 88% at 350°C, 85% at 400°C, 74% at 450°C, to 60% at 500°C (As shown in Figure 4). The recovered permeance before the full recovery at 500°C was found to increase as the temperature was increased from 43% of the original value at 350°C, 65% at 400°C, 74% at 450°C, and ~100% at 500°C (As shown in Figure 4). The results suggested that the higher desorption rate of the H_2S molecules at higher temperatures resulted in less sulfur adsorption on the membranes surface, thereby less poisoning effect of the membrane at higher temperatures. Further, from the linear recovery rate at different temperatures, the activation energy of desorption was estimated to be 71 kJ/mol by the use of Arrhenius relation, which suggested the partial physical and partial chemical adsorption of H_2S. Finally, as Figure 4 shows, the He leak flux showed no significant increase (even decreased) after the sulfur exposures over a lone testing time, indicating that the adsorbed sulfur might have segregated to the grain boundaries and defects in the Pd/Au films to block the passage-way of the non-adsorbing gas, such as He.

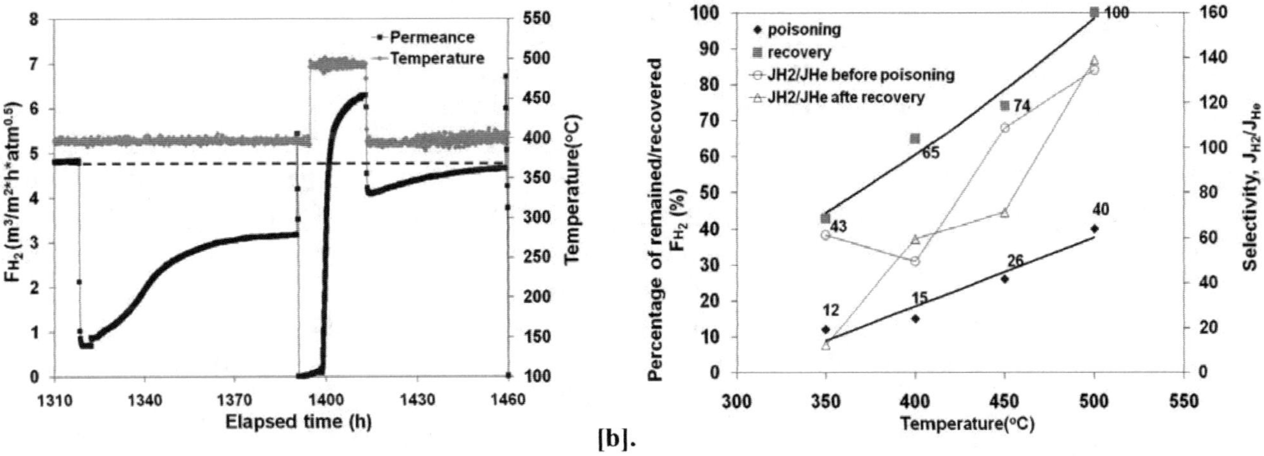

[a]. [b].

Figure 4. [a] Near full permeance recovery at 500°C, and [b] Summary of permeance remained/recovered and selectivity before/after poisoning at different temperatures.

Conclusions

H_2S was found to show severe poisoning effects on pure Pd. A 54.8 ppm (±1ppm) H_2S/H_2 gas mixture caused the Pd to react with, and to form Pd_4S resulting in numerous pin holes formation at the temperature of 350°C to 500°C. The formation of Pd_4S and the effect of poisoning were found to be temperature dependent and lower temperatures resulted in higher degree of poisoning. Further, the Pd membrane showed instantaneous permeance drop after the H_2S/H_2 introduction, and continued to drop over time. No permeance recovery after the re-introduction of pure H_2 has been observed, indicating the irreversible sulfur poisoning effect on pure Pd. The Pd/Au alloy, on the other hand, was found to show better sulfur resistance. No significant morphology changes and no new phase formation after the exposure of H_2S/H_2 have been observed in the entire testing temperature range. Further, although the Pd/Au membrane also showed the instantaneous permeance drop after the H_2S/H_2 introduction, the permeance reached a steady state value within a few minutes. Moreover, the permeance was able to be recovered in H_2 after the poisoning, indicating the reversible poisoning effect. The percentage of permeance recovered in H_2 after the poisoning, as well as the percentage of permeance remained during the poisoning, were found to be higher at higher temperatures, suggesting that the poisoning effect was temperature dependent, and the less poisoning effect at higher temperatures might be resulted from the higher desorption rate of H_2S.

Acknowledgements

The financial support provided by University Coal Research Program, U.S. Department of Energy, Award No. DE-FG26-04NT42194 is gratefully acknowledged.

References

1. McKinley, D.L. and Nitro, W. (1967) "Metal Alloy for Hydrogen Separation and Purification." *US Patent 3,350,845.*
2. Kajiwara, M.; Uemiya, S. and Kojima, T. (1999) "Stability and hydrogen permeation behavior of supported platinum membranes in the presence of hydrogen sulfide." *International Journal of Hydrogen Energy*, pp. 839 - 844.
3. Edlund, D.J. and Pledger, W.A. (1993) "Thermolysis of hydrogen sulfide in a metal-membrane reactor." *Journal of Membrane Science*, pp. 255 – 264.
4. Way, J. D.; Lusk, M and Thoen, P. (2008) "Sulfur-resistance composite metal membranes." *US Patent 20080038567.*
5. Gryaznov, V.M. (2000) "Metal containing membranes for the production of ultrapure hydrogen and the recovery of hydrogen isotopes." *Separation and Purification Methods.*
6. Sonwane, C. G.; Wilcox, J.; Ma, Y. H. (2006) "Achieving optimum hydrogen permeability in PdAg and PdAu alloys." *The Journal of chemical physics*, pp. 184714.
7. Okinaka, Y. (1990) Chapter 15 Electroless plating of gold and gold alloys in Electroless Plating: Fundamentals and Applications. Mallory, G.O. and Hajdu, J.B., ed. Orlando, FL, American Electroplaters and Surface Finishers Society.
8. Parker, E. A) (1974) Chapter 10 in Gold Plating Technology. Reid, F. H. and Goldie, E. ed. Ayr, Scotland, Electrochemical Publications limited.
9. El-Shazly, M. F. and Baker, K. D. (1982) "High build electroless gold process." *Proceeding of First AES electroless plating Symposia*, St Louis.
10. Mardilovich, P.P.; She, Y.; Ma, Y.H. and Rei, M. (1998) "Defect Free Palladium Membranes on Porous Stainless Steel Support." *AIChE Journal*, pp. 310 – 322.

11. Ma, Y.H.; Akis, B.C.; Ayturk, M.E.; Guazzone, F.; Engwall, E.E. and Mardilovich, I.P. (2004) "Characterization of intermetallic diffusion barrier and alloy formation for Pd/Cu and Pd/Ag porous stainless steel composite membranes." *Industrial Engineering and Chemistry Research*. pp. 2936 – 2945.
12. Ayturk, M. E.; Mardilovich, I. P.; Engwall, E. E.; Ma, Y. H. (2006) "Synthesis of composite Pd-porous stainless steel (PSS) membranes with a Pd/Ag intermetallic diffusion barrier." *Journal of Membrane Science,* pp. 385.

RELEVANCE OF CONCENTRATION POLARIZATION ON HYDROGEN FLUX THROUGH A THIN, HIGHLY SELECTIVE Pd_{80}-Ag_{20} CERAMIC SUPPORTED MEMBRANE

Jacopo Catalano, Marco Giacinti Baschetti, and Giulio C. Sarti.
Chemical Engineering, Mining and Environmental Technology, University of Bologna, via Terracini 28, Bologna, Italy

Abstract

Permeation experiments using pure hydrogen and several mixtures of hydrogen, nitrogen, and methane were performed on a Pd-based membrane deposited on a ceramic support. In particular, the permeance of pure hydrogen as well as binary and ternary mixtures containing also nitrogen and methane was measured using different feed flow-rates and H_2 concentrations, in a custom-built, tube-in-shell apparatus. The membranes, supplied by NGK Insulator Ltd Japan, consisted of a 2.5 μm Pd-Ag layer deposited on a tubular ceramic support. They showed extremely high hydrogen permeance and practically infinite selectivity toward hydrogen. Permeability and permeance were calculated from the linear regression of pure hydrogen data, according to Sieverts' law; permeability at 673 K is 3.08 $\times 10^{-6}$ mol s^{-1} m^{-1} $bar^{-0.5}$.

Surprisingly, experiments showed that hydrogen permeability is much lower for gas mixtures experiments than for the pure hydrogen case, and also that the hydrogen flux for mixture experiments deviates significantly from Sieverts' law. The above features reveal the presence of a significant mass transfer resistance in the external gas phase; that is also clearly confirmed by the increase in hydrogen permeability obtained with increasing the feed flow-rate of the gas mixtures.

The data collected were then used to calculate the average mass transfer coefficient in different experimental conditions. The gas phase mass transfer coefficient follows closely a correlation of boundary-layer type, proportional to the square root of the Reynolds number and to the cubic root of the Schmidt number.

Introduction

The idea of a hydrogen economy related to the use of this gas as a clean energy carrier has gained much attention in the last years giving new interest to researches aimed at improvements of the production of pure hydrogen. Thus palladium and palladium based membranes, which are theoretically able to completely purify hydrogen from other gases, have been widely studied in the last years with the focus on the fabrication of stable and long lasting membranes, with high flux and selectivity.

Nowadays, even if not yet at an industrial level, palladium based membranes can be produced, providing a high permeate flux and ultra-pure hydrogen[1,2]. The attention is then progressively moving from the production and characterization to the study of the membrane behavior inside the separation modules or membrane reactors in order to have valuable information for the development of pilot scale processes and for the system scale up. Indeed the peculiar characteristics of these membranes, with high rates of hydrogen permeation, can lead to accumulation of non permeable gases and depletion of hydrogen on the boundary layer close to the membrane[3,4,5]; concentration polarization phenomena have, therefore, to be taken into account for a proper description of the module behavior.

In this study the analysis of such phenomena is considered testing hydrogen permeation behavior of highly permeable and highly selective $Pd_{80}Ag_{20}$ membranes in different operative conditions; different mixture compositions as well as different feed flow-rates were considered in order to reveal the presence, if any, of concentration polarization in the module and to understand their influence on the membrane efficiency as a function of the different operating conditions.

Materials and equipment

The membranes used in the present work were provided by NGK Insulators Ltd. and consist of a dense, 2.5 μm thick, Pd-Ag layer, with silver content of about 20% by weight, deposited via electroless plating on the external surface of a tubular asymmetric α-alumina support[1].

The membranes used have an external diameter of about 1 cm, a total length of 9 cm, and are sealed, on one end, with a stainless steel plug and endowed, on the other end, with a VCR fitting for coupling with the permeation module. The latter was a custom-built stainless-steel shell-in-tube module connected to the gas reservoirs through different pressure and flow controllers which allow to feed the module with mixtures of desired compositions and flow-rates.

The different streams leaving the module were also endowed with pressure controllers and mass flowmeters on both permeate and retentate sides in order to maintain the desired pressure difference across the membrane and to measure the permeating flow. A gas chromatograph connected to all the different streams in the system gave a direct measurement of their compositions.

The module was located inside a tubular furnace and an additional heating element, as well as a thermocouple, were inserted close the membrane to obtain an accurate temperature control, during the experiments[1].

Different permeation experiments were performed testing the membranes at 400 °C±2% and using different pressures (total differential pressures ranging between 0.2 and 6 bar), and different feed flow-rates (1, 2, and 3 NL min^{-1}). Tests were performed with pure hydrogen as well as mixtures with nitrogen and methane; in particular H_2/N_2 mixtures (50 and 88 $\%_{vol}$ of H_2), H_2/CH_4 mixtures (50$\%_{vol}$ of H_2) and the ternary mixtures $H_2/N_2/CH_4$ (50/25/25 $\%_{vol}$) were used for the experiments. All gases used in the experiments were gascromatrographic grade with purity of 99,99% and were available at 11 bar.

During permeation experiments the pressure in the shell side was increased by subsequent steps, measuring the permeate flux and the concentration of the different streams leaving the module until constant steady state was reached, and the permeate, retentate and feed streams remained unchanged for at least one hour.

Theoretical Background

Hydrogen transport through Pd-based membranes is commonly represented by Sieverts' law, which results from the coupling of two processes: i) the hydrogen adsorption and dissociation onto Pd surface ($H_2 \rightarrow 2H$) and ii) the subsequent atomic hydrogen diffusion in the metallic lattice [7, 8]; when the reaction is fast enough for the diffusion to be the controlling step of the process, the hydrogen flux, $\mathbf{N}_{H_2}$, can be written as:

$$\mathbf{N}_{H_2} = \frac{D_H K_S}{2} \cdot \frac{\left(p_{H_2,ret}^{0.5} - p_{H_2,per}^{0.5}\right)}{\delta} \qquad (1)$$

where D_H is the average diffusivity of atomic hydrogen in the Pd-Ag alloy, p_{H_2} is the partial pressure of hydrogen at the retentate (*ret*) or permeate (*per*) side, δ is the membrane thickness and K_S is the Sieverts' constant which represents the equilibrium constant of the adsorption/dissociation reaction considered above.

Consistently with the solution diffusion model for transport in dense membrane an hydrogen permeability, $\bar{\mathbf{P}}_{H_2}$, [$mol\ m^{-1}\ s^{-1}\ bar^{-0.5}$ or $NL\ m^{-1}\ s^{-1}\ bar^{-0.5}$] and permeance, $\bar{\mathbf{K}}_{H_2}$, [$mol\ m^{-2}\ s^{-1}\ bar^{-0.5}$ or $NL\ m^{-2}\ s^{-1}\ bar^{-0.5}$] through the Pd based membrane can be defined as :

$$\overline{\mathbf{K}}_{H_2} = \frac{\overline{\mathbf{P}}_{H_2}}{\delta} = \frac{\mathbf{N}_{H_2}}{\left(p_{H_2,ret}^{0.5} - p_{H_2,per}^{0.5}\right)} \tag{2}$$

which can be used to characterize the transport properties of the membranes under consideration. Eq (2) hold for hydrogen which is undergoing dissociation at the Pd surface, but not for other gases that follow the solution diffusion mechanism and thus, in view of their negligible solubility and diffusivity values in the metallic lattice, cannot practically permeate in the dense palladium layer, giving to this kind of membranes a virtually infinite selectivity.

Due to the presence of other resistances to mass transport, as those related to the porous support or to concentration profile in the gas phase, experimental data often deviate from Sieverts' law and are usually described by modifying the relationship as follows:

$$\mathbf{N}_{H_2} = \overline{\mathbf{P}}_{H_2} \frac{\left(p_{H_2,ret}^{n} - p_{H_2,per}^{n}\right)}{\delta} \tag{3}$$

where n is now an empirical exponent which usually takes values between 0.5 and 1 depending on the dominant resistance in the system.

Such an approach indeed does not give any physical insight to the system behavior and better results are easily achievable considering the different process in series; such a problem was originally treated and solved by Ward and Dao[9] which referred to the works done by Holleck[8] and by King and Wells[10] for the modeling of hydrogen transport in Pd membranes. In particular these authors considered the transport as the result of the following processes:

- Hydrogen transport in the gas phase from the bulk to the membrane surface in the high concentration (and vice versa in the permeate side);
- Adsorprtion (desorption) and dissociation (formation) of atomic (molecular) hydrogen at the high (low) Pd surface;
- Atomic hydrogen diffusion in the metallic layer.

The transport in the porous support was not included in the list because only unsupported membranes were considered in the study but it is easily understood that in the case of supported membranes also the support resistance has to be considered in principle[11].

In the present work the interest was focused on the possible presence of external transport resistances which are responsible of the differences in hydrogen permeability observed between pure hydrogen and mixture experiments. Such resistances indeed are usually neglected for metallic membranes, but different authors have already pointed out that for highly permeable and selective membranes they can have not negligible effects on the separation performances[3,4,5]. The resistance of the gas phase can be easily accounted for through the introduction of an external mass transport coefficient, k_G; the hydrogen flux can therefore be described as:

$$\mathbf{N}_{H_2} - x_{H_2,int}\mathbf{N}_{tot} = k_G\left(x_{H_2,ret} - x_{H_2,int}\right) \tag{4}$$

where $\mathbf{N}_{tot}$ is the total permeate flux and x_{H_2} is the average hydrogen molar fraction at the retentate side (subscript *ret*) and at the interface between the gas phase and the Pd-Ag membrane (subscript *int*).

When the Sieverts' law can be used to describe the transport in the membrane the combination of Eqs. (1) and (4) allows to calculate the interfacial hydrogen concentration and to determine the importance of the mass transport resistance in the gas phase. The relevance of such term depends obviously on the value of k_G which depends on both the material properties and system fluid dynamics, and is related to Schmidt (*Sc*) and Reynolds (*Re*) numbers through a general relationship of the following form[12]:

$$\mathrm{Sh} = \frac{k_G l}{cD} = \alpha \mathrm{Re}^a \, \mathrm{Sc}^b \qquad (5)$$

where *Sh* represents the Sherwood number *l* is a characteristic length of the system, *a* and *b* are constants that depend on the system geometry and fluid dynamics and *c* and *D* represent respectively the molar concentration and the diffusion coefficient of hydrogen in the gas phase.

Results

The membrane behavior was studied extensively considering three different feed flow-rates (1, 2 and 3 NL min^{-1}) at 400 °C, both for the pure hydrogen tests as well as for hydrogen mixtures of different compositions.

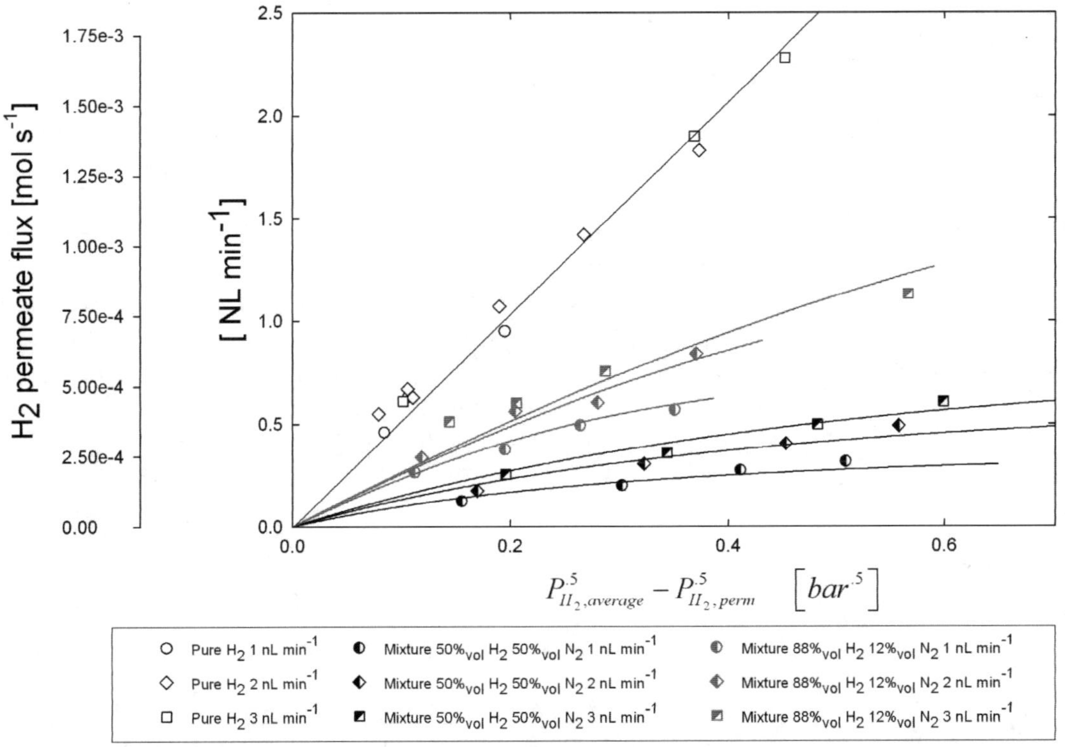

Figure 1 Experiments performed at 400°C at three different flow-rates. Shaded symbols represent pure hydrogen tests, semi-shaded N_2-H_2 mixtures: black symbols 50%$_{vol}$ H_2-50%$_{vol}$ N_2, red 88%$_{vol}$ H_2-12%$_{vol}$ N_2).

The results, reported in compact form in figure 1, show that for pure hydrogen permeation tests permeability follows the Sieverts' law, confirming that the diffusion in the Pd alloy is still controlling the hydrogen transport in the membrane, even if the membranes are very thin. The permeate flux is independent of the feed flow-rate. Permeability and permeance values are calculated through a linear regression of experimental data, according to Eq. (2), in particular hydrogen permeability at 400 °C is 3.08 x10^{-6} mol s^{-1} m^{-1} $bar^{-0.5.}$ and is consistent with the many experimental data reported in previous works[13-19]; on the other hand, due to the small thickness of the metallic layer, the permeance is 1.23 mol s^{-1} m^{-2} $bar^{-.05}$ at 400 °C and is among the highest values found in the literature for similar membranes as it can be observed in table 1.

The hydrogen flux permeating from different gaseous mixtures is also reported in figure 1; it is apparent that hydrogen permeance in pure gas experiments is definitely higher than that measured in the case of mixtures at the same hydrogen partial pressure. In mixture experiments the hydrogen flux across the membrane shows a rather different dependence on the driving force, since it depends on both hydrogen concentration in the feed and on feed flow-rate; it decreases drastically when the feed concentration decreases and increases when the feed flow-rate is increased.

Table 1. Permeante values at several temperatures, nominally 400, 450, and 500 °C and comparison with selected literature data.

Authors	*Ref.*	*Pd*	*T*	$\bar{\mathbf{K}}_{H_2}$
		μm	°C	mol s^{-1} m^{-2} bar$^{-0.5}$
This study		~2.5	400	1.232
Pizzi et Al. (2008).	[6]	~2.5	400	0.853
A. Basile at al. (2001)	[14]	70	400	0.054
S. Uemiya et al (1991)	[15]	5.8	400	0.659
N. Itoh et al.(1993)	[16]	20	400	0.204
L. Yang (2006)	[17]	20	400	0.264
E. Gobina et al(1994)	[18]	6	400	0.017
Y. Guo et al(2003)	[19]	8.6	400	0.864

In addition clear deviations from Sieverts' law are visible since the permeate flux is not linear with the difference of the square root of hydrogen partial pressure at the two sides of the membrane but shows a well defined downward concavity. In any event, the G.C. analysis showed that no nitrogen, or methane, was present in the permeate stream, in all the tests performed, so that the selectivity of the membrane can be assessed to be larger than 10000 (estimated from the accuracy of the gascromatograph) confirming the existence of a continuous, dense and defect free Pd-Ag layer on the external surface of the tested membranes. Thus the behavior observed for mixtures cannot be attributed to the presence of defects on the membrane surface.

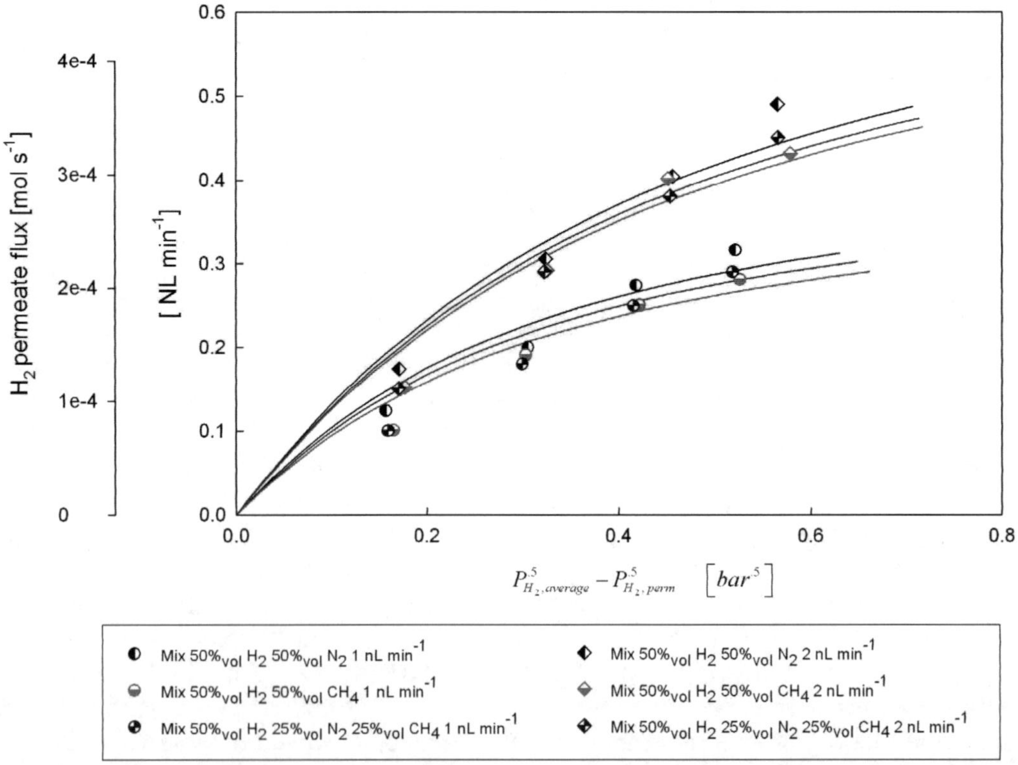

Figure 2. Effect of different non permeable gas in the hydrogen permeation. Tests were carried out with 50%$_{vol}$ H_2 feed content for three different mixtures at 1 and 2 NL min^{-1} feed flow rate.

The same behavior observed for H_2/N_2 mixtures is shown also when hydrogen-methane mixture as well as ternary mixture of the three gases are considered; the data, reported in figure 2, show indeed the general qualitative behavior of the other mixtures with just a slight decrease in the permeate flux when nitrogen was substituted by methane, as the non-permeable gas, in the feed. In mixture permeation tests the clear difference with respect to the pure hydrogen experiments is not directly related to the type of non penetrating gas present in the mixture.

Discussion

The results show that the membranes considered are endowed with an extremely high selectivity towards hydrogen suggesting the presence of a defect free Pd-Ag layer deposited on the porous substrate.

Permeate flux from pure hydrogen feeds follows Sieverts' law, while when mixtures are used in the feed the hydrogen flux shows a different trend and increases when the feed flow-rate increases. Such deviations, observed for gas mixture, can be explained considering the existence of additional resistance to mass transport in the gas phase. Other possible explanations, such as competitive absorption or surface poisoning, which has been reported in some cases[20,21,22], also for inert gas like nitrogen, do not appear applicable to the present case, since two different non permeable gases (methane and nitrogen) lead to very similar behaviors despite of their different chemical nature. Furthermore, surface effects cannot explain the permeate flux dependence on the feed flow-rate shown for gas mixtures.

In order to confirm the role of concentration polarization in the gas phase, the behavior of the system can be described by using Eqs. (2) and (4), thus obtaining the value of k_G as well as of the average interfacial pressure of hydrogen on the membrane surface, for all the experimental conditions inspected. The gas transport coefficient has been considered as the only fitting parameter of the procedure and was used to minimize the deviation between the model predictions and the experimental data. The curves, calculated through such approach are reported, as solid lines in figures 1 and 2, from which the rather good agreement with experimental values is self-evident.

In order to further support the interpretation offered for the behavior of hydrogen flux from gas mixtures, it is important to show that the k_G values obtained have a clear physical meaning and that k_G is not a mere fitting parameter. To that aim Eq. (5) has been used as a guideline to derive a relationship among the Sherwood number, calculated through the fitted k_G values, and Reynolds and Schmidt numbers corresponding to the different operating conditions examined.

The relationship obtained is reported in figure 4, which clearly indicates that there is a nice linearity between the calculated Sherwood number and the product $Re^{0.5}Sc^{0.33}$, showing a dependence very similar to that derived by the well known boundary layer theory over a flat surface; the proportionality constant is different and is about 1.89, nonetheless the existence of such a typical relationship strongly confirms the physical consistence of the interpretation offered. Thus we can conclude that the presence of concentration polarization phenomena cannot be neglected and must be considered in order to simulate and design properly the separation modules.

Conclusions

Permeation experiments have been performed on thin Pd-Ag membranes for hydrogen purification at different feed flow-rates and feed mixture compositions.

The membranes resulted to be extremely selective towards hydrogen and to posses very high permeance values, among the highest found in the open literature for similar membranes. The pure hydrogen data showed excellent agreement with the Sieverts' law while significant deviations from Sieverts' law were found in mixtures experiments, where the permeate flux is lower than for the pure hydrogen case, at a given value of hydrogen partial pressure in the feed side. For gas mixtures, in addition, the hydrogen permeate flux increases both with increasing feed hydrogen concentration and feed flow-rate, indicating the presence of a significant concentration polarization in the separation module.

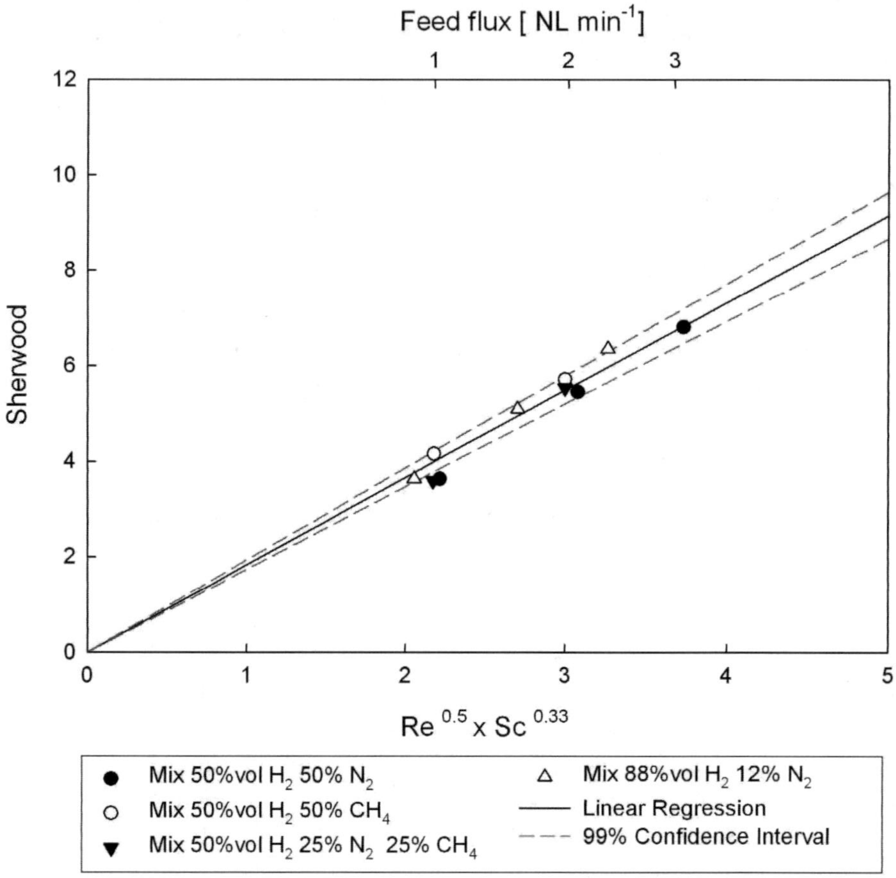

Figure 4. Calculated Sherwood number versus a boundary layer type correlation.

The data collected in the different experimental conditions were modeled accounting for the existence of this additional resistance to mass transport in the gas phase and using the average mass transfer coefficient as a fitting parameter.

The model results are in very good agreement with the experimental data and the Sherwood numbers obtained are related to Reynolds and Schimdt numbers through a boundary layer type correlation, showing the physical consistence of the interpretation offered.

Acknowledgement

This work was performed with financial support of the Italian Ministry of University and Research "Contributo del Fondo Integrativo Speciale Ricerca FISR DM 17/12/2002-anno 2001"

Progetto: Idrogeno puro da gas naturale mediante reforming a conversione totale ottenuta integrando reazione chimica e separazione a membrana".

References

1 F. Gallucci, F. Chiaravalloti, S. Tosti, E. Drioli and A. Basile, The effect of mixture gas on hydrogen permeation through a palladium membrane: Experimental study and theoretical approach, Int. J. Hydr. En., 32 (2007) 1837.

2 J. Zhang, H.Y. Xu, W.Z. Li, S.F. Hou and C.Y. Yu and K. Jiang, CO-free hydrogen production via an ultra permeable palladium membrane, 13th International Congress on Catalysis Paris, (2004) 162.

3 J. Zhang , D. Liu , M. He , H. Xu and W. Li, Experimental and simulation studies on concentration polarization in H_2 enrichment by highly permeable and selective Pd membranes, J. Membr. Sci., 274 (2006) 83.
4 O. Lüdtke, R.-D. Behling and K. Ohlrogge, Concentration polarization in gas permeation, J. Membr. Sci., 146 (1998) 145.
5 G. He, Y. Mi, P.L. Yue and G. Chen, Theoretical study on concentration polarization in gas separation membrane processes, J. Membr. Sci., 153 (1999) 243.
6 D. Pizzi, R. Worth, M. Giacinti Baschetti, G. C.Sarti and K. Noda, Hydrogen permeability of 2.5_m palladium–silver membranes deposited on ceramic supports, J. Membr. Sci. (2008), doi:10.1016/j.memsci.2008.08.020
7 A. Sieverts and W. Danz, Solubilities of D_2 and H_2 in Palladium, Z. Physik. Chem. (B), 34 (1936) 158.
8 G.L. Holleck, Diffusion and Solubility of Hydrogen in Palladium and Silver-Palladium Alloys in Hydrogen in Metals II, ed. by G. Alefeld and J. Volkl (Springer, Berlin Heidelberg) 1978.
9 T.L. Ward and T. Dao, Model of Hydrogen permeation behaviour in palladium membranes, , J. Membr. Sci., 153 (1999) 211.
10 D.A. King.and M.G. Wells, Reaction mechanism in chemisorption kinetics: nitrogen on the {10 0} plane of tungsten, Proceedings of the Royal Society of London, 339 (1974) 245.
11 A. Caravella, G. Barbieri and E.Drioli, Modelling and simulation of hydrogen permeation through supported Pd-alloy membranes with a multicomponent approach., Chem. Eng. Sci, 63 (2008) 2149.
12 R.B. Bird, W.E. Stewart and E.N. Lightfoot, Transport Phenomena, Wiley, New York, 1960.
13 R. C. Raid, J.M Prausnitz and B.E. Poling, The Properties of Gases and Liquids, McGraw Hill, New York, 1987
14 A. Basile, L. Paturzo and F. Laganà, The partial oxidation of methane to syngas in a palladium membrane reactor: simulation and experimental studies, Catal. Today 67 (2001) 65.
15 S. Uemiya, T. Matsuda and E. Kikuchi , Hydrogen permeable palladium-silver alloy membrane supported on porous ceramics, J. Membr. Sci., 56 (1991) 315.
16 N. Itoh and W.-C. Xu, Selective hydrogenation of phenol to cyclohexanone using palladium-based membranes as catalysts, Appl. Catal. A, 107 (1993) 83.
17 L. Yang, Z. Zhang, B. Yao, and X. Gao, Hydrogen Permeance and Surface States of Pd-Ag/Ceramic Composite Membranes, AIChE Journal 52 (2006) 2783
18 E. Gobina, R. Hughes, Ethane dehydrogenation using a high-temperature catalytic membrane reactor, J. Membr. Sci. 90 (1994) 11
19 Y. Guo, G. Lu, Y. Wang, R. Wang, Preparation and characterization of Pd/Ag/ceramic composite membrane and application to enhancement of catalytic dehydrogenation of isobutane, Sep. Purif. Technol. 32 (2003) 271
20 W.Wang, X. Pan, X. Zhang, W. Yang and G. Xiong, The effect of co-existing nitrogen on hydrogen permeation through thin Pd composite membranes, Separ. Purif. Techn., 54 (2007) 267
21 L. Du, S. Ahat, M. Sheng, L. Luo, J. Freytag and W. Senske, Absorption of N atom in Pd wire at low temperature and the influence on its physical properties, J. Alloy. Comp., 311 (2000) 270.
22 L. Du, S. Ahat, M. Sheng, Z. Sun, W. Huang and L. Luo, Inter-diffusion and nitrogen absorption in Pd/Ti/SiO2/Si multi-layer system, J. Mater. Sci. Lett., 20 (2001) 911.

701e Electroless Plating of Palladium Onto a Cordierite Monolith for Portable Hydrogen Extraction

Daejin Kim, Alexander Kellogg, and Benjamin A. Wilhite. Department of Chemical, Materials and Biomolecular Engineering, University of Connecticut, 191 Auditorium Rd, Unit 3222, Storrs, CT 06269

Integrated ceramic minichannel and microchannel membrane networks comprised of cordierite monoliths packaged with appropriate fluid distribution are being developed for hydrogen extraction from hydrocarbons. The final membrane design is envisioned as a catalytic steam reforming layer; over top a hydrogen-permselective palladium layer. By combining these technologies it is expected to produce ultra pure hydrogen in a single integrated device for portable power systems application[1]. In this talk, we present progress to-date in coating thin permselective palladium films within the cordierite minichannel network for hydrogen purification.

In this study, palladium films have been deposited by electroless plating on top of an alumina supporting layer, itself deposited within a cordierite matrix. The alumina supporting layer was coated inside the cordierite channels by wash-coating method to achieve a smooth, uniform coat. This was accomplished in two steps, first laying down a micro-sized layer of γ-alumina to achieve a low-stress deposition surface, followed by deposition of a nano-sized γ-alumina layer to produce a smooth, crack-free, and pinhole-free alumina layer that is ideal for plating Palladium. Two electroless plating methods were compared for depositing thin Palladium films on the resulting substrate. Each palladium film was tested for permeability and permselectivity by flowing hydrogen / helium mixtures through the channels at 350° C.

In this talk we will present details on the coating methods used to prepare the cordierite substrate, plating techniques for realizing permselective membranes, and packaging strategies to ensure stable, leak-free operation. With this there will be a general discussion of how these different variables factor into the selectivity and permeability of the resulting membrane. SEM imaging and EDS analysis were used to examine the thin film morphology. Leak testing and selectivity testing were used to examine the overall performance of the palladium films. To-date, films have been developed that achieve a hydrogen permeation rate of 0.004-0.017moles/m^2s, and hydrogen-to-helium selectivities of 40-100.

[1] B.A. Wilhite, S.E. Weiss, J.Y. Ying, M.A. Schmidt and K.F. Jensen, "Demonstration of 23wt% Ag-Pd Micromembrane Employing 8:1 $LaNi_{0.95}Co_{0.05}O_3/Al_2O_3$ Catalyst for High-Purity Hydrogen Generation," *Advanced Materials*, **18**, 1701 (2006).

701f Development of Group V Based Metal Membranes for Hydrogen Separation

Stephen N. Paglieri, Tritium Science and Engineering, Los Alamos National Laboratory, P.O. Box 1663, MS-C927, Los Alamos, NM 87545, Micheal V. Ciocco, Parsons Project Services Inc., P.O. Box 618, South Park, PA 15129, Robert E. Buxbaum, REB Research and Consulting, 3259 Hilton Rd., Ferndale, MI 48220, and Bryan D. Morreale, Office of Research and Development, U.S. Department of Energy, National Energy Technology Laboratory, 626 Cochrans Mill Rd., P.O. Box 10940, Pittsburgh, PA 15236.

Group IV and V metals of the periodic table that have bcc structure such as titanium, vanadium, and niobium have higher hydrogen permeabilities than palladium but have surface oxides that prevent hydrogen absorption. Therefore, these metals require a thin film (1 micron or less) of catalyst on their surfaces such as palladium. Generally, Group IV and V metals have shown embrittlement upon exposure or cycling in hydrogen. Besides embrittlement, another problem experienced by metal composite membranes is metallic interdiffusion between the substrate foil and the palladium overcoat. However, alloys of these metals may possess some of the qualities desired in a metal membrane including high hydrogen permeability and relatively low cost.

Overall, more than 250 different alloys were investigated. Alloys were screened for potential use as hydrogen separating membranes by Charpy impact testing of striker bars cut from ingots of each alloy prepared by arc melting. Membrane discs were cut from alloys that passed the brittleness test by thinly slicing an ingot using wire electrode discharge machining, and sand blasting to remove bulk oxide. The discs (between 0.3 and 0.5 mm thick) were ion-milled, and 100-nm-thick palladium or palladium-alloy coatings were applied to both sides in situ via electron-beam-evaporation physical vapor deposition. Membranes were tested for hydrogen permeability over a range of temperatures (up to 700° C), pressures (up to 0.3 MPa), and durations (> 500 h). Membrane resilience was often assessed by conducting permeability testing in hydrogen at 400° C and then decreasing the temperature, noting the onset of hydrogen embrittlement as evidenced by crack formation in the membrane.

Several palladium coated titanium and nickel based alloys were tested for hydrogen permeability but these materials exhibited little or no hydrogen flux or broke when exposed to hydrogen. Niobium based materials were permeable to hydrogen but were often too fragile for practical usage. Vanadium based alloys also commonly suffered from embrittlement but were the most ductile of the alloys tested, usually failing in hydrogen at temperatures around 250° C after initial hydrogen permeability testing at 400° C. Most of the vanadium alloys displayed hydrogen permeabilities comparable to pure palladium. The influence of adding various elements to vanadium on the hydrogen permeability, resistance to hydrogen embrittlement, and effect on hydrogen flux stability with time was explored. For example, titanium increases hydrogen permeability but does not appear to lessen embrittlement and may promote atomic diffusion to the surface. Palladium clearly increased ductility, for example, a V-10Pd (at.%) membrane withstood hydrogen permeation testing at temperatures as low as 150° C.

Furthermore, it is essential that the hydrogen dissociating surface catalyst is stable at the temperatures that metal membranes are expected to operate at (up to 450 ° C) and lasts for the membrane lifetime (ideally > 100,000 h). One approach to the problem of metallic interdiffusion between the catalytic coating on the membrane surface and the constituents of the membrane itself is to use palladium alloys. The effect of palladium and palladium-alloy coatings on metallic interdiffusion and the hydrogen permeability of Group V membranes will be described. Interdiffusion between the surface coatings and membranes was characterized using Rutherford backscattering spectrometry (RBS) and Auger electron spectroscopy (AES).

719a Dynamic Adsorption Behaviors of M-Chloronitrobenzene and P-Chloronitrobenzene on Faujasite Zeolites

Benyapa Yensukjit[1], Pramoch Rangsunvigit[1], Santi Kulprathipanja[2], and Akekasit Leardsakulthong[1]. (1) Petrochemical Technology, The Petroleum and Petrochemical College, Chulalongkorn University, 254 Soi Chula 12 Prayathai Rd., Bangkok, 10330, Thailand, (2) UOP LLC, 50 East Algonquin Road, Des Plaines, IL 60017-5016

Chloronitrobenzenes (CNBs) are important organic intermediates that are used practically for azos, sulfur dyes, pesticides, fungicides, and in rubber chemical industries. Many separation and purification techniques, such as distillation, crystallization, and adsorption, have been investigated to achieve the highest purity and conversion of CNB. In this work, adsorption will be explored more in detail for the CNB separation and purification. From our previous study, the static adsorption of *m*- and ,*p*-CNB shows that the selective adsorption of *m*-CNB is higher than that of *p*-CNB for all tested FAU zeolites – LiX, NaX, and NaY. In addition, LiX is found to have adsorption capacities higher than NaX and NaY In this study, the adsorption of *m*- and *p*-CNB mixtures was further investigated for their dynamic adsorption characteristics on LiX, NaX, and NaY zeolites using pulse test and breakthrough techniques. Desorbent and feed composition were varied for the investigation. The effects of temperature were also studied. The results substantiate those from the static adsorption study. A comparison will be made for both adsorption capacity and selectivity. Furthermore, the adsorption behaviors of both CNB isomers deduced from both static and dynamic studies will be presented.

Keyword: Chloronitrobenzene, Zeolite, Adsorption, Separation

719b Generalized Sld-PR Adsorption Model for Pure and Mixed Gases on Coal Matrices

S. Mohammad[1], J. Chen[1], R. L. Robinson[1], and K. a. M. Gasem[2]. (1) Chemical Engineering, Oklahoma State Univ., EN 423, Oklahoma State University, Stillwater, OK 74078, (2) School of Chemical Engineering, Oklahoma State University, 423 Engineering North, Stillwater, OK 74078

The Simplified Local-Density Peng-Robinson (SLD-PR) adsorption model was tested for its generalization capability. In particular, a coal-property, structure-based generalization was employed. Although this SLD-PR model development is similar to our previous efforts with the Simplified Local-Density Hard-Sphere (SLD-HS) model, an important motivation in carrying out this generalization was the readily available mixing rules of the PR equation of state. This considerably simplified the direct extension of this model to mixture adsorption behavior.

Before performing the SLD-PR model generalizations, we have assessed the correlative capability of the model. On average, the model can represent all adsorption data within their experimental uncertainties. Then, simple correlations were developed to generalize the SLD-PR parameters (i.e., matrix surface area, solid-solid interaction parameter, and the slit width) in terms of gas properties and accessible adsorbent characteristics (the proximate and ultimate analyses of the coals). For this analysis, we used our adsorption measurements for methane, nitrogen and CO2 on ten dry and wet coals, including five Argonne Premium coals.

Our results indicate that the generalized SLD-PR model is capable of predicting the adsorption of the coalbed methane gases (methane, nitrogen and CO2) and their mixtures, on average, within twice the expected experimental uncertainties.

719c Optimal Operation and Design of Liquid-Solid Circulating Fluidized Bed Ion Exchange System for Continuous Protein Recovery

Jahirul I. Mazumder, Jesse Zhu, and **Ajay K. Ray**. Department of Chemical & Biochemical Engineering, The University of Western Ontario, 1151 Richmond St., London, ON N6A 5B9, Canada

With the development in the understanding of the life sciences, the demand for various functional proteins is constantly growing. Extraction and purification of protein form various natural sources have continued to evolve during the past fifteen years. Liquid-solid circulating fluidized bed (LSCFB) ion exchange system, which is capable of continuous protein recovery from broth solution, has significant advantages over the other ion-exchange adsorption processes. Two separate operations (adsorption and desorption) are carried out simultaneously in the two columns (downcomer and riser) of LSCFB system and the performance of these operations are mutually dependent. Therefore, optimum design and operation of the LSCFB system for the better overall performance is very critical.

Like most real-life optimization problems, the design and operation of LSCFB system for continuous protein recovery is also associated with several important objectives such as production rate and recovery of protein, and ion exchange resin requirements, which need to be optimized simultaneously. Multi-objective optimization of the LSCFB system at both the operation and the design stages were carried out using an experimentally validated mathematical model to determine the range of optimal solutions. Elitist non-dominate sorting genetic algorithm with its jumping gene adaptation (NSGA-II-aJG) was used to solve a number of two- and three- objective function optimization problems. The optimization resulted in Pareto optimal solutions, which provides a broad range of non-dominated solutions due to conflicting behavior of the operating and design parameters on the system performance indicators. Significant improvements were achieved, for example, for the same recovery level, the production rate at optimal operation increased by 33%, using 11% less solids compared to experimental results. In the design stage optimization, the performance of the system was further improved. This multi-objective optimization study is very general and can be easily extended for the improvement of LSCFB in other applications.

Keywords: Liquid-solid circulating fluidized bed, protein recovery, multi-objective optimization, Pareto sets, genetic algorithm.

719d Polypeptide - Oms Nanocomposite Materials

Daniel F. Shantz and **Jonathan Lunn**. Department of Chemical Engineering, Texas A&M University, TAMU 3122, College Station, TX 77843-3122

Organic-inorganic hybrid materials are of interest for numerous technological problems as ideally they possess the desirable characteristics of both the inorganic and organic constituents. In the current work we will summarize our efforts to grow polypeptide chains directly from the surface of a variety of ordered mesoporous materials once suitably functionalized. Using amine-functionalized OMS materials we have found that it is possible to form nanocomposites of OMS-polypeptide materials using Leuch's NCAs that possess a high density of polypeptide chains covalently attached to the OMS surface. Nitrogen adsorption on these materials indicates a dramatic decrease of porosity as compared to the parent OMS phase. More specifically this talk will summarize our findings on several fronts, namely how 1) the surface amine loading, 2) monomer identity, 3) pore size and 4) pore topology influence the nature of the composite material obtained. Both porosimetry and spectroscopy data will be presented to interrogate the composite material obtained. We will also present preliminary data for these materials regarding their ability to perform chiral separations.

719e Comparison of Water Isotherm Models for 3A, 4A, 5A and 13X Zeolite

Kevin F. Loughlin and Reem Salah Amer. Chemical Engineering, American University of Sharjah, P. O. Box 26666, Sharjah, United Arab Emirates

Abstract: Model isotherms are derived considering a two site hypothesis, one for the alpha cage [or large cage], and one for the beta cage. Two simple model isotherms, one a dual site Toth isotherm and one a dual site Langmuir isotherm are fitted to the data using the optimization package Solver in EXCEL. Both dual site Toth or Langmuir isotherm models fit the data satisfactorily.

The optimized standard enthalpy and entropy of adsorption parameters are surprising for the beta cage, The enthalpy of the beta cage is 1/3rd of that observed calorimetrically, and the standard entropy of adsorption is positive, a physical impossibility. Substituting the calorimetric enthalpy of adsorption corrected the standard differential entropy of sorption values resulting in these values varying significantly with temperature. This variation is postulated to be due to the formation of clusters of water.

KEYWORDS: Adsorption, water, 3A, 4A, 5A or 13X zeolite, Langmuir isotherm, Toth isotherm, dual site isotherms, standard enthalpy of adsorption, standard entropy of adsorption, solver in EXCEL.

719f Water Isotherm Models for 4A (NaA) Zeolite

Kevin F. Loughlin, Chemical Engineering, American University of Sharjah, P.O. Box 26666, Sharjah, United Arab Emirates

ABSTRACT

The adsorption data of Gorbach et al. (1) and Morris (2) for the adsorption of water on 4A zeolite pellets is re-analyzed. Model isotherms are derived considering a two site hypothesis, one for the alpha cage and one for the beta cage. Four simple model isotherms are fitted to the data. Both a dual site Toth or Langmuir isotherm model fit the data adequately.

The optimized standard enthalpy and entropy of adsorption parameters derived from the data are surprising for the beta cage,. The optimized standard enthalpy of the beta cage is 1/3rd of that observed calorimetrically, and the standard entropy of adsorption is positive, a physical impossibility. Substituting the calorimetric enthalpy of adsorption corrected the standard differential entropy of sorption values resulting in the standard entropy of sorption values varying significantly with temperature. This variation is postulated to be due to the formation of clusters of water, such as dimers, trimers etc.

KEYWORDS:

Adsorption, water, 4A (NaA) zeolite, Langmuir isotherm, Toth isotherm, Ruthven isotherm, dual site isotherm, standard enthalpy of adsorption, standard entropy of adsorption,

719g Esiex-Electrical Swing Ion Exchange:a Process Coupling Ion Exchange, Carbonic Acid Elution and Electroregeneration by Electrodialysis

Lu Wei, Laurence Muhr, and **Georges Grevillot**. Laboratoire des Sciences du Génie Chimique CNRS, 1, rue Grandville, Nancy, France

Consider an electrodialysis cell with ion exchange resin between the membranes (this is typically EDI : electrodeionization). The process is in three steps.

In the first step, the mixture to be separated is percolated on the ion exchange bed. Conditions must be chosen in order that only one specie is fixed on the resin. For example, the dipeptide GLYGLY in water is fixed on an anion exchange resin in carbonate initial form.

The second step is elution of GLYGLY with a carbonic acid solution. This solution allows to recover pure GLYGLY in water, contrary to other classical methods in ion exchange where the desorption occurs in a buffer or in a salt solution. During this step, the resin converts to bicarbonate form. The carbonic acid solution was obtained thanks to the third step of the previous cycle.

The third step is to return the resin to the carbonate form. This is done by passing an electrical current trough the cell. An anion exchange membrane is chosen so that hydroxiles anions move toward the resin bed. Half of the bicarbonate anions move trough the opposite anion exchange membrane of the cell toward a compartment where they dissolve in water. In this compartment, protons are produced by a bipolar membrane : thus the elution solution for the second step is regenerated. The second half of bicarbonate anions transform on the resin to carbonate anions. Thus the initial state is recovered.

SOLVENT RECOVERY BY DR-PSA, A SIMULATION STUDY

*Kazuyuki Chihara**, *Keita Sawada*, *Tadahiro Aiko*, *Takashi Kaneko*, *Takuya Chiyoda*

Meiji University, Japan
1-1-1 Higashi-mita, Tama-ku, Kawasaki, Japan
**Corresponding author, TEL +81-44-934-7197, FAX +81-44-934-7197*
e-mail: chihara@isc.meiji.ac.jp

ABSTRACT

Pressure swing adsorption (PSA) has been applied for various mixed gas separation and purification. And process simulation of PSA has become powerful for PSA process design. These days, solvent recovery became one of attractive PSA applications, not only conventional PSA, but also dual reflux PSA, which has middle point feed to the PSA column.

In this study, we investigated optimization about dual reflux PSA for solvent recovery by process simulation, named Stop & Go method. Dichloromethane adsorption on zeolite was assumed as a case study.

Keywords

PSA, Dual Reflux PSA, simulation

INTRODUCTION

Pressure swing adsorption (PSA) method is a cyclic process useful for the separation of gaseous mixtures. Though there were some problems to be solved, there may be some possibility to apply PSA to solvent recovery.

Dual reflux PSA is a new pressure swing adsorption process proposed for treatment of low-VOC-concentration air streams. Feed gas is supplied to the high pressure column at some intermediate position to divide it into an enriching and a stripping sections. A part of air stream leaving the high pressure column is returned to the low pressure column as enriching reflux. And we can get VOC as a liquid by dual reflux PSA process. [1]

In this study, the solvent recovery by DR-PSA with zeolite was tested by process simulation. As a first step, CH_2Cl_2 was chosen as solvent and the following items were examined.

SIMULATION

Figure 1 shows Dual Reflux PSA unit, respectively. Simulation was tried, based on this unit. The Dual Reflux PSA has feed point to the high pressure column at some middle point of the column, which separates the system into the upper enriching section and the lower stripping section. Operation sequence is shown in Figure 2. It was consist of four sequence, pressurization step, adsorption step, evacuation step and purge step. Pressurization, adsorption, evacuation and purge were set to 20, 200, 20 and 200 seconds. Four type of Purge condition was tried. Column conditions were shown in table.2. Adsorbate conditions were shown table.3.

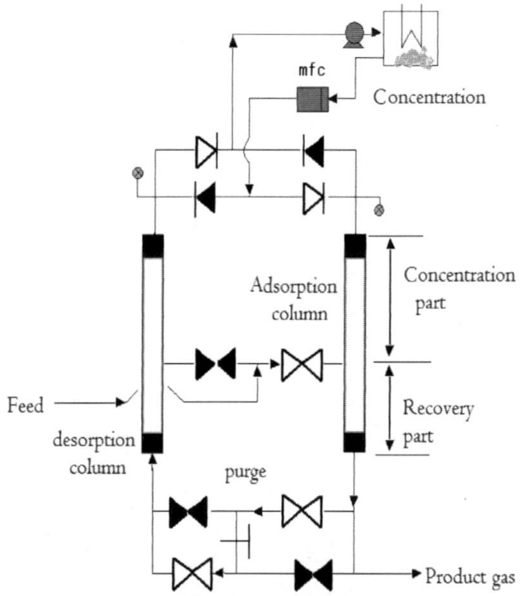

Fig.1 dual reflux PSA unit

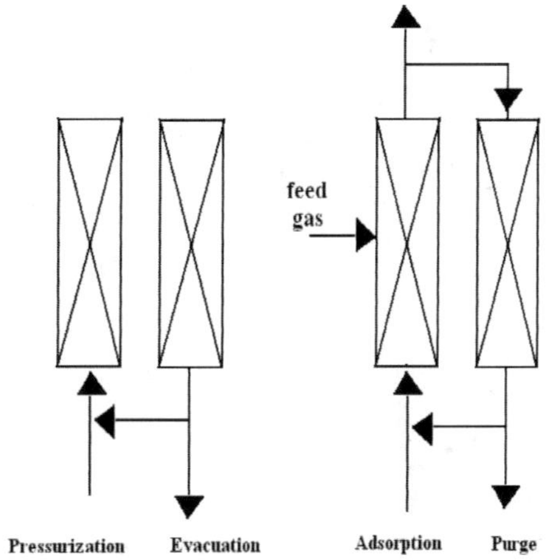

Fig.2 operation seqence

Table.1 operation condision

	feed [m^3/min]	Purge [m^3/min]	product gas [m^3/min]	pressure of saturated vapor [atm@-20deg]	concentration of feed gas [ppm]
A	100	58	99	0.066	10000
B	100	67	99	0.066	10000
C	100	100	99	0.066	10000
D	100	137	99	0.066	10000

Table2. column conditoin

radius[m]	1.02
length[m]	11.34
Amount of adsorbent [kg]	8530
porosity	0.4

Table.3 adsorbent condition

q^{∞}[Nl/g]	0.26
γ/ksav [sec]	150

RESULT AND DISCUSSION

3-1 influence of feed inlet position

Figure 3 shows influence from changing feed inlet position. Changing feed inlet position influence solvent recovery. It means that feed inlet position is very important for dual reflux PSA.

3-2 comparison each optimum feed inlet position

Figure 4 shows influence of amount of purge gas flow. They are each optimum feed inlet position. They are nearly same. It didn't decrease amount of solvent recovery as purge gas flow increasing.

Figure 5 shows concentration solvent in column. About A, CH_2Cl_2 concentration of feed gas equal that in column.

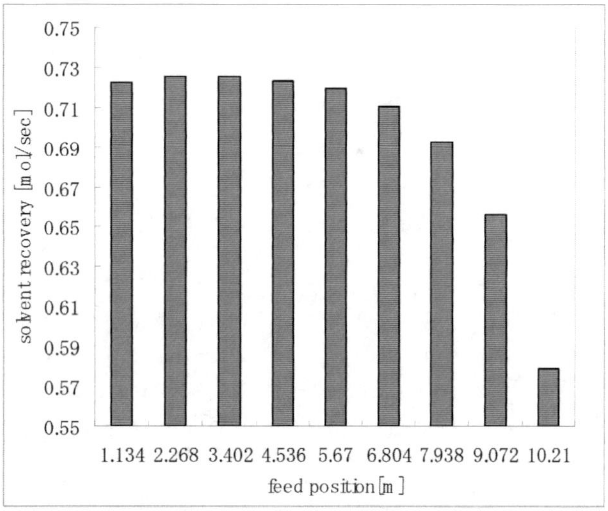

Fig.3 difference of solvent recovery by changing Feed position.

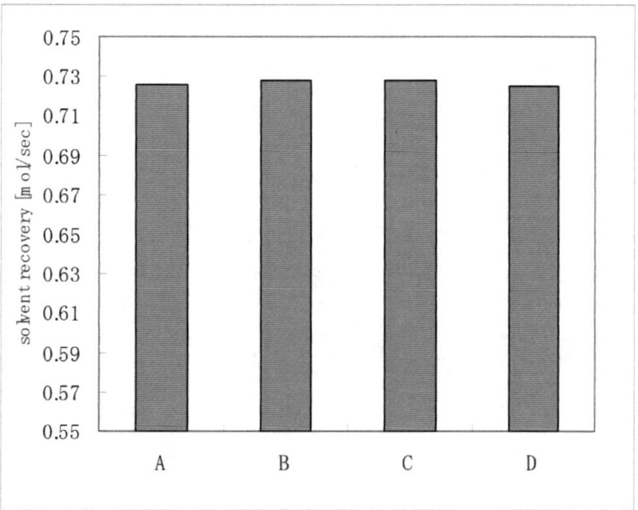

Fig.4 defference of solvent recovery each optimum position

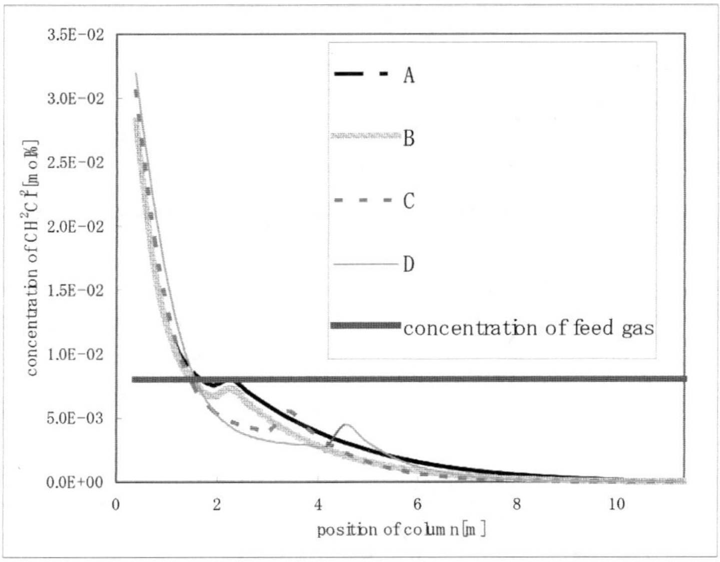

Fig.5 CH_2Cl_2 concentration in column

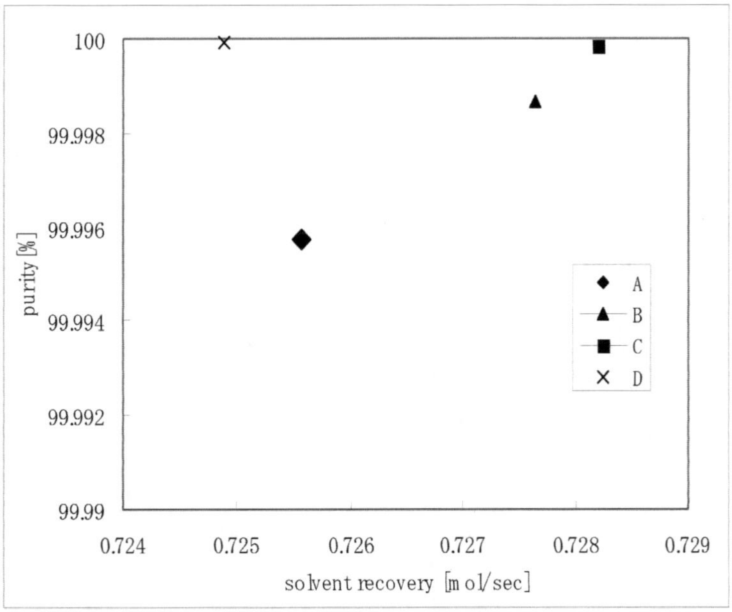

Fig.6 influence of purge gas

CONCLUSION

Numerical calculation for dual reflux PSA were tried using Stop & Go method. Performances were compared as for amount of solvent recovery and purity for different four type of purge gas flow. Running cost will increase as purge gas flow increase. But, amount of solvent recovery isn't change. If purge gas flow was lower than A, it will not be periodic state. So, A is optimum purge gas flow about this unit. When purge gas flow is A, concentration of feed gas equal concentration of gas in column.

REFERENCE

1. Ryoko. Wakasugi et al. (2005), "Dual Reflux PSA Process Applied to VOC Recovery as Liquid Condensate," *Adsorption 11(journal of the international Adsorption Society)*, pp. 561-566
2. Kazuyuki. Chihara et al. (2005), "Solvent Recovery by PSA, an Experimental and a Simulation Study," *Adsorption 11(journal of the international Adsorption Society)*, pp.567-572
3. Kazuyuki. Chihara et al. (2004), "Solvent Recovery by PSA with High Silica Zeolite, an Experimental and a Simulation Study," *14th IZC., Capetown*, pp.2003-2011
4. Diagne D. et al. (1995), "Parametric Studies on CO_2 Separation and Recovery by a Dual Reflux PSA Process Consisting of Both Rectifying and Stripping Sections," *I&EC Research*, 34, pp.3083-3089

Cassava-Based Adsorbent for Removing Water from Ethanol Vapor

Panarat Tomanee, School of Chemical Engineering, Suranaree University of Technology

Abstract

Cassava pulp is a by-product of cassava starch production that contain sufficient amount of fiber and starch. This material, which is normally abundantly produced, is mostly sold very cheaply to animal feed manufacturers. When demand from animal feed industry is low, cassava pulp accumulates, rots and smells badly, especially in a humid rainy season. This causes serious odor pollution, an environmental problem commonly observed around cassava-starch factories.

Starch, starch-based material, cellulose, and hemicelluloses have been reported to have affinity for water and can selectively adsorb water from various organic solutions. It has beed demonstrated the separation of ethanol-water vapor mixture over cracked corn, starch, and carboxymethyl cellulose. Other biomass materials, including corn grits, corn meal, corn cobs, xylan, crystalline cellulose, and wood chips, have been shown to separate water from ethanol. Water was also removed from others alcohols such as methanol, isopropyl alcohol, and tert-butyl alcohol.

This research aims to study the potential of using cassava pulp as adsorbent for removing water from ethanol vapor. Parameters investigated include characteristics in water adsorption of cassava-based adsorbent in packed-bed adsorption column and adsorption capacity. Adsorption isotherm and breakthrough curve for water adsorption over this adsorbent are presented here. It is found that adsorbent prepared from cassava starch and pulp was able to selectively remove water from ethanol solution, and it was possible to obtain anhydrous alcohol, with ethanol concentration higher than 99.5 %wt from this process.

Introduction

Large quantities of water are produced during fermentation process for ethanol production. Anhydrous ethanol, i.e. solution with ethanol concentration of 99.5 wt% or higher, is required for automobile usage (Benson and George, 2005), therefore the water must be removed. Conventional techniques to remove the water are low-pressure distillation and azeotropic distillation with third component addition. However, it has been reported that distillation of ethanol could consume up to 50% of the overall energy used in a typical grain alcohol plant. Distillation of water-ethanol mixture to a concentration near azeotrope, followed by adsorption to remove the remaining water, referred to as dehydration step, has proved to be more energy-efficient alternative. The energy consumption in azeotropic distillation of ethanol-water azeotrope is about 88 kJ per m^3 of ethanol, which is higher than the energy consumption needed for the adsorption process using corn grits by 32 kJ per m^3 of ethanol (Ladisch and Dyck, 1979).

Adsorbents that readily adsorb water have already been studied, and some are being used commercially. These include molecular sieves, chloride or oxide salts, silica gel, and starchy and cellulosic materials from biomass, such as corn grits, ground corn, potato starch, maize starch, wheat straw. It was found that biomass materials were generally less expensive to regenerate when compared to molecular sieves (Ladisch and Tsao, 1982). In addition, if regeneration was not practical and economical, the spent adsorbents could be used as a feedstock of a fermenter to produce ethanol. Deposition of the biobased adsorbents is also environmental friendly as these materials are biodegradable (Benson and George, 2005).

There are two types of starch, amylose and amylopectin (Kyle and Ladisch, 2001). Amylose consists of D-glucos units bound together by α-1,4 O-glucosidic bonds and has about 9-20 α-1,6 O-glucosidic branch point on each amylose chain. Amylopectin is similar to amylose but it has many more α-1,6 O-glucosidic branches. Starch's ability to adsorb water is a result of interaction between free hydroxyl (-OH) groups on the glucose units and the water molecules, as depicted in Figure 1.

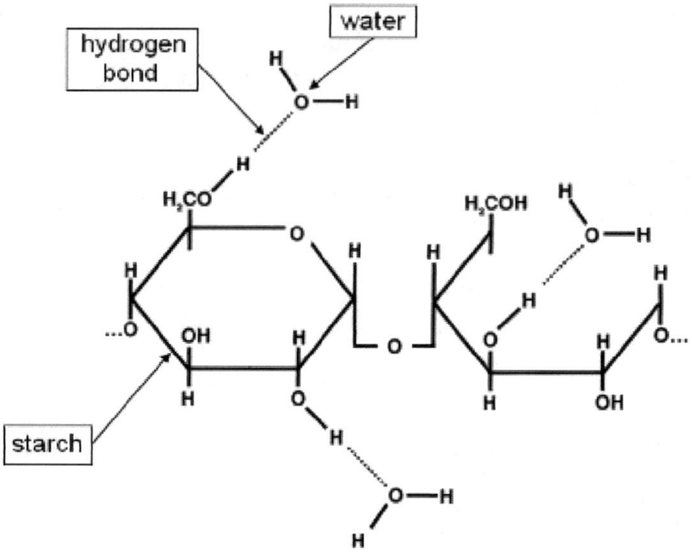

Figure 1. Hydrogen bonding between starch and water

Cassava pulp is a by-product of cassava starch production that contain sufficient amount of fiber and starch. This material, which is normally abundantly produced, is mostly sold very cheaply to animal feed manufacturers. When the demand from animal feed industry is low, cassava pulp accumulates and rots, causing a serious strong odor problem, which is an environmental problem commonly observed around cassava-starch factories. to community near cassava starch plant, especially during a humid rainy season.

In this work, starch-based adsorbent capable of removing water from ethanol vapor was developed. It was found that adsorbent prepared from cassava pulp was able to selectively remove water from ethanol solution. Efficiency of adsorption is affected by particle size of the adsorbent, as well as the starting concentration of ethanol feed. Isotherm for water adsorption over this adsorbent is also presented and discussed here.

Materials and Methods

Adsorbent Preparation

Cassava-based adsorbents were prepared in-house by mixing cassava starch and dried cassava pulp at a weight ratio of 1:1 with a solution of 0.1 M NaOH. The mixture was heated for partially gelatinization of the starch before being dried for 12 h in an air-convection oven at 105°C. The dried mixture was then ground and sieved. The adsorbents used in this study have the particle size in a range of 30-40 mesh.

Bench-Scale Packed Bed Adsorption

A bench-scale packed bed adsorber is shown schematically in Figure 2. Adsorption column with an inside diameter of 2.54 cm and a length of 18 cm was packed with 60 g of adsorbent. Column wall temperature was maintained at 85°C by circulating hot water through the jacket. Ethanol-water mixtures with various water contents were used as starting solutions and were heated from a 500-ml flask surrounded by an electric heating mantle. The vapor was passed through the column, and the exit stream was condensed using water as a cooling medium. The condensate, with an average flow rate of about 0.8 ml/min, was removed every 10 minutes and was analyzed for ethanol concentration using Abbe refractometer.

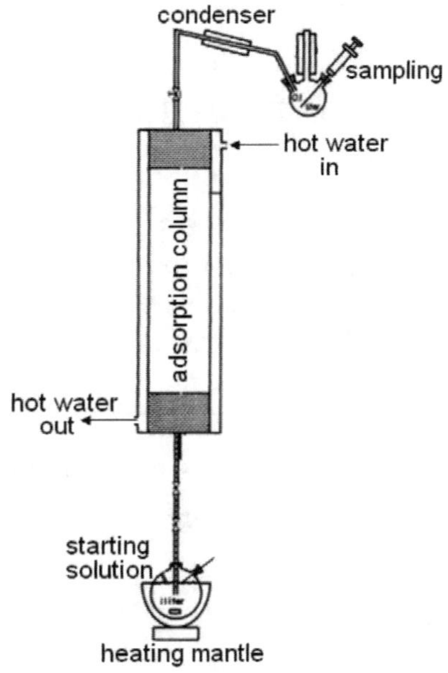

Figure 2. Schematic diagram of bench-scale packed bed.

Scanning Electron Microscopy

Adsorbent samples were fixed on a metal stubs and coated with gold. The morphologies of the samples were then observed using a JEOL JSM-6400 scanning electron microscope operating at 20 kV

Result and Discussion

Breakthrough curve of water from 90 %wt ethanol solution was shown in Figure 3 as an example of breakthrough curves obtained in this study. C/C_0 denotes ratio of water concentration in the condensate to that initially present in starting solution. It can be seen that the adsorbents were able to selectively adsorbed water from ethanol, and it was possible to obtain anhydrous alcohol, i.e. solutions with ethanol concentration higher than 99.5 %wt when this method was used for dehydration of ethanol vapor.

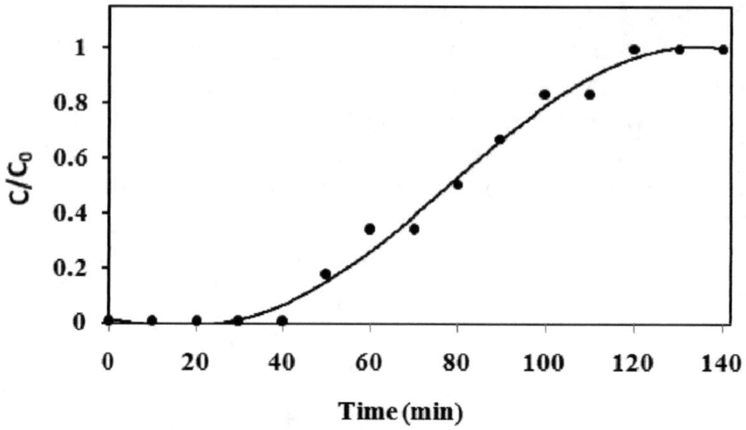

Figure 3. Breakthrough curves demonstrating adsorption characteristics of water over cassava-based adsorbents.

Isotherm for water adsorption was constructed using results obtain from breakthrough curve of water over cassava-based adsorbent. In this study, highest water uptake of about 0.06 $g_{H2O}/g_{adsorbent}$ was achieved. This value, however, could be increased with increasing starch content in the adsorbents

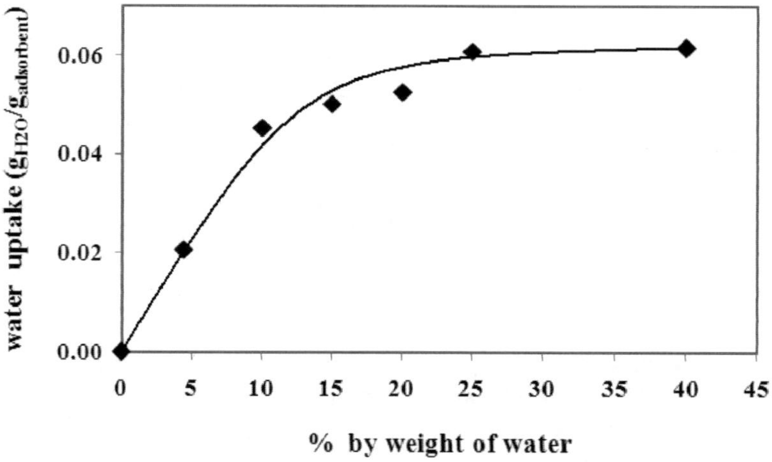

Figure 4. Isotherm of water adsorption on cassava-based adsorbent.

SEM image of cassava-based adsorbent is as shown in Figure 5. Starch can be seen as spherical particles with various sizes. Partially gelatinized starch can also be seen here.

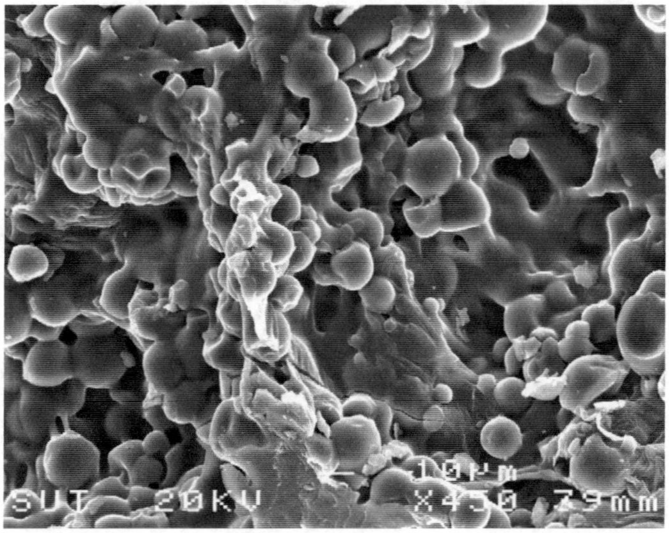

Figure 5. SEM image of cassava-based adsorbent.

Conclusion

Adsorption isotherm of water over cassava-based adsorbents, prepared from cassava starch and cassava pulp, was constructed using a bench-scale packed-bed adsorber. It was found in this study that the adsorbents were able to selectively adsorbed and removed water from ethanol vapor, and it was possible to obtain anhydrous alcohol, with ethanol concentration higher than 99.5 %wt from this process.

References

1. Benson, T.J. and George, C. E. (2005), "Cellulose based adsorbent materials for the dehydration of ethanol using thermal swing adsorption," *Adsorption*, 11: 697-701.
2. Ladisch, M. R. and Dyck, K. K. (1979), "Dehydration of ethanol: new approach gives positive energy balance," *Science*. 205: 898-900.
3. Ladisch, M. R and Tsao, G.T. U.S. Patent No. 4,345,973, 1982.
4. Kyle, E.B. and Ladisch, M. R. (2001) Chemistry and properties of starch based desiccants," *Enz. Microb. Technol.* 28: 573-581.

MULTICOMPONENT GAS ADSORPTION ON MSC5A BY CHROMATOGRAHIC METHOD AND STOP & GO SIMULATION

Kazuyuki CHIHARA[1], Yuki TERAMURA[1], Hidenori NAKAMURA[1] and Yosuke KANEKO[1]*
Meiji University
Higashi-mita, Tama-ku, Kawasaki, Japan

Abstract

Perturbation chromatography with multi component gas carrier and non-equilibrium thermodynamics liner law was applied for discussion of the interference effect and the displacement effect on mass transfer in multi component gas adsorption. Moment analysis method and stop & go simulation method were utilized to obtain each mass transfer parameters of adsorbate gases. Dependency of micropore diffusion on amount adsorbed and correlation of micropore diffusion with chemical potential driving force for microporous adsorbent were confirmed. Cross effect in micropore diffusion was found.

Keywords

chromatograph, stop & go simulation, multi component gas

1. Introduction

The combination of chromatographic method and moment analysis of the response peaks is one of the useful techniques to study adsorption equilibrium and adsorption rate (Chihara et al. 1978). Perturbation chromatography with the mixed multi component adsorbate gas carrier (two adsorbates) has been applied to several studies on adsorption (Ruthven and Kumar 1979, Kumar et al. 1982). In this work, perturbation chromatography with multi component gas carrier (two adsorbates with inert gas) and non-equilibrium thermodynamics liner law was applied for discussion of the interference effect and the displacement effect (those are cross effects) on mass transfer in multi component gas adsorption as previous study (Chihara et al. 2005) for different gas mixture (He, N_2, CH_4). Moment analysis method and stop & go simulation method were utilized to obtain each mass transfer parameters of adsorbate gases. Dependency of micropore diffusion on amount adsorbed and correlation of micropore diffusion with chemical potential driving force for microporous adsorbent were discussed. Also, cross effect were discussed. Ruthven already pointed the dependency of micropore diffusivity on amount adsorbed in single component adsorption (Ruthven 1984). Tondeur et al gave general background on multi component perturbation chromatography for the first moment only (Tondeur 1996).

2. Experimental Method

2.1. Experimental Procedure and Conditions

The experimental apparatus was shown in figure1. The apparatus was similar to a conventional gas chromatograph. Adsorbent particles (molecular sieving carbon 5A, 20/30 mesh, Japan Enviro Chemical Ltd.,) were packed in a column (100cm x 3mm i.d.). Carrier gas was a mixture of two or three components among He, CH_4, CO_2. Perturbation pulse was introduced into the carrier gas stream.

Introduction of pulses was performed by 6-way valve. The pulse size was 1cc, which meant injection period was 1.4 sec. Then pulse response was detected by TCD cell. Output signal of TCD was transmitted to a personal computer through RS232C. This signal was also transmitted to the personal computer. Simulated chromatogram by a personal computer can be overlapped on experimental chromatogram shown in the monitor screen. Further, moment of pulse response, which is shown in the monitor screen, can be automatically calculated by the personal computer. The first absolute moment and the second central moment were evaluated from the effluent peak $C_e(t)$ as follows:

$$\mu_1 = \int_0^\infty C_e(t)tdt \Big/ \int_0^\infty C_e(t)dt \qquad (1)$$

$$\mu_2{}' = \int_0^\infty C_e(t)(t-\mu_1)^2 dt \Big/ \int_0^\infty C_e(t)dt \qquad (2)$$

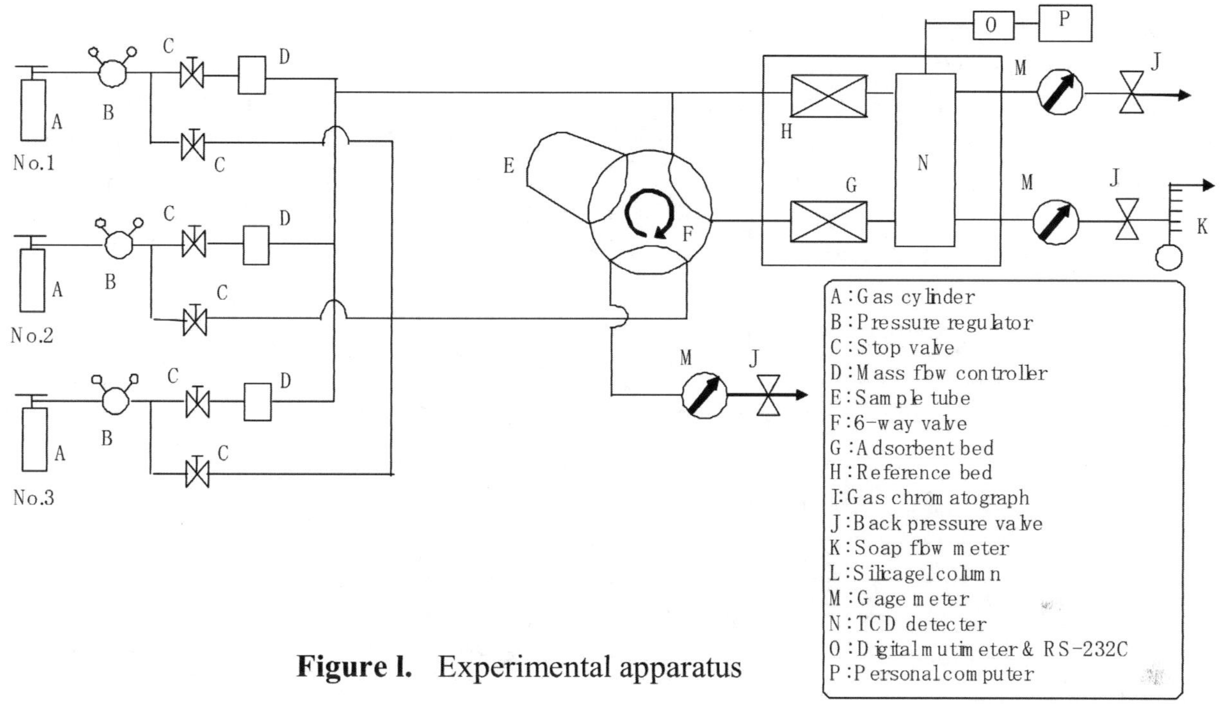

Figure I. Experimental apparatus

Moment Analysis

Basic equations are in ref. (Kumar et al. 1982)

The resultant moment equations of the impulse response are

$$\mu_1 = \frac{z}{u}\left\{1 + \frac{(1-\varepsilon)\rho_p(1-y)K^*}{\varepsilon}\right\} \qquad (3)$$

$$\mu_2 = \frac{2z}{u}\left[Ez\left\{\frac{1+(1-\varepsilon)\rho_p K^*}{\varepsilon u}\right\}^2 + \frac{(1-\varepsilon)\rho_p(1-y)K^*}{\varepsilon K s a v}\right] = \frac{2z}{u}(\delta_d + \delta_f + \delta_a + \delta_i) \qquad (4)$$

where μ_1=first absolute moment of the chromatographic peak[s] μ_2=second central moment[s^2].

$$\delta_d = \frac{Ez}{u^2}\left\{1 + \frac{(1-\varepsilon)\rho_p(1-y)K^*}{\varepsilon}\right\}^2 \qquad (5)$$

$$\delta_f = \frac{1-\varepsilon}{\varepsilon}\frac{R}{3k_f}\rho_p{}^2(1-y)K^{*2} \qquad (6)$$

$$\delta_a = \frac{1-\varepsilon}{\varepsilon}\frac{R^2}{15D_a}\rho_p^{\,2}(1-y)K^{*2} \quad (7)$$

$$\delta_i = \frac{1-\varepsilon}{\varepsilon}\frac{\rho_p(1-y)K^*}{15D}\bar{a}^{-2}\exp(\sigma^2) \quad (8)$$

where K^*=apparent adsorption equilibrium constant, Ez=axial dispersion coefficient based on void spaces in the bed, k_f=external mass transfer coefficient, D_a=diffusivity in macropores and D_i=diffusivity in micropores based on amount adsorbed gradient driving force. The arithmetic average radius $\bar{a}$ is 8.2 μm (Chihara, Suzuki and Kawazoe 1978).

Equation(3) was used along with the experimental μ_1 to obtain the adsorption constant K^*. Equation(4) to (8) were used along with the experimental μ_2' to obtain the diffusivity in micropores D.

Stop & Go Simulation

Numerical solution for multi component chromatogram in time domain could be obtained by appropriate model equations with experimental conditions. This simulated chromatogram can be compared with experimental chromatogram to determine the equilibrium and the adsorption kinetic parameters. Here Markham-Benton equation as for adsorption equilibrium and linear driving force (LDF) approximation as for adsorption kinetics were adapted for numerical calculation, which was based on stop & go method (Chihara et al. 1986, Chihara and Kondo 1986). In particular, LDF model of adsorption kinetics was based on non-equilibrium thermodynamics.

For binary adsorbates, adsorption rate equations are

$$\gamma\frac{\partial q_1}{\partial t} = Ksav_{1,1}(q_1^* - q_1) + Ksav_{1,2}(q_2^* - q_2) \quad (9)$$

$$\gamma\frac{\partial q_2}{\partial t} = Ksav_{2,1}(q_1^* - q_1) + Ksav_{2,2}(q_2^* - q_2) \quad (10)$$

where $Ksav$= Overall mass transfer coefficients.

Overall mass transfer coefficients ($Ksav$) for LDF model were determined. Then, micropore diffusivities were obtained by subtracting other mass transfer effects from overall resistance ($\gamma/Ksav$). Thus obtained micropore diffusivities were correlated with chemical potential driving force by consideration of Fick's diffusion equation, non-equilibrium thermodynamics and extended Langmuir equation (Karger and Bulows 1975).

$$\frac{D_{11}}{a^2} = \frac{D_1'}{a^2}\frac{\partial \ln p_1}{\partial \ln q_1} = \frac{D_1'}{a^2}\frac{1-\theta_2}{1-\theta_1-\theta_2} \quad (11)$$

$$\frac{D_{12}}{a^2} = \frac{D_1'}{a^2}\frac{q_1}{q_2}\frac{\partial \ln p_1}{\partial \ln q_2} = \frac{D_1'}{a^2}\frac{\theta_1}{1-\theta_1-\theta_2} \quad (12)$$

$$\frac{D_{21}}{a^2} = \frac{D_2'}{a^2}\frac{q_2}{q_1}\frac{\partial \ln p_2}{\partial \ln q_1} = \frac{D_2'}{a^2}\frac{\theta_2}{1-\theta_1-\theta_2} \quad (13)$$

$$\frac{D_{22}}{a^2} = \frac{D_2'}{a^2}\frac{\partial \ln p_2}{\partial \ln q_2} = \frac{D_2'}{a^2}\frac{1-\theta_1}{1-\theta_1-\theta_2} \quad (14)$$

For instance, $Ksav$ is related to D in single adsorbate case as

$$\frac{1}{Ksav} = \frac{a^2}{15D} + K^*\left(\frac{R}{3k_f} + \frac{R^2}{15\varepsilon D_a}\right) \quad (15)$$

For binary case, when using Equation (15), apparent K^* could be determined by appropriate slope of adsorption isothermal plane at perturbation point.

3. Result And Discussion

3.1. Single Adsorbate Carrier Mixed with He and The Same Adosorbate Pulse

Figure 2. shows an example of comparison of experimental chromatogram with simulated chromatogram for MSC5A to obtain *Ksav* for LDF model. Experimental conditions were 313 K, column pressure 5 atm, flow rate 25 cm/sec and He+CH_4 mixed gas carrier with CH_4 pulse. Here CH_4 concentration in the carrier gas was changed 10, 30, 50, 70 and 90 %.

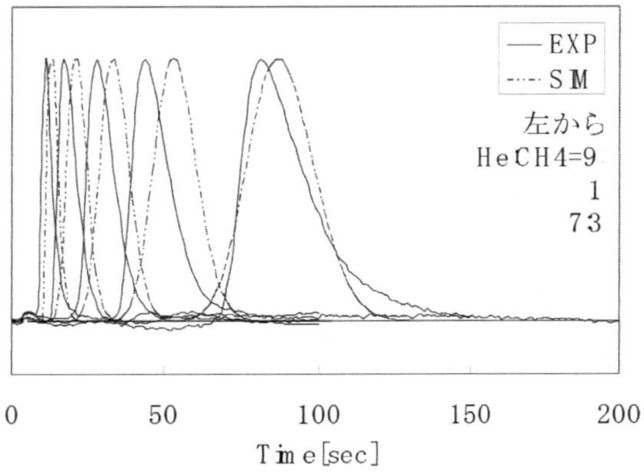

Figure 2. Comparison of experimental peaks with simulation

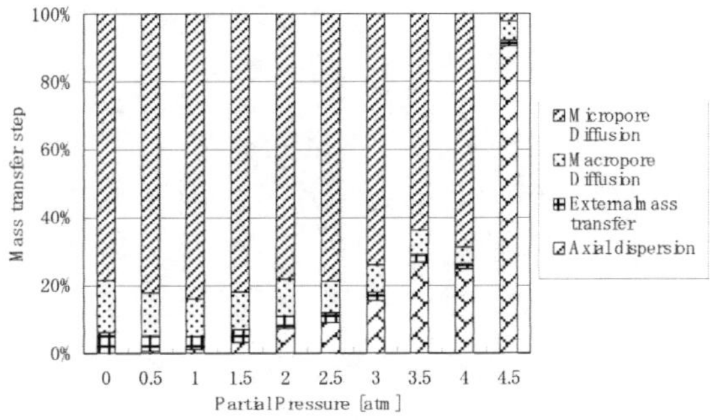

Figure 3. Mass transfer steps of CH_4 on to MSC5A

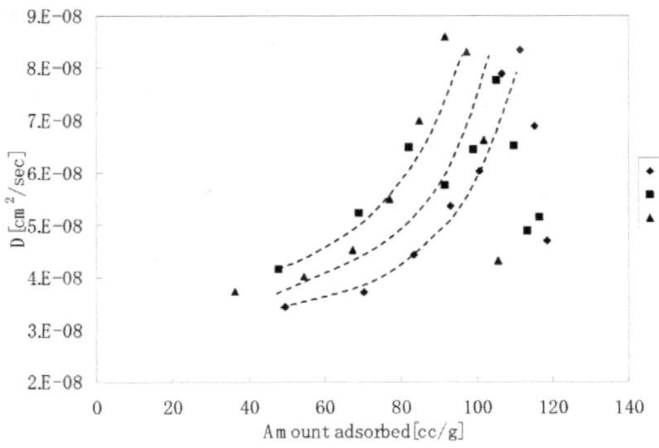

Figure 4. Dependency of micropore diffusivity coefficient on the amount adsorbed

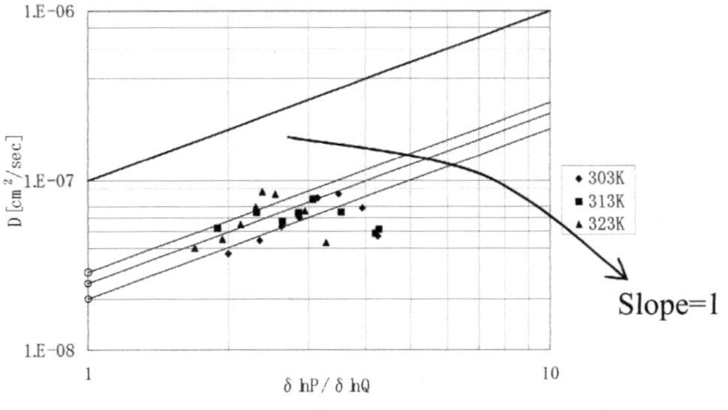

Figure 5. Correlation based on chemical potential driving force

Figure 3. shows contribution of axial dispersion, external mass transfer, macropore diffusion and micropoore diffusion in case of CH_4 for MSC5A at 323 K. This figure illustrated that mass transfer is controlled by micropore diffusion.

Figure 4. shows the dependency of micropore diffusivity coefficient of CO_2 with amount adsorbed at 303, 313 and 323 K for MSC5A.

Figure 5. shows the correlation of micropore diffusivity of CO_2 with chemical potential driving force at 303, 313 and 323 K for MSC5A. Proportional relation was obtained, which means that micropore diffusion is based on chemical potential driving force.

3.2. Binary Adsorbate Carrier Mixed with He and Adsobate Pulse

Figure 6-(a) and (b) show experimental and simulation results in an example case of binary adsorbate carrier mixed with He and an adsorbate pulse for MSC5A. Experimental conditions were 323 K, column pressure 5 atm, flow rate 25 cm/sec and He(60%)+CO_2(10%)+CH_4(30%) mixed gas carrier with CO_2 pulse. Figure6-(a) is the comparison between experimental and simulated TCD peaks. Two simulated TCD peaks are those for accounted and not accounted for cross effect. Figure6-(b) is simulated peaks of each gas component and total peak. The total peak is assumed to be the sum of adsorbate peaks except He and assumed to be as simulated TCD peak, which was found to be conincident with experiment as in Figure6-(a).

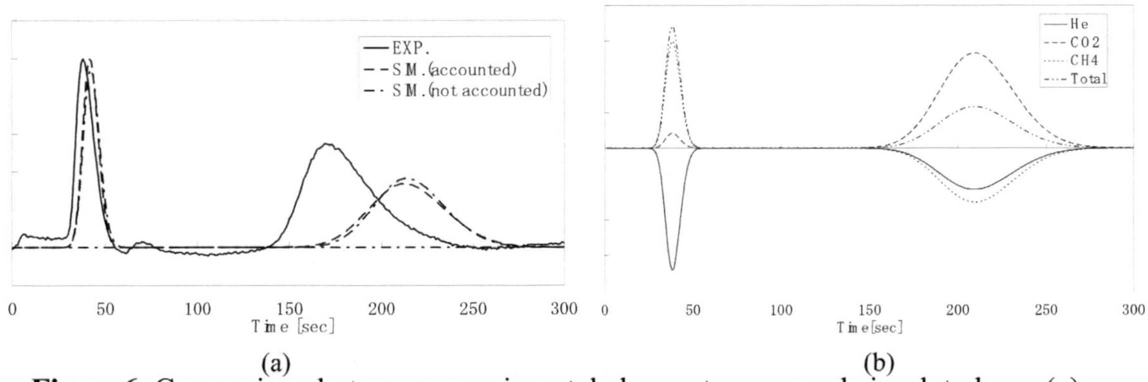

Figure6. Comparison between experimental chromatogram and simulated one (a), Simulation of pulse response of each component (b)

Simulated chromatogram for CO_2 and CH_4 mixture may be regarded as co-diffusion or competitive adsorption for 1st peak and as counter-diffusion or displacement adsorption for 2nd peak.

1st peak shows good agreement between experimental peak and simulated peak. But 2nd peak is not. So it is hard to judge the propriety.

The sum and product of two peaks' retention times can be related with isothems (Glover 1983).

$$(1-y_1^*)\left(\frac{\partial g_1}{\partial y_1}\right)^* + (1-y_2^*)\left(\frac{\partial g_2}{\partial y_2}\right)^* - y_1^*\left(\frac{\partial g_2}{\partial y_1}\right)^*$$

$$- y_2^*\left(\frac{\partial g_1}{\partial y_2}\right)^* = \frac{N}{M_s}(t_{ra}+t_{rb})$$

$$\left(\frac{\partial g_1}{\partial y_1}\right)^*\left(\frac{\partial g_2}{\partial y_2}\right)^* - \left(\frac{\partial g_1}{\partial y_2}\right)^*\left(\frac{\partial g_2}{\partial y_1}\right)^* \tag{16}$$

$$= \left(\frac{N}{M_s}\right)^2 \frac{t_{ra}t_{rb}}{(1-y_1^*-y_2^*)}$$

In this study, eq. (16) is rewrited as follows, assuming extended Langmuir isotherm.

$$\frac{p_0}{(1+K_1p_1+K_2p_2)^2}\{(1-y_1)q_{\infty 1}K_1(1+K_2p_2)$$

$$+(1-y_2)q_{\infty 2}K_2(1+K_1p_1)\}$$

$$+\frac{p_0K_1K_2}{(1+K_1p_1+K_2p_2)^2}(y_1q_{\infty 2}p_2+y_2q_{\infty 1}p_1)$$

$$=\frac{V_r}{M_s}(t_{ra}+t_{rb}) \tag{17}$$

$$\frac{p_0^2q_{\infty 1}q_{\infty 2}K_1K_2}{(1+K_1p_1+K_2p_2)^2}\{(1+K_1p_1)(1+K_2p_2)$$

$$-K_1p_1K_2p_2\}$$

$$=\left(\frac{N}{M_s}\right)^2\frac{t_{ra}t_{rb}}{(1-y_1-y_2)}$$

Where p_0= column pressure (5 atm), K = Langmuir constant, q_∞= saturation amount adsorbed, t_r = retention time, V_r = volume flow rate at 1 atm, M_s = amount of MSC5A in column.
This relation could be used to check experimental and simulated results with assumed isotherm as table 1.

Table 1. The sum and product of two peaks retention times

	Sum [sec]	Product[sec^2]
Exp.	212	6747
Sim.	257	9030
From isothem	242	6664

Sum value of calculation is close to simulation. But, product is close to experiment.

Figure 7. shows Ksav obtained by Eq.(11)~(15) neglecting macro effect for binary adsobates system, corresponding to Figure 6. It was found that *Ksav* obtaind by consideration of chemical potential driving force with cross effect was well applied to get good simulation results for multicomponent pulse response though some discrepancies were founded.

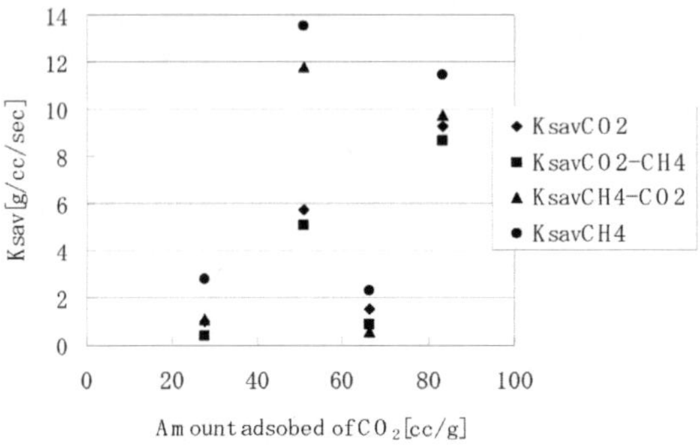

Figure 7. Comparison of *Ksav* of single adsorbate, and *Ksav* of binary adsorbate[He+ CH_4- CO_2]

4. Conclusion

This time, we cannot get good agreements between experimental chromatogram and simulated chromatogram, based on the modeling of Stop & Go method, in case of peturbation chromatography with mixed adsorbate gas carrier. And micropore diffusivities obtained were interpreted by chemical potential driving force consideration based on non-equilibrium thermodynamics law. Some possibilitty was shown for cross effect of micropore diffusion in multicomponent adsorption.

5. References

Chihara K. et al; "Simulation of Pressure Swing Adsorption for Air Separation" Proc. 7th Int. zeolite Conf.,563(1986)

Chihara K. and Kondo A.; "Simulation of Pressure Swing Adsorption –Three Gas Components and Three Adsorption Columns-" 2nd FOA, 165 (1986)

Chihara, K., Suzuki M. and Kawazoe K.; "Adsorption Rate on Molecular Sieving Carbon by Chromatography" AIChE J., 24, 237 (1978)

Chihara K. et al; "Study of Multicomponent Gas Adsorption into MSC5A by Chromatograph" Adsorption, 11, 249-253(2005)

Glober C. J. and Lau W. R.; "Determination of Multicomponent Sorption Equilibria Using Perturbation Gas Chromatograpy" AIChE J., 29, 73-79 (1983).

Karger J. and Bulows M.; Chem. Eng. Sci., 30, 893(1975)

Kumar R., Duncan R.C. and Ruthven D.M.; "A Chromatographic Study of Diffusion of Single Components and Binary Mixtures of Gases in 4A and 5A Zeolites" Can.J.Chem.Eng., 57, 342 (1982)

Tondeur D., Kabir H., Luo L., and Granger J.; "Multicomponent adsorption equilibria from impulse response chromatography" Chem. Eng. Science, 51(N15) 3781-3799 (1996)

Ruthven D.M,; "Principle of Adsorption and Adsorption Process" Wiley, 146 (1984)

Ruthven D.M. and Kumar R.; "A Chromatographic Study of the Diffusion of N_2, CH_4 and Binary CH_4-N_2 Mixtures in 4A Molecular Sieve" Can.J.Chem.Eng., 57, 342 (1979)

ADSORPTION OF MIXED ORGANIC SOLVENT BY FAU TYPE ZEOLITE

Kazuyuki Chihara, Junichi Kabe, Shinji Tomita, Shinji Kondo, Takashi Matsumoto*

Meiji University
Higashi-mita, Tama-ku, Kawasaki, Japan
**Corresponding author, TEL +81-44-934-7197, FAX +81-44-934-7197,*
e-mail: chihara@isc.meiji.ac.jp

ABSTRACT

Instead of the chlorofluorocarbon specified as the ozone-depleting substance, a chlorinated organic compound, hydrocarbon, alcohol, etc. are used as an alternative washing solvent, and they are used as a mixed solvent in many cases. Therefore, adsorption operation of binary systems is needed. Furthermore, since azeotropic may happen, in the case of the design of a solvent recovery system and operation, those adsorption equilibrium are required as basic data. This study considered the adsorption equilibrium relation used as basic data required in order to adsorb various binary mixture solvent vapor by high silica zeolite.

Keywords

azeotropic adsorption, breakthrough curves, gravimetric method

INTRODUCTION

Discharge of organic substances into air has been strongly prohibited since some decades ago, to preserve comfortable natural environment. Though hydrocarbons, alcohols and chlorinated hydrocarbons had been used as the degreasing agent in industries, these solvents may also affect to our environment, and it is necessary to remove them from air as much as possible. This study was performed aiming at presenting useful data for the design of adsorption processes, especially the removal of the solvent vapors as above from air by adsorption. The solvents used were trichloroethylene (TCE) and ethanol (EtOH).

EXPERIMENTAL

2.1 Adsorbent

Fixed-bed adsorption experiment and gravimetric method of laboratory-scale were carried out to remove organic solvent vapors by Y type high silica zeolite. Experimental results for binary component system, which showed azeotropic mixture systems, could be correlated satisfactorily.

2.2 Fixed-bed Adsorption Experiment

Solvents used were ethanol (EtOH) and trichloroethylene (TCE). Adsorbents were packed in a glass column of 0.10m length and 0.0156m i.d. Experimental conditions were as follows: linear flow rate; 0.2 m/s, influent concentration; in the range of 0.004~0.2 mol/m3, and temperature; 298 K. Experimental apparatus are shown in Figure 1.

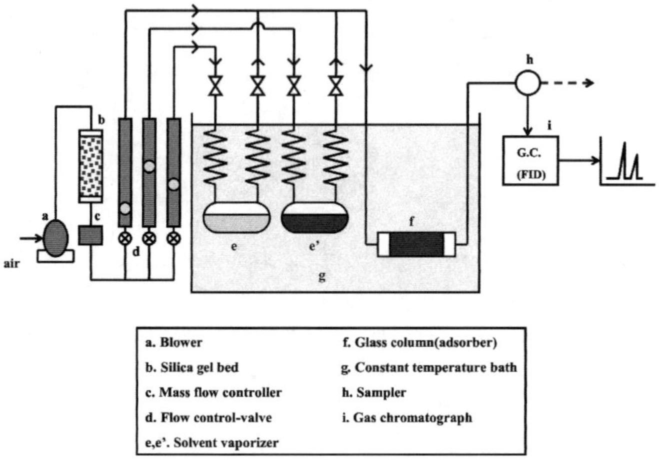

Figure 1: Experimental apparatus to fixed-bed adsorption experiment

2.3 Gravimetric Method

Figure 2 shows experimental apparatus for gravimetric analysis. The zeolite sample (about 0.25g) was placed in a quartz basket (F). Then the adsorbate in flask (B) was fed to adsorption tube (H). The whole apparatus was in a constant temperature air bath. The temperature range was 293-298 K. The amount adsorbed was measured corresponding to the pressure of the vapor in the tube. The pressure was measured by pressure sensor (J) at higher pressure range (> 0.013 atm) and baratron (I) at lower pressure range (<0.013 atm).Gravimetric Method is an experiment that inputs the adsorption quality like the pulse in the faction of the low pressure, and measures the change in the weight in that case. It predicts it from the working curve that measures the amount of adsorption from a pressure decrease in the device by the adsorption phenomenon beforehand. The weight change was used for the curve fitting and the adsorption curve. The prediction of the amount of adsorption by pressure calculates decrease pressure ΔP by adsorption from amount q of adsorption when the adsorption quality is first input. The difference of Pt at P0 and each time to reaching to equilibrium happens because of adsorption. Amount qt of adsorption at each time was requested by the following expressions.

$$qt = \frac{P_0 - P_t}{RT} \times V \times M \qquad (1)$$

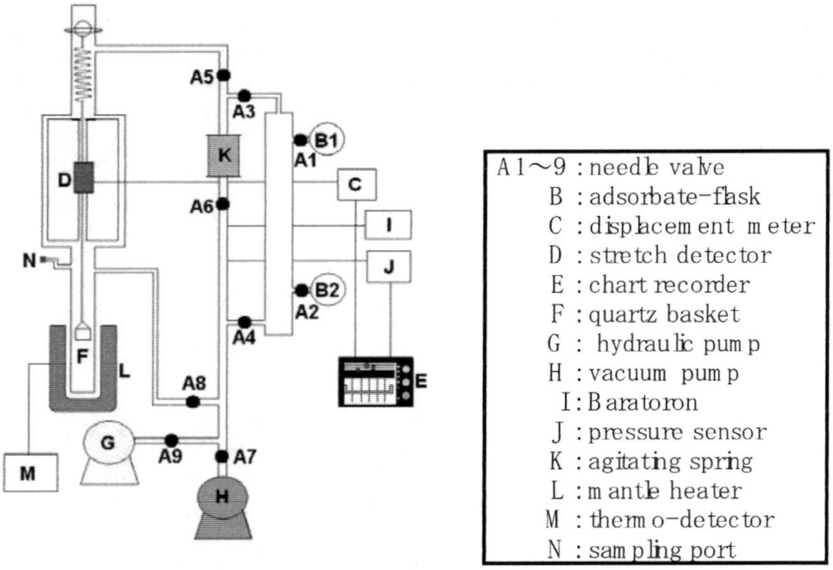

Figure 2: Experimental apparatus to gravimetric method

RESULT AND DISCUSSION

3.1 Breakthrough Curves

Figure 3 show several types of breakthrough curves obtained for EtOH-TCE -Y-type zeolite system. In Figure 3, the range of the total concentration of EtOH and TCE are 2500~3500ppm. For this system reversal of the order of breakthrough (turn over) occurred once at concentrations of 0.222 mole fractions of EtOH. When the mole fractions were 0.222, the mixture of two components behaved as if it was a single component system.

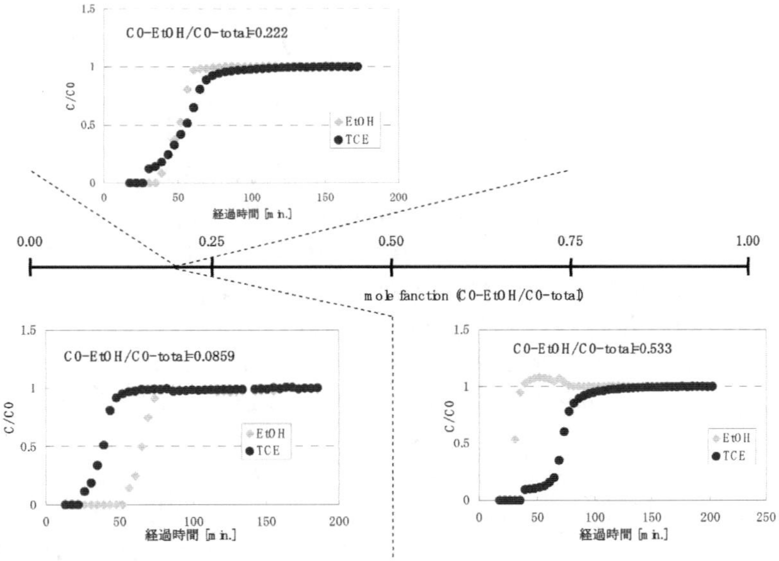

Figure 3: Several types of breakthrough curves observed for EtOH-TCE-Y-type system (2500~3500ppm)

3.2 Adsorption Isotherm

In the gravimetric method, the adsorption isotherm of binary systems was obtained. Figure 4 show several types of adsoption isotherm obtained for EtOH-TCE -Y-type zeolite system. The result of the single element experiment was able to reproduce the result of the flow method. In Figure 4, a ratio of EtOH and TCE is 0.587:0.413 and 0.756:0.244.It was seen that amount of adsorption increased as ratios of TCE increased.

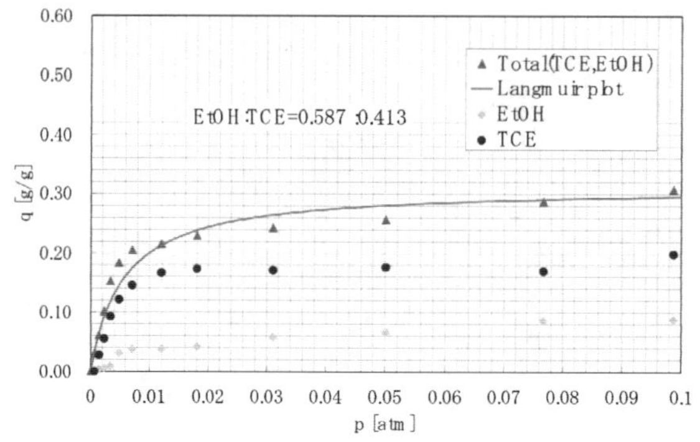

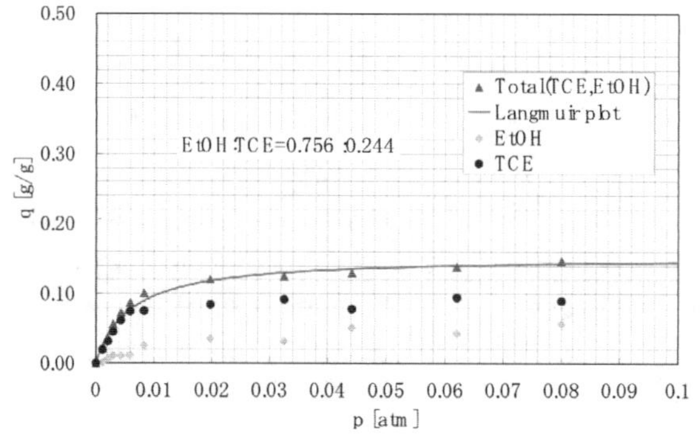

Figure 4: Several types of adsorption isotherm observed for EtOH-TCE-Y-type system

3.3Adsorption Equilibria

Binary adsorption equilibria for the system of EtOH-TCE of several total concentration are shown in Figure 5 and Figure 6 in form of X-Y diagram (X and Y are the mole fractions of gas phase concentration and the amount adsorbed at equilibrium, respectively). Figure 5 is a result of the fixed-bed adsorption experiment, and figure 6 is a result of the gravimetric method. As for the gravimetric method, azeotropic phenomenon was not seen in the low concentration (3000 ppm , 6000 ppm). However the high concentration region that was not able to be measured by the fixed-bed adsorption experiment was able to be measured. Those equilibria have one azeotropic point, which attribute to the reversal of breakthrough order.

For vapor-liquid equilibrium, when the saturation vapor pressures of each component were almost the same, appearance of one azeotropic points were reported. It is thought, therefore, that the phenomenon occurred in adsorption equilibrium for these systems can be ascribed to the fact that the saturation vapor pressure and the boiling points of each component were almost the same.

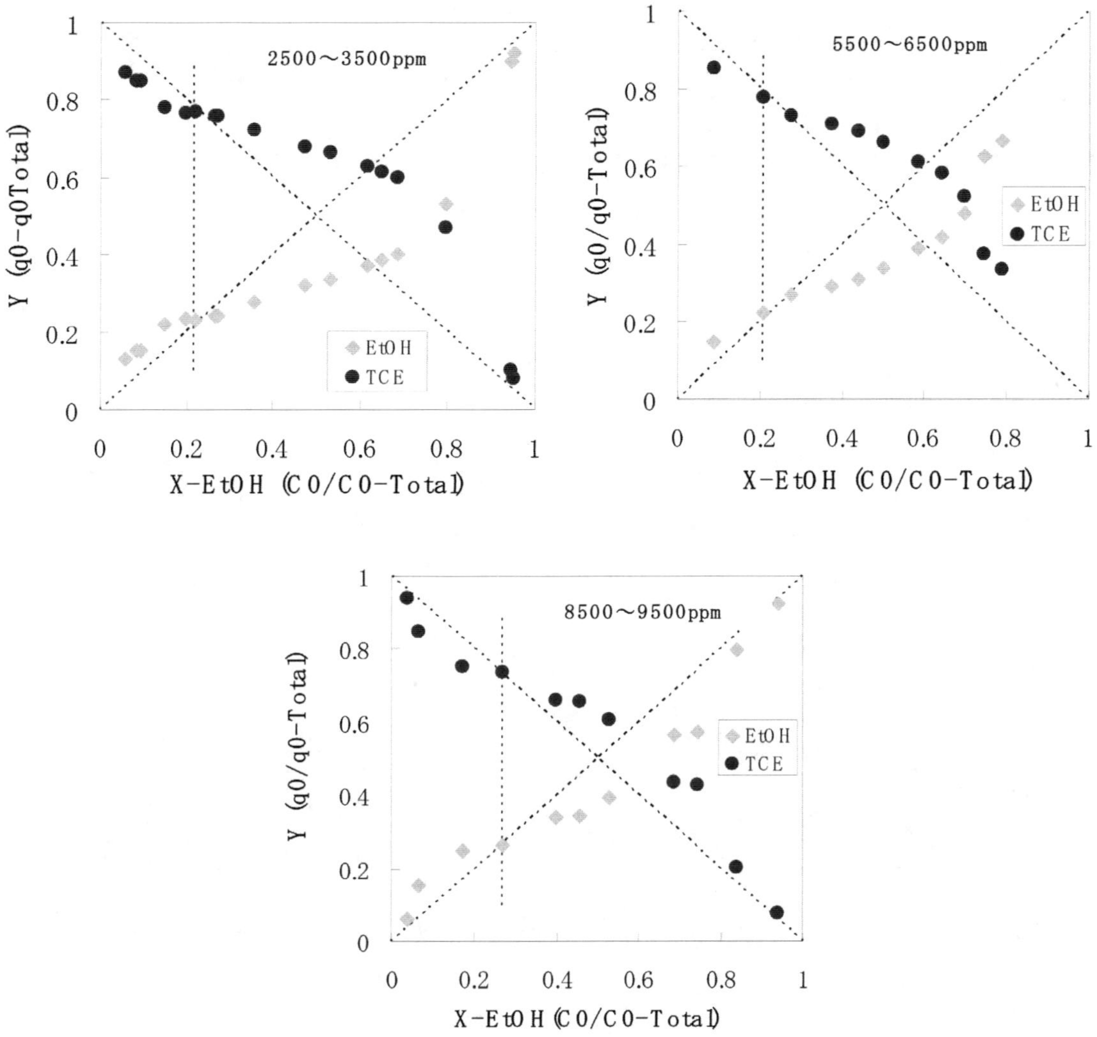

Figure 5: Comparison of adsorption equilibrium for experimental concentration (Fixed-bed adsorption experiment)

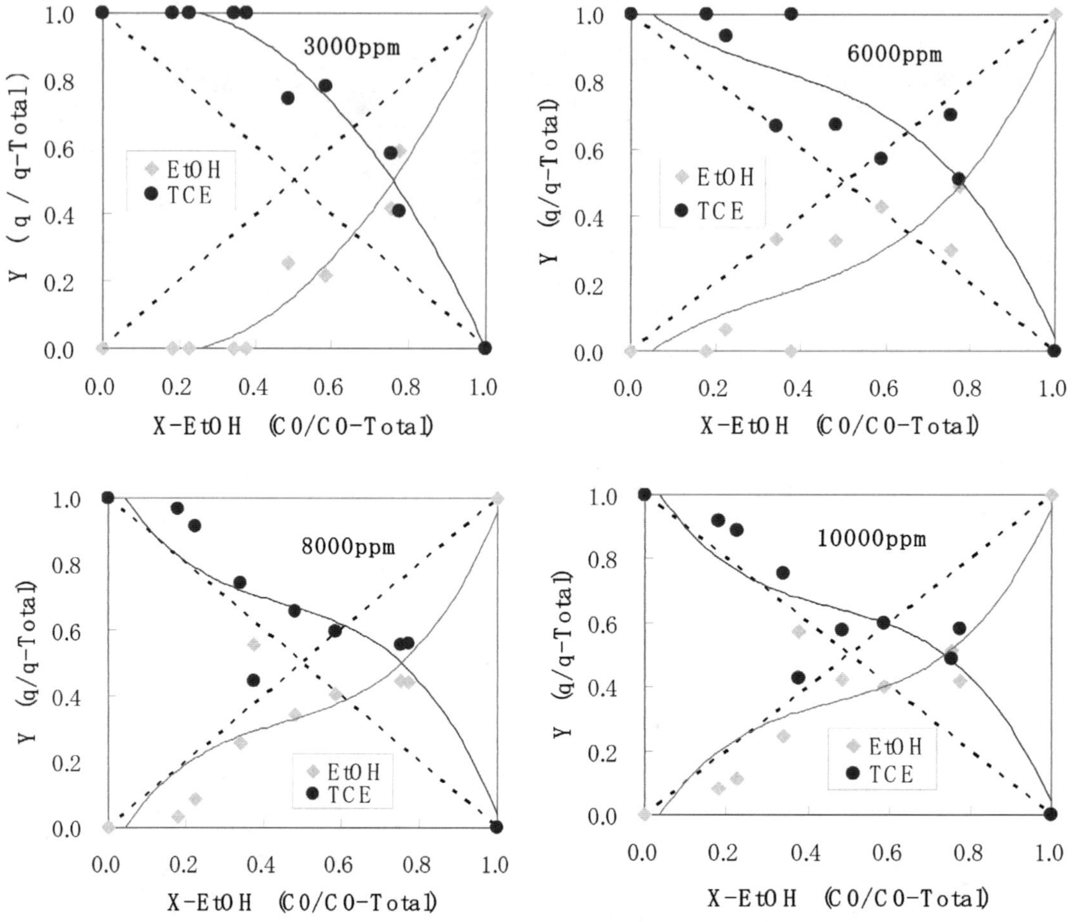

Figure 6: Comparison of adsorption equilibrium for experimental concentration (Gravimetric method)

CONCLUSION

Adsorption of EtOH-TCE vapors onto HSZ was studied. One azeotropic point appeared in the adsorption equilibrium for this EtOH-TCE -Y-type system. As for the gravimetric method, azeotropic phenomenon was not seen in the low concentration (3000 ppm , 6000 ppm). However the high concentration region that was not able to be measured by the fixed-bed adsorption experiment was able to be measured.

REFERENCE

1. K Chihara et al., Proceeding of the 7th International Zeolite Conf, 563-570 (1986)

719l Adsorption Equilibrium and Kinetics of Cesium Ion Using Natural and Synthetic Adsorbents

Tae Young Kim[1], Seon-Gyun Rho[1], Se-Young Jung[1], Seung-Jai Kim[1], and Sung-Yong Cho[2]. (1) Environmental Engineering, Chonnam National University, 300 YongBong-Dong Buk-Gu, Gwangju, 505-757, South Korea, (2) 82-62-530-1969, Chonnam National University, 300 YongBong-Dong Buk-Gu, Gwangju, 505-757, South Korea

There is a great concern about global warming among the scientific and political communities. This inconvenient truth is the impetus driving development of alternative energies with minimal carbon dioxide emissions. For over 50 years, nuclear echnologies have been supplying a better quality in different fields of our life. However, this has resulted in accumulation of substantial amounts of radioactive waste. Americium-241, europium-152 and cesium-137, important sources of radioactivity in the waste, enter the environment mainly through fallout from nuclear weapon tests and controlled releases of industrial nuclear wastes. These radio nuclides pose serious environmental threats and challenge remediation, so they are potentially very hazardous nuclides. They are extremely long lived and do not undergo biodegradation. Their solubility in aqueous solution makes them likely to migrate through ground water to the biosphere. Furthermore, 137Cs, because of its chemical similarity to K and Na, is readily assimilated by terrestrial and aquatic organisms. However, as a cheap energy with no CO2 emissions, nuclear energy plays an important role in Korea's power supply system. The main purpose of this work is to eliminate cesium ion from its aqueous solution using chitosan bead, alginate bead and ion exchange resin and to study its adsorption characteristics experimentally as well as theoretically.

A Simplified Analysis of Temperature Swing Adsorption in Honeycomb Rotor Dehumidifiers

Tsutomu Hirose

Advanced Adsorption Technology, 19-7, 4 Chome, Miwadai, Higashi-ku, Fukuoka, Japan

Abstract

Honeycomb rotor dehumidifiers have been used widely because of large surface area and rapid thermal response. Mass and heat transfer behavior for the performance evaluation is rather complicated and a system of simultaneous partial differential equations of 8 variables must be solved under a cyclic initial condition for the complete simulation.

With high efficiency dehumidifiers, however, a relative heat capacity λ $(=\rho_c c_{pc} L/\rho_g c_{pg} U t_c)$ is smaller than unity and a relative adsorption capacity $l(=\rho_c q^*_{s0} L/\rho_g x_{s0} U t_c)$ is higher than unity. When the cycle time is selected properly under the above condition, each step of adsorption/desorption can be divided into two different time zones, i.e. a transition stage appearing just after switching the step until $t/t_c<\lambda_s$ and a subsequent thermal equilibrium stage during $\lambda_s <t/t_c<1$. Sensible heat transfer is predominant in the transition region and reaches the thermal equilibrium within this stage. Thus, the mathematical solution becomes free from the cyclic condition to save the iteration. In the thermal equilibrium region, heat liberated by adsorption/desorption is balanced to warm/cool the streaming air, and the enthalpy of air is constant everywhere and equal to that of the feed/regeneration air. Since the amount adsorbed changes linearly, the time average values of temperature, humidity etc. can be calculated by a simple integration of the reciprocal of driving force, based on the short cycle time approximation or the continuous countercurrent flow model.

The above simplified model is proposed together with a graphical solution method and the result is compared with the more rigorous simulation.

1. Introduction

An adsorptive dehumidifier or a desiccant rotor adsorbs water vapor at room temperature to produce dry air and desorbs water vapor having been adsorbed at an elevated temperature. It is based on a principle of thermal (or temperature) swing adsorption (TSA) that is a kind of periodical mass transfer operations. This typical example of simultaneous mass and heat transfer process with periodic boundary conditions has been analyzed by a computer simulation by Tsujiguchi and Kodama [1] and others. Heat/mass transfer between wet air and an adsorbent rotor consists of two stages which have different purposes, i. e. cooling/heating for switch of the process step and adsorption/desorption for treatment of water vapor. Each stage is distinguished clearly in a well-designed desiccant rotor. Heat/mass transfer can be analyzed more simply by obtaining an asymptotic solution to each stage. Contributions of each stage to the overall efficiency of dehumidification are discussed based on this asymptotic analysis.

2. Analysis of Heat/Mass Transfer

2.1. Definition of Variables and Heat/Mass Balance

Figure 1 shows a schematic diagram of a 2-step TSA operation in which a desiccant rotor of length of L is rotated slowly and encountered periodically with adsorption (designated with a subscript i=s) and desorption (designated with a subscript i=e). Relevant variables are expressed in terms of dimensionless variables as defined by Eqs. (1)-(7). They are Z for axial position z[m], τ for process time t[s], g for gas phase temperature T_g[K], f for solid phase temperature T_f[K], P for humidity x[kg/kg] and Q for amount adsorbed q[kg/kg]. Subscripts 0 and 1, respectively, refer to the inlet end of feed and the exit end o f product in the adsorption stream..

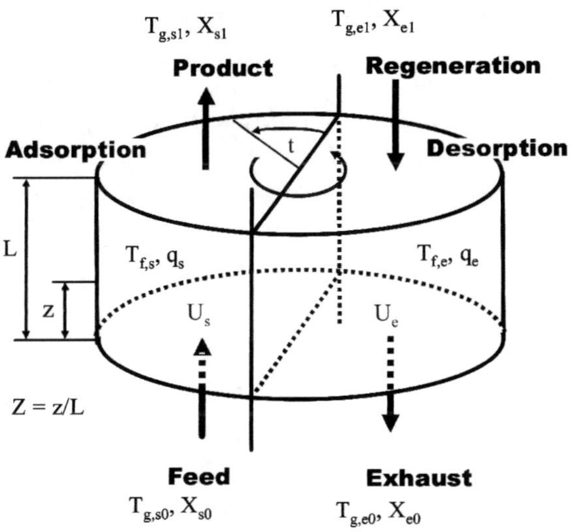

Fig. 1 Schematic diagram of a honeycomb rotor and relevant nomenclature

$$Z = z / L, \qquad (1)$$

$$\tau = t / t_{c,s} \quad (0 < \tau < 1) \qquad (2)$$

$$\tau = 1 + (t - t_{c,s}) / t_{c,e} \quad (1 < \tau < 2) \qquad (2')$$

$$g_i = \frac{T_{g,el} - T_{g,i}}{T_{g,el} - T_{g,s0}} \qquad f_i = \frac{T_{g,el} - T_{f,i}}{T_{g,el} - T_{g,s0}} \qquad (3), (4)$$

$$P_i = x_i / x_{s0} \qquad (5)$$

$$Q_i = q_i / q^*_{s0} \qquad Q^*_i = q^*_i / q^*_{s0} \qquad (6), (7)$$

in which equilibrium amount adsorbed q^*_i is a function of temperature of adsorbent $T_{f,i}$ and humidity of air x_i and given by an adsorption isotherm.

$$q^*_i = q^*_i(T_{f,i}, x_i) \qquad (8)$$

Heat/mass balance and heat/mass transfer rate are expressed in terms of these variables for both adsorption step (subscript i=s) and desorption step (subscript i=e) as follows

$$\partial g_i / \partial Z = \pm \kappa_i (g_i - f_i) \qquad (9)$$

$$\partial P_i / \partial Z = \pm N_i (Q^*_i - Q_i) \qquad (10)$$

$$\lambda_i \, \partial f_i / \partial \tau = \kappa_i (g_i - f_i) - r N_i (Q^*_i - Q_i) \qquad (11)$$

$$l_i \, \partial Q_i / \partial \tau = N_i (Q^*_i - Q_i) \qquad (12)$$

A complex sign $\pm$ takes $-$ for adsorption step (i=s) and + for desorption step (i=e). Ten dimensionless parameters are grouped out in a course of normalization. They are number of heat transfer units κ_i, number of mass transfer units N_i, relative heat capacity λ_i, relative adsorption capacity l_i, relative heat of adsorption r, humidity ratio γ.

$$\kappa_i = \frac{h_i a L}{\rho_g c_{pg} U_i} \qquad N_i = \frac{k_i a L q^*_{s0}}{\rho_g U_i x_{s0}} \qquad (13),(14)$$

$$\lambda_i = \frac{\rho_c c_{pc} L}{\rho_g c_{pg} U_i t_{c,i}} \qquad l_i = \frac{\rho_c q^*_{s0} L}{\rho_g U_i x_{s0} t_{c,i}} \qquad (15), (16)$$

$$r = \frac{x_{s0} \Delta H}{c_{pg}(T_{g,el} - T_{g,s0})} \qquad \gamma = x_{el} / x_{s0} \qquad (17), (18)$$

Boundary and periodic conditions applied are as follows

$$g_s = g_{s0} = 1 \text{ at } Z = 0, \quad g_e = g_{el} = 0 \text{ at } Z = 1 \qquad (19)$$

$$P_s = P_{s0} = 1 \text{ at } Z = 0, \quad P_e = P_{el} = \gamma \text{ at } Z = 1 \qquad (20)$$

$$f_e\big|_{\tau=1} = f_s\big|_{\tau=1}, \qquad f_s\big|_{\tau=0} = f_e\big|_{\tau=2} \qquad (21)$$

$$Q_e\big|_{\tau=1} = Q_s\big|_{\tau=1}, \qquad Q_s\big|_{\tau=0} = Q_e\big|_{\tau=2} \qquad (22)$$

2.2. Transition Stage and Thermal Equilibrium Stage

In a well designed desiccant rotor, the relative heat capacity λ is necessary to be small for a rapid thermal response while the relative adsorption capacity l be large for a delayed breakthrough, i. e. the following equation

$$\lambda_i << 1 << l_i \tag{23}$$

has to be fulfilled.

Figure 2 shows a typical example of a computer simulation for the time variation of humidity P_{s1}, temperature $(1\text{-}g_{s1})$, and the amount adsorbed Q_{s1} at the product end under the above condition of Eq. (23). A transition stage with a rapid change in humidity and temperature starts just after the switch to the adsorption step, and followed by a thermal equilibrium stage with a gentle change in them.

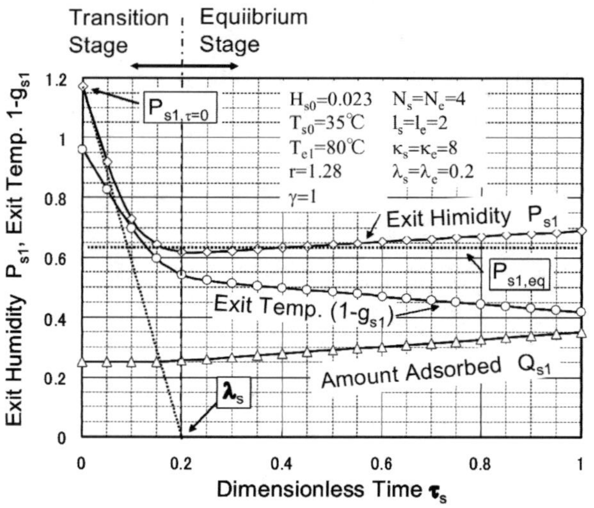

Fig. 2 Typical example of time variation of humidity, temperature and amount adsorbed at the product end

2.3 Thermal Equilibrium Stage

Heat transfer rates are balanced between the sensible heat the adsorption heat and reach a state of the dynamic steady state. Humidity increases with time while temperature decreases with time. Thus the enthalpy of air is kept constant independently of space and time. Amount adsorbed Q changes linearly with time and its slope decreases with decreasing value of N_i/l_i $(=k_i a t_{c,i}/\rho_c)$. at a limit of $N_i/l_i \rightarrow 0$, amount adsorbed Q becomes time independent throughout the whole step of adsorption and desorption and a system of the basic equations, Eqs. (9)-(12) is reduced simply to a system of ordinary differential equations, resulting in a steady state solution.

$$(P_e - \gamma) = (l_e / l_s)(P_s - P_{s1}) \tag{24}$$

$$g_e - rP_e = -r\gamma\ , \tag{25}$$

$$g_s - rP_s = 1 - r \tag{26}$$

$$\kappa_s(g_s - f_s) = rN_O(Q_s^* - Q_e^*) \tag{27}$$

$$\kappa_s(g_s - f_s) = \kappa_e(l_e / l_s)(g_e - f_e) \tag{28}$$

$$Q_s = Q_e = \frac{Q_s^* + (N_e l_s / N_s l_e)Q_e^*}{(1 + N_e l_s / N_s l_e)} \tag{29}$$

$$\mathrm{d}\,P_s / \mathrm{d}\,Z = -N_O(Q_s^* - Q_e^*) \tag{30}$$

ここに $$1/N_O = 1/N_s + (l_e / l_s)/N_e + R \tag{31}$$

$$R = \frac{(1 + l_s / l_e)}{12}\left[\frac{(1 + N_e l_s / N_s l_e)}{l_s(1 + N_e / N_s)}\right]^2 \tag{32}$$

The term R defined by Eq. (32) is an additional resistance to mass transfer caused by a finite value of l_s and l_e. Since every variable of humidity, temperature or amount adsorbed changes linearly with time for a finite but small value of N_i/l_i, time-average values are identical with the above asymptotic solution. This is along with the concept of the short cycle time approximation [2] which has been applied to the analysis of pressure swing adsorption (PSA).

Humidity of the desorption step P_e is calculated by Eq. (24) for a given value of humidity of the adsorption step P_s between values at the inlet (P_s=P_{s0}=1) and the exit (P_s=P_{s1}). Air temperatures g_s and g_e corresponding to P_s and P_e are calculated from Eqs. (25) and (26), respectively. Equilibrium amounts adsorbed Q^*_s and Q^*_e given by an adsorption isotherm are a function of a set of the humidity P and adsorbent temperature f in the adsorption and desorption steps, respectively, and obtained by a trial and error solution of the simultaneous equations of Eqs. (27) and (28) together with f_s and f_e. This can be done straightforward by applying an excel tool of the goalseek. When the reciprocal of thus sought driving force of adsorption ($Q^*_s - Q^*_e$) is integrated,

$$N_O = \int_{Ps1}^{1} \frac{1}{Q_s^* - Q_e^*} \, d P_s \tag{33}$$

the overall number of mass transfer units N_O, therefore the required length of the honeycomb rotor L can be calculated.

2.4. Transition Stage

Sensible heat transfer is predominant in this stage since the temperature difference is large between gas and adsorbent. Time interval of the transition stage is close to the relative heat capacity λ_s,

$$\tau_s = \lambda_s \tag{34}$$

The amount adsorbed $Q_{\tau=0}$ and adsorbent temperature $f_{\tau=0}$ at the start of the adsorption step are known in the above analysis of the thermal equilibrium stage and the exit humidity at the start $P_{s1,\tau=0}$ is given by integrating the equation

$$d P_s / d Z = -N_s (Q_s^* - Q_{\tau=0}) \tag{35}$$

A contribution $\Delta P_{s1,tr}$ of the transition stage to the exit humidity is given by the equation

$$\Delta P_{s1,tr} = (1/2)\lambda_s P_{s1,\tau=0}(1 - P_{s1,eq} / P_{s1,\tau=0})^2 \tag{36}$$

in approximating by a triangle as shown in Fig. 2. In the equation, the term $P_{s1,eq}$ is the exit humidity in the thermal equilibrium stage as obtained in the above section.

3. Performance Evaluation of a Desiccant Rotor

Figure 3 shows an example of the axial distribution of humidity P_s in the thermal equilibrium stage of the adsorption step and the result is compared between the numerical and the asymptotic solutions. Values at the midpoint τ_s = 0.5 are plotted as representative value for the numerical solution since it is time-dependent. The equilibrium amount adsorbed q^* is assumed to be a function of a relative humidity $\phi = p/p_s$ and to be correlated by a Langmuir isotherm as

$$q^* = \frac{q_\infty K \varphi}{1 + K\varphi} \tag{37}$$

with Langmuir parameters of q_∞ = 0.336 kg/kg, K = 2.18. The asymptotic solution is a limiting solution as $N/l = Kat_c/\rho_c \rightarrow 0$ and is in complete agreement with numerical solution for a small value of N/l less than unity. A significant error

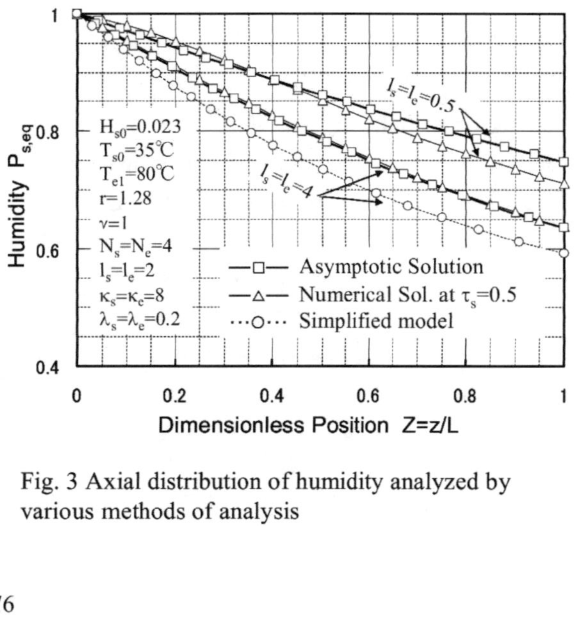

Fig. 3 Axial distribution of humidity analyzed by various methods of analysis

starts to appear for N/l beyond 10.

Figure 4 shows the result of the exit humidity when the relative adsorption capacity l_s is varied in keeping a ratio of relative heat to adsorption capacity l_s/λ_s constant at 10. Each contribution of the transition stage $\Delta P_{s1,tr}$ and the thermal equilibrium stage $P_{s1,eq}$ are plotted together with a sum of them, which is compared with the time average value of the numerical solution P_{s1}. Since the abscissa of this figure is inversely proportional to the adsorption step time t_c, the figure shows the effect of the rotation speed of a desiccant rotor. The contribution of the transition stage $\Delta P_{s1,tr}$ increases with increasing rotation speed while that of the equilibrium stage $P_{s1,eq}$ decreases and approaches to a limiting value with increasing speed. As a result, an optimum value of the rotation speed exists to minimize the exit humidity.

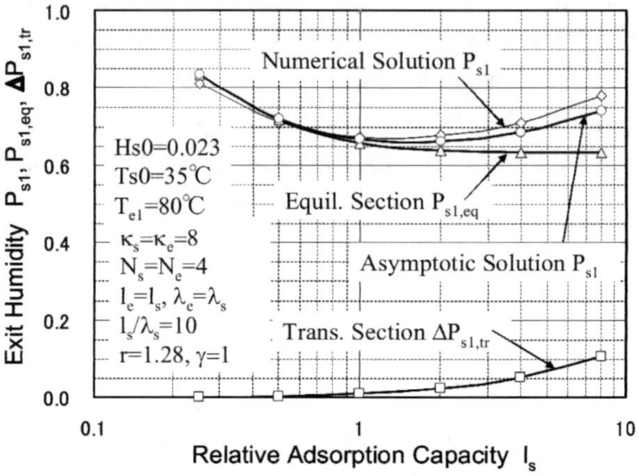

Fig. 4 Dependence of exit humidity on relative adsorption capacity

4. A Simplified Graphical Solution

Since heat transfer resistance ($1/\kappa_s$) in Eq. (27) is smaller than mass transfer resistance ($1/rN_O$), the temperature difference between air and adsorbent may be negligible. Equations (27) and (28) may be replaced with

$$f_s = g_s, \qquad f_e = g_e \qquad (38)$$

and the equilibrium amount adsorbed Q^*_s and Q^*_e can be evaluated simply for air temperature g_s and g_e but not adsorbent temperature. When Eqs. (24)-(26) and (33) are rewritten in a dimensional form as follows

$$x_e - x_{e1} = \frac{U_s t_{c,s}}{U_e t_{c,e}}(x_s - x_{s1}) \qquad \text{Operating line} \qquad (39)$$

$$h_s = h_{s0} = c_{pg} T_{g,s0} + x_{s0} \Delta H \qquad \text{Iso-enthalpic line} \qquad (40)$$

$$h_e = h_{e1} = c_{pg} T_{g,e1} + x_{e1} \Delta H \qquad \text{Iso-enthalpic line} \qquad (41)$$

$$L = \frac{\rho_g U_s}{k_s a} = \int_{xs1}^{xs0} \frac{1}{q_s^* - q_e^*} \mathrm{d}x_s \qquad \text{Rotor length} \qquad (42)$$

A set of Eqs. (39)-(41) can be expressed graphically as in Fig. 5. where the relative humidity ϕ is related to the humidity x with enthalpy h or air temperature T_g as parameter. A diagonal line is drawn for transforming between the ordinate and the abscissa. An adsorption isotherm Eq. (37) is plotted against the abscissa of relative humidity.

The state of air for given values of humidity x_{s0}, x_{s1}, x_{e1} is given by points s0, s1, e1 on an h-line (iso-enthalpy line). The point A1 is a point at which a line turning on the diagonal line above the point e1 intersects a line passing through a point s1. A line passing through the point A1 with a slope of $U_s t_{c,s}/U_e t_{c,e}$ is an operating line given by Eq. (39), which correlates the humidity between the adsorption and the desorption steps. When a state of air in the adsorption step is given by a point s on the h_s-line, the corresponding state of air in the desorption step is given by an intersection e between the h_e-line and a line turning twice on the operating line and the diagonal line. Equilibrium amount adsorbed q^*_s at the position of humidity x_s in the adsorption step is given by a point q^*_s where a line starting the point s and turning twice on the diagonal line and the adsorption isotherm intersects a line above the point s. A corresponding amount adsorbed in the desorption step q^*_e is given by a point q^*_e by a similar graphical operation. A line segment $q^*_s q^*_e$ is the driving force of adsorption (q^*_s - q^*_e) for the given value of the humidity x_s. A graphical integration of Eq. (42) can be carried out to obtain the required length of the rotor L after seeking the driving force for several values of the humidity x_s between the feed x_{s0} and the product x_{s1} by repeating the same operation.

A broken line in Fig. 3 is an example of the axial distribution of the exit humidity obtained by this simplified graphical method. It is useful to find the overall trend in spite of less accuracy.

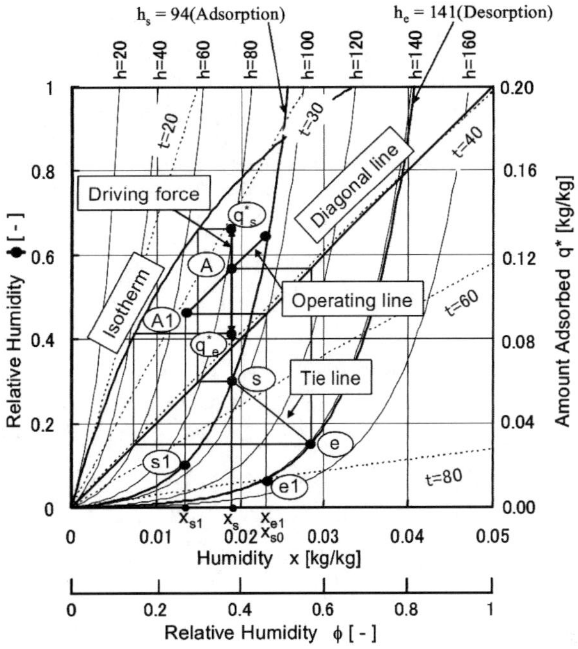

Fig. 5 Simplified graphical method of analysis

5. Summary

Simultaneous transfer of heat and water vapor in a desiccant rotor is analyzed by dividing the whole process into two time zones, i.e. the transition stage and the thermal equilibrium stage. By this simplification, the exit humidity can be calculated by a direct integration of the reciprocal of driving force for adsorption. The above asymptotic model is proposed together with a simplified graphical solution method and the result is compared with the more rigorous numerical simulation.

Nomenclature

c_{pc} : heat capacity of adsorbent, J.kg^{-1}.K^{-1}
c_{pg} : heat capacity of air, J.kg^{-1}.K^{-1}
c_w : heat capacity of water vapor, J.kg^{-1}.K^{-1}
f : dimensionless temperature of adsorbent, Eq. (4)
g : dimensionless temperature of air, Eq.(3)
h : enthalpy of air, J/kg
ha : volumetric heat transfer coefficient, W.m^{-3}.K^{-1}
ΔH : heat of adsorption or vaporization, J.kg^{-1}
K : Langmuir constant
ka : volumetric mass transfer coefficient, kg.m^{-3}.s^{-1}
L : length of rotor, m
l : relative adsorption capacity, Eq.(16)
N : number of mass transfer units, Eq. (14)
N_O : overall number of mass transfer units, Eq. (31)
p : vapor pressure, Pa
p_a : atmospheric pressure, Pa
P : dimensionless humidity, Eq.(5)
ΔP_{s1} : contribution from transition section
q : amount adsorbed, kg/kg
Q : dimensionless amount adsorbed, Eq.(6)
q^* : equilibrium amount adsorbed, kg/kg

Q^* : dimensionless value of q^*, Eq.(7)
q_∞ : Langmuir constant, kg/kg
r : dimensionless heat of adsorption, Eq. (17)
R : additional mass transfer resistance, Eq. (32)
t : time from start of adsorption, s
t_c : step time, s
T_f : temperature of adsorbent, ℃
T_g : temperature of air, ℃
U : superficial velocity of air, m/s
x : humidity, kg/kg
z : distance from feed inlet, m
Z : dimensionless distance, Eq.(1), $= z/L$
γ : ratio of humidity, Eq.(18), $= x_{e1}/x_{s0}$
κ : number of heat transfer units, Eq.(13)
λ : relative heat capacity, Eq.(15)
τ : dimensionless time, Eqs. (2),(2')
ρ_c : bulk density of adsorbent, $kg.m^{-3}$
ρ_g : density of air, $kg.m^{-3}$
ϕ : relative humidity

subscripts
0 : inlet end of feed
1 : exit end of product
e : desorption step
i : i=s(adsorption) or i=e(desorption)
s : adsorption step

References

[1] T. Tsujiguchi and A. Kodama : *Trans. of the JSRAE*, 22(4), 417(2005), 23(4), 467(2006) and 24(3), 205(2007).
[2] T.Nakamura, et al. : *Adsorption,* 11, 609(2005)

719n Separation Characteristics of Chromium Ion Onto Natural Polymer

Tae Young Kim, Jae-hoon Chung, Seung-Shik Park, Byoung-Jun Min Min, Sung-Yong Cho, and Seung-Jai Kim. Environmental Engineering, Chonnam National University, 300 YongBong-Dong Buk-Gu, Gwangju, 505-757, South Korea

Chromium is one of the most strategic and critical materials having a wide range of use in the various industries. Hexavalent chromium is generally found in the chemical industry as an oxidizing agent, metallurgical industry for chrome alloy of chromium metal production, electroplating, corrosion control and photography. On the other hand, trivalent chromium is extensively used in leather tanning industry, paints and pigments industry, fungicides industry and ceramic or glass manufacture. Cr(VI) is known to be toxic to both plants and animals, as a strong oxidizing agent and potential carcinogen. Although Cr (III) is nontoxic or less toxic than Cr(VI), a long-term exposure to trivalent chromium is known to cause allergic skin reactions. Wastewater containing relatively low concentration of Cr(VI) is usually treated with ion-exchange resin which has the advantage of the recovery of chromic acid and also disadvantage of high cost of resin. As an alternative to this, sorption has recently gained more attention due to its possibility to overcome this problem. In several recent publications, various authors have reported the use of different sorbents for trivalent chromium removal using such as: chitosan impregnated microemulsions, sunflower stalks, seaweed biomass. In a similar manner, hexavalent chromium removal was investigated by employing composite chitosan biosorbent, granular activated carbon, immobilized fungal biomass. The sorption of trivalent and hexavalent chromium mixture was effective by few sorbents such as seaweed biosorbent and layer silicates. Recently, the possibility of both chromium species removal from tannery effluents was studied, by the use of calcium alginate beads containing humic acid, in a toxicological approach. The main purpose of this work is to eliminate chromium ion from its aqueous solution using alginate bead and to study its adsorption characteristics experimentally as well as theoretically.

719o Production of Ammonia with Pressure Swing Adsorption Utilized In Key Separation Steps

James A. Ritter[1], Armin D. Ebner[1], Shubhra Bhadra[1], Marjorie A. Nicholson[1], Charles E. Holland[1], and Felix Jegede[2]. (1) Department of Chemical Engineering, University of South Carolina, 301 Main St., Columbia, SC 29208, (2) SmartKoncept Technology, 11999 Katy Frwy,, Houston,, TX 77079

Pressure swing adsorption (PSA) technology enables the energy-efficient recovery of specific compounds from a gas under pressure. In this project, an ammonia production process that incorporates PSA for the recovery of ammonia from the product gas and purification of the process reactants, nitrogen (from air) and hydrogen (from steam methane reforming, SMR) is being developed. The process converts methane, air, and water to ammonia, with carbon dioxide and a small amount of argon as byproducts. Current ammonia processes react air and syngas (gas mixture of carbon dioxide and hydrogen) to form ammonia, which is recovered via energy-intensive refrigeration and condensation. The new approach will increase ammonia yield and reduce natural gas feedstock and fuel consumption. The improved ammonia process will find application in the U.S. chemical industry, which produced 23.7 billion pounds of ammonia in 2004. Ammonia is also used at power plants in selective catalytic reduction (SCR) and selective non-catalytic reduction (SNCR) systems, to reduce NOx emissions. The new process is scaleable and enables on-demand ammonia production for such applications. This presentation will provide an overview of the proposed flow sheet ammonia production flow sheet and give an update on the status of the project with respect to PSA process design for ammonia purification.

ADSORPTION MEASUREMENT OF CHLORINATED HYDROCARBONS INTO HIGH-SILICA ZEOLITE BY CHROMATOGRAPHIC METHOD

Kazuyuki Chihara, Kenta Saito, Hidenori Nakamura and Yosuke Kaneko
Department of Applied Chemistry, Meiji University
1-1-1 Higashi-mita, Tama-ku, Kawasaki Kanagawa 214 - 8571, Japan

Abstract

In this study, adsorption measurement of various chlorinated hydrocarbons on Y-type zeolites were performed using the chromatographic method to examine the adsorption phenomena. Dichloromethane (DCM), chloroform (TCM), trichloroethylene (TCE), and tetrachloroethylene (PCE), were used as the chlorinated hydrocarbons, and PQ-USY, USY-6.18, Na-Y, HSZ-390HUD, Pentasile-2 and Pentacil-1 were used as adsorbent. Each adsorption measurement was performed to get the heat of adsorption and the activation energy of micropore diffusivity. The amount adsorbed and isosteric heats of adsorption became low, when SiO_2/Al_2O_3 ratio of zeolite became high. The isosteric heats of adsorption were roughly 1.4 to 2.0 of the heats of vaporization. The activation energies of diffusion in the micropore of zeolite were found to be almost 40% to 60% of the isosteric heats of adsorption.

1. Introduction

Diffusion as well as adsorption equilibrium in adsorbent particles are important in designing adsorption processes. The possible mechanisms of diffusion in adsorbent particles are classified into three groups in the case of gaseous adsorption. They are diffusion in gas phase of macropores such as molecular diffusion or Knudsen diffusion, surface diffusion which occurs in the adsorbed phase, or diffusion in the micropores whose diameters are comparable with molecular diameter of adsorbates. The engineering estimation of the effective diffusivities for the first and second cases is almost possible, but when the intraparticle diffusion is controlled by the third mechanism, there are still some ambiguities in estimating the order of magnitude of the adsorption rate. Molecular sieving materials such as zeolite have micropores of well-defined structure. Micropores of zeolites have network structures. Study of micropore diffusion in molecular sieving materials is significant since they provide funda-mental characteristics of diffusion in other adsorbents which posses micropores of wide varieties. Though many investigations for intracrystalline diffusion in zeolite have been made, there has been, however, little work on diffusion of chlorinated hydrocarbons in zeolite.

Micropore diffusion of adsorbate molecules occurs in the potential energy field which is due to interactions between adsorbate molecules and adsorbent lattice atoms. This diffusion is generally described as each molecule hopping from one minimum of the potential energy distribution to another, or adsorption site to site, across potential energy barriers, which separate adjacent sites. This type of micropore diffusion is a kind of activated process and is called activated diffusion.

The purpose of this work is to experimentally determine the micropore diffusivities and adsorption equilibrium constants of various chlorinated organics in zeolite at wide temperature ranges and to correlate the obtained activation energies of micropore diffusion and isosteric heats of adsorption with the properties of adsorbate gases.

Chromatographic measurements were made for this purpose. By introducing the concentration pulse of adsorbate to the carrier gas stream at the inlet of the adsorbent column, the concentration elution curve at the outlet of the column was measured. Adsorption equilibrium constant and micropore diffusivity were determined by means of moment analysis of elution curves.

2. Experimental Procedure and Conditions

The experimental apparatus was shown in Figure 1. The apparatus was a conventional gas chromatograph (GC-9A) (Shimadzu Co., LTD). PQ-USY, USY-6.18, Na-Y, HSZ-390HUD, Pentasile-2 and Pentacil-1(crashed and screened to obtain particle size between #20 and #32) were used as adsorbent. These particles were packed in a column. The carrier gas was helium.

Dichloromethane, chloroform, trichloroethylene, and tetrachloroethylene were used as adsorbate. The adsorbate was introduced as pulse into the carrier gas stream by a 6-way valve. The flow was controlled by mass flow controllers (about 25 cm/sec). The outlet signal from the TCD cell was recorded to a chart-recorder and converted into personal computer as pulse response. Column conditions were shown in Table 1.

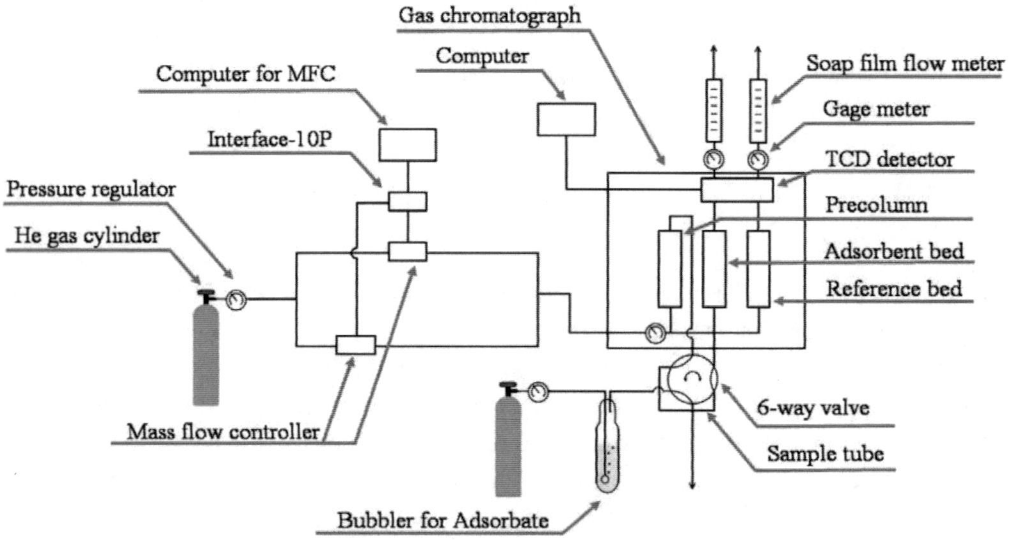

Figure 1 Experimental Apparatus

Table 1 column and adsorbent conditions

		HSZ-390HUD	PQ-USY	USY-6.18	Na-Y	Pentasile-1	Pentasile-2
Zeolite Type		FAU	FAU	FAU	FAU	MFI	MFI
SiO_2/Al_2O_3	[-]	460	70	6.18	5.6	400	30
Particle Density	[g/cm^3]	0.830	0.752	0.730	0.982	0.935	1.32
Macropore fraction	[-]	0.383	0.414	0.438	0.285	0.479	0.265
Bulk Density	[g/cm^3]	0.449	0.423	0.452	0.620	0.637	0.696
Void Fraction	[-]	0.460	0.428	0.385	0.369	0.319	0.473
Lenfth of Column	[cm]	47.21	30.0	30.1	25.0	24.0	18.5
Column Diameter	[cm]	0.30	0.30	0.30	0.30	0.30	0.30
Range of Sieve Openings	[mm]	0.495-0.833	0.495-0.833	0.495-0.833	0.495-0.833	0.495-0.833	0.495-0.833

3. Method of Analysis

The adsorption isotherm is given by a linear equation at near zero surface coverage:

$$q = K^* \cdot c \qquad (1)$$

where q= amount adsorbed per unit mass of adsorbent [g/g], K^*= adsorption equilibrium constant [cm^3/g], c= concentration in the fluid phase [g/cm^3].

The resultant moment equations of the impulse response for a packed bed of spherical particles with bidispersed pore structure and size distribution of the microparticles with helium carrier are (Chihara, Suzuki and Kawazoe 1978 and Kawazoe, Suzuki and Chihara 1973).

$$\mu_1 = \frac{z}{u}(1+\delta_0) \qquad (2)$$

$$\mu_2' = \frac{2z}{u}(\delta_d + \delta_f + \delta_a + \delta_i) \qquad (3)$$

where

$$\delta_0 = \frac{(1-\varepsilon)\varepsilon_a}{\varepsilon}\left(1+\frac{\rho_p K^*}{\varepsilon_a}\right) \qquad (4)$$

$$\delta_d = \frac{Ez}{u^2}(1+\delta_0)^2 \qquad (5)$$

$$\delta_f = \frac{1-\varepsilon}{\varepsilon}\cdot\frac{R}{3K_f}(\varepsilon_a + \rho_p K^*)^2 \qquad (6)$$

$$\delta_a = \frac{1-\varepsilon}{\varepsilon}\cdot\frac{R^2}{15D_a}(\varepsilon_a + \rho_p K^*)^2 \qquad (7)$$

$$\delta_i = \frac{1-\varepsilon}{\varepsilon}\cdot\frac{\rho_p K^*}{15D}\int_{-\infty}^{\infty} a^2 f(a)\, d\ln a \qquad (8)$$

The size distribution of the microparticles is assumed to be the log-normal distribution as

$$f(a) = \frac{1}{\sqrt{2\pi}\cdot\sigma}\exp\left[-\frac{(\ln a - \ln a_0)^2}{2\sigma^2}\right] \qquad (9)$$

By introducing Equation (9) into Equation (8), we get

$$\delta_i = \frac{1-\varepsilon}{\varepsilon}\cdot\frac{\rho_p K^*}{15D}\overline{a^2}\exp(\sigma^2) \qquad (10)$$

δ_f is neglected for its small effects in the following analysis. For instance, by using Carberry's equation (Carberry 1960), k_f can be estimated for the experimental condition. Then, the value of δ_f is found to be one order of magnitude smaller than δ_d or δ_a.

4. Result and Discussion

4.1 Microparticle size distribution

Figure 2 Photomicrograph of Pentasile-2 pellet

The size distribution of the microparticles for zeolite was determined by the scanning electron microscope. By cutting the zeolite pellet, the inner surface was photographed. A typical photomicrograph is shown in Figure 2. It is apparent that the macroparticle is an agglomerate of microparticles, and the microparticles have random shapes and size distribution. The diameters in the same orientation of about 200 microparticles were measured from micrographs.

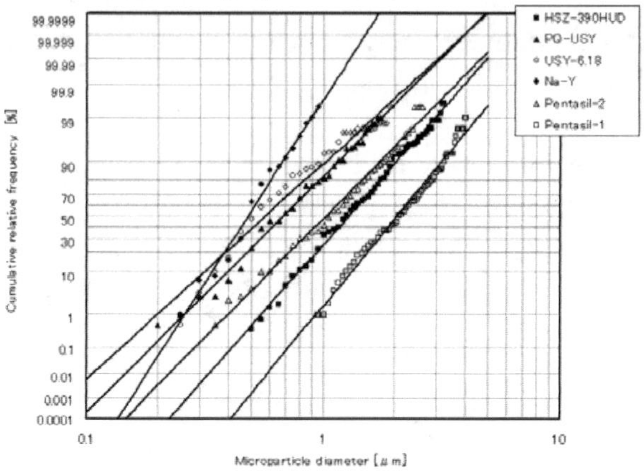

Figure 3 Log-normal size distribution of the microparticles of zeolite

On the assumption that the microparticle is a sphere, the cumulative number fraction was plotted against the microparticle radius $\bar{a}$ on a log-normal probability graph paper; straight lines were obtained as shown in Figure3. It is assumed that microparticles of zeolite have the log-normal size distribution. The radius at the peak of the probability a_0' which corresponds to a number fraction basis, and the standard deviation σ were thus found. In the case of log-normal distribution, the straight lines of distribution both on number fraction basis and on volume fraction basis are parallel on a log-normal probability paper, and the relation between the radius at the peak of the probability on a number fraction basis a_0' and that on a volume fraction basis a_0 and average radius $\bar{a}$ are as follows (Hatch 1933). The value of $\bar{a}$ and σ were shown in Table2.

$$a_0 = a_0' \exp\left(3\sigma^2\right) \tag{11}$$

$$\bar{a} = \frac{\int_{-\infty}^{\infty} af(a)d\ln a}{\int_{-\infty}^{\infty} f(a)d\ln a} = a_0 \exp\left(\frac{\sigma^2}{2}\right) \tag{12}$$

Table 2 Average radius and standard deviation

	HSZ-390HUD	PQ-USY	USY-6.18	Na-Y	Pentasile-1	Pentasile-2
Average Radius [μm]	2.03	1.23	1.17	0.61	3.07	1.84
Standard Deviation [-]	0.365	0.415	0.454	0.266	0.340	0.410

4.2 Moment analysis

The concentration of adsorbate in the effluent from the adsorbent bed was detected by the thermal conductivity cell. Figure 4 shows the effluent chromatographic peaks for dichloromethane pulses at several temperatures for almost the same velocity.

The first absolute moment and the second central moment were evaluated from the effluent peak $C_e(t)$ as follows:

$$\mu_1 = \int_0^\infty C_e(t)tdt \Big/ \int_0^\infty C_e(t)dt \tag{13}$$

$$\mu_2' = \int_0^\infty C_e(t)(t-\mu_1)^2 dt \Big/ \int_0^\infty C_e(t)dt \tag{14}$$

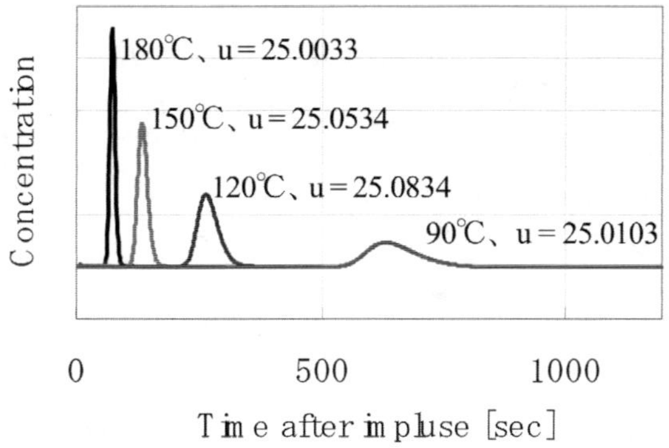

Figure 4 Typical chromatographic peaks for chloroform [HSZ-390HUD]

The analysis of second moment was simplified by first dividing Equation(3) by $[\mu_1/(z/u)]^2$ obtained from Equation(2). The result was

$$H = \frac{\mu_2'}{\mu_1^2}\left(\frac{z}{2u}\right) = H_0 + \frac{Ez}{u^2} \qquad (15)$$

where

$$H_0 = \frac{\delta_a + \delta_i}{(1+\delta_0)^2} \qquad (16)$$

Ez was expected to be proportional to u in the experimental condition. H_0 was able to be obtained as an intercept to the ordinate when H were plotted against $1/u$ according to Equation(15). Since δ_a could be estimated by the established evaluation equation for macropore diffusion and Equation(7), δ_i was obtained by Equation(16).

4.3 Adsorption equilibrium constant

Equation(2) and (4) were used along with the μ_1 to obtain the adsorption constant K^*. The isosteric heat of adsorption q_{st} is related with K^* for each gas by Van't Hoff's equation:

$$\frac{K^*}{T} = \left(\frac{K^*}{T}\right)_{\frac{1}{T}\to 0} \exp\left(\frac{qst}{R_g T}\right) \qquad (17)$$

Here it is assumed that the gas is ideal.

According to Equation(17), the obtained K^* divided by T was plotted against $1/T$ for various gases in Figure5, and the isosteric heat of adsorption q_{st} and $(K^*/T)_{1/T=0}$ were determined and listed in Table3.

Table 3 Properties of gases for adsorption

		HSZ-390HUD	PQ-USY	USY-6.18	Na-Y	Pentasil-1	Pentasil-2
DCM	ΔH [kJ/mol]			28.0			
	q_{st} [kJ/mol]	38.49	38.6	42.2	43.4	59.7	56.3
	E [kJ/mol]	13.39	19.3	23.3	13.7	33.4	43.4
TCM	ΔH [kJ/mol]			29.4			
	q_{st} [kJ/mol]	38.05	35.1	45.2	46.0	57.2	34.4
	E [kJ/mol]	13.33	19.0	39.4	30.3	40.5	8.7
TCE	ΔH [kJ/mol]			31.4			
	q_{st} [kJ/mol]	42.04	43.4	49.2	48.2	62.5	61.1
	E [kJ/mol]	11.22	17.9	20.1	10.8	39.0	30.0
PCE	ΔH [kJ/mol]			34.7			
	q_{st} [kJ/mol]	48.79	47.5	50.5	56.5	52.0	61.5
	E [kJ/mol]	17.87	12.8	21.3	40.1	34.6	38.5

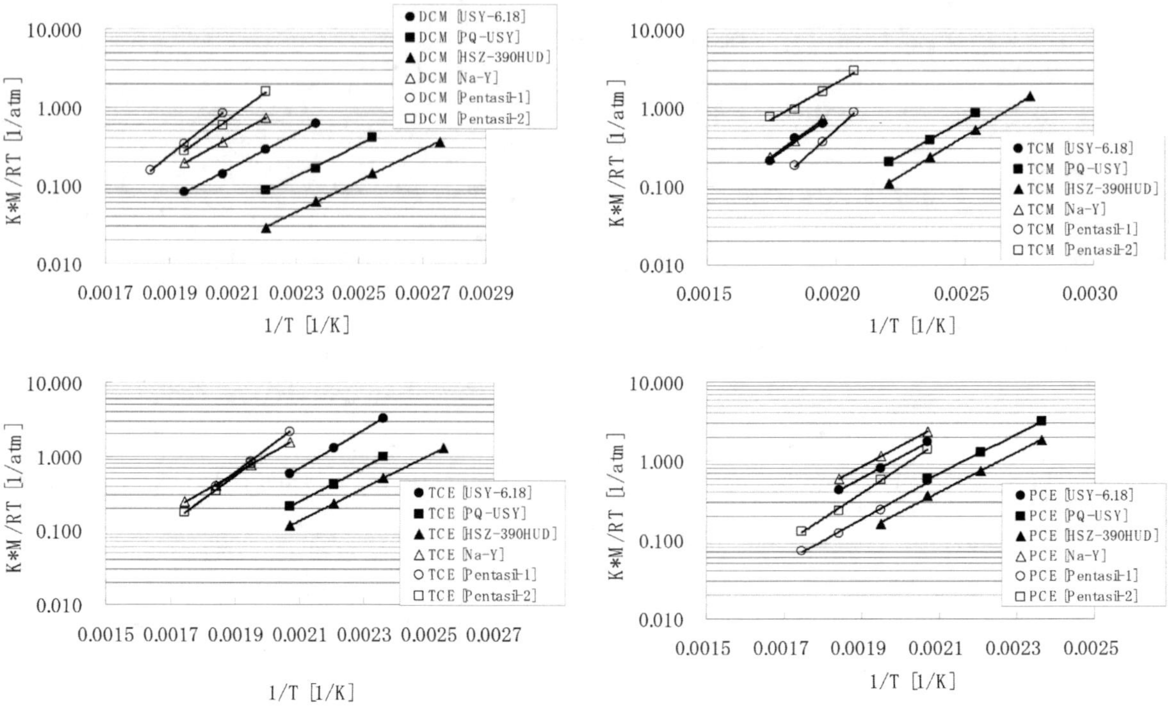

Figure 5 Van't Hoff plot of adsorption equilibrium

4.4 Axial dispersion

Equation(15) shows that the contribution of axial dispersion to H must be evaluated first to estimate H_0 from experimental value of H. In packed beds, Ez is expressed as a sum of a molecular diffusion(D_M) term and a term due to the velocity profile:

$$Ez = \frac{D_M}{\tau_{ext}} + \frac{d_p u}{Pe} \qquad (18)$$

The second term of equation(18) is expected to be dominant (Suzuki and Smith 1971), and H in Equation(15) should depend linearly on $1/u$. Peclet number was evaluated(0.78) from the empirical correlation (Suzuki and Smith 1971) given as:

$$Pe = 1.2dp \; :dp \text{ in mm} \qquad (19)$$

4.5 Diffusion in macropore

H_0 includes the effect of the macropore diffusion and the micropore diffusion.
The following established relation (Satterfield 1970) was used to estimate the diffusivity in the macropore:

$$D_a = \frac{\varepsilon_a}{\tau_p} \cdot D_M \qquad (20)$$

where D_a= diffusivity in macropores D_M= molecular diffusivity ε_a =void fraction of macropore.
Equation (7) was then applied to estimate δ_a. Then, δ_i was obtained by Equation (16).

4.6 Diffusion in micropore

Since the size distribution of the microparticle for zeolite was already obtained by the electron microscope, diffusivities in micropores D were evaluated from δ_i and K^*, a, and σ by using Equation(10). The values of D for different temperatures were plotted according to the following Arrhenius' equation in Figure5.

$$D/\overline{a^2}\exp(\sigma^2)=\left[D_0/\overline{a^2}\exp(\sigma^2)\right]\exp\left(-\frac{E}{RT}\right) \tag{21}$$

From Figure6, the activation energy E was determined from the slope of the straight line corresponding to the date points for each gas: the preexponential factor $D_0\exp(\sigma^2)$ was obtained as the interception to the ordinate at $1/T$=0. These figures were listed in Table3.

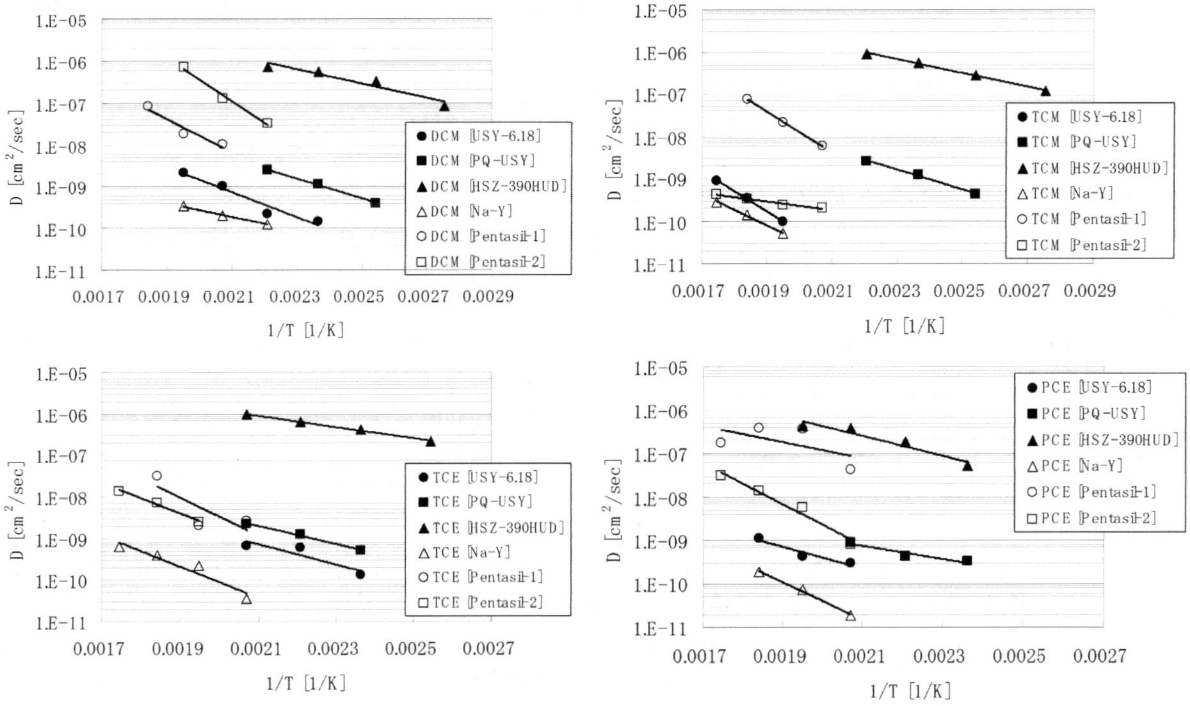

Figure 6 Arrhenius' plot

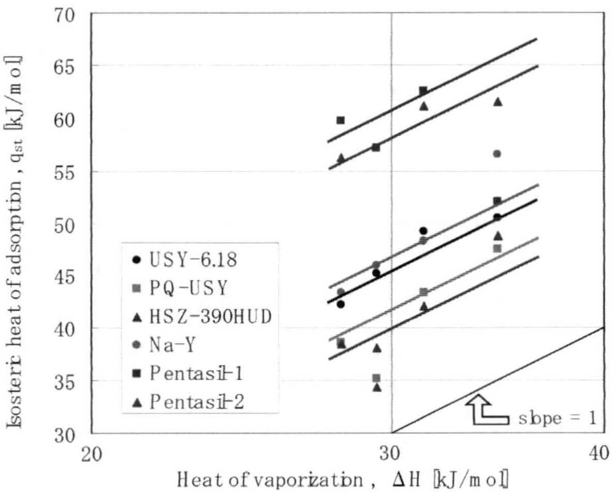

Figure7 Correction of isosteric heat of adsorption with heat of vaporization

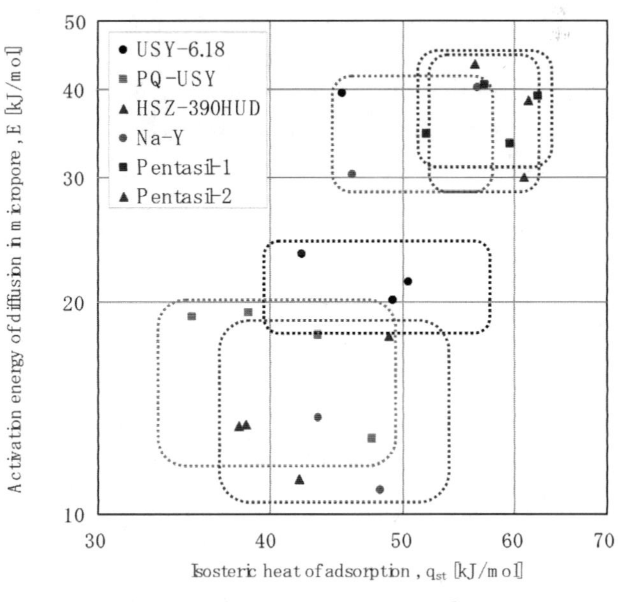

Figure8 Correlation of actibation energy of the micropore diffusion with isosteric heat of adsorption

4.7 Discussion

The isosteric heat of adsorption obtained with the chromatographic method of zeolite in this study is characterized, so that, the estimated values are for zero coverage, or initial isosteric heats of adsorption. Therefore, these values correspond entirely to the interaction between the adsorbate molecule and the adsorbent lattice atoms, excluding the interaction among adsorbate molecules.

Isosteric heat of adsorption q_{st} for five adsorbate with four adsorbent are plotted against the heat of vaporization ΔH of corresponding adsorbate in Figure 7. Here, The amount adsorbed and isosteric heats of adsorption became low, when SiO_2/Al_2O_3 ratio of zeolite became high. The isosteric heats of adsorption were roughly 1.4 to 2.0 of the heats of vaporization. (Barrer and Lee 1968) showed that the ratios $q_{st}/\Delta H$ for zeolite L slightly exceeded 2. Adsorption heat of PQ-USY is the lowest on all adsorbate. Adsorption heat of tetrachloroethylene became highest than others on all adsorbents.

Diffusion in the micropore of the zeolite is so called activated diffusion and is described as the molecules hopping from an adsorption site to another across the energy barrier which separates adjacent sites, which is similar to surface diffusion.

The activation energy was found to be half of isosteric heat of adsorption for the gases examined here.

5. Conclusion

Chromatographic measurements were made for various gas adsorptions on zeolite at three kind of temperatures, for zero coverage. Dichloromethane (DCM), chloroform (TCM), trichloroethylene (TCE), and tetrachloroethylene (PCE), were used as the chlorinated hydrocarbons, and PQ-USY, USY-6.18, Na-Y, HSZ-390HUD, Pentasile-2 and Pentacil-1 were used as adsorbent.

Adsorption equilibrium constants in the Henry's law region and diffusivities in the micropores in zeolite were determined from the first absolute moment and the second central moment where the effects of all the other possible transport processes in the case of bidispersed pore structure and also the log-normal size distribution of the microparticle were taken into account.

The amount adsorbed and isosteric heats of adsorption became low, when SiO_2/Al_2O_3 ratio of zeolite became high. The isosteric heats of adsorption were roughly 1.4 to 2.0 of the heats of vaporization. The activation energies of diffusion in the micropore of zeolite were found to be almost 40% of the isosteric heats of adsorption.

NOTATION

a = radius of microparticle, cm
a_0 = a at the peak of the probability on a volume fraction basis , cm
a_0' = a at the peak of the probability on a number fraction basis, cm
$\bar{a}$ = arithmetic average radius of microparticles defined
C_e = concentration in the effluent gas, mol/cm^3
c = concentration in the fluid phase, mol/cm^3
c = concentration in the particle, mol/cm^3
D = diffusivity in micropores based on amount adsorbed gradient driving force, cm^2/s
D_0 = preexponential factor for the Arrhenius' plot of D, cm^2/s
D_a = diffusivity in macropores, cm^2/s
D_M = molecular diffusivity, cm^2/s
d_p = particle diameter, cm
E = activation energy for micropore diffusion, kJ/mol
E_z = axial dispersion coefficient based on void spaces in the bed, cm^2/s
$f(a)$ = distribution function of the radii of microparticles
H = defined by Equation (13), s
H_0 = defined by Equation (14), s
ΔH = heat of vaporization, kJ/mol
K^* = adsorption equilibrium constant, cm^3/g
k_f = external mass transfer coefficient, cm/s
P_e = Peclet number, $d_p u/E_z f$
q = amount adsorbed per unit mass of adsorbent, mol/g
q_s = isosteric heat of adsorption, kJ/mol
R = radius of particles, cm
R_g = gas constant, Kcal/mol・K
T = absolute temperature, K
t = time, s
u = interstitial velocity of fluid, cm/s
v = volumetric flow rate, cm^3/s
z = longitudinal position in the bed, cm

REFERENCES

Barrer, R. M., Lee, A., Surface Sci., 12 341 (1968)
Carberry, J. J., AIChE, 6, 460 (1960)
Chihara, K., Suzuki, M., Kawazoe, K., AIChE, 24, 237-245 (1978)
Hatch,T.; "Determination of "Average Particles Size" from the Screen-Analysis of Non-Uniform Particulate Substances," J. Franklin Inst., 27, 215 (1933)
Kawazoe, K., Suzuki, M., Chihara, K., SEISAN-KENKYU, Japan, 25, 558 (1973)
Satterfield, C.N., M.I.T Press, London, 33 (1970)
Suzuki, M., Smith,J.M.,Chem., Eng. 3, 256 (1972)

ADSORPTION EQUILIBRIUM OF MIXED ORGANIC SOLVENT TO FAU TYPE HIGH SILICA ZEOLITE

Kazuyuki Chihara, Shinji Tomita, Junichi Kabe, Shinji Kondo, Takashi Matsumoto*
Applied Chemistry, Meiji University, Kawasaki, JAPAN

Introduction

When the solvent collection device like the PSA method is designed, it is very important to know the adsorption equilibrium relations between adsorbent and adsorbates, and movement and adsorption speed of adsorbates in absorbent intergranular. The molecular simulation method is an effective method to the examination of the adsorption mechanism to the zeolite at a molecular level. Especially, the GCMC simulation is used to predict the adsorption equilibrium. The GCMC simulation is a technique for changing a lot of atoms and the positions of the molecule one after another according to the stochastic process by random numbers that generate it, and calculating there by the balanced structure and amount of thermodynamics.
In this study, it made comparative study of the simulation that used Materials Studio with the gravimetric method adsorption experiment.

Experimental

Gravimetric Method

Gravimetric analysis was used to obtain the adsorption curves plotted for the ratio of volumes of solution (q_t/q_∞) against time (t). Micropore diffusivity was obtained by analyzed adsorption curve. The amount adsorbed was measured corresponding to the pressure of the vapor in the tube. The pressure was measured by pressure sensor at higher pressure range (> 0.013 atm) and baratron at lower pressure range (<0.013 atm). A Time with increment value of adsorption was measured by chart recorder. In this way, the adsorption curves were obtained.

In this research, FAU type zeolite (SiO_2/Al_2O_3=460) was used as sorbate and ethanol and trichloroethylene were used as adsorbates. The adsorption equilibrium experiment used the gravimetric method. By using the zeolite of about 0.25g, adsorbates were poured from a high vacuum of pressure 10^{-6}atm and measured the weight change in the absorbent with a differential transformer.

Simulation

Materials Studio

Simulation module in Materials Studio (Accelrys Inc.) was used throughout MC simulations. As for Materials Studio, the adsorption equilibrium can be reproduced in a short time, and two or more calculations are possible at the same time. The power place used cvff that corrected Universal. Ratio of the silica alumina (SiO_2/Al_2O_3) of the FAU type zeolite that is saved in the library of MS was adjusted to 460 and the absorbent was used. It calculated by the Sorption module, and the adsorption isotherm was obtained.

Result And Discussion

The adsorption isothermals that obtained from the gravimetric method experiment and GCMC simulation are shown in Fig.1 and Fig.2.
Result of simulation of two elements is shown in Fig.3. And adsorption equilibrium curve according to total concentration.

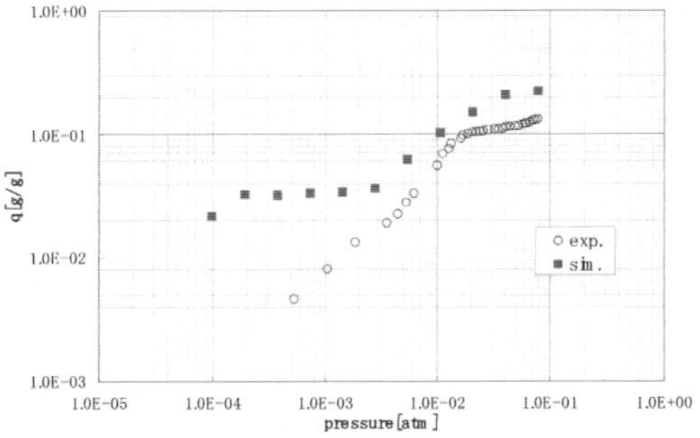

Fig.1 Comparison between experiment and simulation(adsorbate: ethanol)

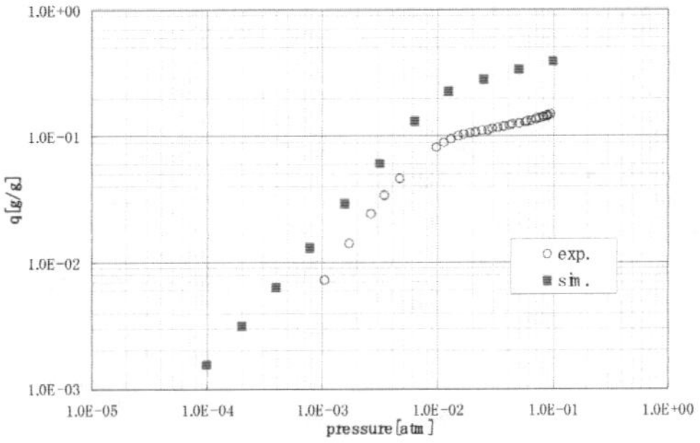

Fig.2 Comparison between experiment and simulation(adsorbate: trichloroethylene)

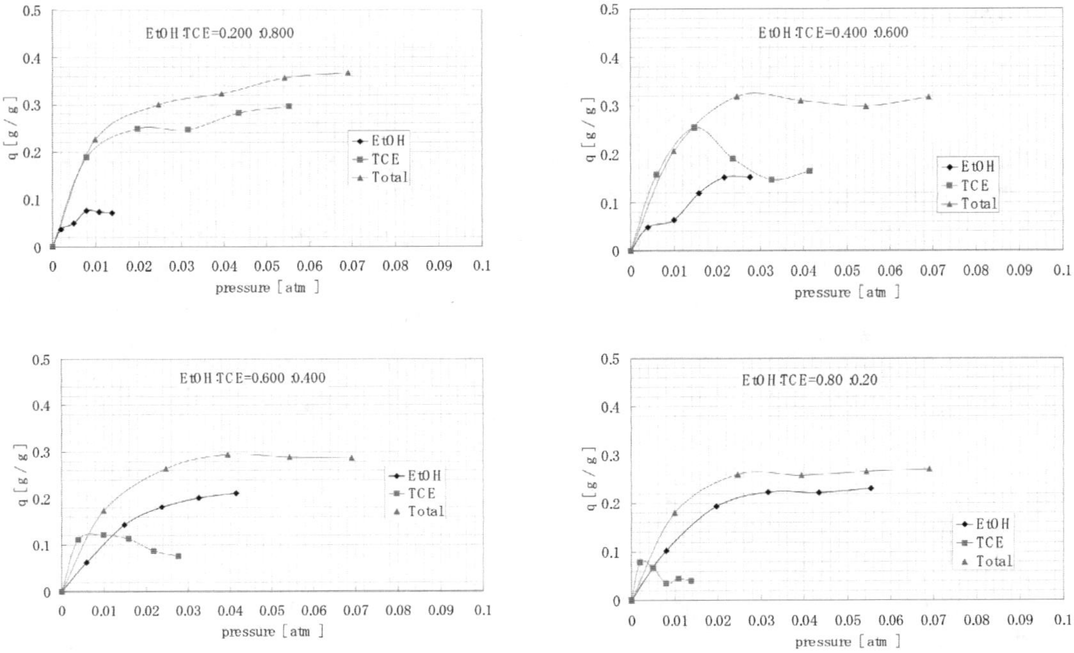

Fig.3 Result of simulation of two elements

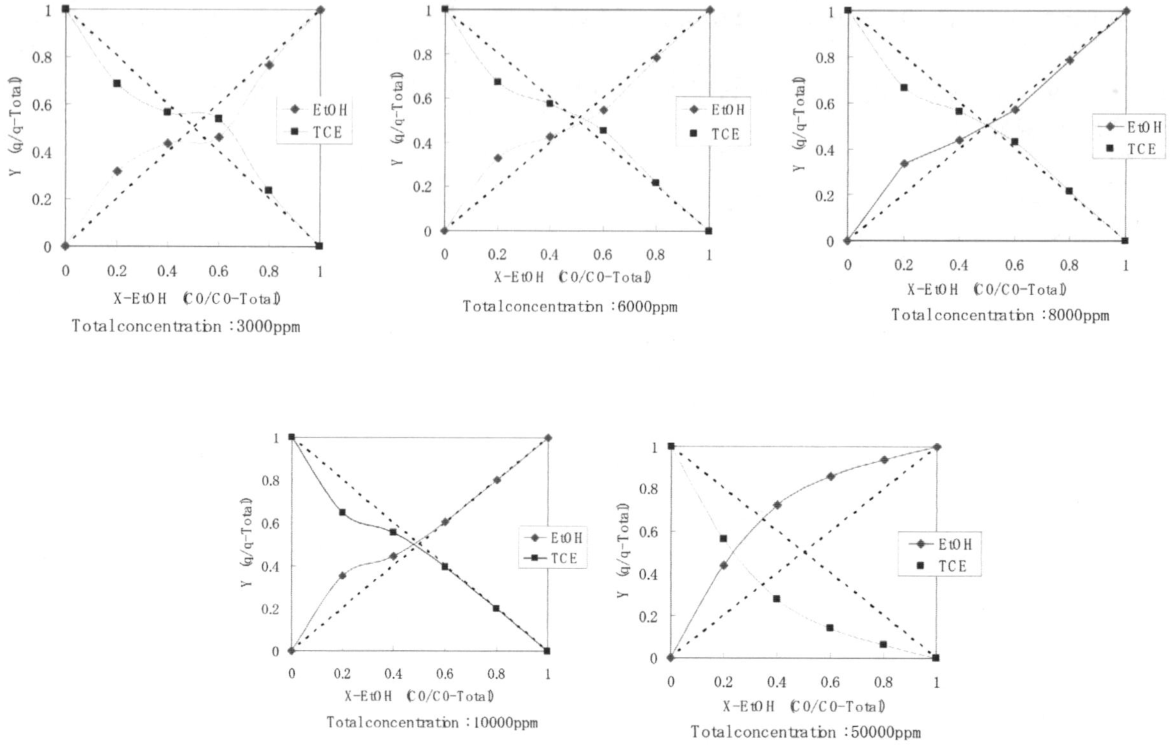

Fig.4 Adsorption equilibrium curve according to total concentration

Conclusions

The value of the amount of the equilibrium adsorption was able to be calculated as a tendency in MS(Fig.1, 2). In accuracy, the amount of adsorption of MS is greatly different compared with the weight method experiment. It will be necessary to improve the crystal structure and the calculation condition of MS in the future. Moreover, the value and the adsorption equilibrium curve of the adsorption equilibrium experiment of two elements of the amount of the equilibrium adsorption were able to be calculated as a tendency in MS(Fig.3, 4).

References

1. K. Chihara, M. Suzuki, K. Kawazoe, Adsorption rate on Molecular Sieving Carbon by Chromatography, AIChE, 24,237(1978)

719r Investigation of the Ternary Mixture of Complex Hydrides for Hydrogen Storage Materials Development

Atsadawuth Siangsai[1], Yindee Suttisawat[1], Pramoch Rangsunvigit[1], Boonyarach Kitiyanan[1], and Santi Kulprathipanja[2]. (1) Petrochemical Technology, The Petroleum and Petrochemical College, Chulalongkorn University, 254 Soi Chula 12 Prayathai Rd., Bangkok, 10330, Thailand, (2) UOP LLC, 50 East Algonquin Road, Des Plaines, IL 60017-5016

Solid hydrogen storage materials have been investigated as hydrogen carriers due to their safety in hydrogen applications. In contrast, they also have some disadvantages, such as a low hydrogen capacity, very high desorption/adsorption temperature, low kinetic rate, and low reversibility. Many attempts have been made to improve and identify materials that can be applied for on-board hydrogen storage for fuel cells. This work investigates the ternary mixture of complex hydrides—$LiNH_2$+$LiAlH_4$+MgH_2—on the desorption/absorption temperature for a set of novel hydrogen storage materials. The ternary mixture is prepared by using mechanical ball-milling. In previous studies, the results from the reaction of the binary mixture $2LiNH_2$+MgH_2 showed that the kinetic rate was slow and created a serious problem in NH_3 emission. The $LiAlH_4$ was also studied and it was reported that the kinetic barrier transformation was easily overcome by mixing it with $LiNH_2$ and its stability of dehydriding was also controlled. One of these reactions, $Mg(NH_2)_2+2LiH \leftrightarrow Li_2Mg(NH)_2+2H_2$ is known as a reversible reaction. Our study demonstrates that by starting from the ternary mixture of complex hydrides—$LiNH_2$+$LiAlH_4$+MgH_2—the kinetic rate is improved and the reversibility is maintained. Phase transformation and the amount of hydrogen desorption are analyzed by XRD and TPD measurements.

Keywords: $LiNH_2$, $LiAlH_4$, MgH_2, Hydrogen Storage, Complex Hydrides

719s Light Olefin/paraffin Separation Using Cu Based Adsorbent

Soon Haeng Cho, S.G.T. Bhat, Sang Sup Han, Jong Nam Kim, and Jong Ho Park. Chemical Process Research Center, Korea Institute of Energy Research, 71-2, Jang-dong, Yusung-gu, Daejoen, 305-343, South Korea

Light olefins are basic raw materials in synthetic chemistry. Cu compound impregnated on solid substrate is prepared, which adsorb selectively light olefins while being amenable to be employed for removing, and recovering substantial quantities of light olefins from the corresponding paraffins. This adsorbent shows high adsorption capacity and fast sorption/desorption rates for light olefin molecules.

The result for propylene/propane separation by pressure swing adsorption process using the above adsorbent is also presented.

ADSORPTION OF ORGANICS ON MSC5A IN SUPERCRITICAL CO2, MOLECULAR SIMULATION

Kazuyuki Chihara, Mai Hikita, Taichi Wada, Ryota Suzuki,*

Meiji University
Higashi-mita, Tama-ku, Kawasaki, Japan
**Corresponding author, TEL +81-44-934-7197, FAX +81-44-934-7197,*
e-mail: chihara@isc.meiji.ac.jp

ABSTRACT

Chromatographic measurements were made for the adsorption of benzene, toluene and m-xylene on molecular sieving carbon (MSC) in supercritical fluid CO2 mixed with organics. Supercritical chromatograph packed with MSC was used to detect pulse responses of organics. Adsorption equilibria and adsorption dynamics parameters for organics were obtained by moment analysis of the response peaks. Dependences of adsorption equilibrium constants, K*, and micropore diffusivity, D, on the amount adsorbed were examined. The dependencies of adsorption equilibrium constants, K*, and micropore diffusivity, D, of toluene, benzene and m-xylene, on molarity of toluene with each parameters of temperature or pressure were obtained.

INTRODUCTION

Supercritical CO2 fluid is attractive solvent whose solubility of organics can be changed with changes of pressure and temperature. New process of adsorptive separation using supercritical fluid might be possible since removal of adsorbate from adsorbent can be performed efficiently.

In this study, chromatographic measurements were made for the adsorption of benzene, toluene and m-xylene on molecular sieving carbon (MSC) in supercritical fluid CO2. Supercritical chromatograph packed with MSC was used to detect pulse responses of organics. Adsorption equilibria and adsorption dynamics for organics were obtained by moment analysis of the response peaks. Dependences of adsorption equilibrium constants, K*, and micropore diffusivity, D, on the amount adsorbed were examined. And, stop & go simulation of multicomponent adsorption equilibria was performed, and potential parameters were determined by comparing the simulation with experimental results.

EXPERIMENTAL

The experimental apparatus (Super 200-type 3; Japan Spectroscopic Co., LTD) was shown in Fig.1. The carrier fluid of the chromatograph was supercritical CO2 (critical temperature 304K, critical pressure 7.3 MPa) and it's mixture with the above-stated organics (benzene, toluene or m-xylene) respectively. The adsorbates used in the form of pulse were the same of or the different from organics mixed with supercritical CO2.

For example, in the case of CO2 mixed with benzene, the organic used in the form of pulse was benzene or, toluene or m-xylene. The volumes of the pulse were fixed to be 10 x 10-9m3 as liquid. MSC 5A (Takeda chemicals Co., HGK882.) was crushed and screened to obtain particle size between 1.49 x 10-4 - 1.77 x. 10-4m (an average particle radius of 8.12 x. 10-5 m). 4.82 x 10-4kg of these particles were packed into the chromatographic column of 15 x 10-2m long and 4.6 x 10-3m in diameter. The void fraction, ε, of the bed was determined to be 0.3256. Flow rate of supercritical CO2 was 1.33 x 10-7m3/s at 268K and at 15.0, 20.0 and 25.0MPa respectively and flow rate of adsorbate (benzene, toluene or m-xylene) was 1.67 x 10-10m3/s, 5.00 x 10-10 m3/s and 1.00 x 10-9 m3/s as liquid at room temperature (298K). The column pressure was kept at 15.0, 20.0 and 25.0 MPa respectively. The pressure drop across the adsorbent bed was estimated to be about 0.1Mpa and was assumed to be negligible. The experimental column temperature was kept at 313, 333 and 353 K respectively. Before experimental runs started, the adsorbent particles were regenerated and stabilized by feeding pure CO2 for 2 hours at the experimental pressure and temperature. Pulse responses were detected using a multi-wave length UV detector (Multi-340; Japan Spectroscopic Co., LTD.) (195-350 nm). Response data were processed by a personal computer. Moment analysis of supercritical fluid chromatogram was tried, and the apparent adsorption equilibrium constant, K* and time constant of micropore diffusivity, D/a2 obtained from first and second moment of response peak, as in references .

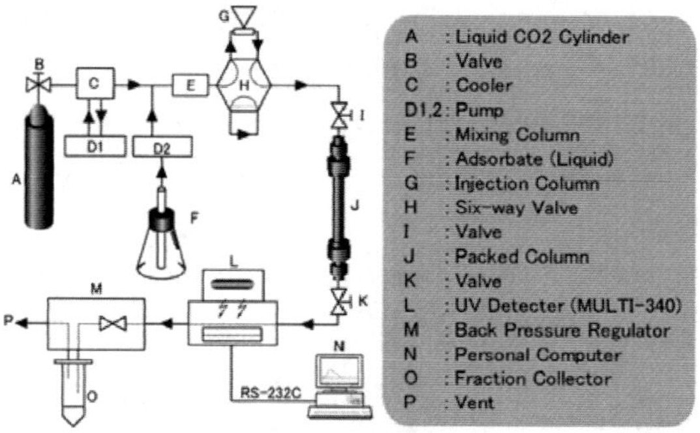

Figure.1 Example of a Sample Table and the Table Caption

RESULT AND DISCUSSION

Figure.2 shows adsorption isotherm of benzene at 313K. According to Fig.2, the amount adsorbed increased with increases of molarity of benzene, and reached to saturation. The amounts adsorbed became larger with decreases of column pressure. It was considered that the situation is competitive adsorption and amount adsorbed of benzene decreases as CO2 adsorption increase with increases of column pressure.

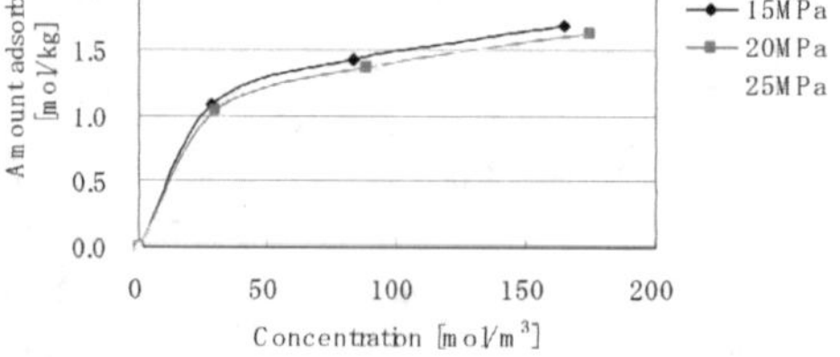

Figure.2 Adsorption isotherm : benzene at 313K

Figure.3 shows dependency of adsorption equilibrium constants, K*, for benzene, toluene, and m-xylene on amount adsorbed of benzene at 20Mpa. This is reasonably decreasing, which corresponds to Figure.2

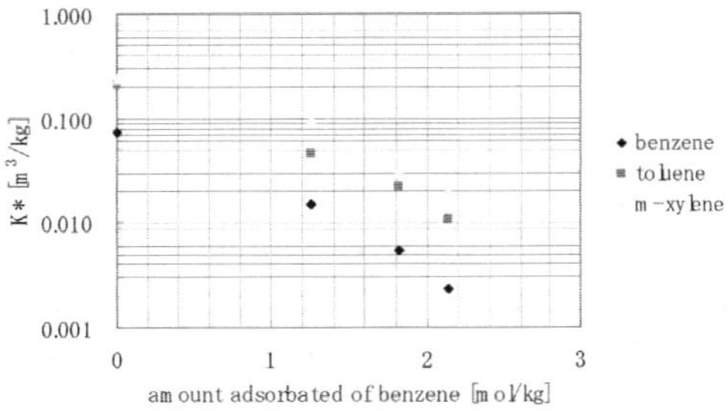

Figure.3 Dependencies of K* on the amount adsorbed of benzene

Figure.4 shows dependency of micropore diffusivity, D/ā2exp(σ2), for benzene, toluene, and m-xylene on amount adsorbed of benzene at 20MPa. The increase of D/ā2exp(σ2) for benzene could be reasonably explained by chemical potential driving force. However, as for dependency of D/ā2exp(σ2) of toluene and m-xylene on amount adsorbed of benzene, further discussion would be necessary.

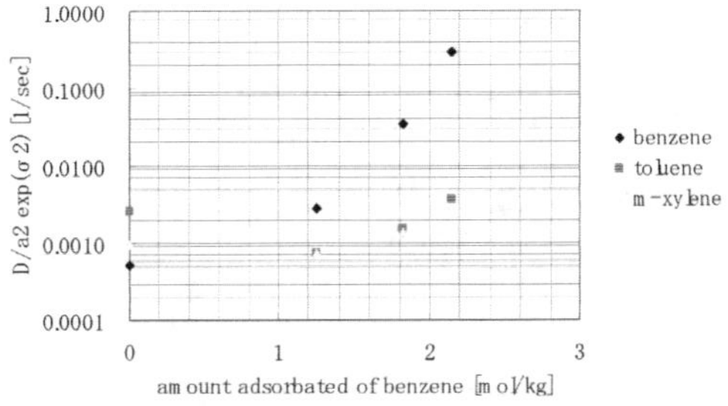

Figure.4 Dependencies of D/a2 exp(σ2) on the amount adsorbed of benzene

Figure.5 shows contribution of four mass transfer steps in pulse response (□'2). The effect of micropore diffusion is dominant in all the conditions examined.

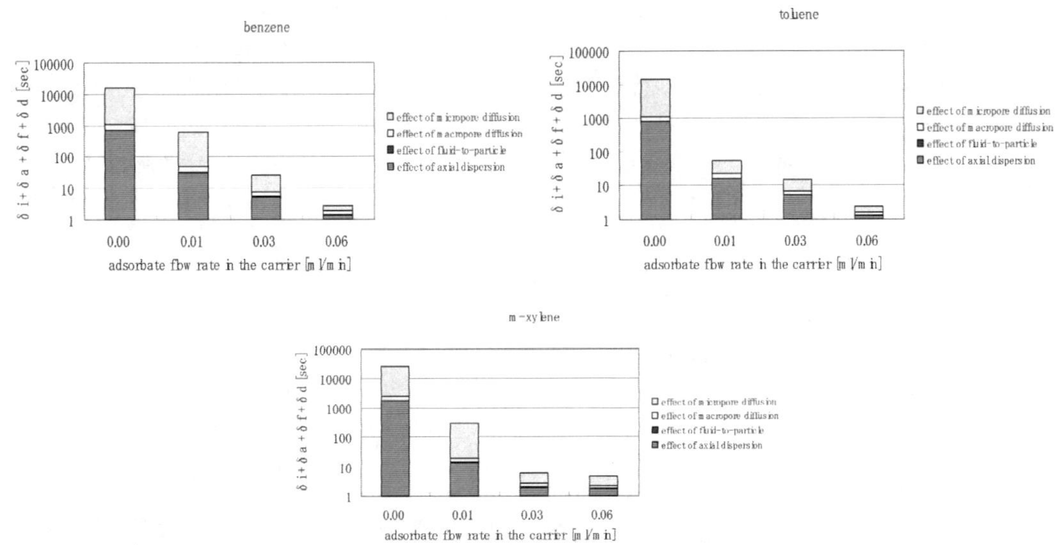

Figure.5 Contribution of four mass transfer steps

STOP & GO SIMULATION

4.1 Adsorption isotherm

stop & go simulation of multicomponent adsorption equilibria was performed, and potential parameters were determined by comparing the simulation with experimental results.

Benzene and CO2 were used for adsorbate. Conditions are 313K and 15MPa. The amount of adsorption increased with increases of molarity of benzene. Figure 6 shows comparison of adsorption isotherm for a stop & go simulation and an experiment.

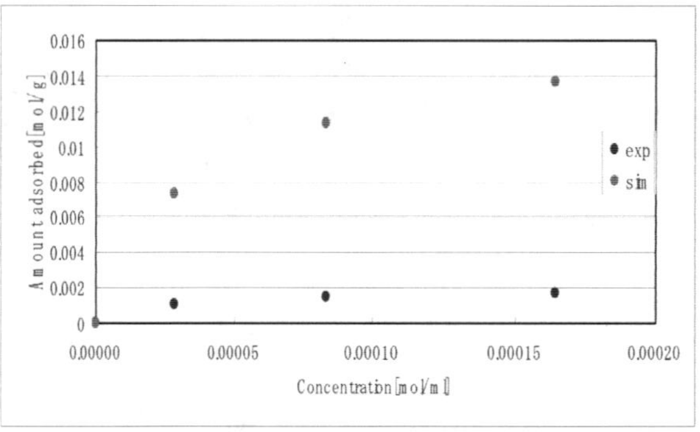

Figure.6 Comparison with experiment and stop & go simulation : Adsorption isothermn

Figure 6.shows comparison of the adsorption isotherm of simulations and experiments. There are the big difference between the simulation result and the experimental result in the amount of adsorption. It was found that the simulation result don't reached amount of saturation adsorbed. More improvement for simulation would be necessary.

5. Molecular simulation

Cerius2 (MSI Inc.) was used throughout the simulations. Three kinds of force field parameters in the Cerius2 library were used. The Grand Caronical Monte Carlo method (under constant chemical potential (□), volume (V), and temperature (T)) was used to get the equilibrium amount adsorbed. The purpose of performing simulation is to elucidate an adsorption mechanism on the molecule level. The simulation was performed on the same conditions as an experiment in order to compare with experiments. MSC68-RC1 model was used as adsorbent. The model has 6.8 A of distance between the centers of two graphitic carbon layers.

Adsorption state

First, we examined how molecules of adsorbate is located. Benzene was used for the adsorbate here. The results are shown in Figure 6. We see from Figure 7 that benzene adsorbs into the adsorption space, which simulates micro pore. Here, benzene adsorbed in parallel to layer in MSC68-RC1 model.

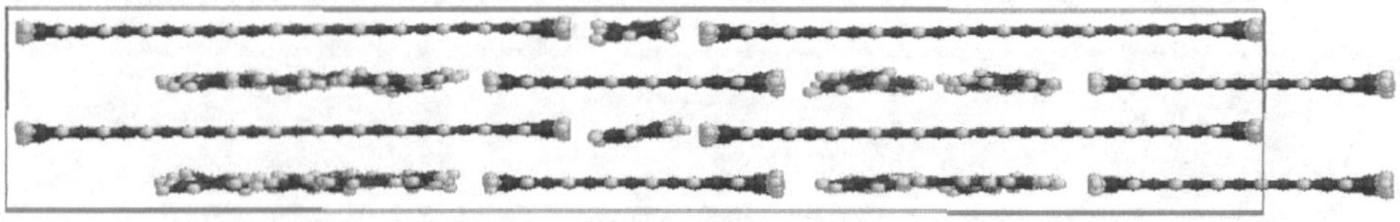

Figure.7 Adsorption state of benzene (MSC68-RC1 model)

Adsorption isotherm

The simulation was carried out for the binary component. Benzene, toluene, m-xylene and CO2 were used for adsorbate. Conditions are 313 K and 15MPa. The amount of adsorption increased with increases of molarity of benzene, toluene, and m-xylene respectively. UNIVERSAL1. 02 was used for the force field. Figure.7 shows adsorption isotherm of benzene, toluene, m-xylene. Figure 8 shows comparison of adsorption isotherm for a molecular simulation the force field and an experiment.

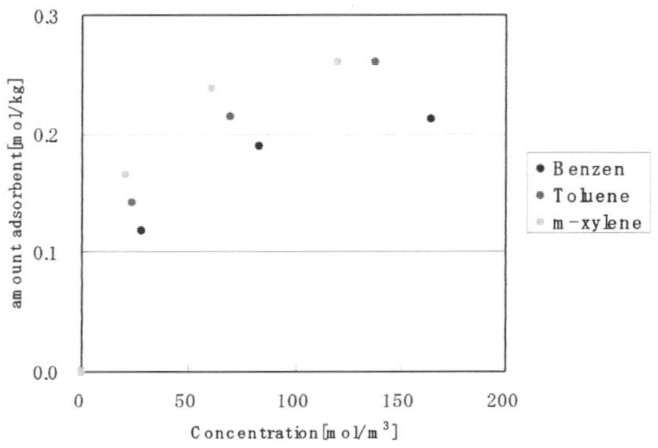

Figure.7 Adsorption isotherm of benzene, toluene, m-xylene at 313K and 15MPa

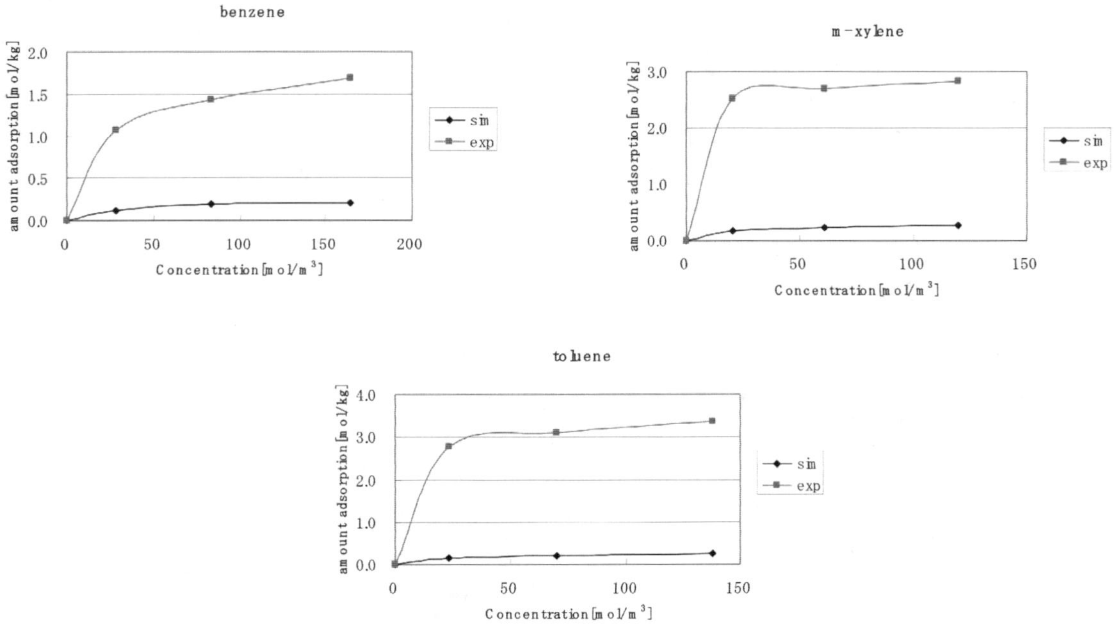

Figure 8 Comparison with experiment and molecular simulation:Adsorption isotherm

CONCLUSION

Adsorption equilibrium and adsorption dynamics on MSC were evaluated for each organics in supercritical CO2 fluid mixed with adsorbate by chromatographic measurement. The dependencies of adsorption equilibrium constants, K*, and micropore diffusivity, D, of toluene, benzene and m-xylene, on molarity of benzene, toluene or m-xylene with each parameters of temperature or pressure were obtained, respectively. It was found that the values of K* and D for an organic substance depended on the amount adsorbed of other organics strongly.

With stop & go simulation and The experimental amount adsorbed could not be simulated yet. So more improvement for simulation w Molecular simulation could be necessary.

719u Partial-Feed Operation for Simulated Moving Bed Chromatography at the Low Selectivity Condition

Kyung-Min Kim, Hyeon-Hui Lee, and Chang-Ha Lee. Department of Chemical and Biomolecular Engineering, Yonsei.university, 262 Sungsanno Seodaemun-Gu, Seoul, 120-749, South Korea

One of the approaches to improve the performance of SMB is using the periodic modulation of operating parameters. Partial-feed operation is also one of these approaches. "Partial-feed" concept was suggested by Zang and Wankat (2002) and compared with total-feed operations. "Partial-feed" means that the feed is introduced into the SMB system during part of a switching period.

In this work, "partial-feed" operation at the low selectivity condition was rigorously studied for the four-zone SMB with one or two column per zone. In the case of SMB with one column per zone, all the performance parameters become better as the feed length decreases even in the extremely low selectivity. The other hand, when the "partial-feed" operation is applied to the four-zone SMB with two columns per zone, the "partial-feed" worsens the performance of SMB in view point of all the performance parameters in low selectivity. However, in high selectivity condition, the "partial-feed" enhanced the purities. Even in the condition of low selectivity, optionally one component purity or recovery can be improved by controlling the feed time while the eluent consumption is worse than the results of "total-feed" operation for all the operating range.

In addition, the "partial-feed" operation was successfully improved the performance of SMB by applying to the "partial-discard" as the "partial-discard with partial-feed".

719v Synthesis of Magnetite-Mesoporous Silica Composites as Adsorbents for Desulfurization from Natural Gas

Seol-Hee Lim, Eun-Ji Woo, and **Chang-Ha Lee**. Department of Chemical and Biomolecular Engineering, Yonsei.university, 262 Sungsanno Seodaemun-Gu, Seoul, 120-749, South Korea

Magnetite-mesoporous silica composites were synthesized as a new adsorbent for desulfurization from natural gas. When nano-sized magnetite was added to the reactant solution prior to a sol-gel reaction, mesoporous silica was formed surrounding nano-sized magnetite, and the silica domain size was reduced significantly comparing with the case that no magnetite was added. This composite shows high capacity for adsorption of methylmercaptane in the presence of methane while mesoporous silica or magnetite only had little adsorption capacity. The sulfur adsorption of this composite show 20 times higher capacity than the mesoporous silica due to the synergetic effect of strong molecular affinity of the magnetite to the sulfur compound and large surface area/pore volume of the mesoporous silica.

719w Surface Modified Superparamagnetic Nanoparticles for Immobilization of Lipase

K.M. Ponvel, Eun-Ji Woo, and **Chang-Ha Lee**. Department of Chemical and Biomolecular Engineering, Yonsei.university, 262 Sungsanno Seodaemun-Gu, Seoul, 120-749, South Korea

It is known that surface modified magnetic nanoparticles have been suggested as a support for the immobilization of enzyme. The surface of magnetic nanopariticle was modified with surfactant such as sodium dodecyl sulfate (SDS) and alkyl benzenesulfonate (ABS) to immobilize lipase using physical adsorption affinity. The surfactant is attached to the magnetite nanoparticles via a chelation of sulfate group of SDS or sulfonate of ABS. Samples are characterized by various techniques. Transmission electron microscope showed that magnetic nanoparticles had an average size of 10,,b2 nm. Magnetic measurement revealed that the magnetic silica nanoparticles were superparamagnetic and the saturation magnetization was about 68-77 emu/g. Porcine Pancreas Lipase (PPL) was immobilized onto the nano particles by physisorption, which was to catalyze hydrolysis of olive oil, showed enhanced durability in the repeated use and catalytic activity was found to be higher than that of free enzyme. Candida Rugosa Lipase (CRL) was also immobilized, however, activity of immobilized lipase was found to be less than that of free enzyme. Since the immobilized superparamagnetic nanoparticles could be easily recovered by magnet, their adsorption affinity and catalytic affinity were compared to each other by using reusability.

719x Comparison of H2 Psa Processes from Low Concentration of Hydrogen Feed

Min-Kyu Kim[1], Dong-Geun Lee[1], Si-Hyung Lee[2], Ki-Hyun Kim[2], and **Chang-Ha Lee**[1]. (1) Department of Chemical and Biomolecular Engineering, Yonsei.university, 262 Sungsanno Seodaemun-Gu, Seoul, 120-749, South Korea, (2) Iron making Research Group of Technical Research Laboratories, POSCO, 1 Goedong-dong Nam-gu, Pohang Gyeongbuk, South Korea

Hydrogen is considered to be one of the most important chemicals used in the chemical and petroleum industries. Furthermore, hydrogen can be used as a general purpose energy source for space heating, electric power generation, and transportation fuel. These potential uses have resulted in a huge increase in hydrogen demand.

The simulation study of PSA processes are studied to produce higher than 90% H_2 from 38% feed with high recovery. The feed contains CO_2, N_2, CO and CH_4 as impurities. The PSA processes are simulated by a non-isothermal dynamic model using gPROMS.

The performance of two-bed PSA is compared to that of four-bed PSA. In addition, the separation performance of the PSA using single adsorbent is also compared to that of the PSA using layered bed. Although the two-bed PSA can produce the target purity, the recovery is low due to the purge gas. However, the four-bed PSA can produce higher than 90% H_2 with very high recovery by minimizing purge gas and twice equalization steps. Furthermore, since the N_2 works as a main impurity in product, the layered bed PSA can control the product quality more efficiently than the single adsorbent bed PSA. To maximize the purity and recovery, the ratio of adsorbent in the bed plays a key role in the PSA process. Moreover, it is important to control the concentration profile of CO_2 in the bed because of its strong adsorption of adsorbent.

719y A Streamlined Two-Column SMB Process for Chiral Separation

João M. M. Araújo[1], Rui C. R. Rodrigues[1], and **José P. B. Mota**[2]. (1) Chemistry, Requimte, Faculdade de Ciências e Tecnologia, Universidade Nova de Lisboa, Quinta Torre, 2829-516 Caparica, Portugal, (2) Chemistry Department, Requimte, Faculdade de Ciências e Tecnologia, Universidade Nova de Lisboa, Quinta Torre, 2829-516 Caparica, Portugal

Simulated moving bed (SMB) chromatography is a powerful tool for chiral separation, because of its many advantages with respect to discontinuous batch chromatography, namely higher product purity, less solvent consumption, and higher productivity per unit stationary phase. Newly developed cyclic operating schemes and column configurations are pushing the trend toward the use of units with fewer columns and more economical set-ups.

Despite the lower productivity and higher solvent consumption of batch processes, they are advantageous in other respects, such as in their simplicity (less design effort is necessary and the scale-up is easier), flexibility (changes in conditions can be easily handled), and versatility (multi- and center-cut separations from multi-component mixtures are easy to implement).

Somewhere in between, and in an attempt to work with the best of "two worlds", we have developed a compact, two-column open-loop SMB process for enantiomeric separation, which is particularly effective when the resolution is limited.

On the other hand, product quality control is mandatory for chiral separations, which requires robust on-line monitoring of the individual enantiomer concentrations. We have successfully implemented an on-line monitoring system, comprising an HPLC set-up with two UV detectors, to measure the individual enantiomeric concentration profiles in the SMB unit. Our monitoring scheme does not use a polarimeter.

The chiral separation considered in this work is the separation of Reboxetine enantiomers on Chiralpak AD, using a mixture of Hexane-Ethanol-DEA as solvent. This system is one in which there is clearly a lack of resolution between the two components. It is thus difficult to obtain both products with high purity and yield by HPLC for the chosen chromatographic conditions. Reboxetine, (RS)-2-[(RS)-*f*Ñ-(2-ethoxyphenoxy)benzyl]-morpholine is an antidepressive NRI drug. Only the (R,R)-, (S,S)-pair is present as a racemic mixture in the active principle and the commercial formulations. Recent studies support the hypothesis that the (S,S)-enantiomer is a more potent inhibitor than the (R,R)- and that it is responsible for the vasomotor and cardiac side effects of Reboxetine.

719z Automated Iterative Refinement of Adsorption Isotherm Models for Simulated Moving-Bed Operation Based on Cyclic Steady State Data

José P. B. Mota, Rui C. R. Rodrigues, and João M. M. Araújo. Chemistry Department, Requimte, Faculdade de Ciências e Tecnologia, Universidade Nova de Lisboa, Quinta Torre, 2829-516 Caparica, Portugal

Enantiomeric separations by simulated moving-bed (SMB) chromatography are usually carried out under strongly nonlinear conditions. The accurate determination of the competitive adsorption equilibrium of the enantiomeric species is thus of fundamental importance to allow computer-assisted optimization or process scale-up. This is especially true for the SMB, because its operating conditions cannot be designed conveniently without knowledge of the adsorption isotherms of the feed components.

The inverse method of isotherm determination is currently becoming very popular as a quick procedure for estimating the adsorption equilibrium data necessary for designing SMB separations. It derives the isotherm from overloaded band profiles of individual solutes or of their mixture. In recent studies, reporting comparisons of frontal analysis and inverse methods, it is concluded that the inverse method gives accurate estimates of the competitive isotherm parameters up to the maximum elution concentration of the overloaded bands. However, it is only moderately accurate from the maximum elution concentration up to the injected concentration.

In this work we describe and validate experimentally a procedure in which the inverse method is applied directly to the cyclic steady-state (CSS) concentration profiles of the running SMB process to update the parameters of the prescribed adsorption isotherm model. The operating conditions are then optimized for the newly determined isotherm parameters and applied to the running SMB process. This process is iterated and automated in our monitoring and control software.

As a proof of concept of the proposed methodology, it has been successfully applied to the chiral separation of Reboxetine enantiomers on Chiralpak AD, using a mixture of Hexane-Ethanol-DEA as solvent. The system was operated near the solubility limit of the the racemic mixture. Reboxetine is an antidepressive NRI drug. Only the (R,R)- and (S,S)-pair is present as a racemic mixture in the active principle and commercial formulations. Recent studies support the hypothesis that the (S,S)-enantiomer is a more potent inhibitor than the (R,R)- and that it is responsible for the vasomotor and cardiac side effects of Reboxetine.

719aa Metal-Organic Frameworks for Adsorption Separations

Jagadeswara R. Karra, Chemical Engineering, Kansas State University, 1005 Durland Hall, Manhattan, KS 66506 and Krista S. Walton, Department of Chemical Engineering, Kansas State University, 1005 Durland Hall, Manhattan, KS 66506.

A critical step in developing adsorption-based technologies is the identification/synthesis of adsorbents that provide the proper adsorption capacities, selectivities, or reactivities for the application. It is well known that metal sites in porous materials have a tremendous influence on the resulting adsorption properties. Metal-organic frameworks (MOFs) are a recent addition to the classes of porous materials and have the potential for providing a flexible platform for developing designer adsorbents. MOFs are synthesized by self assembly of organic ligands and metal oxide clusters and possess pore sizes and chemical functionalities that can be manipulated by modifying the metal group or organic linker. Several MOFs have open metal sites (coordinatively unsaturated) that are built into the pore walls in a repeating, regular fashion. These metal sites have been shown to impart catalytic activity to the materials and also have the potential to enhance general adsorption properties. In this work, we explore the effect of open metal sites in Cu-BTC on adsorption selectivities of carbon monoxide- and carbon dioxide-containing mixtures. Molecular modeling results and experimental results will be discussed.

719ab Molecular Simulation Study of Spontaneous Desorption In Cage-Like Mesoporous Adsorbents

Christopher Rasmussen[1], Aleksey Vishnyakov[1], Matthias Thommes[2], and Alexander V. Neimark[1]. (1) Chemical & Biochemical Engineering, Rutgers University, 98 Brett Road, Piscataway, NJ 08854-8058, (2) Quantachrome Instruments, 1900 Corporate Drive, Boynton Beach, FL 33426

Cavitation of metastable adsorbate in ink-bottle, or blocked mesopores causes a sharp step on the desorption isotherm near the lower closure point of the hysteresis loop [1]. For a long time, the position of this step was considered independent of the pore size. However, recent experimental studies showed that the pressure of cavitation could depend on the pore size and shape [2]. We hypothesize that the position of cavitation depends on the free energy barrier of formation of the critical nucleus. We show that the nucleation barrier can be estimated by comparing simulated and experimental isotherms. In order to establish a relationship between the cavitation pressure and the pore size, a series of argon and nitrogen adsorption-desorption isotherms were calculated in spherical pores ranging from 5.7 to 9.7 nm, which corresponded to the samples of SBA16 materials explored experimentally. The gauge cell Monte Carlo method was employed to calculate the critical nucleus size and the nucleation barrier [3,4]. The grand canonical Monte Carlo simulation was employed to mimic the spontaneous cavitation process. The simulated isotherms were matched to the experimental data. We found that the nucleation barrier of cavitation increases with the pore size. The nucleation barriers vary from 30 kT to 50 kT in the pores of 5.7nm to 9.7nm.

1. P.I. Ravikovitch and A.V. Neimark, Langmuir, 2002, 18, 9830-9837.

2. M. Thommes, B. Smarsly, P.I. Ravikovitch, A.V. Neimark, Langmuir, 2006, 22, 756-764.

3. A. V. Neimark and A. Vishnyakov, Journal of Chemical Physics, 2005, 122, 054707.

4. A. V. Neimark and A. Vishnyakov, Journal of Physical Chemistry, 2006, 110, 9403-9412.

719ac Enhancing Cellulose Accessibility to Cellulase of Various Feedstocks Pretreated by Cellulose-Solvent-Based Lignocellulose Fractionation

Zhiguang Zhu, **Noppadon Sathitsuksanoh**, and Y-H. Percival Zhang. Biological Systems Engineering, Virginia Polytechnic Institute and State University, 200 Seitz Hall (0303), Blacksburg, VA 24061

Effectively overcoming recalcitrant lignocellulose is the largest obstacles to cellulosic ethanol production because natural biomass composite has a limited substrate accessibility to cellulase or catalysts. Quantitative determination of cellulose accessibility to cellulase (CAC) was established based on adsorption of the non-hydrolytic protein TGC containing a cellulose binding domain (CBM) and green fluorescence protein (GFP) (Hong et al. 2007. Langmuir 23: 12535). Here we proposed a new approach to determine the surface area of cellulose and non-cellulose (lignin-rich) fractions of pretreated corn stover, switchgrass, and common reed by cellulose-solvent-based lignocellulose fractionation (CSLF). Quantitative data suggest that CSLF can break lignocellulosic structure efficiently than dilute acid, resulting in much higher surface areas for cellulose and lignin fractionation than does dilute acid pretreatment. High cellulose surface area results in faster hydrolysis rates and high cellulose digestibility. Our results infer that increasing cellulose accessibility is more important than decreasing lignin content or lignin surface area.

719ad Fate of Oxygen Atoms Present as CO Complexes on Activated Carbon Under Heat Treatment In Hydrogen Atmosphere

Atsushi Gabe, Kazunori Kawata, and Yoshinobu Otake. Applied Chemistry, Meiji University, 1-1-1, Higashi-Mita,Tama-ku,, Kawasaki, 2148571, Japan

Activated carbon derived from phenolic resin is oxidized by either HNO3 or atmospheric air at several temperatures for varying oxidation times in order to introduce surface oxygen complexes on the carbon surfaces. The types and amounts of oxygen complexes are estimated by temperature programmed desorption (TPD). The temperature programmed reduction (TPR) is also performed in H2, the concentration of which is ranged from 1.95 to 100vol%. Three different gases of CO2, CO and H2O evolving from the carbons during TPD and TPR runs are analyzed by gas chromatograph. The amount of H2O desorbed in TPR is significantly larger by comparison to that in TPD, but an opposite result is found in CO. The quantities of H2O and CO are affected by the concentration of H2 during TPR, indicating that oxygen atoms in CO complexes are removed partially as H2O by H2 during TPR and the remainder of the CO complexes decomposes as CO without being affected by the presence of H2. No significant change is found in the amount of CO2 among TPD and TPR runs. The TPR runs on the various types of oxygen-containing carbons are performed systematically under a fixed concentration of H2 in order to investigate quantitatively how the CO complexes decompose as H2O and CO during TPR.

719ae Removal of Low Concentration of Pce from Air

Stéphanie Pacault, **Georges Grévillot**, and Cécile Vallières. Chemical Engineering Science Laboratory-CNRS, 1, rue Grandville, Nancy, France

Removal of PCE by adsorption on activated charcoal has been studied. Isotherms has been measured in the range of very low concentrations (5-15 ppmv) using a breakthrough front method. Humidity range from 0 to 85%.Mass transfer length has been also measured in order to design the process.

719af Removal of Pah Compounds by Adsorption

Célio L. Cavalcante Jr.[1], Diana C. S. Azevedo[2], Ivanildo J Silva Jr[2], Francisco Murilo T. Luna[2], Carolina B Veloso[2], and Caio C B Araujo[2]. (1) Department of Chemical Engineering, Universidade Federal do Ceara, Campus do Pici, Bl. 709, Fortaleza, CE, 60455-760, Brazil, (2) Chemical Engineering, Universidade Federal do Ceará, Campus do Pici, bl. 709, Fortaleza, 60.455-900, Brazil

Polycyclic aromatic hydrocarbons (PAHs) are well known for their toxic, carcinogenic, and mutagenic effects. Their removal from commercial products using adsorption has been intensively studied. Because of the dimensions of the typical PAH molecules (e.g., naphthalene, anthracene, pyrene) mesoporous materials seem to be adequate for this purpose. PAH adsorption in non-ordered activated carbons has been reported. In this study, the adsorption of polyaromatic compounds (naphthalene, pyrene and anthracene) on commercial activated carbons was evaluated using batch and chromatographic methods. From these experiments, adsorption equilibria and kinetics were determined and correlated using commonly used models (Langmuir model and pore diffusion model, respectively). The estimated parameters can then be used to scale-up units for PAH removal from heavy oil streams.

719ag Multi-Scale Simulation for Adsorption Process Development. A Case Study of Ortho-Dichlorobenzene on Activated Carbon

Hae-jeong Son[1], Young-il Lim[1], Kyung-soo Nam[2], and Kyung-Seun Yoo[2]. (1) Department of Chemical Engeering, Hankyoung National University, 67 Sokjong-dong, Kyonggi-do, Anseong, South Korea, (2) Department of Envirenmental Engineering, Kwangwoon university, Kwangwoon Road, Nowon-Gu, Seoul, South Korea

This article presents a multi-scale simulation approach starting at the molecular level for the adsorption process development, especially, in o-DCB (ortho-Dichlorobenzene) adsorption on an activated carbon. A grand canonical Monte-Carlo (GCMC) method is realized by using Materials Studio v. 4.2 with the Forcite and the Sorption modules (Accelrys Inc., USA) in the molecular level. In this molecular simulation, adsorption isotherms and kinetic parameters are calculated.

Computational fluid dynamics (CFD) can show the velocity and concentration profiles of the fluid phase within the adsorption column. Using COMSOL Multiphysics v. 3.3 (COMSOL AB., USA) as a CFD code, the characteristics of the fluid is identified according to the column geometry.

After the model parameters of the adsorption process simulation are estimated from, the process simulation is performed for the operating condition optimization or the process design, using these model parameters estimated from the molecular-level and the fluid level. In the process simulation level, the process simulation is performed in FAST-Chrom/SMB (Fast and Accurate Simulation Tools for Chromatography and SMB), which developed by our own laboratory.

The simulation results obtained from the adsorption equilibrium constant predicted at the molecular level are in good agreement with the experimental results of a pulse response. The systematical multi-scale simulation approach addressed in this study may be useful to accelerate the adsorption process development in reducing the number of experiments.

Keywords: Multi-scale simulation, Molecular simulation, Process simulation, Activated carbon, Adsorption isotherms, Computational fluid dynamics (CFD).

Comparison of Odor Transfer Characteristics of Total Heat Exchangers between Ion Exchange Resin and Porous Adsorbent as Desiccant

Hiroshi Okano[1], Tsutomu Hirose[2], and Wei-Li Jin[1].
(1) Research & Development, Seibu Giken Co., Ltd, 3108, Aoyagi, Koga, Fukuoka, Japan.
(2) Advanced Adsorption Technology, 19-7, 4 Chome, Miwadai, Higashi-ku, Fukuoka, Japan

ABSTRACT

A total heat exchanger can recover both sensible heat as temperature and latent heat as humidity lost by the ventilation in the air conditioning of buildings and has been recognized to be an important device to enhance conservation of energy and reduction of CO_2 emission with keeping a high quality of indoor air. Latent heat (ambient water vapor) is exchanged through honeycomb rotors coated on the surface with desiccant material of porous adsorbent etc. but some complaints are made about accumulation of odor substance on adsorbent and accidental transfer of it back to the room. We have developed and commercialized a new type of total heat exchangers by use of ion exchange resin as desiccant. The heat exchange efficiency of it is equivalent to the conventional ones and the accumulation and transfer of odor is reduced to much less extent. Mechanism of the odor transfer is discussed by comparing odor transfer rate in the total heat exchanger between porous adsorbent such as silica gel and ion exchange resin as desiccant.

INTRODUCTION

Recently, residences and offices have been improved in the air-tightness and insulation for energy saving and the ventilation made up with spontaneous draft has been insufficient. Sick building syndrome attracting a lot of attention is closely related to the lack of ventilation and the importance of artificial ventilation is recognized. A total heat exchanger as shown in Fig. 1 is an energy-saving device that recovers and reutilizes energy lost in the ventilation of buildings since latent heat (humidity) as well as sensible heat (temperature) can be recovered there in exchange between indoor and outdoor air streams by a desiccant rotor. Desiccants used to give a latent heat (humidity) exchanging function have been porous adsorbent such as silica gel, alumina and synthetic zeolite. However, because these materials adsorb various kinds of odors as well as humidity, they have become the cause of complaints against offensive odor generation due to odor transfer. To solve this problem, we have successfully developed a novel total heat exchanger with little odor adsorption and transfer by use of ion exchange resin as a desiccant.

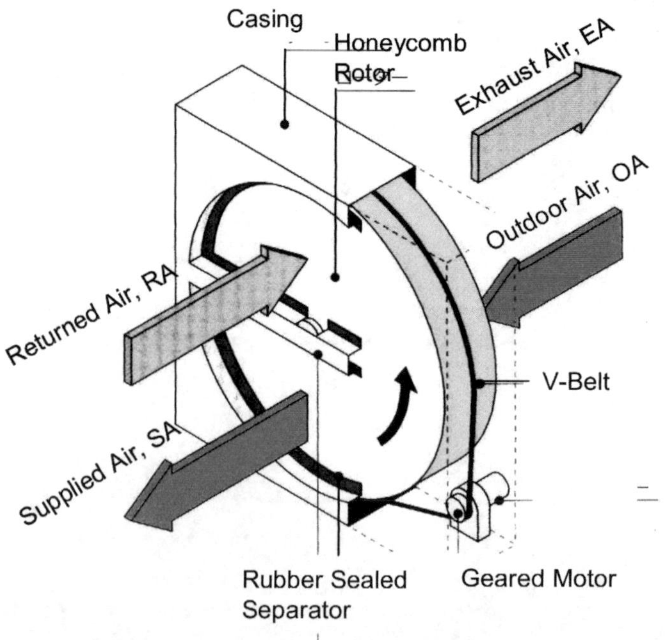

Fig. 1 Schematic diagram of a honeycomb rotor for the total heat exchange

In the present report, odor transfer tests for ammonia and others were conducted using this novel total heat exchanger and odor transfer properties were compared with the conventional types of the total heat exchanger.

1. PRINCIPLE AND ENERGY SAVING EFFECT OF A TOTAL HEAT EXCHANGER

A honeycomb rotor consisting of desiccant matrix is rotated for total heat exchange at the speed of 10-20 rpm in a casing cassette that separates between supply and exhaust air zones, as shown in Fig. 1. For example in summer, the indoor air is cool and dry but contaminated with carbon dioxide from respiration while the fresh outdoor air is warm and humid When the return air from the inside is passed through the rotor in the upper half of the rotor, the rotor is cooled and the desiccant is regenerated by the return air while only the contaminated air is exhausted after warmed and humidified by the exchange of the total heat (temperature and humidity). On the other side of the rotor, the outdoor fresh air being warm and humid is taken in through the lower half of the cassette where it passes through the rotor countercurrently. The outdoor air is cooled and dehumidified by the contact with the cool dry desiccant and the incoming cool and dry air is continuously supplied to the building. Thus the refrigerating load of the air conditioner can be saved by the recovery of the total heat from the exhaust air by as much as 70-80% of energy lost in the ventilation, as explained on a humidity chart given in Fig. 2. The principle is the same in winter as well except that the warm and humid inside air is exhausted and pre-heated and humidified air is supplied to the room. That is to say, a total heat exchanger is energy-saving device that recycles energy lost in ventilation. The total heat exchanger can recover and utilize both sensible heat as temperature and latent heat as humidity and has higher advantage of energy saving to recover much more amount of heat than the conventional sensible heat exchanger.

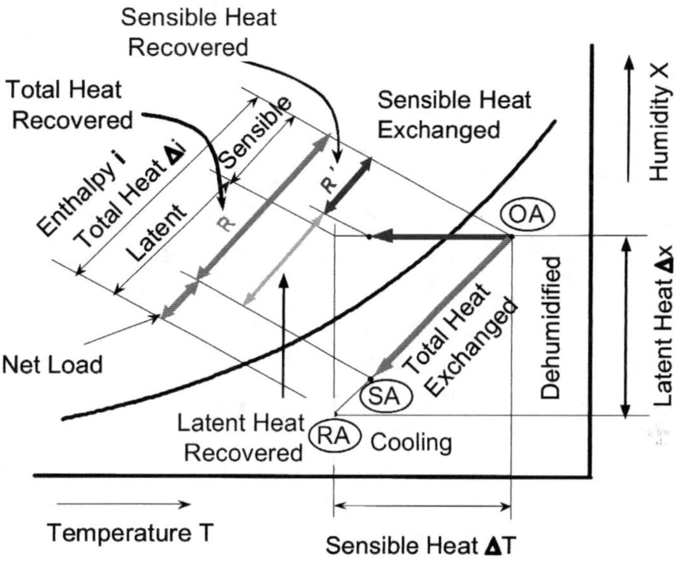

Fig. 2 Presentation of sensible, latent and total heats recovered by a total heat exchanger on a humidity chart

The ability of latent heat exchange in the total heat exchanger is very important in practice. In winter, it prevents troubles of static electricity due to an over-dry atmosphere. Humidity control is important for human health, for example for prevention of flu. In summer, load of latent heat reaches about 50 % of the total load of air conditioning.

The mainstream of total heat exchangers is the rotary type made of aluminum sheet. Metal aluminum itself cannot exchange

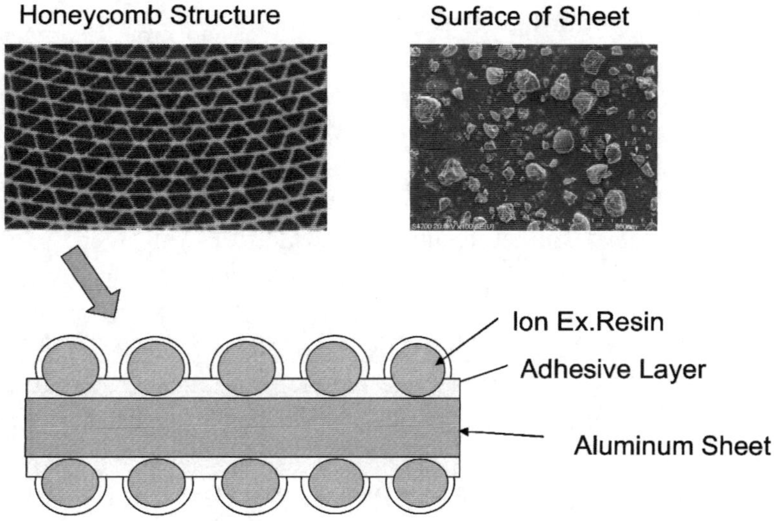

Fig. 3 Detail of honeycomb structure of a total heat exchanger

latent heat (humidity) but is modified to give the desiccating ability in various ways. Thin layer of aluminum oxide is formed by treating the surface of aluminum sheet by some oxidizing agent or aluminum sheet is coated with solid adsorbent such as silica gel, as shown in Fig. 3.

2. RELEASE OF OFFENSIVE ODOR FROM TOTAL HEAT EXCHANGERS

Some odorous matter emitted indoors sometimes is caught in the desiccant rotor and returned back to the room together with the supply air by desorption. Thus, total heat exchangers have not been recommended to be installed in the place encountered with much emission of offensive odor because of this problem of odor transfer. Various kinds of paints, adhesives and construction/interior materials are used even in the ordinary buildings in the case of new construction. A considerable amount of odor substance is emitted from these materials and is expected to be accumulated in total heat exchangers. Transfer of offensive odor was complained in practice in 1-2 years after new construction at the start of the rainy season when the outdoor humidity increased. Situation became serious in some cases although the probability was as low as 0.1-0.4 %.

Mechanism of odor transfer from a total heat exchanger may be explained as follows:

1) Various odor substances and solvent vapors are emitted from construction and interior materials of a new building. Such odor substances are adsorbed and accumulated in adsorbent such as silica gel when the return air from buildings is exhausted through total heat exchangers.

2) In early spring or in the rainy season when the humidity increases rapidly, odor substances are desorbed due to an increased amount adsorbed of water vapor on adsorbent. In other words, accumulated odor substances are desorbed into the room by so called the replacement adsorption of water.

3) When adsorbent of a total heat exchanger is adsorbable to odor substances as well as water vapor, odor released from total heat exchangers is circulated again in the room by transferring from the return air to the supply air. Odor substance cannot be exhausted outdoors endlessly, resulting in customers' complaints.

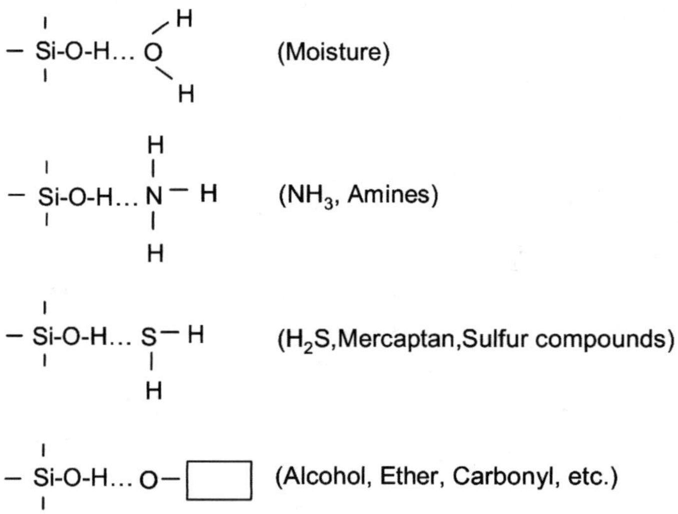

Fig. 4 Mechanism of adsorption of odor substances on a silanol group of porous material

Complaints of the offensive odor transfer are caused by two factors of the replacement desorption of adsorbed odor accumulated and of the endless circulation without exhaust by odor transfer. Total heat exchangers are excellent energy-saving apparatuses and then have been developed by many companies and widespread all over the world. However, this problem of odor transfer has been one of the most difficult problems unsolved.

3. ADSORPTION OF ODOR BY POROUS MATERIAL

Adsorption in porous material such as silica gel is due to either silanol radical (Si-OH) or capillary adsorption. The silanol radical has a very strong affinity to water vapor but also adsorbs odor substances, too, as shown in Fig.4. Also, water-soluble odor substances may be absorbed in the aqueous phase which was adsorbed as water vapor in capillary adsorption. Porous adsorbent such as silica gel adsorbs water vapor in proportional to the relative humidity of air. Many adsorption sites remain vacant and do not contribute to adsorption of water vapor at a low to moderate humidity.

Odor substances are adsorbed on those sites and accumulated there.

When the humidity outdoors is increased in the rainy season, water vapor, too, tends to be adsorbed on sites having been inactive in adsorption/desorption at low to moderate humidity. Odor substances which have occupied such sites are desorbed by the replacement adsorption and released from the total heat exchanger. As a result of these phenomena, adsorption and transfer of offensive odor cannot be avoided when the porous adsorbents such as silica gel are used.

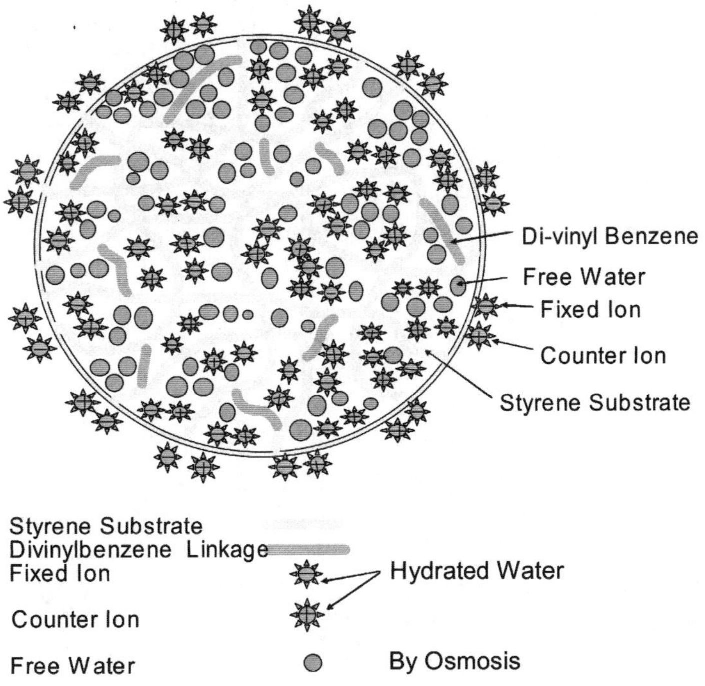

Fig. 5 A structural model of ion exchange resin

4. ION SORPTION TYPE OF TOTAL HEAT EXCHANGERS

We have developed and commercialized a new type of total heat exchangers in which heat exchange efficiency is equivalent to the conventional ones and the accumulation and transfer of odor is reduced to much less extent. The new type is based on the sorption of water vapor into ion exchange resin and the rejection of odor substances and thus it is called an "Ion sorption type" of total heat exchangers in what follows.

Figure 5 shows a structural model of ion exchange resin which is used in the total heat exchanger. Styrene substrates are cross-linked with divinyl benzene in ion exchange resin. A surfonic acid group ($-SO_3H$) is chemically introduced to this polymer structure as a fixed anion and then neutralized with a counter cation ($Na+$). The desiccant ability of ion exchange resin is caused by hydration power and osmotic pressure as shown in Fig. 6 and is similar in principle rather to that of absorbent such as lithium chloride. Unlike such an absorbent, however, fixed ions inside ion exchange resin are bound with cross-linkage and thus the ion exchange resin is not liquefied even in the case of the saturated absorption. Adsorption isotherms of water vapor are shown in Fig. 7 for various adsorbents together with ion exchange resin. The amount adsorbed on ion exchange resin depends on a degree of cross-linkage and the one used in the total heat exchanger has the moisture content between A-type and B-type silica gels.

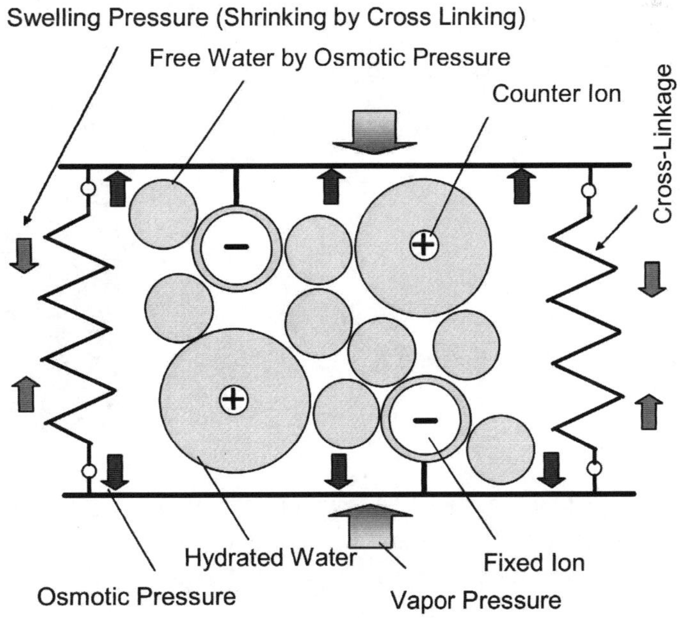

Fig. 6 Pressure balance between swelling and osmosis in ion exchange resin

Ion exchange resin does not adsorb

easily odor substances probably by the following reason. Dry ion exchange resin does not have pores insides unlike porous adsorbents such as silica gel and the void is filled with water sorbed in a hygroscopic state. Thus, odor substances can hardly be caught nor accumulated in vacant sites of adsorption. The shrinking force due to crosslinkage and the swelling force due to hydration and osmotic pressure are balanced inside the resin, and sorption/desorption are caused by the difference between the internal pressure and the external vapor pressure. The interior pressure of the resin is thought to be kept higher than in free liquid solution of adsorbent as shown in Fig. 6. Even a water soluble odor substance is difficult to be solved into water insides resin due to thus high interior pressure and the transfer of odor substance is expected to be depressed correspondingly.

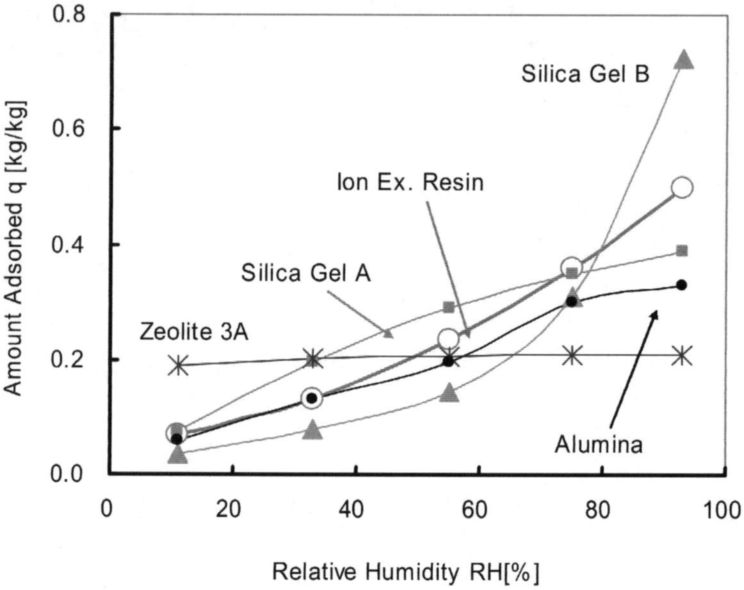

Fig. 7 Adsorption isotherm of water vapor for various desiccant materials

5. HEAT EXCHANGE EFFICIENCY AND ODOR TRANSFER

Humidity, temperature and odor concentration were measured in the return, outdoor and supply air streams in a wind tunnel as shown in Fig. 8, and the result was examined in various ways. The exchange efficiency η of total heat was evaluated in terms of enthalpy I and the water and odor transfer ratios F on the concentration basis C, as follows

Total heat exchange efficiency:

$$\eta = \frac{I_{sa} - I_{oa}}{I_{ra} - I_{oa}}$$

Odor transfer ratio:

$$F = \frac{C_{sa} - C_{oa}}{C_{ra} - C_{oa}}$$

The specific transfer rate R

$$R = (C_{sa} - C_{oa})Vt_c$$

is defined as amount of water or odor substance per unit cross-sectional area in each revolution of the rotor, in which V is the face velocity or superficial velocity of air and t_c is the time required for one revolution.

Fig. 8 Schematic diagram of an experimental apparatus for total heat exchange and odor transfer

The total heat exchange efficiency of the conventional silica gel type and the present ion sorption type were tested and compared in Fig.9. The total heat exchange efficiency η is as high as

80% at the face velocity of V=2m/s and decreases with increasing face velocity although it keeps 70 % at V=4m/s. It was confirmed that the ion sorption type has almost the same total heat exchange efficiency as a silica gel type.

Figure 10 shows a comparison of the transfer ratio F of water vapor and some odor substances between various types of desiccant. The outdoor air was kept at temperature T_{OA}=30 ℃ and relative humidity ϕ_{OA}=50 %

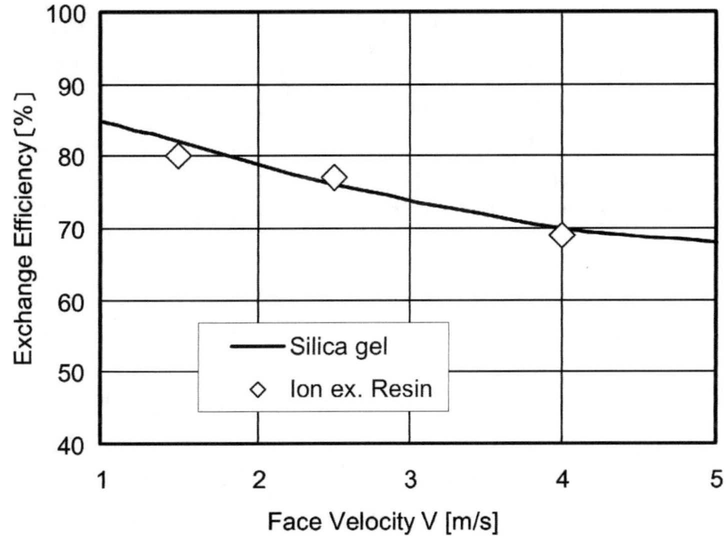

Fig. 9 Comparison of the heat exchange efficiency between silica gel and ion exchange resin

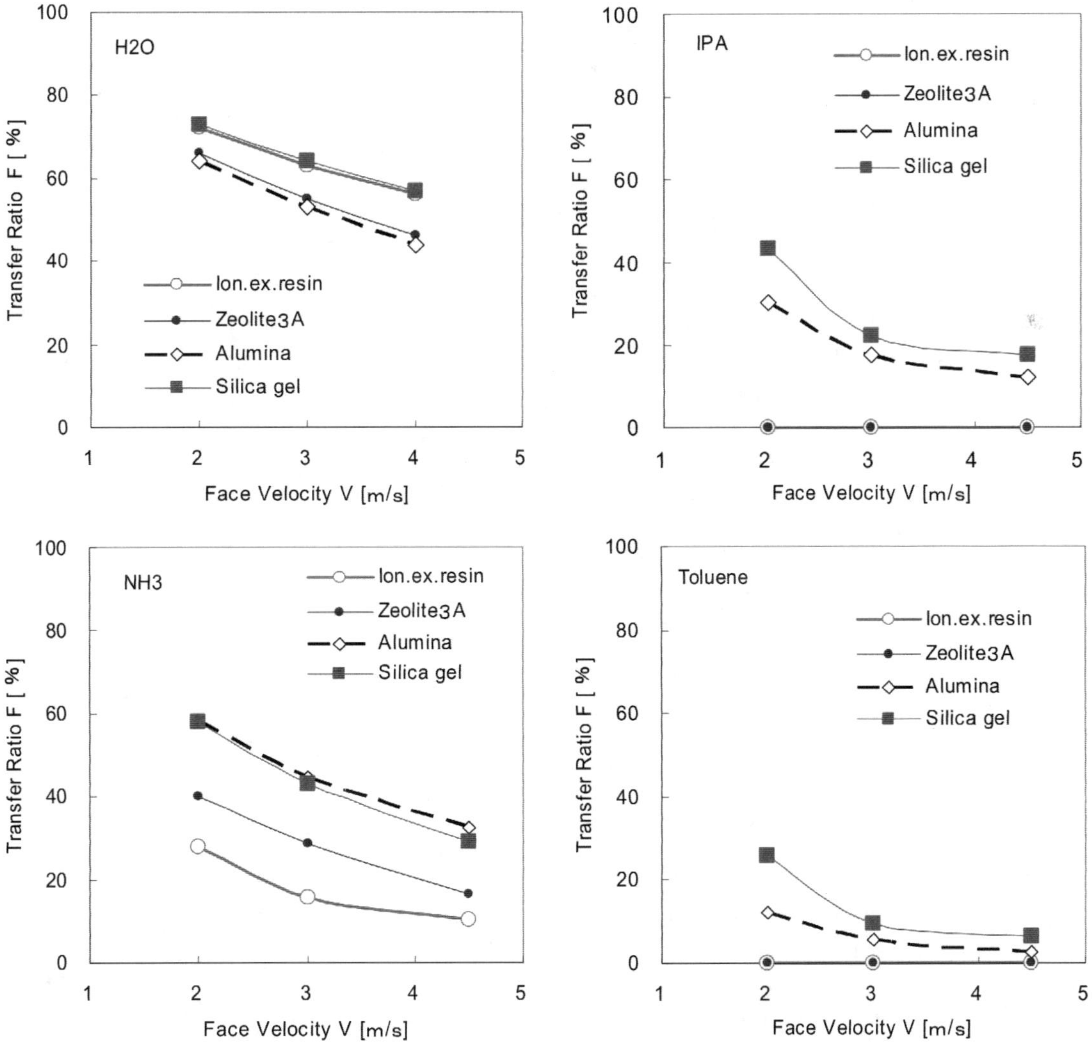

Fig. 10 Comparison of the transfer ratio F of moisture and odor substances between various desiccant materials for a total heat exchanger

while the returned air at T_{RA}= 27℃ and ϕ_{PA}=60 %. The returned air contained the odor substances at concentrations of C_{RA}= 0.016 g/m^3 for NH_3, C_{RA}= 0.13 g/m^3 for IPA and C_{RA}= 0.20 g/m^3 for toluene. The higher value of the transfer ratio F is favorable for water while the lower is favorable for odor substances. The total heat exchanger of ion sorption type has the same transfer ratio F of water to that of the conventional silica gel type and higher ratio than that of zeolite 3A and alumina types. The odor transfer ratio F of ammonium (NH_3) is highest in both silica gel and alumina types, followed by a zeolite 3A type and lowest in the ion sorption type. In the case of IPA (iso-propyl alcohol), no transfer was detected in zeolite 3A and ion sorption types although the odor transfer ratio F is as high as 10-40 % in silica and alumina types. The situation is the same in the toluene case as in IPA although the transfer ratio F for silica gel and alumina types is lower than that in IPA case. The transfer ratio F is influenced by the face velocity V and is decreased with increasing face velocity.

The amount transferred of odor was converted to the specific transfer rate R, Eq.(3), for further discussion from other points of view and the results are shown in Fig. 11. The specific transfer rate R of water vapor per each evolution of the rotor increases with the face velocity V while the transfer ratio F decreases as shown in Fig.10. This is because an increased face velocity, thus increased amount of supplied water vapor, compensates the decrease in humidity in the downstream of the

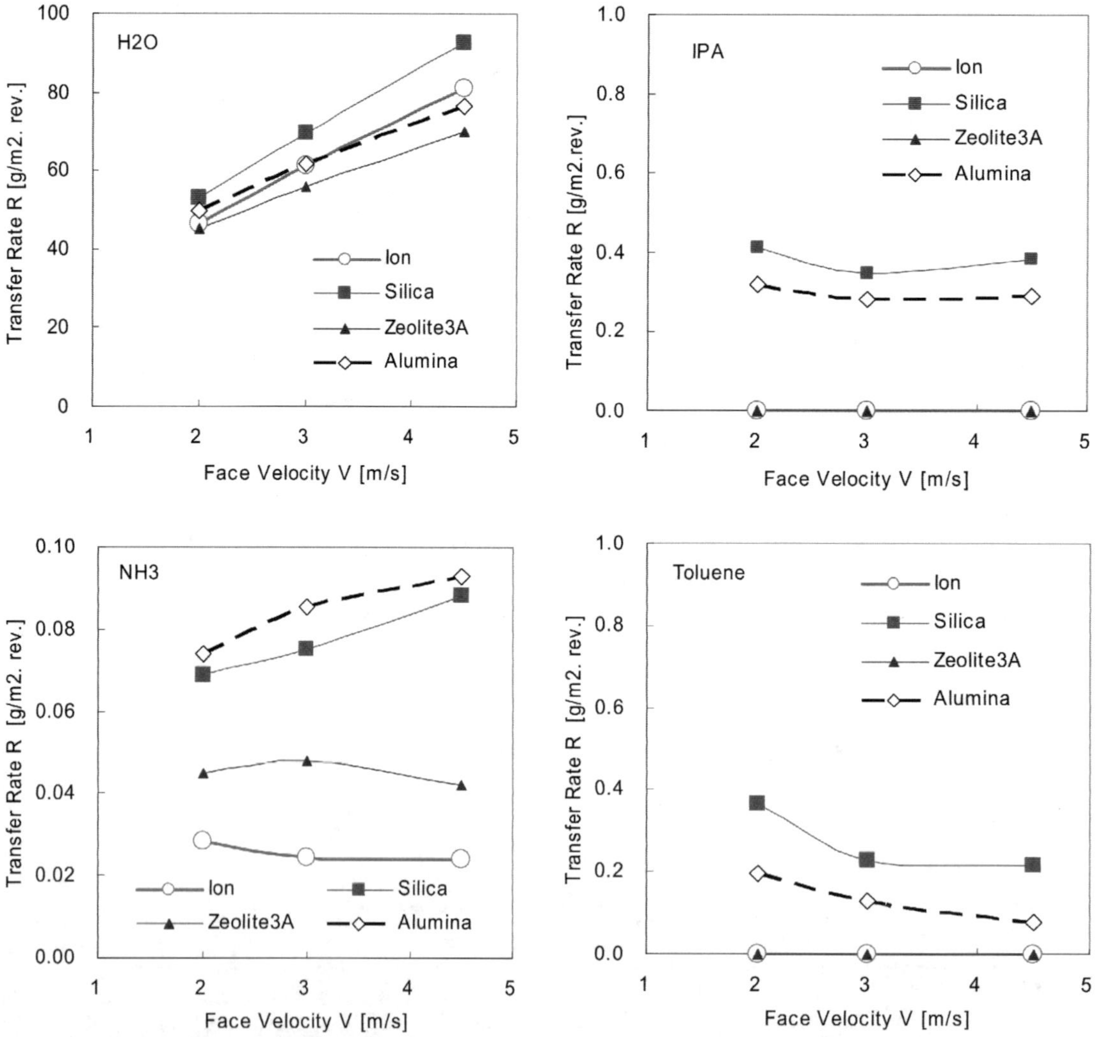

Fig. 11 Comparison of the specific transfer rate R per unit cross-sectional area in each revolution of the rotor between various desiccant materials for a total heat exchanger

outdoor air, resulting in an increase in adsorption.

In the case of ammonium, the specific transfer rate R increases with increasing face velocity V in silica gel and alumina types as in case of water. This is due to the adsorption together with water by a strong affinity to water. On the other hand, it is independent of the face velocity and a constant amount of NH_3 is transferred in zeolite 3A and ion exchange resin types. Ammonium molecules hardly enter ion exchange resin as described previously and the specific transfer R rate is kept at a low value.

The specific transfer rate R of IPA is found to be independent of the face velocity both in silica gel and alumina types. IPA is adsorbed on silica gel and alumina but the rotor reaches the adsorption breakthrough very early because of a low level of equilibrium amount adsorbed. Thus the specific transfer rate of IPA does not change with increase or decrease in the face velocity.

In the case of toluene, the specific rate decreases with increasing face velocity unlike other cases. This unexpected result may not be explained well although toluene once adsorbed may be desorbed by the replacement adsorption with water vapor of which adsorption was enhanced by increased face velocity.

CONCLUSIONS

The mechanical ventilation is important in a recent trend of an air-tight residences and offices and total heat exchanges have been recognized to be very efficient from an energy saving point of view. The most serous problem to be solved has been the odor transfer in which odor substances emitted inside the room is returned back with the supplied air by transferring from the returned-exhaust stream to the outdoor-supplied stream. The authors developed a new "Ion sorption type" of total heat exchangers by use of ion exchange resin as desiccant. The heat exchange efficiency of it is equivalent to the conventional ones and the accumulation and transfer of odor is reduced to much less extent. Mechanism of the odor transfer is discussed by comparing odor transfer rate in the total heat exchanger between porous adsorbent such as silica gel and ion exchange resin as desiccant.

REFERENCES

1) H. Okano, H. Tanaka, T. Hirose, H. Funato and S. Ishihara, " A Novel Total Heat Exchanger with Little Odor Transfer Using Ion Exchange Resin as a Desiccant," ASHRAE Transactions, Vol.107, Part 2, p.864(2001).

2) "DIAION: Manual of Ion Exchange Resins and Synthetic Adsorbent", Vol. 1, Published by Mitsubishi Chemical Corporation, 1995, Tokyo, Japan

719aj ADSORPTION and Diffusion in Flexible Metal Organic Framework Materials : MOLECULAR Simulations Compared to Experiments

G. Maurin[1], A. Ghoufi[1], N. Rosenbach[1], F. Salles[1], H. Jobic[2], S. Bourrelly[3], P. Llewellyn[3], Thomas Devic[4], Christian Serre[4], and Gérard Férey[4]. (1) Université Montpellier 2, Montpellier, France, (2) Institut de Recherches sur la Catalyse et l'Environnement, Villeurbanne, France, (3) Université de Provence, Marseille, France, (4) Institut Lavoisier, Versailles, France

MOF materials have recently emerged as viable candidate adsorbents for a number of different gases that are implicated in industrial processes. In this work, we present an exploration of the adsorption mechanism of CO2 and different alkanes (from C1 to C4) in the flexible MIL-53 (Cr) and its isostructural rigid form MIL-47(V). It is the first attempt to follow an "hybrid" route based on the combination of Grand Canonical Monte Carlo and Molecular Dynamics simulations to capture the breathing of a MOF material along the adsorption process. This strategy relies on newly derived force field which successfully reproduces the structural switching between the narrow and the large pore versions of MIL-53(Cr) during the breathing phénomenom. We thus predict the adsorption isotherms and enthalpies of adsorption and probe the possible adsorption sites of the different adsorbates in both MIL materials. These results are then compared to those obtained using our preliminary approach which consisted of building "composite" isotherms and enthalpies from the contribution of both the narrow and large pore version of MIL-53(Cr)1. We will also report the adsorption behaviours for different CO2/ alkane mixtures. All these simulations are then favorably compared to Microcalorimetry and in situ X-ray diffraction data. The diffusion of H2, CH4 and CO2 in both flexible MIL-53(Cr) and rigid MIL-47 (V) systems has been also successfully characterized by combining QENS and Molecular Dynamics (MD) in MIL-53(Cr) and MIL-47 (V). Different loadings and temperatures were studied. In MIL-53 (Cr), a 1D diffusion model for H2 was found to fit better the experimental spectra than the ordinary 3D one, leading to the observation for the first time of a 1D "supermobility" of H2 along the tunnel of the MIL materials with a self diffusivity Ds around 2.10-7 m2.s-1 at 77K2. Our simulations based on a small reparameterization of the existing forcefield for the H2/MIL framework, were in excellent agreement for both Ds as well as the activation energies. The Ds for CH4 was also successfully compared between experiment and simulation in the whole range of loading. Finally, the transport diffusivity Dt of CO2 in both large and narrow pore structures of MIL-53(Cr) was extracted from QENS and long MD runs since this molecule is a purely coherent scatterer.

Medicinal Herb Extraction Strategy – A Solvent Selection and Extraction Method Study

Xun Yan[1], **Jatinder Rana**[1], **Amitabh Chandra**[1], **Doug Vredeveld**[1], **Haywood Ware**[1], **John Rebhun**[1], **Timothy Mulder**[1], **Kathryn Persons**[1], **Dorothy Zemaitis**[1] and **Yingqin Li**[2], (1)Amway Corp., 7575 Fulton St E, Ada, MI 49301, (2)Nutrilite - Access Business Group, Buena Park, CA 90622

Abstract

Medicinal herbs have a long history of use in improving human health and curing diseases. Herbal materials are naturally an excellent medicine library because they provide unlimited components with complex chemical structures and a wide range of bioactivities. Herbal materials have thus attracted wide interest of researchers in the identification of the active components and verification of their efficacy. Herb preparation and extraction is the first step and in fact one of the most important steps in the chemical identification process. This step makes the down stream processes possible, including bioactive component identification, isolation and characterization.

In this research, using a model herb material, Echinacea, we studied chemical extractability of different solvents with various polarities. Several extraction methods were used and their extraction efficiencies were compared. The extraction methods used in this research were soxhlet extraction (SE), sonication assisted extraction (SAE), and accelerated solvent extraction (ASE). Chemical profile and yield of each extract and the throughput capability of different extraction methods were systematically compared and discussed. Based on the results, a recommendation was made for a consolidated herb extraction-fractionation process.

Introduction

Evidence based activity is the key for the development of modern herbal products. Herb preparation and extraction is the first step and in fact one of the most important steps in the bioactivity and chemical identification process. This step makes the down stream processes possible, including bioactive component identification, isolation and characterization. There are several extraction techniques that are commonly used, including decoction, steam distillation, solvent extraction (soaking and leaching), soxhlet percolation, ultrasound assisted extraction (UAE), microwave assisted extraction (MAE) pressurized solvent extraction (PSE), and the combinations of two or more techniques mentioned above.[1~4]

Considering operational cost and scale-up possibilities, soxhlet extraction (SE), sonication assisted extraction (SAE), and accelerated solvent extraction (ASE, also known as PSE) were selected for comparison in this study. Both sequential and parallel extraction techniques were used in this study. Chemical profiles of each solvent extract, extraction yield and the throughput capability of the different methods were the focus of this study.

Solvents used in herb extraction need to be volatile and leave no residue when dried. This solvent property will make the downstream processing easier, including solvent exchange and bioassay. Hexane, chloroform, ethyl acetate, methanol or ethanol are thus proper candidates and are widely used in herb processing. In this study water and DMSO were also selected due to their popularity in animal studies and bioassay.

Materials and Methods

Materials

Echinacea angustifolia was obtained from Nutrilite, CA. Solvents used for extractions, including hexane, chloroform, ethyl acetate, methanol, reagent alcohol (a mixture of ethanol, methanol and isopropanol), DMSO and water are of HPLC grade. DMSO, reagent alcohol and water were also used as a mixture at 5/3/2 by volume, so called DEW.

ASE Extraction method and procedure

In a "sequential extraction" technique, the residue of the previous extraction is used as the feed for the next extraction. The extraction temperature was 70 °C for hexane, chloroform, and ethyl acetate, 85 °C for methanol, 100 °C for DMSO/methanol/water, and 120 °C for water. The operation pressure was also monitored.

For each extraction step (each solvent), there was a static extraction stage (soaking) and a dynamic extraction stage (flowing through). The static extraction time is about 20 minutes, and the dynamic extraction time is also 20 minutes at a flow rate of 6 Bed Volume/hour. There were two static/dynamic extraction cycles in each solvent extraction step. Before switching to the next solvent, the sample bed was purged with air to avoid carryovers and contaminations. The purged liquid was combined with the extracts.

Soxhlet extraction and sonication assisted extraction (SAE)

Powdered botanicals were extracted at a solvent to mass ratio of 10/1 (v/w). Both sequential and parallel techniques were used in the soxhlet extraction. In SAE process, an individual sample was directly extracted by first soaking it overnight while stirring and followed by sonication for 1 h at room temp. The extract was then filtered out.

Yield measurement

The extract solutions of hexane, chloroform, ethyl acetate, and methanol were vacuum dried with rotary evaporator. The final extract weight was measured in a pre-weighed flask.

HPLC analysis

Two different HPLC conditions were used to analyze the extract samples. The first condition used a Luna PFP column (3 um, 150 mm x 4.6 mm) with a 0.2% phosphoric acid/acetonitrile gradient starting at 5/1. The second condition used a Synergi MAX-RP column (4 um, 150 mm x 4.6 mm) with a water/acetonitrile gradient starting at 6/4.

Experimental Results

Four extraction methods were used in this study, including soxhlet, sonication assisted extraction, and ASE. The extracts were then dried by rotary evaporation. Extraction solvents were arranged in sequential or parallel scheme. The dry mass was measured and the extraction yield was calculated. The results were recorded in Table 1. In this study, yield is defined as ratio of extract dry mass to raw material mass.

Table 1. Extraction yield of different techniques using different solvents. (mg extract /g raw material)

Solvents\Technologies	Soxhlet -sequential	Soxhlet -parallel	SAE -parallel	ASE -sequential
Hexanes	3.26	19.5	10.6	15.6
Chloroform	13.27	31.5	18.8	8.7
Ethyl Acetate	1.94	32.7	17.6	2.6
Methonal	84.8	197.5	115	124.9
DEW(DMSO/ROH/water)	-	-	190	125.4
Water	-	-	-	144.6

Chromatographic profiles

The dried extracts were re-suspended in methanol at a nominal concentration of 5mg/mL. The clear solution part was used for HPLC analysis. The HPLC chromatograms of extracts generated using different technologies were recorded in Figure 1, and 2. Only the chromatograms of chloroform and methanol extracts were shown. When the same solvent was used, the chromatograms were very similar, in spite of the different processes or extraction methods.

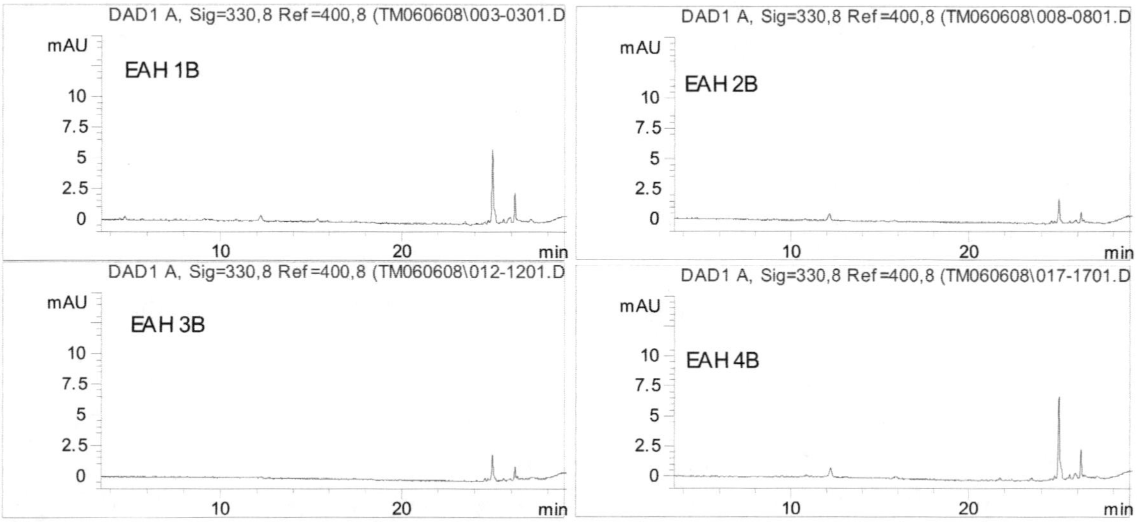

Figure 1. HPLC profiles of chloroform extract. The extraction methods used were, EAH1B, soxhlet-sequential; EAH2B, soxhlet-parallel; EAH3B, vortex-sonication; EAH4B, ASE.

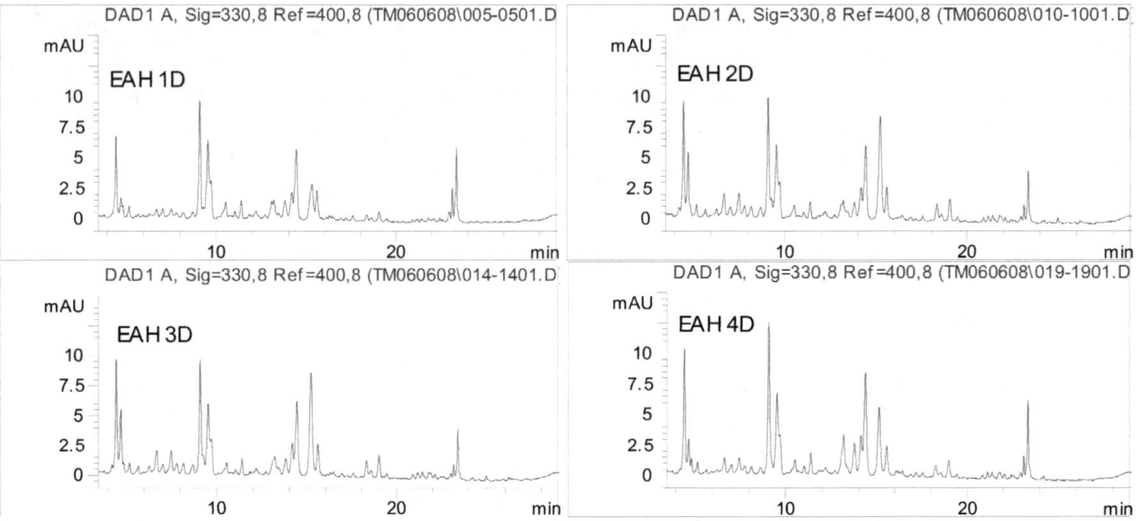

Figure 2. HPLC profiles of methanol extracts. The extraction methods used were, EAH1D, soxhlet-sequential; EAH2D, soxhlet-parallel; EAH3D, vortex-sonication; EAH4D, ASE.

In Figure 3, the chromatograms of ASE extracts with methanol, DEW (DMSO/ROH/water) and water extracts are compared.

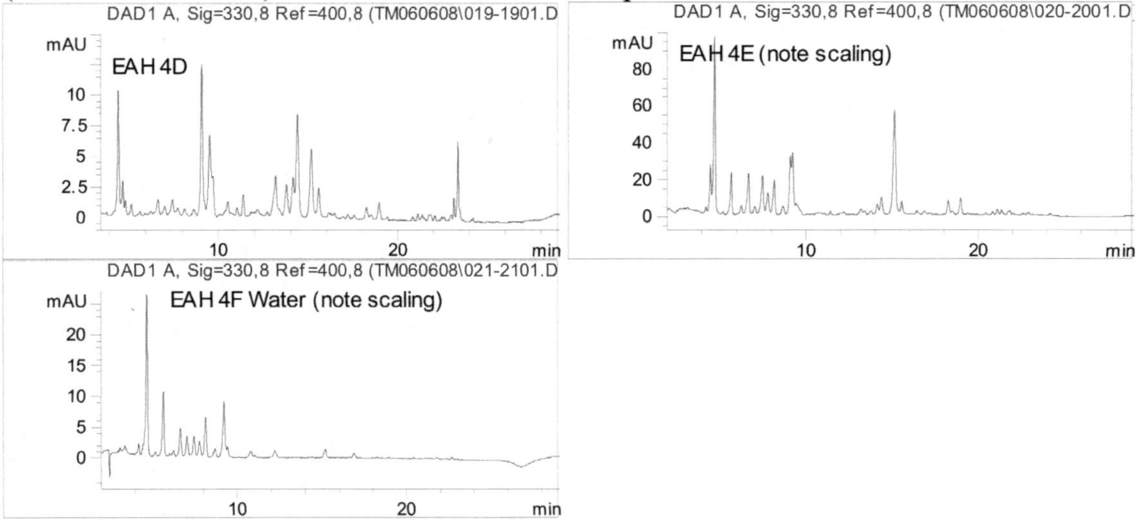

Figure 3. HPLC profiles of ASE extracts using methanol, EAH4D; DMSO/ROH/Water, EAH4E; and water, EAH4F.

Discussion and Conclusion

From the results, both yields and chromatogram profiles of the extracts from the same extraction solvent are fairly close, in spite of the extraction methods. Thus it can be concluded that the extraction techniques are not essential to the chemical extractability or the quality of the extract. The solvents used in the process have far more critical roles on the two extract qualities.

The selection of an extraction technique will be depending on operational requirements. These factors include, but not limited to, instrument cost, labor cost, operational cost, consumable cost. The extraction time is another factor that directly related to the throughput of the method, which is a key consideration in certain laboratory practice. In table 2, some non-technical factors are compared for ASE, SAE and soxhlet.

Table 2 Comparison of several factors that affects the performance of the technologies.

Items\Technologies	ASE	SAE	Soxhlet-parallel
Extraction time	short	long	long
Operation Labor	low (automated)	mid	mid
Operational Cost	mid	low	high
Instrument Cost, $	50,000	1,500	300

Acknowledgements

We are grateful to Dionex scientists for coordinating and carrying out the ASE extractions on Echinacea.

References

1. E. S. Ong, Extraction methods and chemical standardization of botanicals and herbal preparations, J. Chromatography, B, 2004 23-33
2. D. A. Starmans and H.H. Nijhuis, Extraction of secondary metabolites from plant material: A review, Trends in Food Sci and Tech, 1996, 191-197
3. M. Toma et al. Investigation of the effects of ultrasound on vegetal tissues during solvent extraction, Ultrason. Sonochem. 2001, 137-142
4. L. Yang et al. Microwave-assisted extraction technology, a review, J. Guangzhou U., 2004 , 519-522

OPTIMIZATION OF THE EXTRACTION PROCESS OF BETA-GLUCANS FROM BARLEY

Oscar Benito, Esther Alonso and Susana Lucas
Chemical Engineering and Environmental Technology Department.
Prado de la Magdalena s/n. Facultad de Ciencias. Universidad de Valladolid (Spain)

Abstract

β-glucans are non-starchy polysaccharides composed of (1-3), (1-4) mixed linked glucose polymers. They are located mainly in the cell walls of the endosperm and the aleurone layer of different cereal grains, specially oat and barleys. In general, β-glucans are a major component of the soluble dietary fibre and they influence the nutritional values and functional properties of food. Recently, several studies have revealed the benefits of including β-glucans into the diet to the human health, such as cholesterol diminution in blood, and reduction of risk of cancer, coronary heart disease, and diabetes. Although there are several papers related to the extraction and purification of β-glucans from cereals, most of them describe extraction and purification methods at a laboratory scale, which involve a significant number and time-consuming steps. Moreover, since the β-glucan content in barley grains is low (<5%) it is necessary to develop and optimise an extraction/purification method that allow reducing operation times and reactive consumption, to make competitive this product at a big scale.

The aim of our research is focused on three different steps: (1) Raw material pre-treatment, (2) Extraction step, and (3) Purification step. Discussion of experimental results is made in terms of β-glucan and starch content in the extracts, and molecular weight of the extracted β-glucans.

1) Several alternatives have been studied for the pre-treatment of the barley flour: ultrasonic treatment with different energy powers, enzyme deactivation and microwaves.
2) A systematic study of the extraction variables such as: temperature, pH, extraction time, solvent ratio and solvent polarity have been performed in order to optimize the operational extraction conditions. Extraction with subcritical water has also been studied.
3) Two different techniques have been studied and compared for the precipitation step: precipitation with alcoholic antisolvents, and precipitation with supercritical CO_2.

In this work the effect of operating parameters on the extraction of β-glucans from different varieties of barley is analyzed. Several experiences have been carried out in batch extractors, with a reference experiment, in order to check the influence of temperature, pH, time, solvent and flour:solvent ratio. Time and temperature showed to have a strongly marked influence: the higher temperature, the higher extraction rate. Higher temperatures led to improve the extraction yield, increasing the amount of starch co-extracted. Extraction time was a critical factor up to 3 hours. Longer times did not increase the amount extracted. The influence of pH was not very remarkable, just generating a slight increase in extraction yield up to 8; beginning to decrease from that value. The use of alcohols (methanol and ethanol) as solvents, did not improve the extraction rate. On the other hand, it was observed an important decrease on the β-glucan solubility, greater the higher concentration of alcohol employed and the longer carbon chain. Finally, a extraction yield of 40-50% was obtained when operating at solvent:flour ratio from 7 to 12, at 55°C, in almost all range of pH, when lasting the extraction process more than 2,5 hours.

1. Introduction

β-glucans are a kind of non-starchy polysaccharides (formed by the linkage of glucose units via β-(1→3) (approximately 30%) and β-(1→4)-glucosidic linkages) that can be found in several kinds of cereals, such as barley, oat or rye in concentrations from 2 to 12% in dry basis. In barley, β-glucans are minor components (up to 6-7%), located majoritary in the aleurone and endosperm cellwalls of the grain [1]. They form part of the dietetic fibre (the same as other structural compounds of vegetals, such as cellulose and hemicelluloses, nevertheless β-glucans are soluble in water.

It is known that they have good properties for human health, if eaten. Among all these properties it is worth mentioning their implication in process of control of cholesterol and glucose concentration in blood [2,3,4]. These effects have been associated with the molecular weight of the polymers. Recently, FDA (Food and Drug Administration, USA) has included β-glucan in its list of products that contribute to lower the cholesterol level in blood, pointing out how to label the products that contain barley to remark its positive effect on health.

Extraction of β-glucans form cereals has to deal with several difficulties. A typical extraction process involves, at least, three stages [5,6]: inactivation of endogenous enzymes, extraction of β-glucans with a suitable solvent, and finally a purification-isolation stage, because other chemicals present in cereals (proteins, saccharides, or starch) can be co-extracted. All these stages increase the complexity of the overall process, increasing the economical costs, therefore. This is the most limiting factor for the extraction process.

When a solid-liquid extraction process is carried out, temperature, pH, extraction time and polarity of the solvent are some of the operating parameters that are to be taken into account. It is important to characterise the behaviour of the target compound when changing the operating variables, in order to improve the extraction process.

In this work a simple solid-liquid discontinuous extraction process from different waxy-type barley (H13-06 (hull-less) and D24-06) with hot water was proposed. The effect of most important operating parameters such as temperature, pH, extraction time and polarity of the solvent was checked in order to optimize the continuos extraction process. Moreover, a reference experience based on literature revision carried out at 55°C, 3 hours, water as solvent, solid:liquid ratio 1:10 was chosen as a comparative tool [7].

2. Experimental section

Two different kinds of barley were used, both of them grown and harvested in Castilla y León (Spain): H13-06 and D24-06. These varieties correspond to the waxy-type barley, a specie of barley which has higher content in β–glucan and a lower in starch than common species. The main difference between those two kinds of barley is that grains of H13 are hulless, not presenting external hull covering the kernel. Barley was supplied by Instituto Tecnológico Agrario de Castilla y León (ITACYL). Table 1 summarizes the composition of both varieties:

Table 1. Chemical composition of the barley used as raw material

	H13-2006 (%)	D24-2006 (%)
Moisture	11,07	10,98
Fats	2,26	2,48
Ash	1,78	1,86
Proteines	4,9	4,3
Starch	52	44,8
Amilose/Amilopectine	10,22/89,78	9,05/90,95
β-Glucan	4,13	5,33
Dietic Fibre	13,85	18,7

Barley employed was milled and sieved through a 0,5mm sieve.
Initially, temperature, pH, time, solvent, flour:solvent ratio were defined as the factors that had to be tested. According to the literature, basic conditions (of temperature) for the initial experiments were selected: 55ºC, 3 hours, water as solvent, solid:liquid ratio 1:10 (these conditions constitue the basic experiment). In this reference experiment, pH of water was not modified, being around 6.
Barley flour was weighted and added to an erlenmeyer flask. Then water was added, being the flour suspended at high stirring rate. Temperature was kept constant in a water bath (Lauda Ecoline Staredition E100, Lauda, Germany). Stirring was obtained with a multistirrer plate (Fisher Scientific, USA), as shown in figure 1:

Once the extraction process was finished, solid:liquid separation was done by centrifugation (5500rpm, 10 minutes). Solid material was discarded and liquid (supernatant) was stored at 4ºC until the moment of being analysed.
All the experiments were conducted in duplicate.

Figure 1. Experimental set-up used for the extraction of β-glucans from barley

β-glucan determination was made at ITACYL, using the Megazyme Assay Kit for mixed linkage β-glucan analysis (*Megazyme*, Ireland). β-glucan concentration is expressed as % (weight) of the total sample.

Yield results shown in the following figures was calculated by dividing the amount of β–glucan in the extract generated and the total amount of β-glucans in the barley flour used in the extraction process.

3. Results

β-glucan extraction yield was 41,5±1,0% for D24 and 47,2±2,4% for H13 for the reference operating conditions This result, contrary to that expected, can be explained for the presence of interferences. H13 is a hulless barley, with a lower content in β-glucans. D24, despite of its higher content in this compound, is more difficult to extract them, probably due to the presence of the hull of the kernel. This problem can be solved by milling the barley in a industrial mill, that generates three milling fractions, with different compositions of β-glucans.
When the extraction process was divided in three stages (of one hour each), the amount of β-glucan extracted was around 60%, a 20% higher than when developing the process in one single stage.

Several experiments were carried out in order to check the influence of temperature, pH, time, solvent and flour:solvent ratio on β-glucans extraction, changing the study variable and keeping constant the remaining ones. The reference experiment was achieved at 55°C, pH around neutrality (6), 3 hours of extraction time and with 1:10 solid:solvent ratio.

Temperature effect (35-75°C): It was observed a costant increase in the extraction yield when increasing the temperature from 29% at 35°C up to 56% at 75°C. However when temperature is higher than 55°C, operational problems connected with the higher suspension viscosity and the co-extraction of starch in the extract were observed (the amount of starch in the extract was increased dramatically, from 0,12% to 0,71%, which makes more difficult the subsequent purification [8]. For these reasons, the optimal extraction temperature was fixed in 55°C (see Figure 2).

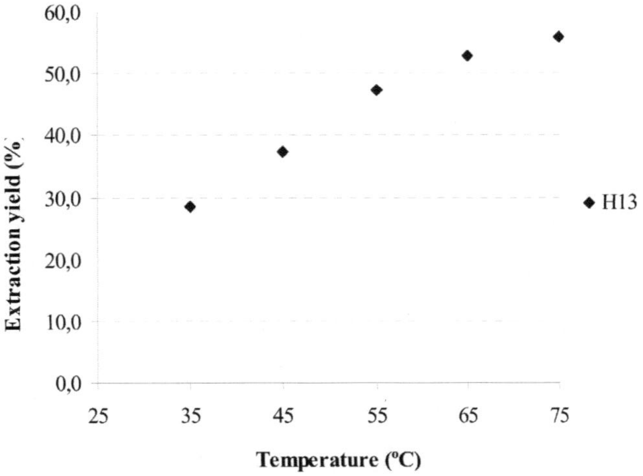

Figure 2. Temperature effect on the β-glucan extraction yield (H13 barley).

Extraction time effect (0.5-5 h). Times under 2,5 hours were found to be too short (yield fell down under 40%), but an extraction time of 5 hours did not entail a dramatic increase in the amount of β-glucan extracted, rather it kept constant (45-47%) after three hours of extraction. Extraction time of 3 hours is enough to ensure the equilibrium was reached and to get a high extraction yield (see Figure 3).

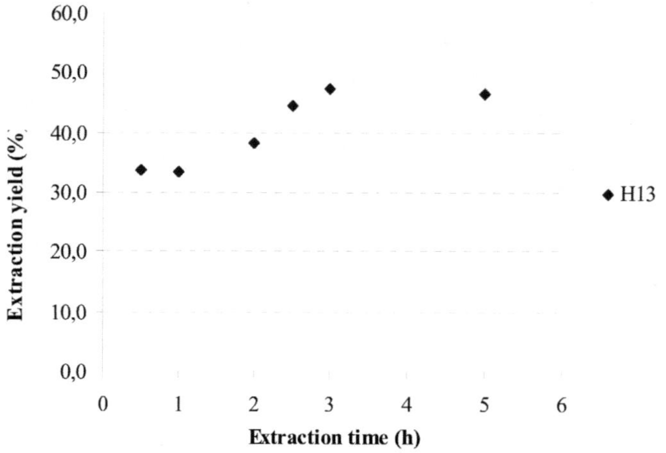

Figure 3. Effect of extraction time on the β-glucan extraction yield (H13 barley).

When the influence of pH (in the range 6-11) was studied (see Figure 4), a slight increase in extraction yield was produced. Water was weakly basified (pH 8) by addition of some drops of NaOH (6N). However, a further increase (pH 11), decreased the amount of β-glucans extracted. This decrease with pH was more marked in the case of H13 type. This could be due to the depolymerisation of our targed compounds. This result reveals that it is not necessary to increase the pH of the original water used as solvent (close to neutral pH).

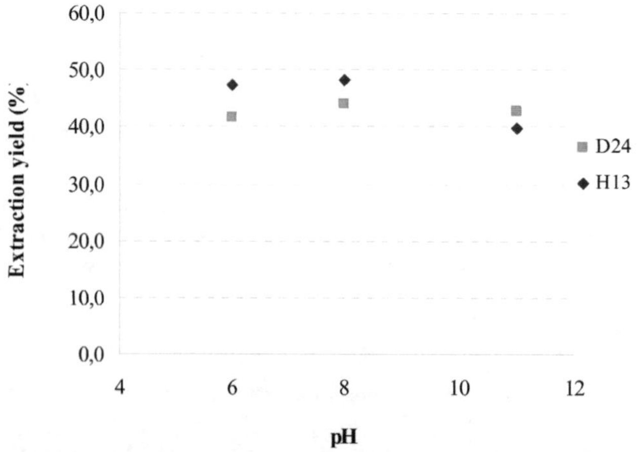

Figure 4. pH effect on β-glucan extraction yield (H13 and D24 varieties of barley).

Solid (flour):liquid ratio (in the range 5-15) was also tested for H13 variety. As it can be observed in Figure 5, an important change in extraction yield (over 50%) could be noticed when this ratio was shift from 5 to 10. The increase of this ratio to 15, did not produce a positive effect in the extraction of β-glucans. When using ratios of 8, 10 or even 12, it was observed a constant yield, around 47%. This result showed that it is possible to operate at lower solid:liquid rate (at 8), which is important taking into account a further purification step. Using less amount of solvent means simplifying the concentration of β-glucans.

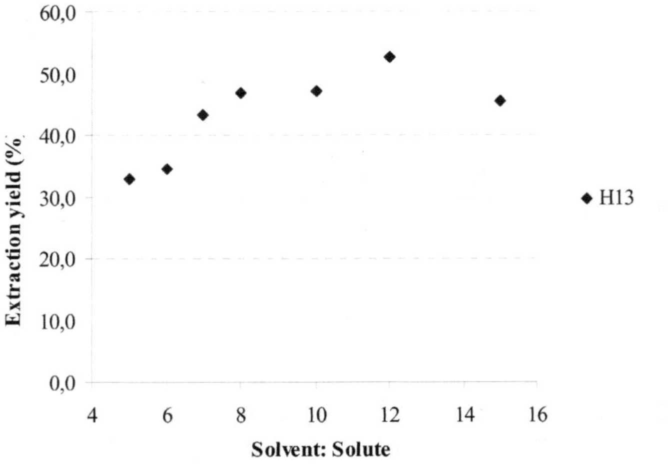

Figure 5. Effect of solvent:flour ratio on β-glucan extraction yield (H13 barley).

Polarity of the solvent can be classified as a critical parameter. This part of the study was developed using the D24 variety. The better results were obtained with water as solvent (yield was 41,5%). When an alcohol was used as solvent (5% in volume), the extraction yield decrease slightly to 41,4% in the case of the use of methanol and to 34,1% when ethanol was used. When the concentration of alcohol was increased (50% or 100%), it was not possible to extract β-glucans. The amount of polymer extracted in those cases was under the detection limit of the analysis method. According to this result, alcohols can not be used as co-solvents. This property would allow to the alcohols to be used to precipitate β-glucans, using the alcohols as anti-solvent.

4. Conclusions

Initial conditions fixed for discontinuous extraction of β-glucans from barley with water (55°C, 3h and flour:solvent ratio of 8) are suitable to get extraction yields in the range 40-50% for both varieties of barley tested. Time and temperature showed to have a strongly marked influence: the higher temperature, the higher extraction rate. Longer times than 3 hours didn't increase the amount of β-glucans extracted. The influence of pH was not very remarkable, just generating a slight increase in extraction yield up to 8; beginning to decrease from that value. The use of alcohols (methanol and ethanol) as solvents, did not improve the extraction rate. The operation in 3 steps of one hour each one, allowed to increase the maximum amount of β-glucan up to 60%. This result shows that β-glucan extraction can be enhanced when increasing the number of extraction steps, which justifies the need to develop a continuos extraction process.

5. References

[1] Irakli M., Biliaderis C.G., Izydorczyk M.S., Papadoyannis I.O. (2004). Isolation, structural features amd rheological properties of water-extractable beta-glucans from different Greek barley cultivars. *Journal of the science of food and agriculture*, 84, 1170-1178.

[2] McIntosh, G.H., Newman, R.K., Newman, C.W. (1995). Barley foods and their influence on cholesterol metabolism plants. *Human Nutrition*, 77, 89-108.

[3] Bhatty R.S. (1999). The potential of hull-less barley. *Cereal Chemistry*, 76(5), 589-599.

[4] Brennan, C.S.; Cleary, L.J. (2005). The potential use of cereal (1→3,1→4)-β-d-glucans as functional food ingredients. *Journal of Cereal Science*, 42(1), 1-13.

[5] Bhatty R.S. (1995). Laboratory and pilot plant extraction and purification of β-Glucans from Hull-less Barley and Oat Brans. Journal *of Cereal Science*, 22, 163-170.

[6] Laroche, C.; Michaud, P. (2007). New developments and prospective applications for beta-(1,3) glucans. *Recent Patents on Biotechnology*, 1, 59-73.

[7] Papageorgiou M, Lakhdara N, Lazaridou A, Biliaderis CG, Izydorczyk MS. (2005). Water extractable (1 → 3,1 → 4)-β-D-glucans from barley and oats: An intervarietal study on their structural features and rheological behaviour. *Journal of Cereal Science*, 42(2), 213-224.

[8] Temelli, F. (1997). Extraction and functional properties of barley β-glucan as affected by temperature and pH. *Journal of Food Science*, 62(6), 1194-1197.

Sugaring-out: a Novel Phase Separation and Extraction System

Bin Wang[a,b], Thaddeus Ezejias[c], Hao Feng[a,b]*, and Hans Blaschek[b,d]

[a] Energy Biosciences Institute, University of Illinois at Urbana-Champaign, Urbana, IL 61801
[b] Department of Food Science and Human Nutrition; University of Illinois at Urbana-Champaign 382F-AESB, 1304 W Pennsylvania Ave, Urbana, IL 61801
[c] Department of Animal Sciences, Ohio State University, OH 43210, USA
[d] Center for Advanced BioEnergy Research, University of Illinois at Urbana-Champaign, Urban, IL 61801, USA

* Corresponding author
Phone: 011-217-244-2571; Fax: 011-217-333-9329; Email: haofeng@uiuc.edu

Keywords: *Sugaring-out; acetonitrile; separation; extraction; phase equilibrium*

The addition of a mass-separating agent (MSA) into a homogeneous solution to create a second immiscible phase has been investigated as an alternative to the traditional liquid-liquid extraction method for the separation of chemical species and recovery of biological products. The concepts that have been exploited include "salting-out" and the aqueous two-phase separation method.

"Salting-out" refers to the separation of a miscible organic compound from its aqueous solution and the formation of a two-phase system caused by addition of a salt as a MSA (Frankforter and Temple, 1915; Kobe and Stone, 1940). The salting-out is attributed to the decreased solubility of organic solvents in water in the presence of electrolytes. Salting-out usually happens at high salt concentrations, which may induce some unwanted chemical reactions during extraction (Jones et al., 1993) and may not be good for the separation of labile biological components. In aqueous two-phase systems (ATPS), the phase separation is triggered by the addition of different polymers or a polymer and a salt into the aqueous solution. In comparison with the salting-out counterpart, the both phases in ATPS are water-based with water contents as high as 85-99%. The major limitations that have hampered the widespread industrial application of ATPS are the high costs associated with the consumption of polymers, the difficulties in predicting the partitioning behavior, and the lack of effective phase-recycling processes (Zijlstra et al., 1998).

Dr. Hao Feng's lab in University of Illinois recently observed that acetonitrile (ACN) that is miscible with water in all proportions can form a two-phase system with water when glucose, xylose, arabinose, fructose, maltose, sucrose or their mixtures was added into the ACN-water solution (Wang et al. 2008). This new phenomena was named "sugaring-out". The phase separation due to sugaring-out is shown in Figure 1, in which Sudan I was added (Fig 1-b) to visualize the phase formation in the colorless ACN and water mixture (Fig 1-a) when 50 g/l glucose or 50 g/l xylose was added. Only 20 seconds after the addition of the sugars, the mixtures were separated into two phases with the dye concentrated in the upper ACN-rich phase (Fig 1-d).

The ACN concentration in the upper phase increased when more sugar was added into the ACN-water mixture, and can reach up to 90% (w/v). The lower concentration limit for the two-phase formation for glucose and xylose at 1°C was 15 g/L and 25 g/L, respectively. At higher temperatures, the concentration needed for phase separation increased. Addition of polysaccharides (starch and dextran) did not result in phase separation.

Figure 1. Sugaring-out phase separation process: ***a.*** Colorless ACN-water (v/v, 1:1) mixture at pH 6.5 and 1°C in both A and B; ***b.*** 0.05 g/L Sudan I was added into the mixture to visualize the subsequent two-phase formation; ***c.*** 5 seconds after 50 g/L glucose (A) and xylose (B) were added; ***d.*** 20 seconds later; ***e.*** 1 minute later; ***f.*** 2 hours later.

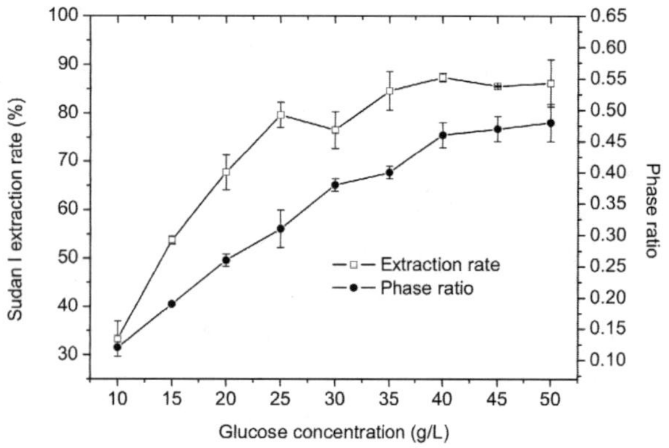

Figure 2. Sudan I extraction by a sugaring-out two-phase system (ACN: water, 1:1 v/v), equilibrated for 5 minutes at 1 °C.

In the Sudan I extraction test with a glucose-triggered ACN-water two-phase system, an extraction rate of > 80% was achieved when the glucose concentration was greater than 25 g/L (Figure 2). The distribution of syringic acid, furfural, para-coumaric acid, ferulic acid and 5-Hydroxymethyl furfural in the upper ACN phase and lower water phase was also examined (results not shown). The distribution coefficient of the five organic compounds in the ACN-water two-phase system was in the range of 1.7-8.9 when the corresponding sugar concentration was 15-50 g/L. The phase ratio of the five organic compounds in the two-phase system was in the range of 0.1-0.5.

The formation of an immiscible ACN phase by the addition of a monomeric sugar or disaccharide as a MSA into an ACN–water solution is a new phenomenon that has not been reported. Uncharged but polar biomolecules such as sugars dissolve readily in water because of the stabilizing effect of hydrogen bonds between the hydroxyl groups or the carbonyl oxygen of the sugars and the polar water molecules. This is also true for compounds containing N-H bonds like ACN (Lehninger et. al. 1993). It is likely that when adding sugars into an ACN and water mixture, some original hydrogen bonds in the mixture are replaced by hydrogen bonds formed between sugar and water molecules, which may force ACN molecules to separate from water molecules and form a new phase. As a result, the ACN volume increased when more sugars were added into the ACN-water mixture. Since ACN is an excellent solvent for both inorganic and organic compounds due to its high dielectric strength and dipole moment (Kroschwitz and Howe-Grant, 1998), it is foreseeable that such an ACN–water two-phase system might find applications in the extraction of compounds such as metal-chelates, ion-pairs, and organic materials, or those that cannot be effectively extracted by conventional liquid-liquid extraction methods.

Acknowledgments

The project was supported by the National Research Initiative of the USDA Cooperative State Research, Education and Extension Service, Grant no. 2006-35504-17419. We are grateful to Dr. Keith Cadwallader and Mr. John Jerrell for their assistance in HPLC analysis. The authors also thank Ms. Lilia Fransisca and Ms. Eun Young Lee for their laboratory assistance during experimentation.

References

Lehninger, A.L., Nelson, D.L., Cox, M.M. Water: its effect on dissolved biomolecules, in: Principle of Biochemistry, 2nd Ed., Worth Publishers Inc. New York, 1993: pp. 81-107.

Frankforter, G.B., Temple, S., (1915). Equilibria in the systems of the higher alcohols, water and salts. Journal of the American Chemical Society, 37, 2697 – 2716.

Jones, L.A., Prabel, J.B., Glennon, J.J., Copeland, M.F., Kavlock, R.J., (1993). Extraction of phenol and its metabolites from aqueous solution. Journal of Agricultural and Food Chemistry, 41, 735-741.

Kroschwitz, J.I., Howe-Grant, M., (1998). Encyclopedia of chemical technology: supplement volume: aerogels to xylylene polymers. Wiley, New York, pp. 437-444.

Wang, B., Ezeji, T., Feng, H., Blaschek, H. 2008. Sugaring-out: A novel phase separation and extraction system. Chemical Engineering Science. 63: 2595– 2600.

Zijlstra, G.M., Gooijer, C.D.D., Tramper, J., (1998) Extractive bioconversions in aqueous two-phase systems. Biochemical Engineering Journal, 9, 191-176.

An Innovative Approach on the Separation of Metal

Yu-Lung Lam, Die Yang, Chi-Yuet Chan, Kwong-Yu Chan, Patrick H. Toy*

Chong Yuet Ming Chemistry Building, The University of Hong Kong, Pokfulam Road, HKSAR

Introduction

There has been increasing environmental concerns on the metallic pollution in recent decades. The conventional technologies used for metal extraction are ion-exchange resins, polymer ultrafiltration and microemulsion. The whole extraction mechanism can be divided into 4 steps: 1) The binding of dissolved metal to active sites, 2) Separation of the metal load extractant from the lean phase, 3) Reverse extraction of loaded metal for recovery 4) Recycling of extractant for repeated extraction. Each of the conventional methods suffers different drawbacks in different steps of the extraction mechanism.

In addition to the above technologies, metal extraction using aqueous biphasic system (ABS) has been reported and investigated. Water-soluble polymers such as polyethylene glycol (PEG) were commonly used in ABS separation. ABS could be applied in separation of various solutes such as biomolecules and particulates. ABS has all advantages of liquid-liquid extraction due to their aqueous properties. Furthermore, polyethylene glycol-based aqueous biphasic systems (PEG-ABS) are virtually nontoxic, nonflammable and inexpensive. However, in ABS, usually a large quantity of salt such as sodium sulfate is required to achieve phase separation. Though most of the salts are inexpensive, the significant amount required salts inevitably introduce additional costs, further post-treatment and corrosion problems.

With respect to the problems mentioned above, a solid phase polymer of PEG with polystyrene (PS) core was firstly used for metal extraction in this work. PS core provides hydrophobic backbone while the PEG has hydrophilic properties. In fact, this solid phase polymer was commonly applied in the biological research for biomolecules such as proteins.

In addition, polyglycidol (PG), a polymer with similar structure with PEG, was synthesized and immobilized on polystyrene (PS) surface. This resin is called PS-PG resin and it is water-compatible and partially soluble in aqueous solution. Because of the PS core, the resin can be readily separated by sedimentation or centrifugation without the need of salt addition.

The synthesized PS-PG resin was tested for precious metal extraction such as gold (I) and silver (I) ions. Reversely extraction of metal ions from the resin was also investigated. Furthermore, since thiol have strong affinity to gold, the extraction capability of PS-PG resin was strengthened by coupling to thiol groups and the resin become thiol-terminated PG (PS-PG-SH). Magnetic aqueous biphasic system was reported for biomolecules and increase in the rate of separation was achieved. Using methodologies well known in the art, magnetite core coated with the PS/PG resin was also synthesized for extraction in order to promote higher separation efficiency.

Results and Discussion

The obtained results indicated that gold can be extracted from the aqueous phase to the polymer phase. The determination of gold in aqueous phase after extraction showed the percentage of gold extracted to the polymer phase was higher than 99%. For silver extraction, the percentage of silver extracted to the polymer phase was 99%.

In fact, the silver solution had to be treated with diluted hydrochloric acid to remove free cyanide anions. Since $[Au(CN)_2]^-$ is the only complex between gold and cyanide anions while $[Ag(CN)_2]^-$, $[Ag(CN)_3]^{2-}$ and $[Ag(CN)_4]^{3-}$ were all been found. $[Ag(CN)_3]^{2-}$ and $[Ag(CN)_4]^{3-}$ may interfere the complex formation between the polymer and the complexes. Therefore, the percentage extraction of silver was lower than that of gold.

$[Ag(CN)_3]^{2-}$ and $[Ag(CN)_4]^{3-}$ can be removed by adding HCl. The purpose of adding HCl is to remove free cyanide (KCN). Free cyanide, if present, reacts with HCl to form HCN which can escape from the solution as it formed. If HCl was added, it react with $[Ag(CN)_2]^-$ and white precipitates AgCN is formed and can be removed by filtration.

The percentage extraction of silver using such treated silver solution was higher than 99%.

Recovery of gold and silver

After extraction, gold can be extracted reversely and recovered by means of electro-deposition as illustrated in Figure 1.

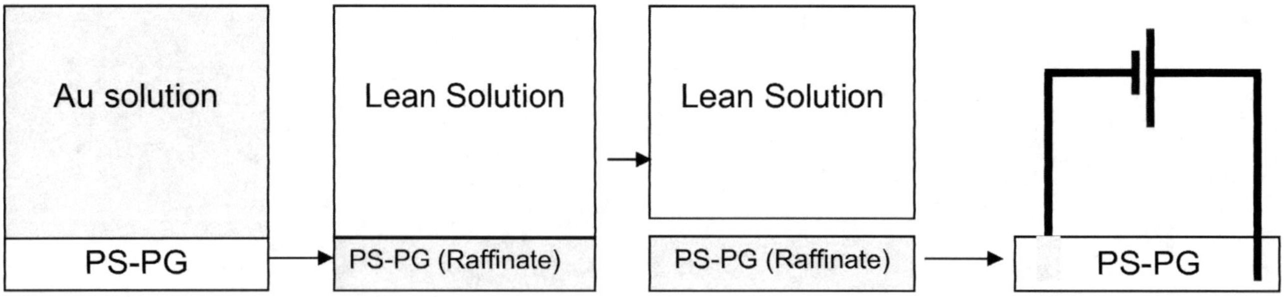

Figure 1. Schematic illustration of gold recovery

First, the gold ions in the water phase were extracted to the polymer phase so that resin saturated with gold was obtained. Secondly, the aqueous phase was discarded and only the resin phase (raffinate) was remained. Finally, a platinum wire as anode, the gold was plated on a clean nickel or copper wire. The electrical resistance in the polymer was very high and this caused high electrical voltage. The voltage could be reduced by adding small amount of salt such as sodium chloride or sodium sulfate.

The nickel wire was weighted before and after electro-deposition so that the mass of gold plated could be calculated.

Similarly, silver recovery was also performed. From the experiment, though only about 40% to 50% of gold or silver were recovered from the resins, no deterioration of extraction capability on the resins was observed.

Enhancement for the extraction capability

In order to enhance the extraction performance of the polymer, a functional group was introduced into the PS/PG copolymer. The functional group incorporated is thiol group in which some of hydrogen atoms in the polymer are replaced by sulfur atoms. The lone pair electrons in the sulfur atoms have strong affinity towards metal ions as metal ions are generally electron deficiency. Therefore, metal-sulphide bond has been regarded as strong bonding and promote the extraction performance.

The thiolated PG-PS was used for silver and gold extraction. It was found that the thiolated resin showed higher extraction capability than the normal one as shown in Figure 3 and 4. With 5 consecutive extractions, the thiolated resin still showed good extraction performance while the normal one was declined in third or fourth extractions.

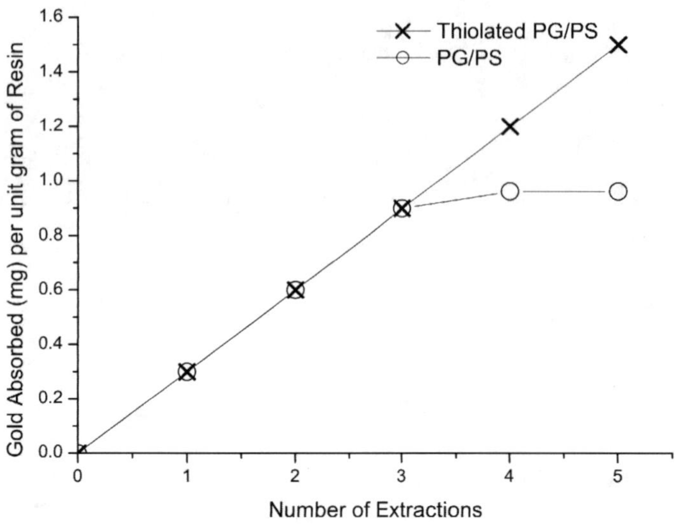

Figure 3. Comparsion between thiolated PS-PG and normal PS-PG on gold extraction capability

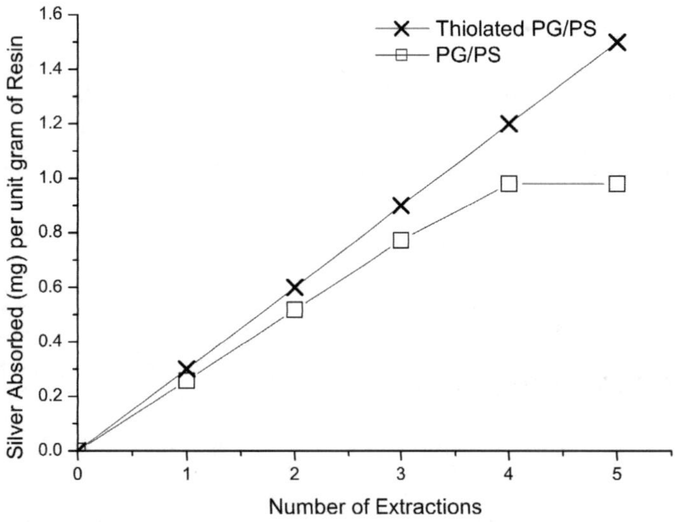

Figure 4. Comparsion between thiolated PS-PG and normal PS-PG on silver extraction capability

Gold could be reversely extracted from the thiolated polymer by means of sodium cyanide. A range of sodium cyanide with different concentrations was used to extract the gold from the thiolated resin saturated with gold reversely.

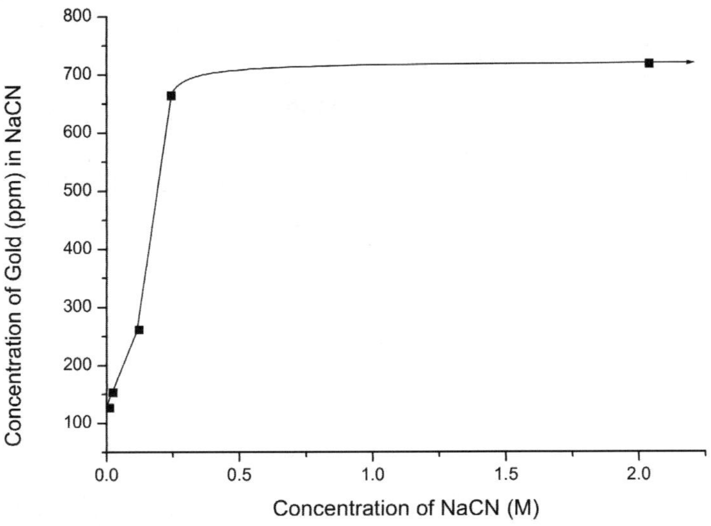

Figure 5. Effect of concentration of sodium cyanide on the gold reversely extracted from the thiolated PS-PG resins.

From the Figure 5, the optimum sodium cyanide concentration for gold reverse extraction is about 0.5M as higher concentration didn't provide further improvement.

Conclusions

In conclusion, a polystyrene-grafted-polyglycidol (PS-PG) resin synthesized and was used for gold and silver extraction. The resin can be recycled without deterioration in extraction capability. In addition, significant increase in loading could be accomplished by introduction of thiol groups to the polymer. The metal re-extraction from the polymer phase is possible and can be made by electro-deposition or sodium cyanide leaching.

PS-PG can be a good candidate or replacement for polyethylene glycol in polyethylene glycol-based aqueous biphasic systems (PEG-ABS) as no addition of salt is required. Therefore, the system becomes less complicated and more convenient and easier for investigation and study.

Acknowledgement

This research is a joint project between the University of Hong Kong and Kenlap PGC Manufacturer Company Limited and has been supported by an award from the Innovation and Technology Fund. (UIM/166)

References

1. Zhang, T.X.; Li, W.J.; Zhou, W.J.; Gao, H.C.; Wu, J.G.; Xu, G.X.; Chen, J.; Liu, H.Z.; Chen, J.Y. Extraction and separation of gold (I) cyanide in polyethylene glycol-based aqueous biphasic systems. *Hydrometallurg.* 2001, *62*, 41–46

2. Fontana, D.; Ricci, D. Poly(ethylene glycol)-based aqueous biphasic systems: effect of temperature on phase equilibria and on partitioning of 1,10-phenanthroline-copper(II) sulphate complex. *Journal of Chromatography B*. 2000, 743, 231-234

3. Chen, J.; Spear, S.K.; Huddleston, J.G.; Rogerse, R.D. Polyethylene glycol and solutions of polyethylene glycol as green reaction media. *Green Chemistry.* 2005, 7, 64-82

4. Du, Y.J.; Brash, J.L. Synthesis and Characterization of Thiol-Terminated Poly(ethylene oxide) for Chemisorption to Gold Surface. *Journal of Applied Polymer Science*. 2003, 90, 594-607

720f Atomistic Simulation of Uranyl Ion Extraction by Tbp/dodecane Solution

Xianggui Ye[1], Shengting Cui[1], Valmor de Almeida[2], and Bamin Khomami[1]. (1) Chemical and Biomolecular Engineering, University of Tennessee, Knoxville, TN 37996, (2) Nuclear Science and Technology Division, Oak Ridge National Laboratory, Oak Ridge, TN 37831

In the UREX-based (Uranium Extraction) process, tri-butyl-phosphate (TBP) in a diluent dodecane is used as the extraction agent. The aqueous phase containing the spent nuclear fuel, namely uranyl, is in contact with the organic phase containing TBP. Based on experimental findings, the uranium in the form of uranyl is extracted into the organic phase as $UO_2(NO_3)_2$•2TBP. One of the main chemical processes giving rise to the extraction is believed to be the following interfacial reaction:

$UO_2^{2+} + 2NO_3^- + 2TBP \leftrightarrow UO_2(NO_3)_2 \bullet 2TBP$ (1)

Basically, it is envisioned that the extractant molecules bind to the cation and form a complex at the aqueous-organic interface. The complex, $UO_2(NO_3)_2$•2TBP, as a neutral entity, then migrates to the organic phase. This simple picture, however, only gives the initial reactants and final products. Hence, many molecular processes including the migration of reactants to the interface, the formation of various intermediates, the binding kinetics of the extractant and the cation; and the migration of the extractant-cation complex from the interface to the organic phase need to be elucidated before a mechanistic understanding of the extraction process is attained. To this end, we have carried out a comprehensive molecular dynamics study. Specifically, the aqueous phase consists of: water, uranyl ions, nitrate ions, and in some cases hydronium ions. The organic phase consists of TBP in dodecane diluent. Our atomistic simulations demonstrate the formation of the $UO_2(NO_3)_2$•2TBP•H_2O complex under acidic condition. In particular, most uranyl complexes are in the form of $UO_2(NO_3)_n$•mTBP•kH_2O (n+m+k=5) with a coordination number 5 under these conditions while under neutral conditions we predominantly observe UO_2^{2+}•5TBP . Overall, our simulations clearly demonstrate that the extraction efficiency and even the types of complexes formed are a strong function of the acidity of solution or equivalently the nitrate concentration.

721a Separation of the Benzene/cyclohexane Mixture by a Distillative Crystallization Technology

Lie-Ding Shiau, Department of Chemical and Materials Engineering, Chang Gung University, 259 Wen-Hwa 1st Road, Taoyuan, 333, Taiwan and Chia-Chen Yu, Chemical and Materials Engineering, Chang Gung University, 259 Wen-Hwa 1st Road, Taoyuan, 333, Taiwan.

Cyclohexane (Cx) is mainly produced by catalytic hydrogenation of benzene (Bz) in the petrochemical industry. Due to close boiling points (Bz:80.1°C, Cx:80.7°C) and azeotrope formation, it is very difficult to separate Cx and Bz by conventional distillation. Although azeotropic distillation and extractive distillation can be used for this separation, these two processes suffer from high capital and operation cost. Therefore, searching for new alternatives to separation of the Bz/Cx mixture has been an important research issue.

A distillative crystallization technology, called distillative freezing (DF) or stripping crystallization (SC), has been applied to separate xylene mixtures with close boiling temperatures. In principle, DF is operated at a triple point condition, in which the liquid mixture is simultaneously vaporized and crystallized due to the three phase equilibrium. Thus, DF is a distillative crystallization technology, which combines distillation and crystallization to produce pure crystals. By lowering temperature and reducing pressure during the operation, DF results in the formation of pure crystals, and liquid phase and vapor phase of mixtures.

Distillative freezing (DF) will be applied in this work to separate the mixture of benzene (Bz) and cyclohexane (Cx). The experiments show that three DF operations are required to purify Cx (B) from to in the Cx-rich mixture. On the other hand, only one DF operation is required to purify Bz (A) mixture from to in the Bz-rich mixture. Thus, it is easier to separate Bz from the Bz-rich mixture than to separate Cx from the Cx-rich mixture. These experimental results are consistent with the DF simulations predicted by the proposed model.

In essence, the DF operation is a combination of distillation and crystallization. There are several unique features of the DF operation: (1) The DF process is conducted under an adiabatic condition at a three-phase equilibrium, where the latent heat released in crystallization is removed by vaporization. Thus, it is an energy-conserving separation method compared to distillation. (2) The addition of solvent is not needed in DF compared to azeotropic or extractive distillation in the separation of the close-boiling mixture. Thus, it is a clean separation technique. (3) The DF process is continued until the liquid phase is completely eliminated and only the pure crystals remain in the feed. Compared to crystallization, subsequent filtration or centrifugation is not needed since no mother liquor is present with the pure crystals. (4) Crystal washing is not required since only the pure crystals remain in the feed and no impurities are adhered on the crystal surfaces at the end of the DF operation.

[1] L.D. Shiau, C.C. Wen, B.S. Lin, Application of distillative freezing in the separation of o-xylene and p-xylene, AIChE J. 52 (2006) 1962-1967.

721b Semi-Continuous Processing In the Pilot Scale Purification of Monoclonal Antibodies

Beth McCooey, Alison Chow, Craig Emery, and Brian Kluck. Genentech, Inc, 1 Antibody Way, Oceanside, CA 92056

Semi-continuous downstream processing of monoclonal antibodies can be achieved by the integration of two chromatography skids, allowing two columns, or a membrane and a column, to be run without an intermediate hold step. To accomplish this, dilution and pH adjustment are performed in-line prior to the second unit operation. Implementation of semi-continuous processing in manufacturing facilities will allow for significant savings by reducing processing time, product hold tanks, and cleaning. An evaluation of semi-continuous processing was performed at pilot scale with the intent of scaling the process up to clinical and manufacturing scales. To ensure consistent process performance, pH, conductivity, and UV absorbance were monitored. Our results demonstrate that product quality, impurity clearance, and yield are comparable to standard processing.

Sulfur Dioxide Separation and Recovery with Ionic Liquid Absorbents

Chang Soo Kim*, Ki Yong Lee, Gyeong Taek Gong, Honggon Kim, Byung Gwon Lee, and Gwang-Deog Jung*

Energy & Environment Research Division, Korea Institute of Science Technology
39-1 Hawolgok-dong Seongbuk-gu, Seoul, South Korea 136-791

ABSTRACT

In Sulfur-Iodine and Hybrid Sulfur cycles, sulfur dioxide (SO_2) as a product of sulfuric acid decomposition at high temperature needs to be separated from gaseous mixtures and recycle to the next process. Depending on the hydrogen production cycles, pure sulfur dioxide is recycled into Bunsen reaction or electrolysis after oxygen separation. Absorption process with ionic liquids was suggested and studied to confirm the feasibility of continuous SO_2 separation process without an absorbent loss during absorption and desorption. High SO_2 solubility by ionic liquids were examined and [EMIm]$EtSO_4$ was tested as a candidate. Cyclic absorption and desorption by temperature swing was confirmed without any loss of solubility of 0.95 gmol SO_2/gmol IL at various desorption temperatures. Lab-scale SO2/O2 separation process was realized and resulted in pure sulfur dioxide recycle into following process was confirmed with preliminary data.

Corresponding Author
* Chang Soo Kim: mizzou333@kist.re.kr
* Gwang-Deog Jung: jkdcat@kist.re.kr

INTRODUCTION

Hydrogen is a promising game changer to replace fossil fuels which have been used for over 2 centuries leading to industrial development in spite of environmental problems and global warming issues. Hydrogen should be produced from environment-friendly resources to avoid hydrogen being regarded as pollution site shift from fossil fuel at power generation site to fossil fuel at the hydrogen production sites. To be completely naive to environment, hydrogen should be produced from water with clean energy resources.

These days in order to produce hydrogen from water without emitting any carbon dioxide during production process, nuclear hydrogen energy was suggested as a heat source for the hydrogen production cycle such as Sulfur-Iodine thermochemical cycle and Hybrid Sulfur cycle [1-4]. In Fig.1 Sulfur-Iodine and Hybrid Sulfur cycles are illustrated with sulfuric acid decomposition in common. During whole cycle, water is supplied and decomposed into hydrogen and oxygen in a closed loop.

To complete a full closed loop, the sulfur dioxide decomposed from sulfuric acid should be separated and recycled into next process without loss. Moreover, pure sulfur dioxide after separation process is required for the next process such as electrolysis and Bunsen reaction. In this study, an absorption process with novel absorbents is suggested to satisfy two requirements of complete SO_2 recovery with high purity.

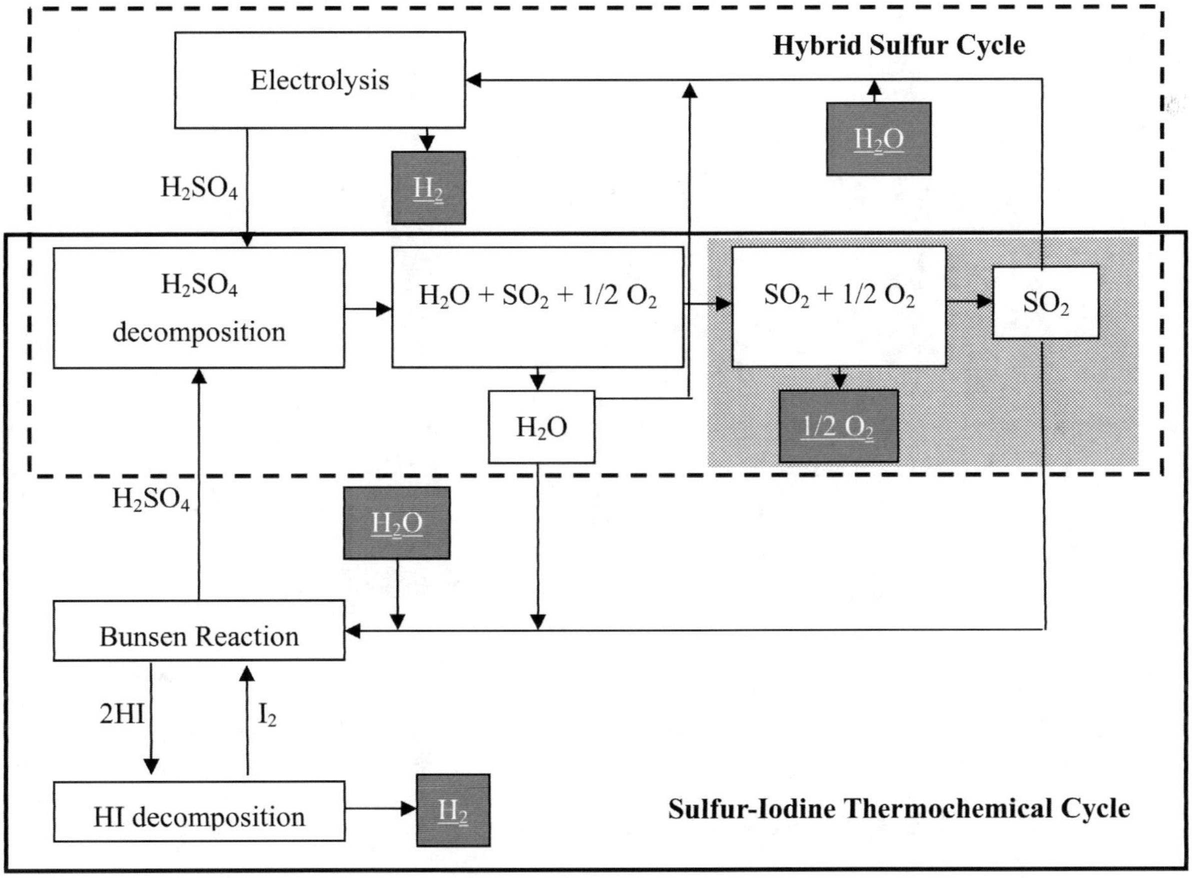

Figure 1. SO_2/O_2 separation process in Sulfur-Iodine Thermochemical Cycle and Hybrid Sulfur cycle

NOVEL ABSORBENTS

Ionic liquids are characterized as room temperature liquids with extremely low vapor pressures, high thermal and chemical stability. These characteristics are the most appropriate for SO_2 separation and recovery due to low vapor pressure to high SO_2 purity as well as high thermal and chemical stability for temperature swing absorption process. The only requirement to be confirmed is high SO_2 absorbing capability and reversible absorption during temperature swing. From various candidating ionic liquids, after screening process with criteria of thermal stability, cost, and absorption capability [EMIm]EtSO$_4$ was selected and examined in detail.

In Fig.2 reversibly SO_2 absorption and high SO_2 solubility by [EMIm]EtSO$_4$ was verified even with different desorption temperatures. At 50 °C, [EMIm]EtSO4 absorbed 0.95 gmol SO_2 per gmol IL and desorbed SO_2 fully at high temperature. Depending on desorption temperature, the desorption speed is changed while full desorption was observed.

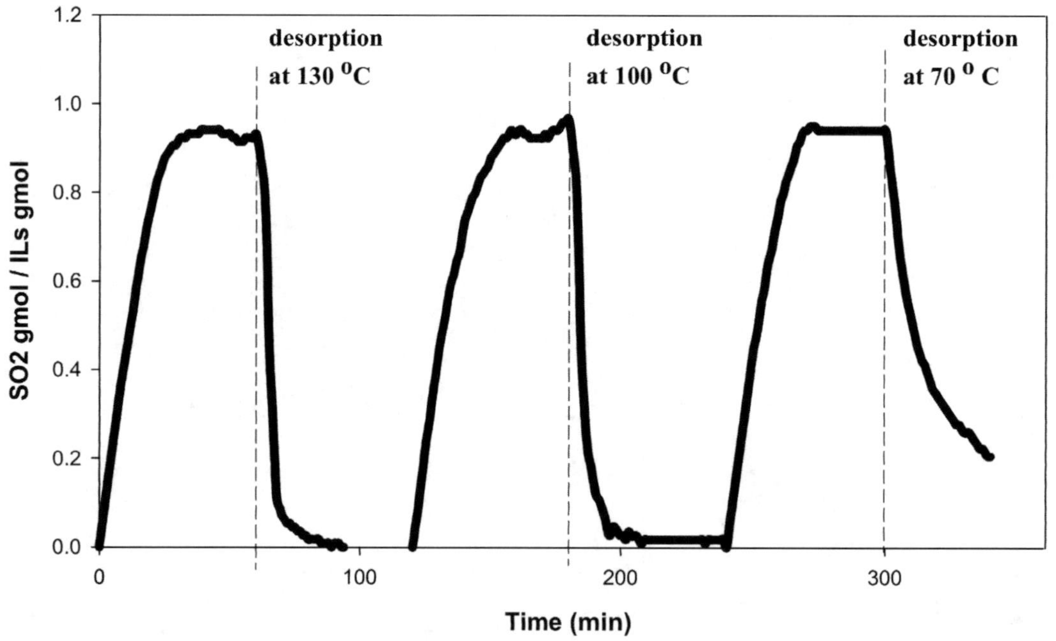

Figure 2 Cycle test of SO_2 absorption with [EMIm]EtSO$_4$ absorbed at 50 °C and 0.7 atm.

CONTINUOUS SO_2 SEPARATION FROM SO_2/O_2 MIXTURE

Additional requirement for SO_2 separating absorbents is the inertness with oxygen including negligible absorption and no reaction even at high temperature. To verify additional requirement and continuous process, bench-scale SO_2/O_2 separation apparatus was set up and operated. In Fig.3, the operation results were reported based on preliminary data. Absorption tower was operated at room temperature and temperature rise was observed from the heat of absorption reported as temperature range 20~40 °C.

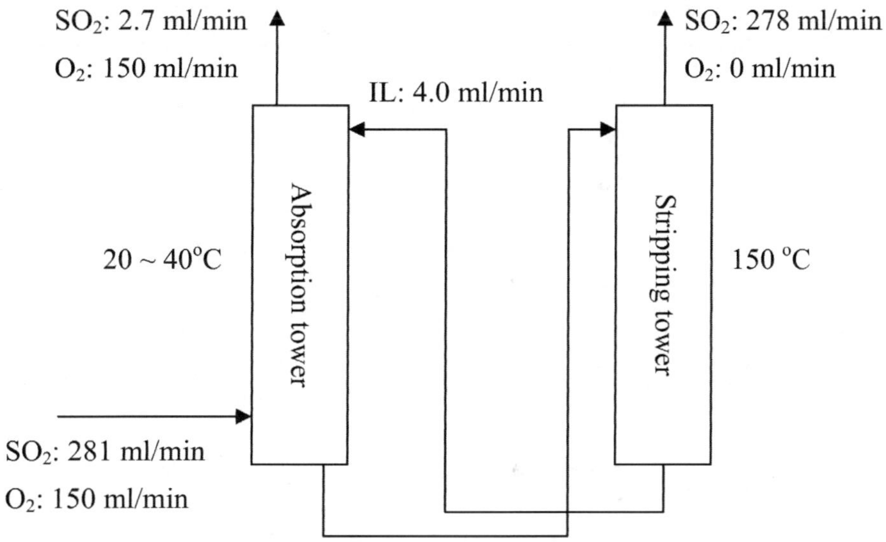

Figure 3 Lab-scale continuous SO_2/O_2 separation process with ionic liquid

After titration of gaseous outlet in the absorption and stripping towers, 278 ml/min SO_2 can be recovered even with 4.0 ml/min ionic liquid flow rate at 150 °C. During collecting stripped gas for titration, negligible oxygen content could be concluded from the observation of no bubble in the titration trap. Continuous SO_2/O_2 separation was demonstrated for a couple of hours without any disruption in the operation.

CONCLUSIONS

Novel absorbent for SO_2/O_2 separation process was evaluated from material properties to bench-scale process operation. Ionic liquids were appropriate SO_2 absorbents to satisfy pure SO_2 recovery for Sulfur-Iodide Cycle and Hybrid Sulfur Cycle which are suggested for hydrogen production by water splitting. From the absorption and desorption test, [EMIm]EtSO$_4$ was applicable to SO_2 separation process in that it is stable during temperature swing with constant SO_2 absorption capability. In bench-scale continuous process, 278 ml/min SO_2 could be separated and recovered in high purity with 4.0 ml/min ionic liquid flow rate. Moreover, the inertness of [EMIm]EtSO$_4$ with oxygen even at high temperature was confirmed from constant SO_2 separation capability.

REFERENCES

1. K. Onuki, S. Shimizu, H. Nakajima, Y Ikezoe, and S. Sato, "Coupling of Thermochemical hydrogen production processes with an HTGR," Int. J. Hydrogen Energy (1984) 9(5) pp.391-396.
2. D. O'Keefe, C. Allen, G. Besenbruch, L. Brown, J. Norman, R. Sharp, K. McCorkle, "Preliminary results from bench-scale testing of a sulfur-iodine thermochemical water-splitting cycle," Int. J. Hydrogen Energy (1982) 7(5) pp.381-392.
3. M. Dokiya, T. Kameyama, K. Fukuda, "Thermochemical hydrogen preparation-part V. A feasibility study of the sulfur iodine cycle," Int. J. Hydrogen Energy (1979) 4(4) pp.267-277.
4. D. Henderson, "Nuclear hydrogen initiative overview. DOE Hydrogen Program Review," Philadelphia. PA. May 2004.

721d Separation of Chiral Amine by Capillary Electrophoresis

Lin Yang, Rongji Dai, Wei Liu, and Yulin Deng. School of Life Science and Technology, Beijing Institute of Technology, 5 South Zhongguancun Street, Haidian District, Beijing, China

Stereochemistry can have a significant effect on the biological activity and the side effects of a drug, different from those of the optically pure drugs. Therefore, the development of a chiral separation method for the determination of optical purity is desirable in many areas.

In the more conventional chromatographic procedures such as high performance liquid chromatography (HPLC), gas chromatography (GC) and thin-layer chromatography, chiral separations were achieved by the use of chiral separations were achieved by the use of chiral additives in the mobile phase. Capillary electrophoresis provided a highly efficient separation method for chiral compounds by the use of a chiral selector in the running buffer. These chiral selectors used in chiral separation mainly include bile salts, chiral surfactants, chiral crown ethers, cyclodextrin derivatives and so on. In the chiral separation for capillary electrophoresis, ¦Â-CD has been used extensively as chiral selector.

2-HPA is a new drug to cure the Parkinson's disease and only the R-isomer is pharmacological activity [1]. In order to separate 2-HPA we have experienced some methods and found that separation of 2-HPA by capillary electrophoresis was efficient[2]. Further more the precursor 2-HA and outgrowth 2-HPPA were separated. Because there were not ultraviolet absorbing groups in 2-HPA and 2-HA, dansyl chloride was chosen to derivatize 2-HPA and 2-HA. 2-HPPA was separated directly.

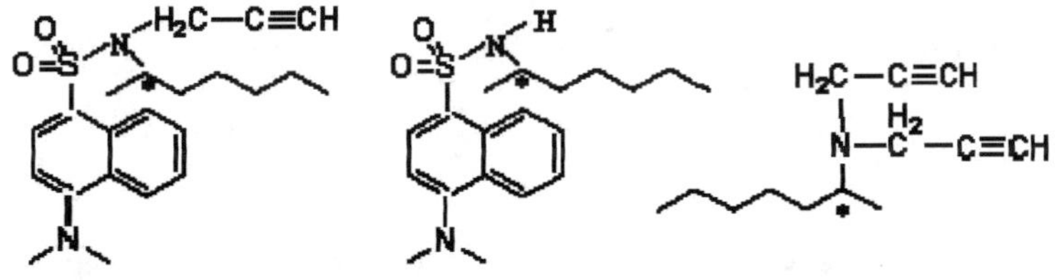

2-HPA-Dns 2-HA-Dns 2-HPPA

FIG.1 Structures of three chiral compounds

Experimental:

Reagents and Instrumentation: 2-HA (Adrich, 99%), PBS (Adrich, 96%), ¦Â-CD (Sigma, 99%), others are chemical pure. CE apparatus (BINDA), capillary columns (50¦Ìm¡Á70cm).

Derivation: 1ml 2-HPA, 1ml dansyl chloride (9g/l) and 15ml Na_2CO_3 aqueous solution (pH=11) were stirred for 15 minutes at 60¡æ in water bath, and then cooled to the ambient temperature. 15ml of ethyl acetate was used to extract the derivatized compounds. The extract was blew with the nitrogen under 40¡æ.

Chromatographic conditions: A suitable chiral analysis conditions in capillary electrophoresis was as following, 10mmol/l ¦Â-cyclodextrin dissolved in 20mmol/l PBS, pH=4.34; samples concentration 0.1g/l; temperature 25¡æ; separation voltage: 10kV; samples were injected into the capillary by 5kV for 5 seconds. The detection wavelengths were 210 nm and 254 nm. The capillary column was first flushed 10 minutes with the 0.1 mol/l NaOH, then with distilled water for 10 minutes, balanced with buffer liquid for about 10 minutes. In this experiment the samples and the buffer were degassed ultrasonically before use. Each component of the mobile phase was filtered through a 0. 45¦Ìm membrane.

Results and discussion

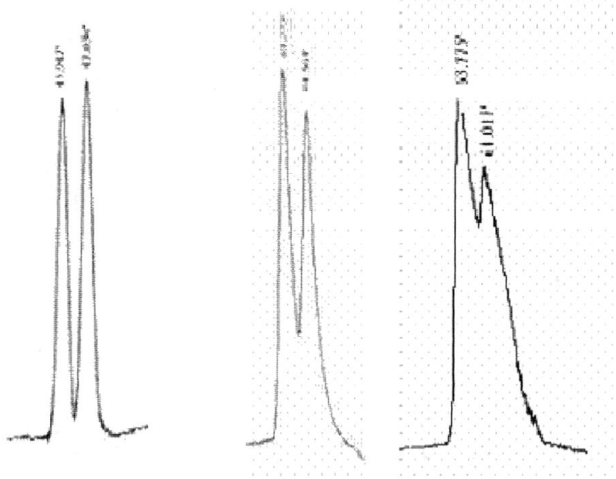

2-HA-Dns 2-HPA-Dns 2-HPPA

FIG.2 Electropherograms of the enantioseparation of three chiral compounds by Capillary Electrophoresis

(10mmol/l β-cyclodextrin dissolved in 20mmol/l PBS, pH=4.34; samples concentration 0.1g/l; temperature 25℃; separation voltage: 10kV; samples were injected into the capillary by 5kV for 5 seconds.)

The result showed that pH of buffer played a key role in separation. The experiment proved that the speed of isomer moving toward to cathode decelerated under the low pH condition, so the coalescent time between ¦Â-cyclodextrin and isomer were prolonged. It was advantageous to separate. On the contrary, higher pH was disadvantageous in separation .We confirmed that the best separate pH condition was 4.34. The effect of chiral selector was important also. According to the Stephen's formula [3]: ¡°Â¡Ì={[C](¦Ì$_1$-¦Ì$_2$)(K_2-K_1)}/{1+[C](K_1+K_2)+K_1K_2[C]}, we choose the ¦Â-CD concentration 10mmol/l. Under the optimal condition 2-HPA-Dns was separated by based-line, 2-HA-Dns and 2-HPPA were separated well.

1. Effect of pH on ¡°Â¡Ì: We found that the pH was the significant factor affecting the chiral separation. And the ¡°Â¡Ì increased with a decrease in buffer pH. When the buffer pH was 4.34, ¡°Â¡Ì tended to a constant value. It is well known to us that the degree of protonation of enantiomers increases as the pH value of the buffer decreases, on the basis of the acid base equilibria of the R- and S-enantiomers. The effective mobilities are related to the degree of protonation of the enantiomers. The effective mobility of basic compounds increases with an increase in the hydrogen ion concentration of the buffer, which leads to an increase in ¡°Â¡Ì,. Additionally, the electrosmotic flow decreased with an increase in hydrogen ion concentration. The low pH of the buffer caused an improvement in the chiral separation.

2. Effect of ¦Â-CD concentration on ¡°Â¡Ì: The chiral recognition of ¦Â-CD is based on the difference in the formation constants of ¦Â-CD complexes with enantiomers, which reflect the difference in hydrophobic complex-forming interaction between the ¦Â-CD cavity and the guest molecule. The ¦Â-CD concentration is an important factor affecting the chiral separation. When the ¦Â-CD concentration was higher than 10 mmol/l, ¡°Â¡Ì tended towards a constant value.

3. Effect of electrolyte concentration on the chiral separation: The chiral separation of enantiomers improved with an increase in electrolyte concentration. This was because an increase in electrolyte concentration decreased the effects which were caused by the differences of the pH and electrolyte concentrations of the sample zone from that of the buffer. Moreover, the increase in electrolyte concentration made the electroomotic flow decrease, and led to an improvement in the resolution of enantiomers. However, the high concentration of electrolyte in buffer was not beneficial to chiral separation, due to a Joule heating effect.

Keyword: dansylchloride; chiral separation; capillary electrophoresis; 2-HPA

References

[1] David A£¬Durden, Lillian E. Drug Metabolism and Disposition, 28(2):147-154 (2000).

[2] Blomberg LG, Wang H., Electrophoresis, 21(10): 1940-1952 (2000).

[3] Wren SAC, Rowe R. C., J. Chromatogr, 603:235(1992).

A continuous reactive distillation process for the production of cyclohexanol from cyclohexene

Rakesh Kumar[1,*], Amit Katariya[1], Hannsjoerg Freund[1], Kai Sundmacher[1,2]
[1]Max Planck Institute for Dynamics of Complex Technical Systems, Magdeburg, Germany
[2] Process Systems Engineering, Otto von Guericke University, Magdeburg, Germany
Email: kumar@mpi-magdeburg.mpg.de

1 Introduction

Cyclohexanol is one of the important bulk chemicals used widely in the chemical industries as an intermediate for the production of polymers (Nylon 6,6 and Nylon 6). Conventionally, cyclohexanol is produced by the hydrogenation of benzene to cyclohexane and its subsequent oxidation by air to cyclohexanol/cyclohexanone. This process has a high energy demand; a further disadvantage is the risk associated with the operation of the process. In recent years, a promising alternative process has been developed that includes hydrogenation of benzene to cyclohexene and further hydration of cyclohexene to cyclohexanol. Cyclohexanol production by direct hydration of cyclohexene is a challenging task due to its low solubility with water. Asahi Chemical Co. made an attempt in this direction to improve the reaction rate using a zeolite catalyst of HZSM5 type[1]. However, still a large amount of catalyst is required to obtain significant reaction rates of cyclohexene hydration.

We propose a novel process scheme for the indirect hydration of cyclohexene to cyclohexanol in a reactive distillation column using formic acid as a reactive entrainer. The proposed process includes the continuous esterification of cyclohexene with formic acid in a reactive distillation column followed by hydrolysis of cyclohexyl formate in a second reactive distillation column to produce pure cyclohexanol[1]. Previously, our group performed thermodynamic, kinetic and process feasibility studied regarding this system [2,3,4]. In the present study, the feasibility of the first step of this process scheme (i.e. the esterification of cyclohexene with formic acid) is investigated experimentally in a vacuum reactive distillation column.

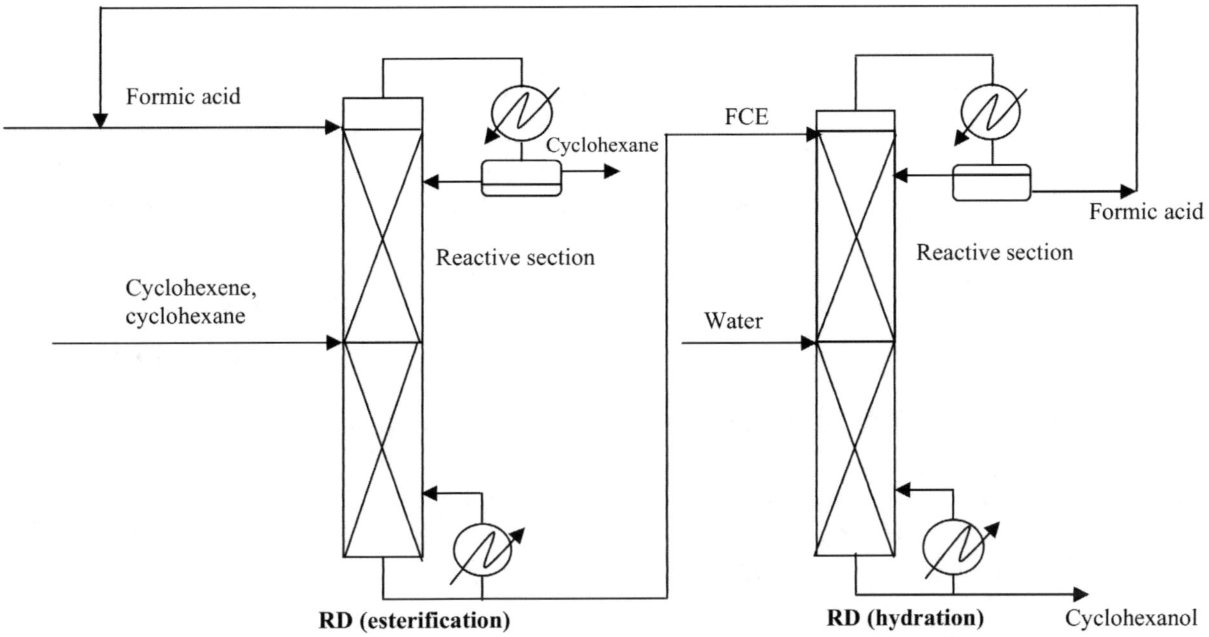

Figure 1: Process flowsheet for the indirect hydration of cyclohexene to cyclohexanol

2. Experimental

2.1 Chemicals:

Cyclohexene (99%) and 1,4 dioxane (99.5%) was obtained from Sigma Aldrich. Formic acid was supplied by Merck (Germany).Cyclohexyl formate (>97%w/w) was prepared in the laboratory. Detail of the method can be found elsewhere[3].

2.2 Analytical method

Analysis of samples obtained from the different position of column was performed on a Gas chromatograph equipped with FID and TCD as a combination of detectors. INNOVAX column (30m, 250 μm, 0.25 μm) was used for the quantification of the component. GC was calibrated using 1,4 dioxane as an internal standard. In case of biphasic mixture, samples were centrifuged before analysis on GC.

2.3 Experimental Setup and procedure

Esterification of cyclohexene with formic acid was conducted in a continuous reactive distillation column as shown in Figure 2. The reactive distillation column consists of two different sections: non reactive stripping section and reactive section. Reactive section is packed with the structured packing Sulzer-KATAPAK filled with Amberlyst-15 cationic ion exchange resin as acid catalyst. Non reactive stripping section of the column is packed with Sulzer-BX packing. The temperature sensors were (Pt-100) provided at different sections in the column. To prevent the heat loss from surrounding, the column wall was covered with the insulating materials and external heating was provided to column. Reboiler was equipped with electrical heating power supply (2.8 kW). Low volatile component (Formic acid) was fed from the top of the column and cyclohexene from the middle of stripping section. Since experiments were performed with pure cyclohexene, overhead product of the column was condensed and total refluxed

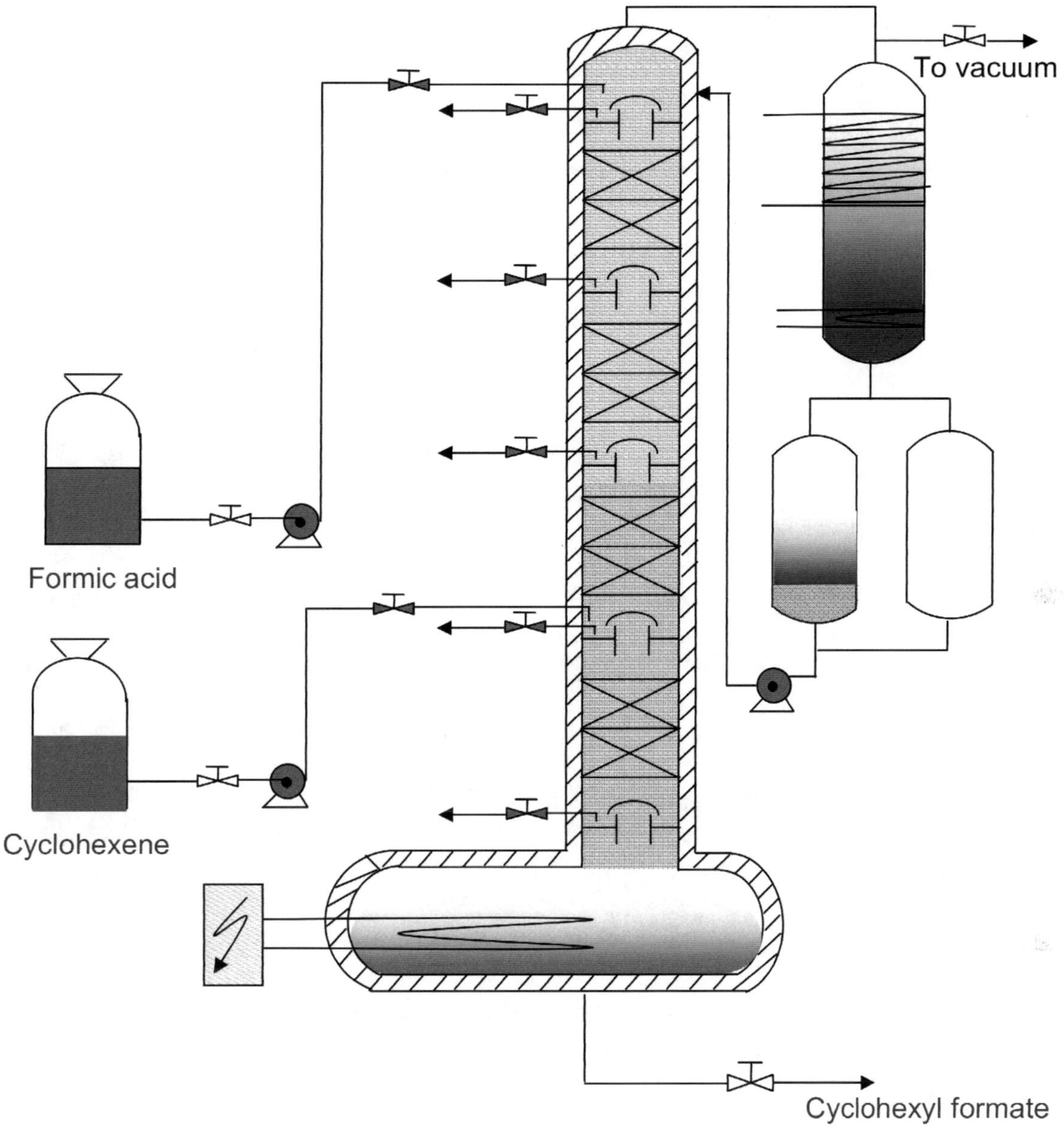

Figure 2: Experimental setup for the esterification of cyclohexene with formic acid

to reactive section of the column. Reboiler holdup was maintained constant by removing of bottom product continuously. In order to minimize the decomposition of formic acid, experiments were performed at low pressure conditions (<0.7 bar). Samples were collected from different position along the height of column and analyzed on GC. At steady state, flow rates of the bottom stream were measured and overall material balance was verified. Effect of feed mole ratio was studied to see the effect on column performance

3. Results and discussion

Initially, experiments were performed to obtain the steady state compositions and temperature profile of column. It took more than 72 hrs to get steady because of large hold up of liquid inside the reboiler. Representative results of concentration and temperature profile obtained under steady state condition are shown in Figure 3. Liquid-Liquid phase splitting take place in whole column except top of reactive section and reboiler part of column. Due to the presence of high concentration of the formic acid in top part of reactive section, it leads to an increase in reaction temperature. Towards the bottom of reactive section and in stripping section, temperature in the column decreases because of increase in concentration of cyclohexene. In bottom part of stripping section and reboiler part of column, there is a steep rise in temperature with increase in concentration of cyclohexyl formate. With 2.19:1 feed molar ratio of the reactants (Formic acid: cyclohexene), 90% conversion of cyclohxene was observed in column. Bottom stream contains about 58 %w/w cyclohexyl formate, 37 %w/w formic acid and unreacted cyclohexene. In the bottom streams of column, presence of water and cyclohexanol was also detected because of the decomposition of formic acid in presence of catalyst in the reactive section.

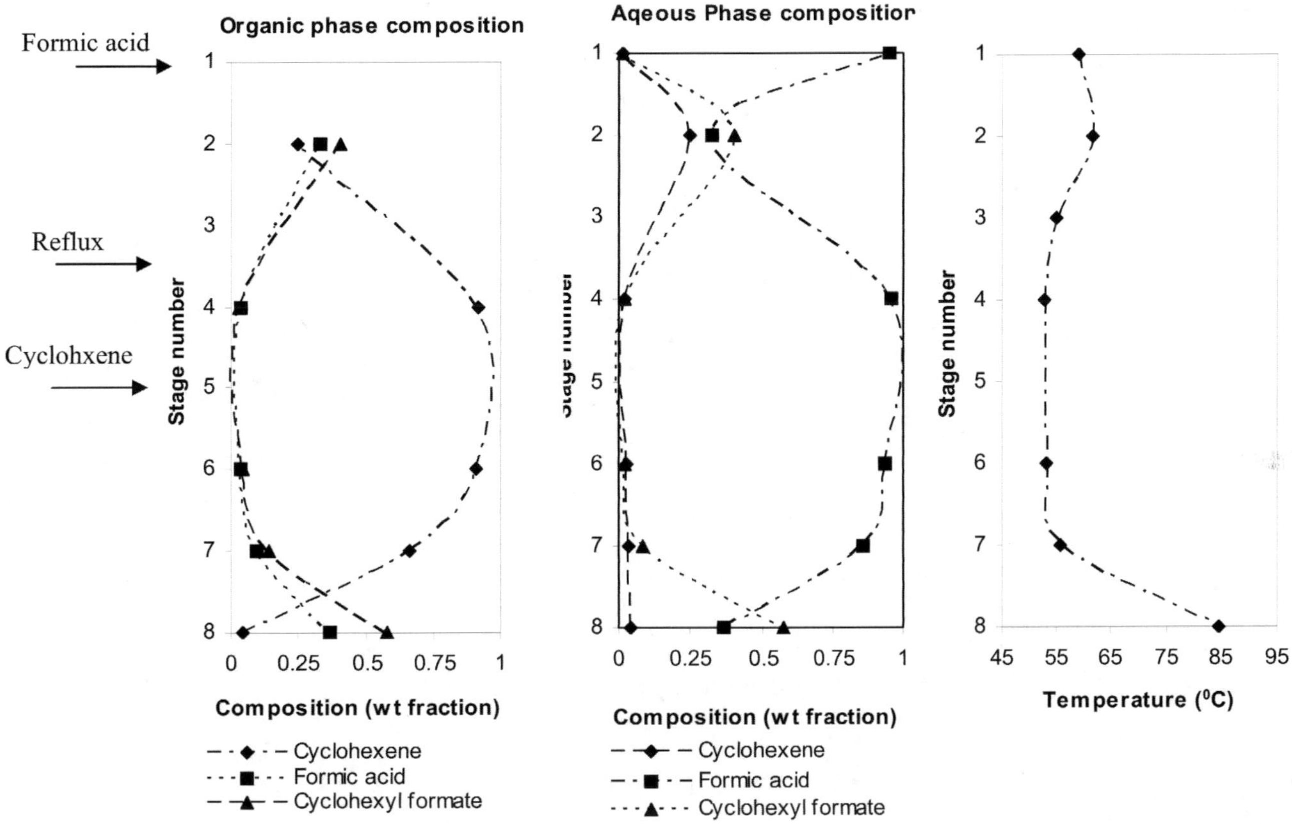

Figure 3: Composition and temperature profile of column. Feed mole ratio (FA: Cyclohexene =2.19), Reboiler duty =700 W, Pressure = 0.544 bar, Feed position: Formic acid from top, Cyclohexene from middle of stripping section

3.1 Effect of feed mole ratio

Feed flow rate of cyclohexene was changed to observe its effect on the column performance by keeping other parameter constant. At high feed mole ratio of the reactants (formic acid: cyclohexene), reaction rate of cyclohexene esterification increases that lead to increase in the conversion. However, excess amount of formic acid in column, reduces the concentration of cyclohexyl formate in the bottom product (Figure

4a). Temperature in the reboiler part of column increases because of the increase in formic acid and cyclohexyl formate concentration (Figure 4b).

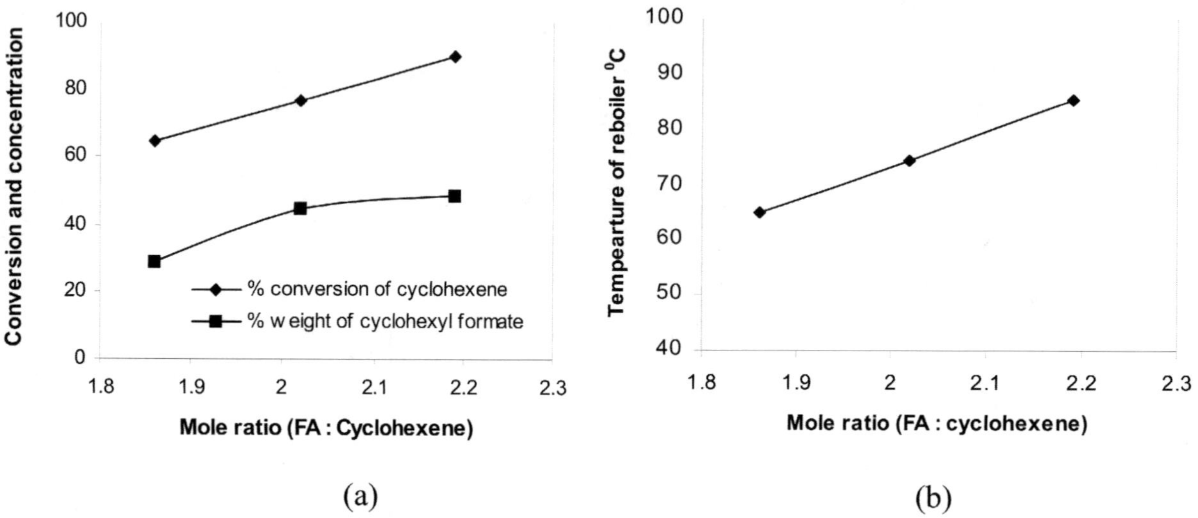

Figure 4: Effect of feed mole ratio of the reactants

4 Conclusions

Esterification of cyclohexene was studied in a continuous reactive distillation column using the formic acid as a reactive entrainer. Experiments were conducted in a pilot scale column using amberlyst-15 as a catalyst in KATAPAK structured packing. About 90% conversion of cyclohxene was obtained with 2:19 mole ratio of the reactants (Formic acid: cyclohexene) under steady state conditions. Effect of feed mole ratio was investigated to evaluate the column performance.

References

[1] O. Mitsui, Y. Fukuoka, Process for producing cyclic alcohol, *United state patent* 4588846 (1986).

[2] F. Steyer and K. Sundamcher, Cyclohexanol production via esterification of cyclohexene with formic acid and subsequent hydration of the ester- reaction kinetics, *Ind. Eng Chem. Res.,* 46 (2007) 1099-1104.

[3] F. Steyer, K. Sundamcher, VLE and LLE data set for the system Cyclohexane + Cyclohexene + Water + Cyclohexanol + Formic acid + Formic Acid Cyclohexyl Ester *J. Chem Eng. Data,* 50 (2005) 1272-1282.

[4] F. Steyer, H. Freund and K. Sundamcher, A novel reactive distillation process for the indirect hydration of cyclohexene using a reactive entrainer, *Ind. Eng. Chem. Res.,* (Submitted)

Operating Strategy of Heteroazeotropic Batch Distillation Systems

Min Jhong Wei and I-Lung Chien, Department of Chemical Engineering, National Taiwan University of Science and Technology, Taipei 106, Taiwan

Mixtures containing azeotrope or close-boiling can be separated by adding an entrainer into the system to form an additional heterogeneous minimum-boiling azeotrope. This heterogenous azeotrope can be separated in a decanter at the top of the column into aqueous phase and organic phase. In this paper, the operating strategy of heteroazeotropic batch distillation for two mixture systems having different residue curve maps will be investigated. The purpose of the batch distillation operation is to recover near pure components in the mixture and also to recover near pure entrainer for further batch usage. The goal of the operating strategy is to minimize total batch time and also to obtain high recovery percentage of the three components in the system.

The first system is to separate acetic acid and water. This mixture does not form azeotrope, however because of tangent pinch on the pure water end it is customary in industry to add an entrainer into this mixture to aid the separation via heterogeneous azeotropic distillation. This investigation shows that although iso-butyl acetate is a good entrainer for acetic acid dehydration operating at continuous mode, it is not suitable for heteroazeotropic batch distillation. Instead, a much better entrainer was found to be vinyl acetate when acetic acid dehydration system is operated in batch mode. Extremely simple operating strategy of this system is developed to simultaneous accumulate aqueous and organic phase materials in the decanter and recover near pure acetic acid at the end of the batch. From dynamic simulation results, it is found that the proposed operating strategy is very robust in terms of the setting of the operating variables.

The second system is to separate Propylene glycol Monomethyl ether (PM) and water. There is an azeotrope for this mixture, thus using simple batch distillation will not obtain two pure components. However, by adding isopropyl acetate into this mixture, a heterogeneous minimum-boiling azeotrope of isopropyl acetate-water is formed. With the same simple operating strategy used in acetic acid dehydration system, near pure water product can be accumulated in the aqueous phase of the decanter and simultaneously organic product (mostly isopropyl acetate) can be accumulated in the organic phase of the decanter. The pure PM is collected at the end of the batch from bottom still. The organic product, although not in pure entrainer form, is demonstrated to be successfully used in the following batch run.

The key control loop in the proposed operating strategy is a middle tray temperature loop which can easily be implemented in industry. Rigorous dynamic modeling from the start-up of the batch run until the end of the batch can be simulated using Aspen Dynamics^{TM}.

721g Using Surface Plasmon Resonance Spectroscopy to Understand Protein-Multimodal Ligand Interactions

Srinavya Vutukuru, Sridhar R. Bethi, and Ravi S. Kane. Chemical and Biological Engineering, Rensselaer Polytechnic Institute, 110 8th Street, Troy, NY 12180

Ion-exchange chromatography offers cost-effective, rapid, and efficient separations and is widely used for the concentration and purification of proteins. The use of classical ion-exchange resins, however, is not always optimal, particularly under high salt conditions. We describe the use of surface plasmon resonance (SPR) spectroscopy and self-assembled monolayers (SAMs) to understand the characteristics of surfaces that promote the adsorption of proteins at high ionic strengths (high-salt conditions). We synthesized SAMs presenting different multi-modal ligands, and determined the influence of surface composition, solution composition, and the nature of the protein on the extent of protein adsorption onto the SAMs. Our results confirm that hydrophobic interactions can contribute significantly to protein adsorption under high-salt conditions. The combination of SPR and SAMs is well-suited for elucidating the fundamental physics underlying the interaction of proteins with complex surfaces of relevance to chromatography.

Optimization of Stepwise Elution Chromatography on the Basis of Iso-Resolution Curve

Shuichi Yamamoto, Mareto Hosono and Noriko Yoshimoto

Bioprocess Engineering Laboratory,
School of Engineering and Graduate School of Medicine
Yamaguchi University, Tokiwadai, Ube, 755-8611, Japan

E-mail shuichi@yamaguchi-u.ac.jp

We have proposed a concept "iso-resolution curve" for linear (salt concentration) gradient elution of proteins in electrostatic interaction chromatography (ion-exchange chromatography, IEC).

In this study this concept was extended to stepwise elution chromatography. The distribution coefficient was determined from linear gradient elution experiment by using our model. The iso-resolution curves were similar to those for linear gradient elution. However, compared with the curves for linear gradient elution where no distinct optimum conditions were found, the curves for stepwise elution showed optimum conditions where the separation can be performed with the shortest separation time and the minimum elution volume consumption.

Figs.1-4 show the concept of iso-resolution curve for step-wise elution.

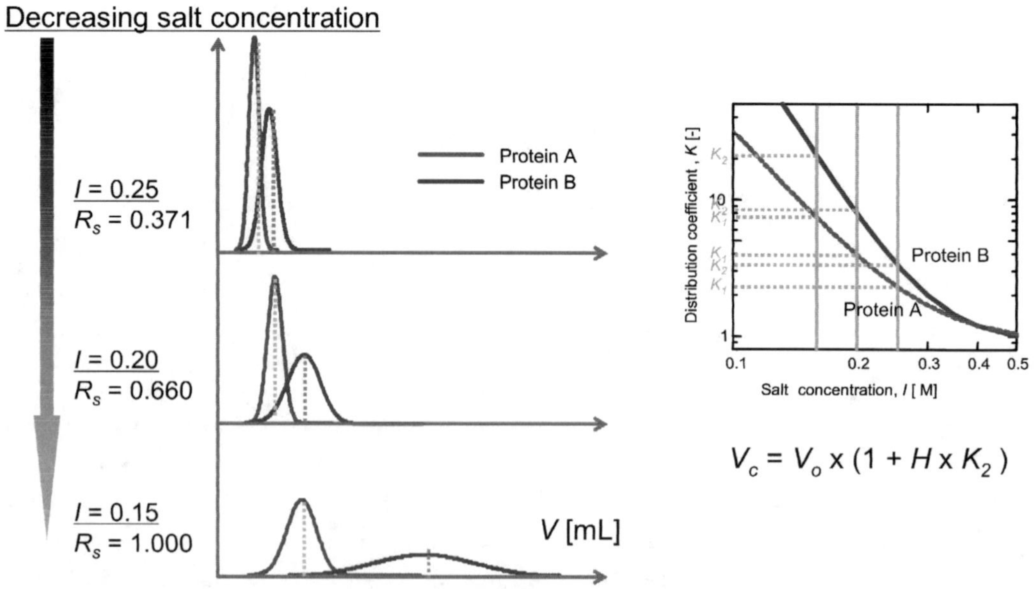

Fig.1 Effect of mobile phase salt concentration on the resolution at a fixed mobile phase velocity;. The right panel shows the distribution coefficient as a function of salt concentration.

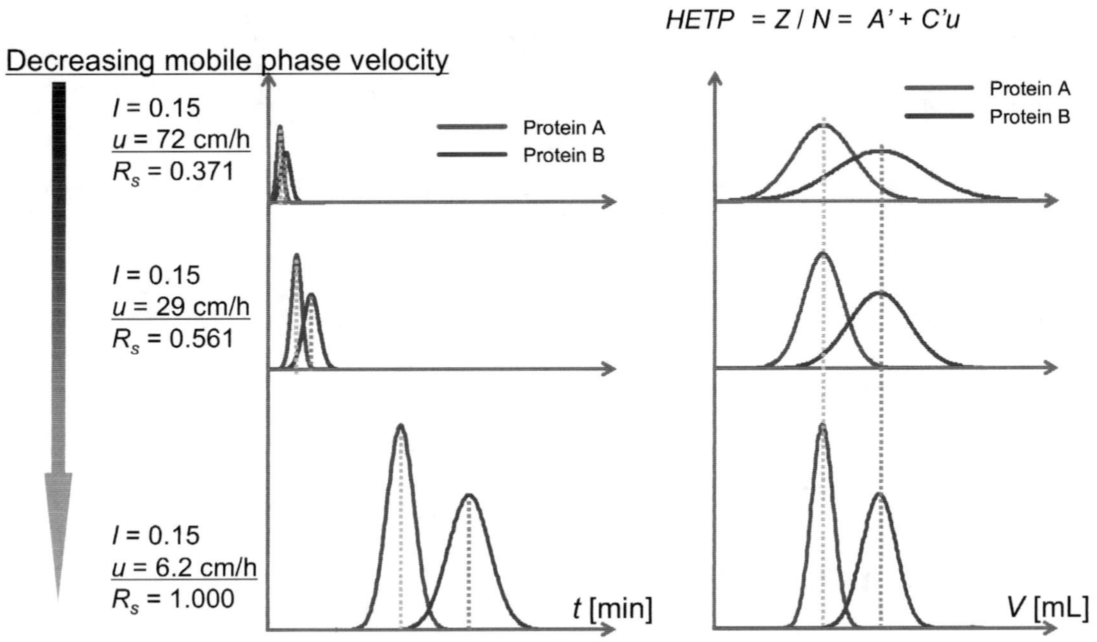

Fig.2 Effect of mobile phase velocity on the resolution at a fixed mobile phase salt concentration.

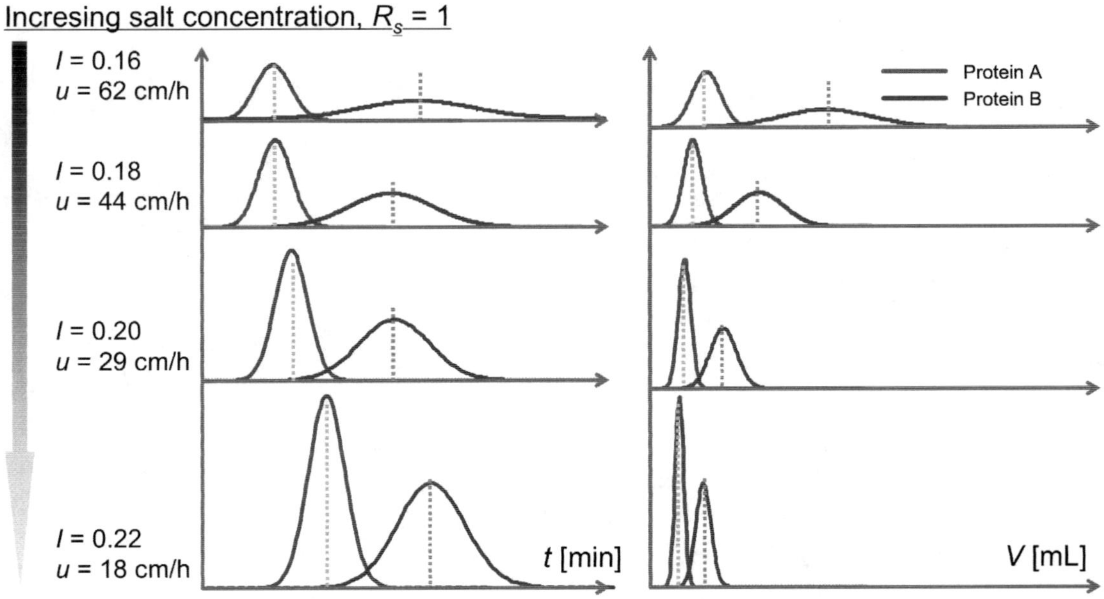

Fig.3 Separation behavior at a fixed resolution (R_s=1) at different combinations of mobile phase salt concentration and velocity.

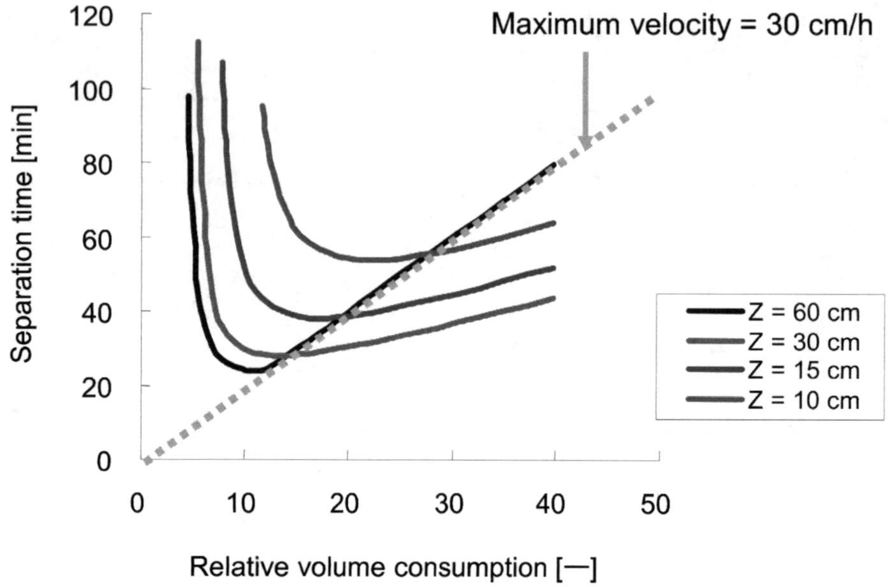

Fig. 4 Iso-resolution curves – Separation time vs. relative volume consumpation.

References

Shuichi Yamamoto and Ayako Kita

Theoretical background of short chromatographic layers. Optimization of gradient elution in short columns,

Journal of Chromatography A, 1065, 45-50(2005)

721i Effect of Pore Size and Particle Diameter on Adsorption and Elution Performance of Hydroxyapatite Chromatography

Shuichi Yamamoto, Yuuki Sakaguchi, Tomohiko Yoshitake, and Noriko Yoshimoto. Chemical and Biomolecular Engineering, Yamaguchi University, Tokiwadai, Ube, 755-8611, Japan

Effects of pore size and particle diameter on adsorption (dynamic binding capacity) and elution (peak salt concentration and peak width) in hydroxyapatite chromatography were investigated. The retention volume of proteins decreased with increasing pore size irrespective of particle diameter. The binding site value was not markedly affected by pore size. Dynamic binding capacity (DBC) of immunoglobulin G depended strongly on particle size and pore diameter. We established a new correlation between DBC and a normalized residence time with respect to particle diameter.

721j Structured Adsorbent Bed for Rapid Adsorption and Regeneration Process

Wei Liu, Energy and Environmental Directorate, Pacific Northwest National Lab, 902 Battelle Boulevard, MSIN:K2-12, Richland, WA 99352

In the conventional fixed bed adsorption process, adsorbent materials are typically made into pellets of various shapes such as sphere and cylinder in a hydraulic diameter in the order of 1mm level, and those pellets are packed inside a large adsorbent bed vessel. A major limitation to use such an adsorbent bed for flue gas CO2 capture is pressure drop. Due to huge gas flow rate of the flue gas, the large pressure drop through the pellet-packed bed is directly associated with high energy consumption. A structured adsorbent bed design is presented here to substantially reduce capital cost and energy consumption for CO2 capture and separation from a gas mixture.

In the structured bed, fine adsorbent powder ranged from a few hundred nm to a few um is confined by a gas-permeable medium to form straight flow channels. As the gas mixture flows through the channel, CO2 diffuses across the gas-permeable medium, into the adsorbent layer, and further adsorbs inside individual adsorbent particles. The structured bed provides following advantages over the conventional packed bed:

• Fast mass and heat transfer rate between the adsorbent material and convective fluid by minimizing all the transport resistance from bulk fluid onto channel surface, from the channel surface into the adsorbent layer, and from the external surface of individual adsorbent particles into the adsorbent.

• 102 – 103 lower DP than the pellet-packed bed at comparable hydraulic diameter under the same flow conditions

• Little dead volume

Those critical attributes enable adsorption/regeneration process be conducted at exceptionally fast cycle speed with low energy consumption. Concept of this new adsorbent bed design is demonstrated by using certain zeolite adsorbent powder and conducting multiple adsorption/regeneration cycles in a flow system. The breakthrough curves of N2, H2O, O2, and CO2 during adsorption and desorption for the conventional packed bed and for the present structured bed are measured. A characteristic time, tau, which corresponds time duration for 2/3 change of the CO2 concentration in the adsorption or desorption profile, is used to compare the global adsorption and desorption kinetics. For the adsorption process, the smaller tau means a sharper adsorption front and thus better utilization of the adsorption bed. For the desorption process, tau number measures the regeneration time. Thus, for an efficient adsorbent bed design, small tau number is needed for better utilization and rapid regeneration.

The testing results show that both adsorption and regeneration time is reduced by a few times with the structured adsorbent bed. It is noted that in addition to the adsorption and regeneration kinetics advantage, the structured bed provides much lower pressure drop and other benefits as well. With some optimization of the structured bed, even higher level of kinetics rate enhancement is obtainable.

Fast adsorption/regeneration cycling increases throughput of the adsorbent material and low pressure drop reduces parasitic power consumption during constant adsorption/regeneration cycles. The structured adsorbent bed is a potential breakthrough adsorption technology to reduce the capital cost and energy consumption of an adsorption process.

723a Optimization of Synthesized Sodium a (NaA) Zeolite Membranes

Dacho Kunnakorn[1], Sujitra Wongkasemjit[1], Thirasak Rirksomboon[1], Pavadee Aungkavattana[2], Neeranut Kuanchertchoo[1], and Santi Kulprathipanja[3]. (1) The Petroleum and Petrochemical college, Chulalongkorn University, Bangkok, Thailand, (2) National Metal and Materials Technology Center (MTEC), Patumthani, Thailand, (3) UOP LLC, 50 East Algonquin Road, Des Plaines, IL 60017-5016

In general, hydrophilic NaA zeolite membranes are used for dehydration of organic solvent. The membranes can be synthesized on alumina support by three different heating methods, namely, microwave, electrophotoresis and autoclave (or conventional heating). Performance and the formation of this zeolite membrane were analyzed using pervaporation technique and XRD/SEM, respectively. From our previous study, the lab-scaled NaA zeolite membrane synthesized possessed the total flux and separation factor of around 0.4-1.3 kg/m^2.hr and 6,000, respectively.

In this work, the optimum conditions for synthesizing the lab-scaled NaA membranes on tubular alumina support by the aforementioned techniques are investigated to improve the total flux and separation factor of the membranes to higher than 1.3 kg/m^2.hr and 10,000, respectively.

723b Mixed Matrix Membrane for Gas Separation

Thanakorn Suntiworawut[1], Thirasak Rirksomboon[1], and Santi Kulprathipanja[2]. (1) Petrochemical Technology, The Petroleum and Petrochemical College, Chulalongkorn University, 254 Soi Chula 12 Prayathai Rd., Bangkok, 10330, Thailand, (2) UOP LLC, 50 East Algonquin Road, Des Plaines, IL 60017-5016

Mixed matrix membranes (MMMs) have been considered as an alternative gas separation technique due to their low energy consumption and high selectivity. In previous studies, it was found that silicon rubber MMMs incorporated with polyethylene glycol (PEG) and some diol isomers such as 1,2-, 1,3-, 1,4- and 2,3-butanediol can improve the selectivity and permeation rate of some gases. In this study, solid-liquid-polymer MMMs are prepared by a solution-casting method and then tested for permeation rate of various pure gases (CO2, N2, CH4 and H2) in order to ascertain the selectivity of membranes. The Ultem (polyimide) polymer is selected as a polymer base due to the higher selectivity of gases when compared to the silicon rubber. Activated carbon is selected for stabilizing PEG and diol isomers into its pores avoiding undesirable liquid leakage. The effects of diol isomers and pressure on Ultem-MMM performance have been investigated and will be reported at the meeting.

723c Transport Phenomena In Polymer Electrolyte Membranes Using Molecular Dynamics

Pil Seung Chung, Parag Jain, Lorenz T. Biegler, and Myung S. Jhon. Chemical Engineering, Carnegie Mellon University, 5000 Forbes Ave., Pittsburgh, PA 15213

Recently, fuel cells have gained considerable attention as an alternative energy source exhibiting attractive features of higher efficiency (more than twice the internal combustion engines), near zero greenhouse emissions, as well as simple and modular design. Amongst the various fuel cells, polymer electrolyte membranes PEM fuel cells (PEMFCs) are especially promising due to their higher efficiencies, low temperature of operation leading to short start-up times, and various applications including automobiles and portable electronics. Here PEM, which plays a critical role in determining fuel cell cost and performance, has been examined via molecular dynamics (MD) simulations.

Based on our previous studies [1, 2], Nafion® molecules were modeled via a coarse-grained, bead-spring model, which neglects the detailed atomic structural information while keeping the essence of the molecular structure. Lennard Jones and finitely extensible nonlinear elastic models were used to describe potential energies among the beads in the system. In addition, to examine the water uptake in PEM, long-range Coulombic interactions between water molecules and the beads were explicitly incorporated in our model. With the total potential energy constructed, MD simulations were performed using the Langevin equation.

Using these simulations, we examined the static and dynamic properties of PEM to obtain a relationship between the PEM nano/micro-structure, its water uptake, and the macroscopic properties such as the conductivity and water diffusion. In order to examine the PEM swelling mechanism and its effect, we calculated PEM density and the radius of gyration of the molecules for different amounts of water uptake. We also investigated the transport processes of protons and water molecules under various water uptake conditions. Specifically, we calculated the self-diffusion coefficient of protons to examine the diffusion process, and the proton conductivity as a function of water uptake. In addition, different types of PEMs were examined by modifying the molecular structure.

The findings from this study, which relate the PEM diffusion coefficient to its structure and composition, can serve to model the non-homogeneity in PEM diffusion processes. This information can be employed to build nano/microscopic models for the PEM water management, as well as to construct an integrated modeling and optimization framework for PEMFCs [3].

References:

1. Q. Guo, H. G. Chen, B. C. Smith, and M. S. Jhon, J. Appl. Phys., 97, 10P301 (2005).

2. Q. Guo, P. S. Chung, H. G. Chen, and M. S. Jhon, J. Appl. Phys., 99, 08N105 (2006).

3. P. Jain, L.T. Biegler, and M.S. Jhon, "Parametric Study and Estimation in CFD-based PEM Fuel Cell Models," AIChE J., to appear (2008).

723d Pure Gas Permeability Measurements on Bis(phenoxyethoxy) and Bis (tert-butylphenoxyethoxy) Polyphosphazenes

Christopher J. Orme and Frederick F. Stewart. Chemical Sciences, Idaho National Laboratory, P.O. Box 1625, Idaho Falls, ID 83415-2208

Polyphosphazenes are hybrid polymers having organic pendant groups attached to an inorganic backbone. The inorganic backbone provides these materials with good chemical and mechanical stability. In addition, phosphazene materials can be tailored to specific applications through the placement of organic pendant groups on the phosphazene backbone. Tailoring phosphazene polymers can yield materials with widely varying physical characteristics, such as, glass transitions that range from -80 „aC to 50 „aC. Polymer morphology that ranges from amorphous rubbery to glassy can be created with the careful selection of pendant groups.

In this study two different homo-polymers were synthesized using the following organic pendant groups; 2-phenoxyethanol and 4-tert-butyl-2-phenoxyethanol. The effects of these pendant groups on polymer morphology and on the gas transport properties are presented. Pure gas permeability measurements were made using the time-lag method. Permeabilities for six gases (He, H2, O2, N2, CO2, and CH4) were determined over a range of temperatures from -20 „aC to 75 „aC, thus providing gas transport information both above and below the polymer glass transition. Activation energies will be presented. Ideal selectivity for the gas pairs: O2/N2, H2/O2, H2/CO2, and CO2/CH4 are addressed.

723e Evaluating the Impacts of Fouling Index Values on Reverse Osmosis Membrane Fouling

Young Haeng Lee[1], Chanhyuk Park[1], Sanghyup Lee[1], Jaehoon Kim[2], Seockheon Lee[1], Dooil Kim[1], and Seongpil Jeong[1]. (1) Center for Environment Technology Research, Korea Institute of Science and Technology, 39-1 Howelgok-Dong, Sungbuk-Gu, Seoul, 130-650, South Korea, (2) Environmental & Process Technology Division, Korea Institute of Science and Technology, 39-1 Howelgok-Dong, Sungbuk-Gu, Seoul, 130-650, South Korea

In reverse osmosis (RO) applications, silt density index (SDI) and modified fouling index (MFI) have been most widely applied for assessing a fouling potential and evaluating the requirement of pre-treatment. However, their limitations have been reported by many researchers so far.

In the present study, the impacts of membrane fouling indices on reverse osmosis membrane fouling were investigated systematically. Bench-scale fouling experiments with thin-film composite (TFC) RO membranes were performed at various combinations of particle size, particle concentration, and membrane type. Mono-dispersed silica suspensions with mean diameters ranging from 0.1 to 20 μm were used as model colloids.

Existing fouling indices such as SDI and MFI showed the impact of various particle and membrane characteristics on their measurements. Specifically, fouling indices significantly increased as feed particle size decreased under identical particle loading, however, the effect of membrane materials on MFI were not markedly observed when using particles larger than 3.0 μm although MFI value for the MCE membranes was higher than PVDF membrane due to increased membrane resistance.

Prediction of MFI by using Happel cell model based on the hydraulic resistance of the particle cake layer was performed. Modeling of experimental data revealed no noticeable changes in cake layer structure when the RO membrane was fouled by inorganic particles larger than 1.0 μm particles. However, the model predicted MFI values did not exhibit good agreement with the experimental results for particles smaller than 1.0 μm.

Overall, based on the experimental results, when the inorganic particles were the primary cause of RO membrane fouling, both the SDI and the MFI values were not closely related to the true fouling potential of feed water. Therefore, fouling index values of SDI and MFI could not predict RO membrane fouling accurately and were not absolute numbers for the design of RO membrane processes.

723f Surface Modified Nonwoven Fabrics for Bioseparations

Haiyan Liu[1], Patrick V. Gurgel[2], Yong Zheng[2], Omon Herigstad[1], Benham Pourdeyhimi[3], and Ruben G. Carbonell[1]. (1) Chemical and Biomolecular Engineering, North Carolina State University, College of Engineering 1, 911 Partners Way, Raleigh, NC 27695, (2) Department of Chemical and Biomolecular Engineering, ProMetic Life Sciences and NC State University, 8168 chemin Montview, 911 Partners Way, Mont-Royal, QC H4P 2L7, Canada, (3) College of Textiles, North Carolina State University, Raleigh, NC 27695

The isolation and purification of biotherapeutic protein products from genetically engineered cell culture or from human plasma contributes anywhere from 50 to 80% of the cost of these drugs. One of the major challenges of the industry is to reduce operating separation costs by minimizing the number of purification steps. One approach to this is the use of affinity adsorption methods to recognize the desired protein product. The other is to do a preliminary isolation or concentration step early in the process, even in the presence of whole cells from the cell culture medium.

In this project, we are investigating the use of small, robust and inexpensive peptide affinity ligands attached to nonwoven fabrics to effect the separation. This might provide an attractive alternative to traditional chromatographic resin columns or porous membranes. The advantages of this nonwoven separation system are the low cost, negligible diffusional limitations, low pressure drop and the controllable pore size and fiber diameter which makes it possible to pass complex media such as a homogenized cell culture medium through the fabric matrix. A commercially available PBT nonwoven fabric was used to create a new affinity membrane in this study. Hydroxylated polyacrylates were grafted onto the nonwoven surface, together with a spacer arm to decrease nonspecific binding and to allow for convenient ligand coupling. The physical and chemical properties of this modified material were studied by SEM, FTIR, XPS and EA (elemental analysis). A conformal hydrophilic layer around each fiber was achieved by optimizing the surface treatment process. The results of preliminary studies on the removal of prion protein from red blood cell concentrate and plasma, as well as the removal of viruses from plasma samples will be presented.

723g Preparation of Superhydrophobic Pvdf Membranes by Electrospinning of Fluorine-Silane Functionalized Pvdf

Yingbo Chen, Qifeng Liu, and Hern Kim. Department of Environmental Engineering and Biotechnology, Myongji University, San 38-2 Nam-dong Cheoin-gu, Yongin, Kyonggi-do, 449-728, South Korea

Superhydrophobic surface can be obtained by either lowering the surface energy of the chemical composition through introduction of low surface energy groups or increasing the toughness of the surface through formation of hierarchical structures. Fluorine-silane functionalized PVDF is synthesized by graft polymerization of 3-trimethoxylpropyl methylacrylate from PVDF and followed with coupling of fluorine silanes. Flat membrane prepared using this fuctionalized PVDF has a water contact angle of 140o. Superhydrophobic PVDF membrane with a contact angle larger than 150o is prepared by electrospinning of the fluorine-silane functionalized PVDF and collecting on a stainless steel plate. The morphologies of the membranes are characterized using scanning electron microscopy. The surface composition of the membranes is analyzed using FTIR and water drops on the surface of the membrane are observed using a microscopy.

723i Supported Liquid Membranes with Strip Dispersion for the Recovery of Cephalexin

Michael E. Vilt, Chemical and Biomolecular Engineering Department, The Ohio State University, 125 Koffolt Laboratories, 140 West 19th Ave., Columbus, OH 43210 and W. S. Winston Ho, Ohio State University, 140 west 19th. Ave., Columbus, OH 43210.

Cephalexin is an important and widely used semisynthetic cephalosporin. Cephalosprorins along with penicillins are Beta-lactam antibiotics, which account for the majority of the antibiotic world market. Cephalexin is traditionally produced by a ten-step chemical synthesis. An enzymatic synthesis for Cephalexin has been developed, and offers several advantages over the classical route. The enzymatic synthesis reduces energy and solvent waste, but has been used in industrial production on a limited basis due to its difficult separation. Liquid membranes, in particular supported liquid membranes (SLMs), are a promising solution to the separation. SLMs are not used commercially, as they are still plagued with problem of long term instability. SLM with strip dispersion has been a recent development to solve the issue of stability.

SLMs with strip dispersion are formed when an aqueous strip solution is dispersed in an organic membrane solution by a mixer, and passed onto one side of a membrane support. When a microporous hydrophobic support is used, the organic phase of the dispersion becomes imbedded in the pores of the support, forming a stable SLM. Stability is maintained by having a constant supply of organic membrane solution to the pores. An aqueous feed solution is passed on the other side of the membrane support, and the target species is extracted into the organic solution by a selective carrier. The target species is then stripped by the aqueous strip solution. For final recovery of the target species, the mixer is turned off or a settler is used, and the dispersion quickly separates.

In this study, Cephalexin has been separated and concentrated from an aqueous solution using SLMs with strip dispersion. The following key parameters were investigated: feed and strip dispersion flowrate, strip dispersion mixing rate, carrier concentration, counter ion concentration, pH, and the volume of aqueous strip solution. Experiments used a Liqui-Cel® hollow fiber module as a microporous support. The organic membrane solution of the SLM consisted of Aliquat 336, Isopar L (isoparaffinic hydrocarbon solvent), and 1-decanol. The aqueous strip solution was composed of potassium chloride and citrate buffer. An enrichment factor of 3.2 was observed in the aqueous strip solution while achieving over 99% extraction and 96.2% total recovery. In this case, the aqueous feed solution of 5500 ppm (15 mM) was lowered to 30 ppm when using an organic membrane solution containing 2.5% Aliquat 336. The resulting overall mass transfer coefficient was 1.6 x 10-5 cm/sec. The mass flux of Cephalexin for this system was found to be independent of aqueous feed and strip dispersion flowrates, suggesting a major mass transfer resistance due to chemical reaction kinetics, which is supported by calculated individual mass transfer resistances. The pH of the aqueous strip phase was found to play a more significant role when trying to achieve higher enrichment ratios. It was observed that the highest stripping efficiency occurs when the pH of the aqueous strip phase is between the values of 5 and 6.

Synthesis and Characterization of Cross-linked Polyvinylalcohol/Polyethyleneglycol Blend Membranes for CO_2/CH_4 Separation

Rong Xing, W.S. Winston Ho
Department of Chemical and Biomolecular Engineering
Department of Materials Science and Engineering
The Ohio State University, Columbus, OH, 43210

Abstract:

Polymeric membrane technology has received increasing attention in the field of gas separation, e.g., natural gas sweetening [1], hydrogen recovery and purification [2], and flue gas separation [3], due to its low cost, energy efficiency and ease of operation. Natural gas sweetening aims to separate acid gas, mainly CO_2, from CH_4 and other mixed gas components, to meet pipeline specifications (2% CO_2 in the U.S.A.). Currently, the polymer membranes, either commercially or reported for CO_2/CH_4 separation in the literature, show either a high permeability with a low selectivity [4] or a high selectivity with a low permeability [5,6], which potentially restrict their scale-up application. Recently, polymer membranes containing polar ether segments have been investigated to separate CO_2 from CH_4 due to the strong affinity between CO_2 and polar ether segments. For example, pure acrylated PEG has been reported to fabricate cross-linked polymer membranes via UV-irradiation, and the resulting membranes showed good performance in CO_2/CH_4 separation [1,7,8]. However, it was difficult to control the formation of large, homogeneous and defect-free ultra thin films. Also, the membranes have relatively low mechanical strength and high cost.

The objective of this research, therefore, is to develop an economically and technically viable polymer membrane for CO_2/CH_4 separation. The target membrane should have desirable separation performance, good mechanical strength and low cost. A cross-linked polyvinyl alcohol (PVA)/polyethylene glycol (PEG) blend membrane was developed to meet the above requirements. The ethylene groups in PEG enhances CO_2 transport through membrane while PVA provides mechanical strength of polymer matrix. The cross-linking and entanglements of PVA and PEG create robust and homogeneous network. In order to improve separation performance, zeolite 5A was chosen as the inorganic filler due to its stronger adsorption of CO_2 than CH_4 [9].

The cross-linking time, molecular weight of PEG and PEG (MW 200) content were found to significantly affect the transport properties of the zeolite unfilled PVA/PEG membrane. The optimal cross-linking time was found to be 84 minutes in order to obtain simultaneously good CO_2 permeability and CO_2/CH_4 selectivity. PEG (MW 200) showed the maximum CO_2 permeability among the PEGs with different average molecular weight investigated. Figure 1 shows the effect of PEG (MW 200) content on the separation performance of CO_2/CH_4, and it was found that, at 64 wt.% PEG (MW 200), the membranes showed good CO_2 permeability of 80.2 Barrers (1Barrer = 10^{-10} cm^3 (STP)•cm/cm2•s•cmHg) and CO_2/CH_4 selectivity of 33.

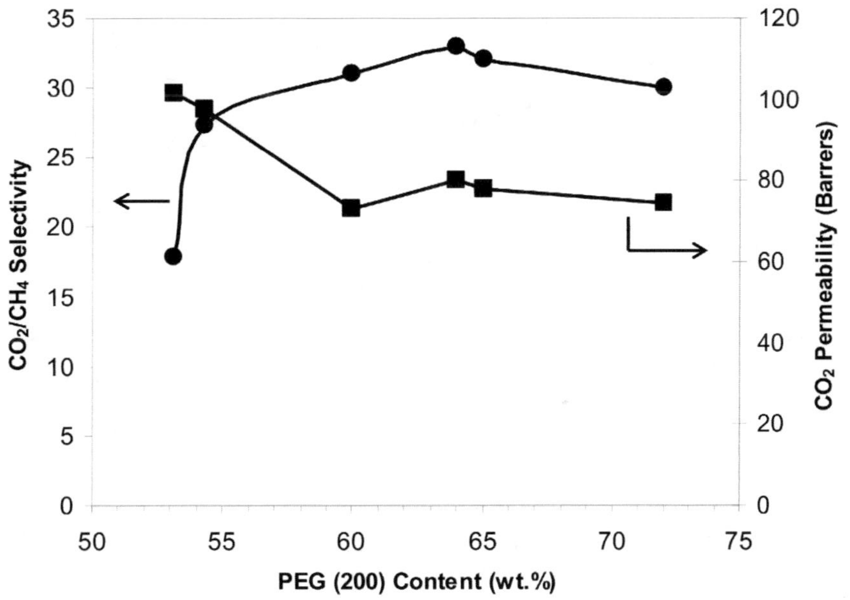

Figure 1. Effect of PEG(MW 200) content on the separation performance of CO_2/CH_4 for the PVA/PEG membrane at 30°C and 150 psi.

The effects of zeolite 5A loading, temperature and feed pressure on the gas separation performance of both zeolite-filled and unfilled PVA/PEG membranes were investigated. Figure 2 illustrates the effect of zeolite 5A loading on CO_2 permeability and CO_2/CH_4 selectivity for the PVA/PEG membrane. As can be seen that the CO_2/CH_4 selectivity decreased as the amount of zeolite 5A loading increased; while the CO_2 permeability first decreased, then drastically increased. The drop in CO_2/CH_4 selectivity could be ascribed to the formation of interfacial cracking at the polymer-zeolite 5A interfaces, which have been evidenced by SEM pictures. The initial decrease of CO_2 permeability was mostly likely caused by the partial pore blockage of the zeolite by polymer chains; while the drastic increase of CO_2 permeability suggested that, with the increase of zeolite 5A loading, CO_2 molecules permeated not only through the polymer matrix, but an additional network created by interconnected zeolite 5A particles. The effect of temperature on the separation performance of both zeolite 5A filled and unfilled PVA/PEG membranes was investigated in the range of 16.4 - 40°C at the feed pressure of 150 psi. On the one hand, the CO_2/CH_4 selectivities of all the membranes decreased linearly as temperature increased due to an increase in polymer chain mobility and thereby, diffusivity at the increased temperatures. However, compared to the zeolite-unfilled membranes, the CO_2/CH_4 selectivity could be influenced to a lesser extent from the incorporation of zeolite particles into the polymer matrix. On the other hand, the permeability of both CO_2 and CH_4 increased as temperature increased for all the membranes. The trade-off relationship exhibited by the CO_2/CH_4 selectivity and CO_2 permeability suggested that overall gas permeation in the PVA/PEG membranes follows the solution-diffusion mechanism. Thus, the gas permeability could be approximately described using the Van't Hoff-Arrhenius approach. Based on the calculation results, we could conclude that (1) the zeolite-filled PVA/PEG membrane had higher activation energy for both CO_2 and CH_4 than that of the unfilled PVA/PEG membrane, and (2) CO_2 always showed a lower pre-exponential value and activation energy than CH_4. These results indicated that the

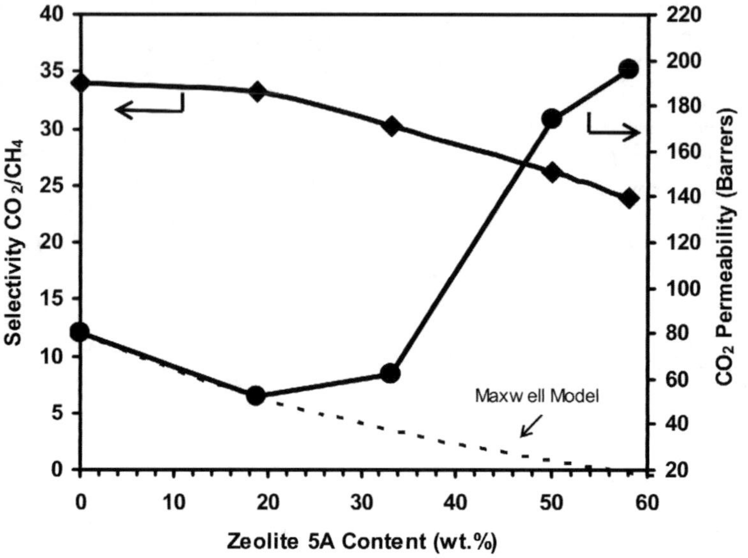

Figure 2. Effect of zeolite 5A loading content on the separation performance of CO_2/CH_4 for the PVA/PEG membrane at 30°C and 150 psi.

temperature effect on the zeolite-filled PVA/PEG membrane was more significant than the unfilled ones, and that CO_2 had a lower enthalpy of sorption in the PVA/PEG membrane than CH_4, possibly due to the strong interaction of CO_2 with ethylene oxide segments. It is noteworthy that the unfilled PVA/PEG membrane could perform beyond the Robeson's upper bound as temperature decreased. As the feed pressure increased, the CO_2/CH_4 selectivity and CO_2 permeability decreased for the unfilled PVA/PEG membranes, while the filled ones were not affected significantly, indicating that the incorporation of zeolitic particles could significantly improve the performance stability of membranes at high pressures.

The separation performance results of some zeolite-filled and unfilled PVA/PEG membranes were summarized with reference to the Robeson's upper bound line for the CO_2/CH_4 separation as shown in Figure 3. The performance of the zeolite-filled membranes approached the upper bound line with the increase of zeolite 5A loading. The performance of unfilled membranes gradually approached the bound line, and finally exceeded the boundary as the temperature decreased from 40 to 16.4°C. In comparison with the commercialized CA and Matrimid®, both the zeolite-filled and unfilled PVA/PEG membranes showed potentially to be industrially attractive.

Future work is required to eliminate or at least minimize the interfacial voids in order to improve the CO_2/CH_4 selectivity, and reduce membrane thickness to obtain high CO_2 permeance at low cost.

Acknowledgement: This report is based on work funded by Shell Oil Company, the National Science Foundation, and The Ohio State University. Part of this material is based upon work

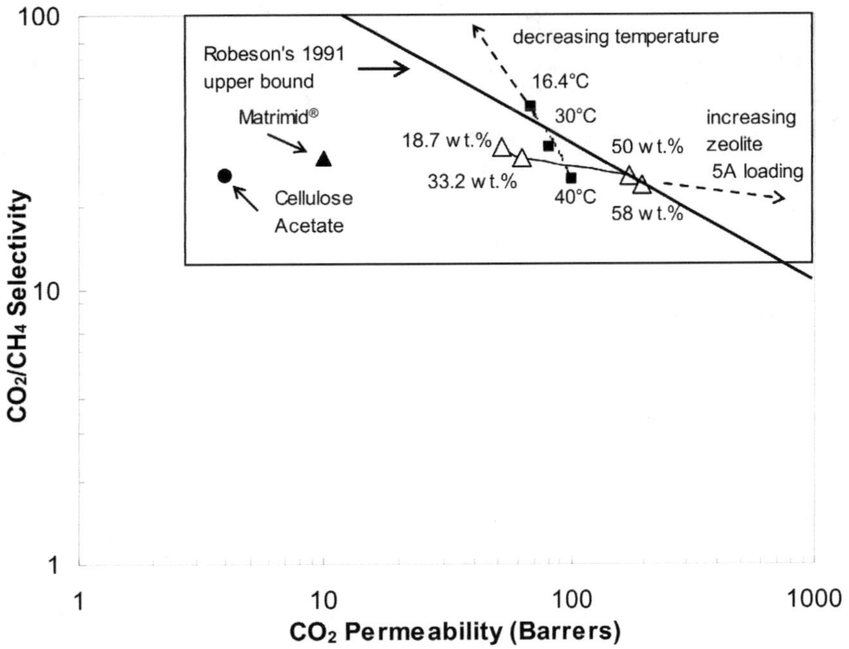

Figure 3. CO_2/CH_4 selectivity versus CO_2 permeability for zeolite-filled and unfilled cross-linked PVA/PEG blend membranes. The solid line represents Robeson's 1991 "upper bound". Solid squares refer to zeolite-unfilled membranes and open triangles represent zeolite-filled membranes. The right-top shaded area is commercially attractive.

supported by the National Science Foundation under Grant No. 0625758. We would like to thank Chi Yen for taking the SEM pictures.

Reference:

[1] H.Lin, E.V. Wagner, R. Raharjo, B.D. Freeman and I. Roman, High performance polymer membranes for natural-gas sweetening, Adv. Mater. 18 (2006) 39-44.

[2] J. Zou, J. Huang, W.S.W. Ho, CO_2-selective water gas shift membrane reactor for fuel cell hydrogen processing, Ind. Eng. Chem. Res. 46 (2007) 2272-2279.

[3] A. Chakma, Separation of CO_2 and SO_2 from flue gas streams by liquid membranes, Energy Convers. Mgmt. 36 (1995) 405-410.

[4] M.P. Chenar, M. Soltanieh, T. Matsuura, A. Tabe-Mohammadi, K.C. Khulbe, The effect of water vapor on the performance of commercial polyphenylene oxide and Cardo-type polyimide hollow fiber membranes in CO_2/CH_4 separation applications, J. Membr. Sci. 285 (2006) 265-271.

[5] Y. Li, T.S. Chung, Highly selective sulfonated polyethersulfone (SPES)-based membranes with transition metal counterions for hydrogen recovery and natural gas separation, J. Membr. Sci. 308 (2008) 128-135.

[6] C. Staudt-Bickel, W.J. Koros, Improvements of CO_2/CH_4 separation characteristics of polyimides by chemical crosslinking. J. Membr. Sci. 155 (1999) 145-154.

[7] H. Lin, E.V. Wagner, J.S. Swinnea, B.D. Freeman, S.J. Pas, A.J. Hill, S. Kalakkunnath, D.S. Kalika, J. Membr. Sci. 276 (2006) 145-161.

[8] N.P. Patel, A.C. Miller, R.J. Spontak, Highly CO_2-permeable and selective membrane derived from crosslinked poly(ethylene glycol) and its nanocomposites, Adv. Funct. Mater. 14 (2004) 699-707.

[9] S. Pakseresht, M. Kazemeini, M.M. Akbarnejad, Equilibrium isotherms for CO, CO_2, CH_4, and C_2H_4 on the 5A molecular sieve by a simple volumetric apparatus, Separation and Purification Technology, 28 (2002), 53-60.

723k Preparation of Composite Nanofibers and Membranes by Electrospinning of Pvdf Solution Containing Hpa

Yingbo Chen and Hern Kim. Department of Environmental Engineering and Biotechnology, Myongji University, San 38-2 Nam-dong Cheoin-gu, Yongin, Kyonggi-Do, 449-728, South Korea

Heteropoly acids (HPA), such as phosphotungstic acid, are widely used as homogeneous and heterogeneous catalysts because of their good thermal stability, high acidity and high oxidizing ability. In addition, heteropoly acids have high proton conductivity even at a temperature higher than 100 oC. Poly(vinylidene fluoride) (PVDF) is used generally for the advantages of high strength, resistance to solvents, acids, bases. It is easy to process due to its relatively low melting point, low density and low cost compared to the other fluoropolymers. In this study, a composite nanofiber containing PVDF and heteropoly acid is prepared by electrospinning of their mixtures. The nanofibers are characterized by scanning electron microscopy, FTIR, TGA, and XRD. The nanofibers can be used a heterogeneous catalysts for organic reactions. Membranes prepared from the nanofibers can be used as a membrane reactor or a proton conductivity membrane for fuel cells application.

723l Preparation of Poly(ethylene glycol)-Grafted Polycaprolactone Porous Membranes by Using Oxygen Plasma to Prevent Fibroblast Adhesion

Chi Yen[1], Hongyan He[2], Zhengzheng Fei[3], W.S. Winston Ho[2], and L. James Lee[1]. (1) Chemical and Biomolecular Engineering, The Ohio State University, 140 WEST 19TH AVE, Columbus, OH 43210, (2) Department of Chemical and Biomolecular Engineering, The Ohio State University, 140 WEST 19TH AVE, Columbus, OH 43210, (3) Center for Affordable Nanoengineering of Polymeric Biomedical Devices, Ohio State University, 140 west 19th. Ave., Columbus, OH 43210

The adhesion and proliferation of fibroblast onto the surface of an implant biomedical device can lead to fibrous tissue encapsulation. The fibrous encapsulation can be detrimental to various implant applications, such as drug delivery devices. Therefore, the development of a nonfouling surface is necessary for the implantable drug delivery devices.

Poly(ethylene glycol) (PEG) is a nontoxic and nonimmunogenic polymer. PEG has the ability to inhibit cell adhesion on a surface due to its hydrophilicity, steric hindrance, and chain mobility. To produce a nonfouling surface, a new method was proposed for grafting PEG on the polycaprolactone (PCL) porous membrane surface. In this study, PEG(400)-monoacrylate was pre-coated on the PCL membrane surface. By using oxygen plasma, PEG could be successfully bonded on the PCL membrane.

The operating condition of the oxygen plasma was optimized to prevent fibroblast adhesion. When the membrane was treated with soaking in a 0.1M PEG (Mw 400)-monoacrylate solution and the oxygen plasma of 25W power for 5 seconds, there was around 90% less number of fibroblast attached onto the PEG-grafted PCL porous membrane than the untreated membrane surface. Attenuated total reflectance-Fourier transform infrared spectroscopy (ATR-FTIR) showed that the intensity of a PEG characteristic peak (~1100 cm-1) increased significantly for the (Mw 400)-grafted membrane. Moreover, the contact angle for the PEG (Mw 400)-grafted surface was decreased significantly in comparison with the untreated membrane. Therefore, the anti-biofouling surface of the PEG-grafted PCL porous membrane can be obtained for the implant drug delivery devices.

SUPRAMOLECULAR FUEL CELL MEMBRANES

A. Fuchs[1], J. Sutrisno[1], C. Evrensel [2], F. Gordaninejad [2], B. Kavlicoglu [2], Y. Liu [2]

1) *Chemical Engineering Department, University of Nevada, Reno – Nevada, USA*
2) *Mechanical Engineering Department, University of Nevada, Reno – Nevada, USA*

Introduction

The development of fuel cell technology depends on the availability and low cost of high quality proton exchange membranes (PEMs). The trend in PEMs research includes: high temperature operation, low humidity, and high proton conductivity. A variety of polymers are being considered to improve the performance of the membranes so that they meet the expectations of DOE's 2020 projections.

The currently available PEMs are suitable for use at maximum 80^0C because of the requirement of the presence of water for proton conductivity. However, for emerging applications, a fuel cell operating at temperatures as high as 120 ^{0}C is required. This can possibly be achieved by creating novel PEMs. Supramolecular polymers are being investigated extensively and may be conjugated with heteropolyacids (HPAs) to prepare high performance PEMs. Metal coordinated polymers, such as: ruthenium or zinc bipyridine polybenzimidazole (PBI) will be synthesized during this project. Self assembled supramolecular structures are formed by the interaction of metal (Ru or Zn) and the nitrogen heterocycle from bipyridine PBI [1]. In addition, hydrophilic – hydrophobic blocks of polymer backbones also create supramolecular interactions, resulting in bi-directional proton conductivity.

Membrane characterization techniques including: differential scanning calorimetry (DSC), dynamic mechanical analysis (DMA), electrochemical impedance spectroscopy (EIS), rheology and single membrane test cell studies will be done as part of this project.

Nanostructured Composite PEMs

Nanostructured polymers are synthesized from macromolecules demonstrating supramolecular interactions. Examples are: hydrogen bonding, hydrophobic interactions, organometallics, electrostatic interactions, liquid crystal polymers and van der Waals [2]. Composite membranes are prepared using polyimide (PI) and polybenzimidazole (PBI) with heteropolyacids (HPAs) [3]. Nanostructured PEMs may be prepared from supramolecular polymers and HPAs.

Surface Polymerization – covalently bonded polymers and supramolecular polymers

Most work on surface polymerization has involved covalent bonding of polymers to a particle surface. It is also of interest to investigate the use supramolecular polymers for surface polymerization.

Surface polymerization has been widely done in with covalent bonding for silicon dioxide and Si- H groups. Germanium has also been surface polymerized by Xu et al [4] by covalently immobilized polymer brushes from surface-initiated ATRP. Well-defined polymer-Ge hybrids, consisting of covalently tethered polymer brushes of pentafluorostyrene (PFS), (2-dimethylamino)ethyl methacrylate (DMAEMA), and PFS-DMAEMA diblock copolymers, were prepared. This is a process for covalent bonding of an ATRP initiator monolayer on the Ge-H surface. Well-defined polymer-Ge hybrids could be readily prepared via surface-initiated ATRP on the substrates.

Surface polymerization with supramolecular polymers has been used for the fabrication of thin films with metallo-supramolecular polyelectrolytes (MEPE) for stimuli responsive layers as well as electrochromic windows. Investigation of the self-assembly of three-dimensional MEPE networks on the surface as a means to control the amount deposited were conducted by adjusting the cross-linking degree. Self assembled iron(II) and 1,4-bis(2,2':6',2"-terpyridin-4'-yl)benzene (tpy-ph-tpy) on quartz slides have resulted in homogeneous thin layers. A base layer of polyethyleneimine (PEI) / poly (styrene-sulfonate) (PSS) was used for a well defined starting condition [5].

Supramolecular PEMs - Polybenzimidazole

PBI (5 wt.%) was dissolved in N,N-dimethylacetamide (DMAc) or 1-Methyl-2-pyrollidone (NMP), and mixed together with sulfonated polymer. 3-pentadecylphenol (PDP, 1-5% of PBI-sulfonated polyamide) was added and mixed to dissolve PDP. Then, solution was evaporated and dried in a vacuum oven at 80 oC for 24 hours to form a membrane.

Results and Discussion

The focus of our efforts is the development of a nanocomposite membrane with HPA and supramolecular polymers. The first report consisted of differential scanning calorimeter (DSC) results of a bypyridine PBI. The results indicated that the bipyridine PBI had a thermal transition of 357.2^0 C. The synthesis method of the bipyridine PBI was described in the earlier bimonthly report.

During the course of this research dissolving the bipyridine PBI was investigated using various solvents including: dimethyl acetamide (DMAc), n- methyl pyrrolidone (NMP) and dimethyl formamide (DMF). DMAc was found to be most commonly used in the literature and was the most effective, although high concentrations of PBI could not be achieved.

Characterization of HPA

The diameter of silicotungstic acid particles was measured using Nikon Eclipse LV100 microscope. Figure 2 shows that the particulate sizes ranged from 2.7 to 3.8 microns.

PBI and Inorganic Proton Conductor Composite Membrane

In order to enhance the proton conductivity, inorganic proton conductors were used to prepare composites with the supramolecular PEMs. The inorganic proton conductor also improved the mechanical strength of the composite PEM [6- 14]. In addition to the curve shown on the left hand side, the peak on the right side represents the Warburg impedance behavior. The conductivity for this system was 1.4 x 10-5 S/cm which is substantially lower than the Nafion™ membrane. In addition, these membranes were very brittle.

Supramolecular PEMs - Polybenzimidazole

The synthesized polymer is supramolecular because of hydrophobic / hydrophilic interactions between the sulfonated polymer (hydrophilic) and the PBI (hydrophobic) segments. These polymers are expected to form controlled "lameller" or layered structures by adding the amphiphile (PDP). This is anticipated to provide an additional level of hierarchical design.

There is also hydrogen bonding between the amphiphile (PDP) and the sulfonic acid group. Proton transfer occurs between the sulfonic acid group and the nitrogen heterocycle in PBI. We believe that phase separation takes place in the hydrophobic / hydrophilic system, providing the opportunity for enhanced directional proton conductivity. Mechanical shearing provides orientation of within the membrane. Channels of hydrophobic and hydrophilic regions are formed during shearing.

These regions provide different proton conductivities in the radial and tangential directions. We speculate that because of the long chained nature of the polymers that a lameller or "onion-skin" structure of both the hydrophobic and hydrophilic regions is developed during shearing. Therefore enhanced proton conductivity results tangentially. It is more difficult for protons to cross over the hydrophobic bands formed within the membrane. The proton conductivites measured were: 3.30.10-3 S/cm for the tangential direction and 1.03.10-3 S/cm for the radial direction.

Composite Membranes - Polyamic Acid and Organoclay

Composite membranes were synthesized from polyamic acid and sulfuric acid treated organoclay. This system was investigated in order to improve the proton conductivity of the composite membrane. Organoclay was chosen as reinforcement because it is known to provide high quality composite PEM membranes [8-16]. The sample of composite membranes is shown in Figure 10. In our case, the organoclay was first treated with sulfuric acid in order to increase the conductivity. In imidization process the first step involves reaction between the dianhydride and diamine. The second step is the thermal imidization step and ring closing. This approach was used for synthesis of the composite membrane.

The proton conductivity as a function of temperature from 25 – 80 ^{o}C. The proton conductivity increases from 1.59.10-2 S/cm (25 ^{o}C) to 2.56.10-2 S/cm (80 ^{o}C). This is the highest proton conductivity achieved in our experiments to date. This approach may have applications for anhydrous operation.

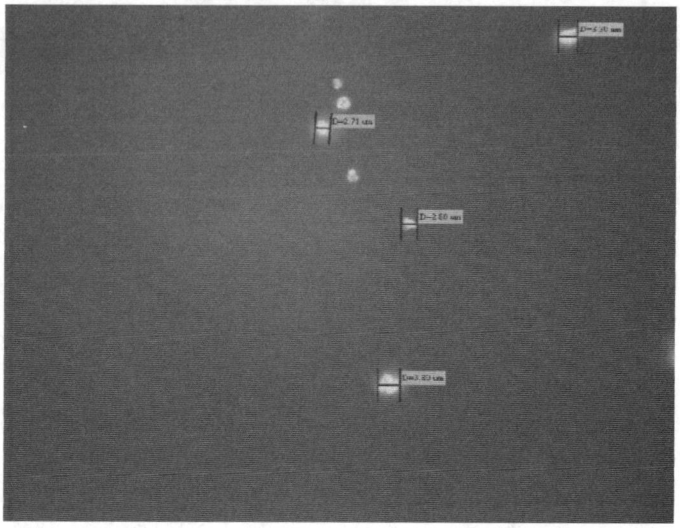

Figure1: Microscopic image of finely ground silicotungstic acid

The surface initiator will be immobilized on the HPA using silane through the formation of silanol group that is covalently bonded to the HPA surface. In the ATRP technique, copper based metal and sparteine will be used as metal catalyst and ligand respectively. The goal of this surface polymerization is to prevent the HPA from being washed out with water during fuel cell operation.

References

1. Sutrisno, J, “Supramolecular proton exchange membrane for fuel cells”, In progress.
2. H. Jiang et al., ‘Infinite chains constructed from flexible bis(benzimidazole)-based ligands’, Polyhedron (2008), in press.
3. He et al, ‘Proton conductivity of phosphoric acid doped polybenzimidazole and its composites with inorganic proton conductors’ , Journal of membrane sciences, 226,169-184, 2003.
4. Xu et al, ‘Well-Defined Polymer-Germanium Hybrids via Surface-Initiated Atom Transfer Radical Polymerization on Hydrogen-Terminated Ge(100) Substrates’, Organometallics, 24, 1768, 2005.
5. Sievers et al, ‘Thin Films of Cross-Linked Metallo-Supramolecular Coordination Polyelectrolytes’, Langmuir, 23, 12179, 2007.
6. R. He, Q. Li, A. Bach, J. O. Jensen, and N. J. Bjerrum, Journal of Membrane Science, 277, 38 – 45 (2005).
7. A. Gültek, M. G. ˙Içduygu, and T. Seçkin, Materials Science and Engineering, B107, 166–171 (2004).
8. S. Liu, Q. Pu., L. Gao, C. Korzeniewski, and C. Matzke, Nano Letters, 5, 1389 – 1393 (2005).
9. Y. F. Lin, C. Y. Yen, C. H. Hung, Y. H. Hsiao, and C. C. M. Ma, Journal of Power Sources, 168, 162–166 (2007).

10. D. H. Jung, S.Y. Cho, D. H. Peck, D. R. Shin, and J. S. Kim, J. Power Sources, 118, 205 – 211 (2003).
11. M. K. Song, S. B. Park, Y. T. Kim, K. H. Kim, S. K. Min, and H. W. Rhee, Electrochim. Acta, 50, 639 – 643 (2004).
12. J. M. Thomassin, C. Pagnoulle, D. Bizzari, G. Caldarella, A. Germain, and R. Jérôme, Solid State Ionics, 177, 1137 – 1146 (2006).
13. C. H. Rhee, H. K. Kim, H. Chang, and J. S. Lee, Chem. Mater., 17, 1691 – 1697 (2005).
14. J. M. Thomassin, C. Pagnaulle, G. Caldarella, A. Germain, and R. Jérôme, Polymer, 46 11389 – 11395 (2005).
15. S. L. C. Hsu, and K. C. Chang, Polymer, 43, 4097 – 4101 (2002).
16. S. H. Kwak, T. H. Yang, C. S. Kim, and K. H. Yoon, Solid State Ionics, 160, 309 – 315 (2003).

723o Membrane Extraction in Cross-Flow Rectangular Modules with Reflux for Improved Performance

Ho-Ming Yeh, Tamkang University, Department of Chemical and Materials Engineering, Tamsui, 251, Taiwan

Application of recycle to the design and operation of a mass or heat transfer equipment can effectively enhance the effects on mass or heat transfer, leading to improved performance . The effect of internal reflux in the raffinate phase for membrane extraction through a cross-flow rectangular module was investigated. Theoretical analysis of mass transfer in cross-flow devices with or without recycle was analogous to heat transfer in cross-flow heat exchangers. In contrast to a device without reflux, considerable improvement in mass transfer is achievable if cross-flow membrane extractors are operated with reflux which provides the increase of fluid velocity, resulting in reduction of mass-transfer resistance. It is concluded that reflux can enhance mass transfer, especially for large flow rate and feed-concentration solution operated under high reflux ratio.

723p Ultrafiltration In Turbular Membranes Inserted Concentrically with a Ring Rod of Varying Ring Distances

Ho-Ming Yeh, Tamkang University, Department of Chemical and Materials Engineering, Tamsui, 251, Taiwan

Raising fluid velocity or creating turbulent behavior in the cross-flow membrane modules has two conflicting effects on ultrafiltration: They are the desirable effect of decrease in concentration polarization resistance and the undesirable effect of increase in frictional pressure loss. Since, along the flow channel in an ultrafilter, transmembrane pressure decreases while concentration polarization resistance increases. It appears therefore that gradually increasing the turbulent strength along the flow channel might suitably suppress the increasing resistance to permeation due to concentration polarization while properly maintaining the decreasing tranemembrane pressure. It is the purpose of present study to investigate the effect of varying baffled-ring distance on ultrafiltration and to derive the prediction equation for estimating the permeate flux, in a tubular membrane of a ring-rod insert with decreasing buffled-ring distance. It was found that the concentration polarization resistance could be reduced in a ring-rod tubular membrane ultrafilter by the turbulent behavior. The performance will be further improved if the baffled-ring distances gradually and properly decrease along the cross-flow channel with fixed number of baffled rings. Theoretical analysis was based on the mass and momentum balances coupled with the application of resistance-in-series model. Correlation predictions are rather acceptable for higher feed concentration comparing with the experimental results for ultrafiltration of dextran T500 aqueous solution in a tubular ceramic membrane of a nine-ring rod insert with gradually decreasing the baffled-ring distances.

723q Preparation of Zr(IV)-Loaded Membranes for Decontamination of as(V) from Aqueous Solutions

Shuaiwen Zou[1], Yuming Zheng[1], Soh Fong Lim[2], Yu-Ting Wei[3], and **J. Paul Chen**[4]. (1) Division of Environmental Science and Engineering, National University of Singapore, 10 kent ridge, Singapore, Singapore, (2) Chemical and Biomolecular Engineering Department, National University of Singapore (NUS), 1 Engineering Dr 2, E2-05-05, Singapore, 117576, Singapore, (3) Division of Environmental Science & Engineering, National University of Singapore, Block E2, #05-05, 5 Engineering Drive 2, National University of Singapore, Singapore, Singapore, (4) Environmental Science and Engineering, Natioanl University of Singapore, Singapore, Singapore

A zirconia-loaded PVDF composited membrane was developed in this study for removal of arsenic from aqueous solutions. The XRD analysis shows that the Zr(IV) is successfully encapsulated in the membrane. The SEM study illustrates that the higher Zr(IV) loading results in higher surface pore size and porosity of the membranes. The optimal pH range of adsorption is from 3.0 to 8.0. The maximum adsorption capacity can be achieved as high as 21.5 mg/g. The complete adsorption can be achieved in 30 h. The membrane filtration experiment shows that it effectively removes arsenic from synthesized waste stream. No leaching of Zr(IV) from the membrane is observed. The FT-IR and XPS studies show that the adsorption mechanism is a ligand-exchange process between As(V) ions and the hydroxyl ions present on the surface of Zr(IV)-membrane.

723r Fabrication of Polysulfone Membrane for the Treatment of Oil-in-Water Emulsion

Bandana Chakrabarty, Mihir K. Purkait, and **Aloke Kumar Ghoshal**. Indian Institute of Technology Guwahati, Guwahati, India

This paper focuses on the fabrication of polysulfone (PSf) membrane for the treatment of oily wastewater coming out from the post-treatment unit of petroleum industries where finely divided oil droplets are uniformly dispersed in large volumes of water. Flat sheet asymmetric polymeric membranes were prepared from homogeneous solution of PSf by phase inversion method. N-methyl-2-pyrrolidone (NMP) and Dimethyl Acetamide (DMAc) were used as solvents separately. Different molecular weights of PVP and PEG were used as the polymeric additives in the casting solution. The morphology and structure of the resulting membranes were observed by scanning electron microscope (SEM). The pore number, pore permeability and their distribution and average pore size of the membranes were determined by the liquid displacement method. The permeation performances of the membranes were evaluated in terms of pure water flux (PWF), equilibrium water content (EWC) and hydraulic permeability. Results show that with increase in molecular weight of PVP and PEG, the pore number as well as pore area in membranes increases. With increase in molecular weight of PEG from 400 to 20000, the PWF increases from 15.3 to 2713.4 $L.m^{-2}.hr^{-1}$with NMP as solvent while with DMAc as solvent, the PWF increases from 24.5 to 555.6 $L.m^{-2}.hr^{-1}$. The performances of different PSf membranes were evaluated by treating with pure water as well as with laboratory made oil-in-water (o/w) emulsion. Experiments were carried out with different membranes in a semi batch filtration cell made of Teflon and the influence of operating conditions such as transmembrane pressure and feed properties such as initial oil concentration and pH of feed solution on membrane performance were investigated. Results show that the membrane properties such as porosity and hydrophilicity are found to have a significant influence on the permeate flow rate and hence subsequent oil removal. The experimental results showed that oil retentions of almost all the membranes were over 90% and oil concentration in the permeate was below 10 mg/L, which met the requirement for discharge. It was concluded that the ultrafiltration (UF) membranes developed in the study were reasonably resistant to fouling and hence the developed PSf membranes may be considered feasible in treating oily wastewater.

1. Introduction

For producing nitrogenous fertilizers a variety of feed stocks may be used such as – lignite, coke, coke oven gas, electrolysis of water, natural gas, associated gas, naphtha, fuel oil etc. Nomally fertilizer plants are located at the point nearest to the feed stocks considering the variety of the resources [1]. Pollution control of fertilizer industry is a very complex problem. Industries are following the three pronged approach of pollution abatement at source, recycle/reuse and treatment of effluents in dealing with problem [2]. The central board for prevention and control of water pollution has laid down minimal national standards, (MINAS) and industries follow these standards [1, 2].

Like any other chemical industry, fertilizer industry produces some gaseous/liquid/solid effluents depending on the technology adopted, feed stock used and location of the industry etc (Table 1). The two major effluent streams are from ammonia production and phosphoric acid production plants. Effluent from ammonia production is highly alkaline and contains excess ammonia from gas scrubbing and gas cleaning operations. The phosphoric acid effluent is acidic and contains high amount of phosphates and suspended solids. The characteristics are different for different effluents according to their feed stocks.

Table 1 Effluent components in fertilizer industry

Air Emissions	Liquid Effluents	Solid Wastes
Oxides of sulphur (SO_X)	Ammonical nitrogen	Carbon sludge
• Sulphur dioxide (SO_2	• Free ammonia	
• Sulphur trioxide (SO_3)	• Ammonium salts	
• Hydrogen sulphide (H_2S)		
Oxides of nitrogen (NOx)	Oxidized nitrogen	Hard coke
• Nitrogen dioxide (NO_2)	• Nitrite nitrogen	
• Nitric oxide (NO)	• Nitrate nitrogen	
Ammonia	Organic nitrogen	Lime sludge
	• Urea	
	• Arsenic	
	• Methanol	
	• Oil and grease	
	• Chromate	
	• Phosphate	
	• Sulphide	
Fluoride compounds		Arsenic sludge
Acid mist		Gypsum

The impacts of discharge of these wastes to the environment causes varied adverse conditions which mainly depend on the nature and the quantum of discharge of pollutants. Due to discharge of effluent the pollution level increases and the increased pollution levels both in atmosphere and in water causes serious environmental problems. When excessive amount of ammonia is being emitted to the atmosphere or discharged with wastewater effluent streams, it is considered as one of those harmful pollutant constituents that damage seriously the environment. Due to its high solubility in water and to its chemical reactivity, ammonia is considered cause for eutrophication of surface water (i.e. the enrichment of water by nutrients causing an accelerated growth of algae for example). Due to this algal boom, whole stretch of water becomes choked, plants rot, all oxygen used up and fish die [3]. As active chemical species, ammonia in atmosphere oxidizes and when it is washed out by rains, it forms nitric acid. As a consequence, acidic rains increase acidification problem in soil and ground water. A decayed alga imparts disagreeable odors and tastes to water [4, 5]. Ammonia in free-state is an irritant and highly toxic to fish. Oily wastewater generated by various industries and subsequently discharged into the natural environment not only creates a major ecological problem but also wastes water resources throughout the world, particularly in arid and semi-arid countries [6].

There are many methods of treatment of effluents to control the pollutants. Unlike other industrial effluents, fertilizer effluent stream should be segregated and treated separately and combined together for final disposal. Wastes utilization programme basically involves three phases namely reduction of pollution at source, recycling /reuse/recovery and waste treatment and ultimate disposal.

The key focus of this study is to prevent pollution and sustain industrial development by removing ammonia from wastewater. Removal of ammonia from liquid effluent by advanced methods of membrane distillation or pervaporation is considered. The effects of different operating parameters on ammonia removal from aqueous solutions have been investigated. Experimental results compared with simulated results of ammonia removal by steam stripping method using PRO II for same ammonia feed composition.

2. Methodologies

In the present work ammonia bearing effluent was collected in a pit. The wastewater specification used for the present work considered on the basis of average data of a local industry, as in Table 1. The pilot plant (Permionics India Ltd., Baroda, India) pervaporation set up was used in this experiment for removing ammonia from wastewater (shown elsewhere). A circular flat sheet membrane was used in the flat plate module of the experimental setup. The effect of feed temperature exposure on the removal efficiency of ammonia was investigated. The effect of downstream pressure on ammonia removal efficiency was observed by varying the down stream pressure in the range of 63 mbar to 103 mbar. The feed pressure, feed temperature and pH maintained constant. The thermodynamic experiments were studied over a temperature range of 49 – 55 °C at pH 9. Permeate samples were collected at one hour interval.

A steam stripper was simulated using PRO/II simulation software and the flow conditions like feed temperature, feed pressure, steam flow rate etc. optimized to acquire target ammonia concentrations (25 ppm) in bottom of stripper.

Table 1 Feed and steam specifications

Feed specifications		Steam specifications	
Feed flow rate	10 m^3/h	Steam temperature	148 °C
NH_3	3000 ppm	Steam pressure	4.5 kg/cm^2
H_2O	Remaining		

3. Results and discussions

Ammoniacal wastewater containing ammonia and water with flow rate 10 m^3/hr, having ammonia composition 3000 ppm was treated by pervaporation method to obtain less than 25 ppm ammonia in treated water. Experiments were performed at 49 °C and 55 °C to investigate the effect of temperature of the feed solution on ammonia removal. Fig.1. showed that the ammonia removal was affected by temperature. Higher feed temperature resulted in higher ammonia removal efficiencies.

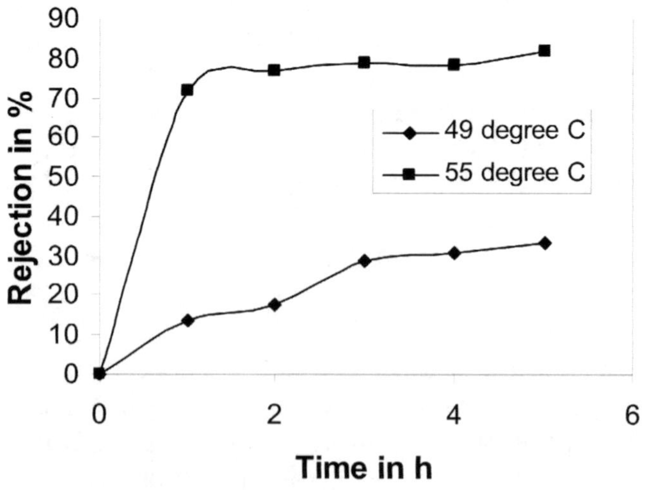

Fig.1. Plot of time vs. rejection on ammonia removal by pervaporation, influence of temperature. Operating parameters: Feed Pressure 1 kg/cm^2, pH 9, downstream pressure 63 mbar, feed concentration 3000 ppm

Experiments were conducted at 83 mbar to 103 mbar downstream pressure to find out the influence of downstream pressure on ammonia removal efficiency. Fig.2. showed that the ammonia removal was affected by downstream pressure.

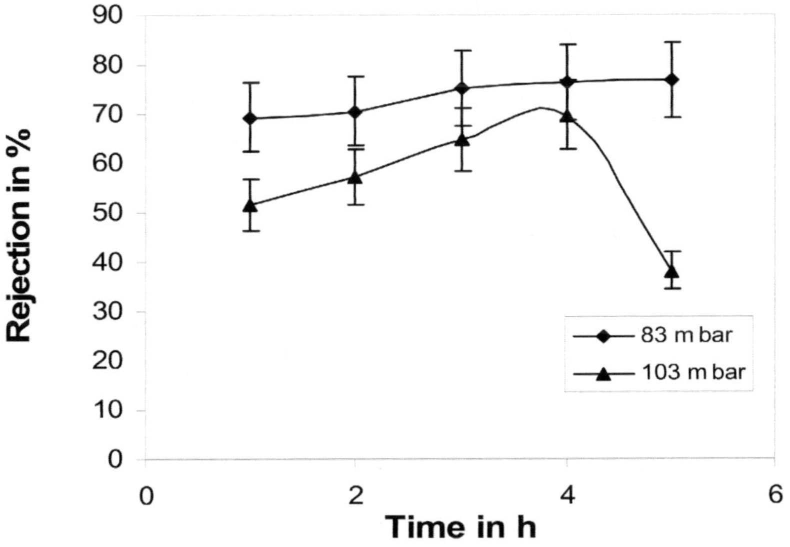

Fig.2. Plot of time vs. rejection on ammonia removal by pervaporation, Influence of downstream pressure. Operating parameters: Temperature 49 °C, pH 9, feed pressure, 1 kg/cm^2, feed concentration 3000 ppm

To compare removal of ammonia from wastewater by conventional steam stripper, theoretical simulation was performed at same feed and steam conditions. On increasing the temperature of the feed i.e. waste water stream at constant steam pressure, steam flow rate and steam temperature after simulation it was found that at 100°C simulation converges. So 100°C was assumed as optimum temperature. On increasing the steam pressure on top tray the removal efficiency of ammonia decreases at constant feed temperature, steam flow rate and steam temperature (Table 2). The steam flow rate affected the ammonia removal efficiency also. For same feed conditions and constant steam pressure and temperature, the ammonia removal efficiency increases with increasing steam flow rate (Table 3).

Table 2. Effect of pressure on ammonia concentration in the stripper bottom

Pressure(kg/cm^2)	Ammonia Concentration in bottom (ppm)
1	149.8
1.2	209.8
1.3	241.9
1.4	276.6
1.5	313.1

Table 3. Effect of Steam flow rate on ammonia concentration in the stripper bottom

Steam Flow Rate (kg/h)	Ammonia Concentration in Bottom (ppm)
500	1458.2
750	533.0
800	413.3
900	240.8
1000	137.4
1100	78.5
1200	45.4
1300	26.8
1325	23.6
1350	20.7

5. Conclusion

The effluents from fertilizer industry are treated by the above techniques according to their composition present in the effluent. At constant feed concentration, feed pressure, feed temperature and down stream pressure the concentration in permeate side decreases with increase in time. An experimental result shows that high feed temperature and low downstream pressures enhanced ammonia removal. From the simulation results it is concluded that 100°C is the optimum temperature for ammonia removal from fertilizer waste water. On increasing the feed pressure or pressure at the top of the column the removal of ammonia decreases at fixed steam flow rate. It has been observed that as steam flow rate increases the ammonia concentration in bottom of the stripper decreases.

References

[1]. S.C. Bhatia, Handbook of industrial pollution and control (V-1), CBS publications and distributor, New Delhi (2001) 353-373.

[2]. N. Manivasakaram, Industrial effluents origin, characteristics, effects, analysis and treatment, Sakthi publications, Coimbatore (1987)151-165.

[3].Department for Environment, Food & Rural Affairs (DEFRA) publications, Ammonia in the UK, 2002, ISBN: PB6865.

[4]. P.H. Liao, A. Chen, K.V. Lo, Removal of nitrogen from swine manure wastewaters by ammonia stripping, Bioresource. Technol. 54 (1995) 17–20.

[5]. A. Bonmat, X. Flotats, Air stripping of ammonia from pig slurry: characterization and feasibility as a pre or post-treatment to mesophilic anaerobic digestion, Waste Manage. 23 (2003) 261–272.

[6]. F.L. Hua, Y.F. Tsang, Y.J. Wang, S.Y. Chan, H. Chua, S.N. Sin, Performance study of ceramic microfiltration membrane for oily wastewater treatment, Chem. Eng. J. 128 (2007) 169–175.

731a New Picture of First-Order Phase Transitions In Metastable Fluids: Homogeneous Nucleation and Growth

Mark J. Uline, Korosh Torabi, and **David S. Corti**. School of Chemical Engineering, Purdue University, 480 Stadium Mall Drive, West Lafayette, IN 47907-2100

Homogeneous nucleation is the activated process by which the new phase (vapor, liquid or solid) is formed from a bulk metastable fluid (superheated liquid, supercooled vapor or liquid) in the absence of impurities or solid surfaces. According to classical nucleation theory (CNT)[1], if an embryo of the new phase is less than some critical size, the embryo collapses back into the metastable fluid; if the embryo exceeds this critical size, the nucleus "reversibly" grows to macroscopic size. The critical nucleus corresponds to the saddle point in free energy space (if the embryo is not treated as incompressible): a maximum in the radius and a minimum in the number of particles, *n*, inside the embryo. The free energy surface continues on indefinitely beyond the saddle point, serving to channel the embryo toward the much lower lying minimum corresponding to the new, bulk phase. Though this subsequent growth of nuclei may be rapid, the post-critical embryos nevertheless follow well-defined pathways that describe the reversible change of n and radii. CNT also invokes another feature of the free energy surface, namely that the region about the saddle point is sharply peaked. Hence, only a small area centered about the saddle point describes the most likely transition paths between a pre-critical embryo and the new phase.

Recently, we adapted density-functional theory (DFT) to calculate the free energy surface, or reversible work $W(n,v)$, of both homogeneous bubble formation within the pure-component superheated Lennard-Jones (LJ) liquid[2] and homogeneous droplet formation within the pure-component supercooled LJ vapor. (Here, n is the number of particles inside an embryo of a given spherical volume v.) The DFT calculations, which constrain the number of particles located inside the bubble for a fixed radius, indicate that $W(n,v)$ is quite different from what is predicted from CNT. For example, DFT reveals that liquid-to-vapor liquid nucleation is more appropriately described by an "activated instability". As the free energy barrier is surmounted, $W(n,v)$ abruptly ends along a locus of instabilities. Further growth of the post-critical bubbles must necessarily proceed via a mechanism appropriate for an unstable system. DFT also suggests that the saddle point, which still corresponds to the critical bubble, is not the only pathway an embryo may take in order to cross the activation barrier. The ridge corresponding to the maximum free energy for each n that leads to the critical bubble is not steep, which indicates that an embryo will more likely than not surmount the barrier along pathways that do not pass through the saddle point. Some of these conclusions also apply to vapor-to-liquid nucleation, though key differences between both free energy surfaces are noted. In contrast to the bubble surface, $W(n,v)$ for droplet formation does not abruptly end after the ridge is surmounted. Nevertheless, limits of stability do appear when the liquid clusters become too dense, such that the surrounding supercooled vapor can no longer maintain its vapor-like density. Furthermore, unlike what was seen in previous theoretical approaches [3], a valley no longer appears beyond the saddle point. The surface beyond the ridge merely serves to channel the post-critical embryos towards the locus of instabilities. In the end, the DFT results highlight the important differences between droplet and bubble nucleation, indicating that any future descriptions of bubble formation cannot solely rely on ideas that have emerged from the study of droplet formation.

For simplicity, the above DFT studies only considered the case of spherical embryos, a not unreasonable assumption for liquid droplets that may not be strictly applicable to bubble formation. Bubbles forming within superheated liquids are not constrained to maintain a spherical shape as previous molecular simulation studies of bubble nucleation[4,5] have demonstrated (although they do not deviate significantly from a sphere). We therefore explore the effects of the shape of the bubble on the physics of liquid-to-vapor nucleation. We present an extension of our DFT method to the study of the growth of bubbles of various non-spherical shapes. Specifically, we present results for the work of prolate and oblate spheroid bubbles. The density profiles surrounding these embryos are now a function of two spatial dimensions as opposed to the single one needed when spherical symmetry is imposed. As a result, the aspect ratio and not just the volume of these spheroids is another important parameter in constructing the free energy surface, i.e., the free energy surface is properly described by $W(n,v,c/a)$, where c is the length of the major axis and a is the length of the minor axis. We find that a locus of instabilities still exists for these non-spherical shapes, so that the previous conclusion regarding an activated instability as the appropriate mechanism describing bubble nucleation and growth remains unchanged.

We also consider the simultaneous growth of two bubbles within the pure-component superheated LJ liquid. By varying both the sizes of and center-to-center separations between the two bubbles, we discuss how the resulting density profiles and free energy surfaces provide insights into the importance of coalescence during the unstable growth phase of bubbles. A recent simulation study of Zahn[6] on the boiling of a water-like fluid indicates that phase separation is initiated by the formation of several tiny cavities, which in turn coalesce to form a larger cavity that appears to then grow further into a macroscopic bubble. We also speculate on what the appropriate signature is needed to identify these unstable growth stages and discuss how some current expressions for the rate of nucleation need to be modified to incorporate the different properties of the free energy surface that emerge from our studies.

Finally, we extend our constrained DFT method to the study of crystallization in the pure-component supercooled LJ liquid. The free energy surface for a given crystal structure is determined and compared to predictions of CNT. Our analysis of the work of formation of crystalline embryos also provides interesting insights into the ongoing debate concerning the appearance of spinodal (or barrierless) nucleation in deeply quenched LJ liquids[7].

[1] L. Gunther, *Am. J. Phys.*, **71**, 351 (2003).

[2] M. J. Uline and D. S. Corti, *Phys. Rev. Letters*, **99**, 076102 (2007).

[3] P. Schaaf, B. Senger, and H. Reiss, *J. Phys. Chem. B*, **101**, 8740 (1997).

[4] T. Kinjo and M. Matsumoto, *Fluid Phase Equilibria*, **144**, 343 (1998).

[5] B. R. Novak, E. J. Maginn and M. J. McCready, *Phys. Rev. B*, **75** 085413 (2007).

[6] D. Zahn, *Phys. Rev. Letters*, **93**, 227801 (2004).

[7] L. S. Bartell and D. T. Wu, *J. Chem. Phys.*, **127**, 174507 (2007).

731b Role Played by Cross-Nucleation Events during Crystal Nucleation and Growth

Jerome Delhommelle and Caroline Desgranges. Department of Chemistry, University of North Dakota, 151 Cornell Street Stop 9024, Grand Forks, ND 58202

In many cases, a process yields crystals of more than one crystalline structure (or polymorph) at the same time, making the control of polymorphism extremely challenging. Several mechanisms have been proposed to account for this phenomenon, known as concomitant polymorphism. It has been attributed either to competing processes of homogeneous nucleation of different polymorphs or to solvent-mediated conversion of one polymorph into another or, more recently, to the heterogeneous nucleation (or cross-nucleation) of one polymorph on another. In recent work [1-4], we used Hybrid Monte Carlo (HMC) simulations as well as molecular dynamics simulations to study the early stages of crystallization in a supercooled liquid of spherical particles. We carry out two different simulation methods corresponding to the two mechanistic steps of crystal nucleation and growth. To simulate the formation of a nucleus of a critical size, we combine HMC simulations together with a method suited to study rare events such as the umbrella sampling technique. The growth step is simply simulated using unconstrained molecular dynamics simulations. We observed the onset of concomitant polymorphism and demonstrated that this phenomenon resulted from the cross-nucleation of a metastable polymorph on the stable polymorph. Our simulations provide detailed insights into the molecular mechanism underlying concomitant polymorphism and cross-nucleation. We discuss the significance of this phenomenon and of its effects on both crystal nucleation and growth.

[1] C.Desgranges and J. Delhommelle, J. Am. Chem. Soc. 128, 15104 (2006).

[2] C.Desgranges and J. Delhommelle, J. Am. Chem. Soc. 128, 15104 (2006).

[3] C.Desgranges and J. Delhommelle, J. Phys. Chem. B 111, 1465 (2007).

[4] C.Desgranges and J. Delhommelle, Phys. Rev. Lett. 98, 235502 (2007).

The homogeneous nucleation of a crystal phase is one of the most interesting phenomena of molecular fluids. Although it has been the subject of extensive experimental and theoretical study, the microscopic mechanism of crystal nucleation still remains poorly understood. It is even more a mystery in chain molecule systems because the chain connectivity could produce very different crystal nucleus conformations, which are important factors in determining the subsequent crystal growth process and the properties of the final product. For example, people observed anomalously low undercooling for homogeneous nucleation of *n*-alkane from C_{16} to C_{32} [1], which has been proposed to be a result of the change in nucleus conformation. However, experiments were not able to examine the exact conformations of the crystal nuclei, especially for the critical nuclei.

In this work we report the results of molecular simulations to study crystal nucleation of n-octane from the melt. A realistic united atom force field was employed for n-octane. The melting behavior was first determined by ramping temperature in a set of molecular dynamics simulations. The adiabatic nucleation trajectory was then sampled using the Monte Carlo umbrella sampling technique. The surface energy of the crystal nuclei was calculated assuming a spherical nucleus model and compared with previous studies to validate our numerical definition of a crystal nucleus. We were also able to calculate the end and side surface free energies of a cylinder nucleus model from the simulation data without making further assumption. This method can be extended to study longer n-alkane molecules and the change of nucleus conformation as n increases.

[1]. Turnbull, D. and R.L. Cormia, *Kinetics of Crystal Nucleation in Some Normal Alkane Liquids.* The Journal of Chemical Physics, 1961. **34**(3): p. 820-831.

731d Minimum Free Energy Pathways for the Nucleation of Molecular Crystals

Erik E. Santiso, Department of Chemical Engineering, Massachusetts Institute of Technology, Cambridge, MA 02139 and Bernhardt L. Trout, Chemical Engineering, Massachusetts Institute of Technology, Room E19-502B, 77 Massachusetts Ave, Cambridge, MA 02139.

Crystallization is extensively used in the pharmaceutical and food industries, among others, to purify chemicals intended for human consumption. Many molecules, however, can crystallize into different polymorphs depending on the precipitation conditions (e.g. solvent, degree of supersaturation). These polymorphs often exhibit different physical properties, which affects their suitability for various applications. For example, different polymorphs of a drug may have different dissolution profiles, and hence bioavaliabilites, which makes the control of the crystallization process critical in drug manufacturing.

In this work, we combine a novel method to characterize the degree of order in molecular crystals with the Finite Temperature String Method in collective variables [1] to obtain minimum free energy pathways for the nucleation of crystals from the melt and from aqueous solution. Our results shed light on the mechanisms by which molecules come together and organize into crystalline structures, which is a first step in understanding the mechanisms of polymorph selection. These results can be used to develop an improved, rational methodology for the manufacturing of drugs with a desired crystal structure.

[1] L. Maragliano, et al., J. Chem. Phys. 125, 024106 (2006)

731e Systematic Investigation of the Nucleation Mechanism of Lysozyme Crystallization In the Liquid-Liquid Co-Existence Region

Yingxin Liu[1], Xiujuan Wang[2], and Chi Bun Ching[2]. (1) Division of Chemical and Biomolecular Engineering, Nanyang Technological University, N1.2-B3-15, 62 Nanyang Drive, SCBE, NTU, Singapore, Singapore, (2) School of Chemical and Biomedical Engineering, Nanyang Technological University, 62 Nanyang Drive, Singapore, 637459, Singapore

Previous work within this group has highlighted that both liquid-liquid phase separation and crystallization occur in protein solutions. The liquid-liquid phase separation and the crystal nucleation and growth processes observed and recorded by an optical microscope suggested that two different nucleation pathways may exist. The current research aims to assess the nucleation of lysozyme crystals in the liquid-liquid co-existence region systematically. The nucleation kinetics has been investigated by the method of initial rates. Micro-batch crystallization experiments were conducted at constant temperature, 10 degree Celsius, with different supersaturations. An optical microscope with a heating/ cooling stage was applied to determine the liquid-liquid co-existence curve, measure the initial nucleation rate and observe the liquid-liquid phase separation and subsequent crystallization process. It was found that the nucleation was arrested in the protein-rich phase. The crystals nucleated from the protein-lean phase. As the crystal nucleated from the solution, the concentration of the protein-lean phase dropped, the protein rich phase dissolved to replenish the supersaturation. By considering the buffering zone effect caused by liquid-liquid phase separation, the nucleation kinetics data were modeled using classical nucleation theory. The nucleation rate was found to increase with the protein concentration as expected.

In addition, the conformation changes and the related protein-protein interactions of lysozyme in aqueous solutions, protein-rich phase, protein-lean phase and crystalline state were studied by Raman microscope. The Raman spectra of lysozyme molecules in different states proved the proposed nucleation mechanism from the view of protein conformation. There was a good agreement among the Raman spectra of lysozyme molecules in different states in amide I and amide III regions, which suggested that lysozyme molecules retained their backbone structure upon aggregation or crystallization. However, the changes in the tyrosine doublet positions and relative intensity revealed that the side chain conformation changed under different protein-protein interactions. The doublet ratio increases when approaching the condition of crystallization with the highest doublet ration observed in crystal. The high doublet ratio was attributed to the exposure of the originally buried tyrosine residue to the outer surface of protein molecule to involve in intermolecular hydrogen bonding, which may increase the protein-protein interaction. The doublet ratio of the protein-lean phase was similar to that of the protein crystallization solution without liquid-liquid phase separation. The doublet ratio of the protein rich phase was larger than that of the protein crystallization solution, which suggested a stronger protein-protein interaction. In protein-lean phase, the relative weaker interactions enabled the protein molecules to adapt their orientations as they did in the crystal more easily. On the contrary, in protein-rich phase, the stronger interactions may take a certain period of time for the protein molecules to change their conformation and orientation, which may be the reason of the arrestment of the nucleation in the protein-rich phase.

In summary, this detailed experimental and theoretical study provides further understanding of the mechanism of protein nucleation in the liquid-liquid co-existence region.

731f Understanding the Molecular Mechanisms Underlying the Nucleation and Growth of Metal Nanoparticles

Jerome Delhommelle and Caroline Desgranges. Department of Chemistry, University of North Dakota, 151 Cornell Street Stop 9024, Grand Forks, ND 58202

Understanding the crystallization of metals is of key importance for many applications e.g. for metal nanoparticles and catalysts. In particular, it is crucial to control the morphology as well as the structure of the crystallites formed during the crystallization process. When and how the selection of a specific structure (or polymorph) occurs remains a long-standing issue. This is a very complex problem, resulting from a subtle interplay between thermodynamics and kinetics. Solving this issue has remained elusive so far, even on simple model systems composed of spherical particles.

In this talk, we use molecular simulations to shed light on the molecular mechanisms underlying the crystallization from liquid metals. We model the metals with a many-body embedded-atom model (EAM) potential [1,2]. The ability of the model to predict the thermodynamic properties of the bulk (including the melting temperature) was assessed in previous work. We consider here liquid metals, supercooled at temperatures 15% and 20% below the melting point, and study the crystallization process in those systems. For both degrees of supercooling, we simulate the homogeneous nucleation event and the subsequent growth of the critical nuclei. Our simulation results shed light on the molecular mechanisms underlying the structure selection process during the crystallization process.

[1] C. Desgranges and J. Delhommelle, J. Am. Chem. Soc. 129, 7012 (2007).

[2] C. Desgranges and J. Delhommelle, J. Chem. Phys. 127, 144509 (2007).

AUTHOR INDEX

AUTHOR INDEX

AUTHOR INDEX

AUTHOR INDEX

AUTHOR INDEX

AUTHOR INDEX

AUTHOR INDEX

AUTHOR INDEX

AUTHOR INDEX

AUTHOR INDEX

AUTHOR INDEX

AUTHOR INDEX

AUTHOR INDEX

AUTHOR INDEX

AUTHOR INDEX

AUTHOR INDEX

American Institute of Chemical Engineers
3 Park Avenue
New York, New York 10016-5991